Heavy Metal Toxicity and Tolerance in Plants

Heavy Metal Toxicity and Tolerance in Plants

A Biological, Omics, and Genetic Engineering Approach

Edited by

Mohammad Anwar Hossain
Bangladesh Agricultural University, Mymensingh, Bangladesh

AKM Zakir Hossain
Bangladesh Agricultural University, Mymensingh, Bangladesh

Sylvain Bourgerie
Université d'Orléans, Orléans, France

Masayuki Fujita
Kagawa University, Kagawa, Japan

Om Parkash Dhankher
University of Massachusetts Amherst, Massachusetts, USA

Parvez Haris
De Montfort University, Leicester, UK

WILEY

This edition first published 2023

Registered Offices
John Wiley & Sons, Inc., 111 River Street, Hoboken, NJ 07030, USA
John Wiley & Sons Ltd, The Atrium, Southern Gate, Chichester, West Sussex, PO19 8SQ, UK

For details of our global editorial offices, customer services, and more information about Wiley products visit us at www.wiley.com.

Wiley also publishes its books in a variety of electronic formats and by print-on-demand. Some content that appears in standard print versions of this book may not be available in other formats.

Library of Congress Cataloging-in-Publication Data

Names: Hossain, Mohammad Anwar, editor. | Hossain, AKM Zakir, editor. | Bourgerie, Sylvain, editor. | Fujita, Masayuki, editor. | Dhankher, Om Parkash, editor. | Haris, P. I. (Parvez I.), editor.
Title: Heavy metal toxicity and tolerance in plants : a biological, omics, and genetic engineering approach / edited by Mohammad Anwar Hossain, AKM Zakir Hossain, Sylvain Bourgerie, Masayuki Fujita, Om Parkash Dhankher, Parvez Haris.
Other titles: Biological, omics, and genetic engineering approach
Description: First edition. | Hoboken, NJ : Wiley, 2023 | Includes index.
Identifiers: LCCN 2023024444 (print) | LCCN 2023024445 (ebook) | ISBN 9781119906469 (cloth) | ISBN 9781119906476 (adobe pdf) | ISBN 9781119906483 (epub)
Subjects: LCSH: Plants–Effect of heavy metals on. | Heavy metals. | Plants–Heavy metal content.
Classification: LCC QK753.H4 H42 2023 (print) | LCC QK753.H4 (ebook) | DDC 581.7–dc23/eng/20230721
LC record available at https://lccn.loc.gov/2023024444
LC ebook record available at https://lccn.loc.gov/2023024445

Cover Design: Wiley
Cover Image: © pingphuket/Shutterstock, Courtesy of the Editors, Denis Belitsky/Shutterstock, Timofeev Vladimir/Shutterstock

Set in 9.5/12.5pt STIXTwoText by Straive, Pondicherry, India

Contents

List of Contributors *xix*
Preface *xxix*
Editor Biographies *xxxi*

1 Plant Response and Tolerance to Heavy Metal Toxicity: An Overview of Chemical Biology, Omics Studies, and Genetic Engineering *1*
Lovely Mahawar, Sakshi Pandey, Aparna Pandey, and Sheo Mohan Prasad
1.1 Introduction *1*
1.2 Plant–Metal Interaction *2*
1.3 Effect of Heavy Metals on Plants *3*
1.3.1 Morphoanatomical Responses *3*
1.3.2 Physiological Responses *8*
1.3.3 Biochemical Responses *8*
1.3.4 Molecular Responses *9*
1.4 Mechanisms to Tolerate Heavy Metal Toxicity *10*
1.4.1 Avoidance *10*
1.4.1.1 Mycorrhizal Association *10*
1.4.1.2 Root Exudates *12*
1.4.2 Sequestration *12*
1.5 Important Strategies for the Enhancement of Metal Tolerance *15*
1.5.1 Omics *15*
1.5.1.1 Genomics *15*
1.5.1.2 Transcriptomics *15*
1.5.1.3 Proteomics *17*
1.5.1.4 Metabolomics *17*
1.5.1.5 Ionomics *18*
1.5.1.6 miRNAomics *19*
1.5.1.7 Metallomics *19*
1.5.2 Genetic Engineering *20*
1.5.2.1 CRISPR Technology *20*
1.5.2.2 Plastid Transformation *21*
1.5.2.3 Gene Silencing *22*
1.6 Conclusion and Future Prospects *22*
References *23*

2 Advanced Techniques in Omics Research in Relation to Heavy Metal/Metalloid Toxicity and Tolerance in Plants *35*
Ali Raza, Shanza Bashir, Hajar Salehi, Monica Jamla, Sidra Charagh, Abdolkarim Chehregani Rad, and Mohammad Anwar Hossain
2.1 Introduction *35*
2.2 An Overview of Plant Responses to Heavy Metal Toxicity *36*
2.3 How the Integration of Multi-omics Data Sets Helps in Studying the Heavy Metal Stress Responses and Tolerance Mechanisms? 39
2.3.1 The Contribution of State-of-the-Art Genomics-Assisted Breeding *39*
2.3.1.1 Quantitative Trait Locus (QTL) Mapping *39*
2.3.1.2 Genome-Wide Association Studies *41*
2.3.2 Transcriptomics *42*
2.3.3 Proteomics *44*
2.3.4 Metabolomics *46*
2.3.5 miRNAomics *47*
2.3.6 Phenomics *49*
2.4 Conclusion and Perspectives *50*
References *50*

3 Heavy Metals/Metalloids in Food Crops and Their Implications for Human Health *59*
Shihab Uddin, Hasina Afroz, Mahmud Hossain, Jessica Briffa, Renald Blundell, and Md. Rafiqul Islam
3.1 Introduction *59*
3.2 Arsenic *60*
3.2.1 Sources and Forms *60*
3.2.2 Food Chain Contamination *62*
3.2.3 Pharmacokinetic Processes *62*
3.2.4 Toxicology Processes *62*
3.2.5 Remedial Options *63*
3.3 Cadmium *63*
3.3.1 Sources and Forms *64*
3.3.2 Food Chain Contamination *64*
3.3.3 Pharmacokinetic Processes *66*
3.3.4 Toxicology Processes *66*
3.3.5 Remedial Options *67*
3.4 Lead *67*
3.4.1 Sources and Forms *68*
3.4.2 Food Chain Contamination *68*
3.4.3 Pharmacokinetic Processes *68*
3.4.4 Toxicology Processes *70*
3.4.5 Remedial Options *71*
3.5 Chromium *72*
3.5.1 Sources and Forms *72*
3.5.2 Food Chain Contamination *74*

3.5.3 Pharmacokinetic Processes *74*
3.5.4 Toxicology Processes *74*
3.5.5 Remedial Options *75*
3.6 Mercury *76*
3.6.1 Sources and Forms *76*
3.6.2 Food Chain Contamination *77*
3.6.3 Pharmacokinetic Processes *79*
3.6.4 Toxicology Processes *79*
3.6.5 Remedial Options *80*
3.7 Conclusions *81*
References *81*

4 Aluminum Stress Tolerance in Plants: Insights from Omics Approaches *87*
Richa Srivastava, Ayan Sadhukhan, and Hiroyuki Koyama
4.1 Introduction *87*
4.2 Exploration of Al Tolerance QTLs *89*
4.3 Unraveling the Genetic Architecture of Al Tolerance from Natural Variation *91*
4.4 Identification of Novel Al Tolerance Genes Through Genome-Wide Association Studies *91*
4.5 Exploring Expression Level Polymorphisms to Identify Upstream Al Signaling *92*
4.6 Comparative Transcriptome Analyses Identify Novel Al Tolerance Genes *93*
4.7 Identification of Al Tolerance Genes from Proteomics *95*
4.8 Conclusion and Future Perspectives *99*
References *99*

5 Breeding Approaches for Aluminum Toxicity Tolerance in Rice and Wheat *105*
Buu Chi Bui and Lang Thi Nguyen
5.1 Introduction *105*
5.2 Plant Signaling *107*
5.3 Rice Genetic Mapping *107*
5.3.1 Linkage Mapping *107*
5.3.2 Association Mapping *108*
5.4 Root Transcriptome *109*
5.5 Wheat Genetic Mapping *114*
5.5.1 Wheat MATE Gene Family *116*
5.6 Wheat Proteomics *117*
5.7 Conclusion *118*
References *118*

6 Chromium Toxicity and Tolerance in Plants: Insights from Omics Studies *125*
Sonali Dubey, Manju Shri, and Debasis Chakrabarty
6.1 Introduction *125*
6.2 Chromium Sources and Bioavailability *126*
6.3 Chromium Uptake, Translocation, and Sub-cellular Distribution in plants *127*

6.4 Detoxification Mechanisms for Cr *129*
6.5 Omics Approaches Used by Plants to Combat Cr Toxicity *130*
6.5.1 Transcriptomics *130*
6.5.2 Chromium-Induced miRNAs in Plants *132*
6.5.3 Metabolomics *133*
6.5.4 Proteomics *133*
6.6 Phytoremediation of Cr Metal by Plants *134*
6.6.1 Phytoremediation Approach for Cr Detoxification *134*
6.6.2 Other Strategies Involved in Cr Remediation *135*
6.6.3 Phytostabilization/Phytoextraction for Cr Decontamination *136*
6.7 Conclusion *136*
References *136*

7 Manganese Toxicity and Tolerance in Photosynthetic Organisms and Breeding Strategy for Improving Manganese Tolerance in Crop Plants: Physiological and Omics Approach Perspectives *141*
Daisuke Takagi
7.1 Introduction *141*
7.2 The Change in Mn Availability Within the Soil *143*
7.3 Why Should We Consider the Occurrence of Mn Toxicity in Plants? Possible Threats of Mn Toxicity in Agricultural Land *144*
7.4 The History of Mn Toxicity *146*
7.5 The Features of Mn Toxicity in Terrestrial Plants and Possible Molecular Mechanisms *147*
7.5.1 The Mechanisms of Emergence of Brownish Patchy Spots in Leaves: The Apoplastic Mn Toxicity *147*
7.5.2 The Mechanisms of Foliar Chlorosis Under Excess Mn: Symplastic Mn Toxicity *150*
7.6 Breeding Strategy for Overcoming the Future Threat of Excess Mn Conditions *154*
7.6.1 Limiting Mn Absorption from Soil to Root *155*
7.6.2 Sequestration of Mn from Cytosol to the Vacuole or Apoplast *156*
7.6.3 Maintenance of Auxin Homeostasis *157*
7.6.4 The Reinforcement of Silicon Uptake and Its Distribution *157*
7.7 Conclusion and Future Prospects *158*
Acknowledgments *158*
References *158*

8 Iron Excess Toxicity and Tolerance in Crop Plants: Insights from Omics Studies *169*
May Sann Aung and Hiroshi Masuda
8.1 Iron Uptake and Translocation Mechanism in Plants *169*
8.1.1 Importance of Iron in Living Organisms *169*
8.1.2 Fe Acquisition Systems in Plants *170*
8.1.3 Fe Translocation Mechanisms in Plants *171*
8.2 Fe Excess Toxicity in Plants *171*

8.2.1 Fe Excess Toxicity in Global Agriculture *171*
8.2.2 Causes of Fe Excess Toxicity in Soils and Its Interaction with Plants *172*
8.2.2.1 State of Fe in Soils and Soil pH Effects on Fe Excess Toxicity *172*
8.2.2.2 Soil Improvement Methods to Ameliorate Fe Excess Toxicity *173*
8.2.2.3 Soil Water and Drainage Effects on Fe Excess Toxicity *173*
8.2.3 Effects of Fe Excess Toxicity on Plant Growth *174*
8.3 Crop Defense Mechanisms Against Excess Fe and Genes Regulating Fe Excess *175*
8.3.1 Defense I: Fe Exclusion from Roots *175*
8.3.1.1 Genes Involved in Defense I *176*
8.3.2 Defense II: Fe Retention in Roots and Suppression of Fe Translocation to Shoots *177*
8.3.3 Defense III: Fe Compartmentalization in Shoots *177*
8.3.3.1 Genes Involved in Defense II and III *178*
8.3.3.2 Role of YSL4 and YSL6 Transporters in Preventing Fe Excess in Early Plant Development *179*
8.3.4 Defense IV: ROS Detoxification *179*
8.3.4.1 Genes Involved in Defense IV *180*
8.3.4.2 GLY1 as a Detoxifying Agent *180*
8.4 Research Outlook on Fe Excess Response of Plants *180*
8.4.1 Regulation of Fe homeostasis in Plants in Response to Fe Excess Stress *180*
8.4.2 Transcription Factors *181*
8.4.3 Cis-Regulatory Elements *182*
8.5 Conclusion and Future Prospects *183*
Acknowledgments *183*
Author Contributions *183*
Disclosures *183*
References *183*

9 Molecular Breeding for Iron Toxicity Tolerance in Rice (*Oryza sativa* L.) *191*
Dorothy A. Onyango, Mathew M. Dida, Khady N. Drame, Benson O. Nyongesa, and Kayode A. Sanni
9.1 Introduction *191*
9.2 Role of Iron in Plants and Rice *192*
9.3 Iron Toxicity and Its Effects on Rice *192*
9.4 Iron Toxicity Tolerance Mechanisms in Rice Plants *193*
9.4.1 Fe Exclusion from Roots *193*
9.4.2 Fe Retention in Roots and Suppression of Fe Translocation to Shoots *194*
9.4.3 Fe Compartmentalization in Shoots *194*
9.4.4 ROS Detoxification *195*
9.4.5 Candidate Genes Involved in the Mechanisms of Fe Toxicity *196*
9.4.6 Genetic Variants for Iron Toxicity Tolerance in Rice Germplasm *197*
9.5 Molecular Breeding for Fe Toxicity Tolerance in Rice *197*
9.6 Conclusion *200*
References *202*

10 Cobalt Induced Toxicity and Tolerance in Plants: Insights from Omics Approaches *207*
Abdul Salam, Muhammad Siddique Afridi, Ali Raza Khan, Wardah Azhar, Yang Shuaiqi, Zaid Ulhassan, Jiaxuan Qi, Nu Xuo, Yang Chunyan, Nana Chen, and Yinbo Gan
10.1 Introduction *207*
10.2 Plant Response to Cobalt Stress *208*
10.2.1 Uptake and Translocation of Cobalt in Plants *209*
10.3 Cobalt-Induced ROS Generation and Their Damaging Effects *211*
10.3.1 ROS-Induced Lipid Peroxidation *211*
10.3.2 ROS-Induced Damage to Genetic Material *212*
10.4 Cobalt-Induced Plant Antioxidant Defense System *213*
10.4.1 Enzymatic Antioxidants *213*
10.4.1.1 Superoxide Dismutase (SOD) *213*
10.4.1.2 Catalases (CAT) *213*
10.4.1.3 Glutathione Peroxidases (GPX) *214*
10.4.1.4 Glutathione Reductase (GR) *214*
10.4.2 Nonenzymatic Antioxidants *215*
10.4.2.1 Ascorbic Acid *215*
10.4.2.2 Tocopherols *215*
10.4.2.3 Reduced Glutathione (GSH) *216*
10.5 Omics Approaches in Cobalt Stress Tolerance *216*
10.5.1 Transcriptomic *216*
10.5.2 Metabolomics *218*
10.5.3 Proteomics *219*
10.6 Conclusion and Future Prospects *220*
Acknowledgments *221*
References *221*

11 Nickel Toxicity and Tolerance in Plants *231*
Sondes Helaoui, Marouane Mkhinini, Iteb Boughattas, Noureddine Bousserrhine, and Mohamed Banni
11.1 Introduction *231*
11.2 Sources of Ni *232*
11.2.1 Natural Sources of Ni *232*
11.2.2 Anthropogenic Sources of Ni *233*
11.3 Role of Ni in Plants *233*
11.4 Ni Uptake and Accumulation in Plants *233*
11.5 Ni Toxicity in Plants *234*
11.5.1 Growth Inhibition *234*
11.5.2 Photosynthesis Inhibition of Ni *236*
11.5.3 Induction of Oxidative Stress *236*
11.6 Tolerance Mechanisms *237*
11.7 Omics Approaches in Ni Stress Tolerance *238*
11.7.1 Transcriptomics *238*
11.7.2 Proteomics *239*

11.7.3 Metabolomics *240*
11.8 Conclusion *240*
References *241*

12 Copper Toxicity and Tolerance in Plants: Insights from Omics Studies *251*
Moreira A, Moraes LAC, Delfim JJ, and Moreti LG
12.1 Introduction *251*
12.2 Copper in Plants *253*
12.2.1 Functions of Copper *253*
12.2.2 Uptake, Transport, Distribution, and Remobilization Mechanisms *255*
12.2.3 Deficient, Sufficient, and Toxic Levels of Copper in Plants *255*
12.2.4 Copper Sources: Fertilizers and Fungicides *256*
12.3 Omics Approaches for Cu Responses and Tolerance in Plants *259*
12.3.1 Genomics *259*
12.3.2 Transcriptomics *259*
12.3.3 Proteomics *261*
12.3.4 Metabolomics *263*
12.3.5 miRNAomics *264*
12.4 Concluding Remarks *266*
Acknowledgments *266*
References *267*

13 Zinc Toxicity and Tolerance in Plants: Insights from Omics Studies *275*
Imran Haider Shamsi, Qichun Zhang, Zhengxin Ma, Sibgha Noreen, Muhammad Salim Akhter, Ummar Iqbal, Muhammad Faheem Adil, Muhammad Fazal Karim, and Najeeb Ullah
13.1 Introduction *275*
13.1.1 Zinc Uptake and Translocation Mechanisms in Plants *275*
13.1.2 Transporters and Metal-Binding Compounds Involved in Zinc Homeostasis *277*
13.2 Impact of Excess Zinc on Physio-genetics Aspects of Plants *277*
13.2.1 Effect of Zinc Toxicity on Seed Germination and Growth of Plants *278*
13.2.2 Effect of Zinc Toxicity on Oxidative Metabolism in Plants *279*
13.2.3 Effect of Zn Toxicity on Physiology and Biochemistry of Plants *280*
13.3 Plants Stress Adaptation to Zinc Toxicity *281*
13.4 Multi-omics Approaches for Zinc Toxicity and Tolerance in Plants *281*
13.4.1 Genomics and Metabolomics *281*
13.4.2 Proteomics and Transcriptomics *283*
13.4.3 miRNA Omics and CRISPR/Cas9 System *284*
13.4.4 Quantitative Trait Locus Mapping and Genome-Wide Association Study *286*
13.5 Conclusion and Future Prospective *286*
Acknowledgments *286*
References *287*

14 Arsenic Toxicity and Tolerance in Plants: Insights from Omics Studies *293*
Barsha Majumder, Palin Sil, and Asok K. Biswas
14.1 Introduction *293*
14.2 Occurrence and Distribution of As in the Environment *295*

14.3 Arsenic Uptake, Accumulation, and Detoxification in Plants *296*
14.3.1 Uptake of Inorganic Arsenic *296*
14.3.2 Uptake of Methylated Arsenic *297*
14.3.3 Arsenic Accumulation and Detoxification *297*
14.3.4 Arsenic Methylation and Volatilization *298*
14.4 Influence of Arsenic on Phytotoxicity *298*
14.4.1 Germination and Growth *298*
14.4.2 Nutrient Uptake *299*
14.4.3 Oxidative Stress and Antioxidative Defense *299*
14.4.4 Ascorbate–Glutathione Cycle *300*
14.4.5 Photosynthesis *300*
14.4.6 Respiration *301*
14.4.7 Carbohydrate Metabolism *302*
14.4.8 Nitrogen Metabolism *302*
14.5 Modulation in “Omics” Profiling Under As Challenged Environment *303*
14.5.1 Genomic Profiling *303*
14.5.2 Transcriptomic Profiling *304*
14.5.3 Proteomic Profiling *307*
14.5.4 Metabolomic Profiling *308*
14.6 Progress in Molecular Biotechnology to Evolve As-Tolerant Plants *308*
14.7 Conclusion and Future Perspective *311*
Acknowledgment *311*
Author Contributions *312*
References *312*

15 Selenium Toxicity and Tolerance in Plants: Insights from Omics Studies *323*
Ali Kıyak, Selman Uluısık, Ertugrul Filiz, and Firat Kurt
15.1 Introduction *323*
15.2 Selenium Toxicity in Plants *324*
15.2.1 Se-Induced Protein Malformation *324*
15.2.2 ROS-Induced Se Phytotoxicity *325*
15.3 Selenium Tolerance in Plants *326*
15.4 Selenium Biofortification in Plants *328*
15.5 Conclusion *329*
References *330*

16 Breeding for Rice Cultivars with Low Cadmium Accumulation *335*
Li Tang, Yaokui Li, Yan Peng, Bigang Mao, Ye Shao, Zhongying Ji, and Bingran Zhao
16.1 Introduction *335*
16.2 Molecular Mechanisms of Cd Accumulation in Rice *335*
16.2.1 Cd Uptake *336*
16.2.2 Radial Transport and Xylem Loading *338*
16.2.3 Distribution of Cd in Shoots *338*
16.3 Transgenic Approach for Breeding Low-Cd Rice *339*
16.3.1 Traditional Transgenic Technology *339*

16.3.2 Genome-Editing Technology *340*
16.4 Mutation Breeding for Low-Cd Rice Cultivars *341*
16.5 Molecular Marker-Assisted Breeding for Low-Cd Rice Cultivars *342*
16.6 Future Perspectives *343*
References *344*

17 Mercury Toxicity: Plant Response and Tolerance *349*
Arifin Sandhi, Abu Bakar Siddique, and Meththika Vithanage
17.1 Introduction *349*
17.2 Global Mercury Pollution *350*
17.3 Mercury Uptake and Toxicity in Plants *352*
17.4 Existence of Differential Plant Response to Hg Stress *353*
17.4.1 Plant Morphological Responses *353*
17.4.2 Plant Anatomical Responses *354*
17.4.3 Cellular Responses *354*
17.4.4 Plant Photosynthetic Response *355*
17.4.5 Enzymatic and Metabolic Responses *355*
17.4.6 Plant Hormonal Responses *356*
17.4.7 Reactive Oxygen Species and Oxidative Responses *356*
17.5 Plant Tolerance Mechanisms *357*
17.5.1 Chelation *357*
17.5.2 Enzymatic and Antioxidative Tolerance *358*
17.5.3 Hormonal Regulations *359*
17.5.4 miRNA-Mediated Tolerance *360*
17.6 Phytoremediation Prospects *360*
17.7 Conclusion *361*
References *362*

18 Lead Toxicity and Tolerance in Plants: Insights from Omics Studies *373*
Sayyeda Hira Hassan, Yassine Chafik, Manhattan Lebrun, Gabriella Sferra, Sylvain Bourgerie, Gabriella Stefania Scippa, Domenico Morabito, and Dalila Trupiano
18.1 Introduction *373*
18.2 Omics' Contribution in Uncovering Molecular Alterations in Plants Under Pb Exposure *375*
18.3 Genetics and Epigenetics Regulations of Pb Toxicity and Tolerance *380*
18.4 The Role of Plant Cell Wall, Cell Signaling, and Transduction *382*
18.5 Pb-Binding Proteins/Transporters and Their Involvement in Tolerance *384*
18.6 Pb-Induced Oxidative Stress and Antioxidative Mechanisms *385*
18.7 Metabolic Pathways Associated with Pb Tolerance *388*
18.7.1 Sugar/Carbohydrate and Energy Metabolic Pathway *388*
18.7.2 Phenylpropanoid Pathway *389*
18.7.3 Sulfur-Related Pathway and Phytohormones *390*
18.8 Conclusion and Future Perspective *392*
References *394*

19 Interaction of Heavy Metal with Drought/Salinity Stress in Plants *407*
Ziqian Li, Wentao Chen, Qianlong Tan, Yuanyuan Hou, Taimoor Hassan Farooq, Baber Iqbal, and Yong Li
19.1 Introduction *407*
19.2 Plant Physiology and Biochemistry *409*
19.2.1 Zinc (Zn) *409*
19.2.2 Cadmium (Cd) *410*
19.2.3 Aluminium (Al) *411*
19.2.4 Other Metals *412*
19.3 Photosynthesis *413*
19.4 Antioxidant System *414*
19.5 Conclusions and Prospects *415*
Acknowledgments *416*
References *416*

20 Hormonal Regulation of Heavy Metal Toxicity and Tolerance in Crop Plants *425*
Éderson Akio Kido, Gizele de Andrade Luz, Valquíria da Silva, Maria Fernanda da Costa Gomes, and José Ribamar Costa Ferreira Neto
20.1 Introduction *425*
20.2 General Aspects of Plants Under HM Stress *426*
20.3 Phytohormone-Mediating Plant Response to HM Stress *427*
20.3.1 Abscisic Acid *430*
20.3.2 Auxin *432*
20.3.3 Brassinosteroid *434*
20.3.4 Cytokinin *435*
20.3.5 Ethylene *437*
20.3.6 Gibberellin *438*
20.3.7 Jasmonate *439*
20.3.8 Melatonin (MT) *440*
20.3.9 Salicylic Acid (SA) *442*
20.3.10 Strigolactone (SL) *444*
20.4 Crosstalk of Phytohormones in Plants Responding to Heavy Metals *445*
20.5 Final Considerations *447*
References *448*

21 Heavy-Metal-Induced Reactive Oxygen Species and Methylglyoxal Formation and Detoxification in Crop Plants: Modulation of Tolerance by Exogenous Chemical Compounds *461*
Beatrycze Nowicka, Tahsina Sharmin Hoque, Sheikh Mahfuja Khatun, Jannatul Naim, Ahmed Khairul Hasan, and Mohammad Anwar Hossain
21.1 Introduction *461*
21.2 Heavy-Metal-Induced ROS and Methylglyoxal Production in Plant Cells *464*
21.3 Detoxification of ROS and Methylglyoxal in Plant Cells *468*
21.4 Exogenous Chemical-Compounds-Mediated Heavy Metal/Metalloid Tolerance in Crop Plants *473*
21.5 Conclusions and Future Perspectives *484*
References *486*

22 Biochar Amendments in Soils and Heavy Metal Tolerance in Crop Plants *493*
Agnieszka Medyńska-Juraszek and Bhakti Jadhav
22.1 Introduction *493*
22.2 Heavy Metal Immobilization Mechanisms on Biochar *495*
22.2.1 Heavy Metal Immobilization Through Soil pH Modification *496*
22.3 Biochar Interactions Through Rhizosphere *496*
22.3.1 Effect on Plant Root Development *497*
22.3.2 Changes in Elements Uptake from Rhizosphere *498*
22.4 Biochar-Induced Plant Respond to Metal Stress *499*
22.4.1 Biochar Induces Changes in Photosynthetic Activity *499*
22.4.2 Biochar Induces Changes in Antioxidant and Phytohormone Activity *499*
22.4.3 Biochar as a Source of Specific Chemical Compounds Affecting Heavy Metal Uptake By Plants *501*
22.5 Effect of Biochar on Heavy Metal Concentrations in Different Crops *503*
22.6 Effect of Biochar Type on Heavy Metal Immobilization *503*
References *504*

23 Plant-Growth-Promoting Rhizobacteria and Their Metabolites: Clean and Green Approaches to Deal with Heavy Metal Toxicity *513*
Imtinen Sghaier, Ameur Cherif, and Mohamed Neifar
23.1 Introduction *513*
23.2 Chemical Fertilizers and Their Impacts *515*
23.2.1 Impacts of Chemical Fertilizers on Atmospheric Ecosystem *515*
23.2.2 Impacts of Chemical Fertilizers on Aquatic Ecosystem *515*
23.2.3 Impacts of Chemical Fertilizers on Soil *515*
23.2.4 Impacts of Chemical Fertilizers on Plants *516*
23.3 PGPR and Biofertilization Traits *516*
23.3.1 Acquisition of Nutrients *516*
23.3.2 Production of Siderophores *517*
23.3.3 Production of Exopolysaccharides *517*
23.4 Resistance to Abiotic Stress *518*
23.5 Biostimulation Potential and PGPR *519*
23.6 Biocontrol Potential and PGPR *520*
23.7 PGPR and Heavy Metal Bioremediation *521*
23.8 Conclusion and Future Prospects *524*
Acknowledgments *525*
References *525*

24 Applications of Nanotechnology for Improving Heavy Metal Stress Tolerance in Crop Plants *533*
Meng Jiang, Yue Song, Mukesh Kumar Kanwar, and Jie Zhou
24.1 Introduction *533*
24.2 Impacts of NPs on the HM Stress in Plants *535*
24.2.1 Silicon *535*
24.2.2 Selenium *535*
24.2.3 Iron *536*
24.2.4 Zinc Oxide *537*

24.2.5 Titanium Dioxide *537*
24.2.6 Cerium Dioxide *538*
24.3 Mechanisms of NPs to Mitigate the Toxicity of HM *539*
24.4 Summary and Prospect *543*
References *545*

25 The Dynamics of Phytoremediation of Heavy Metals: Recent Progress and Future Perspective *553*
Imran Haider Shamsi, Xiaoli Jin, Xin Zhang, Qidong Feng, Zakir Ibrahim, Muhammad Faheem Adil, Muhammad Fazal Karim, and Najeeb Ullah
25.1 Introduction *553*
25.1.1 Types of Phytoremediation *554*
25.1.1.1 Phytostabilization *554*
25.1.1.2 Phytovolatalization *554*
25.1.1.3 Phytoextraction *554*
25.1.2 Modified Concept *555*
25.1.2.1 Chemical-Assisted Phytoremediation Employing Non-hyperaccumulator Plants *556*
25.1.2.2 Biochar-Assisted Phytoremediation *556*
25.1.2.3 Microbial-Assisted Phytoremediation *557*
25.2 Importance of Phytoremediation *557*
25.3 Role of Phytoremediation as a Sustainable Solution *558*
25.4 Biophilic Design as Phytoremediation in Urban Sustainability *559*
25.4.1 Eco-Design *559*
25.4.2 Biophilic Design *559*
25.4.2.1 Hypothesis of Biophilic *562*
25.4.2.2 Dimensions of Biophilic Design *562*
25.4.2.3 Direct Experience of Nature *562*
25.4.2.4 Indirect Experience of Nature *563*
25.4.2.5 Experience of Place and Space *563*
25.4.2.6 Sustainable Biophilic Cities *563*
25.4.3 Health Benefits *564*
25.4.4 Biophilic as an Antidepressant in Urban Environment *565*
25.4.5 Economic Benefits *566*
25.4.6 Sustainability and Resilience *566*
25.5 Conclusion *567*
25.6 Future Perspective *568*
Acknowledgment *569*
References *569*

26 Genetic Engineering for Heavy Metal/Metalloid Stress Tolerance in Plants *573*
Mohammad Anwar Hossain, Md. Tahjib-Ul-Arif, Sopnil Ahmed Jahin, Abu Bakar Siddique, Mumtarin Haque Mim, Sharif-Ar-Raffi, Muhammad Javidul Haque Bhuiyan, and Beatrycze Nowicka
26.1 Introduction *573*
26.2 Mechanisms of Heavy Metal/Metalloid Tolerance in Plants *574*

26.3 Strategies for Improving Metal/Metalloid Stress Tolerance in Plants *576*
26.4 Transgenic Plants and Heavy Metal/Metalloid Stress Tolerance in Plants *577*
26.4.1 Sulfur Metabolism Engineering and Heavy Metal Tolerance *577*
26.4.2 Glyoxalase Pathway Genes and Heavy Metal Stress Tolerance *577*
26.4.3 Enhanced Antioxidant Defense and Heavy Metal Tolerance *579*
26.4.4 Phytochelatin and Metallothionein Genes and Heavy Metal Tolerance *579*
26.4.5 Metal Ion Transporter Genes/Proteins and Heavy Metal Stress Tolerance *579*
26.5 CRISPR/Cas System and Heavy Metal Tolerance Development *585*
26.6 Conclusions and Future Prospects *585*
Acknowledgment *586*
References *586*

Index *593*

List of Contributors

Muhammad Faheem Adil
Department of Agronomy
College of Agriculture and Biotechnology
Key Laboratory of Crop Germplasm
Resource, Zhejiang University
Hangzhou, China

Muhammad Siddique Afridi
Department of Plant Pathology
Federal University of Lavras (UFLA)
Lavras, MG, Brazil

Hasina Afroz
Department of Soil Science
Bangladesh Agricultural University
Mymensingh, Bangladesh

Muhammad Salim Akhter
Department of Botany, Institute of Pure
and Applied Biology, Bahauddin Zakariya
University, Multan, Pakistan

May Sann Aung
Department of Biological Production
Faculty of Bioresource Sciences
Akita Prefectural University
Akita City, Japan

Wardah Azhar
Zhejiang Key Lab of Crop Germplasm
Department of Agronomy
College of Agriculture and Biotechnology
Zhejiang University, Hangzhou, China

Mohamed Banni
Department of plant protection
Laboratory of Agrobiodiversity and
Ecotoxicology
Higher Institute of Agronomy
Chott-Meriem, Sousse University
Sousse, Tunisia

Higher Institute of Biotechnology
Monastir University, Monastir, Tunisia

Shanza Bashir
Institute of Environmental Sciences
and Engineering (IESE), School of Civil
and Environmental Engineering (SCEE)
National University of Sciences and
Technology (NUST)
Islamabad, Pakistan

Muhammad Javidul Haque Bhuiyan
Department of Biochemistry and
Molecular Biology
Bangladesh Agricultural University
Mymensingh, Bangladesh

Asok K. Biswas
Department of Botany, Plant Physiology
and Biochemistry Laboratory
Centre of Advanced Study
University of Calcutta, Kolkata, India

Renald Blundell
Department of Physiology and Biochemistry, Faculty of Medicine and Surgery, University of Malta, Msida, Malta

Centre for Molecular Medicine and Biobanking, University of Malta, Msida, Malta

Iteb Boughattas
Department of plant protection, Laboratory of Agrobiodiversity and Ecotoxicology, Higher Institute of Agronomy Chott-Meriem, Sousse University, Sousse, Tunisia

Department of agronomy, Regional Field Crops Research Center of Beja, University of Jendouba, Beja, Tunisia

Sylvain Bourgerie
Laboratoire de Biologie des Ligneux et des Grandes Cultures, Université d'Orléans, INRAE USC 1328, Orléans, France

Noureddine Bousserrhine
Faculty of Sciences and Technology Laboratory of Water Environment and Urban systems, University Paris-Est Créteil, Créteil, France

Jessica Briffa
Department of Physiology and Biochemistry
Faculty of Medicine and Surgery
University of Malta, Imsida, Malta

Buu Chi Bui
Department of Biotechnology, Institute of Agriculture Science for Southern Vietnam (IAS), Ho Chi Minh City, Vietnam

Yassine Chafik
Laboratoire de Biologie des Ligneux et des Grandes Cultures, Université d'Orléans, Orléans, France

Département de Biologie
Faculté des Sciences
Université Mohammed Premier
Oujda, Morocco

Debasis Chakrabarty
Molecular Biology & Biotechnology Division, Council of Scientific and Industrial Research - National Botanical Research Institute (CSIR-NBRI), Lucknow, Uttar Pradesh, India

Sidra Charagh
State Key Laboratory of Rice Biology
China National Rice Research Institute
Chinese Academy of Agricultural Sciences (CAAS), Hangzhou, China

Nana Chen
Zhejiang Key Lab of Crop Germplasm
Department of Agronomy
College of Agriculture and Biotechnology
Zhejiang University, Hangzhou, China

Wentao Chen
Life and Science Department, National Engineering Laboratory for Applied Technology of Forestry & Ecology in South China, Laboratory of Urban Forest Ecology of Hunan Province, College of Life Science and Technology, Central South University of Forestry and Technology, Changsha, Hunan, China

Ameur Cherif
BVBGR-LR11ES31
Higher Institute of Biotechnology of Sidi Thabet (ISBST)
University of Manouba, Ariana, Tunisia

Yang Chunyan
Zhejiang Key Lab of Crop Germplasm
Department of Agronomy
College of Agriculture and Biotechnology
Zhejiang University, Hangzhou, China

Delfim JJ
Department of Crop Science,
UEL–Londrina State University,
Londrina, Paraná State, Brazil

Mathew M. Dida
Department of Applied Plant Sciences,
School of Agriculture and Food Security,
Maseno University, Kisumu, Kenya

Khady N. Drame
Centre d'Etude Régional pour
l'Amélioration de l'Adaptation à la
Sécheresse (CERAAS)/Institut Sénégalais
de Recherches Agricoles (ISRA)
Thiès, Sénégal

Sonali Dubey
School of Biosciences, IMS Ghaziabad
University Courses Campus
Ghaziabad Uttar Pradesh, India

Taimoor Hassan Farooq
Bangor College China, A Joint Unit of
Bangor University and Central South
University of Forestry and Technology
Changsha, Hunan, China

Qidong Feng
Department of Agronomy
College of Agriculture and Biotechnology
Key Laboratory of Crop Germplasm
Resource, Zhejiang University
Hangzhou, China

Ertugrul Filiz
Department of Crop and Animal
Production, Cilimli Vocational School
Duzce University, Duzce, Turkey

Yinbo Gan
Zhejiang Key Lab of Crop Germplasm
Department of Agronomy
College of Agriculture and Biotechnology
Zhejiang University, Hangzhou, China

Maria Fernanda da Costa Gomes
Department of Genetics, Biosciences
Center, Federal University of
Pernambuco (UFPE),
Recife, Pernambuco, Brazil

Ahmed Khairul Hasan
Department of Agronomy
Bangladesh Agricultural University
Mymensingh, Bangladesh

Sayyeda Hira Hassan
Department of Biosciences and Territory
University of Molise, Pesche, Italy

Laboratoire de Biologie des Ligneux et des
Grandes Cultures, Université d'Orléans,
INRAE USC 1328, LBLGC EA 1207,
Orléans, France

Sondes Helaoui
Department of plant protection, Laboratory
of Agrobiodiversity and Ecotoxicology,
Higher Institute of Agronomy Chott-
Meriem, Sousse University, Sousse, Tunisia

Tahsina Sharmin Hoque
Department of Soil Science, Bangladesh
Agricultural University, Mymensingh
Bangladesh

Mahmud Hossain
Department of Soil Science
Bangladesh Agricultural University
Mymensingh, Bangladesh

Mohammad Anwar Hossain
Department of Genetics and Plant
Breeding, Bangladesh Agricultural
University, Mymensingh, Bangladesh

Yuanyuan Hou
Life and Science Department,
National Engineering Laboratory for Applied Technology of Forestry & Ecology in South China
Laboratory of Urban Forest Ecology of Hunan Province, College of Life Science and Technology, Central South University of Forestry and Technology
Changsha, Hunan, China

Zakir Ibrahim
Department of Agronomy
College of Agriculture and Biotechnology
Key Laboratory of Crop Germplasm Resource, Zhejiang University
Hangzhou, China

Faculty of Agriculture
Lasbela University of Agriculture Water and Marine Sciences
Uthal, Pakistan

Baber Iqbal
School of Environment and Safety Engineering, Jiangsu University
Zhenjiang, China

Ummar Iqbal
Department of Botany, The Islamia University of Bahawalpur, Bahawalpur
Pakistan

Md. Rafiqul Islam
Department of Soil Science, Bangladesh Agricultural University, Mymensingh
Bangladesh

Bhakti Jadhav
Institute of Soil Science, Plant Nutrition and Environmental Protection
Wroclaw University of Environmental and Life Sciences, Wroclaw, Poland

Sopnil Ahmed Jahin
Department of Genetics and Plant Breeding, Bangladesh Agricultural University, Mymensingh, Bangladesh

Monica Jamla
Department of Biotechnology
Modern College of Arts, Science and Commerce, Savitribai Phule Pune University, Pune, India

Zhongying Ji
State Key Laboratory of Hybrid Rice
Hunan Hybrid Rice Research Center
Changsha, China

Meng Jiang
College of Agriculture and Biotechnology
Zhejiang University, Hangzhou, PR China

Hainan Institute, Zhejiang University
Sanya, PR China

Institute of Crop Sciences
National Key Laboratory of Rice Biology
Zhejiang University, Hangzhou, PR China

Xiaoli Jin
Department of Agronomy
College of Agriculture and Biotechnology
Key Laboratory of Crop Germplasm Resource, Zhejiang University
Hangzhou, China

Agnieszka Medyńska-Juraszek
Institute of Soil Science, Plant Nutrition and Environmental Protection, Wroclaw University of Environmental and Life Sciences, Wroclaw, Poland

Mukesh Kumar Kanwar
College of Agriculture and Biotechnology
Zhejiang University, Hangzhou, PR China

Department of Horticulture
Zhejiang Provincial Key Laboratory of Horticultural Plant Integrative Biology
Zhejiang University, Hangzhou, PR China

Muhammad Fazal Karim
Department of Agronomy, Pir Mehar Ali Shah Arid Agriculture University
Rawalpindi Pakistan

Ali Raza Khan
Zhejiang Key Lab of Crop Germplasm
Department of Agronomy, College of
Agriculture and Biotechnology
Zhejiang University, Hangzhou, China

Sheikh Mahfuja Khatun
Department of Genetics and Plant
Breeding, Bangladesh Agricultural
University, Mymensingh, Bangladesh

Éderson Akio Kido
Department of Genetics, Biosciences
Center, Federal University of
Pernambuco (UFPE),
Recife, Pernambuco, Brazil

Ali Kıyak
Research Center for Scientific and
Technology Applications, Burdur Mehmet
Akif Ersoy University, Burdur, Turkey

Hiroyuki Koyama
Faculty of Applied Biological Sciences
Gifu University, Gifu, Japan

Firat Kurt
Department of Plant Production and
Technologies, Faculty of Applied Sciences
Mus Alparslan University, Mus, Turkey

Manhattan Lebrun
Laboratoire de Biologie des Ligneux et des
Grandes Cultures, Université d'Orléans,
INRAE USC 1328, LBLGC EA 1207
Orléans, France

Department of Environmental
Geosciences, Faculty of Environmental
Sciences, Czech University of Life Sciences
Prague, Suchdol, Czech Republic

Yaokui Li
State Key Laboratory of Hybrid Rice
Hunan Hybrid Rice Research Center
Changsha, China

Yong Li
Life and Science Department,
National Engineering Laboratory for
Applied Technology of Forestry &
Ecology in South China, Laboratory
of Urban Forest Ecology of Hunan
Province, College of Life Science and
Technology, Central South University
of Forestry and Technology, Changsha
Hunan, China

Ziqian Li
Life and Science Department, National
Engineering Laboratory for Applied
Technology of Forestry & Ecology in
South China, Laboratory of Urban Forest
Ecology of Hunan Province, College of
Life Science and Technology, Central South
University of Forestry and Technology
Changsha, Hunan, China

Gizele de Andrade Luz
Department of Genetics, Biosciences
Center, Federal University of
Pernambuco (UFPE),
Recife, Pernambuco, Brazil

Zhengxin Ma
Department of Agronomy
College of Agriculture and Biotechnology
Key Laboratory of Crop Germplasm
Resource, Zhejiang University
Hangzhou China

Lovely Mahawar
Ranjan Plant Physiology and Biochemistry
Laboratory, Department of Botany,
University of Allahabad, Prayagraj, India

Barsha Majumder
Department of Botany, Plant Physiology
and Biochemistry Laboratory
Centre of Advanced Study
University of Calcutta, Kolkata, India

Bigang Mao
State Key Laboratory of Hybrid Rice
Hunan Hybrid Rice Research Center
Changsha, China

Mumtarin Haque Mim
Department of Genetics and Plant Breeding, Bangladesh Agricultural University, Mymensingh, Bangladesh

Domenico Morabito
Laboratoire de Biologie des Ligneux et des Grandes Cultures, Université d'Orléans INRAE USC 1328, LBLGC EA 1207 Orléans, France

Hiroshi Masuda
Department of Biological Production Faculty of Bioresource Sciences Akita Prefectural University Akita City, Japan

Marouane Mkhinini
Department of plant protection, Laboratory of Agrobiodiversity and Ecotoxicology, Higher Institute of Agronomy Chott-Meriem, Sousse University, Sousse, Tunisia

Moraes LAC
Department of Soil Science and Ecophysiology, Embrapa Soja Brazilian Agricultural Research Corporation (Embrapa), Londrina Paraná State, Brazil

Moreira A
Department of Soil Science and Ecophysiology, Embrapa Soja Brazilian Agricultural Research Corporation (Embrapa), Londrina Paraná State, Brazil

Moreti LG
Department of Crop Science São Paulo State University Botucatu São Paulo State, Brazil

Jannatul Naim
Department of Genetics and Plant Breeding, Bangladesh Agricultural University, Mymensingh, Bangladesh

Mohamed Neifar
BVBGR-LR11ES31, Department of Biotechnology, Higher Institute of Biotechnology of Sidi Thabet (ISBST), University of Manouba, Ariana, Tunisia

APVA-LR16ES20, Department of Biology National School of Engineers of Sfax (ENIS) University of Sfax, Sfax, Tunisia

José Ribamar Costa Ferreira Neto
Department of Genetics, Biosciences Center, Federal University of Pernambuco (UFPE), Recife, Pernambuco, Brazil

Lang Thi Nguyen
Department of Rice Breeding, High Agricultural Technology Research Institute (HATRI) Can Tho City, Vietnam

Sibgha Noreen
Department of Botany, Institute of Pure and Applied Biology, Bahauddin Zakariya University, Multan, Pakistan

Beatrycze Nowicka
Department of Plant Physiology and Biochemistry, Faculty of Biochemistry Biophysics and Biotechnology Jagiellonian University, Kraków, Poland

Benson O. Nyongesa
Department of Biological Sciences School of Science, University of Eldoret Eldoret, Kenya

Dorothy A. Onyango
Department of Product Development and Commercialization, African Agricultural Technology Foundation (AATF) Nairobi, Kenya

Aparna Pandey
Ranjan Plant Physiology and Biochemistry Laboratory, Department of Botany University of Allahabad, Prayagraj, India

Sakshi Pandey
Ranjan Plant Physiology and Biochemistry Laboratory, Department of Botany, University of Allahabad, Prayagraj, India

Yan Peng
State Key Laboratory of Hybrid Rice
Hunan Hybrid Rice Research Center
Changsha, China

Sheo Mohan Prasad
Ranjan Plant Physiology and Biochemistry Laboratory, Department of Botany
University of Allahabad, Prayagraj, India

Jiaxuan Qi
Zhejiang Key Lab of Crop Germplasm
Department of Agronomy
College of Agriculture and Biotechnology
Zhejiang University, Hangzhou, China

Abdolkarim Chehregani Rad
Department of Biology, Laboratory of Plant Cell Biology, Bu-Ali Sina University, Hamedan, Iran

Sharif-Ar-Raffi
Department of Genetics and Plant Breeding, Bangladesh Agricultural University, Mymensingh, Bangladesh

Ali Raza
College of Agriculture, Fujian Agriculture and Forestry University (FAFU)
Fuzhou, China

Ayan Sadhukhan
Department of Bioscience and Bioengineering, Indian Institute of Technology Jodhpur, Jodhpur India

Abdul Salam
Zhejiang Key Lab of Crop Germplasm
Department of Agronomy, College of Agriculture and Biotechnology
Zhejiang University, Hangzhou, China

Hajar Salehi
Department for Sustainable Food Process
Università Cattolica del Sacro Cuore
Piacenza, Italy

Arifin Sandhi
Department of Biology and Environmental Science, Faculty of health and life sciences
Linnaeus University, Kalmar, Sweden

Kayode A. Sanni
Department of Product Development and Commercialization, African Agricultural Technology Foundation (AATF),
Nairobi, Kenya

Gabriella Stefania Scippa
Department of Biosciences and Territory
University of Molise, Pesche, Italy

Gabriella Sferra
Department of Biosciences and Territory
University of Molise, Pesche, Italy

Imtinen Sghaier
BVBGR-LR11ES31, Department of Biotechnology, Higher Institute of Biotechnology of Sidi Thabet (ISBST), University of Manouba, Ariana, Tunisia

Department of Biology, Faculty of Sciences of Tunis, University of Tunis El Manar
Tunis, Tunisia

Imran Haider Shamsi
Department of Agronomy
College of Agriculture and Biotechnology
Key Laboratory of Crop Germplasm Resource, Zhejiang University
Hangzhou China

Ye Shao
State Key Laboratory of Hybrid Rice
Hunan Hybrid Rice Research Center
Changsha, China

Manju Shri
School of Applied Sciences and Technology
Gujarat Technological University
Ahmedabad, Gujarat, India

Yang Shuaiqi
Zhejiang Key Lab of Crop Germplasm,
Department of Agronomy
College of Agriculture and Biotechnology
Zhejiang University, Hangzhou, China

Abu Bakar Siddique
Department of Plant Physiology
Umeå Plant Science Centre (UPSC)
Umeå University, Umeå, Sweden.

Palin Sil
Department of Botany, Plant Physiology
and Biochemistry Laboratory, Centre of
Advanced Study, University of Calcutta
Kolkata, India

Valquíria da Silva
Department of Genetics, Biosciences
Center, Federal University of Pernambuco
(UFPE), Recife, Pernambuco, Brazil

Yue Song
College of Agriculture and Biotechnology
Zhejiang University, Hangzhou, PR China

Hainan Institute, Zhejiang University
Sanya, PR China

Institute of Crop Sciences, National Key
Laboratory of Rice Biology, Zhejiang
University, Hangzhou, PR China

Richa Srivastava
Department of Bioscience and
Bioengineering, Indian Institute of
Technology Guwahati, Guwahati, India

Md. Tahjib-Ul-Arif
Department of Biochemistry and
Molecular Biology, Bangladesh
Agricultural University, Mymensingh
Bangladesh

Daisuke Takagi
Faculty of Agriculture, Setsunan
University, Hirakata, Osaka, Japan

Qianlong Tan
Life and Science Department
National Engineering Laboratory for
Applied Technology of Forestry & Ecology
in South China, Laboratory of Urban
Forest Ecology of Hunan Province
Changsha, Hunan, China

Li Tang
State Key Laboratory of Hybrid Rice
Hunan Hybrid Rice Research Center
Changsha, China

Dalila Trupiano
Department of Biosciences and Territory
University of Molise, Pesche, Italy

Shihab Uddin
Department of Soil Science, Bangladesh
Agricultural University, Mymensingh
Bangladesh

Zaid Ulhassan
Zhejiang Key Lab of Crop Germplasm
Department of Agronomy, College of
Agriculture and Biotechnology
Zhejiang University, Hangzhou, China

Najeeb Ullah
Agricultural Research Station, Office of VP
for Research and Graduate Studies, Qatar
University, Doha, Qatar.

Selman Uluısık
Burdur Food Agriculture and Livestock Vocational School, Burdur Mehmet Akif Ersoy University, Burdur, Turkey

Meththika Vithanage
Ecosphere Resilience Research Center
Faculty of Applied Sciences
University of Sri Jayewardenepura
Nugegoda, Sri Lanka

Nu Xuo
Zhejiang Key Lab of Crop Germplasm
Department of Agronomy, College of Agriculture and Biotechnology
Zhejiang University, Hangzhou, China

Qichun Zhang
MOE Key Laboratory of Environment Remediation and Ecosystem Health
College of Environmental and Resource Sciences, Zhejiang University
Hangzhou, China

Xin Zhang
School of Marxism
Zhejiang University
Hangzhou, China

Bingran Zhao
State Key Laboratory of Hybrid Rice
Hunan Hybrid Rice Research Center
Changsha, China

Jie Zhou
College of Agriculture and Biotechnology
Zhejiang University, Hangzhou, PR China

Department of Horticulture
Zhejiang Provincial Key Laboratory of Horticultural Plant Integrative Biology
Zhejiang University
Hangzhou, PR China

Shandong (Linyi) Institute of Modern Agriculture, Zhejiang University, Linyi
PR China

Preface

Unrestricted anthropogenic activities, industrialization, urbanization, and unorganized waste disposal have significantly increased the environmental pollution in recent times. In addition, factors like continuous use of phosphatic fertilizers, industrial waste, leaded gasoline and paints, dust from smelters, pesticides, sewage sludge application, wastewater irrigation, and the weathering of the minerals are the major causes of heavy metals (HMs)/metalloids contamination of soil and environmental pollution worldwide. In soil, HMs either exist freely as ionic form or complexed with inorganic/organic ligands or bound with soil organic matter content. Bioavailability and chemical speciation of HMs is extensively governed by soil geochemical properties. Accumulation of HMs in soil poses a pervasive hazard to the agricultural ecosystems, which in turn affects the human population also through the transmission via food chain. Heavy metals are categorized as essential and nonessential metals considering the biological importance and their impacts on the plant growth and yield. Heavy metals/metalloids such as arsenic (As), cadmium (Cd), lead (Pb), and mercury (Hg) are nonessential for the plant metabolism and can greatly reduce the crop productivity after their concentration reaches to the supraoptimal levels. These metals are also classified as most toxic due to their toxicity, occurrence frequency in the environment, and potential human exposure. On the other hand, HMs such as iron (Fe), copper (Cu), zinc (Zn), boron (B), manganese (Mn), nickel (Ni), cobalt (Co), and molybdenum (Mo) are involved in the essential functioning and hence display positive impacts on the crop productivity when present at optimum levels. Though the optimal amount of these essential elements is required for the ideal plant growth and development, their excess availability severely impact plant growth and productivity. Climate change is likely to have a greater influence on HM/metalloid contamination based on significant impact on all factors related to HM bioavailability, fate, and toxicity including their eco-physiological properties.

Plants have developed unique strategies to respond to HM/metalloid stress, which enable them to monitor their surroundings and adjust their metabolic systems to maintain metal homeostasis. Plants respond to HM stresses by activating a cascade or network of events that starts with stress perception and ends with the expression of a battery of stress-associated genes conferring cellular, biochemical, physiological, and molecular plant processes to cope with lethal effects of HM toxicity. Recently, a substantial interest has developed in HM tolerance mechanisms, especially the ones enabling plants to thrive well in environments having high metal concentrations. Recent advancement and approaches in crop improvement in toxic HM/metalloid stress is predominantly focused on multidimensional regulatory network at molecular level. Meanwhile, current developments in various disciplines of biology, for example genomics, transcriptomics, metabolomics, phenomics, and metallomics have aided in the characterization of genomes, RNA biology, transcription factors, metabolites, and phenome gene products involved in metal tolerance in crop plants. In comparison with practicing techniques, using omics technology is greatly helpful, pragmatic, and feasible approach for improving plant systems.

Recently, a significant proportion of the crop genetic research is focused on establishing and finding the elusive blocks of knowledgeable links between the physiological significance of metal integration and relative associated toxicity of the transient flow of the metal from the roots to shoots thus affecting the plant biomass. Although significant progress has been made in HM toxicity and tolerance in plants, in-depth understanding of the molecular mechanisms-associated HM stress tolerance in plants is important to establish several lines of genetic research to advance the understanding of the metal translocation and the involvement of the metal in several physiological responses. In addition, to date most of the information obtained on tolerance mechanisms has been obtained from experiments where plants have been exposed to a single form of HMs and the mechanisms associated with the tolerance of plants' to a mixture of HMs is not fully understood.

In this book, ***Heavy Metal Toxicity and Tolerance in Plants: A Biological, Omics, and Genetic Engineering Approach,*** we represent a collection of 26 chapters contributed by the leading expert engages with HM/metalloid toxicity and tolerance in crop plants. The aim of this book is to provide a comprehensive overview of the latest understanding of the physiological, biochemical, and molecular basis of HM/metalloid tolerance and functional *omics* that will allow for a deeper understanding of the HM/metalloid tolerance for deliberate manipulation of plants to increase tolerance to HM/metalloid toxicity/deficiency, crop quality improvement, as well as phytoremediation. This would help researchers to develop strategies to enhance metal toxicity/deficiency tolerance as well as crop productivity under stressful conditions and to better utilize natural resources to ensure future food security. Finally, this book will be a valuable resource for promoting future research into plant HM/metalloid tolerance and aims to be a reference book for researchers working on developing plants tolerant to metal stress and effective strategies for reducing the risk of health hazards. We believe that the information presented in this book will make a sound contribution to this fascinating area of research.

Mohammad Anwar Hossain
AKM Zakir Hossain
Sylvain Bourgerie
Masayuki Fujita
Om Parkash Dhankher
Parvez Haris

Editor Biographies

Dr. Mohammad Anwar Hossain is serving as a Professor in the Department of Genetics and Plant Breeding, Bangladesh Agricultural University (BAU), Mymensingh, Bangladesh. He received his BSc in Agriculture and MS in Genetics and Plant Breeding from BAU, Bangladesh. He also received an MS in agriculture from Kagawa University, Japan, in 2008 and a PhD in abiotic stress physiology and molecular biology from Ehime University, Japan, in 2011 through Monbukagakusho scholarship. As a JSPS postdoctoral researcher, he has worked on isolating low phosphorus stress-tolerant genes from rice at the University of Tokyo, Japan, during the period of 2015–2017. His current research program focuses on understanding physiological, biochemical, and molecular mechanisms underlying abiotic stresses in plants and the generation of stress-tolerant and nutrient-efficient plants through breeding and biotechnology. He has over 75 peer-reviewed publications and has edited 15 books, including this one, published by CRC Press, Springer, Elsevier, Wiley, and CABI.

Dr. AKM Zakir Hossain is working as a Professor in the Department of Crop Botany, Bangladesh Agricultural University (BAU), Mymensingh, Bangladesh. He is currently appointed as a Vice-Chancellor of Kurigram Agricultural University, Kurigram, Bangladesh, by the Honorable President of the Government of the People's Republic of Bangladesh on May, 2022. Dr. Zakir received his BSc in agriculture and MS in crop botany (plant physiology) from BAU, Bangladesh. He achieved his PhD from Gifu University, Japan, in 2004 and got the degree on "Plant Cell Physiology in Cereal Crops" under Monbukagakusho

Scholarship (Japan Government), Japan. He also got postdoctoral research scholarship and worked on "Biological Nitrification Inhibition (BNI) in Cereals and its Molecular Characterization" from Japan International Research Center for Agricultural Sciences (JIRCAS) during the period of 2006–2008. His research specialization confined in plant physiology, molecular biology, and plant environmental chemistry. He is currently doing researches on plant stress physiology, plant nutritional physiology, abiotic stress tolerance in plants, characterization of secondary metabolites in medicinal plants. He published about 50 peer-reviewed research articles in national and reputed international journals with high impact factor and edited 3 books including this one published by Science, Springer, Elsevier, Wiley, Taylor and Francis, MDPI etc.

Dr. Sylvain Bourgerie is an Associate Professor in biochemistry and molecular biology at the University of Orleans (France). He received a PhD in biochemistry from the University of Limoges (France). His current research, done at the Laboratory of biology of wood and crops, in the team "Trees and Responses to Water and Environmental Constraints," since 2009, focuses on the fate of metallic trace elements (MTE) in the different abiotic and biotic compartments of contaminated soils and for associated plants. He seeks to define the mechanisms that condition MTE transfer, bioaccumulation capacities, and, finally, their toxic and ecotoxicological effects on the different biological levels of integration. He also tries to better understand the interactions between trees and contaminated soils, particularly the influence of rhizospheric processes on the mobility, availability and toxicity of MTE, in order to better understand the phytoremediative capacities of woody species to vegetate soils that are unsuitable for plant development. In details, he develops alternative approaches in aided phytostabilization using biochar as amendment. He has over 60 peer-reviewed publications.

Dr. Masayuki Fujita is a Professor in the Department of Plant Science, Faculty of Agriculture, Kagawa University, Kagawa, Japan. He received his BSc in chemistry from Shizuoka University, Shizuoka, and his MAgr and PhD in plant biochemistry from Nagoya University, Nagoya, Japan. His research interests include physiological, biochemical, and molecular biological responses based on secondary metabolism in plants under biotic (pathogenic fungal infection) and abiotic (salinity, drought, extreme temperatures, and heavy metals) stresses, phytoprotectants and biostimulants, phytoalexin, cytochrome P-450, glutathione *S*-transferase, phytochelatin, and redox reaction and antioxidants. He has over 200 peer-reviewed publications and has edited 32 books and special issues of journals.

Dr. Om Parkash Dhankher is a Professor of agriculture biotechnology in the Stockbridge School of Agriculture, University of Massachusetts, Amherst (USA). He received his MSc and MPhil in Botany from Kurukshetra University (India) and PhD in plant molecular biology from Durham University (United Kingdom). He was the recipient of the prestigious Commonwealth Scholarship by the Commonwealth Commission London. He developed the first transgenic plant-based approach for arsenic phytoremediation by combining the expression of two bacterial genes and translating this research from model plant Arabidopsis to high biomass nonfood field crops. His major research focus is phytoremediation, bioenergy production, and developing climate-resilient crops. Along with this, his laboratory is developing arsenic-free and arsenic-tolerant food crops in order to improve human health using both forward and reverse genetic approaches. Prof. Dhankher has published more than 120 referred publications and book chapters in high-impact journals including *Nature, Nature Biotechnology, PNAS, Plant Cell, Plant Biotechnology, New Phytologist, Plant Physiology, Environmental Science & Technology*, ACS Nano, etc., four edited books, Guest Edited five special issues for several journals, and six international patents were awarded to him. He is an elected fellow of Crop Science Society of America (CSSA), Agronomy Society of America (ASA), International Society of Environmental Biologist (ISEB), Indian Society of Plant Physiology and a member of the executive committee of the American Society of Plant Biologists (ASPB), elected Vice President for the International Society for Phytotechnologies (IPS, 2015–2022). Prof. Dhankher is also a serving as the senior associate editor for the *International Journal of Phytoremediation*, editor for the *Plant Cell Reports, International Journal of Plant & Environment*; *Plant Physiology Reports;* and the Associate Editor for the *Crop Science, The Plant Genome*, MDPI *Plants*, and the *Food and Energy Security* journal, etc. Prof. Dhankher has supervised over two dozen PhD and MSc students, and postdoctoral research associates. Prof. Dhanker has established widespread national and international collaborations with researchers in Australia, India, China, Italy, Egypt, UK, and USA.

Professor Parvez Haris currently holds the Chair of Biomedical Science at De Montfort University (Leicester, United Kingdom) and served as the Head of Research for the School of Allied Health Sciences for many years. He is engaged in research at the interface of chemistry, life, health, and environmental sciences. After gaining a first class BSc honors degree, Parvez was awarded a scholarship by the UK Science & Engineering Research Council in 1985 to study PhD in Biochemistry at the Royal Free Hospital School of Medicine (University of London). The scholarship involved link with industry (SmithKline Beecham) and Parvez conducted research on a medically important drug molecule and its interaction with enzymes. His project involved application of spectroscopic techniques for molecular analysis and drug discovery. His PhD supervisor was one of the leading British Scientists, Professor Dennis Chapman FRS (Fellow of the Royal Society), who was the founder of Biocompatibles International Plc (a medical design technology company). After completion of his PhD, Parvez carried out research for seven years as a post-doctoral research fellow (partly funded by the Wellcome Trust), at the Royal Free Hospital School of Medicine. At the end of his post-doctoral fellowship, he joined De Montfort University as a lecturer in 1996. His research includes the analysis of elements and molecules in living systems and the environment with particular emphasis on addressing global challenges. He is engaged in highly interdisciplinary research, which involves the use of diverse biochemical and spectroscopic methods for improving our understanding of biochemical processes, food systems, and the environment. Parvez has published hundreds of scientific articles and has co-edited five books.

1

Plant Response and Tolerance to Heavy Metal Toxicity

An Overview of Chemical Biology, Omics Studies, and Genetic Engineering

Lovely Mahawar, Sakshi Pandey, Aparna Pandey, and Sheo Mohan Prasad

Ranjan Plant Physiology and Biochemistry Laboratory, Department of Botany, University of Allahabad, Prayagraj, India

1.1 Introduction

As plants are sessile organisms, they are forced to experience several environmental stresses throughout their life cycle (Bhat et al. 2019). Among various plant stresses, heavy metals (HMs) have emerged as the serious threat in the recent decades. Heavy metals are group of inorganic elements having atomic mass >20 and specific gravity >5 (Rascio and Navari-Izzo 2011). Among the 90 naturally occurring elements in the environment, 53 of them are considered as HMs (Jamla et al. 2021a), but only few play beneficial roles in plant metabolism. Considering their biological importance on plant system, HMs are categorized as essential and nonessential metals (Jamla et al. 2021a). Metals including Ni, Cu, Mo, Zn, Fe, and Mn are required in trace quantity by plants and considered as essential micronutrients, while others such as Cd, Pb, Hg, As, and Cr are included in nonessential HMs as they are toxic at almost all concentrations and do not play any crucial role in plants (Edelstein and Ben-Hur 2018). All HMs (essential and nonessential) are imperishable in cells and cannot be naturally eliminated from the environment. Remarkably, plants can naturally extract and accumulate them irrespectively of their requirements for proper growth and development (Ozturk et al. 2017). The accumulation rate and toxicity vary from species to species. Toxic effects of HMs on plants depend not only on their amount available in the environment but also on their reactivity and oxidation capacity. Plant interactions with the HMs alter their metabolism, mainly photosynthesis (by disrupting photosynthetic apparatus), and disturb protein synthesis. Other results of HM-induced stress are hampering of stomatal conductance, as well as the disturbance of plant–water relationship and mineral uptake via roots. By interactions with the cellular compartments that form reactive oxygen species (ROS), HMs disrupt cellular redox homeostasis, which ultimately leads to cell death/oxidative burst (reviewed in Hossain et al. 2012). Besides, alteration of plant metabolism by HMs has an impact on agricultural yield (both qualitative and quantitative) and food safety (Rai et al. 2019).

Heavy Metal Toxicity and Tolerance in Plants: A Biological, Omics, and Genetic Engineering Approach, First Edition. Edited by Mohammad Anwar Hossain, AKM Zakir Hossain, Sylvain Bourgerie, Masayuki Fujita, Om Parkash Dhankher, and Parvez Haris.

To defend from the detrimental effects of HMs, plants have developed several adaptive mechanisms including compartmentalization and sequestration of HMs in the cellular structures; their neutralization by organic ligands, namely metallothioneins (MTs) and phytochelatins (PCs); and exclusion *via* transporters (Khan et al. 2021). Moreover, several signaling molecules such as phytohormones (auxins, cytokinins, ethylene, gibberellic acid, and abscisic acid) (Khan et al. 2021), gasotransmitters (nitric oxide, carbon monoxide, methane, and hydrogen sulfide), calcium (Ca^{2+}), and ROS prevent plants from HM toxicity (Nazir et al. 2022). In addition, HM toxicity promotes surplus production of ROS that oxidize biomolecules (proteins, lipids, nucleic acids, pigments, etc.) of the cells and enhance the activity of antioxidants (Mahawar et al. 2018a). Plant antioxidant defense system comprises of enzymes participating in ROS detoxification and antioxidant recycling (superoxide dismutase, catalase, ascorbate peroxidase, glutathione reductase, etc.) and other antioxidant molecules (ascorbate, thiols, glutathione, proline, etc.) that neutralize ROS (Mahawar and Shekhawat 2016). Beside these physiological and biochemical alterations, recent reports demonstrated the use of advanced omics tools (genomics, transcriptomics, proteomics, and metabolomics) to study the changes at gene, protein, and metabolite levels under HM stress (Jamla et al. 2021a; Raza et al. 2021). Recently, Cao et al. (2019) analyzed the transcriptome of rice mutant for low cadmium accumulation (*lcd1*) in response to cadmium stress. Similarly, Kintlova et al. (2017) studied the *Hordeum vulgare* L. transcriptome in response to Cu, Zn, and Cd. In other studies, the effects of HM exposure on plant species were examined by metabolic profiling (Navarro-Reig et al. 2015). In addition to transcriptomics and metabolomics, there is also research on micro-RNA (miRNA) that has evolved as an effective epigenetic regulator for the HM responses in several crop plants (Jamla et al. 2021b). Though there are several reports on the impact of HMs on plants, this chapter provides the consolidated information on HM–plant interactions encompassing the metal uptake, their translocation mechanisms, negative impacts, and potential remediation strategies to mitigate HM toxicity in plants. The study also decodes the current knowledge on HM-related changes in crops using omics tools to develop metal-tolerant crop varieties so that sustainable production without HM load in food chain can be achieved. The identification of HM tolerance genes and metabolites in plants by employing omics tools will provide solid support to develop transgenic plant varieties with enhanced tolerance to HMs for global food safety.

1.2 Plant–Metal Interaction

Heavy metals are present in various forms (ionic, complex with organic/inorganic ligands, and attached with soil organic matter) in the terrestrial environment (Jamla et al. 2021a). Mobility and bioavailability of metals mainly depend on the physiochemical properties of soil such as structure, soil topography, pH, organic matter, binding of metal with colloidal substances, cation exchange capacity, temperature, and electrical conductivity (Sharma et al. 2022). Beside soil properties, various other factors including formation of insoluble metal complexes, type of metal ions, plant species, and diverse microbial community in the terrestrial environment affect HM mobility and its absorption by plant roots (Rahman and Singh 2019). Mostly, HMs are absorbed to the rhizosphere region by passive mechanism and oozed out to the soil via roots (Jamla et al. 2021a). Uptake of HMs by plants occurs by two key pathways – apoplast and symplast. In apoplastic pathway, non-cationic metal chelates are absorbed in exchange of apoplastic cations, while in symplastic movement cationic metal ions are uptaken by roots and translocated to foliar tissues through the xylem via ion channels and specific/nonspecific metal transporters (Rabêlo et al. 2017). Besides, several divalent and trivalent metal ions bind with the cell wall, particularly pectin that serves as a significant site to concentrate large amount of metals/metalloids ion in hypodermal and epidermal root cells (Raza et al. 2021). Pectin–HM binding depends on metal ions, for example $Al^{3+} > Cu^{2+} > Pb^{2+} > Zn^{2+} = Ca^{2+}$ or $Cu^{2+} = Pb^{2+} > Cd^{2+} = Zn^{2+} > Ca^{2+}$, $Pb^{2+} \gg Cu^{2+} > Co^{2+} > Ni^{2+} \gg Zn^{2+} > Cd^{2+}$ or $Pb^{2+} = Cd^{2+}$ (Debbaudt et al. 2004; Sharma et al. 2022).

When metal is accumulated into the cells/tissues, plant utilizes its detoxification mechanism to regulate ion homeostasis by either sequestering them into the cellular compartments or oozing out from the tissues (Jamla et al. 2021a). Depending on their ability to exclude or accumulate HMs, plants can be classified into excluders (that accumulate large amount of metals in roots but limit their translocation to shoots), accumulators (containing higher level of metals in foliar tissues than in the soil), and indicators (metals accumulated in the foliar tissues reflect the metal concentration in the terrestrial environment). Among these, accumulators play a key role to remediate metal contamination from the environment due to their unique mechanism of tolerating HM toxicity without any damage (Mishra and Chandra 2022). Therefore, understanding the molecular mechanism of HM tolerance/avoidance using genetic engineering and omics approaches is valuable to develop metal resilience crop plants for sustainable agricultural improvement.

1.3 Effect of Heavy Metals on Plants

Heavy metal pollution in the terrestrial environment (through increased anthropogenic activities such as excessive use of fertilizers, pesticides, swift urbanization, industrialization, and irrigation by wastewater) results in the contamination of food and fodder grown in such environmentally degraded lands (Rai et al. 2019). As plants are the primary level of food chain, they become route for various hazardous metals to enter the food cycle endangering human and animal health (Rai et al. 2019). Several HMs are required in trace amounts for regulating numerous physiological responses in plants (Soumya et al. 2022). But surplus concentration of these metals alters morphoanatomical, biochemical, and molecular processes in plants (Jalmi et al. 2018). In the following section, we are describing various morphological, physiological, and molecular changes in plants in response to HM toxicity.

1.3.1 Morphoanatomical Responses

Uptake and accumulation of HMs in plant tissues results in several morphological changes (Table 1.1), which are easily visualized in plants growing in metal-contaminated soil. These alterations in plant morphology are evident in all tissues at all developmental

Table 1.1 Morphoanatomical, biochemical, physiological, and molecular effects of different heavy metals on plants.

S. No.	Responses	Heavy metals	Plants	Types of effect	References
1.	**Morphological**	Cu and Cd	*Solanum melongena*	Reduced seed germination and lateral root formation	Neelima and Reddy (2003)
		Cu, Pb, Cr, Zn, and Cd	*Arabidopsis thaliana, Triticum aestivum, Sesbania rostrata, Sesbania cannabina, and Lupinus luteus*	Decreased root elongation and increased lateral root formation	Yang et al. (2004), Pasternak et al. (2005), and Sofo et al. (2017)
		Pb	*Vigna unguiculata*	Swelling observed behind the root tip	Kopittke et al. (2007) and Emamverdian et al. (2015)
		Zn	*Solanum lycopersicum* L.	Decreased root and shoot length, reduced leaf area	Emamverdian et al. (2015)
		Cd	*Triticum* sp.	Inhibition of seed germination, decreased nutrient content and shoot and root length	Yourtchi and Bayat (2013)
			Allium sativum	Reduced shoot growth	Asati et al. (2016)
			Zea mays	Decreased root and shoot growth	Asati et al. (2016)
		Hg	*Oryza sativa*	Reduced tiller height, panicle formation and plant yield, enhanced Hg bioaccumulation in seedlings	Asati et al. (2016)
			Lycopersicon esculentum	Reduced plant height, flowering, and fruit weight	Asati et al. (2016)
2.	**Anatomical**	Cu	*Brassica juncea* L.	Tissue lignification, reduced biomass accumulation and trichomes formation on areal tissue of the leaves and stems	Martins et al. (2020)
		Ni	*Plantago lanceolate*	Reduced meristem mitosis	Pavlova (2017)
		Cr	*Calendula officinalis, Taraxacum officinale, Solanum lycopersicum* L.	Reduced root size, damaged root anatomy (root cap, epidermal cells, trichomes)	Bini et al. (2012)
			Oryza sativa	Disrupted mitotic cell division	Sundaramoorthy et al. (2010)
			Solanum lycopersicum	Disorganization of chloroplast	Bini et al. (2012)

3.	**Physiological**	Cr	*Saccharum officinarum*	Leaf chlorosis at low concentration, necrosis at high concentration	Radha et al. (2000)
			Salvia sclarea L.	Chlorosis, reduced carotenoid content	Corradi et al. (1993)
		Zn, Cd, and Cu	*Beta vulgaris*, *Sorghum vulgaris*	Reduced stomatal index	Kasim (2006) and Sagardoy et al. (2010)
		Cu	*Brachiaria decumbens*	Reduced formation of parenchymatous tissues and xylem vessels	Gomes et al. (2011)
		As	*Helianthus annuus* L.	Reduced chlorophyll content	Yadav et al. (2014)
		Mn	*Pisum sativum* L., *Glycine max* L.	Reduced chlorophyll content	Emamverdian et al. (2015)
		Ni and Co	*Cicer arietinum* L.	Poor nodulation, reduced yield	Emamverdian et al. (2015)
		Ni	*Zea mays* L.	Altered photosynthetic protein complexes, decreased rate of Hill reaction	Ghasemi et al. (2012)
		Cu	*Chloris gayana* Knuth, *Syzygium aromaticum* L., *Cucumis sativus*, *Eucalyptus* species	Chlorosis, necrosis, leaf discoloration	Emamverdian et al. (2015)
			Brassica napus L.	Reduced content of chlorophylls (Chl*a* and Chl*b*) and carotenoids	Peško and Kráľová (2013)
		Zn	*Vigna subterranean*	Anomalous chromosomes, sticky metaphase, and premature separation of chromosomes	Emamverdian et al. (2015)
			Cyamopsis tetragonoloba	Reduced germination and growth, reduced chlorophyll, carotenoid, sugar, and amino acid content	Asati et al. (2016)
			Pisum sativum	Reduced chlorophyll, photosynthesis, and plant growth	Asati et al. (2016)
			Lolium perenne	Reduced growth, nutrient content, and photosynthetic energy conversion	Bonnet et al. (2000)

(*Continued*)

Table 1.1 (Continued)

S. No.	Responses	Heavy metals	Plants	Types of effect	References
4.	**Biochemical**	Cu	*Hibiscus cannabinus* L.	Damaged root and shoot tissues (excessive production of malondialdehyde (MDA) and hydrogen peroxide (H_2O_2), and electrolyte leakage)	Saleem et al. (2020)
		Fe	*Vigna unguiculata* L.	Increased MDA content	Ifie et al. (2020)
		Cr	*Vigna sinensis* (L.) savi ex Hassk	Reduced amylase activity, sugar content, and seed germination	Emamverdian et al. (2015)
5.	**Molecular**	Cd	*Amaranthus cruentus*, *A. hypochondriacus* × *A. hybridus*, and *A. hypochondriacus* × *A. hybridus*	Enhanced expression of *Chit5* (one of the chitinase encoding genes) in roots compared to shoots	Lancíková et al. (2020)
			Populus trichocarpa	*PtrMTP* (metal tolerance proteins) gene induces tolerance toward metal contamination in roots, stems, and leaves (through homeostasis and detoxification mechanisms)	Gao et al. (2020)

stages, but the HM toxicity effects are more noticeable during seed germination and root growth (Adrees et al. 2015). Previous research on the effect of lead (Pb) stress on various crop plants demonstrated that exogenous application of Pb negatively affected the seed germination, morphology, physiology, and early growth of crops (Riyazuddin et al. 2021). Similarly, Neelima and Reddy (2003) studied the differential effect of cadmium (Cd) and mercury (Hg) on growth and metabolism of *Solanum melongena* L. and noticed that both metals decreased seed germination, seedling growth and metabolism, number of lateral roots, and fresh and dry weights of seedlings when applied individually to crop (Neelima and Reddy 2003). Furthermore, root structure alteration is the other symptom detected in plants growing in HM-contaminated soils. Several plants such as *Triticum aestivum* (Sofo et al. 2017), *Arabidopsis thaliana* (Pasternak et al. 2005), *Sesbania cannabina,* and *Sesbania rostrata* (Yang et al. 2004) showed inhibition in root length and improved lateral root formation when exposed to different concentrations of HMs (Cu, Pb, Cr, Zn, and Cd) (Riyazuddin et al. 2021). Kopittke et al. (2007) observed confined swelling in roots behind root tip (due to initiation of lateral roots) and reduced root growth and fresh biomass in *Vigna unguiculata* grown under Pb stress. The first sign of HM toxicity is the initiation of lateral root development. Terzano et al. (2008) indicated that deposition of zinc (Zn) in the endodermis and pericycle stimulates lateral root formation in plants. The increased formation of lateral roots disturbs plant water relations and sugar translocation from leaves to roots, which leads to reduced growth of primary and secondary roots and ultimately results in decreased root length and increased root/shoot area ratio (Rucińska-Sobkowiak 2016; Wang et al. 2020). Additionally, HM toxicity reduces the plant height in numerous plants by hindering the cellular metabolism of shoots when metal ions are translocated (from roots to shoots) and accrued in shoots. Several research reported that HMs, mainly arsenic (As) and Hg, promote cell flaccidity, prevent cell elongation and mitotic activity, thus lead to stunted plant growth (Riyazuddin et al. 2021). Study on the effect of As on *Helianthus annuus* L. seedlings showed that As disassembled the chlorophyll molecule by binding with Mg^{2+}, which prevents photosynthesis, thus reduces crop growth (Yadav et al. 2014). Furthermore, copper (Cu) stress increases lignin deposition in both underground and overground tissues, which in turn decreases biomass accumulation due to the disturbance of cell development (Martins et al. 2020). To tolerate/sequester surplus Cu, *Brassica juncea* produces large number of appendages, trichomes on the aerial tissues (Martins et al. 2020). Effect of HM toxicity has also been apparent in leaves. Heavy metals decrease the number, area, thickness, and pigmentation of leaves by damaging absorption and conduction of water, which affects key physiological processes (photosynthesis and transpiration) in plants (Kopittke et al. 2007). Numerous studies reported the negative impact of HMs on plants, for example browning of leaf tips has been observed in *Phaseolus vulgaris* under Hg stress (Sharma and Shrivastava 2014). Similarly, in *Jatropha curcas*, Pb toxicity causes growing of small brittle purple leaves (Islam et al. 2007). Moreover, HMs alter stomatal index in certain plants. The decrease in stomatal index was observed in *Beta vulgaris* exposed to Zn (Sagardoy et al. 2010) and in *Sorghum vulgaris* exposed to Cu and Cd (Kasim 2006). Conversely, increase in stomata number has been observed in *Vigna radiata* exposed to initial period of As toxicity. This effect was associated with the development of fused, seized, and abnormal stomata (Gupta and Bhatnagar 2015).

1.3.2 Physiological Responses

Various physiological responses in plants are influenced by HM toxicity. Experimental evidence of Hg stress on *J. curcas* demonstrated that increase in concentration of Hg^{2+} (from 5 to 80 $\mu g\,ml^{-1}$) reduces the number of roots, shoots, and leaves (Marrugo-Negrete et al. 2016). In *H. annuus*, nickel (Ni) inhibits seed germination by decreasing the protease and α-amylase activities (Ashraf et al. 2011). Cd toxicity induces chlorosis in *P. vulgaris* and affects leaf area in *Solanum lycopersicum* (Raza et al. 2022). Moreover, Cd and Cu stress limits pollen germination in *Pisum sativum* (Raza et al. 2022). Roots are the primary site that comes in contact with HMs in the terrestrial environment. Exposure of roots to metals results in restricted root growth, decreased root hair surface, and increased root dieback (Rucińska-Sobkowiak 2016). To avoid intrusion of organelle functioning by metals/metalloid ions, root cells promote their movement into the vacuole. Eleftheriou et al. (2015) observed the sequestration of chromium (Cr) ions in vacuoles and cell walls of *A. thaliana*. In spite of the presence of above mentioned protective mechanisms, adverse effect of Cd on structure and function of cell membrane has been reported, which generates varied negative consequences including shortage in nutrient and water supply (Raza et al. 2022).

Though certain metals (Fe, Zn, Cu, Mn, Mo, and Co) are regarded as key elements for photosynthesis because they serve as prosthetic groups for several photosynthetic enzymes, their higher concentrations hinder normal activity (Chandra and Kang 2016) (Table 1.1). For example, in *Citrus aurantium* Cu decreases content of photosynthetic pigments and hinders photosynthesis (Giannakoula et al. 2021). Similarly, Zn causes the decline in the content of photosynthetic pigments in *Brassica chinensis* L. (Yang et al. 2012). Heavy metal toxicity affects the entire photosynthetic machinery resulting in decline in chlorophyll content, mesophyll thickness, and photosynthetic rate (Schmidt et al. 2020). Study conducted by Mathur et al. (2016) on *T. aestivum* revealed that Cr decreases the electron transport rate and the number of active reaction centers of photosystem II (PSII) that hampers the activity of light-harvesting complex II in crops. Moreover, metal toxicity negatively affects mass flow, transpiration rate, and stomatal movement in plants (Wani et al. 2018). In poplar hybrids, reduced transpiration rate was observed in response to Cu, Cd, Zn, and Cr (Chandra and Kang 2016). In consistency with the above mentioned results, Zn stress resulted in lower stomatal activity in *Pistacia vera* (Tavallali 2017). Furthermore, toxic metals/metalloids harmed the function of vascular tissues in plants, for example in *Salix caprea* Zn and Cd reduced the xylem area in roots (Vaculík et al. 2012).

1.3.3 Biochemical Responses

Heavy metals generally inactivate biochemical processes either by substituting the essential metal ions or by inhibiting the functional groups of enzymes (Table 1.1). They compete with essential elements to bind in the active sites of enzymes and hinder their activity or denature them (Ghori et al. 2019). *Vigna radiata* seeds planted in Cr-contaminated soil showed a considerable reduction in protein, starch, and chlorophyll contents (Rath et al. 2019). Correspondingly, reduction in root activity, net photosynthetic rate, and chlorophyll content was observed in various ornamental plants cultivated in Pb polluted soil (Song et al. 2020). However, improvement in proline, soluble sugar content, and membrane stability index indicated that plants develop tolerance strategies toward metal stress by maintaining osmotic balance and producing metal–proline complexes (Song et al. 2020). Modification in the biochemical responses leads to several phytotoxicity symptoms. Study on *Hibiscus cannabinus* L. treated with Cu revealed the Cu-induced oxidative damage in kenaf seedlings, identified by measuring electrolyte leakage, and the levels of malondialdehyde (MDA, product of lipid peroxidation used as an indicator of membrane damage) and hydrogen peroxide (ROS) in the stressed tissues (Saleem et al. 2020). In another study, increased concentration of MDA was observed in *V. unguiculata* exposed to iron (Fe) stress (Ifie et al. 2020). To offset the damage caused by metal-induced oxidative stress, plants enhance the activity of antioxidants (Mahawar et al. 2021). Study on *Sesuvium portulacastrum*, halophytic phytoremediator of Cu, showed that surplus MDA production due to the exposure to Cu stress was neutralized by plant antioxidant defense system (Lokhande et al. 2020).

Antioxidants function by locating, deactivating, and eliminating reactive species (Hasanuzzaman et al. 2020). Report on *Medicago sativa* treated with Cu and nano Cu showed increased activity and expression of superoxide dismutase (SOD) (Cota-Ruiz et al. 2020). Correspondingly, exposure to Ni led to the enhanced peroxidase (POD) and glutathione-*S*-transferase (GST) activity in *M. sativa*, which is comparable with the results on expression of *Prx1C* mRNA, indicating the role of the enhancement of antioxidant defense in gaining of tolerance to HM-induced stress (Helaoui et al. 2020).

Additionally, HM toxicity significantly affect carbon dioxide (CO_2) assimilation, activity of Rubisco enzyme, and nitrogen metabolism in plants (Raza et al. 2022). For example, surplus concentration of Cd and Zn, hindered CO_2 assimilation in *T. aestivum* (Paunov et al. 2018), and excessive Cr reduced the activity of nitrate and ammonia absorption enzymes in *Cyamopsis tetragonoloba* (Sangwan et al. 2014).

1.3.4 Molecular Responses

Reactive oxygen species produced in response to metal stress oxidize plasma membrane and cause harm to cellular biomolecules such as proteins, lipids, and nucleic acids (Hasanuzzaman et al. 2020). Heavy metal toxicity denatures DNA, causes breakage of DNA strands, mutations, and cross-linking of DNA and proteins (Emamverdian et al. 2015). Moreover, toxic metals/metalloids disturb cellular homeostasis and hinder electron transport chains and ATP synthesis that ultimately leads to programmed cell death (Parmar et al. 2013). To protect from the damaging effects of metal ions, plants limit their root growth by ceasing the mitotic activity in response to a metal stimulus (Campos et al. 2018). Study by Pavlova (2017) suggested that Ni toxicity hinders root elongation by reducing mitotic divisions in root meristems of *Plantago lanceolata*. Similarly, in *Oryza sativa*, Cr disrupts mitosis by slowing down the cell cycle (Sundaramoorthy et al. 2010). Furthermore, upregulation of several proteins that play a role in preventing entry of metal ions into the nucleus protects plants from molecular damage (Raza et al. 2022). Improved expression of *Chit5* (gene encoding chitinase enzyme) was observed in roots of Amaranth grain varieties subjected to Cd stress (Lancíková et al. 2020). In *Populus trichocarpa*, upregulation of gene encoding metal-tolerant protein, namely *PtrMTP* (*P. trichocarpa* metal-tolerance proteins), provided tolerance toward metal toxicity in this plant via HM detoxification and cellular homeostasis

(Gao et al. 2020). Moreover, the expression of antioxidant genes was upregulated in response to metal stress, which provided tolerance toward metal toxicity in plants (Bashri and Prasad 2015). Recently, Alaraidh et al. (2018) studied the transcription profiles of the genes encoding antioxidant enzymes (CAT, POD, and APX) in fenugreek seedlings treated with Cd, Cr, and Pb and demonstrated that genes of all evaluated antioxidant enzymes showed a substantial increase in mRNA expression levels in response to HMs tested in comparison to the control seedlings.

Additionally, microRNAs have been shown to regulate metal chelation, antioxidant response, and hormonal signaling (auxin and cytokinin) in metal-treated plants (Table 1.1) (Ding et al. 2020). Experimental evidence on *Carthamus tinctorius* subjected to Cd stress showed expression of several microRNAs (miR156, miR162, miR164, miR166, miR172, miR398, and miR408) that regulate their corresponding gene transcripts in response to metal stress (Kouhi et al. 2020). MicroRNA expression related to the biosynthesis of secondary metabolites suggested their significant contribution in increased tolerance to Cd and aluminum (Al) toxicity in early growth stages of transgenic tobacco (*Nicotiana tabacum* cv. Xanthi nc) (Cedillo-Jimenez et al. 2020). Recent study conducted by Casarrubia et al. (2020) revealed the combined role of microRNA with mycorrhiza in providing tolerance toward metal ions. The study suggested that microRNA in association with mycorrhizal network substantially improves Cd tolerance in *Vaccinium myrtillus* (Casarrubia et al. 2020).

1.4 Mechanisms to Tolerate Heavy Metal Toxicity

1.4.1 Avoidance

Avoidance is the first line of defense that inhibits the entry of HMs into plants. It involves inhibition or prevention of metal uptake by roots (Figure 1.1) (Yan et al. 2020). Roots adapt various methods to prevent mobility of metals into plants, which include (i) mycorrhizal associations; (ii) secretion of root exudates (amino acids); (iii) release of metal-binding carboxyl group-containing organic acids such as oxalic acid and malic acid that bind metal ions; and (iv) alteration in pH of rhizosphere, thus forming redox barrier to decrease the metal uptake (Yan et al. 2020). Moreover, cell walls of roots play a role in HM avoidance by binding them to polygalacturonic acid present in the pectin layer of cell wall (Yan et al. 2020).

1.4.1.1 Mycorrhizal Association

Mycorrhizal association is a mutual symbiotic relation between fungi and roots of higher plants, which acts as a barrier and effectively excludes metal uptake in plants in contaminated soil (Figure 1.1). Ectomycorrhizas (ECM, fungal associations that form a sheath around roots without penetrating into the roots) and arbuscular mycorrhizas (AM, fungal hyphae penetrate into roots) are the two major forms of mycorrhizae involved in metal tolerance (Dalvi and Bhalerao 2013). Mycorrhizae enhance plants' ability to tolerate metal ions either directly, by improving phytostabilization and phytoextraction strategies (immobilization, adsorption, chelation, and absorption of metal

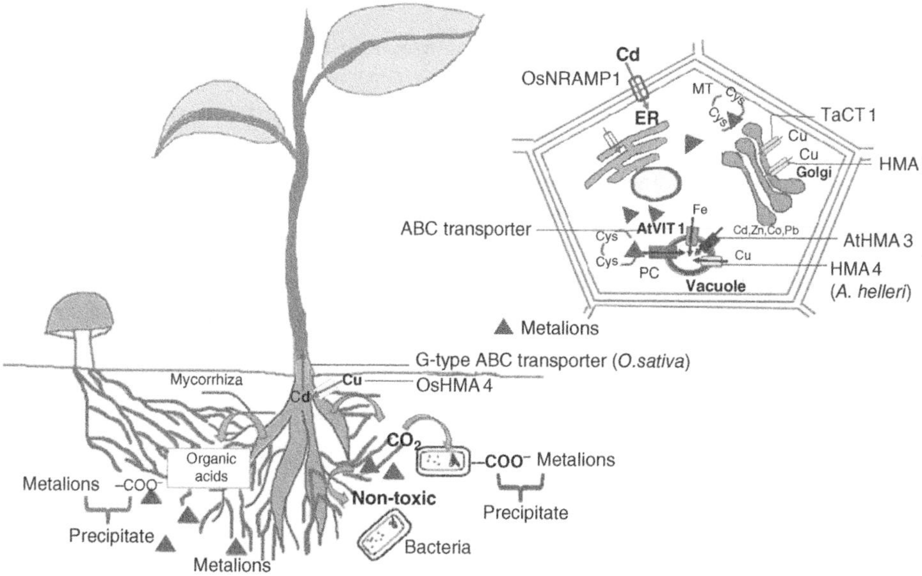

Figure 1.1 Illustrates different tolerance strategies adopted by plants to overcome metal toxicity-avoidance (through mycorrhizal association, secretion of organic acids) and sequestration (of excess metals into cell organelles through various transporters present at the membranes) (MT, metallothioneins; PC, phytochelatins).

ions by the hyphae), or indirectly by modifying the plant morphology and physiology (improved plant growth, mineral and water uptake, augmented antioxidants activities, metabolites accumulation, and reduced root biomass) (Janeeshma and Puthur 2020). Study by Chen et al. (2003) in *Trifolium pratense* associated with arbuscular mycorrhiza growing under Zn stress suggested that mycorrhiza prevents the translocation of Zn^{2+} into the aerial tissues by increase in metal retention in the roots. Regvar et al. (2006) observed the gradual substitution of vegetation with mycorrhizal plants in metal-contaminated site. Arbuscular mycorrhiza exudes organic acids that have an impact on the bioavailability of metals in the contaminated environment through acidification, chelation, and redox reactions (Teotia et al. 2016). Experimental evidence of *Glomus* (AM)-associated *Eucalyptus globulus* exposed to Pb stress showed the potential of AM in the improvement of Pb^{+} uptake and accumulation in plant shoots (Arriagada et al. 2005). Moreover, plants associated with AM can tolerate metal toxicity due to the presence of a glycoprotein, glomalin (that possess metal-binding ability), synthesized by external mycelium. Recently, Zhang et al. (2019) demonstrated the direct involvement of mycorrhiza in metal retention. Mycorrhizal fungi were shown to accumulate HMs in their mycelium, which decreases the bioavailability of these metals to plants. Likewise, Hrishikesh et al. (2010) suggested the role of extraradical hyphae in the adsorption of HMs in the contaminated soil. Arbuscular mycorrhiza also improves adsorption of HMs to the chitin of hyphal cell wall that decreases the metal ion content both in plant and rhizosphere (Janeeshma and Puthur 2020). Saravanan et al. (2013) studied the uptake of chromium ions by chitin and chitosan.

1.4.1.2 Root Exudates

As an adaptive mechanism against metal contamination in soils, roots release variety of metabolites called root exudates such as diffusates (organic acids, amino acids, inorganic ions, sugars, etc.), excretions (protons, bicarbonates, carbon dioxide, etc.) (Figure 1.1), and secretions (allelopathic compounds, siderophores, mucilage, etc.) (Dalvi and Bhalerao 2013). The exudates limit the availability of HMs to roots by forming stable metal–metabolite complexes (Figure 1.1). It is an important avoidance strategy by roots (Chen et al. 2017a). These excreted metabolites increase the pH of the rhizosphere to alkaline, which leads to the precipitation of the metal ions and make them less available to the roots (Dalvi and Bhalerao 2013). Bali et al. (2020) studied the role of different root exudates in ameliorating Cd tolerance in plants. Similarly, Salt et al. (2000) examined the role of root exudates in tolerance to Ni^{2+} in *Thlaspi* species.

1.4.2 Sequestration

Metal sequestration is one of the vital methods adapted by plants to tolerate excessive concentrations of HMs by accumulating high amounts of metal ions in the cellular compartments/shoots without any damage to plant cell (Jogawat et al. 2021). It is mediated by several biochemical and molecular mechanisms at the subcellular level. To sequester metal ions in cellular structures in nontoxic forms, plants employ different transporters to uptake, translocate, and relocate HMs. Metals are translocated and redistributed in either chelated or ionic forms. In chelated form, metals are bound by metal-binding proteins/metal chelators to form complexes that after translocation accumulate mostly into vacuoles (Figure 1.1). Metal chelators that are mainly involved in the sequestration process are phytochelatins (PCs) and metallothioneins (MTs). Phytochelatins are small cysteine-rich metal-binding peptides synthesized by phytochelatin synthase (PCS). The PCS protein is usually synthesized under HM stress in plants and marine algae (Navarrete et al. 2019). The downregulation of phytochelatin synthase gene results in reduced accumulation of HMs. The complexes of phytochelatins with HMs are transported into the vacuoles via ABC-type transporters located on the tonoplast (Figure 1.1) (Song et al. 2014). Wang et al. (2012) expressed PCS gene from *T. aestivum* (TaPCS1) in *O. sativa* exposed to Cd stress and demonstrated that transgenic *O. sativa* displayed enhanced tolerance toward Cd^{2+} by increasing Cd^{2+} accumulation in crop tissues (roots and shoots) (Wang et al. 2012).

Metallothioneins (MTs) are cysteine-rich, low-molecular-weight (6.5 kDa) metal-binding proteins universally present in all organisms. These proteins bind to both essential and nonessential metal ions and play an important role in providing tolerance against them (Figure 1.1) (Rai et al. 2021). Metallothioneins are produced in elevated concentration in response to metal toxicity and are known to detoxify HMs by scavenging the ROS formed due to oxidative stress. The expression of metallothionein genes from *P. sativum* (*PsMT1* and *PsMT2*) in *Rhizobium leguminosarum* strain efficiently improved tolerance and accumulation of Cd^{2+} in pea plants by increasing nodule formation (Table 1.2) (Tsyganov et al. 2020). Moreover, nicotianamine (NA), a non-proteogenic amino acid, is other important metal chelator that alleviates metal stress by translocating and accreting HMs in plant tissues. Nicotianamine mainly alleviates Fe stress; the Fe-NA complexes are translocated in the phloem through YSL (YSL1 and YSL3) transporters. It regulates Fe concentration in the phloem by sensing the root–shoot Fe content (Mahawar et al. 2022).

Table 1.2 Illustrates the improved metal tolerance in transgenic plants achieved by insertion or enhanced expression of foreign genes.

S. No.	Metal type	Genes from isolated organism	Transgenic plants	Function of overexpressed genes	References
1.	Cd	*VaP5CS* – *Vigna aconitifolia*	*Medicago truncatula*	Improved proline accumulation, phytochelatin synthesis, and enhanced Cd tolerance	García de la Torre et al. (2022)
		gshl – *Escherichia coli*	*Brassica juncea*	γ-ECS transgenic seedlings showed higher concentrations of phytochelatins, γ-GluCys, glutathione, and total nonprotein thiols, and increased tolerance to cadmium	Zhu et al. (1998)
		PsMT1 and *PsMT2* – *Pisum sativum*	*Rhizobium leguminosarum*	Overexpression of metallothioneins biosynthetic genes, enhanced tolerance toward Cd in pea plant by enhanced nodule formation	Tsyganov et al. (2020)
2.	As and Cd	*OsNRAMP1* – *Oryza sativa*	*Arabidopsis thaliana*	Enhanced tolerance by accumulation of As and Cd in roots and shoots	Tiwari et al. (2014)
3.	Hg	*merA* and *merB* – *Escherichia coli* plasmid	*Nicotiana tabacum, Arabidopsis thaliana, Oryza sativa,* and *Solanum tuberosum*	Tolerance to high Hg concentration in soil, reduced Hg content in leaves, seeds, and grains	Li et al. (2020b)
4.	As	*PvACR3* – *Pteris vitata*	*Arabidopsis thaliana* and *Nicotiana tabacum*	Increased resistance to As(III) and As(V), decreased As accumulation in shoots	Chen et al. (2017b)
5.	Cd, Cu, Fe, Mn, Zn, and Ni	*HvNAS1* – *Hordeum vulgare*	*Arabidopsis thaliana* and *Nicotiana tabacum*	Enhanced NAS (nicotinamine synthase) gene expression conferred tolerance by homeostasis of metals through NA (nicotianamine). However, transgenic tobacco plants showed tolerance against Ni, only.	Kim et al. (2005)
6.	Cd	*OsMTP1* – indica rice (*Oryza sativa* L. cv. IR64)	*Nicotiana tabacum*	Enhanced vacuolar sequestration of Cd and reduced Cd toxicity symptoms such as cell death and growth inhibition	Das et al. (2016)

(Continued)

Table 1.2 (Continued)

S. No.	Metal type	Genes from isolated organism	Transgenic plants	Function of overexpressed genes	References
7.	Pb, Cd, and Zn	*ZntA* – *Escherichia coli*	*Arabidopsis thaliana* (Col–1 type)	*ZntA* gene encodes P-type ATPase metal pump, which pumps Pb and Cd outside of the cell, thus its expression induced tolerance to Pb, Cd, and Zn and decreased their content in shoots	Lee et al. (2003)
8.	Cd, Cu, and Zn	*TdSHN1* – Durum wheat (*Triticum turgidum* L. subsp *durum*)	*Nicotiana tabaccum*	Transgenic tobacco exhibited improved tolerance to heavy metal stress, reduced oxidative stress, increased photosynthetic pigment content	Djemal and Khoudi (2022)

Beside metal chelators, activated transporters play a crucial role in HM detoxification by translocation of metal ions to distinct parts of the plant what enables to avoid their toxic accumulation in plant tissues, particularly in the roots. In some plants, redistribution of metal ions and its storage is a tolerance strategy toward HMs. In *O. sativa*, upregulation of *OsNRAMP1* (NATURAL RESISTANCE-ASSOCIATED MACROPHAGE PROTEIN 1), biomembrane-localized Fe transporter improved uptake and storage of Cd^{2+} in leaves of the crop plant (Jogawat et al. 2021). Moreover, Takahashi et al. (2011) observed that expression of *OsNRAMP1* was two-fold higher in Cd hyperaccumulating *O. sativa* (Habataki cultivar) than in hypoaccumulating crop variety Sasanishiki. Additionally, plants employ extrusion/metal efflux process by ATP-driven metal transporters such as ABC (ATP-binding Cassettes) and HMAs (HEAVY METAL ATPases) that efflux metal ions from roots to other plant parts/external environment (Jogawat et al. 2021). Recently, Fu et al. (2019) studied the role of G-type ABC transporter (ABCG36) in Cd translocation in *O. sativa* from roots to shoots and observed that loss of function of ABCG36 (mutant type) results in increase in Cd content in crop roots. Similarly, Eren and Argüello (2004) examined the role of AtHMA2 in efflux of surplus cytosolic Zn^{2+}, Cd^{2+}, and other metal ions to the external environment to reduce the intracellular metal concentration in *A. thaliana*. In *O. sativa*, mutation of *OsHMA2* alters the Zn^{2+} to Cd^{2+} ratio, which is an efficient way to decrease Cd concentration in plants (Satoh-Nagasawa et al. 2012). Furthermore, chemical modification of metal ions by changing their oxidation state and converting them into less toxic forms reduces the metal toxicity in plants. The example of such protection mechanism is a conversion of Cr(VI) into Cr(III) (less toxic form). Fe^{3+} is reduced into Fe^{2+} via enzyme FRO (ferric reductase oxidase) before uptake by plants (Mahawar et al. 2022).

1.5 Important Strategies for the Enhancement of Metal Tolerance

1.5.1 Omics

Different molecular components are necessary for the regulation of HM toxicity responses in plants. As a result, combined omics methods have been widely employed to understand the biological relations and molecular mechanisms of tolerance to hazardous HMs (Raza et al. 2022). Irrespective of recent developments and genomic methods for improving crops under HM stress, a strong emphasis on examination of multidimensional regulatory systems is essential to extend the molecular knowledge (Raza et al. 2022). Omics advances are practical approaches to identify candidate genes and their roles and discover vital molecular regulators. These techniques enable to study differential gene expression, distinguish molecular targets and signaling pathways, and to improve metal tolerance in crop plants (Jamla et al. 2021a). Commonly used omics techniques such as genomics (studying of the gene functions), transcriptomics (focused on coding RNA and its types), proteomics (examining protein profiles), and metabolomics (profiling of metabolites), as well as innovative technologies like ionomics (micro and macro ions profiling), miRNAomics (examining noncoding RNA), phenomics (studying of plant phenotypes), and multi-omics approaches have gained significant importance in enhancing plant resilience and crop improvement (Jamla et al. 2021a; Khan et al. 2021; Raza et al. 2022). In the following section, a detailed account on the use of several omics' approaches is provided in plant response to HM stress.

1.5.1.1 Genomics

Genomics is the study of genes and genetic interactions under various physiological conditions (Gilliham et al. 2017). Several omics approaches, namely transcriptomics, transgenomics, epigenomics, metabolomics, and phenomics are interconnected to genomics. These technologies in combination with high-throughput (HTP) methods have made remarkable improvement in plant genomics. They enabled to gain wide knowledge about several important agricultural crops (Varshney et al. 2021), which paved the way for new openings for genome editing and multi-omics technologies. Genomics can assist the identification of key genes, enzymes, and other molecules involved in metal stress response (particularly those involved in transport, homeostasis, chelation, and sequestration of HMs), using analytical technologies linked with computational tools (Figure 1.2). Analysis of metal-tolerant genes in hyperaccumulator plants by comparative genomics suggested various groups of genes, namely metal-tolerance protein (MTP), HM-associated proteins (HMPs), HM ATPase (HMA), MATE (multidrug and toxic compound extrusion), YSL (yellow stripe-like), and ZIP (ZRT-IRT-like proteins family) transporters (Khalid et al. 2019). In their genome-wide study, Li et al. (2020a) detected several HM stress-associated proteins (HMPs) in *A. thaliana* (55 proteins) and *O. sativa* (46 HMPs) that neutralize HMs by binding them with their cysteine residues. In similar study, He et al. (2020) identified 36 genes from 6 subfamilies in *Solanum tuberosum* encoding HMA that significantly contribute to uptake, translocation, and sequestration of HMs.

1.5.1.2 Transcriptomics

Genomics suggests the presence or absence of a candidate gene in an organism, but transcriptomics determines the functional role of those genes. The word "transcriptomics" signifies a

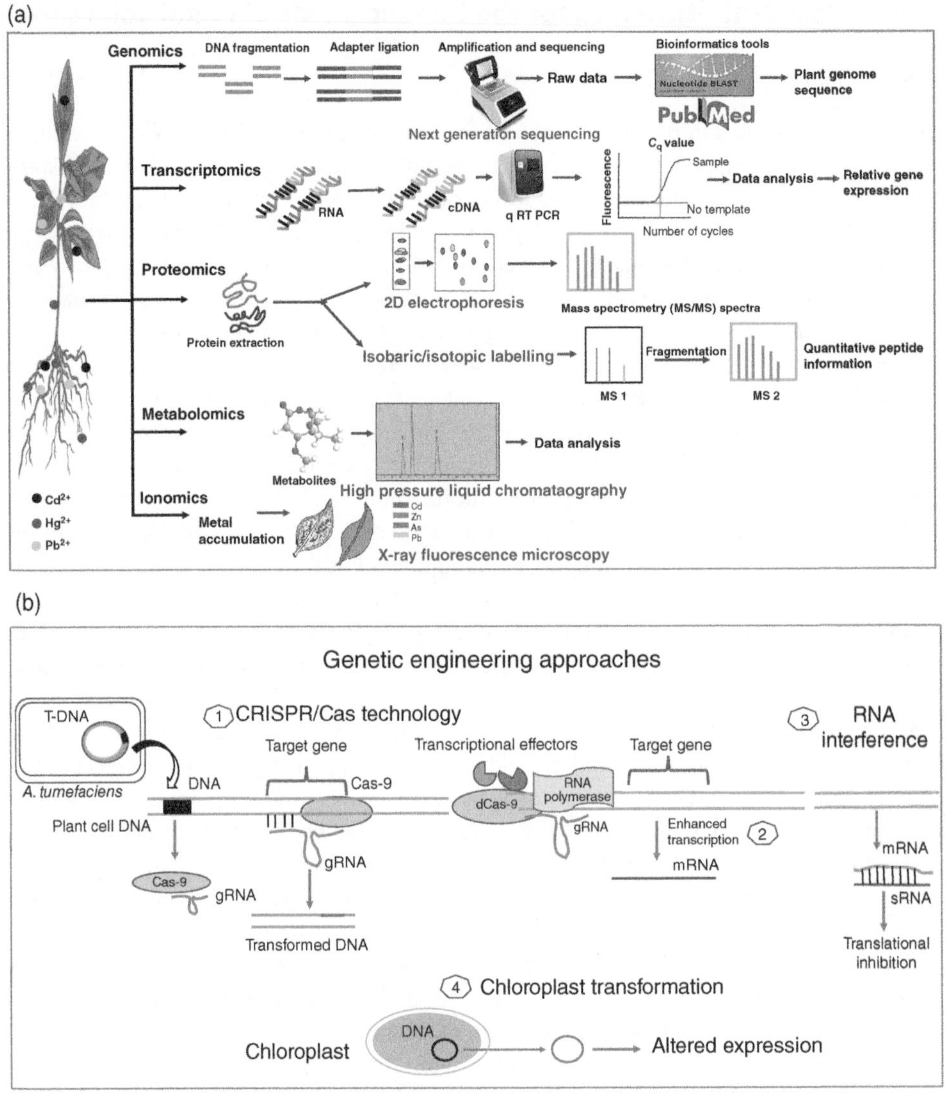

Figure 1.2 Represents the advanced technologies – various omics (a) and genetic engineering (b) approaches for the improvement of metal stress tolerance in plants.

group of methods used to examine the RNA transcripts in an organism including plants. Numerous transcriptomic approaches have been extensively utilized to quantify the abundance level of target transcripts (Figure 1.2) (Mehmood et al. 2021; Raza et al. 2021). Transcriptomic profiling enhances our understanding of plant responses under changing conditions, because RNA transcript levels alter among plant varieties, tissues, during developmental stages, and depending on the external environment (Jamla et al. 2021b). Stress-induced alterations in the RNA transcription result in the production of novel proteins, transcription factors, and metabolic compounds that control stress-mediated genes. With the progression in transcriptomic approaches such as RNA sequencing, next-generation sequencing (NGS), microarrays, differential display reverse transcription polymerase chain reaction, and serial analysis of gene expression (SAGE), it is feasible to improve plant tolerance toward toxic metals. RNA-sequencing analysis on *T. aestivum* exposed to treated with Cd stress showed transcriptional alterations in the crop roots accumulating low level of Cd^{2+} (Xiao et al. 2019). Comparative transcriptomic profiling of *B. juncea* L. under As and Cd stress uncovered the mechanisms of metal uptake and tolerance in this plant (Thakur et al. 2019). Similarly, transcriptome analysis by Chen et al. (2019) in Cd-treated *Gossypium hirsutum* revealed possible strategies for the improvement of tolerance toward Cd stress. Microarray study conducted by Jobe et al. (2021) in *A. thaliana* exposed to As toxicity showed the function of transcription factor SLIM1 in the regulation of As sensitivity in plants. Another study suggested the role of MYB4 transcription factor in the regulation of Cd tolerance in *A. thaliana* by improving the expression of metal chelator proteins, namely metallothionein 1C (MT1C) and phytochelatin synthase 1 (PCS1), and by enhancing the tolerance to metal-induced oxidative stress (Agarwal et al. 2020). Differential expression analysis in Pb-treated *Zea mays* resulted in the detection of 262/464 transcription factor (TFs) genes that respond toward metal treatment, among which bZIP TFs are the most notably enriched TFs associated with Pb tolerance (Zhang et al. 2017).

1.5.1.3 Proteomics

Proteomics deals with the study of complete set of proteins produced in organisms, their structure, subcellular location, and physiological functions (Figure 1.2) (Raza et al. 2022). At molecular level, proteomics plays an important role to study the cellular processes (Mehmood et al. 2021). In recent decades, proteomics has emerged as a vital way to understand plant adaptation and tolerance in metal-contaminated environment. For example, in *Brassica napus*, Luo and Zhang (2019) elucidated the mechanism of plant response to metal toxicity by studying the effect of Cd^{2+} on xylem sap proteome using label-free shotgun proteomic approach. The study suggested that hazardous metals (Cd) transported through xylem sap along with water and minerals from roots to overground tissues are neutralized by proteome alteration in sap. The metabolic pathways mainly affected during this process were associated with metabolism of proteins and lipids, stress/oxidoreductase, and changes in the cell wall. BnPDFL, a defensin-like protein, belonging to the category of stress/oxidoreductase was found to play a major role in Cd tolerance in crop plant (Luo and Zhang 2019). Similarly, proteomic analysis conducted by Farooq et al. (2021) using absolute quantification (iTRAQ) proteomics approach in As-treated *B. napus* cultivars (ZS758 and ZD622) revealed the multilevel coordination of defense mechanism and energy metabolism in crop. In another study, differential expression of proteins in Se and non-Se-treated *O. sativa* seedlings was analyzed using iTRAQ proteomics approach (Zeng et al. 2019). Proteomic analysis of *Eucalyptus camaldulensis* seedlings exposed to Cu stress identified expression of 26 protein targets (11 proteins upregulated and 15 down-regulated) in response to metal. The detected proteins were related to plant metabolism, photosynthesis, ROS detoxification, and RNA protein synthesis (Alotaibi et al. 2019).

1.5.1.4 Metabolomics

Metabolomics refers to the identification and quantification of all low-molecular-weight metabolites that organisms need to function during developmental stages (Arbona et al. 2013). Among several omics' advances, metabolomics is relatively more innovative, well-known approach to understand the functioning and response of plants under changing environment. To tolerate and survive under stress environment, plants modulate their growth and physiology, which leads to alteration in gene expression, affecting the type and quantity of metabolite production (Jamla et al. 2021a). Proline is one of the vital plant metabolites that plays an important role in osmotic balance, metal chelation, and activation of antioxidant enzymes. Therefore, the manipulation of its level seems a promising approach for improved crop production due to obtaining of varieties/transgenic lines with enhanced tolerance toward HM stress (Mahawar et al. 2018b). Similarly, the concentrations of α-tocopherol and putrescine increase in response to HMs, providing tolerance to metal stress. Experimental evidence in *A. thaliana* exposed to Cd stress suggested that improved α-tocopherol production is associated with the mitigation of stress in crop plants (Khalid et al. 2019). In another study, augmented concentration of α-tocopherol in Cd-treated *B. napus* prevents absorption of Cd^{2+} by roots and its further translocation to shoots (Ali et al. 2020). Recent research by Zhang et al. (2017) in *Lemna minor* subjected to Hg stress demonstrated that Hg^{2+} exposure results in accretion of tocopherol and glutathione, which reduce metal accumulation and oxidative damage in plants. Metabolites such as ascorbate and glutathione also contribute during response to metal toxicity by neutralizing ROS produced as a result of metal-induced oxidative stress. Moreover, certain secondary metabolites, namely phenols, carotenoids, tannins, and flavonoids alleviate stress effects by either acting as antioxidants or improving the activity of antioxidant enzymes. In addition to mitigate ROS levels, these metabolites protect photosynthetic machinery of plants (Khalid et al. 2019). Combination of metabolomics with other advanced technologies such as chromatography (gas chromatography, high-performance thin-layer chromatography, ultra-performance liquid chromatography, etc.) and mass spectrometry have further advanced the research in metabolite profiling (Figure 1.2) (Jamla et al. 2021a). Metabolic profiling of *Raphanus sativus* treated with Pb and Cd using GC-MS showed the alteration in key metabolic pathways (carbohydrate, glutathione and amino acid metabolism, ATP synthesis, and oxidative phosphorylation pathways) and metabolite synthesis (sugar and secondary metabolites) (Wang et al. 2015). Likewise, metabolomics coupled with comparative transcriptomic analysis in Cd-stressed *Festuca arundinacea* enabled the detection of the vital elements involved in nitric-oxide-regulated metal stress responses/adaptation in plants (Zhu et al. 2020). Another study by Xie et al. (2019) aimed at the identification of metabolites and thiol compounds in *Amaranthus hypochondriacus* treated with Cd, using LC–MS/MS, and HPLC demonstrated the increased content of phytochelatins (PCs) in leaves of stressed plants that accumulate 40 times more Cd^{2+} than the control plants.

1.5.1.5 Ionomics

Ionomics is the study of mineral nutrients and trace elements present in living organisms at a particular time and their alteration in response to an external stimulus (Stich et al. 2020). Plants need the mineral nutrients in minute concentration for their proper functioning and development. Nitrogen is a vital element for plant function and a key component of proteins, nucleic acids, and hormones that play an essential role in HM toxicity (Raza et al. 2022). Lin et al. (2011) observed that nitrogen alleviates Cd stress in *Populus yunnanensis* by producing N-containing antioxidants, chlorophylls, and metabolites. Similarly, N fertilizer decreases Cd toxicity in *Sedum* and *Solanum nigrum* (Yang et al. 2019).

To understand phenotyping, gene cloning, and screening, Punshon et al. (2013) used a multi-elemental imaging system (synchrotron X-ray fluorescence microscopy, SXRF) in *A. thaliana* (Figure 1.2). Recently, Su et al. (2017) used Fourier transform infrared spectroscopy (FTIR) to study the subcellular distribution, accumulation, and metabolic and physiological changes in Cd-treated *Spirodela polyrhiza*. Júnior et al. (2014) employed ICP-MS ionomic approach to examine the effect of Cd stress in *H. annuus* L. Furthermore, the role of Mg^{2+} in alleviation of Cd stress was determined (Kashem and Kawai 2007). Recent report on the effect of As and Si in *Z. mays* revealed that Si mitigates the toxic effect of As^{3+} in crop plant by augmenting antioxidant activity and lowering ROS level (Mišúthová et al. 2021). Further, the molecular mechanism of alleviation of As stress was studied by ionomic profiling (ICP-MS) and transcriptomic variations in *O. sativa* (Xiao et al. 2021).

1.5.1.6 miRNAomics

miRNAs are small noncoding RNAs that regulate gene expression by base pairing with the target mRNA, resulting in translation repression or gene silencing. In recent decades, function of miRNAs as important post-transcriptional regulators to confer abiotic stress tolerance in plants has been revealed. Several studies reported the involvement of miRNAs in regulating TFs needed during metal stress adaptation and tolerance in plants (Khalid et al. 2019). These investigations suggested that at high metal concentration, miRNAs alter their expression patterns, which in turn regulate the uptake and translocation of metals provoking oxidative stress and antioxidant defense mechanisms in plants (Jamla et al. 2021b). Numerous research let scientists to effectively identify the metal responsive miRNAs in plants; for example Khalid et al. (2019) detected miR156, miR171, and miR396 in *B. napus* treated with Cd stress. Similarly, miR156j from miR156 family was identified to confer As tolerance in *O. sativa* (Pandey et al. 2020). Three miRNAs, namely miR160, miR167, and miR397, that regulate auxin synthesis in plants provide Cd tolerance in *A. thaliana* (Pegler et al. 2021). Besides providing tolerance toward HM toxicity, miRNAs play a key role in gene interaction and protein folding. For instance, miR390, miR319, miR396, and miR160 were observed to interact with stress-responsive genes under Al stress in *M. sativa* and *Glycine max* (Noman and Aqeel 2017). Further, Ding et al. (2020) identified interaction of miR159, miR167, and miR268 with ABC transporters and NRAMP in *B. napus*. miR159, miR397, and miR408 were reported to interact with ion-binding proteins, scaffold proteins, laccase, ABC transporters, and TFs (MYB, TCP) to confer As stress tolerance (Noman and Aqeel 2017). Thus, it is evident from above studies that miRNAs play a vital role in the alleviation of the negative effects of metal stress.

1.5.1.7 Metallomics

Metallomics is the study of metals present in living organisms and their interactions with the biomolecules (metallome). Metals are vital components of the living systems as they serve as cofactors for structural arrangements, optimum functioning, and stability of majority of proteins (metalloproteins). Metallomics is a novel, less explored omics approach in comparison to other omics technology (Jamla et al. 2021a). During HM stress, variation in metal concentrations affects the function of metalloproteins; thus, metallomics provides a better understanding of uptake, translocation, and accumulation of metal ions in plants by analyzing different metallomes at distinct time periods and exposure levels

(López-Barea and Gómez-Ariza 2006). There are four ways for generating a metallome of a biological system – mass spectrometry (MS), X-ray fluorescence (XRF), synchrotron radiation X-ray absorption spectroscopy (XAS), and metalloproteomic and computational. Metallomics techniques have been employed in various plants to study the metal interactions in them, for example the "S"- or "O"-based Cd ligands have been identified in leaves of Cd hyperaccumulators using XAS technique. Furthermore, Kashiwabara et al. (2021) detected As in root section of *Pteris vittata* using synchrotron radiation micro XRF imaging and demonstrated that method of As uptake by plant roots changes with the direction of root growth.

1.5.2 Genetic Engineering

The role of genes in the regulation of metal uptake, translocation, and accumulation paves the way for the improvement of stress tolerance and the development of metal-tolerant plant varieties by modulating gene expression (Arora and Narula 2017). Genetic engineering has immense potential to enhance plants' ability to mobilize, avoid, and accumulate/sequester HMs from the contaminated environment by using effective gene transfer methods (Yan et al. 2020). Genetic engineering can lead to enhanced phytoremediation carried out by transgenic plants, which efficiently detoxify metal ions (Rai et al. 2021) as a result of increased expression of genes encoding metal transporters and enzymes (which reduce metals into less toxic forms) (Fasani et al. 2018). Due to the benefits of heterologous gene expression, it is possible to improve remediation potential of agriculturally important crop plants by expressing phytoremediation genes (metal uptake, translocation, and detoxification genes) in them (Table 1.2) (Yan et al. 2020). Therefore, the practical genetic engineering approaches are provided in the present section to overwhelm the negative effects of metal ions on yield and physiology of plants.

1.5.2.1 CRISPR Technology

CRISPR–Cas9 system is a recently developed gene editing technique. It can be employed to knockout genes and in genome editing. Clustered regulatory interspaced short palindromic repeats (CRISPR) consist of an assembly of spacer sequences (flanking sequences), which upon transcription generate precursor CRISPR RNA/crRNA (Rai et al. 2021). This crRNA together with transactivating crRNA (tracrRNA) form crRNA/tracrRNA complex, which works in association with Cas9 nuclease (Cas9). The guide RNA/gRNA (transcribed spacer sequences) targets Cas9 nuclease to a specific region and induces its binding to protospacer-adjacent motif (PAM). Then, nuclease creates double-strand break in gene of interest. CRISPR-mediated DNA fragmentation alters molecular make-up of organisms by modifying gene sequence (Figure 1.2) (Arora and Narula 2017; Zhang and Showalter 2020). It was first observed working as a defensive strategy against pathogenic nucleic acids in *Streptococcus pyogenes* (Le Rhun et al. 2019). Currently, CRISPR/Cas and related techniques are being efficiently employed in plants to engineer crop genes against several environmental stresses (Rai et al. 2021). Furthermore, the application of caspases with altered activity, i.e. dead caspase enables to regulate the gene expression, holds promising results in metal tolerance of plants. CRISPRa (CRISPR Activation) does not delete a dsDNA, but rather regulates transcription. Dead Cas9 increases transcription by binding to domains containing sites for binding of transcription factors or proteins (Figure 1.2). It can be therefore employed to alter the expression of the genes involved in metal uptake and sequestration. The application of this genome-editing technology in plant metal tolerance is a promising endeavor in improving the remediation efficiency of plants. Moreover, due to the accessibility of complete genome sequences of phytoremediators, it is feasible to identify and characterize metal-tolerant genes and genes whose products are involved in metal transport and binding (metal ligands) in plants (Basu et al. 2018; Kumar and Trivedi 2018). Overexpression of metallothioneins (MT1 and MT2) and other metal responsive genes were observed in several plants subjected to HM toxicity (Basu et al. 2018). Additionally, CRISPR/Cas systems can be implemented to stimulate numerous processes such as phytoaccumulation (uptake of metals by roots), phytodegradation (uptake and metabolization of metals into less toxic forms by roots), phytostabilization (reduction of the mobility of HMs in soil), and rhizofiltration (filtration of polluted water through hydroponic cultivation of plants) (Rai et al. 2019). In comparison to other genome-editing tools such as zinc finger nucleases (ZFNs) and transcription activator-like effector nucleases (TALENs), CRISPR/Cas can ease the production of transgenic plants having desired set of genes (Lino et al. 2018). These days, CRISPR technology is employed to develop transgenic *O. sativa* lines by knocking down the expression of metal transporter genes. Inactivation of K^+ transporter, OsHAK1 in Cs-stressed *O. sativa* results in the production of low Cs crop that strongly tolerate Cs stress by reducing the uptake of Cs^+ (Nieves-Cordones et al. 2017). Similarly, mutation in *OsNRAMP5* gene produces low Cd-accumulating *O. sativa* indica (Tang et al. 2017) and *O. sativa* japonica (Yang et al. 2019). Thus, CRISPR-Cas enables to develop transgenic crops by altering the expression of metal uptake genes providing metal tolerance without even insertion of foreign genes. Despite several advancements in CRISPR/Cas technology, there are still certain limitations, and the major drawback in CRISPR/Cas-mediated transformation in plants is that only few transformed plant colonies show gene of interest (Rai et al. 2021). Basu et al. (2018) reported that the rate of transformed colonies with desired mutation formed via CRISPR/Cas system is only 20–50%. Overexpression of Cas9 gene in several plant species produces toxic effects, which might limit transformation efficiency in them (Shan et al. 2013). However, CRISPR/Cas system-mediated transformation in plants has shown high differences in transgenic colonies for the targeted alteration; further work such as bypassing Cas9 to reduce off-target transformation is needed for efficient transformation in plants (Zhang et al. 2020). Circumventing Cas9 with Cpf1 system (CRISPR from *Prevotella* and *Francisella* 1) improves transformation efficiency by correctly regulating the transcription of guide mRNA so that it binds to specific complementary base sequence and induces cleavage at target site (Chen et al. 2019). Recently, in several plant species CRISPR/Cpf1 system has been effectively employed to generate point mutations and introducing knockins and knockouts without any damage to plant cells (Xu et al. 2017).

1.5.2.2 Plastid Transformation

Chloroplast is a type of plastid commonly present in all photosynthetic organisms and known to play a major role in carbohydrate metabolism. Besides photosynthesis, chloroplast is also involved in genetic transformation due to its small genome size (100–200 kb genes) (Kumar and Ling 2021). Transformation of chloroplast genome is an alternative method of genetic engineering that has been extensively used in the development of transgenic plants (Figure 1.2). It is an eco-friendly approach that has several advantages over nuclear transformation such as improved transgene expression, multigene expression in a single transformation event, regulation of transgene via maternal inheritance, lack of position effects, gene silencing, and undesirable foreign DNA (Grevich and Daniell 2005). Due to the ability of chloroplast to express multiple desired genes, chloroplast transformation has become a promising tool to develop transgenic plants that show enhanced tolerance toward various environmental stresses (Rascón-Cruz et al. 2021). Chloroplast transformation is an ideal method for Hg phytoremediation in plants as chloroplast is the primary site of damage due to Hg^{2+}. Expression of bacterial genes, namely *merA* and *merB*, in chloroplast genome of *N. tabacum* conferred tolerance toward high concentration of phenylmercuric acetate (Hussein et al. 2007). Study conducted by Jin and Daniell (2014) suggested that overexpression of γ-tocopherol methyltransferase (γ-TMT) gene in *N. tabacum* plastome increased α-tocopherol content, thereby improving Cu tolerance in transgenic plants by augmenting the concentration of antioxidants. Similarly, Kim et al. (2007) successfully targeted *Brassica rapa* type-1 metallothionein (BrMT1) gene in the chloroplast of *A. thaliana*. The transplastomic plant showed improved Cd detoxification and decrease in chlorosis and H_2O_2 content (Kim et al. 2007).

1.5.2.3 Gene Silencing

Gene silencing is a process of inhibition or reduction of the protein synthesis from its corresponding gene. Each gene synthesizes protein in two steps – transcription (formation of mRNA from DNA) and translation (synthesis of protein from mRNA). In gene silencing, the gene expression is modified before translation by using small RNAs (sRNA) that participate in specific degradation of target gene transcripts, thus preventing the subsequent protein formation (Figure 1.2) (Rai et al. 2021). In plants there are two classes of small RNAs (~20–30) – short interfering RNAs (siRNAs) and microRNAs (miRNAs) that inactivate homologous sequences (Moazed 2009). Besides the regulation of gene transcription, sRNAs have been shown to improve abiotic stress tolerance in plants (Dai and Zhao 2011). Regulation of abiotic stress response by sRNAs as a genetic tool was initially observed in *A. thaliana* (Borges and Martienssen 2015). Several reports demonstrated the use of gene silencing technology in the improvement of metal tolerance in plants (Ghosh et al. 2017). Sunkar et al. (2012) reported the role of various miRNAs, namely miR397, miR398, miR408, miR857, in providing tolerance toward Cu stress in plants by improving the antioxidant gene expression. Similarly, upregulation of miR95 caused an increase in tolerance against As and Cu stress in *A. thaliana* (Liang et al. 2012). Recently, with the assistance of several innovative technologies (deep RNA sequencing, MALDI-TOF analysis, in-silico-based approaches, next-generation sequencing), it is feasible to detect metal tolerance inducing sRNAs in those plants whose complete genome is unavailable (Kruszka et al. 2012).

1.6 Conclusion and Future Prospects

Ever-increasing anthropogenic activities, swift urbanization, industrialization, and excessive use of synthetic fertilizers have resulted in the contamination of the environment with toxic HMs. These HMs are uptaken by the plants along with mass flow and hamper plant growth and development by adversely affecting their physiological and molecular processes. Toxic concentration of HMs not only hinders plant growth but also impedes human and animal health as plants are the primary level of food chain and they readily uptake metal ions and translocate them to edible parts, where HMs can accumulate in high concentrations. Plants, like other organisms, have developed several mechanisms such as induction of stress responsive genes, proteins and metabolites, sequestration of surplus metal ions using metal chelators (MTs, PCs), and stimulation of antioxidants activity to tolerate/adapt to toxic concentrations of these HMs. Several researchers studied these mechanisms at laboratory level, but very few reports demonstrated HM detoxification methods in the field due to poor understanding of metal uptake from rhizosphere to roots and translocation and accumulation processes, as well as lack of knowledge on HM responsive genes, transcription factors, gene products, their expression profiles, and miRNAs. By using novel omics and genetic engineering approaches, it is feasible to identify specific physiological and molecular changes in plant cells/tissues in response to HM stress. The chapter furnishes current studies on noxious metals, their adverse effects on plant growth and responses (morphoanatomical, physiological, and molecular responses), tolerance strategies adapted by plants (avoidance, sequestration) to overcome metal toxic effects, and the use of genetic engineering (CRISPR/Cas system) and omics approaches in enhancing plant avoidance/tolerance toward HM stress. Overexpression of genes encoding phytochelatin synthase, metallothioneins, and vacuolar transporters through genetic engineering can be effectively employed these days to develop transgenic plants with improved tolerance toward metal ions and other environmental stress factors (Table 1.2). The omics approaches such as genomics, transcriptomics, proteomics, metabolomics, miRNAomics, metallomics, and ionomics have helped researchers to get better understanding of novel transporters and molecular pathways involved in metal stress tolerance at DNA, RNA, proteins, metabolites, or miRNA levels. Besides the above discussed omics technologies, innovative omics tools such as next-generation sequencing platform-driven genomics and transcriptomics and high-throughput multi-omics approaches are evolving as promising technologies for decoding the gene expression mechanisms (in plant tissues) of the plant–metal interaction. Amalgamation of described knowledge about combined approaches of genetic engineering and omics tools would be very effective to develop genetically engineered plants for the removal of toxic metals from the contaminated site or to enhance HM stress regulation in different crop plants for sustainable agriculture. Nevertheless, complete investigations of metal metabolism, alteration of retrograde signaling pathways, and effects of co-uptake of metals are needed before explorations of omics approaches at the field level to develop hyperaccumulator or HM resilient plants.

References

Adrees, M., Ali, S., Rizwan, M. et al. (2015). The effect of excess copper on growth and physiology of important food crops: a review. *Environ. Sci. Pollut. Res.* 22: 8148–8162.

Agarwal, P., Mitra, M., Banerjee, S. et al. (2020). MYB4 transcription factor, a member of R2R3-subfamily of MYB domain protein, regulates cadmium tolerance via enhanced protection against oxidative damage and increases expression of PCS1 and MT1C in Arabidopsis. *Plant Sci.* 297: 110501.

Alaraidh, I.A., Alsahli, A.A., and Razik, E.A. (2018). Alteration of antioxidant gene expression in response to heavy metal stress in *Trigonella foenum-graecum* L. *S. Afr. J. Bot.* 115: 90–93.

Ali, E., Hassan, Z., Irfan, M. et al. (2020). Indigenous tocopherol improves tolerance of oilseed rape to cadmium stress. *Front. Plant Sci.* 11: 547133.

Alotaibi, M.O., Mohammed, A.E., Almutairi, T.A. et al. (2019). Morpho-physiological and proteomic analyses of *Eucalyptus camaldulensis* as a bioremediator in copper-polluted soil in Saudi Arabia. *Plants* 8: 43.

Arbona, V., Manzi, M., Ollas, C.D. et al. (2013). Metabolites as a tool to investigate abiotic stress tolerance in plants. *Int. J. Mol. Sci.* 14: 4885–4911.

Arora, L. and Narula, A. (2017). Gene editing and crop improvement using CRISPR-Cas9 system. *Front. Plant Sci.* 8: 1932.

Arriagada, C.A., Herrera, M.A., and Ocampo, J.A. (2005). Contribution of arbuscular mycorrhizal and saprobe fungi to the tolerance of *Eucalyptus globulus* to Pb. *Water Air Soil Pollut.* 166: 31–47.

Asati, A., Pichhode, M., and Nikhil, K. (2016). Effect of heavy metals on plants: an overview. *Int. J. Appl. Innov. Eng. Manag.* 5: 56–66.

Ashraf, M.Y., Sadiq, R., and Hussain, M. (2011). Toxic effect of nickel (Ni) on growth and metabolism in germinating seeds of sunflower (*Helianthus annuus* L.). *Biol. Trace Elem. Res.* 143: 1695–1703.

Bali, A.S., Sidhu, G.P.S., and Kumar, V. (2020). Root exudates ameliorate cadmium tolerance in plants: a review. *Environ. Chem. Lett.* 18 (4): 1243–1275.

Bashri, G. and Prasad, S.M. (2015). Indole acetic acid modulates changes in growth, chlorophyll a fluorescence and antioxidant potential of *Trigonella foenum-graecum* L. grown under cadmium stress. *Acta Physiol. Plant.* 37: 1745.

Basu, S., Rabara, R.C., Negi, S. et al. (2018). Engineering PGPMOs through gene editing and systems biology: a solution for phytoremediation? *Trends Biotechnol.* 36 (5): 499–510.

Bhat, J.A., Shivaraj, S.M., Singh, P. et al. (2019). Role of silicon in mitigation of heavy metal stresses in crop plants. *Plants (Basel)* 8: 71.

Bini, C., Wahsha, M., Fontana, S. et al. (2012). Effects of heavy metals on morphological characteristics of *Taraxacum officinale* web growing on mine soils in NE Italy. *J. Geochem. Explor.* 123: 101–108.

Bonnet, M., Camares, O., and Veisseire, P. (2000). Effects of zinc and influence of *Acremonium lolii* on growth parameters, chlorophyll a fluorescence and antioxidant enzyme activities of ryegrass (*Lolium perenne* L. cv Apollo). *J. Exp. Bot.* 51: 945–953.

Borges, F. and Martienssen, R.A. (2015). The expanding world of small RNAs in plants. *Nat. Rev. Mol. Cell Biol.* 16: 727–741.

Campos, N.V., Arcanjo-Silva, S., Freitas-Silva, L. et al. (2018). Arsenic hyperaccumulation in *Pityrogramma calomelanos* L. (Link): adaptive traits to deal with high metalloid concentrations. *Environ. Sci. Pollut. Res.* 25: 10720–10729.

Cao, Z.Z., Lin, X.Y., Yang, Y.J. et al. (2019). Gene identification and transcriptome analysis of low cadmium accumulation rice mutant (lcd1) in response to cadmium stress using MutMap and RNA-seq. *BMC Plant Biol.* 19: 250.

Casarrubia, S., Martino, E., Daghino, S. et al. (2020). Modulation of plant and fungal gene expression upon Cd exposure and symbiosis in ericoid mycorrhizal *Vaccinium myrtillus*. *Front. Microbiol.* 11: 341.

Cedillo-Jimenez, C.A., García-Ortega, L.F., Martínez-Reséndiz, M. et al. (2020). Response of transgenic tobacco overexpressing the CchGLP gene to cadmium and aluminium: phenotypic and microRNAs expression changes. *Physiol. Mol. Biol. Plant.* 26: 3–13.

Chandra, R. and Kang, H. (2016). Mixed heavy metal stress on photosynthesis, transpiration rate, and chlorophyll content in poplar hybrids. *Forest Sci. Technol.* 12: 55–61.

Chen, B.D., Li, X.L., Tao, H.Q. et al. (2003). The role of arbuscular mycorrhiza in zinc uptake by red clover growing in a calcareous soil spiked with various quantities of zinc. *Chemosphere* 50: 839–846.

Chen, Y., Zhi, J., Zhang, H. et al. (2017a). Transcriptome analysis of *Phytolacca americana* L. in response to cadmium stress. *PLoS One* 12: e0184681.

Chen, Y., Hua, C., Jia, M. et al. (2017b). Heterologous expression of *Pteris vittata* arsenite antiporter PvACR3;1 reduces arsenic accumulation in plant shoots. *Environ. Sci. Technol.* 51: 10387–10395.

Chen, H., Li, Y., Ma, X. et al. (2019). Analysis of potential strategies for cadmium stress tolerance revealed by transcriptome analysis of upland cotton. *Sci. Rep.* 9: 86.

Corradi, M., Bianchi, A., and Albasini, A. (1993). Chromium toxicity in *Salvia sclarea*. Effects of hexavalent chromium on seed germination and seedling development. *Environ. Exp. Bot.* 33: 405–413.

Cota-Ruiz, K., Ye, Y., Valdes, C. et al. (2020). Copper nanowires as nanofertilizers for alfalfa plants: understanding nano-bio systems interactions from microbial genomics, plant molecular responses and spectroscopic studies. *Sci. Total Environ.* 742: 140572.

Dai, X. and Zhao, P.X. (2011). psRNATarget: a plant small RNA target analysis server. *Nucleic Acids Res.* 39: W155–W159.

Dalvi, A.A. and Bhalerao, S.A. (2013). Response of plants towards heavy metal toxicity: an overview of avoidance, tolerance and uptake mechanism. *Ann. Plant Sci.* 2: 362–368.

Das, N., Bhattacharya, S., and Maiti, M.K. (2016). Enhanced cadmium accumulation and tolerance in transgenic tobacco over expressing rice metal tolerance protein gene OsMTP1 is promising for phytoremediation. *Plant Physiol. Biochem.* 105: 297–309.

Debbaudt, A.L., Ferreira, M.L., and Gschaider, M.E. (2004). Theoretical and experimental study of M^{2+} adsorption on biopolymers. III. Comparative kinetic pattern of Pb, Hg and Cd. *Carbohy. Polym.* 56 (3): 321–332.

Ding, Y., Ding, L., Xia, Y. et al. (2020). Emerging roles of microRNAs in plant heavy metal tolerance and homeostasis. *J. Agric. Food Chem.* 68: 1958–1965.

Djemal, R. and Khoudi, H. (2022). The ethylene-responsive transcription factor of durum wheat, TdSHN1, confers cadmium, copper, and zinc tolerance to yeast and transgenic tobacco plants. *Protoplasma* 259: 19–31.

Edelstein, M. and Ben-Hur, M. (2018). Heavy metals and metalloids: sources, risks and strategies to reduce their accumulation in horticultural crops. *Sci. Hortic.* 234: 431–444.

Eleftheriou, E.P., Adamakis, I.D.S., Panteris, E. et al. (2015). Chromium-induced ultrastructural changes and oxidative stress in roots of *Arabidopsis thaliana*. *Int. J. Mol. Sci.* 16: 15852–15871.

Emamverdian, A., Ding, Y., Mokhberdoran, F., and Xie, Y. (2015). Heavy metal stress and some mechanisms of plant defense response. *Sci. World J.* 2015: 756120.

Eren, E. and Argüello, J.M. (2004). Arabidopsis HMA2, a divalent heavy metal-transporting P1B-type ATPase, is involved in cytoplasmic Zn^{2+} homeostasis. *Plant Physiol.* 136: 1–12.

Farooq, M.A., Hong, Z., Islam, F. et al. (2021). Comprehensive proteomic analysis of arsenic induced toxicity reveals the mechanism of multilevel coordination of efficient defense and energy metabolism in two *Brassica napus* cultivars. *Ecotoxicol. Environ. Saf.* 208: 111744.

Fasani, E., Manara, A., Martini, F. et al. (2018). The potential of genetic engineering of plants for the remediation of soils contaminated with heavy metals. *Plant Cell Environ.* 41: 1201–1232.

Fu, S., Lu, Y., Zhang, X. et al. (2019). The ABC transporter ABCG36 is required for cadmium tolerance in rice. *J. Exp. Bot.* 70 (20): 5909–5918.

Gao, Y., Yang, F., Liu, J. et al. (2020). Genome wide identification of metal tolerance protein genes in *Populus trichocarpa* and their roles in response to various heavy metal stresses. *Int. J. Mol. Sci.* 21: 1680.

García de la Torre, V.S., Coba de la Peña, T., Lucas, M.M. et al. (2022). Transgenic *Medicago truncatula* plants that accumulate proline display enhanced tolerance to cadmium stress. *Front. Plant Sci.* 13: 829069.

Ghasemi, F., Heidari, R., Jameii, R. et al. (2012). Effects of Ni^{2+} toxicity on Hill reaction and membrane functionality in maize. *J. Stress Physiol. Biochem.* 8: 55–61.

Ghori, N.H., Ghori, T., Hayat, M. et al. (2019). Heavy metal stress and responses in plants. *Int. J. Environ. Sci. Technol.* 16: 1807–1828.

Ghosh, S., Singh, K., Shaw, A.K. et al. (2017). Insights into the miRNA-mediated response of maize leaf to arsenate stress. *Environ. Exp. Bot.* 137: 96–109.

Giannakoula, A., Therios, I., and Chatzissavvidis, C. (2021). Effect of lead and copper on photosynthetic apparatus in citrus (*Citrus aurantium* L.) plants. The role of antioxidants in oxidative damage as a response to heavy metal stress. *Plants* 10: 155.

Gilliham, M., Able, J.A., and Roy, S.J. (2017). Translating knowledge about abiotic stress tolerance to breeding programmes. *Plant J.* 90: 898–917.

Gomes, M.P., Marque, T.C.L.L.D.S., Nogueira, M.D.O.G. et al. (2011). Ecophysiological and anatomical changes due to uptake and accumulation of heavy metal in *Brachiaria decumbens*. *Sci. Agricola.* 68: 566–573.

Grevich, J.J. and Daniell, H. (2005). Chloroplast genetic engineering: recent advances and future perspectives. *Crit. Rev. Plant Sci.* 24: 83–107.

Gupta, P. and Bhatnagar, A.K. (2015). Spatial distribution of arsenic in different leaf tissues and its effect on structure and development of stomata and trichomes in mung bean, *Vigna radiata* (L.) Wilczek. *Environ. Exp. Bot.* 109: 12–22.

Hasanuzzaman, M., Bhuyan, M., Zulfiqar, F. et al. (2020). Reactive oxygen species and antioxidant defense in plants under abiotic stress: revisiting the crucial role of a universal defense regulator. *Antioxidants* 9: 681.

He, G., Qin, L., Tian, W. et al. (2020). Heavy metal transporters-associated proteins in *Solanum tuberosum*: genome-wide identification, comprehensive gene feature, evolution and expression analysis. *Genes* 11: 1269.

Helaoui, S., Boughattas, I., Hattab, S. et al. (2020). Physiological, biochemical and transcriptomic responses of *Medicago sativa* to nickel exposure. *Chemosphere* 249: 126121.

Hossain, M.A., Piyatida, P., da Silva, J.A.T., and Fujita, M. (2012). Molecular mechanism of heavy metal toxicity and tolerance in plants: central role of glutathione in detoxification of reactive oxygen species and methylglyoxal and in heavy metal chelation. *J. Bot.* 2012: 1–37.

Hrishikesh, U., Panda, S.K., Bhattacharjee, M.K. et al. (2010). Role of arbuscular mycorrhiza in heavy metal tolerance in plants: prospects for phytoremediation. *J. Phytol.* 2: 16–27.

Hussein, H.S., Ruiz, O.N., Terry, N. et al. (2007). Phytoremediation of mercury and organomercurials in chloroplast transgenic plants: enhanced root uptake, translocation to shoots, and volatilization. *Environ. Sci. Technol.* 41: 8439–8446.

Ifie, J., Ifie-Etumah, S., and Ikhajiagbe, B. (2020). Physiological and biochemical responses of selected cowpea (*Vigna unguiculata* (L.) Walp.) accessions to iron toxicity. *Acta Agric. Slovenica* 115: 25–38.

Islam, E., Yang, X., Li, T. et al. (2007). Effect of Pb toxicity on root morphology, physiology and ultrastructure in the two ecotypes of *Elsholtzia argyi*. *J. Hazard. Mater.* 147: 806–816.

Jalmi, S.K., Bhagat, P.K., Verma, D. et al. (2018). Traversing the links between heavy metal stress and plant signaling. *Front. Plant Sci.* 9: 12.

Jamla, M., Khare, T., Joshi, S. et al. (2021a). Omics approaches for understanding heavy metal responses and tolerance in plants. *Curr. Plant Biol.* 27: 100213.

Jamla, M., Patil, S., Joshi, S. et al. (2021b). MicroRNAs and their exploration for developing heavy metal-tolerant plants. *J. Plant Growth Regul.* 1–17.

Janeeshma, E. and Puthur, J.T. (2020). Direct and indirect influence of arbuscular mycorrhizae on enhancing metal tolerance of plants. *Arch. Microbiol.* 202: 1–16.

Jin, S. and Daniell, H. (2014). Expression of γ-tocopherol methyltransferase in chloroplasts results in massive proliferation of the inner envelope membrane and decreases susceptibility to salt and metal-induced oxidative stresses by reducing reactive oxygen species. *Plant Biotechnol. J.* 12: 1274–1285.

Jobe, T.O., Yu, Q., Hauser, F. et al. (2021). The SLIM1 transcription factor regulates arsenic sensitivity in *Arabidopsis thaliana*. *bioRxiv* 426316. https://doi.org/10.1101/2021.01.12.426316.

Jogawat, A., Yadav, B., and Narayan, O.P. (2021). Metal transporters in organelles and their roles in heavy metal transportation and sequestration mechanisms in plants. *Physiol. Plant.* 173: 259–275.

Júnior, C.A.L., Mazzafera, P., and Arruda, M.A.Z. (2014). A comparative ionomic approach focusing on cadmium effects in sunflowers (*Helianthus annuus* L.). *Environ. Exp. Bot.* 107: 180–186.

Kashem, M.A. and Kawai, S. (2007). Alleviation of cadmium phytotoxicity by magnesium in Japanese mustard spinach. *Soil Sci. Plant Nutr.* 53: 246–251.

Kashiwabara, T., Kitajima, N., Onuma, R. et al. (2021). Synchrotron micro-X-ray fluorescence imaging of arsenic in frozen-hydrated sections of a root of *Pteris vittate*. *Metallomics* 13: mfab009.

Kasim, W.A. (2006). Changes induced by copper and cadmium stress in the anatomy and grain yield of *Sorghum bicolor* (L.) Moench. *Int. J. Agric. Biol.* 8: 123–128.

Khalid, N., Aqeel, M., and Noman, A. (2019). System biology of metal tolerance in plants: an integrated view of genomics, transcriptomics, metabolomics, and phenomics. In: *Plant Metallomics and Functional Omics* (ed. G. Sablok), 107–144. Cham: Springer International Publishing.

Khan, M.I.R., Chopra, P., Chhillar, H. et al. (2021). Regulatory hubs and strategies for improving heavy metal tolerance in plants: chemical messengers, omics and genetic engineering. *Plant Physiol. Biochem.* 164: 260–278.

Kim, S., Takahashi, M., Higuchi, K. et al. (2005). Increased nicotianamine biosynthesis confers enhanced tolerance of high levels of metals, in particular nickel, to plants. *Plant. Cell. Physiol.* 46: 1809–1818.

Kim, S.H., Lee, H.S., Song, W.Y. et al. (2007). Chloropiasto targeted BrMT1 (*Brassica rapa* Type-1 Metallothionein) enhances resistance to cadmium and ROS in transgenic *Arabidopsis* plants. *J. Plant Biol.* 50 (1): 1–7.

Kintlova, M., Blavet, N., Cegan, R. et al. (2017). Transcriptome of barley under three different heavy metal stress reaction. *Genom. Data.* 13: 15–17.

Kopittke, P.M., Asher, C.J., Kopittke, R.A. et al. (2007). Toxic effects of Pb^{2+} on growth of cowpea (*Vigna unguiculata*). *Environ. Pollut.* 150: 280–287.

Kouhi, F., Sorkheh, K., and Ercisli, S. (2020). MicroRNA expression patterns unveil differential expression of conserved miRNAs and target genes against abiotic stress in safflower. *PLoS One* 15: e0228850.

Kruszka, K., Pieczynski, M., Windels, D. et al. (2012). Role of microRNAs and other sRNAs of plants in their changing environments. *J. Plant Physiol.* 169: 1664–1672.

Kumar, A.U. and Ling, A.P.K. (2021). Gene introduction approaches in chloroplast transformation and its applications. *J. Genet. Eng. Biotechnol.* 19: 148.

Kumar, S. and Trivedi, P.K. (2018). Glutathione-*S*-transferases: role in combating abiotic stresses including arsenic detoxification in plants. *Front. Plant Sci.* 9: 751.

Lancíková, V., Tomka, M., Žiarovská, J. et al. (2020). Morphological responses and gene expression of grain Amaranth (*Amaranthus* spp.) growing under Cd. *Plants* 9: 572.

Le Rhun, A., Escalera-Maurer, A., Bratovič, M. et al. (2019). CRISPR-Cas in *Streptococcus pyogenes*. *RNA Biol.* 16: 380–389.

Lee, J., Bae, H., Jeong, J. et al. (2003). Functional expression of a bacterial heavy metal transporter in Arabidopsis enhances resistance to and decreases uptake of heavy metals. *Plant Physiol.* 133: 589–596.

Li, J., Zhang, M., Sun, J. et al. (2020a). Heavy metal stress-associated proteins in rice and Arabidopsis: genome-wide identification, phylogenetics, duplication, and expression profiles analysis. *Front. Genet.* 11: 477.

Li, R., Wu, H., Ding, J. et al. (2020b). Transgenic merA and merB expression reduces mercury contamination in vegetables and grains grown in mercury-contaminated soil. *Plant Cell Rep.* 39: 1369–1380.

Liang, G., He, H., and Yu, D. (2012). Identification of nitrogen starvation-responsive microRNAs in *Arabidopsis thaliana*. *PLoS One* 7: e48951.

Lin, T., Zhu, X., Zhang, F. et al. (2011). The detoxification effect of nitrogen on cadmium stress in *Populus yunnanensis*. *Bot. Res. J.* 4: 13–19.

Lino, C.A., Harper, J.C., Carney, J.P. et al. (2018). Delivering CRISPR: a review of the challenges and approaches. *Drug Deliv.* 25: 1234–1257.

Lokhande, V.H., Patade, V.Y., Srivastava, S. et al. (2020). Copper accumulation and biochemical responses of *Sesuvium portulacastrum* (L.). *Mater. Today Proc.* 31: 679–684.

López-Barea, J. and Gómez-Ariza, J.L. (2006). Environmental proteomics and metallomics. *Proteomics* 6: S51–S62.

Luo, J.S. and Zhang, Z. (2019). Proteomic changes in the xylem sap of *Brassica napus* under cadmium stress and functional validation. *BMC Plant Biol.* 19: 280.

Mahawar, L. and Shekhawat, G.S. (2016). Salt induces oxidative stress and its tolerance mechanism in plant: morphological, biochemical and molecular perspective. *Biotech Today* 6: 80–87.

Mahawar, L., Kumar, R., and Shekhawat, G.S. (2018a). Evaluation of Heme oxygenase 1 (HO 1) in Cd and Ni induced cytotoxicity and cross talk with ROS quenching enzymes in two to four leaf stage seedlings of *Vigna radiata*. *Protoplasma* 255: 527–545.

Mahawar, L., Khator, K., and Shekhawat, G.S. (2018b). Role of proline in mitigating NaCl induced oxidative stress in *Eruca sativa* Miller: an important oil yielding crop of Indian Thar Desert. *Vegetos* 31: 55–63.

Mahawar, L., Popek, R., Shekhawat, G.S. et al. (2021). Exogenous hemin improves Cd^{2+} tolerance and remediation potential in *Vigna radiata* by intensifying the HO-1 mediated antioxidant defence system. *Sci. Rep.* 11: 2811.

Mahawar, L., Ramasamy, K.P., Pandey, A. et al. (2022). Iron deficiency in plants: an update on homeostasis and its regulation by nitric oxide and phytohormones. *Plant Growth Regul.* https://doi.org/10.1007/s10725-022-00853-6.

Marrugo-Negrete, J., Durango-Hernández, J., Pinedo-Hernández, J. et al. (2016). Mercury uptake and effects on growth in *Jatropha curcas*. *J. Environ. Sci.* 48: 120–125.

Martins, J.P.R., Vasconcelos, L.L.D., Braga, P.D.C.D.S. et al. (2020). Morphophysiological responses, bioaccumulation and tolerance of *Alternanthera tenella* Colla (Amaranthaceae) to excess copper under in vitro conditions. *Plant Cell Tissue Organ Cult.* 143: 303–318.

Mathur, S., Kalaji, H., and Jajoo, A. (2016). Investigation of deleterious effects of chromium phytotoxicity and photosynthesis in wheat plant. *Photosynthetica* 54: 185–192.

Mehmood, S.S., Lu, G., Luo, D. et al. (2021). Integrated analysis of transcriptomics and proteomics provides insights into the molecular regulation of cold response in *Brassica napus*. *Environ. Exp. Bot.* 187: 104480.

Mishra, B. and Chandra, M. (2022). Evaluation of phytoremediation potential of aromatic plants: a systematic review. *J. Appl. Res. Med. Aromat. Plants* 1–37.

Mišúthová, A., Slováková, Ľ., Kollárová, K., and Vaculík, M. (2021). Effect of silicon on root growth, ionomics and antioxidant performance of maize roots exposed to As toxicity. *Plant Physiol. Biochem.* 168: 155–166.

Moazed, D. (2009). Small RNAs in transcriptional gene silencing and genome defence. *Nature* 457 (7228): 413–420.

Navarrete, A., González, A., Gómez, M. et al. (2019). Copper excess detoxification is mediated by a coordinated and complementary induction of glutathione, phytochelatins and metallothioneins in the green seaweed *Ulva compressa*. *Plant Physiol. Biochem.* 135: 423–431.

Navarro-Reig, M., Jaumot, J., García-Reiriz, A. et al. (2015). Evaluation of changes induced in rice metabolome by Cd and Cu exposure using LC-MS with XCMS and MCR-ALS data analysis strategies. *Anal. Bioanal. Chem.* 407: 8835–8847.

Nazir, M.M., Li, Q., Noman, M. et al. (2022). Calcium oxide nanoparticles have the role of alleviating arsenic toxicity of barley. *Front. Plant Sci.* 13: 843795.

Neelima, P. and Reddy, K.J. (2003). Differential effect of cadmium and mercury on growth and metabolism of *Solanum melongena* L. seedlings. *J. Environ. Biol.* 24: 453–460.

Nieves-Cordones, M., Mohamed, S., Tanoi, K. et al. (2017). Production of low-Cs^+ rice plants by inactivation of the K^+ transporter Os HAK 1 with the CRISPR-Cas system. *Plant J.* 92: 43–56.

Noman, A. and Aqeel, M. (2017). miRNA-based heavy metal homeostasis and plant growth. *Environ. Sci. Pollut. Res.* 24: 10068–10082.

Ozturk, M., Altay, V., and Karahan, F. (2017). Studies on trace elements distributed in *Glycyrrhiza taxa* in Hatay-Turkey. *Int. J. Plant Envron.* 3 (2): 1–7.

Pandey, A.K., Gedda, M.R., and Verma, A.K. (2020). Effect of arsenic stress on expression pattern of a rice specific miR156j at various developmental stages and their allied co-expression target networks. *Front. Plant Sci.* 11: 752.

Parmar, P., Dave, B., Sudhir, A. et al. (2013). Physiological, biochemical and molecular response of plants against heavy metals stress. *Int. J. Curr. Res.* 5: 80–89.

Pasternak, T., Rudas, V., Potters, G. et al. (2005). Morphogenic effects of abiotic stress: reorientation of growth in *Arabidopsis thaliana* seedlings. *Environ. Exp. Bot.* 53: 299–314.

Paunov, M., Koleva, L., Vassilev, A. et al. (2018). Effects of different metals on photosynthesis: cadmium and zinc affect chlorophyll fluorescence in durum wheat. *Int. J. Mol. Sci.* 19: 787.

Pavlova, D. (2017). Nickel effect on root-meristem cell division in *Plantago lanceolata* (Plantaginaceae) seedlings. *Aust. J. Bot.* 65: 446–452.

Pegler, J.L., Oultram, J.M.J., Nguyen, D.Q. et al. (2021). MicroRNA mediated responses to cadmium stress in *Arabidopsis thaliana*. *Plants* 10: 130.

Peško, M. and Kráľová, K. (2013). Physiological response of *Brassica napus* L. plants to Cu(II) treatment. *Proceedings of ECOpole* 7: 155–161.

Punshon, T., Ricachenevsky, F.K., Hindt, M.N. et al. (2013). Methodological approaches for using synchrotron X-ray fluorescence (SXRF) imaging as a tool in ionomics: examples from *Arabidopsis thaliana*. *Metallomics* 5: 1133–1145.

Rabêlo, F.H.S., Jordão, L.T., and Lavres, J. (2017). A glimpse into the symplastic and apoplastic Cd uptake by Massai grass modulated by sulfur nutrition: plants well-nourished with S as a strategy for phytoextraction. *Plant Physiol. Biochem.* 121: 48–57.

Radha, J., Srivastava, S., and Madan, V.K. (2000). Influence of chromium on growth and cell division of sugarcane. *Indian J. Plant Physiol.* 5: 228–231.

Rahman, Z. and Singh, V.P. (2019). The relative impact of toxic heavy metals (THMs) (arsenic (As), cadmium (Cd), chromium (Cr)(VI), mercury (Hg), and lead (Pb)) on the total environment: an overview. *Environ. Monit. Assess.* 191: 419.

Rai, P.K., Lee, S.S., Zhang, M. et al. (2019). Heavy metals in food crops: health risks, fate, mechanisms, and management. *Environ. Int.* 125: 365–385.

Rai, K.K., Pandey, N., Meena, R.P. et al. (2021). Biotechnological strategies for enhancing heavy metal tolerance in neglected and underutilized legume crops: a comprehensive review. *Ecotoxicol. Environ. Saf.* 208: 111750.

Rascio, N. and Navari-Izzo, F. (2011). Heavy metal hyperaccumulating plants: how and why do they do it? And what makes them so interesting? *Plant Sci.* 180: 169–181.

Rascón-Cruz, Q., González-Barriga, C.D., and Iglesias-Figueroa, B.F. (2021). Plastid transformation: advances and challenges for its implementation in agricultural crops. *Electron. J. Biotechnol.* 51: 95–109.

Rath, B.P., Hota, S., Subhadarshini, S. et al. (2019). Consequence of chromium tainted soil on physical and biochemical responses of *Vigna radiata* L. *J. Appl. Biol. Biotechnol.* 7: 35–41.

Raza, A., Su, W., Hussain, M.A. et al. (2021). Integrated analysis of metabolome and transcriptome reveals insights for cold tolerance in rapeseed (*Brassica napus* L.). *Front. Plant Sci.* 12: 721681.

Raza, A., Tabassum, J., Zahid, Z. et al. (2022). Advances in "omics" approaches for improving toxic metals/metalloids tolerance in plants. *Front. Plant Sci.* 12: 794373.

Regvar, M., Vogel-Mikuš, K., Kugonič, N. et al. (2006). Vegetational and mycorrhizal successions at a metal polluted site: indications for the direction of phytostabilisation? *Environ. Pollut.* 144: 976–984.

Riyazuddin, R., Nisha, N., Ejaz, B. et al. (2021). A comprehensive review on the heavy metal toxicity and sequestration in plants. *Biomolecules* 12: 43.

Rucińska-Sobkowiak, R. (2016). Water relations in plants subjected to heavy metal stresses. *Acta Physiol. Plant.* 38: 257.

Sagardoy, R., Vázquez, S., Florez-Sarasa, I.D. et al. (2010). Stomatal and mesophyll conductances to CO_2 are the main limitations to photosynthesis in sugar beet (*Beta vulgaris*) plants grown with excess zinc. *New Phytol.* 187: 145–158.

Saleem, M.H., Fahad, S., Rehman, M. et al. (2020). Morpho-physiological traits, biochemical response and phytoextraction potential of short-term copper stress on kenaf (*Hibiscus cannabinus* L.) seedlings. *Peer J.* 8: e8321.

Salt, D.E., Kato, N., Krämer, U. et al. (2000). The role of root exudates in nickel hyperaccumulation and tolerance in accumulator and non-accumulator species of Thlaspi. In: *Phytoremediation of Contaminated Soil and Water* (ed. N. Terry and G. Bañuelos), 189–200. Boca Raton: CRC Press.

Sangwan, P., Kumar, V., and Joshi, U. (2014). Effect of chromium(VI) toxicity on enzymes of nitrogen metabolism in clusterbean (*Cyamopsis tetragonoloba* L.). *Enzyme Res.* 2014: 784036.

Saravanan, D., Gomathi, T., and Sudha, P.N. (2013). Sorption studies on heavy metal removal using chitin/bentonite biocomposite. *Int. J. Biol. Macromol.* 53: 67–71.

Satoh-Nagasawa, N., Mori, M., Nakazawa, N. et al. (2012). Mutations in rice (*Oryza sativa*) heavy metal ATPase 2 (OsHMA2) restrict the translocation of zinc and cadmium. *Plant Cell Physiol.* 53 (1): 213–224.

Schmidt, S.B., Eisenhut, M., and Schneider, A. (2020). Chloroplast transition metal regulation for efficient photosynthesis. *Trends Plant Sci.* 25: 817–828.

Shan, Q., Wang, Y., Li, J. et al. (2013). Targeted genome modification of crop plants using a CRISPR-Cas system. *Nat. Biotechnol.* 31 (8): 686–688.

Sharma, J. and Shrivastava, S. (2014). Physiological and morphological responses of *Phaseolus vulgaris* caused by mercury stress under lab conditions. *Recent Adv. Biol. Med.* 1: 136.

Sharma, A., Kapoor, D., Gautam, S. et al. (2022). Heavy metal induced regulation of plant biology: recent insights. *Physiol. Plant.* 174 (3): e13688.

Sofo, A., Bochicchio, R., Amato, M. et al. (2017). Plant architecture, auxin homeostasis and phenol content in *Arabidopsis thaliana* grown in cadmium-and zinc-enriched media. *J. Plant Physiol.* 216: 174–180.

Song, W.Y., Mendoza-Cózatl, D.G., Lee, Y. et al. (2014). Phytochelatin-metal(loid) transport into vacuoles shows different substrate preferences in barley and Arabidopsis. *Plant Cell Environ.* 37: 1192–1201.

Song, X., Zhang, C., Chen, W. et al. (2020). Growth responses and physiological and biochemical changes in five ornamental plants grown in urban lead-contaminated soils. *Plant Environ. Int.* 1: 29–47.

Soumya, V., Kiranmayi, P., and Kumar, K.S. (2022). Morpho-anatomical responses of *Catharanthus roseus* due to combined heavy metal stress observed under scanning electron microscope. *Plant Sci. Today.* 9 (3): 623–631.

Stich, B., Benke, A., Schmidt, M. et al. (2020). The maize shoot ionome: its interaction partners, predictive power, and genetic determinants. *Plant Cell Environ.* 43: 2095–2111.

Su, C., Jiang, Y., Li, F. et al. (2017). Investigation of subcellular distribution, physiological, and biochemical changes in *Spirodela polyrhiza* as a function of cadmium exposure. *Environ. Exp. Bot.* 142: 24–33.

Sundaramoorthy, P., Chidambaram, A., Ganesh, K.S. et al. (2010). Chromium stress in paddy: (i) nutrient status of paddy under chromium stress; (ii) phytoremediation of chromium by aquatic and terrestrial weeds. *Comptes Rendus Biol.* 333: 597–607.

Sunkar, R., Li, Y.F., and Jagadeeswaran, G. (2012). Functions of microRNAs in plant stress responses. *Trends Plant Sci.* 17: 196–203.

Takahashi, R., Ishimaru, Y., Senoura, T. et al. (2011). The OsNRAMP1 iron transporter is involved in Cd accumulation in rice. *J. Exp. Bot.* 62 (14): 4843–4850.

Tang, L., Mao, B., Li, Y. et al. (2017). Knockout of OsNramp5 using the CRISPR/Cas9 system produces low Cd-accumulating indica rice without compromising yield. *Sci. Rep.* 7: 1–12.

Tavallali, V. (2017). Interactive effects of zinc and boron on growth, photosynthesis, and water relations in pistachio. *J. Plant Nutr.* 40: 1588–1603.

Teotia, P., Kumar, V., Kumar, M. et al. (2016). Rhizosphere microbes: potassium solubilization and crop productivity–present and future aspects. In: *Potassium Solubilizing Microorganisms for Sustainable Agriculture* (ed. V. Meena, B. Maurya, J. Verma, and R. Meena), 315–325. India: Springer.

Terzano, R., Al Chami, Z., Vekemans, B. et al. (2008). Zinc distribution and speciation within rocket plants (*Eruca vesicaria* L. Cavalieri) grown on a polluted soil amended with compost as determined by XRF microtomography and micro-XANES. *J. Agric. Food Chem.* 56: 3222–3231.

Thakur, S., Choudhary, S., Dubey, P. et al. (2019). Comparative transcriptome profiling reveals the reprogramming of gene networks under arsenic stress in Indian mustard. *Genome* 62: 833–847.

Tiwari, M., Sharma, D., Dwivedi, S. et al. (2014). Expression in Arabidopsis and cellular localization reveal involvement of rice NRAMP, OsNRAMP1, in arsenic transport and tolerance. *Plant Cell Environ.* 37: 140–152.

Tsyganov, V.E., Tsyganova, A.V., Gorshkov, A.P. et al. (2020). Efficacy of a plant-microbe system: *Pisum sativum* (L.) cadmium-tolerant mutant and *Rhizobium leguminosarum* strains, expressing pea metallothionein genes *PsMT1* and *PsMT2*, for cadmium phytoremediation. *Front. Microbiol.* 11: 15.

Vaculík, M., Konlechner, C., Langer, I. et al. (2012). Root anatomy and element distribution vary between two *Salix caprea* isolates with different Cd accumulation capacities. *Environ. Pollut.* 163: 117–126.

Varshney, R.K., Barmukh, R., Roorkiwal, M. et al. (2021). Breeding custom-designed crops for improved drought adaptation. *Adv. Genet.* 2: e202100017.

Wang, F., Wang, Z., and Zhu, C. (2012). Heteroexpression of the wheat phytochelatin synthase gene (TaPCS1) in rice enhances cadmium sensitivity. *Acta Biochim. Biophys. Sin.* 44 (10): 886–893.

Wang, Y., Xu, L., Shen, H. et al. (2015). Metabolomic analysis with GC-MS to reveal potential metabolites and biological pathways involved in Pb & Cd stress response of radish roots. *Sci. Rep.* 5: 18296.

Wang, Y., Zhang, L., Wang, J. et al. (2020). Identifying quantitative sources and spatial distributions of potentially toxic elements in soils by using three receptor models and sequential indicator simulation. *Chemosphere* 242: 125266.

Wani, W., Masoodi, K.Z., Zaid, A. et al. (2018). Engineering plants for heavy metal stress tolerance. *Rend. Lincei. Sci. Fis. Nat.* 29: 709–723.

Xiao, Y., Du, Z., Qi, X. et al. (2019). RNA-sequencing analysis reveals transcriptional changes in the roots of low-cadmium-accumulating winter wheat under cadmium stress. *Acta Physiol. Plant.* 41: 13.

Xiao, W., Liu, P., Wang, K. et al. (2021). Relationship between ionomics and transcriptomics of rice plant in response to arsenite stress. *Environ. Exp. Bot.* 189: 104565.

Xie, M., Chen, W., Lai, X. et al. (2019). Metabolic responses and their correlations with phytochelatins in *Amaranthus hypochondriacus* under cadmium stress. *Environ. Pollut.* 252: 1791–1800.

Xu, R., Qin, R., Li, H. et al. (2017). Generation of targeted mutant rice using a CRISPR-Cpf1 system. *Plant Biotechnol. J.* 15 (6): 713–717.

Yadav, G., Srivastava, P.K., Singh, V.P. et al. (2014). Light intensity alters the extent of arsenic toxicity in *Helianthus annuus* L. seedlings. *Biol. Trace Elem. Res.* 158: 410–421.

Yan, A., Wang, Y., Tan, S.N. et al. (2020). Phytoremediation: a promising approach for revegetation of heavy metal-polluted land. *Front Plant Sci.* 11: 359.

Yang, Z., Chen, F., Yuan, J. et al. (2004). Responses of *Sesbania rostrata* and *S. cannabina* to Pb, Zn, Cu and Cd toxicities. *J. Environ. Sci.* 16: 670–673.

Yang, H.F., Zhang, J., and Li, J.L. (2012). Physiological response to zinc pollution of rape (*Brassica chinensis* L.) in paddy soil ecosystem. *Adv. Mat. Res.* 356: 39–43.

Yang, W., Dai, H., Dou, X. et al. (2019). Effect and mechanism of commonly used four nitrogen fertilizers and three organic fertilizers on *Solanum nigrum* L. hyperaccumulating Cd. *Environ. Sci. Pollut. Res.* 26: 12940–12947.

Yourtchi, M.S. and Bayat, H.R. (2013). Effect of cadmium toxicity on growth, cadmium accumulation and macronutrient content of durum wheat (Dena CV.). *Int. J. Agric. Crop Sci.* 6: 1099–1103.

Zeng, R., Farooq, M.U., Wang, L. et al. (2019). Study on differential protein expression in natural selenium-enriched and non-selenium enriched rice based on iTRAQ quantitative proteomics. *Biomolecules* 9: 130.

Zhang, Y. and Showalter, A.M. (2020). CRISPR/Cas9 genome editing technology: a valuable tool for understanding plant cell wall biosynthesis and function. *Front. Plant Sci.* 11: 589517.

Zhang, T., Lu, Q., Su, C. et al. (2017). Mercury induced oxidative stress, DNA damage, and activation of antioxidative system and Hsp70 induction in duckweed (*Lemna minor*). *Ecotoxicol. Environ. Saf.* 143: 46–56.

Zhang, X.F., Hu, Z.H., Yan, T.X. et al. (2019). Arbuscular mycorrhizal fungi alleviate Cd phytotoxicity by altering Cd subcellular distribution and chemical forms in *Zea mays*. *Ecotox. Environ. Safe.* 171: 352–360.

Zhang, N., Roberts, H.M., Eck, J.V. et al. (2020). Generation and molecular characterization of CRISPR/Cas9-induced mutations in 63 immunity-associated genes in tomato reveals specificity and a range of gene modifications. *Front. Plant Sci.* 11: 10.

Zhu, Y., Pilon-Smits, E., Tarun, A. et al. (1998). Cadmium tolerance and accumulation in Indian mustard is enhanced by overexpressing gamma-glutamyl-cysteine synthetase. *Plant Physiol.* 121: 1169–1178.

Zhu, H., Wang, S., Zhang, Y. et al. (2020). QTL excavation and analysis of candidate genes in contents of As, Cu, Fe, Hg and Zn in rice grain. *Sci. Sin. Vitae.* 50: 623–632.

2

Advanced Techniques in Omics Research in Relation to Heavy Metal/Metalloid Toxicity and Tolerance in Plants

Ali Raza[1,], Shanza Bashir[2], Hajar Salehi[3], Monica Jamla[4], Sidra Charagh[5], Abdolkarim Chehregani Rad[6], and Mohammad Anwar Hossain[7]*

[1] *College of Agriculture, Fujian Agriculture and Forestry University (FAFU), Fuzhou, China*
[2] *Institute of Environmental Sciences and Engineering (IESE), School of Civil and Environmental Engineering (SCEE), National University of Sciences and Technology (NUST), Islamabad, Pakistan*
[3] *Department for Sustainable Food Process, Università Cattolica del Sacro Cuore, Piacenza, Italy*
[4] *Department of Biotechnology, Modern College of Arts, Science and Commerce, Savitribai Phule Pune University, Pune, India*
[5] *State Key Laboratory of Rice Biology, China National Rice Research Institute, Chinese Academy of Agricultural Sciences (CAAS), Hangzhou, China*
[6] *Department of Biology, Laboratory of Plant Cell Biology, Bu-Ali Sina University, Hamedan, Iran*
[7] *Department of Genetics and Plant Breeding, Bangladesh Agricultural University, Mymensingh, Bangladesh*

2.1 Introduction

Environmental contamination is progressively rising every day at an alarming pace. During the past numerous decades, fast-rising industrial mechanization and expansion, modern agricultural operations, anthropogenous activities, and sweltering and dumping of fossil fuels have augmented the verge amount of numerous basic heavy metals (HMs)/metalloids in the plant, agricultural soil, and marine milieu, which is triggering noxiousness to the diversity of living creatures (Rai et al. 2019; Adimalla 2020; Jamla et al. 2021a; Raza et al. 2021a, b). Heavy metals, including metals and metalloids with an atomic density greater than 5 g cm^{-3}, are nonbiodegradable and toxic environmental pollutants persistent in the ecosystem and causing substantial damage to the sessile plants. They are required for plant growth and developmental processes. However, above the permissible limit, they become toxic (https://www.atsdr.cdc.gov/) and cause plant death. Additionally, human settlements and industrialization are causing the disturbance of the required amount of HMs in the soil–water–air ecosystem. HM toxicity (including (i) light metals like aluminum [Al], (ii) HMs like cadmium [Cd], mercury [Hg], lead [Pb], chromium [Cr], manganese [Mn], nickel [Ni], zinc [Zn], and metalloids like selenium [Se], arsenic [As], boron [B], etc.) has been considered as a challenging concern worldwide. Much scientific and objective

*https://orcid.org/0000-0002-5120-2791

Heavy Metal Toxicity and Tolerance in Plants: A Biological, Omics, and Genetic Engineering Approach, First Edition. Edited by Mohammad Anwar Hossain, AKM Zakir Hossain, Sylvain Bourgerie, Masayuki Fujita, Om Parkash Dhankher, and Parvez Haris.

evidence exists implying HMs to be common environmental pollutants. Plants growing in the contaminated areas, including terrestrial and aquatic ecosystems, are continuously exposed to these toxic materials, leading to tremendous yield losses. For this, understanding the comprehensive responses of plants to HMs is critical to improving their performance. Basically, HMs interrupt the intracellular homeostasis, causing oxidation in macromolecules, including lipids, proteins, genetic material, and enzyme inactivation mainly due to the overproduction of free radicals (Rai et al. 2019; Adimalla 2020; Jamla et al. 2021a; Raza et al. 2021a, b).

Like other organisms, plants have also evolved various strategies to maintain their homeostasis to minimize the harmful effects of HMs. For the past decades, extensive research has been carried out to explore terrestrial plants' growth, developmental and physiological responses, and molecular mechanisms when exposed to HMs. Techniques established by plants for surviving in HM-polluted environments can be divided into two categories: (i) A restriction mechanism in which plant cells exudate specific small molecules to the rhizosphere to keep metal ions from entering into cellular parts by chelating them and probably transforming them to be less toxic and (ii) a detoxification mechanism, where plants absorb high levels of HM ions to allocate them in specific internal tissues and sequestrate them (Gallo-Franco et al. 2020). As a toxicity mechanism, certain unnecessary HMs are likely to substitute the important ones in enzymes/pigments, interrupting their implementation (Erickson et al. 2019). Consequently, plants want these HMs to execute their metabolic events at a minimal quantity.

Different biotechnological techniques are being used to boost our knowledge of the key processes and ways of participating in plant tolerance to HM toxicity. Among different tools, "omics" tools, such as genomics, transcriptomics, proteomics, metabolomics, miRNAomics, and phenomics, are commonly used for this objective (Raza et al. 2020, 2021b; Jamla et al. 2021a; Khan et al. 2021; Raza et al. 2022). Therefore, this chapter presented an overview of plant responses and tolerance mechanisms against HM toxicity and recent advances in omics tools in identifying the novel genes, proteins, metabolites, and metabolic pathways. The data obtained from multi-omics tools can be further engineered to develop HM-tolerant future crop plants for a safe food supply to feed the growing population worldwide.

2.2 An Overview of Plant Responses to Heavy Metal Toxicity

Understanding the HM-induced toxicity and their effects on plants is as vital as HM-stimulated responses to manage and improve crop productivity in polluted areas. Studies have shown that visible symptoms can emerge on almost all plant parts and at all the developmental stages ranging from germination to senescence. However, since seeds are the first part exposed to HMs, thus seed germination, seedling, and root development are greatly influenced. Regarding this, impaired seed germination and lower vigor, and, therefore, a decline in subsequent plant growth have been reported in almost all studied cases (Rai et al. 2019; Adimalla 2020; Jamla et al. 2021a; Raza et al. 2020, 2021a, b). Heavy metal exposure, e.g., has influenced early seed development, such as exterior appearance, internal condition, and germination in a variety of crop plants (Sengar et al. 2010; Zulfiqar et al. 2019). Likewise, Sanjosé et al. (2021) reported a significantly reduced germination of *Salsola vermiculata* when exposed to 4000 μM Cu and Zn. Since what happens to the seed morphology is hidden behind the alteration at the biochemical and molecular level, the protein modification and thus the inhibition of enzymatic activities have been suggested as the main reasons for seed development-related changes (Cui et al. 2014). Aspects of toxicity, frequency in terms of existence, and mainly the exposure potential of HMs to different parts of plants have been considered for this classification.

Furthermore, HMs directly or indirectly affect physicochemical properties, resulting in growth changes. Membrane lipid peroxidation, which can directly cause biomembrane deterioration, and modified enzyme activity are among the direct effects. Particularly, HM toxicity (like Ni, Cu, Hg, and Zn) induces malondialdehyde (MDA) production as a reliable indicator of peroxidation, and oxidative stress, with increasing concentration (Georgiadou et al. 2018; Sorrentino et al. 2018). The hormesis response, known as an adaptive response of triphasic dose to a stress-characterizing factor, has been reported in some stressed plants with various environmental agents, including HMs (Molina and Segura 2021). For instance, Li et al. (2022b) showed that atmospheric Pb significantly induces hormesis in *Tillandsia usneoides* (an accumulator plant). Other biomarkers, including hormone balance distribution, photosynthesis inhibition, indispensable nutrient deficiency, and changes in photoassimilate translocation can be categorized as indirect effects of HM toxicity resulting in HM-induced growth decline (Rai et al. 2019; Jamla et al. 2021a; Raza et al. 2021a, b).

On the other hand, physiological mechanisms involving cellular and biochemical changes to induce HM responses follow several strategies. Organic acids, root reductase, phytochelatins, metallothioneins, mycorrhizae, and heat shock proteins have been engaged in HM-induced responses in plants (Adimalla 2020; Raza et al. 2021b). First, these mechanisms mobilize the involving components to avoid HM translocation into the cellular cytoplasm by ions binding to cell wall components or pumping out the HMs through active efflux. Cell wall exudates and surrounding mycorrhizae serve as key defenders against HMs in this way, as evidenced by studies of various HM-induced toxicity (Shi et al. 2019; Dhalaria et al. 2020). Besides, the capacity of the cell wall to bind metal ions has been attributed to its composition, mainly those compounds containing functional groups such as –COOH, –OH, and –SH, which are found primarily in polysaccharides (Ghorbani et al. 2022). However, it is believed this protective capability varies among different plant species due to having different cell wall structures and compositions, which might affect the binding affinity and result in different protection levels (Parrotta et al. 2015).

Plants have also evolved two types of peptide metal-binding ligands in response to HMs: phytochelatins (PCs) and metallothioneins (MTs) (Raza et al. 2020, 2021b). Phytochelatins, defined as a class of HM-binding Cys-rich peptides involved in detoxification, are enzymatically synthesized by plants from sulfur-rich glutathione in response to HM exposure, and were first discovered as Cd-binding components (Grill et al. 1985; Hossain et al. 2012; Raza et al. 2020). It has been reported that the level of PCs synthesized in cell cultures correlates with the metal ions decline from the medium, interpreting the valuable role of PCs in the homeostasis balance of ionomic metabolism (Sauge-Merle et al. 2003). Heavy metal detoxification mechanism by PCs is much more complicated than simply producing the PC-metal ion chelation. In the first place, HM ions must effectively trigger PC synthase, producing the PC-metal ion chelation, and finally be sequestrated to the vacuole and possibly be aggregated with other components, such as organic acids (Raza et al. 2020).

At the level of histological response, secondary thickened cell wall and lignification have also been reported to play a protective role in overcoming the HM toxicity (Mehrabanjoubani et al. 2019). HM exposure causes an increase in the root vascular system and parenchymal cells along with increased lignification. For example, a study reported that Cd stress upregulated GH3 activity in poplar, downregulated auxin in wood cells, and thus switched the metabolic pathways to increased lignin synthesis (Elobeid et al. 2012). Consistence with these findings, excessive synthesis of monolignols, a precursor of lignin, in soybean is triggered by Cd stress, leading to cell wall solidification in roots and thus restricted root growth (Finger-Teixeira et al. 2010).

Besides these mechanisms, plants have also evolved some other molecular strategies involved in epigenetic, genetic, and gene expression regulation under HM exposure. It has been demonstrated that toxic HMs can modify epigenetic regulatory features and thus affect gene expression mainly by binding to different ion metal response elements located in the promoters of the targeted genes (Jing et al. 2022). DNA methylation and histone modifications are among epigenetic mechanisms reported for alleviating the HM-induced effects in plants (Gallo-Franco et al. 2020; Saeed et al. 2022). All these chromatin modifications are strongly correlated to transcriptional activation. Methylation has been considered a rapid response strategy for improving plant survival under HM exposure (Greco et al. 2019; Jing et al. 2022). Studies showed that methylation protects DNA from the HM-induced damage and mismatch probably by excision of single-strand breaks or multi-stranded transposition from genomic locations (Greco et al. 2019; Gallo-Franco et al. 2020).

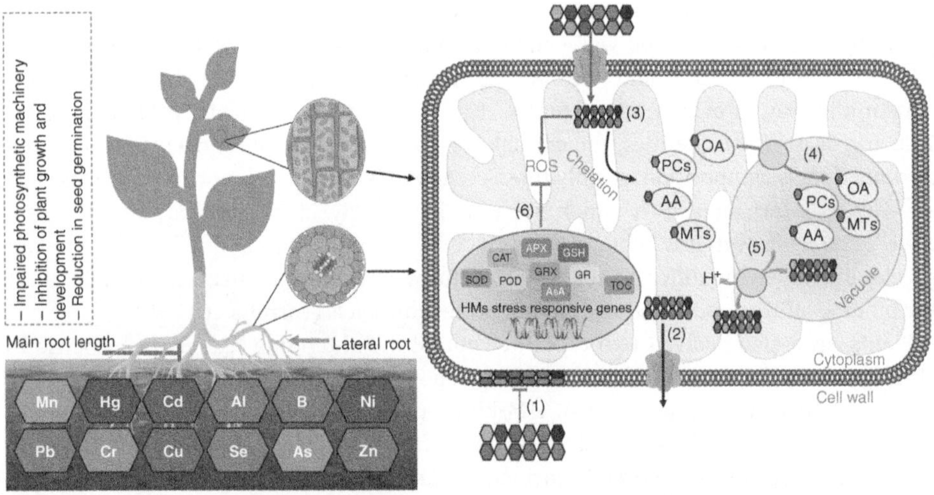

Figure 2.1 A putative schematic diagram describes overall plant responses to heavy metal (HM) exposure in plants. In short, plants evolved several mechanisms to cope with HMs: (1) HM–cell wall components binding; (2) the efflux of HM ions from cytoplasm to apoplast; (3) HM–ligand chelation; (4) accumulation of HM–ligand complexes in the vacuole; (5) vacuolar sequestration by tonoplast-located transporters; (6) triggering the antioxidant defense system and scavenging the ROS. Abbreviations: ROS, reactive oxygen species; OA, organic acid; AA, amino acid; PCs, phytochelatins; MTs, metallothioneins; SOD, superoxide dismutase; CAT, catalase; APX, ascorbate peroxidase; POD, peroxidase; GR, glutathione reductase; GRX, glutaredoxins; AsA, ascorbic acid; GSH, reduced glutathione; TOC, tocopherol.

According to the findings mentioned earlier, a schematic highlight of the plant responses to HM toxicity on a whole scale is presented in Figure 2.1. The worth mentioning question here is which strategies are common and can be adopted? Based on the mentioned overview, the chelation process is the most commonly used strategy against HM toxicity, which improves metal excretion. However, the induced tolerance in plants against HMs combines several mechanisms. Apart from the induction of metal chelators, boosting the antioxidant defense system and DNA methylation/demethylation processes involved in gene expression regulation play a critical role in HM detoxification.

2.3 How the Integration of Multi-omics Data Sets Helps in Studying the Heavy Metal Stress Responses and Tolerance Mechanisms?

Plant responses to HM toxicity depend on the management of molecular components. Thus, multi-omics approaches have been comprehensively explored to understand the molecular processes and interactions against HM toxicity. Notwithstanding the extraordinary innovations in genomics, it is essential to appraise other omics approaches for extensive information at the molecular point (Figure 2.2). Systematic examinations and current data from omics approaches focus on signaling and molecular mechanisms to improve tolerance to toxic HMs for better crop advancement.

2.3.1 The Contribution of State-of-the-Art Genomics-Assisted Breeding

Genomics is the area of "omics" science that studies the complete genome, its structure, functions, expressions, and correlations in any given organism, through its gene sequencing (Varshney et al. 2018, 2021). In recent years, genomic-assisted breeding (GAB) has been discovered and acknowledged to have great prospects in various aspects of life, especially in crop enhancement, improvement, and optimization, given the current intensifying complications of all-time high climate change impacts and environmental contamination, changing the dynamics of agriculture in forms previously unknown (Varshney et al. 2018, 2021; Sandhu et al. 2021).

GAB, since its inception, has provided many tools and strategies that have revolutionized modern genome characterization, development, and manipulation for various sorts of objectives (Varshney et al. 2018, 2021; Aglawe et al. 2021; Sandhu et al. 2021). Among various GAB tools, the prominent ones being quantitative trait loci (QTL) and genome-wide association studies (GWAS) will be the focus of attention in this study. With environmental levels of HM contamination rising all around the globe and food security concerns being at an all-time high, there is a need for more desperate than ever to apply modern biotechnology tools such as GAB for optimizing plant species to be able to tackle through and bear the ever-high HM toxicity.

2.3.1.1 Quantitative Trait Locus (QTL) Mapping

QTL can be defined as the specific locus or polygene of the genome responsible for a phenotype varying continuously due to environmental factors (Varshney et al. 2018, 2021).

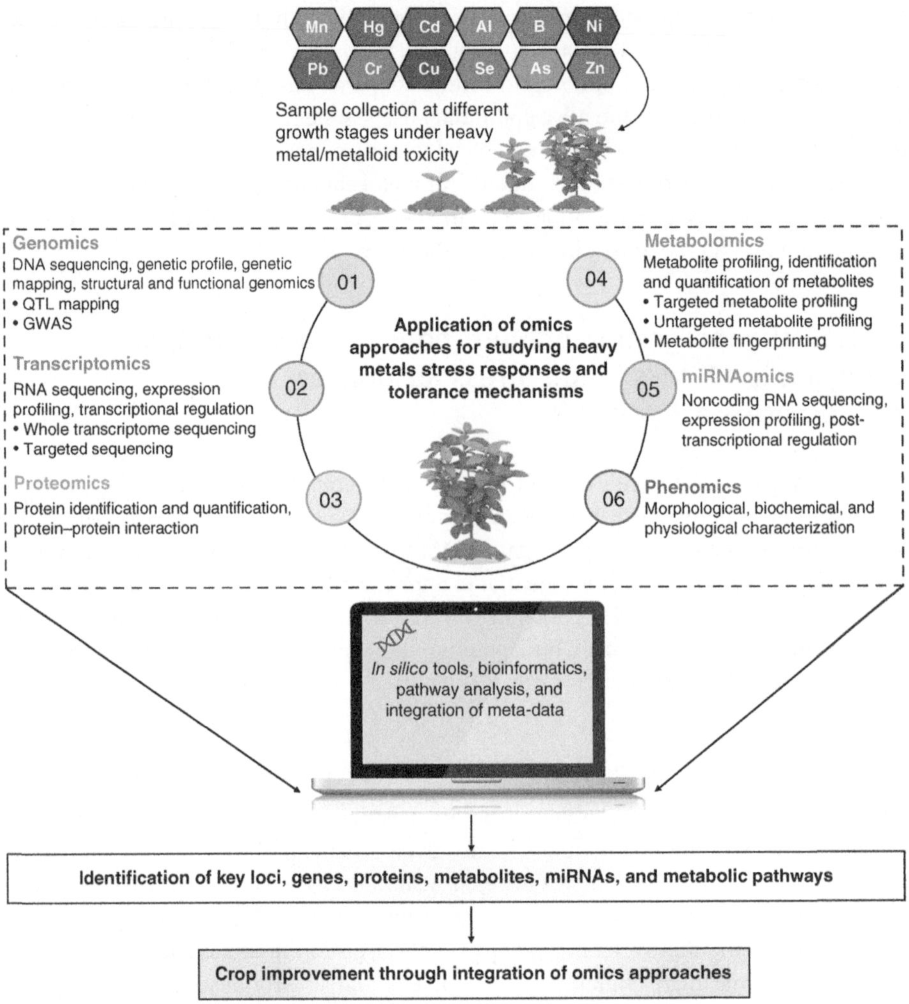

Figure 2.2 A model for integrated six-omics approaches for rising heavy metal (HM)-tolerant plants. The application of multi-omics tools can aid in disclosing stress-responsive mechanisms at the molecular level.

Quantitative traits cannot be separated following the Mendelian standards; thus, studying them is comparatively more complex; therefore, high-efficiency advanced tools, such as the QTL mapping, are of great help in understanding and exploiting them. For this purpose, three main ingredients are considered as requisites, including a study population that contains linkage disequilibrium, a gene/molecular (DNA) marker(s), and an appropriate statistical tool (Varshney et al. 2018, 2021).

QTL mapping is being utilized in identifying the QTLs that promote tolerance against HM toxicity among various crop plants (Jamla et al. 2021a; Khan et al. 2021; Raza et al. 2022). For instance, Velu et al. (2017) mapped QTLs for high Fe and Zn accumulation in wheat plants. Zn increasing genes were found to be on chromosomes 1B and 6B, along with a gene on 2B co-located with Fe expanding gene. They concluded that Zn accumulation is governed by a genetic basis. African wild rice (*Oryza longistaminata*) was reported to have 29 new QTLs for some micronutrient contents of Fe, Zn, Se, Cd, As, and Pb, using QTL mapping analysis of the genotype and phenotypic data. Among these QTLs, Fe, Zn, Se, and Cd contents were mainly regulated by the *qFe5.2/qZn5.2* and *qSel/qCd1*, respectively (Liu et al. 2020). QTL interval mapping helped identify QTLs and their respective molecular markers for the bioavailability of Zn, Fe, and Se in wheat, which was then used for MAS to biofortify these micronutrients (Gupta et al. 2021).

Cd accumulation in rice has been linked to negative impacts on the plant's physical and physiological health, particularly seed germination and growth, as well as chlorosis and necrosis (Chen et al. 2020). Therefore, understanding the genetic basis can help combat the issue at root levels. Indica and japonica rice varieties in China were studied for QTL identification of their high and low Cd accumulation alleles, respectively, to provide information that could prove helpful in breeding rice varieties of good quality with low Cd accumulation capability to prevent Cd toxicity in the consumers (Pan et al. 2020). Another study identified QTLs *qCDSL1.1* and *qCDSL1.2* responsible for Cd tolerance in rice at the seedling stage in two cities in China, Hangzhou and Lingshui, respectively (Shilin et al. 2021).

A light metal, aluminum (Al), is another toxicant for plants that can modify physiological processes in the organism, causing stunted root and shoot growth, decreased yield, and entering the food chain, affecting human health. Rice populations have been analyzed to identify Al tolerance genes, resulting in the location of 48 QLTs on chromosomes 1, 3, 9, and 12 (Gallo-Franco et al. 2020). RIL of soybean obtained by cross-breeding of varieties Zhonghuang 24 (ZH 24) and Huaxia 3 (HX 3) were mapped for high-density genetic linkage. The results expressed five main QLTs, namely *qAAC_04, qRRE_04, qRRE_16, qRRE_17,* and *qAAC_19,* to account for trivalent Al tolerance. The QLTs could be found on chromosomes Gm04, Gm16, Gm17, and Gm19 and caused a 39.65% comprehensive genetic effect (Wang et al. 2019). In another study, mapping of rice characterized 17, 21, and 22 QTLs for Cd, Pb, and As accumulation, respectively (Liu et al. 2019a). Overall, it has been observed that QLT mapping techniques have been proven to be a precursor to genetic characterization, leading to avenues for breeding combinations and tools and strategies that can revolutionize the crop as we know them today.

2.3.1.2 Genome-Wide Association Studies

GWAS is a robust tool commonly employed in breeding programs because it can quickly examine complicated features in various environmental settings. It is incredibly efficient and precise and is conducted on a genome-wide scale, allowing several alleles to be detected simultaneously. Moreover, it also eliminates the need for repeated population establishment, therefore considerably reducing gene sequencing expenses (Varshney et al. 2018, 2021; Xiao et al. 2022).

QTLs linked to HM accumulations have been identified using GWAS. Accumulation of microelements including Zn, Fe, Cu, and Mn in maize was interpreted using GWAS and principal component analysis (PCA) (Ma et al. 2021). The authors found 11 genes highly related to 27, 1, 5, and 3 SNPs, which were associated with *Zn-PC1, Zn-PC2, Cu-PC1,* and *Mn-PC2,* respectively. The gene *GRMZM2G142870* was determined to highly impact the accumulation of Zn, and *GRMZM2G045531* affects Mn accumulation in the plant (Ma et al. 2021). In another study, rice tolerance against Mn excess was studied through GWAS. Seven selected traits were run through association mapping with 416741 SMP markers implementing a linear model. The results discovered six significant associations for those traits' shoot concentration of Mn and their relative shoot length. Genes relating to the transportation of HM, peroxidase precursor, and Mn ion-binding proteins were located in the proposed candidate region. SNP-2.22465867 marker was determined to induce a change in one of the amino acids in *LOC_Os02g37170* gene (Shrestha et al. 2018).

A study conducted by Chen et al. (2018) detected 419 rapeseed (*Brassica napus*) accessions and inbred lines genotyped with a 60K Illumina Infinium SNP array of *Brassica*. They discovered a wide range of genetically regulated Cd accumulation and translocation variations. In another case, GWAS analysis performed on *B. napus* identified four QTLs responsible for tolerance to Pb. It was concluded that the expression of *GSTUs, BCATs, UBP13, TBR,* and *HIPP01* genes was affected and increased by Pb toxicity (Zhang et al. 2020a). Another GWAS was executed by implementing 200510 SNPs developed by SLAF-seq (specific-locus amplified fragment sequencing) technology on 254 inbred lines of *B. napus* varying in their Al tolerance (Zhou et al. 2022). They determined the presence of 43 SNPs relating to Al tolerance in the organism on 14 chromosomes at the seedling stage. Furthermore, 777 candidate genes were identified at the flanking 100 kb region of the SNPs. Research showed 64 candidate genes to have been differentially expressed at least once in Al-tolerant or Al-sensitive lines (Zhou et al. 2022).

A study of 312 distinct rice accessions found that grain's Cd values change significantly between subgroups, with Cd concentrations decreasing from indica through temperate japonica to tropical japonica (Zhao et al. 2018). Furthermore, no common QTL was found between indica and japonica, implying that Cd accumulation has a different genetic basis or that some QTLs for Cd accumulation in rice are subspecies-specific. *OsNRAMP2* was also suspected of being a novel functional gene linked to Cd accumulation in rice (Zhao et al. 2018). GWAS analysis of *Medicago truncatula*'s Cu and Hg accumulation and relative root growth was carried out. Trait associations of Cu and Hg were determined to be scattered throughout the genome, suggesting that these traits are regulated by several QTLs and are thus polygenic. P-type ATPase transporters and membrane-associated ATP-binding cassette transporters, oxidative stress response genes, and stress-related UDP-glycosyltransferases were all found in candidate genes (Paape et al. 2021). Thus, GWAS as a tool for GAB is of high significance and provides information on the genes of interest, and their phenotypic association and functions, leading pathways to efficacious breeding practices and combinations.

2.3.2 Transcriptomics

Recent advances show that the genomic/genetic level studies can assist in identifying and characterizing the genes of interest and are of great value in understanding and exploiting the issues at the genetic level, but further manipulation can only be carried out if the expression of those genes and their functioning can be understood. This is where transcriptomics comes into play (Jamla et al. 2021a; Khan et al. 2021; Raza et al. 2021d, e, 2022). Transcriptomics provides ways to understand gene expression in terms of mRNA formation, composition, and processing and ties the journey from the genetic level to the proteome functioning (Jamla et al. 2021a; Khan et al. 2021; Raza et al. 2021d, e, 2022). Like other stresses, HM toxicity in plants is associated with regulatory and functional genes responsible for survival mechanisms and pathways. The regulatory genes produce various transcription factors, which play a vital role in regulating a network of stress genes. In contrast, the functional genes control the release of amines, sugars, alcohols, and other compounds (Khalid et al. 2019; Jamla et al. 2021a; Khan et al. 2021; Raza et al. 2022).

With the help of transcriptome analysis, As accumulation and toxicity were observed to be decreased in rice by expression of As transporter genes: *NIP1;1, NIP2;1, ABCG5, NRAMP1, NRAMP5,* and *TIP2;2* (Chauhan et al. 2020). In response to As stress in paddy rice, 3812 differentially expressed genes (DEGs) were identified in the plant shoots and were mainly associated with transmembrane transporters and ion binding. The release of ABC transporters and aquaporins was reported as a detoxification mechanism in rice plants (Xiao et al. 2021). In another study, Di et al. (2021) observed 2983 genes to be triggered as a result of As(III) and 1844 genes as a resultant of As(V) exposure. Upregulation of transcription factors such as NAC, including NAM (no apical meristem), ATAF (*Arabidopsis thaliana* activating factor), and CUC (cup-shaped cotyledon) were reported, whereas WRKY4, bHLH, and AP2-EREB were downregulated.

Mungbean (*Vigna radiata*) responded to Cd exposure by detoxification and antioxidative defense mechanisms regulated by 6737, 10,279, and 9672 DEGs on the first, fifth, and ninth day of exposure, respectively. Whereas genes that regulate DNA replication and repair, cell division and cell wall synthesis, and photosynthesis were downregulated (Leng et al. 2020). Chen et al. (2020) identified 3926 DEGs (1206 upregulated and 2720 downregulated) in Kenaf (*Hibiscus cannabinus*) as a Cd exposure-response. Analysis showed these genes regulate Cd toxicity through various processes such as carbohydrate and energy metabolism, replication and repair, signal transduction, HM transport and catabolism, catalytic activity, and detoxification activities. In a recent case, a comparative transcriptomic analysis of *Phytolacca americana* confirmed 5054 DEGs, of which genes for lignin production were found to have been upregulated. In contrast, genes regulating protein production for photosynthesis and flavonoid synthesis were downregulated (Zhao et al. 2021).

Cross-species comparison in response to a metalloid, boron (B) exposure was performed on *Schrenkiella parvula* and *A. thaliana* (Wang et al. 2021a). The authors concluded that *S. parvula*'s transcriptome is pre-adapted to combat B toxicity. In contrast *A. thaliana* experiences higher stress upon B exposure and alters its RNA metabolism, and the cell wall is adapted to sequester excess B. Recently, Yu et al. (2020) reported that cytochrome gene *osCYP* in rice was more differentially expressed in roots as compared to that in the shoots upon exposure to two variants of Cr, i.e., Cr(III) and Cr(VI). A comparative analysis of transcriptome in pepper (*Capsicum annuum*) expressed upregulation of 3846 genes with downregulation of 2841 genes in response to Cr(VI) exposure with 24-EBR supplementation. While 1223 genes were upregulated and only 649 were downregulated for Cr(VI) exposure with 24-EBR supplementation (Mumtaz et al. 2022).

Under Cd stress, comparative transcriptomic analysis was performed for two cultivars of sweet potato (*Ipomoea batatas*), namely N88 and X16. N88 had 29 genes expressed differentially compared to the control, while X16 had 20. Analysis showed X16 to have higher number of DEGs, with 2649 being upregulated, while 3173 being downregulated. DEGs were confirmed to be controlling Cd detoxification, cell wall biosynthesis, hormone signal transduction, and glutathione metabolism, with prominent DEGs identified as AuX_1, *CAT*, CAX_3, *CCR*, $COPT_5$, *DR*, *GAUT*, *GSR*, *GST*, HMA_3, and *SOD* (Yin et al. 2022). *Solanum nigrum*'s exposure to Cd triggered 3908 DEGs, of which 1049 were upregulated, while the rest, 2859, were downregulated. The analysis presented these DEGs associated with various biological processes, cellular components, and molecular functions. Upregulated DEGs were found to be regulating catalytic activity and catabolic processes, extracellular region and ribosome and its structural constituents, peptide metabolic process and translation, and isoprenoid biosynthetic process. However, the downregulated DEGs mainly controlled structural molecule activity, intracellular ribonucleoprotein complex, and non-membrane-bounded organelles and ribosomes (Wang et al. 2022). These examples show how transcriptome analysis can aid in gaining a better knowledge of HM exposure, genetic implications and responses, detoxification, mitigation, and adaptability mechanisms. Therefore, it can help modify species capable enough to cope with the modern times of extreme environmental contamination and climate change.

2.3.3 Proteomics

Proteins, which are functionally categorized as versatile macromolecules, are involved in most biological processes (in some cases along with other biomolecules) in living cells. Proteins have been identified as the main and primary targets of HMs where they disturb protein function either by forming an HM–protein functional side chain groups complex or by displacing the essential ions in MTs (Tamás et al. 2018; Jamla et al. 2021a; Khan et al. 2021; Raza et al. 2022). Proteomics is a cutting-edge technology that deals with the full-fledged screen of proteins quantitatively, their structure, and functional role in a biological system under specific conditions (Liu et al. 2019b).

Over the last decade, extensive proteomics studies on HM–plant responses have been run to untangle the molecular mechanism against HM toxicity by mining proteins' differential regulation and proposing the putative targets for genetic engineering to improve plant tolerance and productivity (Jamla et al. 2021a; Raza et al. 2022). As previously mentioned, induced cellular ROS production is raised after HM stress by inducing cellular redox reaction or disturbing the thiol-containing antioxidants and proteins and inducing lipid peroxidation (Hasanuzzaman et al. 2020; Raza et al. 2021c). The excessive accumulation of ROS leads to oxidation in both amino acid side chains and thus protein structure, causing protein fragmentation and affecting their cross-linking (Hasanuzzaman et al. 2020; Raza et al. 2021c). For instance, a targeted proteomic approach showed a downregulation of chloroplastic-related proteins upon exposure to Cd, indicating photosynthesis disturbance (Gutsch et al. 2018). Apart from the antioxidant defense system, omics analyses confirmed that the regulation of HM ion transporters, regulatory elements, photosynthesis and energy production, and metabolism components are effectively involved in HM tolerance (Jamla et al. 2021a; Raza et al. 2022).

A combination of molecular approaches, in particular proteomics, has identified a series of proteins that contributed to HM influx, transport, and detoxification, such as natural resistance-associated macrophage protein (NRAMP), ZRT, IRT-like protein (ZIP), metallothionein (MT), plant defensins (PDF), cation/proton exchanger (CAX), ATP-binding cassette C subfamily (ABCC), metal tolerance protein (MTP), oligopeptide transporter (OPT), heavy metal ATPase (HMA) plant Cd resistance (PCR), and ABCG (Li et al. 2022a, b). The transport-related proteins are responsible for maintaining the optimal dosages of non-essential HMs below their toxicity threshold levels by ions' absorption and compartmentalization. The various transporter proteins were regulated and expressed by a wide range of transduction signaling under environmental stress. The expression of *OsHMA9*, identified as a member of HMAs in rice, was triggered by an excess dosage of Cd, Zn, and Cu and its role in Pb efflux from internal parts of cells (Lee et al. 2007). A similar finding was observed for *OsHMA6* expression under excessive Cu^{+2} in rice plants (Wenli et al. 2020). These findings demonstrate a different response of plants to individual HMs at the protein level, triggering a compensatory response for some HMs and inhibiting almost all metabolic pathways for others.

In response to Cd toxicity, Jian et al. (2020) reported the upregulation of glutathione metabolism-related proteins in insensitive wheat variety compared to sensitive one, indicating higher detoxification. However, impaired protein synthesis and degradation were inferred from the proteome profile. Additionally, proteomics studies have proved the upregulation of molecular chaperones upon exposure to HMs and reported their role in preventing the aggregation of nascent peptide chains or stress-unfolded protein by re-establishing and folding of the functional proteins contributed to redox homeostasis maintenance and finally, plant survival (Tamás et al. 2018). A wide range of studies has shown that various HSPs and protein disulfide isomerase (PDIs) were differentially expressed in response to HMs. For example, the upregulation of DnaK-type molecular chaperone gene has been documented in response to Cd toxicity (Sahr et al. 2005). Similarly, different types of low-molecular-mass HSPs were markedly expressed under HMs such as Cu, Cd, Ni, Pb, and so on (Ezemaduka et al. 2017; Cui et al. 2019; Abassi et al. 2019).

A label-free proteome technique identified 6345 proteins in Cu and citric acid (CA)-treated plants, of which 426 were differentially expressed in rapeseed. The interpretation of gene ontology pathways analysis showed that most of the identified genes involved in primary metabolism including energy and carbohydrate metabolism, protein metabolism, and also secondary metabolism, such as stress and defensive system along with metal detoxification, were found to be Cu-responsive mechanisms (Ju et al. 2021). Other regulatory mechanisms such as post-translational modifications (PTMs), protein folding, stability and localization, and protein–protein interactions can be contributed to protein and genome differences. Therefore, proteomics would offer a powerful platform to identify these modified changes at the proteome level. In plants, PTMs occur in almost all proteins after their biosynthesis in response to internal or external cues and have a significant role in extensive biological pathways by modifying their structure and dynamics (Audagnotto and Dal Peraro 2017; Greco et al. 2019; Gallo-Franco et al. 2020). The efficiency of stress-induced adaptive responses of plants, in particular HMs toxicity, depends on the intricate coordination of multiple signaling pathways in which protein PTMs are involved too (Audagnotto and Dal Peraro 2017). The regulation of PC biosynthesis has been pointed out at the post-translational level, which HMs activate. For instance, a study reported that PCs were regulated by Cd-dependent phosphorylation (at a threonine residue near its catalytic site), acting as a Cd sensor and causing hypersensitivity to Cd (Wang et al. 2009).

To summarize, various proteomics techniques have aided in exploring potential proteins implicated in HM-induced responses and, as a result, cellular homeostasis and tolerance.

Consequently, subcellular proteome-wide identification provides novel insights into detecting HM-specific markers.

2.3.4 Metabolomics

Plant metabolome profile gives the qualitative and quantitative data of the entire metabolite pool under specific conditions. Metabolomics is relatively a more knowledgeable and comprehensive omics approach encompassing the information of plant metabolites (low-molecular-weight molecules; <1 kDa) and critical pathways of the system. The complex communicating network profile of metabolites gives an insight into the changes at genotype and phenotype levels, which eventually can be analyzed and explored for sustainable agriculture programs (Jamla et al. 2021a; Khan et al. 2021; Raza et al. 2021d, 2022; Raza 2022). During stress, plants undergo several molecular, biochemical, cellular, and physiological changes to maintain homeostasis. Thus, metabolite-linked metabolic pathways play a crucial role as they act as a networking link governing the status of growing plants under stress conditions, such as HM toxicity.

The metabolomic response was studied between Cd-tolerant (CB671) and Cd-sensitive (ZD622) genotypes of rapeseed (Mwamba et al. 2020). Carbon flux was rearranged by increased production of solutes, sugar storage forms, jasmonates, ethylene, and vitamin B6 in CB671 compared to ZD622 genotype. Apart from this detoxification mechanism, the production of secondary metabolites and secretion of cell wall polysaccharides was escalated in the CB671 genotype (Mwamba et al. 2020). In another study, MS-based metabolomics analysis of low-Cd-accumulating and high-Cd-accumulating rice grains was studied. With the increasing Cd stress, most of the amino acids and carbohydrates were downregulated, except myoinositol, which was upregulated. Organic acids involved in alpha-linolenic acid metabolism and jasmonic acid hormone production were also activated (Zeng et al. 2021). Furthermore, Li et al. (2021) hypothesized that the application of zinc-oxide nanoparticles (ZnO NPs) in Cd-stressed fragrant varieties of rice results in improved tolerance to Cd. The metabolic profile shows possible networking of amino acid (alanine, aspartate, and glutamate) synthesis pathways, besides phenylpropanoid biosynthesis, taurine, and hypotaurine metabolism pathways.

Wang et al. (2021b) characterized 145 and 133 metabolites in leaves and roots of maize (*Zea mays*) under Cd stress and exogenously applied GSH hormone. Tricarboxylic acid cycle (TCA cycle) metabolism was decreased, and amino acid contents were increased in leaves compared to roots. Also, increased sugar storage and shikimic acid metabolism were observed in root tissues. GSH alleviates the Cd stress by changing the production of antioxidant enzymes and alleviating lipid peroxidation (Wang et al. 2021b). Soybean (*Glycine max*) was subjected to nanomolar concentration of Cd stress, and metabolomics analysis was performed. Amino acid and carbohydrate metabolism was decreased during Cd stress in soybean plants (Andresen et al. 2020). Metabolomic analysis of root exudates of *Sedum plumbizincicola* was done under Cd stress (Sun et al. 2020). Out of 155 metabolites, 33 showed substantial differences, including amino acids, polyols, lipids, and organic acids. Sixteen metabolic pathways involved (purine metabolism, glutathione metabolism, arginine, and proline metabolism, valine, leucine, and isoleucine biosynthesis, glycine, serine, and threonine metabolism) were found correlated to Cd stress resistance (Sun et al. 2020).

In another study, physiology status as well as metabolic profile of shoots and cytosol of the macrophyte (*Elodea nuttallii*) was investigated in response to three stresses, Cd, methylmercury, and inorganic mercury, respectively. Cd-mediated metabolic changes were more than the other stresses. Aminoacyl-tRNA biosynthesis and branched-chain amino acid pathways were regulated in response to Cd stress (Cosio and Renault 2020). Transcriptomic, metabolomics, and cytology studies were done in potato (*S. nigrum* L.) in response to Cd stress. The profile of identified metabolites depicts their association in several pathways like cell wall biosynthesis; alanine, aspartate, and glutamate metabolism; pantothenate and CoA biosynthesis; glutathione metabolism; tyrosine metabolism; glyoxylate and dicarboxylate metabolism; pyrimidine metabolism; and citrate cycle (Wang et al. 2022). Roots of sunflower (*Helianthus annuus*) were exposed to Cr(IV) stress, and metabolome analysis was observed. Metabolites such as oxylipins, linoleic acid, and alpha-linolenic acid were identified in response to Cr stress (Ibarra et al. 2019).

Pidatala et al. (2018) performed comparative metabolic studies of *Chrysopogon zizanioides* and maize in response to Pb toxicity. Metabolite (amino acids, organic acids, and coenzymes) profile was significantly different for both plants. Inhibition of Cd and Pb HM uptake by wheat via HM immobilizing bacteria, *Enterobacter bugandensis* TJ6 was investigated using metabolomics and proteomics profiles. The synthesis of indole-3-acetic acid, arginine, betaine, abscisic acid, and jasmonic acid was improved in the presence of bacterium (Han et al. 2021). Toxic effects of Hg and Se were investigated in a halophyte (*Suaeda salsa*) at gene, protein, and metabolic levels. The plants treated only with Hg stress show metabolite profile involved in ATP synthesis and binding, tolerance, sugar-phosphatase activity, photosynthesis, citrate cycle, glycolysis, substance metabolism (cysteine and methionine, amino acid, pyrimidine), and many more. In comparison, combined treatment (Se+Hg) revealed a different metabolomics profile (Liu et al. 2021b).

2.3.5 miRNAomics

MicroRNAs (miRNAs) belong to a class of small endogenous noncoding RNA molecules of 19–24 nucleotides in length. They are ubiquitous and are present in prokaryotic and eukaryotic systems. They are reported as key regulators/modulators in plants' post-transcriptional and post-translational gene expression (Jamla et al. 2021b). They bind via complementary base pairing to its single/multiple target messenger RNA(s) and result in differential expression patterns of single/myriad of genes involved in either single/multiple developmental, biological, physiological pathways, and more (Jamla et al. 2021b; Patil et al. 2021). Recently, HM stress-responsive miRNAs have been identified in various crops and plant types (Jamla et al. 2021b). The target genes are involved in balancing the cell's redox status by regulating metal uptake and transport, sequestration of metals in cellular organelles, metal chelation, switching antioxidant enzymatic pool, and production of ROS. Current trends and key findings of miRNA in response to five toxic HMs: Pb, As, Cr, Cd, and Hg are mentioned below.

Zhang et al. (2020b) showed that overexpression of miR156 in *A. thaliana* results in enhanced tolerance to Cd stress, and knock-down data showed sensitivity to Cd stress. The overexpressing lines have higher biomass and chlorophyll content, longer primary root length, increased activity of antioxidant enzymes, and low levels of ROS production. The expression profile analysis reveals the role of Cd uptake transporters (*AtZIP1* and *AtZIP2*), vacuole segregated transporter (*AtABCC1*), and Cd transporter from root to shoot (*AtHMA4*) in stressed plants (Zhang et al. 2020b). A collective study of miRNA and mRNA expression against Cd stress response was studied in rapeseed root and shoot (Fu et al. 2019). Data showed that 399 DEGs in the root and 389 DEGs in the shoot were identified after Cd treatment. Differential expression patterns of three novel miRNAs and five miRNA families (miR395, miR397, miR398, miR408, and miR858) were observed in both tissues. The target genes were involved in key stress-responsive pathways such as detoxification of superoxide radicals, TFs activation, antioxidative response pathways and secondary metabolic pathways (Fu et al. 2019).

Yang et al. (2021) performed an integrated evaluation of miRNA–mRNA in potato (*Solanum tuberosum* L.) in response to Cd stress. The investigation identified 161 existing miRNAs, 383 known miRNAs, and 7361 novel miRNAs; and 18 miRNAs were differentially expressed. Additionally, Cd stress response also involves regulation of glutathione metabolism, plant hormone signal transduction, and phenylpropanoid biosynthesis pathways. In another study, role of miR166 in response to Cd accumulation and tolerance was studied in rice (Ding et al. 2018). Overexpression of miR166 reduced Cd accumulation in the grains and improved Cd tolerance. The target gene expression analysis showed overexpression of *OsHB4*, a class of HD-Zip family proteins, responsible for increased Cd sensitivity and Cd accumulation in the leaves and grains of transgenic rice lines. Previously, Ding et al. (2017) showed that miR268 acts as a negative regulator responding to Cd stress in rice. The expression of its target gene, *NRAMP325*, was decreased by Cd treatment. Enhanced levels of H_2O_2 and MDA were also observed in the transgenic rice lines compared to the wild types.

Cultivar-dependent expression of miRNAs was observed in *Ipomea aquatica* Forsk. in response to Cd stress (Shen et al. 2017). Low-shoot-Cd-cultivar (QLQ) and high-shoot-Cd-cultivar (T308) were selected for the study. The expression of five miRNAs was regulated in QLQ only. Among these, miR395 was upregulated, leading to Cd retention and detoxification in root tissue. Differential expression patterns of miRNA5139, miRNA1511, and miRNA8155 take part in the Cd reduction in shoot tissues. On the other hand, the pattern was more complex in the T308 cultivar. Here, the expression of miRNA397 was exclusively regulated in T308 via MATE efflux family of the target gene (Shen et al. 2017). Recently, role of miRNA in response to Cd stress was further studied in *A. thaliana* (Pegler et al. 2021). It was reported that the expression level of analyzed six miRNAs (miR160, miR167, miR393, miR395, miR399, and miR408) was regulated by double-stranded RNA binding (DRB) protein 1, and miR396 was regulated by *DRB2* target gene, respectively.

In a previous study, Jian et al. (2018) identified 44 existing miRNAs and 103 novel miRNA in rapeseed against Cd stress. Targeted genes were involved in the biotic stress response, ion homeostasis, secondary metabolism, and transcription factor regulation. Downregulated (miR156d, miR1140, and miR2111b-3p) and upregulated (miR6035) expression was reported in response to stress. Roots of the B73 and Mo17 inbred maize lines were treated at different intervals. Downregulated (miR156, miR166, miR167, and miR171) and upregulated (miR393) expression pattern was observed in maize against Cd stress (Gao et al. 2019). The target genes involved belong to the TF families (SBP-box, bZIP, ARFs, and GRAS), F-box-like protein, and some metabolic pathways.

miRNA-mediated Cr response was studied in *Miscanthus sinensis* (Nie et al. 2021). A total of 104 conserved miRNAs and 158 non-conserved miRNAs were identified in different tissues. Based on clustering analysis, tissue-specific miRNA expression was also observed. Additionally, 833 DEGs targeting 45 miRNAs (roots) and 280 DEGs targeting 13 miRNA (leaves) were identified. Expression profile analysis showed potential involvement of miR167a, novel_miR15, and novel_miR22 target genes in Cr transportation and chelation. Besides, miR156a, miR164, miR396d, and novel_miR155 were also identified for their participation in the physiological and biochemical metabolisms, and the Cr detoxification in plants (Nie et al. 2021). Moreover, Dubey et al. (2020) studied the role of miRNAs in response to Cr stress in rice. Notably, 512 miRNAs were identified in Cr-treated samples. Differential expression patterns of 13 conserved were observed. Target gene prediction reveals the role of HSPs, ABC transporters, TFs, auxin hormone, and metal ion transport in Cr stress (Dubey et al. 2020).

In short, these findings reveal that miRNA substantially controls gene expression and plays a crucial part in alleviating the harmful effect of HM toxicity. Therefore, miRNAomics must be further used to find HM-associated miRNAs in various major crop plants.

2.3.6 Phenomics

Phenomics is a term that refers to a collection of biological, physical, and metabolic processes that an organism expresses as phenotypes in a given environment. At the same time, a phenotype encompasses a set of qualities that may be observed through direct observation or analytical methods and can be defined as a phenotype–environment interaction (Pratap et al. 2019). Changes in environmental conditions, such as abiotic stressors, can cause the plant phenotype to show differences within the same genotype (Walter et al. 2015; Pratap et al. 2019). Plant phenomics uses recordings and analytical techniques to quantify various phenotypes, resulting in high-dimensional phenotypic data for organisms (Walter et al. 2015; Pratap et al. 2019). In the post-genomic era, the phenomics implications are significant. It is clear after employing GWAS, QTL mapping, MAS, and genomic selection, all of which rely heavily on HTP for crop development. HTP approaches use automated digital analysis to compile data samples, evaluate the data, and treat it using appropriate statistical processing (Awada et al. 2018).

Identifying plant phenotypic expression is possible because of new approaches for genetic dissection, knowledge of plant structure, and the study of all functional activities (Pratap et al. 2019). Falk et al. (2020) used the phenome technique to evaluate genetic diversity in soybean root system architecture (RSA) features, as it is challenging to analyze RSA attributes for many accessions manually. Clusters of accessions on phenotype and genotype demonstrated similarities, with genotype-based groups showing regional co-relation. This research shows how functional genetic diversity can be exploited to develop root features that can be used to benefit crops in the future (Falk et al. 2020). A dataset of stress-responsive genes in *Arabidopsis* has been created under biotic and abiotic conditions such as Fe and Al toxicity, osmotic, light, temperature, and oxidative stress. Numerous shared and unshared molecular functions, phenomic traits, biological functions, and metabolic pathways were identified in this work, which could aid in the development of superior varieties using genome editing techniques (Naika et al. 2013). Liu et al. (2021a) recently investigated the rice root system's phenomic and metabolomic response to Cd stress. According to the researchers, plants cope with HM stress by reprogramming root architecture and upregulating phenylethanoid glycosides (Liu et al. 2021a). Because just a few phenomics applications against HM toxicity in plants have been reported, there is a need to bridge this gap by employing phenomics to develop HM-tolerant crop plants.

2.4 Conclusion and Perspectives

The last century's intensive and extensive anthropogenic activities have produced an unparalleled level of environmental contamination, producing adverse consequences for all living organisms. Heavy metal contamination emerging as one of the significant contaminants affects plant's vital physiological and molecular characteristics, leading to massive impacts on food security and human health. Prominent among the HM-induced effects include damage to cellular components and interruptions with cell cycle, generation of ROS, impairment of DNA, nutrient imbalances, and up- and downregulation of genomes, transcriptomes, and proteases. In such times, advanced and rapid omics approaches have provided an array of hope in understanding plant responses to HM exposure. Innovations and improvements in genomics, transcriptomics, proteomics, metabolomics, miRNAomics, phenomics, etc., have provided tools to comprehend pathways and mechanisms adapted by plants to optimize reactions to HMs. They provide the molecular, transcriptomic, and genetic profiling of the plant body in response to HM toxicity, thus providing opportunities to develop and enhance various mechanisms to combat the issue at hand. The application of various omics techniques individually and in combination has also opened avenues for the production and optimization of plant species/breeds. High-quality and quantity yields are underway and are predicted to further enhance sustainably to meet the global development of the environmental scenario. Moving forward, researchers have high hopes for the various omics techniques to unravel complex nature of plant and HM interactions.

References

Abassi, S., Wang, H., Ponmani, T., and Ki, J.S. (2019). Small heat shock protein genes of the green algae *Closterium ehrenbergii*: cloning and differential expression under heat and heavy metal stresses. *Environmental Toxicology* 34: 1013–1024.

Adimalla, N. (2020). Heavy metals pollution assessment and its associated human health risk evaluation of urban soils from Indian cities: a review. *Environmental Geochemistry and Health* 42: 173–190.

Aglawe, S.B., Verma, A.K., and Upadhyay, A.K. (2021). Bioinformatics tools and databases for genomics-assisted breeding and population genetics of plants: a review. *Current Bioinformatics* 16: 766–773.

Andresen, E., Lyubenova, L., Hubáček, T. et al. (2020). Chronic exposure of soybean plants to nanomolar cadmium reveals specific additional high-affinity targets of cadmium toxicity. *Journal of Experimental Botany* 71: 1628–1644.

Audagnotto, M. and Dal Peraro, M. (2017). Protein post-translational modifications: In silico prediction tools and molecular modeling. *Computational and Structural Biotechnology Journal* 15: 307–319.

Awada, L., Phillips, P.W., and Smyth, S.J. (2018). The adoption of automated phenotyping by plant breeders. *Euphytica* 214: 1–15.

Chauhan, R., Awasthi, S., Indoliya, Y. et al. (2020). Transcriptome and proteome analyses reveal selenium mediated amelioration of arsenic toxicity in rice (*Oryza sativa* L.). *Journal of Hazardous Materials* 390: 122122.

Chen, L., Wan, H., Qian, J. et al. (2018). Genome-wide association study of cadmium accumulation at the seedling stage in rapeseed (*Brassica napus* L.). *Frontiers in Plant Science* 9: 375.

Chen, P., Chen, T., Li, Z. et al. (2020). Transcriptome analysis revealed key genes and pathways related to cadmium-stress tolerance in Kenaf (*Hibiscus cannabinus* L.). *Industrial Crops and Products* 158: 112970.

Cosio, C. and Renault, D. (2020). Effects of cadmium, inorganic mercury and methyl-mercury on the physiology and metabolomic profiles of shoots of the macrophyte *Elodea nuttallii*. *Environmental Pollution* 257: 113557.

Cui, L., Feng, X., Lin, C.J. et al. (2014). Accumulation and translocation of 198Hg in four crop species. *Environmental Toxicology and Chemistry* 33: 334–340.

Cui, Y., Wang, M., Yin, X. et al. (2019). OsMSR3, a small heat shock protein, confers enhanced tolerance to copper stress in *Arabidopsis thaliana*. *International Journal of Molecular Sciences* 20: 6096.

Dhalaria, R., Kumar, D., Kumar, H. et al. (2020). Arbuscular mycorrhizal fungi as potential agents in ameliorating heavy metal stress in plants. *Agronomy* 10: 815.

Di, X., Zheng, F., Norton, G.J. et al. (2021). Physiological responses and transcriptome analyses of upland rice following exposure to arsenite and arsenate. *Environmental and Experimental Botany* 183: 104366.

Ding, Y., Wang, Y., Jiang, Z. et al. (2017). MicroRNA268 overexpression affects rice seedling growth under cadmium stress. *Journal of Agricultural and Food Chemistry* 65: 5860–5867.

Ding, Y., Gong, S., Wang, Y. et al. (2018). MicroRNA166 modulates cadmium tolerance and accumulation in rice. *Plant Physiology* 177: 1691–1703.

Dubey, S., Saxena, S., Chauhan, A.S. et al. (2020). Identification and expression analysis of conserved microRNAs during short and prolonged chromium stress in rice (*Oryza sativa*). *Environmental Science and Pollution Research* 27: 380–390.

Elobeid, M., Göbel, C., Feussner, I., and Polle, A. (2012). Cadmium interferes with auxin physiology and lignification in poplar. *Journal of Experimental Botany* 63: 1413–1421.

Erickson, T.B., Brooks, J., Nilles, E.J. et al. (2019). Environmental health effects attributed to toxic and infectious agents following hurricanes, cyclones, flash floods and major hydrometeorological events. *Journal of Toxicology and Environmental Health, Part B* 22: 157–171.

Ezemaduka, A.N., Wang, Y., and Li, X. (2017). Expression of CeHSP17 protein in response to heat shock and heavy metal ions. *Journal of Nematology* 49: 334.

Falk, K.G., Jubery, T.Z., O'Rourke, J.A. et al. (2020). Soybean root system architecture trait study through genotypic, phenotypic, and shape-based clusters. *Plant Phenomics* 2020: 1925495.

Finger-Teixeira, A., Ferrarese, M.L.L., Soares, A.R. et al. (2010). Cadmium-induced lignification restricts soybean root growth. *Ecotoxicology and Environmental Safety* 73: 1959–1964.

Fu, Y., Mason, A.S., Zhang, Y. et al. (2019). MicroRNA-mRNA expression profiles and their potential role in cadmium stress response in *Brassica napus*. *BMC Plant Biology* 19: 1–20.

Gallo-Franco, J.J., Sosa, C.C., Ghneim-Herrera, T., and Quimbaya, M. (2020). Epigenetic control of plant response to heavy metal stress: a new view on aluminum tolerance. *Frontiers in Plant Science* 11: 2025.

Gao, J., Luo, M., Peng, H. et al. (2019). Characterization of cadmium-responsive MicroRNAs and their target genes in maize (*Zea mays*) roots. *BMC Molecular Biology* 20: 1–9.

Georgiadou, E.C., Kowalska, E., Patla, K. et al. (2018). Influence of heavy metals (Ni, Cu, and Zn) on nitro-oxidative stress responses, proteome regulation and allergen production in basil (*Ocimum basilicum* L.) plants. *Frontiers in Plant Science* 9: 862.

Ghorbani, E., Nowruzi, B., Nezhadali, M., and Hekmat, A. (2022). Metal removal capability of two cyanobacterial species in autotrophic and mixotrophic mode of nutrition. *BMC Microbiology* 22: 1–15.

Greco, M., Sáez, C.A., Contreras, R.A. et al. (2019). Cadmium and/or copper excess induce interdependent metal accumulation, DNA methylation, induction of metal chelators and antioxidant defences in the seagrass *Zostera marina*. *Chemosphere* 224: 111–119.

Grill, E., Winnacker, E.L., and Zenk, M.H. (1985). Phytochelatins: the principal heavy-metal complexing peptides of higher plants. *Science* 230: 674–676.

Gupta, P., Balyan, H., Sharma, S., and Kumar, R. (2021). Biofortification and bioavailability of Zn, Fe and Se in wheat: present status and future prospects. *Theoretical and Applied Genetics* 134: 1–35.

Gutsch, A., Zouaghi, S., Renaut, J. et al. (2018). Changes in the proteome of *Medicago sativa* leaves in response to long-term cadmium exposure using a cell-wall targeted approach. *International Journal of Molecular Sciences* 19: 2498.

Han, H., Zhang, H., Qin, S. et al. (2021). Mechanisms of *Enterobacter bugandensis* TJ6 immobilization of heavy metals and inhibition of Cd and Pb uptake by wheat based on metabolomics and proteomics. *Chemosphere* 276: 130157.

Hasanuzzaman, M., Bhuyan, M., Zulfiqar, F. et al. (2020). Reactive oxygen species and antioxidant defense in plants under abiotic stress: revisiting the crucial role of a universal defense regulator. *Antioxidants* 9: 681.

Hossain, M.A., Piyatida, P., da Silva, J.A.T., and Fujita, M. (2012). Molecular mechanism of heavy metal toxicity and tolerance in plants: central role of glutathione in detoxification of reactive oxygen species and methylglyoxal and in heavy metal chelation. *Journal of Botany* 2012: 872875.

Ibarra, A.A.G., Wrobel, K., Barrientos, E.Y. et al. (2019). Impact of Cr (VI) on the oxidation of polyunsaturated fatty acids in *Helianthus annuus* roots studied by metabolomic tools. *Chemosphere* 220: 442–451.

Jamla, M., Khare, T., Joshi, S. et al. (2021a). Omics approaches for understanding heavy metal responses and tolerance in plants. *Current Plant Biology* 27: 100213.

Jamla, M., Patil, S., Joshi, S. et al. (2021b). MicroRNAs and their exploration for developing heavy metal-tolerant plants. *Journal of Plant Growth Regulation* 41: 2579–2595.

Jian, H., Yang, B., Zhang, A. et al. (2018). Genome-wide identification of microRNAs in response to cadmium stress in oilseed rape (*Brassica napus* L.) using high-throughput sequencing. *International Journal of Molecular Sciences* 19: 1431.

Jian, M., Zhang, D., Wang, X. et al. (2020). Differential expression pattern of the proteome in response to cadmium stress based on proteomics analysis of wheat roots. *BMC Genomics* 21: 1–13.

Jing, M., Zhang, H., Wei, M. et al. (2022). Reactive oxygen species partly mediate DNA methylation in responses to different heavy metals in pokeweed. *Frontiers in Plant Science* 13: 845108.

Ju, Y.-H., Roy, S.K., Roy Choudhury, A. et al. (2021). Proteome changes reveal the protective roles of exogenous citric acid in alleviating Cu toxicity in *Brassica napus* L. *International Journal of Molecular Sciences* 22: 5879.

Khalid, N., Aqeel, M., and Noman, A. (2019). System biology of metal tolerance in plants: an integrated view of genomics, transcriptomics, metabolomics, and phenomics. In: *Plant Metallomics and Functional Omics* (ed. G. Sablok), 107–144. Springer.

Khan, M.I.R., Chopra, P., Chhillar, H. et al. (2021). Regulatory hubs and strategies for improving heavy metal tolerance in plants: chemical messengers, omics and genetic engineering. *Plant Physiology and Biochemistry* 164: 260–278.

Lee, S., Kim, Y.-Y., Lee, Y., and An, G. (2007). Rice P1B-type heavy-metal ATPase, *OsHMA9*, is a metal efflux protein. *Plant Physiology* 145: 831–842.

Leng, Y., Li, Y., Wen, Y. et al. (2020). Transcriptome analysis provides molecular evidences for growth and adaptation of plant roots in cadimium-contaminated environments. *Ecotoxicology and Environmental Safety* 204: 111098.

Li, Y., Liang, L., Li, W. et al. (2021). ZnO nanoparticle-based seed priming modulates early growth and enhances physio-biochemical and metabolic profiles of fragrant rice against cadmium toxicity. *Journal of Nanobiotechnology* 19: 1–19.

Li, D., He, T., Saleem, M., and He, G. (2022a). Metalloprotein-specific or critical amino acid residues: perspectives on plant-precise detoxification and recognition mechanisms under cadmium stress. *International Journal of Molecular Sciences* 23: 1734.

Li, P., Zhang, J., Sun, X. et al. (2022b). Atmospheric Pb induced hormesis in the accumulator plant *Tillandsia usneoides*. *Science of The Total Environment* 811: 152384.

Liu, X., Chen, S., Chen, M. et al. (2019a). Association study reveals genetic loci responsible for arsenic, cadmium and lead accumulation in rice grain in contaminated farmlands. *Frontiers in Plant Science* 10: 61.

Liu, Y., Lu, S., Liu, K. et al. (2019b). Proteomics: a powerful tool to study plant responses to biotic stress. *Plant Methods* 15: 1–20.

Liu, X., Fan, F., Liu, M. et al. (2020). Quantitative trait loci mapping of mineral element contents in brown rice using backcross inbred lines derived from *Oryza longistaminata*. *Frontiers in Plant Science* 11: 1229.

Liu, C., Lan, M.-m., He, E.-k. et al. (2021a). Phenomic and metabolomic responses of roots to cadmium reveal contrasting resistance strategies in two rice cultivars (*Oryza sativa* L.). *Soil Ecology Letters* 3: 220–229.

Liu, T., Chen, Q., Zhang, L. et al. (2021b). The toxicity of selenium and mercury in *Suaeda salsa* after 7-days exposure. *Comparative Biochemistry and Physiology Part C: Toxicology & Pharmacology* 244: 109022.

Ma, L., Qing, C., Zhang, M. et al. (2021). GWAS with a PCA uncovers candidate genes for accumulations of microelements in maize seedlings. *Physiologia Plantarum* 172: 2170–2180.

Mehrabanjoubani, P., Abdolzadeh, A., Sadeghipour, H.R. et al. (2019). Silicon increases cell wall thickening and lignification in rice (*Oryza sativa*) root tip under excess Fe nutrition. *Plant Physiology and Biochemistry* 144: 264–273.

Molina, L. and Segura, A. (2021). Biochemical and metabolic plant responses toward polycyclic aromatic hydrocarbons and heavy metals present in atmospheric pollution. *Plants* 10: 2305.

Mumtaz, M.A., Hao, Y., Mehmood, S. et al. (2022). Physiological and transcriptomic analysis provide molecular insight into 24-epibrassinolide mediated Cr(VI)-toxicity tolerance in pepper plants. *Environmental Pollution* 306: 119375.

Mwamba, T., Islam, F., Ali, B. et al. (2020). Comparative metabolomic responses of low-and high-cadmium accumulating genotypes reveal the cadmium adaptive mechanism in *Brassica napus*. *Chemosphere* 250: 126308.

Naika, M., Shameer, K., and Sowdhamini, R. (2013). Comparative analyses of stress-responsive genes in *Arabidopsis thaliana:* insight from genomic data mining, functional enrichment, pathway analysis and phenomics. *Molecular BioSystems* 9: 1888–1908.

Nie, G., Liao, Z., Zhong, M. et al. (2021). MicroRNA-mediated responses to chromium stress provides insight toward to tolerance characteristics of *Miscanthus sinensis*. *Frontiers in Plant Science* 12: 1019.

Paape, T., Heiniger, B., Santo Domingo, M. et al. (2021). Genome-wide association study reveals complex genetic architecture of cadmium and mercury accumulation and tolerance traits in *Medicago truncatula*. *bioRxiv* https://doi.org/10.1101/2021.10.29.465169.

Pan, X., Li, Y., Liu, W. et al. (2020). QTL mapping and candidate gene analysis of cadmium accumulation in polished rice by genome-wide association study. *Scientific Reports* 10: 1–11.

Parrotta, L., Guerriero, G., Sergeant, K. et al. (2015). Target or barrier? The cell wall of early-and later-diverging plants vs cadmium toxicity: differences in the response mechanisms. *Frontiers in Plant Science* 6: 133.

Patil, S., Joshi, S., Jamla, M. et al. (2021). MicroRNA-mediated bioengineering for climate-resilience in crops. *Bioengineered* 12: 10430–10456.

Pegler, J.L., Oultram, J.M., Nguyen, D.Q. et al. (2021). MicroRNA-mediated responses to cadmium stress in *Arabidopsis thaliana*. *Plants* 10: 130.

Pidatala, V.R., Li, K., Sarkar, D. et al. (2018). Comparative metabolic profiling of vetiver (*Chrysopogon zizanioides*) and maize (*Zea mays*) under lead stress. *Chemosphere* 193: 903–911.

Pratap, A., Gupta, S., Nair, R.M. et al. (2019). Using plant phenomics to exploit the gains of genomics. *Agronomy* 9: 126.

Rai, P.K., Lee, S.S., Zhang, M. et al. (2019). Heavy metals in food crops: health risks, fate, mechanisms, and management. *Environ International* 125: 365–385.

Raza, A. (2022). Metabolomics: a systems biology approach for enhancing heat stress tolerance in plants. *Plant Cell Reports* 41: 741–763.

Raza, A., Habib, M., Kakavand, S.N. et al. (2020). Phytoremediation of cadmium: physiological, biochemical, and molecular mechanisms. *Biology* 9: 177.

Raza, A., Charagh, S., Najafi-Kakavand, S., and Siddiqui, M.H. (2021a). The crucial role of jasmonates in enhancing heavy metals tolerance in plants. In: *Jasmonates and Salicylates Signaling in Plants*, 159–183. Springer.

Raza, A., Habib, M., Charagh, S., and Kakavand, S.N. (2021b). Genetic engineering of plants to tolerate toxic metals and metalloids. In: *Handbook of Bioremediation* (ed. M. Hasanuzzaman and M.N.V. Prasad), 411–436. Elsevier.

Raza, A., Hussain, S., Javed, R. et al. (2021c). Antioxidant defense systems and remediation of metal toxicity in plants. In: *Approaches to the Remediation of Inorganic Pollutants* (ed. M. Hasanuzzaman), 91–124. Springer.

Raza, A., Su, W., Hussain, M.A. et al. (2021d). Integrated analysis of metabolome and transcriptome reveals insights for cold tolerance in rapeseed (*Brassica napus* L.). *Frontiers in Plant Science* 12: 721681.

Raza, A., Tabassum, J., Kudapa, H., and Varshney, R.K. (2021e). Can omics deliver temperature resilient ready-to-grow crops? *Critical Reviews in Biotechnology* 41: 1209–1232.

Raza, A., Tabassum, J., Zahid, Z. et al. (2022). Advances in 'omics' approaches for improving toxic metals/metalloids tolerance in plants. *Frontiers in Plant Science* 12: 794373.

Saeed, F., Chaudhry, U.K., Bakhsh, A. et al. (2022). Moving beyond DNA sequence to improve plant stress responses. *Frontiers in Genetics* 13: 874648.

Sahr, T., Voigt, G., Paretzke, H.G. et al. (2005). Caesium-affected gene expression in *Arabidopsis thaliana*. *New Phytologist* 165: 747–754.

Sandhu, N., Yadav, S., Catolos, M. et al. (2021). Developing climate-resilient, direct-seeded, adapted multiple-stress-tolerant rice applying genomics-assisted breeding. *Frontiers in Plant Science* 12: 637488.

Sanjosé, I., Navarro-Roldán, F., Infante-Izquierdo, M.D. et al. (2021). Accumulation and effect of heavy metals on the germination and growth of *Salsola vermiculata* L. seedlings. *Diversity* 13: 539.

Sauge-Merle, S., Cuiné, S., Carrier, P. et al. (2003). Enhanced toxic metal accumulation in engineered bacterial cells expressing *Arabidopsis thaliana* phytochelatin synthase. *Applied and Environmental Microbiology* 69: 490–494.

Sengar, R., Gautam, M., Sengar, K. et al. (2010). Physiological and metabolic effect of mercury accumulation in higher plants system. *Toxicological & Environmental Chemistry* 92: 1265–1281.

Shen, C., Huang, Y.-Y., He, C.-T. et al. (2017). Comparative analysis of cadmium responsive microRNAs in roots of two *Ipomoea aquatica* Forsk. cultivars with different cadmium accumulation capacities. *Plant Physiology and Biochemistry* 111: 329–339.

Shi, W., Zhang, Y., Chen, S. et al. (2019). Physiological and molecular mechanisms of heavy metal accumulation in nonmycorrhizal versus mycorrhizal plants. *Plant, Cell & Environment* 42: 1087–1103.

Shilin, D., Chaolei, L., Lianguang, S. et al. (2021). Identification of QTLs for cadmium tolerance during seedling stage and validation of qCDSL1 in rice. *Rice Science* 28: 81–88.

Shrestha, A., Dziwornu, A.K., Ueda, Y. et al. (2018). Genome-wide association study to identify candidate loci and genes for Mn toxicity tolerance in rice. *PloS One* 13: e0192116.

Sorrentino, M., Capozzi, F., Amitrano, C. et al. (2018). Performance of three cardoon cultivars in an industrial heavy metal-contaminated soil: effects on morphology, cytology and photosynthesis. *Journal of Hazardous Materials* 351: 131–137.

Sun, L., Cao, X., Tan, C. et al. (2020). Analysis of the effect of cadmium stress on root exudates of *Sedum plumbizincicola* based on metabolomics. *Ecotoxicology and Environmental Safety* 205: 111152.

Tamás, M.J., Fauvet, B., Christen, P., and Goloubinoff, P. (2018). Misfolding and aggregation of nascent proteins: a novel mode of toxic cadmium action in vivo. *Current Genetics* 64: 177–181.

Varshney, R.K., Singh, V.K., Kumar, A. et al. (2018). Can genomics deliver climate-change ready crops? *Current Opinion in Plant Biology* 45: 205–211.

Varshney, R.K., Bohra, A., Yu, J. et al. (2021). Designing future crops: genomics-assisted breeding comes of age. *Trends in Plant Science* 26: 631–649.

Velu, G., Tutus, Y., Gomez-Becerra, H.F. et al. (2017). QTL mapping for grain zinc and iron concentrations and zinc efficiency in a tetraploid and hexaploid wheat mapping populations. *Plant and Soil* 411: 81–99.

Walter, A., Liebisch, F., and Hund, A. (2015). Plant phenotyping: from bean weighing to image analysis. *Plant Methods* 11: 1–11.

Wang, H.-C., Wu, J.-S., Chia, J.-C. et al. (2009). Phytochelatin synthase is regulated by protein phosphorylation at a threonine residue near its catalytic site. *Journal of Agricultural and Food Chemistry* 57: 7348–7355.

Wang, X., Cheng, Y., Yang, C. et al. (2019). QTL mapping for aluminum tolerance in RIL population of soybean (*Glycine max* L.) by RAD sequencing. *PloS One* 14: e0223674.

Wang, G., DiTusa, S.F., Oh, D.H. et al. (2021a). Cross species multi-omics reveals cell wall sequestration and elevated global transcript abundance as mechanisms of boron tolerance in plants. *New Phytologist* 230: 1985–2000.

Wang, R., Lin, K., Chen, H. et al. (2021b). Metabolome analysis revealed the mechanism of exogenous glutathione to alleviate cadmium stress in maize (*Zea mays* L.) seedlings. *Plants* 10: 105.

Wang, J., Chen, X., Chu, S. et al. (2022). Comparative cytology combined with transcriptomic and metabolomic analyses of *Solanum nigrum* L. in response to Cd toxicity. *Journal of Hazardous Materials* 423: 127168.

Wenli, Z., Chang, L., Yajun, Z. et al. (2020). Rice heavy metal P-type ATPase *OsHMA6* is likely a copper efflux protein. *Rice Science* 27: 143–151.

Xiao, W., Liu, P., Wang, K. et al. (2021). Relationship between ionomics and transcriptomics of rice plant in response to arsenite stress. *Environmental and Experimental Botany* 189: 104565.

Xiao, Q., Bai, X., Zhang, C., and He, Y. (2022). Advanced high-throughput plant phenotyping techniques for genome-wide association studies: a review. *Journal of Advanced Research* 35: 215–230.

Yang, X., Kang, Y., Liu, Y. et al. (2021). Integrated analysis of miRNA-mRNA regulatory networks of potato (*Solanum tuberosum* L.) in response to cadmium stress. *Ecotoxicology and Environmental Safety* 224: 112682.

Yin, A., Shen, C., Huang, Y. et al. (2022). Transcriptomic analyses of sweet potato in response to Cd exposure and protective effects of K on Cd-induced physiological alterations. *Environmental Science and Pollution Research* 1-15.

Yu, X.-Z., Lu, C.-J., Tang, S., and Zhang, Q. (2020). Transcriptomic analysis of cytochrome P450 genes and pathways involved in chromium toxicity in *Oryza sativa*. *Ecotoxicology* 29: 503–513.

Zeng, T., Fang, B., Huang, F. et al. (2021). Mass spectrometry-based metabolomics investigation on two different indica rice grains (*Oryza sativa* L.) under cadmium stress. *Food Chemistry* 343: 128472.

Zhang, F., Xiao, X., Xu, K. et al. (2020a). Genome-wide association study (GWAS) reveals genetic loci of lead (Pb) tolerance during seedling establishment in rapeseed (*Brassica napus* L.). *BMC Genomics* 21: 1–12.

Zhang, L., Ding, H., Jiang, H. et al. (2020b). Regulation of cadmium tolerance and accumulation by miR156 in Arabidopsis. *Chemosphere* 242: 125168.

Zhao, J., Yang, W., Zhang, S. et al. (2018). Genome-wide association study and candidate gene analysis of rice cadmium accumulation in grain in a diverse rice collection. *Rice* 11: 1–15.

Zhao, L., Zhu, Y.-H., Wang, M. et al. (2021). Comparative transcriptome analysis of the hyperaccumulator plant *Phytolacca americana* in response to cadmium stress. *3 Biotech* 11: 1–23.

Zhou, H., Xiao, X., Asjad, A. et al. (2022). Integration of GWAS and transcriptome analyses to identify SNPs and candidate genes for aluminum tolerance in rapeseed (*Brassica napus* L.). *BMC Plant Biology* 22: 1–17.

Zulfiqar, U., Farooq, M., Hussain, S. et al. (2019). Lead toxicity in plants: impacts and remediation. *Journal of Environmental Management* 250: 109557.

3

Heavy Metals/Metalloids in Food Crops and Their Implications for Human Health

Shihab Uddin[1], *Hasina Afroz*[1], *Mahmud Hossain*[1], *Jessica Briffa*[2], *Renald Blundell*[3,4], *and Md. Rafiqul Islam*[1]

[1] *Department of Soil Science, Bangladesh Agricultural University, Mymensingh, Bangladesh*
[2] *Department of Physiology and Biochemistry, Faculty of Medicine and Surgery, University of Malta, Imsida, Malta*
[3] *Centre for Molecular Medicine and Biobanking, University of Malta, Msida, Malta*
[4] *Department of Physiology and Biochemistry, Faculty of Medicine and Surgery, University of Malta, Msida, Malta*

3.1 Introduction

In general, metals are elements that are hard, lustrous, fusible, malleable, ductile, and have good electrical and thermal conductivity. An element without the characteristics of a metal is referred to as a nonmetal. An element with properties halfway between those of metals and nonmetals is said to be a metalloid. Heavy metals are ubiquitous and naturally occurring elements that cannot be degraded or destroyed. Heavy metals and/or metalloids are generally referred to any naturally occurring metallic chemical element having an atomic number greater than 20, a relatively high atomic weight in the range of 63.5–200.6 $g\,mol^{-1}$ and the density at least five times of water (5 $g\,cm^{-3}$) (Koller and Saleh 2018). Heavy metals and/or metalloids include arsenic (As), cadmium (Cd), mercury (Hg), chromium (Cr), thallium (Tl), lead (Pb), aluminum (Al), etc. Heavy metals can be broadly classified into two categories: essential and nonessential. Essential heavy metals are required by living organisms for carrying out fundamental processes viz. growth, metabolism, and development of different organs. Essential heavy metals like copper (Cu), iron (Fe), manganese (Mn), cobalt (Co), zinc (Zn), and nickel (Ni) are required by plants for their functions in forming cofactors that are structurally and functionally vital for enzymes and other proteins (LeFauve and Connaughton 2017). Nonessential heavy metals and metalloids like As, Cd, Pb, Hg, Cr, and Al are not required by plants. They are considered as contaminants and toxic substances even at very low concentrations, risk to the human health and environment (Koller and Saleh 2018). Human activities like industrial, mining, and agricultural activities have led to a widespread distribution of heavy metals in the ecosystems, posing risks to eco-physiological systems and human health (Yin et al. 2017).

Heavy Metal Toxicity and Tolerance in Plants: A Biological, Omics, and Genetic Engineering Approach, First Edition. Edited by Mohammad Anwar Hossain, AKM Zakir Hossain, Sylvain Bourgerie, Masayuki Fujita, Om Parkash Dhankher, and Parvez Haris.

Heavy metals are nonbiodegradable and thus cannot be broken down. Organisms may purify metal ions through a number of ways such as hiding the active heavy metal inside a protein or placing the insoluble form in intracellular granules, which will be either stored long term or excreted in the feces. Bioaccumulation of the heavy metals causes both biological and physiological impediments. Heavy metals also interact with nucleic acids causing deoxyribonucleic acid (DNA) damage and DNA repair mechanism, regulatory molecular inactivation, which will lead to carcinogenesis. Heavy metals can also react with lipids causing cell membrane damage and thus causing cell damage. The metal ions also react with protein, causing misfolding, aggregation, enzyme activation, and conformational change, all leading to cellular function loss. Heavy metal ions also interact with the nervous system causing neuron damage and neurotransmitter inhibition causing neurotoxicity (Engwa et al. 2019; Briffa et al. 2020). Owing to their bioaccumulation potential along the food chain, high toxicity, prevalence, and persistence in the environment, As, Cd, Pb, Cr, and Hg have become a major public health concern across the world (Suhani et al. 2021). The purpose of this chapter is to provide an overview of the key heavy metals and metalloids that pose a threat to human health, including their sources, forms, potential for contaminating the food supply, pharmacokinetic processes, toxicological processes, and possible remedial measures.

3.2 Arsenic

Arsenic is a notorious toxin classified as a group-I carcinogen by the International Agency for Research on Cancer (IARC) (Pandey et al. 2018). Albertus Mannus (1250) initially isolated this substance as arsenic sulfide in the early Bronze Age (2500 BCE). Arsenic is the third element after phosphorus (P) in Group VA on the periodic table. It is a semi-metallic compound (metalloid) having atomic number 33, atomic weight 74.92, boiling point 613 °C at sublimes, and density 5.75 $g\,cm^{-3}$ at 25 °C. There are a number of recognized isotopes, for example ^{73}As. It has no beneficial role in plants and humans rather exposure of As induces adverse effects on human health and causing cancer.

3.2.1 Sources and Forms

Arsenic is the 21 most abundant element in the earth crust with an average soil concentration of 5–13 $mg\,kg^{-1}$. Volcanoes are important natural sources of As, soil As level near the volcanoes is around 20 $mg\,kg^{-1}$ (Chung et al. 2013). Arsenic is mostly found in association with sulfide and oxide complexes in soil and rock formations, and it is concentrated in magnetic sulfides and iron ores. Arsenopyrite (FeAsS), scorodite ($FeAsO_4{\cdot}2H_2O$), realgar (As_2S_3), and orpiment (As_2S_3) are a few of the well-known As minerals (James et al. 2017). The extraction and smelting of nonferrous metals, the use of pesticides, the burning of coal, wood, and garbage are the anthropogenic sources of As (Figure 3.1).

Arsenic can be found in nature in the oxidation states of arsenic (0), arsenate (+V), arsenite (+III), and arsine (−III) (James et al. 2017). Arsenite (H_3AsO_3, $H_2AsO_3^-$, $HAsO_3^{2-}$, and AsO_3^{3-}) and arsenate (H_3AsO_4, $H_2AsO_4^-$, $HAsO_4^{2-}$, and AsO_4^{3-}) species dominate in the aquatic environment. Arsenate species are more usually found in under oxidized conditions, whereas arsenite species typically predominate under reduced conditions. Inorganic As species can be methylated by bacteria, fungi, yeasts, and mammals to produce

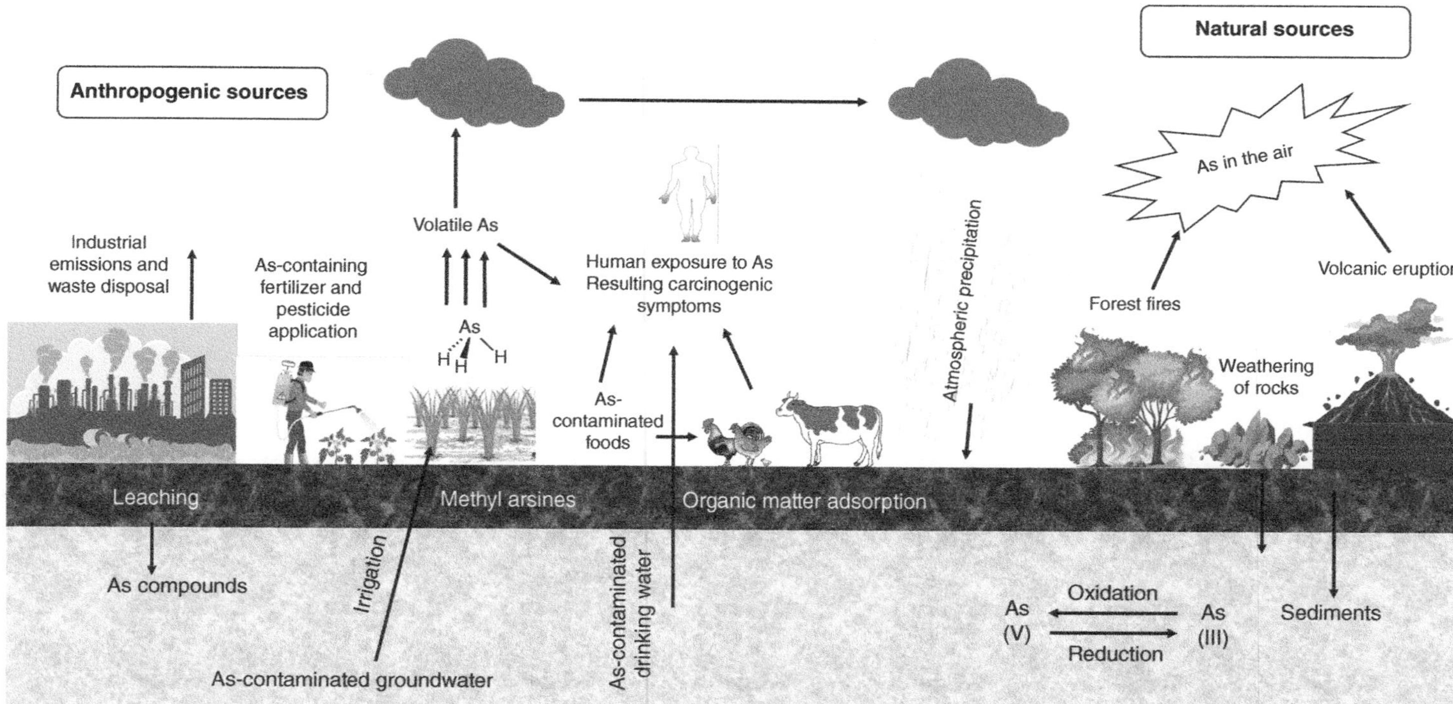

Figure 3.1 Arsenic contamination in the environment and its exposure to human.

monomethyl As (MMA), dimethyl As (DMA), and trimethyl As (TMA). As compounds are classified according to their increasing toxicity in the following order: organic arsenicals < As0 < inorganic species (As^{5+} < As^{3+}) < arsine (Kuivenhoven and Mason 2019).

3.2.2 Food Chain Contamination

Groundwater and food continue to be the main sources of exposure for the majority of the people in the world (Chung et al. 2013) (Figure 3.1). Elevated inorganic As exposure comes mostly from groundwater due to the extensive natural pollution of aquifers around the world. Although the particular mechanisms of As mobilization are not fully known, oxidation, reduction, and/or carbonation mechanisms are used to release As from rock deposits into groundwater. Besides drinking water, food is a significant source of As exposure (Figure 3.1). Rice irrigated with As-tainted water is the main source of inorganic As exposure in humans contributing >60% of total exposure. However, exposure to inorganic As in vegetables, fruits, and meat can represent another major part of the As exposure through food. Arsenic can also be found in foods like fish, shellfish, pork, poultry, and dairy products. Besides, seafood is a major source of methylated As. Young children may also be exposed to modest quantities of As through routine hand-to-mouth activities, a significant route of exposure. Smokers may also be exposed to the naturally occurring inorganic As found in tobacco.

3.2.3 Pharmacokinetic Processes

Arsenic is absorbed orally according to its chemical form, where three fourths of the metal are found in the form of MMA, As^{3+}, As^{5+}, and DMA in humans. Absorption occurs through passive diffusion and the pathway is through inhalation, the oral route, and at a lesser absorption though the dermal route. There is no superior distribution of As and its metabolites in any of the bodies' organs according to human tissue autopsies. Urine and feces are the general route to excrete As from human body, though minimal excretion also occurs through the nails and hair. Methylation of the inorganic As is the primary metabolic pathway. The chief factors of As toxicity are the dose and the period of its exposure.

3.2.4 Toxicology Processes

Animal studies have shown carcinogenic activity in several organs, which are similar to the As-induced cancer that is seen in humans. The organs affected were noted to be the skin, lungs, urinary ladder, and the liver. Hence, there are an involvement of mutual mechanisms of action. Other organs affected by As poisoning are the prostate together with the Kupffer cells. The mechanisms involved in carcinogenesis are (i) epigenetic alterations, which include alteration to DNA methylations, histones, and micro-ribonucleic acid (miRNA), which have been observed to cause malignant growth; (ii) impairment to the maintenance system of the DNA; and (iii) generation of reactive oxygen species (ROS). *In vitro* studies have shown the observation of P53 expression to be altered causing a lowered expression in the p21 protein. Humans have also been noted to be affected by arsenic compounds in their leukocytes, where genotoxicity has been observed. Methylated arsenic inhibits DNA repair processes and the generation of ROS as metabolic products, found in the spleen and liver. Arsenic can bind to the DNA-binding proteins and can also disrupt the DNA repair processes, which increases the carcinogenesis risk (Engwa et al. 2019; Briffa et al. 2020).

Effects caused by inorganic arsenic toxicity include irritation of the gastro-intestinal system; lung irritation and heart problems; brain damage and skin alterations; red blood cells and white blood cells produced at a lesser volume; DNA damage; infertility and miscarriages; increased risk of cancer. Moreover, organic arsenic toxicity causes stomach distress and nerve damage (does not involve the DNA and is not carcinogenic) (ATSDR 2007a; Sabine and Wendy 2009; Marcus 2013; Briffa 2020; Briffa et al. 2020).

3.2.5 Remedial Options

Arsenic contamination in soil–plant–water systems make it complicated to design the remedial measures. Drinking water and foods are the main sources of As in the human body. Filtration of water using adsorption, ion exchange, and membrane filters can reduce the As concentration to an allowable level. Few of these filtration systems, viz. Sono filter and Shapla filter, are already in use in the As-contaminated areas in Bangladesh and many other countries to supply clean drinking water. Arsenic safe rice production is another challenge due to its excessive demand of irrigation water in the contaminated hotspots. International Rice Research Institute has developed alternate Wetting and Drying technique for reducing the water use as well as reducing As concentration in rice. Intermittent irrigation technique is another achievement in that end (Honma et al. 2016). Arsenic tolerant variety development is considered to be the best remedial option but yet to get any successful variety. Genetic and biotechnological studies reported some arsenic tolerant genes (*OsHAC1;1, OsHAC1;2, OsABCC1*), phosphate transporters (*OsPT1, OsPT8*), and aquaporins (NIP1;1), etc., for the development of As-tolerant varieties (Kamiya and Fujiwara 2009; Kamiya et al. 2013; Wang et al. 2016). A number of paddy management techniques like silicon, P, biochar, and zero valent iron application have potentials to reduce As in rice to some extent. There is a scope to remove As from rice during rice processing and cooking. Arsenic-contaminated sites can be corrected by hyper-accumulating plants, viz. *Pteris vittata, Pityrogramma calomelanos, Mimosa pudica, Melastoma malabrathricum* (Bancewicz 2021). Arsenic is also released in the atmosphere under the reduced environment with high organic matter content in soil (Mestrot et al. 2011). In addition, remediation techniques based on nanotechnology have become more widespread in the twenty-first century. Complexation and ligand exchange on the surfaces of nanomaterials are the main mechanisms of As adsorption (Siddiqui et al. 2019). Arsenic removal from water has been successfully accomplished using a variety of media, including nanoparticles made of cupric oxide, iron oxide, titanium dioxide, zirconium dioxide, and zero-valent iron (ZVI) (Bhardwaj et al. 2019).

3.3 Cadmium

Cadmium is an inorganic hazardous heavy metal belonging to the group IIB of the periodic table, having atomic number 48, atomic weight 112.41, boiling point 765 °C (1410 °F), and density 8.65 g at 25 °C. The eight stable isotopes of Cd are ^{106}Cd, ^{108}Cd, ^{110}Cd, ^{111}Cd, ^{112}Cd,

^{113}Cd, ^{114}Cd, and ^{116}Cd. ^{112}Cd and ^{114}Cd are the most prevalent isotopes (Rahimzadeh et al. 2017). For plants and animals, Cd is regarded as a nonessential element, yet it is extremely damaging to humans and the environment (Wang et al. 2021). As there is enough proof that Cd causes cancer in both human and animal, IARC (2012) has classified it as a human carcinogen.

3.3.1 Sources and Forms

The earth crust contains large amounts of Cd, which occurs naturally and has a concentration between 0.1 and 0.2 ppm. Although Cd can also be found in ocean water, it is mainly found in sedimentary rocks (approximately 500 ppm), phosphorites, and marine phosphate (Saini and Dhania 2020). The weathering of Cd-rich rocks (greenockite), Zn deposits as cadmium sulfide, volcanic activity, and forest fires are other natural sources of Cd. About 100–500 tons of Cd are released during volcanic eruptions, and movement of the released Cd toward the sea is the significant natural source of Cd in aquatic bodies (Unsal et al. 2020). Moreover, environmental Cd pollution has been worse in recent decades as a result of the rapid rise of heavy industries. Important anthropogenic causes of soil Cd pollution in terrestrial and aquatic environments include atmospheric deposition from mining, smelting, coal and fossil fuel combustion, fertilizer manufacture, waste incineration, and sewage sludge (Srivastava et al. 2017; Wang et al. 2021) (Figure 3.2). According to Böhlandt et al. (2012), environmental smoking is the primary cause of Cd contamination in indoor air.

Divalent Cd^{2+} ions make up 55–90% of the total soluble Cd in groundwater, depending on the composition of the groundwater. The remaining Cd is present as organic and inorganic complexes, such as $CdCl^+$, $CdCl_2$, $CdCl_3^-$, $Cd(SO4)_2^{2-}$, $CdSO_4^0$, $CdHCO_3^-$, $CdCO_3^0$, $Cd(CO_3)_2^{2-}$, $CdOH^+$, $Cd(OH)_2^0$, $Cd(OH)_3^-$, $Cd_2OH_3^+$, and $CdNO_3^+$ (Wilkin 2007). With increasing pH, Cd complexation with inorganic carbon becomes more prevalent, and precipitation of $CdCO_3$ (otavite) is favorable at pH > 8. Otavite has a low solubility for conventional groundwater inorganic carbon concentrations around pH 9–10. Above this pH, solubility improves due to the stability of the $Cd(CO_3)_2^{2-}$ complex. Cd complexes appear as $Cd(HS)_2^0$, $Cd(HS)_3^-$, $Cd(HS)_4^{2-}$, and $CdHS^+$ in anoxic and sulfidic environments. Generally, CdS (greenockite) has a limited solubility, although equilibrium Cd concentrations rise with rising dissolved sulfide levels (Kubier et al. 2019).

3.3.2 Food Chain Contamination

Through multiple pathways, including soil and water, Cd has the ability to enter the food chain. When Cd is released into the atmosphere by both natural and man-made means, it travels via water, air, and soil where it can be ingested by plants, animals, and shellfish from the soil and water. A substantial amount of Cd is absorbed and accumulated in agricultural products due to its fast rate of transfer from soil to plant and relatively high bioavailability of Cd (Zhu et al. 2018). According to Bernhoft (2013), Cd substitutes for calcium cation (Ca^{2+}) during root uptake because of similarities in their chemical characteristics (Meng et al. 2018). Similar to other macro- and micronutrients (Mg, Ca, Fe, Zn, and Cu), Cd enters plant cells through the same transport systems (EFSA 2009). Depending on the soil pH, effective soil-to-plant transfer enables Cd to go to the aerial parts of the plant with other

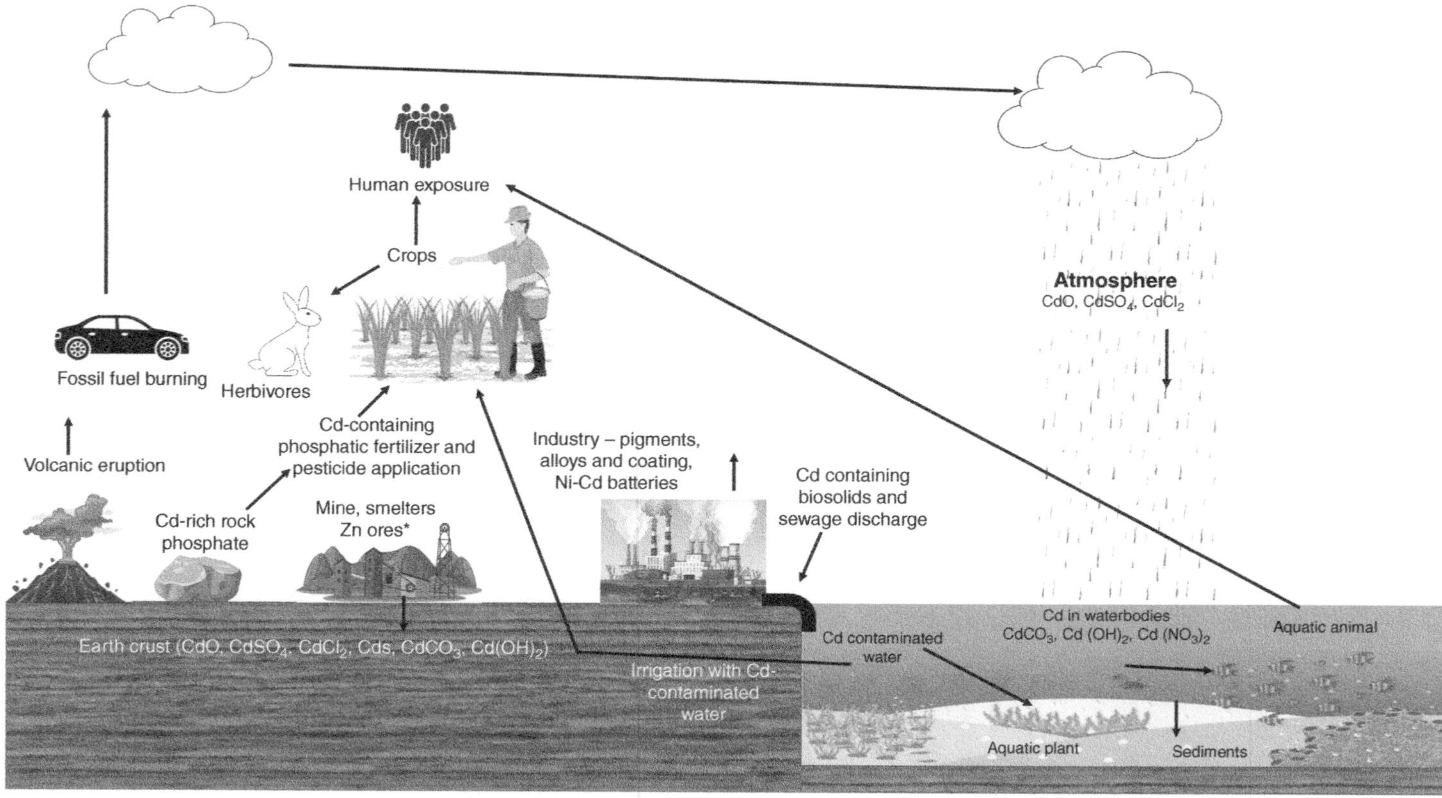

Figure 3.2 Cadmium contamination in the environment and its exposure to human.

micronutrients, resulting in substantial Cd accumulation in the edible parts of plants (Khan et al. 2017) (Figure 3.2). The largest source of ambient Cd exposure in the population, aside from certain occupational activities like smoking, is foodstuff (Jean et al. 2018). The most significant sources of dietary Cd are in specific grains like rice and wheat, leafy salad vegetables, marine goods, chicken meat, and preparing diet (Soisungwan 2018). Moreover, half of the world's population consumes rice regularly, making it one of the most significant sources of dietary Cd transfer from soil to human (Zhao and Wang 2020).

3.3.3 Pharmacokinetic Processes

Cadmium has been observed to be absorbed generally through the inhalation route, though absorption depends on the particle size of the metal. If the particle size of Cd is large, it will be deposited onto the nasopharyngeal and the tracheobronchial through impaction. These are removed through the mucociliary processes. Some of the soluble Cd salts are acetate, nitrate, chloride, and sulfate. Insoluble Cd salts on the other hand are sulfides. The Cd sulfides react with the carbon dioxide that is present in the lungs. After absorption of Cd, the organs found with the highest concentration of Cd are the liver and the kidneys. Cadmium has been observed to have a high affinity for metallothionein, together with the sulfhydryl groups present in albumin. From the lungs, Cd is then excreted through both the fecal and urinary route in equal amounts. If Cd is absorbed through the gastrointestinal tract, it would be excreted through the fecal route (ATSDR 2012a; Briffa et al. 2020).

3.3.4 Toxicology Processes

Apoptosis promotion has been linked with Cd, together with the promotion of oxidative stress, DNA damage, and methylation of DNA (Engwa et al. 2019). The main targets for Cd toxicity are the bones, kidneys, and lungs (ATSDR 2012a). Its carcinogenicity affects the kidneys, prostate, pancreas, and the lungs (Valko et al. 2005).

Primarily, the cause of Cd toxicity is thought to be due to the deactivation of an important DNA repair activity. The MMR is disrupted due to a mutation of the MutS homolog (MSH) protein. When this disruption occurs due to the Cd, the mismatch repair system (MMR) is inactivated, leading to an increase in the instability of the genome, thus increasing the incidence of a number of human cancer (Valko et al. 2005; Briffa et al. 2020). An indirect generation of free radicals from Cd occurs, which are the superoxide radical, hydroxyl radical together with the nitric oxide. Through the Fenton reaction, the non-radical hydrogen peroxide is also generated as seen in studies (Valko et al. 2005). Lipid composition was disturbed due to Cd toxicity, while the lipid peroxidation was observed to be enhanced by Cd. Depletion of antioxidant enzymes, including superoxide dismutase and glutathione peroxidase, causes cardiotoxic effects. Glucose levels in the blood have seen to be increased by Cd due to the hepatic glycogen stores being affected. Other affects caused by Cd toxicity are focal mononuclear infiltrates, intralobular fibrosis, cirrhosis, and smooth endoplasmic reticulum explosion (ATSDR 2012a; Briffa et al. 2020).

Cadmium causes an alteration to the metabolism of several metals such as Fe, Cu, Zn, and Se, which bring about the toxicity in testes due to the interference of Cd with the Zn-protein complexes, which control the DNA transcription. This then leads to cell apoptosis (ATSDR 2012a; Briffa et al. 2020). Bone fractures are observed due to Cd, which decreases the density of the bone mineral. Effects caused by Cd toxicity include nephrotoxicity, infertility, decreased immune system, calcium metabolism alterations, bone fractures and osteoporosis, renal dysfunction, central nervous system complications and psychological disorders, gastrointestinal disorders, DNA impairment, cancer, and Itai-Itai disease (IARC and WHO 1993; ATSDR 2012a; Briffa et al. 2020).

3.3.5 Remedial Options

Cadmium contamination renders the soil unfit for use in agriculture and other activities. Therefore, rehabilitation of certain soil types is crucial. It is challenging to forecast or manage the dangers associated with Cd exposure. However, there are a variety of mitigation strategies that can lessen the transfer of Cd, including increasing soil organic matter (which helps control Cd solubility), adding lime to the soil (which raises soil pH and reduces Cd bioavailability), choosing and developing low-Cd accumulating crop varieties, biofortification (applying soil amendments based on soil characteristics), and using Zn-based fertilizers (which lessens Cd accumulation in crops), avoid using phosphate-based fertilizer, and checking of irrigation water for salinity and chloride levels (reduces Cd uptake in crops) (Schaefer et al. 2020; Kumar et al. 2021). Application of clay and modified clay minerals to areas with high concentrations of Cd and other heavy metals can also lower the amount of bioavailable Cd in the rhizosphere (Biswas et al. 2015). Additionally, a wide range of engineering techniques are currently in use (such as excavation, landfilling, thermal treatment, leaching, and electro-reclamation), but they are not entirely satisfactory because they destroy the biotic and abiotic components of the soil. They are also technically challenging and expensive to use. The need for alternative low-cost solutions has been driven by the high cost and failure or inadequate removal of heavy metals by the physico-chemical and biological approaches. Phytoremediation, often known as the use of plants to degrade, sequester, and remove hazardous chemicals from the environment, is a noble, cost-effective, and environmentally friendly remediation approach. *Eleusine indica* L. (KB-1), *Ageratum conyzoides* L. (KB-2), and *Euphorbia hirta* L. (KB-3) are three native plants used in earlier studies to phytoremediate for Cd. As a result, concentrations of Cd were reduced by 58.8%, 52.2%, and 51.8%, respectively (Lata et al. 2019). However, there are a lot of issues with this approach as well (Verma et al. 2017). In light of this, phytoremediation including rhizospheric bacteria has become a viable agronomic remediation technique (Helmisaari et al. 2007). Due to their extensive uptake of Cd, bacteria including *Bacillus licheniformis*, *Bacillus biosubtyl*, and *Bacillus thuringiensis* in combination with *Indian mustard* can lessen the toxicity of Cd (Hussein 2008). Other than *Bacillus* sp., *Phormidium valderianum*, *Chlorella vulgaris*, *Aspergillus niger*, *Pleurotus ostreatus* sp., and *Rhizopus arrhizus* are also employed as a method of bioremediation and have the capacity to digest and detoxify Cd (Lata et al. 2019).

3.4 Lead

Lead is an inorganic metal found in Group 14 (IV A) periodic table of elements having atomic number 82, atomic weight 207.2, melting point 327.5 °C (621.5 °F), and density 11.34 g cm^{-3} at 25 °C (Külahc 2020). Natural Pb is a mixture of 49 isotopes, with the four most common being ^{208}Pb (52%), ^{206}Pb (24%), ^{207}Pb (22%), and ^{204}Pb (less than 2%). Lead is sometimes referred to as a microelement with no known physiological function that is present in trace amounts in all biotic resources, such as soil, water, plants, and animals. The Agency for Toxic Substances and Disease Registry rates Pb as the second-most dangerous contaminant in the environment after As (ATSDR 2015).

3.4.1 Sources and Forms

Lead is a poisonous element that occurs naturally with background concentrations in the solar system, the earth's surface, and seawater are 1.03×10^{-8}%, 14 ppm, and 0.003 ppb, respectively, making it the 37th most plentiful element in the crust of the planet (Sarkar et al. 2011). Volcanic eruptions, wildfires, and different rock/soil erosional processes are all natural ways that Pb enters the environment (Natasha et al. 2020). In addition, substantial levels of Pb are released into the environment as a result of human activities including Pb mining, smelting, industrial processes, batteries, color paints, electronic waste, thermal power plants, ceramics, electroplating, atmospheric deposition from petrol use, pesticides, and fertilizers, and bangle manufacturing (Singh et al. 2015) (Figure 3.3). Nearly 98% of the stable Pb found in the environment comes from paints, petrochemicals, pipes, and other supply systems. Large amounts of Pb are released into the air when leaded fuel is burned. These Pb particles fall on the soil's surface and are then absorbed into the ground by precipitation and irrigation techniques.

Lead is mostly found in the crystalline forms of galena (PbS), cerussite ($PbCO_3$), anglesite ($PbSO_4$), and minum (Pb_3O_4). The two most prevalent forms of Pb found in soil are Pb-oxides and hydroxides. Lead compounds primarily exist as free ions (Pb^{2+}), which is the most poisonous, bioavailable, and stable form found in water. Other forms that are easily accessible to plants include Pb-chlorides and carbonates, and its complexes with organic matter in freshwater or marine environments.

3.4.2 Food Chain Contamination

Lead exposure can occur in both the workplace and the environment in several processes that include using unclean or adulterated hands or face in situations where personal hygiene and health are at risk, inhaling Pb particles from the atmosphere contaminated by cigars and automobile exhaust, ingesting Pb-contaminated soil dust, drinking Pb-contaminated water, eating Pb-contaminated vegetables and crops, and so on (Figure 3.3). Lead is typically not absorbed via the skin (Mudgal et al. 2010). Food plants like roots and vegetables that are cultivated in contaminated soil absorb a considerable amount of Pb from the environment. When people or other animals consume these contaminated foods, Pb is biomagnified. Additionally, fish consumption from tainted freshwater or saltwater contributes to food chain pollution.

3.4.3 Pharmacokinetic Processes

Inorganic Pb present as small particles are taken up through the respiratory tract, while the larger particles are taken away by the mucociliary cells, which then transport the particles to the oropharynx then swallowed. When the particles arrive in the gastrointestinal tract, a

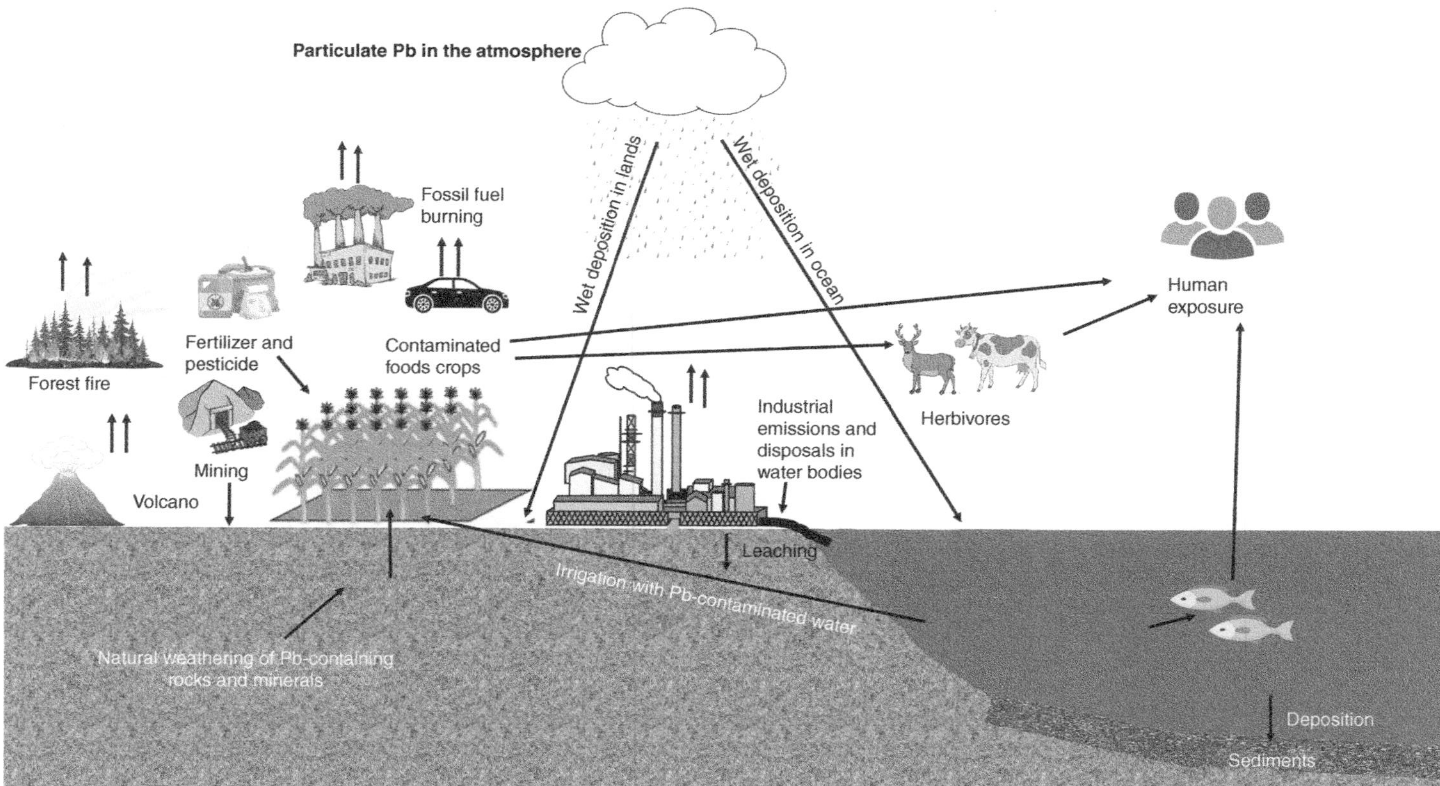

Figure 3.3 Lead contamination in the environment and its exposure to human.

number of factors need to be considered regarding absorption, such as age, nutrition, and diet, physiological characteristics of Pb, and in which medium it is absorbed. Absorption of inorganic Pb mostly occurs in the duodenum via saturable mechanisms. When it comes to inorganic Pb dermal absorption, it is not efficient though it can be increased due to hand to mouth behavior (ATSDR 2012a; Briffa et al. 2020). When absorbed, it is distributed all over the body and Pb is transported independently, where most is distributed to the bones. Conditions where resorption of bone occurs are pregnancy and lactation, osteoporosis, and menopause. Since Pb is mostly distributed to bones, these conditions will have an increased Pb. Lead is also found in the red blood cells; it can be easily transferred from mother to fetus and can also be transferred through lactation (ATSDR 2012a; Briffa et al. 2020).

Organic Pb such as alkyl Pb compounds can be observed in the liver to be actively metabolized as a result of an oxidative dealkylation and are also catalyzed by the P-450 chromosome. When exposure of tetraethyl Pb occurs in people, ethyl Pb, diethyl Pb, together with inorganic Pb were observed to be excreted via the urinary route. Workers exposed to tetraalkyl Pb illustrated the presence of trialkyl Pb metabolites in the brain, kidney, and livers. Nonworkers were also noted to have trialkyl metabolites. The principal route for inorganic Pb excretion is the urinary and fecal route though other routes can include sweat, saliva, hair, nails, breast milk, and seminal fluid (ATSDR 2012a; Briffa et al. 2020).

3.4.4 Toxicology Processes

High blood pressure is seen when persistent exposure to Pb occurs. Lead affects vital hormones together with neuronal systems in the body; these play an essential role for the regulation of cardiac output, heart rate, and peripheral vascular resistance. Hypertension due to Pb also causes depleted nitric oxide (NO), as observed in rats. Nitric oxide is imperative in the control of blood pressure via the exterior and central mechanisms. Lead is known to cause hypertension and is associated with several adrenergic system abnormalities. These abnormalities include a reduced concentration of the vascular β-adrenergic receptors, a raised plasma norepinephrine, and an elevated activity in the central sympathetic nervous system. The renin–angiotensin–aldosterone system is stimulated both directly and indirectly via the sympathetic nervous system stimulation. Chronic exposure to Pb elevates the plasma renin activity, aldosterone concentrations, in addition to the angiotensin-converting enzyme (ACE).

Lead toxicity has been noted due to Pb binding with several proteins in renal cells. In the proximal renal tubule, the formation of intra-nuclear inclusion bodies, made up of Pb complexed with protein, occurs which is linked in the modification of the Pb compartmentalization beginning with the cytosol up to the nuclear fraction, generating Pb nephron-toxicity. Cytosolic proteins (e.g. 2-microglobulin cleavage products) can aid as Pb carriers or other transitional ligands with regards to the uptake of Pb into the nucleus and can also be implicated in the ligand exchange reaction to further cytosolic binding sites (e.g. δ-aminolevulinic dehydratase), which binds and the metal Pb inhibits. Lead can furthermore bind to metallothionein but does not stimulate the protein when compared to the Zn and Cd inducers. In the renal proximal tubule, the mitochondria present can be found with structural abnormalities due to Pb, thus abnormal respiratory functions are noted alongside a reduced respiratory control ration. In the isolated renal mitochondria, the Pb inhibits the calcium uptake, and in the mitochondria, Pb is noted to enter as a substrate for a calcium transporter. Cellular degeneration and transport insufficiencies are contributed due to the oxidative metabolism impairment. Oxidative stress can similarly be caused through exposure of Pb inducing a secondary response of NO synthase induction, transketolase, together with glutathione *S*-transferase present in the kidney. It was noted that Pb reduces the glomerular filtration rate in *in vitro* studies in rats, in conjunction with the reduction of renal blood flow. Increase in blood pressure might also affect the proximal tubule causing injury in conjunction with impairment or affect the glomerulus sclerosis, which will cause the discharge of renin and/or renal insufficiency (ATSDR 2007b; Briffa et al. 2020).

Patients who suffer from Parkinson's disease have been noted to have an impairment in the dopamine synthesis regulation due to Pb, together with dopamine release and consequently revealing a presynaptic site of action. Divergent effects of Pb on the mesolimbic and nigrostriatal dopamine systems have been reported involving the receptor binding, turnover and uptake, and synthesis of dopamine. Since the mesolimbic system is affected by Pb, lesions were noted to occur on the frontal cortex including the nucleus accumbens, where perseverative deficits were detected. Learning is not influenced by Pb though alterations in the dopaminergic system have been noted, which can cause fixed-interval (FI) schedule reinforcement to have a change in response rate, producing the inhibition of an inappropriate response to fail (ATSDR 2007b; Briffa et al. 2020).

Effects caused by Pb toxicity include hypertension, signs of encephalopathy, peripheral nerve damage, injury to the brain, cognitive impairment, abdominal pain, renal impairment, pica, anemia due to disruption of hemoglobin synthesis causes iron deficiency, problems in reproductive organs (miscarriages, premature and low births, stillbirths, sperm damage), problems for children (brain development alteration, central nervous system development altered, reduced intelligence, a regression in educational accomplishments, a decrease in the attention span, antisocial behavior increased) (ATSDR 2007b, 2019; Clinic 2016; Briffa 2020; Briffa et al. 2020).

3.4.5 Remedial Options

The most popular decontamination techniques for Pb-contaminated sites include *in situ* vitrification, excavation and landfilling, soil incineration-washing-flushing-reburial, solidification, stabilization of electrokinetic systems, pump and treat systems, ion exchange chemical precipitation, ultrafiltration, adsorption, electrodialysis, flocculation, and others. Of these, ion exchange, adsorption, ultrafiltration, chemical precipitation, and electrodialysis are more useful. Electrokinetic remediation (EKR) is another technique that includes a mix of electrolytes to bind Pb and render it immobile in soil by influencing soil conductivity (H^+ and Fe^{2+}) and current by substituted soil ions from EKR ions (e.g. KNO_3, $NaNO_3$, Na_2CO_3, K_2HPO_4, KH_2PO_4, etc.) (Li et al. 2014).

In last two decades, the use of microorganisms to control and destroy Pb has grown attention. Microorganisms like bacteria (*B. licheniformis, Escherichia coli, Pseudomonas fluorescens, Salmonella typhi*), algae (*C. vulgaris, Cladophora glomerata, Oedogonium rivulare, Oscillatoria quadripunctulata*), fungi (*Aspergillus flavus, Aspergillus terreus, Saccharomyces cerevisiae*), yeast, and some other microbes have been found to absorb and detoxify Pb (Jagetiya and Kumar 2020). Green plants are also used to remove Pb from contaminated soil. *Jatropha curcas*, *Populus deltoides*, *Populus nigra*, *Populus trichocarpa*, and *Trifolium alexandrinum* are some higher plants used in the phytoremediation of Pb. There are some hyperaccumulator plant species like *Arabis paniculata*, *Nora mucronate*, *Baccharis latifolia*, *Sonchus oleraceus*, *Bidens triplinervia*, *Brassica juncea*, Buckwheat, *Cynara cardunculus*, *Helianthus annuus*, *Hemidesmus indicus*, *Lepidium bipinnatifidum*, *I. mustard*, *Poa pratensis*, *Pisum sativum*, *Plantago orbignyana*, *Sedum alfredii*, *Sesuvium portulacastrum*, *Tagetes minuta* L., *Zea mays*, *Phaseolus vulgaris*, and *Raphanus sativus* are used for phytoextraction of Pb (Jagetiya and Kumar 2020). Production of genetically modified plant can also be successfully employed to promote phytoextraction of Pb.

3.5 Chromium

Chromium (Cr) is the first element in the d-block of group (VI) in the periodic table (Owlad et al. 2009) having atomic number 24, atomic weight 51.99, boiling point 2672 °C, and density 7.19 g cm^{-3} at 25 °C. There are four stable isotopes of Cr present in nature, and their abundances are as follows: ^{50}Cr (4.31%), ^{52}Cr (83.76%), ^{53}Cr (9.55%), and ^{54}Cr (2.38%) (Chen et al. 2021). Chromium is the number one carcinogen and has been listed the seventh among the top 20 hazardous compounds by the IARC and the ATSDR (Sharma et al. 2021). Chromium can occur in any of the oxidation states from II to VI because it is a typical transition metal; however, it is less often in oxidation levels other than 0, III, and VI. In the diet, Cr(III) is a crucial element that supports the metabolism of lipids and glucose (Sharma et al. 2021). However, consuming significant quantities of Cr may result in health issues, such as lung cancer. On the other hand, it has been calculated that Cr(VI) is 300 times more hazardous than Cr(III) and is a powerful, exceedingly toxic carcinogen (Chen et al. 2021).

3.5.1 Sources and Forms

The majority of the Cr on the earth is found in the core and mantle, making it the seventh most abundant element. Chromium is the 21st most abundant element in the earth's crust, with an average concentration of 185 mg kg^{-1}, substantially lower than the planet's overall concentration of 3700 mg kg^{-1} (Sperling 2005). The single and most significant commercial source of Cr is chromite ore, which typically has a chromic oxide content of 15–65%. Elevated Cr levels in the environment are caused by anthropogenic activities such as electroporation, industrial waste disposal, wood preservatives, textile dying, and leather tanning (Figure 3.4).

Chromium exists in a variety of oxidation states (from 2 to +6), but the most prevalent and stable forms in the natural world are hexavalent chromate Cr(VI) and trivalent chromite Cr(III). When compared to Cr(III), Cr(VI) is highly mobile in soil, exceedingly hazardous to living things, and potentially mutagenic, carcinogenic, and teratogenic. However, Cr(III) is considerably less harmful and, in low concentrations, is a useful element. In natural waters, species such as $CrOH^{2+}$, $Cr(OH)_2^{+}$, $Cr(OH)_3^{0}$, $Cr(OH)_4^{-}$, $Cr_2(OH)_2^{4+}$, and $Cr_3(OH)_4^{5+}$ are anticipated to predominate. Chromate (CrO_4^{2-}), dichromate ($Cr_2O_7^{2-}$), hydrogen chromate ($HCrO_4^{-}$), dihydrogen chromate (chromic acid, H_2CrO_4), hydrogen dichromate ($HCr_2O_7^{-}$), trichromate ($Cr_3O_{10}^{2-}$), and tetrachromate ($Cr_4O_{13}^{2-}$) are possible forms of Cr(VI) in aqueous solutions (Sperling 2005).

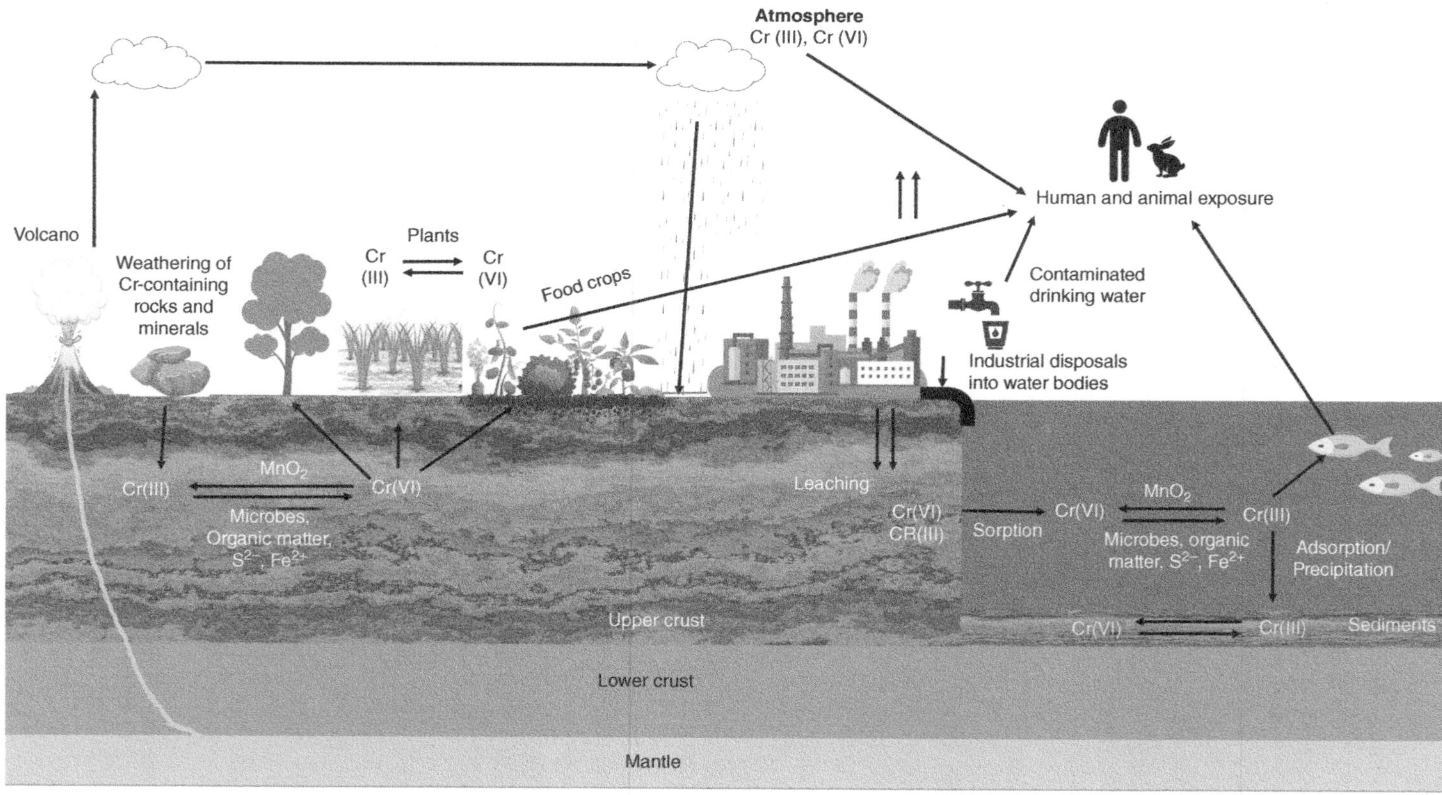

Figure 3.4 Chromium contamination in the environment and its exposure to human.

3.5.2 Food Chain Contamination

Humans are primarily exposed to Cr through skin contact, ingestion of contaminated food, drinking contaminated water, and breathing of contaminated air. Various natural and man-made processes, such as weathering of rocks and mudstone, microbial interaction with mafic and ultramafic rocks, and geogenic processes, which naturally release Cr(VI) in the environment by chromite oxidation, can cause Cr to frequently enter surface and groundwater bodies (Tumolo et al. 2020). Through trophic transfer, contaminated water bodies may lead to Cr bioaccumulation in aquatic life and later in people. Anthropogenic activities, especially the disposal of liquid and solid wastes and mining byproducts that include Cr, as well as atmospheric deposition, may increase the concentration of Cr in the soil (Figure 3.4) (Yadav et al. 2017). The bioaccumulation of Cr in foods is brought on by the cultivation of food crops in contaminated soil and irrigation with contaminated water (Figure 3.4). When people eat contaminated foods, the amount of Cr in the body is biomagnified. Additionally, the most direct pathway for Cr ecotoxicity that contaminates the food chain is the production of feeds and fertilizers from browned skin-cut wastes (SCW). The protein-rich SCW is unscientifically employed to make protein concentrates for fish and poultry feeds as well as organic fertilizer. Given the facts, a significant migration of Cr into poultry products, fish, and vegetables may occur, and this migration may further biomagnify into the food chain (Hossain et al. 2007). In a contaminated area, high levels of Cr (0.53–0.96 ppm) were found in milk, poultry eggs (23.3809 g per egg), fish (1.46 ppm), and vegetables (21.9–22.9 ppm) (Naser et al. 2012; Mandal and Ahmed 2013; Hezbullah 2014).

3.5.3 Pharmacokinetic Processes

The absorption of Cr on inhalation is affected according to both the chemical and physical properties of the Cr particle and due to the macrophage's activity in the alveoli. Chromium (Cr^{6+}) is absorbed at a more rapid rate in the blood stream than Cr^{3+}. The mucociliary clearance may pass Cr to the gastrointestinal tract, which is then absorbed primarily in the jejunum. The absorption of the metal is affected by its formulation and oxidation state (ATSDR 2012b; Briffa et al. 2020). The Cr^{6+} is absorbed much more than Cr^{3+} though it is reduced to the Cr^{3+} state by the gastric juices. In the gastrointestinal tract, inorganic Cr^{3+} is not absorbed well. This reduction to Cr^{3+} occurs in the liver, lungs, blood, erythrocytes, epithelial fluid, alveolar macrophages, and the peripheral parenchyma cells. Once Cr^{3+}reaches the blood stream, it binds to transferrin. Organs that absorb the Cr are the spleen, bone marrow, lymph nodes, lungs, kidneys, and the liver, and it is excreted through urine and bile excretion, and to a lesser extent, hair and nails, sweat, and milk (Valko et al. 2005; Briffa et al. 2020).

3.5.4 Toxicology Processes

Several DNA lesions are caused by groups that are made up of Cr^{3+} and peptides, DNA, and proteins present inside the cell (ATSDR 2012b). These lesions include DNA–protein crosslinks, DNA–DNA crosslinks, Cr-DNA adducts, DNA strand breaks, and cellular signaling pathway changes. All these may be a cause for the occurrence of toxicity and carcinogenicity by the compounds of Cr (Valko et al. 2005; ATSDR 2012b). When comparing the oxidative states of Cr, it can be seen that Cr^{6+} is more toxic state due to having the ability to enter the cells more and having a higher redox potential. At physiological pH, Cr^{6+} is present in a tetrahedral chromate anion, while Cr^{3+} is present in an octahedral complex structure. Once Cr^{6+} enters the cell, it becomes reduced to Cr^{3+} through a redox reaction, also producing Cr^{4+} and Cr^{5+} as intermediates. The Cr^{4+} has been seen to have a higher toxic potency than Cr^{3+}. Due to Cr being found in the liver and in the blood, free radicals have been detected *in vivo* studies in animals. The Cr^{5+} generates one-electron reduction through intermediates (Engwa et al. 2019). Studies have shown that DNA damage is induced by chromate compounds that caused production of DNA adducts, chromosomal aberrations, transcription of DNA, and variations in replication sister chromatid exchanges (Engwa et al. 2019; Briffa et al. 2020).

The Cr yields a variety of detrimental effects on the cells attributed to the oxidative cells affected by the creation of the oxygen radical species. This process causes other effects such as lipid peroxidation, cellular cytoskeleton communication modification, cellular cytoskeleton modification, and signaling pathways. An amount of functional damage is caused to the body due to Cr. The damage caused include RNA and DNA polymerase arrest, mutagenesis, together with modified gene expression. DNA double strand breaks are noticed to initiate modifications to the cellular communication, consequences on signaling pathways, and cytoskeleton. The DNA double strand breaks are the result of repair errors and distorted replication forks, which are attributed by Cr-DNA ternary adducts. Carcinogenicity is witnessed because of Cr-induced chromosome instability, which encompasses the mediation by means of the spindle and centrosome assembly checkpoint bypass. Furthermore, another fact is the lung particle deposition location and the extracellular Cr^{6+} dissolution (ATSDR 2012b; Briffa et al. 2020).

Effects caused by inorganic Cr toxicity are Cr(VI) inhalation, or chronic contact can cause poisoning, allergic contact dermatitis and eczema, chrome holes which will occur mostly in the forearms, hands, fingers and nose (these chrome holes can lead to secondary infections), sinusitis, bronchitis and pneumonia, lung cancer, irritation of mucous membranes, gingivitis, liver and kidney disease, oral intake of Cr(VI) usually causes acute poisoning, gastrointestinal ulceration, nausea and vomiting, diarrhea, fever, vertigo, toxic nephritis and liver damage, coma, death can occur usually at 1–3 g (Dayan and Paine 2001; Sabine and Wendy 2009; ATSDR 2012b; Briffa 2020; Briffa et al. 2020).

3.5.5 Remedial Options

Physical, chemical, and biological methods are available for treating Cr. The physical approach for removing Cr uses procedures including membrane filtrations, adsorption, ion exchange, photocatalysis, electrodialysis, etc. These methods make use of Cr's physicochemical properties (Sharma et al. 2021). Hexavalent Cr is converted to trivalent Cr using a chemical remediation process that uses sulfur dioxide, ferrous sulfate, and $CaCO_3$ (Madhavi et al. 2013). Along with the adsorption of trivalent Cr, organic acids can diminish hexavalent Cr.

Utilizing the capacity of microorganisms or plants to transform or adsorb heavy metals are regularly used technologies called biotransformation and biosorption. When choosing between these two options, biotransformation – the conversion of hazardous, mobile Cr(VI) to less toxic, less mobile Cr(III). Cr(VI), which has a positive charge, forms a chemical bond with bacteria's negatively charged surface. The attachment of the metal ion is aided by many functional groups, including glycoproteins, glycolipids, polysaccharides, and proteins, on the surface of the microbial cell. Hexavalent Cr precipitates on bacterial surfaces during bio-sorption, when it is converted to trivalent Cr (Sharma et al. 2021). Another potential strategy that makes use of these helpful bacteria to reduce metal toxicity in Cr-polluted soils is rhizoremediation. The use of Cr-tolerant plant growth-promoting bacteria (PGPB) as beneficial inoculants, which reduce metal toxicity by increasing plant nutrient uptake and also improve plant vigor and biomass production under unfavorable environmental conditions, may also increase the efficacy of the remediation process (Francisco et al. 2018). Due to their effectiveness in removing Cr(VI) from aqueous solutions, zero valent iron nanoparticles (nZVI) have received a great attention recently.

3.6 Mercury

Mercury is a liquid silver or quicksilver, has atomic number 80, standard atomic weight 200.59, boiling point 356.7 °C, density 13.534 g cm^{-3} at 25 °C (Eisler 2006; Meija et al. 2010; Zheng et al. 2012). It is quite flammable; a saturated atmosphere of Hg vapor at 24.0 °C contains roughly 18 ml Hg cm^{-3} (Satoh 2013). The only metal that is liquid at room temperature is Hg, which is heavy, glittering, silvery white with a slight bluish tinge. It is thick, has a high surface tension, and is comparatively stable. White solid Hg exists below the melting point. It exists as a colorless vapor in its gaseous state. Mercury is a fair conductor of electricity but a fairly poor one for heat (Beckers and Rinklebe 2017). It has seven stable isotopes with the percentages of abundance listed below that correspond to the atomic weight previously mentioned: ^{196}Hg (0.16%), ^{198}Hg (10.04%), ^{199}Hg (16.94%), ^{200}Hg (23.14%), ^{201}Hg (13.17%), ^{202}Hg (29.74%), and ^{204}Hg (6.82%) (Meija et al. 2010). The health of people can be seriously affected by Hg and its compounds. It has been considered an immunotoxin and a neurotoxin, and the World Health Organization has named it one of the ten substances that pose the greatest threat to human health (WHO 2017).

3.6.1 Sources and Forms

The atmosphere, terrestrial habitats (soil), and aquatic ecosystems (ocean) are the three main worldwide Hg reservoirs. Generally, anthropogenic discharge and naturally occurring release are the principal causes of Hg pollution in soil. Due to the uneven distribution of Hg in the soil caused by the geological history of high Hg, the concentration of Hg in various soils varies. In addition, Hg is still being released into the soil via natural weathering process in the earth crust. However, in recent years, manmade emissions have been a significant source of Hg pollution. According to the UNEP's worldwide assessment records, the rate of Hg deposition has increased 1.5–3 times over the previous century as a result of an increase in industrial and agricultural activities as well as pollutant emissions from medical procedures (Jiang et al. 2006). Mercury contamination in soil is caused by the intentional or unintentional direct release of substantial amounts of Hg into the environment during daily living and manufacturing. The primary human-made processes that release Hg are burning coal, incinerating trash, making metals, using chlorine alkali in certain industries, mining for and using Hg in metallurgy, operating landfills, and other similar processes (Wang et al. 2012; UNEP 2018). Anthropogenic Hg can enter the soil through wet and dry deposition, waste discharge, and usage of pesticides and fertilizers that contain Hg (Figure 3.5). Mining activities, particularly Hg mining, can result in serious Hg contamination of mines and surrounding area, as it promotes the mobilization of Hg, leading to increased levels of Hg in soil (Feng and Qiu 2008). Metal smelting seriously pollutes the environment with Hg, especially when smelting nonferrous metals and Zn. Around 75% of the Hg in the atmosphere is caused by the combustion of fossil fuels, particularly coal (Jiang et al. 2006). A terrestrial ecosystem can absorb the majority of the Hg created by combustion, which builds up in the soil. The area surrounding a coal-fired power plant is typically the most seriously affected by Hg emissions (Ali et al. 2016, 2017).

Mercury is ubiquitous toxic element, which is found in air, water, and soil and there are both organic and inorganic forms of Hg, including elemental or metallic Hg (Hg^0), inorganic Hg (Hg^+, Hg^{2+}), alkylated compounds, and organic Hg (He et al. 2015; Al-Saleh et al. 2016; Li et al. 2017). Among the Hg species, the most prevalent Hg compounds are mercuric sulfide (HgS), mercuric chloride ($HgCl_2$), mercuric oxide (HgO), methylmercury (MeHg), dimethylmercury ($(CH_3)_2Hg$), ethylmercury (C_2H_5Hg), and phenylmercury (C_6H_5Hg) (Raj and Maiti 2019). At normal temperature, elemental Hg is a liquid that easily turns into vapor. The vaporized form of Hg is more hazardous than the liquid version. When a Hg container breaks, Hg pours out, and breathing in this massive Hg vapor can be deadly. Organic Hg compounds like methylmercury (Me-Hg) and ethylmercury (Et-Hg) are more toxic than inorganic Hg compounds. The hierarchy of different forms of Hg regarding increasing Hg toxicity is defined as $Hg^0 < Hg^{2+}$, $Hg^+ <$ Me-Hg (Balali-Mood et al. 2021).

3.6.2 Food Chain Contamination

Mercury enters the food chain through anthropogenic and natural sources, as well as soil and water that have been exposed to atmospheric deposition. The ease with which Hg travels over distances and through air, water, and soil is often determined by its chemical form. In soil or sediments of lakes or streams, microorganisms (mainly sulfate-reducing bacteria) are capable of converting inorganic Hg to methylmercury, which is the most physiologically relevant change of Hg (CRS 2006). Methylation is important because, compared to inorganic Hg, methylmercury is far more poisonous to animals and easier to absorb by living tissues. It is also more likely to be consumed through food (Figure 3.5). When consumed more quickly than it can be eliminated, methylmercury is easily absorbed by the digestive system and builds up in the bodies of fish and other animals. Animals higher on the food chain typically have higher levels of exposure because methylmercury tends to be deposited in muscle tissue (i.e. the edible meat of fish and other animals). The longest food chains, which include predatory fish such as walleye, large-mouth bass, and tuna, fish-eating birds such as loons, ospreys, and eagles, and fish-eating mammals including as raccoons, otters, and mink, acquire the highest quantities of methylmercury. Methylmercury concentrations in fish can be up to 10 million times higher than those in water (CRS 2006). Additionally, food crops like rice that are grown in damp conditions have the potential to

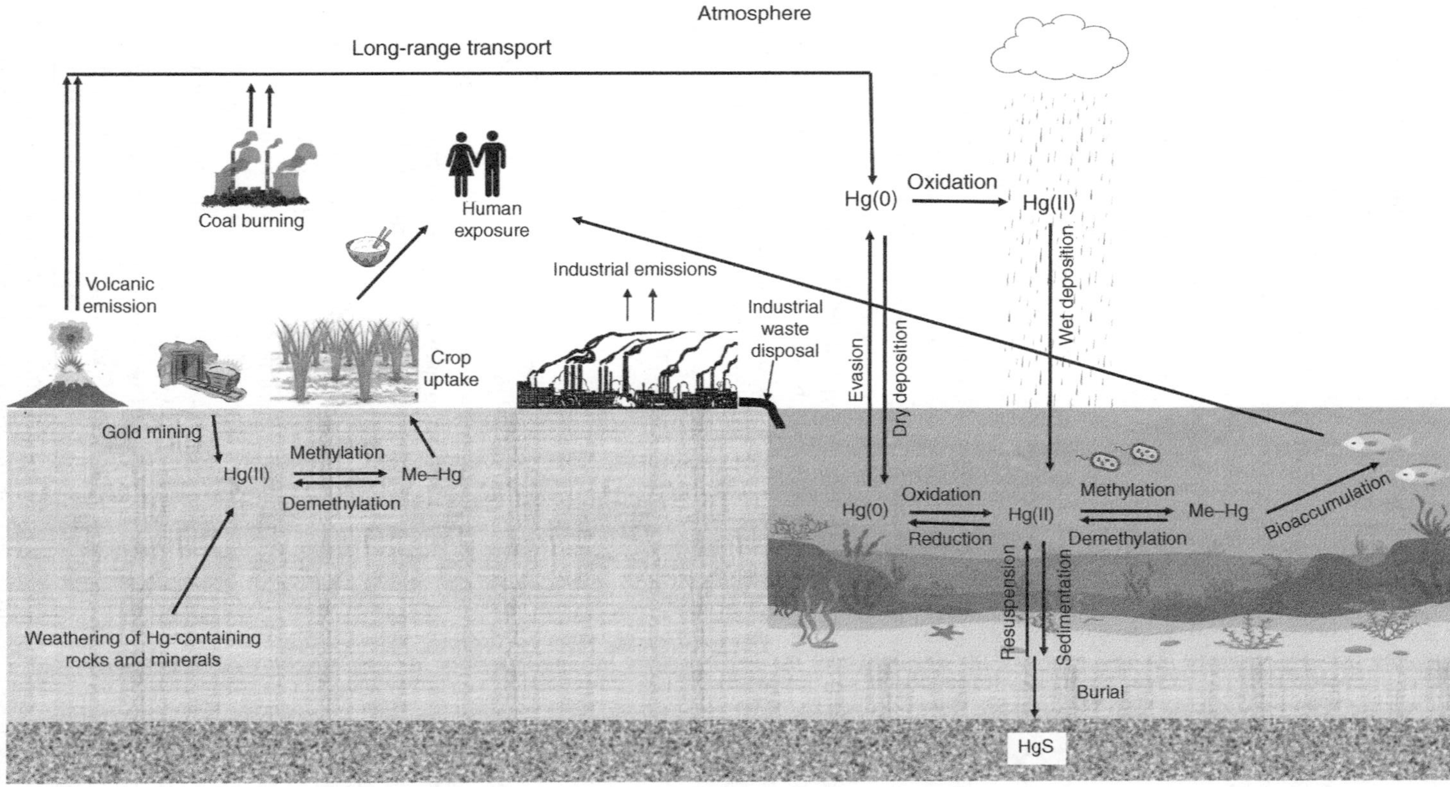

Figure 3.5 Mercury contamination in the environment and its exposure to human.

accumulate more Hg in their grains. After irrigation water delivery, other root crops and vegetables can also absorb a substantial amount of Hg.

3.6.3 Pharmacokinetic Processes

Inhalation is one of the major routes that metallic Hg passes through, followed by ingestion. By attaching to the red blood cells, metallic Hg can pass efficiently through the blood–brain barrier and through the placenta. The central nervous system takes up some of the Hg though the majority is oxidized quickly to mercuric Hg. Other organs and tissues where metallic Hg is deposited are the thyroid, breast, myocardium, muscles, liver, adrenals and kidneys, skin, pancreas, salivary glands, sweat glands, lungs, prostate, testes, enterocytes, and breast milk. Metallic Hg has a great affinity for T cell surfaces and the sulfhydryl group. When it binds to the T cells, together with the sulfhydryl group, it influences the T cell function. The Hg is then excreted in majority as mercuric Hg (Bernhoft 2012; Briffa et al. 2020).

Mercuric Hg is corrosive, and when absorbed by the intestine, its absorbency will be increased due to this effect. Like metallic Hg, mercuric Hg also binds to sulfhydryl groups, erythrocytes, glutathione, metallothionein, or it is pendulous in plasma. It aggregates in the placenta, amniotic fluid together with the fetal tissues. Even though mercuric Hg does not pass through the blood–brain barrier adequately, amino acid transporters like cysteine aid the accumulation of it in the brain. Mercuric Hg is mainly found in the convoluted renal tubule, where it is seen to bond to metallothionein. Apart from depositing on the renal tube, it is also deposited on the periportal of the liver and to a lesser extent in the testes, epithelial tissues, and the choroidal plexus. Excretion of mercuric Hg occurs to a high degree through the urine and feces and a lesser degree from the sweat, saliva, tears, and breast milk (Bernhoft 2012; Briffa et al. 2020).

Organic Hg is absorbed through inhalation, ingestion, and through the skin, especially methylmercury. Organic Hg also adheres to the sulfhydral group when in the blood stream, especially to cysteine. Methylmercury is seen to be absorbed by the brain, liver and kidneys, fetus, placenta, fetal brain, peripheral nerves, and bone marrow. Slow demethylation occurs to the methylmercury, into inorganic Hg. Excretion of methylmercury occurs primarily through the fecal route and then by the breast milk (Bernhoft 2012; Briffa et al. 2020).

3.6.4 Toxicology Processes

Mercury has been observed to start an oxidative stress when both forms are exposed to animals. A reduction of intracellular thiols such as glutathione occurs as a result of mercuric ions, which demonstrates a high affinity to the thiol binding side. This causes a direct and an indirect cause or susceptibility for the proximal tubular cells to oxidative stress (Valko et al. 2005). Modification of intracellular thiol by Hg causes oxidative stress, dysfunction of the mitochondria, lipid peroxidation, together with heme metabolism alterations. Hydrogen peroxide formation is increased due to the depolarization of the mitochondrial inner membrane by Hg^{2+} (ATSDR 1999; Valko et al. 2005).

Impairments to the kidneys are epithelial cell damage, which is caused by lipid peroxidation and augmented free radical formation. When treated with Hg, it is observed that there will be a decrease in the cellular defense mechanism against oxidative damage (ATSDR 1999). Lipid peroxidation is augmented due to Hg, and this is seen mostly in the kidneys, epididymis, and testes (Valko et al. 2005). Neurotoxicity is seen as the early symptoms of methylmercury poisoning as it targets the mitochondria. In the brain, at the electron transport chain in the mitochondria, methylmercury poisoning causes an excess of the reactive oxygen species transpiration, which triggers an induction of the oxidative stress (Valko et al. 2005; Briffa et al. 2020).

Inorganic Hg in high levels is observed to cause nonlethal effects in the cells of kidneys, which are linked to the redistribution of intracellular calcium stores. Depending on the extracellular calcium existence, the escalation of the intracellular calcium causes cytotoxicity. The non-mitochondrial intracellular stores produce more calcium in the cytosol due to the endoplasmic reticulum. Buffering processes causes a reduction in the intracellular calcium and an uptake may be seen through the microsomal Ca^{2+}, the Mg^{2+}-ATPase, or the mitochondrial uniporter (Valko et al. 2005; Briffa et al. 2020).

Effects caused by Hg toxicity when inhaled (ATSDR 1999; Bernhoft 2012; Li et al. 2017; Briffa 2020; Briffa et al. 2020): eye inflammation (inhalation), lung irritation (inhalation), nausea and vomiting (inhalation), diarrhea (inhalation), rashes (inhalation). Effects caused by Hg toxicity through diet such as in fish if taken in excess amounts: myocardial infarction, coronary heart disease, cardiovascular disease. Other effects caused by Hg toxicity: genotoxic, DNA and chromosomes impairment, Mongolism (Down's syndrome), impact on reproductive system (miscarriage, sperm impairment in men, congenital disabilities, neurological disorders).

3.6.5 Remedial Options

Mercury seriously contaminates the soil as a persistent and flammable element that can enter the soil through a variety of routes and persist there in a variety of forms. As a result, remediation of Hg in contaminated soil has drawn much more attention. For Hg-contaminated soils, there are two remediation mechanisms: eliminating harmful compounds to lower the total concentration of Hg and altering the mobility and biological availability of Hg in the soil to lower the risk of pollution (Koptsik 2014).

Physical remediation primarily entails the use of soil replacement, physical separation, soil vapor extraction, fixed/stabilized soil, vitrification, thermal desorption, and electrokinetic remediation technologies. In general, for physical remediation to be as effective as possible, chemical reagents must be added to the soil (Wang et al. 2012). Chemical reagents, reactions, and principles are employed in chemical remediation to remove Hg (Song et al. 2017); the aforementioned procedures often cause pollutants to decompose, which either eliminates or lessens the toxicity of the soil. Soil washing, chemical stabilization, oxidation, reduction, and reduction dechlorination, solvent extraction, and soil performance enhancement remediation technology are currently the key chemical remediation techniques. Biological remediation/bioremediation lowers, eliminates, or immobilizes harmful pollutants from the soil. The two primary categories of bioremediation are plant remediation and microbial remediation, with plant remediation receiving far more attention. A viable option for removing Hg from contaminated soil includes *J. curcas*, *Cyrtomium macrophyllum*, or *Miscanthus sinensis*. Bioremediation agents might be microorganisms that are resistant to metals. Microbes with the capacity to remove Hg from a site include *Bacillus thuringiensis* and *Paenibacillus* sp. (Teng et al. 2020).

3.7 Conclusions

Heavy metals/metalloids are naturally occurring toxins and widely distributed in the soil-plant system and the environment. In addition to the natural abundance, rapid industrialization is contributing in the amplification of the heavy metal concentration in the environment in the recent decades. Arsenic, Cd, Pb, Cr, and Hg are few of the more toxic heavy metals having worldwide concerns for their environmental contamination and human health implications. Therefore, a detail dissection of the sources and forms, food chain contamination, pharmacokinetic processes, toxicology processes and remedial options of these five heavy metals/metalloids gives a comprehensive understanding. These metals/metalloids tend to bioaccumulate in the edible parts of the food crops and fishes that pose a serious threat to human health. There are unbreakable ties between heavy metals/metalloids, food security, and human health. These heavy metals cause hematologic and immunologic illnesses, diabetes, hearing loss, cardiovascular diseases, developmental abnormalities, neurologic and neurobehavioral disorders, as well as many types of cancer. In order to stop metallic pollutants from entering the food chain and to create effective remediation plans, it is necessary to quickly and accurately map soil pollution. For contaminated soils, heavy metal/metalloid tolerant varieties, good soil and water management, biological remediation including phytoremediation and PGPR can be more cost- and environment-friendly alternatives.

References

Ali, J., Kazi, T.G., Afridi, H.I. et al. (2016). The evaluation of sequentially extracted mercury fractions in Thar coal samples by using different extraction schemes. *Int. J. Coal Geol.* 156: 50–58.

Ali, J., Kazi, T.G., Tuzen, M., and Ullah, N. (2017). Evaluation of mercury and physicochemical parameters in different depths of aquifer water of Thar coalfield Pakistan. *Environ. Sci. Pollut. Res.* 24: 17731–17740.

Al-Saleh, I., Nester, M., Abduljabbar, M. et al. (2016). Alterations in biochemical markers due to mercury (Hg) exposure and its influence on infant's neurodevelopment. *Int. J. Hyg. Environ. Health* 219: 898–914.

ATSDR (1999). Toxicological profile for mercury. https://www.atsdr.cdc.gov/toxprofiles/tp46.pdf (accessed 29 June 2018).

ATSDR (2007a). Toxicological profile for arsenic. https://www.atsdr.cdc.gov/toxprofiles/tp2.pdf (accessed 29 June 2018).

ATSDR (2007b). Toxicological profile for lead. https://www.atsdr.cdc.gov/toxprofiles/tp13.pdf (accessed 29 June 2018).

ATSDR (2012a). Toxicological profile for cadmium. Georgia. https://www.atsdr.cdc.gov/toxprofiles/tp.asp?id=48&tid=15 (accessed 29 June 2018).

ATSDR (2012b). Toxicological profile for chromium. https://www.atsdr.cdc.gov/csem/chromium/docs/chromium.pdf (accessed 29 June 2018).

ATSDR (2015). Agency for toxic substances and disease registry toxicological profile for lead. U.S. Department of Health and Human Services. https://wwwn.cdc.gov/TSP/ToxProfiles/ToxProfiles.aspx?id=96&tid=22 (accesses 25 July 2022).

ATSDR (2019). Toxicological profile for lead. U.S. Department of Health and Human Services. https://www.atsdr.cdc.gov/ToxProfiles/tp13-c2.pdf (accesses 25 July 2022).

Balali-Mood, M., Naseri, K., Tahergorabi, Z. et al. (2021). Toxic mechanisms of five heavy metals: mercury, lead, chromium, cadmium, and arsenic. *Front. Pharmacol.* 12: 643972.

Bancewicz, B.K. (2021). A call to action: incentivizing arsenic remediation. In: *Arsenic Monitoring, Removal and Remediation* (ed. M. Stoytcheva and R. Zlatev), 1–18. United Kingdom: IntechOpen Limited https://doi.org/10.5772/intechopen.99376.

Beckers, F. and Rinklebe, J. (2017). Cycling of mercury in the environment: sources, fate, and human health implications – a review. *Crit. Rev. Environ. Sci. Technol.* 47 (9): 693–794.

Bernhoft, R. (2012). Mercury toxicity and treatment: a review of the literature. *J. Environ. Public Health* 2012: 460508. https://doi.org/10.1155/2012/460508.

Bernhoft, R. (2013). Cadmium toxicity and treatment. *Sci. World J.* 2013: 394652.

Bhardwaj, A., Rajput, R., and Misra, K. (2019). Status of arsenic remediation in India. In: *Advances in Water Purification Techniques* (ed. S. Ahuja), 219–258. Netherlands: Elsevier.

Biswas, B., Sarkar, B., Mandal, A., and Naidu, R. (2015). Heavy metal immobilizing organoclay facilitates polycyclic aromatic hydrocarbon biodegradation in mixed-contaminated soil. *J. Hazard Mater.* 298: 129–137.

Böhlandt, A., Schierl, R., Diemer, J. et al. (2012). High concentrations of cadmium, cerium and lanthanum in indoor air due to environmental tobacco smoke. *Sci. Total Environ.* 414: 738–741.

Briffa, J. (2020). Heavy metals in Maltese agricultural soil. PhD Thesis. University of Malta.

Briffa, J., Sinagra, E., and Blundell, R. (2020). Heavy metal pollution in the environment and their toxicological effects on humans. *Heliyon* 6 (9): e04691. https://doi.org/10.1016/j.heliyon.2020.e04691.

Chen, Y., Wu, H., Sun, P. et al. (2021). Remediation of chromium-contaminated soil based on *Bacillus cereus* WHX-1 immobilized on biochar: Cr(VI) transformation and functional microbial enrichment. *Front. Microbiol.* 12: 641913. https://doi.org/10.3389/fmicb.2021.641913.

Chung, C.J., Huang, Y.L., Huang, Y.K. et al. (2013). Urinary arsenic profiles and the risks of cancer mortality: a population-based 20-year follow-up study in arseniasis-endemic areas in Taiwan. *Environ Res.* 122: 25–30.

Clinic, M. (2016). Lead poisoning. https://www.mayoclinic.org/diseases-conditions/lead-poisoning/symptoms-causes/syc-20354717 (accessed 29 June 2018).

CRS (2006). Mercury in the Environment: Sources and Health Risks. Congressional Research Service Report for Congress (19 January 2006). http://www.crs.gov_RL32420 (accesses 27 July 2022).

Dayan, A.D. and Paine, A.J. (2001). Mechanisms of chromium toxicity, carcinogenicity and allergenicity: review of the literature from 1985 to 2000. *Hum. Exp. Toxicol.* 20 (9): 439–451. https://doi.org/10.1191/096032701682693062.

EFSA (2009). Scientific opinion of the panel on contaminants in the food chain on a request from the European Commission on cadmium in food. *EFSA J.* 7 (3): 1831–4732.

Eisler, R. (2006). *Mercury Hazards to Living Organisms*. USA: CRC Press.

Engwa, G.A., Ferdinand, P.U., Nwalo, F.N., and Unachukwu, M.N. (2019). Mechanism and health effects of heavy metal toxicity in humans. In: *Poisoning in the Modern World – New Tricks for an Old Dog?* (ed. O. Karcioglu and B. Arslan). United Kingdom: IntechOpen Limited https://doi.org/10.5772/intechopen.82511.

Feng, X. and Qiu, G. (2008). Mercury pollution in Guizhou, Southwestern China – an overview. *Sci. Total Environ.* 400: 227–237.

Francisco, R., Branco, R., Schwab, S. et al. (2018). Impact of plant-associated bacteria biosensors on plant growth in the presence of hexavalent chromium. *World J. Microbiol. Biotechnol.* 34: 12.

He, F., Gao, J., Pierce, E. et al. (2015). In situ remediation technologies for mercury-contaminated soil. *Environ. Sci. Pollut. Res. Int.* 22: 8124–8147.

Helmisaari, H.S., Salemaa, M., Derome, J. et al. (2007). Remediation of heavy metal contaminated forest soil using recycled organic matter and native woody plants. *J. Environ. Qual.* 36: 1145–1153.

Hezbullah, M. (2014). Scenario of food adulteration in Bangladesh: a ringing alarm for health security. *Bangladesh J. Med.* (Supplement issue) 25: 16.

Honma, T., Ohba, H., Kaneko-Kadokura, A. et al. (2016). Optimal soil Eh, pH, and water management for simultaneously minimizing arsenic and cadmium concentrations in rice grains. *Environ. Sci. Technol.* 50 (8): 4178–4185.

Hossain, A.M.M.M., Monir, T., Rezwan-Ul-Haque, A.M. et al. (2007). Heavy metal concentration in tannery solid wastes used as poultry feed and the ecotoxicological consequences. *Bangladesh J. Sci. Indust. Res.* 42: 397–416.

Hussein, H.S. (2008). Optimization of plant–bacteria complex for phytoremediation of contaminated soils. *Int. J. Bot.* 4: 437–443.

IARC (2012). Cadmium and cadmium compounds. In: *IARC Monographs on the Evaluation of Carcinogenic Risks to Humans. Arsenic, Metals, Fibres, and Dusts*, vol. 100. France: International Agency for Research on Cancer, C. Lyon.

IARC and WHO (1993). Beryllium, cadmium, mercury, and exposures in the glass manufacturing industry. *IARC*. http://publications.iarc.fr/Book-And-Report-Series/Iarc-Monographs-On-The-Evaluation-Of-Carcinogenic-Risks-To-Humans/Beryllium-Cadmium-Mercury-And-Exposures-In-The-Glass-Manufacturing-Industry-1993 (accessed 29 June 2018).

Jagetiya, B. and Kumar, S. (2020). Phytoremediation of lead: a review. In: *Lead in Plants and the Environment* (ed. D.K. Gupta, S. Chatterjee, and C. Walther), 171–202. Switzerland: Springer https://doi.org/10.1007/978-3-030-21638-2_9.

James, K.A., Meliker, J.R., and Nriagu, J.O. (2017). Arsenic. In: *International Encyclopedia of Public Health* (ed. S.R. Quah), 170–175. London: Academic Press.

Jean, J., Sirot, V., Hulin, M. et al. (2018). Dietary exposure to cadmium and health risk assessment in children – results of the French infant total diet study. *Food Chem. Toxicol.* 115: 358–364.

Jiang, G.B., Shi, J.B., and Feng, X.B. (2006). Mercury pollution in China: an overview of the past and current sources of the toxic metal. *Environ. Sci. Technol.* 40: 3672–3678.

Kamiya, T. and Fujiwara, T. (2009). Arabidopsis NIP1;1 transports antimonite and determines antimonite sensitivity. *Plant Cell Physiol.* 50 (11): 1977–1981.

Kamiya, T., Islam, M.R., Duan, G. et al. (2013). Phosphate deficiency signaling pathway is a target of arsenate and phosphate transporter OsPT1 is involved in As accumulation in shoots of rice. *Soil Sci. Plant Nutri.* 59 (4): 580–590.

Khan, M., Khan, S., Khan, A., and Alam, M. (2017). Soil contamination with cadmium, consequences and remediation using organic amendments. *Sci. Total Environ.* 601–602 (1): 1591–1605.

Koller, M. and Saleh, H.M. (2018). Introducing heavy metals. In: *Heavy Metals* (ed. H.E.M. Saleh and R.F. Aglan). United Kingdom: IntechOpen Limited https://doi.org/10.5772/intechopen.74783.

Koptsik, G.N. (2014). Problems and prospects concerning the phytoremediation of heavy metal polluted soils: a review. *Eurasian Soil Sci.* 47: 707–722.

Kubier, A., Wilkin, R.T., and Pichler, T. (2019). Cadmium in soils and groundwater: a review. *Appl. Geochem.* 108: 104388. https://doi.org/10.1016/j.apgeochem.2019.104388.

Kuivenhoven, M. and Mason, K. (2019). Arsenic toxicity. In: *StatPearls, 31082169*. USA: StatPearls Publishing.

Külahcı, F. (2020). Environmental distribution and modelling of radioactive lead (210): a Monte Carlo simulation application. In: *Lead in Plants and the Environment, Radionuclides and Heavy Metals in the Environment* (ed. D.K. Gupta, S. Chatterjee, and C. Walther), 15–32. Switzerland: Springer https://doi.org/10.1007/978-3-030-21638-2_2.

Kumar, A., Subrahmanyam, G., Mondal, R. et al. (2021). Bio-remediation approaches for alleviation of cadmium contamination in natural resources. *Chemosphere* 268: 128855.

Lata, S., Kaur, H.P., and Mishra, T. (2019). Cadmium bioremediation: a review. *Int. J. Pharm. Sci. Res.* 10 (9): 4120–4128.

LeFauve, M.K. and Connaughton, V.P. (2017). Developmental exposure to heavy metals alters visually-guided behaviors in zebrafish. *Curr. Zool.* 63 (2): 221–227.

Li, D., Tan, X.Y., Wu, X.D. et al. (2014). Effects of electrolyte characteristics on soil conductivity and current in electrokinetic remediation of lead-contaminated soil. *Sep. Purif. Technol.* 135: 14–21.

Li, R., Wu, H., Ding, J. et al. (2017). Mercury pollution in vegetables, grains and soils from areas surrounding coal-fired power plants. *Sci. Rep.* 7 (1): 46545. https://doi.org/10.1038/srep46545.

Madhavi, V., Reddy, A.V.B., Reddy, K.G. et al. (2013). An overview on research trends in remediation of chromium. *Res. J. Recent Sci.* 2277: 2502.

Mandal, S. and Ahmed, A.T.A. (2013). Copper, cadmium, chromium and lead bioaccumulation in Stinging Catfish, *Heteropneustes fossilis* (Bloch) and freshwater mussel, *Lamellidens corrianus* Lia and to compare their concentration in sediments and water of Turag river. *J. Asiatic Soc. Bangladesh Sci.* 39 (2): 231–238.

Marcus, J.B. (2013). Vitamin and mineral basics: the ABCs of healthy foods and beverages, including phytonutrients and functional foods. In: *Culinary Nutrition* (ed. J.B. Marcus), 279–231. Netherlands: Elsevier.

Meija, J., Yang, L., Sturgeon, R.E., and Mester, Z. (2010). Certification of natural isotopic abundance inorganic mercury reference material NIMS-1 for absolute isotopic composition and atomic weight. *J. Anal. Atom. Spectrom.* 25: 384–389.

Meng, D., Li, J., Liu, T. et al. (2018). Effects of redox potential on soil cadmium solubility: insight into microbial community. *J. Environ. Sci.* 75: 224–232.

Mestrot, A., Feldmann, J., Krupp, E.M. et al. (2011). Field fluxes and speciation of arsines emanating from soils. *Environ. Sci. Technol.* 45 (5): 1798–1804.

Mudgal, V., Madaan, N., and Mudgal, A. (2010). Heavy metals in plants: phytoremediation: plants used to remediate heavy metal poll. *Agric. Bio. J. North. Amer.* 1: 40–46.

Naser, H.M., Sultana, S., Mahmud, N.U. et al. (2012). Heavy metal levels in vegetables with growth stage and plant species variations. *Bangladesh J. Agric. Res.* 36 (4): 563–574.

Natasha, Dumat, C., Shahid, M. et al. (2020). Lead pollution and human exposure: forewarned is forearmed, and the question now becomes how to respond to the threat! In: *Lead in Plants and the Environment* (ed. D.K. Gupta, S. Chatterjee, and C. Walther), 33–65. Switzerland: Springer https://doi.org/10.1007/978-3-030-21638-2_3.

Owlad, M., Aroua, M.K., Daud, W.A.W., and Baroutian, S. (2009). Removal of hexavalent chromium-contaminated water and wastewater: a review. *Water Air Soil Pollut.* 200: 59–77.

Pandey, D., Zoomi, I., Akhtar, O. et al. (2018). Approaches for remediation of arsenic contamination from soil and water: a review. *Int. J. Life Sci. Res.* 6 (3): 146–162.

Rahimzadeh, M.R., Rahimzadeh, M.R., Kazemi, S., and Moghadamnia, A. (2017). Cadmium toxicity and treatment: an update. *Caspian J. Intern. Med.* 8 (3): 135–145.

Raj, D. and Maiti, S.K. (2019). Sources, toxicity, and remediation of mercury: an essence review. *Environ. Monit. Assess.* 191: 566.

Sabine, M. and Wendy, G. (2009). Human health effects of heavy metals. *Environmental Science and Technology Briefs for Citizens*, (15). http://www.engg.ksu.edu/CHSR/outreach/resources/docs/15_HumanHealthEffectsofHeavyMetals2013.pdf (accessed 29 June 2018).

Saini, S. and Dhania, G. (2020). Cadmium as an environmental pollutant: ecotoxicological effects, health hazards, and bioremediation approaches for its detoxification from contaminated sites. In: *Bioremediation of Industrial Waste for Environmental Safety* (ed. R.N. Bharagava and G. Saxena), 357–387. Switzerland: Springer.

Sarkar, D., Datta, R., and Hannigan, R. (2011). *Concepts and Applications in Environmental Geochemistry*. Netherlands: Elsevier.

Satoh, H. (2013). Mercury. In: *Aging and Vulnerability to Environmental Chemicals: Age-related Disorders and Their Origins in Environmental Exposures* (ed. B. Weiss), 125–150. London: Royal Society of Chemistry.

Schaefer, H.R., Dennis, S., and Fitzpatrick, S. (2020). Cadmium: mitigation strategies to reduce dietary exposure. *J. Food Sci.* 85: 260–267.

Sharma, N., Sodhi, K.K., Kumar, M., and Singh, D.K. (2021). Heavy metals eco-toxicity with major concern to chromium and recent advancement in remediation technologies. *Environ. Nanotechnol. Monitor. Manag.* 15: 100388. https://doi.org/10.1016/j.enmm.2020.100388.

Siddiqui, S.I., Naushad, M., and Chaudhry, S.A. (2019). Promising prospects of nanomaterials for arsenic water remediation: a comprehensive review. *Process Saf. Environ. Prot.* 126: 60–97.

Singh, S., Srivastava, P.K., Kumar, D. et al. (2015). Morphoanatomical and biochemical adapting strategies of maize (*Zea mays* L.) seedlings against lead and chromium stresses. *Biocatal. Agric. Biotechnol.* 4: 286–295.

Soisungwan, S. (2018). Dietary cadmium intake and its effects on kidneys. *Toxics* 6: 1.

Song, B., Zeng, G., Gong, J. et al. (2017). Evaluation methods for assessing effectiveness of in situ remediation of soil and sediment contaminated with organic pollutants and heavy metals. *Environ. Int.* 105: 43–55.

Sperling, M. (2005). Chromium. In: *Encyclopedia of Human Nutrition* (ed. L. Allenand and A. Prentice), 113–126. Oxford: Elsevier.

Srivastava, V., Sarkar, A., Singh, S. et al. (2017). Agroecological responses of heavy metal pollution with special emphasis on soil health and plant performances. *Front. Environ. Sci.* 5: 64.

Suhani, I., Sahab, S., Srivastava, V., and Singh, R.P. (2021). Impact of cadmium pollution on food safety and human health. *Curr. Opin. Toxicol.* 27: 1–7.

Teng, D., Mao, K., Ali, W. et al. (2020). Describing the toxicity and sources and the remediation technologies for mercury-contaminated soil. *RSC Adv.* 10: 23221–23232.

Tumolo, M., Ancona, V., Paola, D.D. et al. (2020). Chromium pollution in European water, sources, health risk, and remediation strategies: an overview. *Int. J. Environ. Res. Publ. Health* 17: 5438.

UNEP (2018). *Global Mercury Assessment 2018*. Geneva: United Nations Environment Programme.

Unsal, V., Dalkõran, T., Çiçek, M., and Kölükçü, E. (2020). The role of natural antioxidants against reactive oxygen species produced by cadmium toxicity: a review. *Adv. Pharmaceut. Bull.* 10: 184.

Valko, M., Morris, H., and Cronin, M. (2005). Metals, toxicity and oxidative stress. *Curr. Medic. Chem.* 12 (10): 1161–1208.

Verma, C., Das, A.J., and Kumar, R. (2017). PGPR-assisted phytoremediation of cadmium: an advancement towards clean environment. *Curr. Sci.* 113: 715–724.

Wang, J., Feng, X., Anderson, C.W. et al. (2012). Remediation of mercury contaminated sites – a review. *J. Hazard. Mater.* 221–222: 1–18.

Wang, P., Zhang, W., Mao, C. et al. (2016). The role of OsPT8 in arsenate uptake and varietal difference in arsenate tolerance in rice. *J. Exp. Bot.* 67: 6051–6059.

Wang, M., Chen, Z., Song, W. et al. (2021). A review on cadmium exposure in the population and intervention strategies against cadmium toxicity. *Bullet. Environ. Contamin. Toxicol.* 106: 65–74.

WHO (2017). *Ten Chemicals of Major Health Concern*. World Health Organization.

Wilkin, R.T. (2007). Cadmium. In: *Monitored Natural Attenuation of Inorganic Contaminants in Ground Water* (ed. R.G. Ford, R.T. Wilkin, and R.W. Puls), 1–9. Oklahoma: Environmental Protection Agency.

Yadav, K.K., Gupta, N., Kumar, V., and Singh, J.K. (2017). Bioremediation of heavy metals from contaminated sites using potential species: a review. *Indian J. Environ. Protect.* 37: 65–84.

Yin, X.X., Wang, L.H., Yu, X.J. et al. (2017). Arsenic accumulation and speciation of $PM_{2.5}$ and relevant health risk assessment in Jinan, China. *Polish J. Environ. Stud.* 26 (2): 949–954.

Zhao, F.J. and Wang, P. (2020). Arsenic and cadmium accumulation in rice and mitigation strategies. *Plant Soil* 446: 1–21.

Zheng, Y., Jensen, A.D., Windelin, C., and Jensen, F. (2012). Review of technologies for mercury removal from flue gas from cement production processes. *Prog. Energ. Combust.* 38: 599–629.

Zhu, G., Xiao, H., Guo, Q. et al. (2018). Effects of cadmium stress on growth and amino acid metabolism in two Compositae plants. *Ecotoxicol. Environ. Saf.* 158: 300–308.

4

Aluminum Stress Tolerance in Plants

Insights from Omics Approaches

Richa Srivastava[1]*, Ayan Sadhukhan*[2]*, and Hiroyuki Koyama*[3]

[1] *Department of Bioscience and Bioengineering, Indian Institute of Technology Guwahati, Guwahati, India*
[2] *Department of Bioscience and Bioengineering, Indian Institute of Technology Jodhpur, Jodhpur, India*
[3] *Faculty of Applied Biological Sciences, Gifu University, Gifu, Japan*

4.1 Introduction

Acidification affects 40% of global agricultural soils, severely affecting the productivity of essential crops (Kochian et al. 2015). Aluminum (Al) is converted to Al^{3+} ions in acid soils having pH < 5.5, which are toxic to plants even at micromolar concentration. Al causes DNA damage leading to severe root growth inhibition by preventing cell division and elongation (Nezames et al. 2012; Sjogren et al. 2015; Sjogren and Larsen 2017). To counter the toxic effects of Al, plants have adopted various strategies, including alkalization of the root surface, preventing the deposition of Al^{3+} by altering cell wall, chelation of Al^{3+} by secretion of organic acids in the rhizosphere, and sequestration of Al in vacuoles and detoxification of reactive oxygen species (ROS) to prevent cytotoxicity (Yan et al. 2022) (Figure 4.1).

Classically, essential Al tolerance genes were identified by mutant screening and subsequent positional cloning from Arabidopsis and rice. The *sensitivity to proton rhizotoxicity 1* (*stop1*) mutant, having drastically reduced root length under both low pH and Al, was identified in a forward genetic screen from mutagenized Arabidopsis and was eventually mapped to the C_2H_2-typezinc finger transcription factor (TF) *STOP1* (*sensitive to proton rhizotoxicity 1*: Iuchi et al. 2007). The *stop1* mutant had repressed transcript levels of *aluminum-activated malate transporter1* (*ALMT1*), which encodes a highly Al-inducible malate transporter extruding the organic acid (OA) malate to chelate Al^{3+} in the rhizosphere (Sasaki et al. 2004; Kobayashi et al. 2007). Mutants of *ALMT1* were severely impaired in root growth under Al suggesting critical involvement of this gene in Al tolerance (Hoekenga et al. 2006). Later, STOP1 was found to regulate the transcription of many genes through microarray analysis. Moreover, the overexpression of *ALMT1* alone could not complement the Al sensitivity of the *stop1* mutant, suggesting various other Al tolerance mechanisms (Sawaki et al. 2009). A paralog of *STOP1*, *STOP2* functions downstream and regulates most of the STOP1-regulated genes in Arabidopsis (Kobayashi et al. 2014). Upstream phosphatidylinositol (PI)

Heavy Metal Toxicity and Tolerance in Plants: A Biological, Omics, and Genetic Engineering Approach, First Edition. Edited by Mohammad Anwar Hossain, AKM Zakir Hossain, Sylvain Bourgerie, Masayuki Fujita, Om Parkash Dhankher, and Parvez Haris.

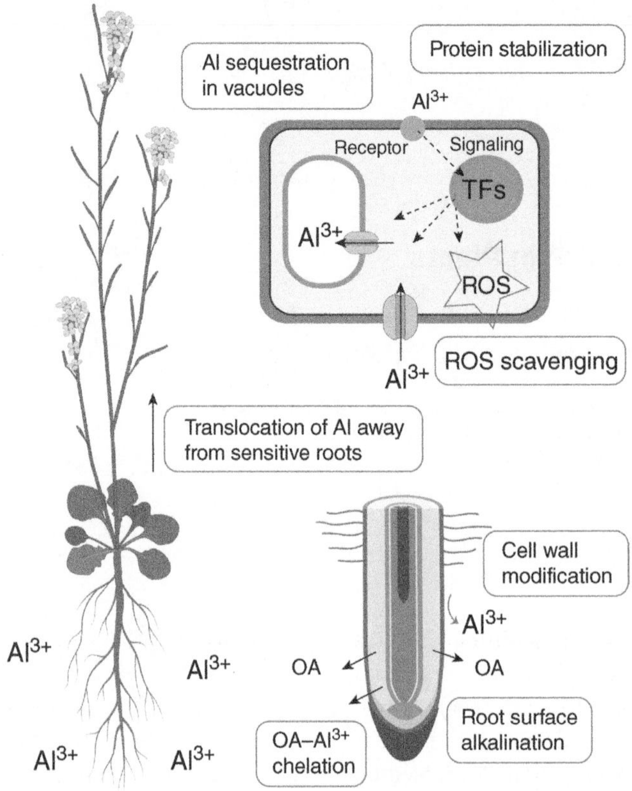

Figure 4.1 Mechanisms of plant tolerance to aluminum stress. Plants counter the toxic Al^{3+} by both external and internal detoxification mechanisms. The roots release various types of organic acid (OA), viz. malate, citrate, and oxalate, to chelate Al^{3+} in the rhizosphere. The root surface pH is also increased, and cell wall properties are modified to prevent the binding of Al^{3+}. The internal detoxification mechanisms involve the diversion of Al^{3+} away from the sensitive root tissues through the phloem. Al^{3+} is sequestered in the vacuoles to protect photosynthesizing leaf cells from damage. Al signaling pathways activate transcription factors (TFs), which induce Al tolerance genes leading to reactive oxygen species (ROS) scavenging and protein stabilization. *Source:* Figure was created using the BioRender web application (https://biorender.com/).

signaling directs a prompt nuclear localization of the STOP1 protein upon initial exposure to Al stress leading to the rapid upregulation of *STOP2*, *ALMT1*, *GDH1* (*glutamate dehydrogenase 1*), and *GDH2* genes having STOP1-binding sites (Tsutsui et al. 2011) in their promoters. Again, the roles of *GDHs* in Al tolerance via regulation of the intracellular pH were discovered only recently (Tokizawa et al. 2021). Other STOP1-regulated Al tolerance genes, *multidrug and toxic compound exclusion protein* (*MATE*) transporting another OA, citrate, into the rhizosphere (Magalhaes et al. 2007; Liu et al. 2009), and *aluminum-sensitive 3* (*ALS3*), a phloem-localized transporter redistributing the toxic Al^{3+} away from the sensitive root tissues (Larsen et al. 2004), are induced more slowly, due to lower affinity of STOP1 binding and indirect regulation, respectively (Tokizawa et al. 2021). On the other hand, low phosphate and iron serve as additional factors to control the nuclear localization of STOP1 under Al stress in acid soil (Godon et al. 2019).

An ortholog of *STOP1*, *aluminum resistance transcription factor 1* (*ART1*), regulates several Al tolerance genes in rice, including a bacterial ATP-binding cassette-type transporter, encoded by *sensitive to Al-rhizotoxicity 1* (*STAR1*) and *sensitive to Al-rhizotoxicity 2* (*STAR2*), which transports UDP-glucose to the apoplast for altering the cell wall to prevent binding of Al^{3+} (Huang et al. 2009; Yamaji et al. 2009). Rice has multiple copies of STOP1 orthologs in its genome, and no wonder it is one of the most Al-tolerant species (Yamaji et al. 2009; Fan et al. 2016). STOP1-like proteins have been found almost in all terrestrial plants, indicating the ubiquitous nature of the STOP1-regulating system for Al tolerance (Ohyama et al. 2013; Ito et al. 2019). STOP1 levels are kept in check under control conditions to reduce the deleterious effects on plant growth by proteasomal degradation (Fang et al. 2021).

Bioinformatic analysis of the *cis*-elements of the *ALMT1* promoter led to the observation that other TFs, apart from STOP1, may bind to the *ALMT1* promoter in an Al-regulated manner. Similarly, another Al tolerance TF *calmodulin binding transcription activator 2* (*CAMTA2*) was identified, which binds to the *ALMT1* promoter to activate transcription under Al (Tokizawa et al. 2015).

In recent years, the powerful technologies of genomics and transcriptomics have led to the identification of new Al tolerance genes and mechanisms. Mapping bi- or multi-parental populations with backgrounds contrasting in Al-tolerance-identified quantitative trait loci (QTLs) associated with Al tolerance in Arabidopsis and rice (Ikka et al. 2007; Meng et al. 2017). These QTLs often overlapped with candidate genes for Al tolerance identified using genome-wide association studies (GWAS). The forthcoming sections summarize the discovery of novel Al tolerance genes from recent genomic and transcriptomic analyses.

4.2 Exploration of Al Tolerance QTLs

In the last two decades, three to five dominant loci that control the variation in Al-tolerance levels have been mapped in several models and edible crops, such as Arabidopsis (Kobayashi and Koyama 2002), rice (Famoso et al. 2011), barley (Coelho et al. 2019), maize (Guimaraes et al. 2014), and soybean (Qi et al. 2008; Wang et al. 2019) using root growth as the phenotypic index (Table 4.1). Over the last few years, next-generation sequencing (NGS) techniques, viz. restriction-site-associated DNA sequencing (RAD-seq), have produced high-density genetic maps consisting of thousands of single-nucleotide polymorphism (SNP), as opposed to the traditional molecular markers, restriction fragment length polymorphism (RFLP), simple sequence length polymorphism (SSLP), simple sequence repeat (SSR), and amplified fragment length polymorphism (AFLP), having poor mapping resolutions and uneven genomic distribution. The development of SNP markers has improved QTL mapping efficiency, promoting the identification of the genetic basis of Al tolerance in complex polyploid crops (Guimaraes et al. 2014; Wang et al. 2019). But QTL studies by linkage mapping are time-consuming and laborious as they require the development of polymorphic markers and mapping populations such as recombinant inbred lines (RILs) using bi-parental crosses, enabling the study of only two alleles at any given locus. Indeed, association mapping using natural genetic diversity could overcome the limitations of QTL mapping.

Table 4.1 QTLs associated with Al tolerance identified in edible crops and model plants.

Crop	Marker type	QTL ID or locus	Al-tolerant × Al-sensitive	Phenotypic index	References
Arabidopsis	SSLP	apx1A-ATTS0477 and mi51-mi204	RIL derived from Landsberg erecta × Columbia	Root length	Kobayashi and Koyama (2002)
Rice	SNP	RM265, RM6333, RM221, L688, RM242, RM247, and R2708	RIL derived from Azucena X IR64; Nipponbare × Kasalath	Total root growth, primary root growth, and longest root growth	Famoso et al. (2011)
Wheat	SSR	Xbarc164 and Xwmc331	RIL derived from Atlas 66 × Chisholm	Relative root growth and root response to hematoxylin stain	Zhou et al. (2007)
Soybean	RFLP	O103-1n, Cr207-1, A847-1, K644-1, B122-1	RIL derived from Young × PI 416937	Root extension	Bianchi-Hall et al. (2000)
	SSR	GMKF046-GMKF080, satt172-satt703, GMKF058-Satt397, Satt397-satt669, satt278-sat_195	RIL derived from Kefeng No.1 × Nannong 1138-2	Plant height, number of leaves, shoot and root dry weight	Qi et al. (2008)
	SNP	Gm04_bin93-bin94, Gm16_bin8, Gm17_bin93, and Gm19_bin19	RIL derived from Zhonghuang 24 × Huaxia 3	Relative root elongation and apical Al^{3+} content	Wang et al. (2019)
Maize	SNP	S2_212940514, S3_187460236, S5_30301926, ZmMATE1, and S8_22681622	RIL derived from Cateto Al237 × L53	Relative net root growth	Guimaraes et al. (2014)
Barley	SSR AFLP	E32M6091-E32M48207, E35M51241-E32M50299, E42M51165-E32M4862, ZmMATE2-E40M48214, E42M48137-E40M5072, E40M48715-E42M51190, E44M5187-Bnlg1176, E32M60139-E35M51107, E35M5191-E44M5184	F2 population derived from LT 99-4 × LS 04-2	Differential root length	Coelho et al. (2019)

4.3 Unraveling the Genetic Architecture of Al Tolerance from Natural Variation

Ikka et al. (2007) observed heritable variation in Al tolerance among 260 worldwide Arabidopsis ecotypes, determined by differences in root length after seven days of hydroponic culture with Al. Soon, sequence variations were mapped to the Al tolerance genes previously identified by mutant screening, providing genetic evidence of their role in Al tolerance. For example, amino acid polymorphisms within ALMT1 were associated with the Al tolerance variation in Arabidopsis ecotypes (Hoekenga et al. 2006). Although the AtSTOP1 protein remained conserved across Arabidopsis ecotypes, extensive amino acid sequence polymorphisms in the rice STOP1 ortholog, ART1, shaped the natural variation in Al tolerance in rice genotypes (Arbelaez et al. 2017), and variations in CsSTOP1 led to the natural selection of Al tolerance in tea (Zhao et al. 2018). Variation in the regulatory and coding sequences of the root cell membrane-localized *Nramp aluminum transporter* (*NRAT1*) determined the transcript levels and Al transport capacities of the NRAT1 protein, contributing to the Al tolerance variation in rice. NRAT1 helps in Al^{3+} transport into the cytoplasm to protect the apoplast from damage (Xia et al. 2014; Li et al. 2014). The toxic Al^{3+} is subsequently sequestered inside vacuoles by another tonoplast-localized Al transporter ALS1 (Huang et al. 2012). (Figure 4.2). Apart from

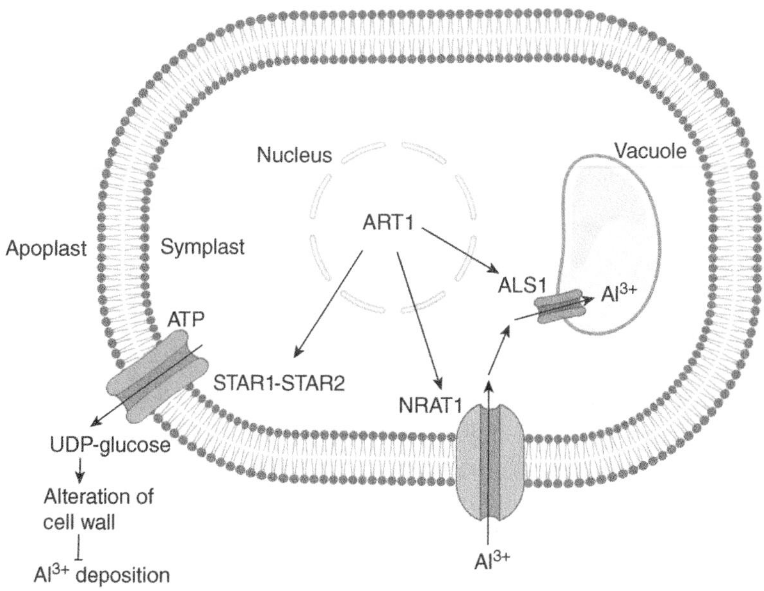

Figure 4.2 Natural variation in rice aluminum tolerance genes. *Oryza sativa* genes having polymorphisms associated with aluminum (Al) tolerance are shown. A rice STOP1-like protein, ART1 is a C_2H_2-type zinc finger transcription factor regulating many Al tolerance genes. These include a bacterial-type ATP-binding cassette UDP-glucose transporter on the plasma membrane encoded by *STAR1* and *STAR2*. UDP-glucose leads to cell wall modifications preventing Al^{3+} deposition. Another transporter, NRAT1, transports Al^{3+} into the cytoplasm to protect the apoplast from Al. Al^{3+} is subsequently sequestered into the vacuole through a tonoplast transporter ALS1. ALS1, Aluminum sensitive 1; ART1, Aluminum resistance transcription factor 1; NRAT1, Nramp aluminum transporter 1; STAR1, Sensitive to aluminum rhizotoxicity 1; STOP1, Sensitivity to proton rhizotoxicity 1. *Source:* Figure was created using the BioRender web application (https://biorender.com/).

these genes, Zhang et al. (2016) identified several new loci by association mapping of different rice landraces contrasting in Al tolerance, using simple sequence repeat (SSR) markers.

4.4 Identification of Novel Al Tolerance Genes Through Genome-Wide Association Studies

GWAS using dense single nucleotide polymorphism (SNP) markers is one of the most potent forward genetic tools to identify genes for abiotic stress tolerance in plants (Zhu et al. 2022). Famoso et al. (2011) identified the candidate genes: *NRAT1* (*Nramp aluminum transporter 1*), *ART1* (*aluminum resistance transcription factor 1*), and *STAR2* by a GWAS using natural accessions of rice. Introgression of the *japonica*-type alleles of *NRAT1* into the sensitive *aus* subpopulations of rice led to improvement of Al tolerance. GWAS of Arabidopsis root growth under Al led to identifying the well-known candidate *ALMT1* and several novel candidate genes conferring Al tolerance (Nakano et al. 2020a). Reverse genetic analysis demonstrated drastically reduced root length in the mutants of the GWAS-identified genes, viz. *TON1 recruiting motif 28* (*TRM28*), *thioredoxin H-type 1* (*TRX1*), and *sensitive to freezing 6* (*SFR6*), under Al. TRX1 is involved in ROS signaling for cellular protection and cell wall biosynthesis (Navrot et al. 2007). TRM28 helps polymerize microtubules required in cell wall synthesis (Höfte and Voxeur 2017). SFR6 is also implicated in the cell wall cellulose biosynthesis (Sorek et al. 2015). These genes highlight the importance of cell wall modifications to protect the cells from toxic Al^{3+}.

4.5 Exploring Expression Level Polymorphisms to Identify Upstream Al Signaling

GWAS on the expression level polymorphism (ELP) across natural ecotypes or expression GWAS (eGWAS) can unravel upstream signaling loci regulating the gene of interest. Two genes of the STOP1-signaling pathway, *ALS3* and *polygalacturonase-inhibiting protein 1* (*PGIP1*), are strongly induced in the Arabidopsis shoot under Al stress (Sawaki et al. 2016). GWAS of *ALS3* in Arabidopsis shoots led to the identification of phosphatidylinositol (PI)-derived signaling cascades involving phospholipase C9 (PLC9) and calcium-dependent protein kinase (CDPK32) upstream to *ALS3*. This pathway regulated *ALS3* via the direct binding of the TF ANAC071 to the *ALS3* promoter under Al stress (Sadhukhan et al. 2021). These results supported the findings of a different pharmacological study. The application of human kinase inhibitors suppressed malate release, unraveling the regulation of phosphatidylinositol 4-kinase (PI4K)-PLC signaling for early Al-induced activation of *ALMT1* (Wu et al. 2019).

PGIP1 is implicated in cell wall pectin stabilization under low pH conditions (Kobayashi et al. 2014). A candidate gene-based eGWAS of *PGIP1* levels in the shoot unraveled

STOP1-dependent and independent signaling pathways associated with the ELP of *PGIP1*. The eGWAS-identified R-R-type MYB protein and NAC027-regulated *PGIP1* independently of STOP1 via nitric oxide (NO) signaling. By contrast, a TRX superfamily protein, under direct STOP1 regulation, was another regulating factor, possibly through ROS signaling (Agrahari et al. 2021). The genes identified through eGWAS were Al inducible and participated in upstream Al signaling, although their mutants did not show root length suppression under Al (Agrahari et al. 2021; Sadhukhan et al. 2021).

GWAS on *TRX1* expression levels under Al led to the detection of the most significant SNPs in its promoter, pointing to the roles of *cis*-variants in the expression level variation and Al tolerance (Nakano et al. 2020a). On the other hand, eGWAS of *ALMT1* led to detecting promoter *cis*-variants, including transposon insertions in the higher *ALMT1* expression-haplotypes from west Europe with predominant acid soils. Several other *trans* loci regulating ALMT1 were also identified, whose functions remain to be explored (Nakano et al. 2020a). Similar transposon insertions in *MATE* orthologs from Arabidopsis and other plants lead to higher gene expression and citrate release in Al-tolerant varieties (Magalhaes et al. 2007; Fujii et al. 2012; Tovkach et al. 2013; Yokosho et al. 2016; Nakano et al. 2020b) (Figure 4.3).

4.6 Comparative Transcriptome Analyses Identify Novel Al Tolerance Genes

Comparative transcriptomics in varieties contrasting in Al resistance has identified a diverse class of genes governing Al tolerance via disparate signaling pathways and metabolic processes in pulses, cereals, and model species. A recent study on rice indicated energy conservation in roots as a crucial mechanism for adapting Al tolerance. Transcriptome analysis revealed the suppression of glycolytic genes in the Al-tolerant landrace, while genes involved in exergonic reactions were activated in the sensitive landrace. This study identified 40 new genes as potential molecular breeding tools regulating Al tolerance, classified into signaling factors (*Os12g0581900*, *Os02g0445600*), cell-wall components (*EP_003637074*, *EXC01915*, and *AP014959*), enzymes (*Os11g0112200* and *Os10g01044*), and translational machinery components (*Os03g0798600* and *Os05g0541900*) (Tyagi et al. 2020).

Induction of two cellulose synthase encoding genes, *Glyma.02G205800* and *Glyma.04G153700,* in roots of Al-tolerant soybean variety and corresponding suppression in the sensitive variety indicated cellulose production as the critical determinant of Al tolerance in soybean (Zhao et al. 2020). In comparison, another soybean study showed genes encoding the antioxidant defense system, such as *GmCAT1* (Glycine max *catalase 1*), *GmPOD1* (Glycine max *peroxidase 1*), *GmGST1* (Glycine max *glutathione S-transferase*), *GmAPX* (Glycine max *ascorbate peroxidase*), *GmGSH1* (Glycine max *gamma glutamyl-cysteine synthetase*), and *GmSOD* (Glycine max *super oxide dismutase*) play a critical role in tolerant cultivars to mitigate Al stress (Li et al. 2020). Similarly, in sugarcane, comparative transcriptome analysis identified genes involved in detoxification and transport (*Zinc-induced facilitator-like 1*, *ZIFL1*; *Tonoplast dicarboxylate transporter*, *TDT*; *Heavy metal-associated isoprenylated plant protein*, *HIPP39*; *ATPases*), lateral root formation (*Auxin*

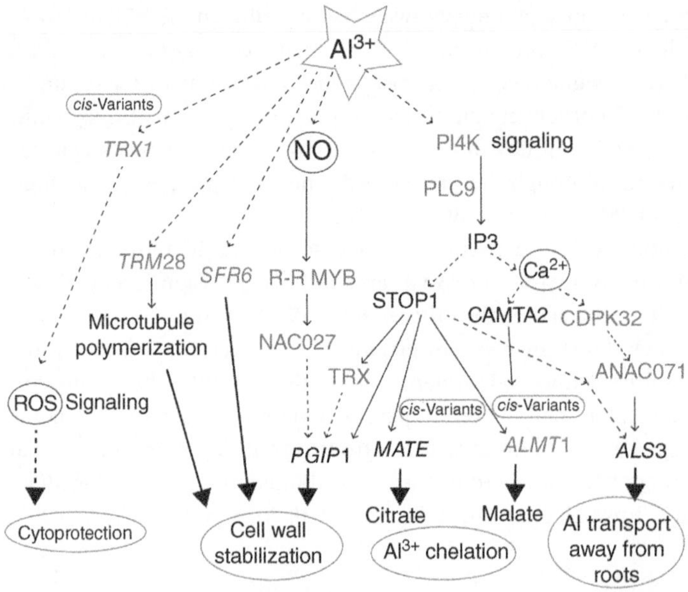

Figure 4.3 Aluminum tolerance and signaling genes identified by genome-wide association studies (GWAS) in Arabidopsis. Al stress signals are transduced by the phosphatidylinositol signaling via PI4K and PLC9 to the second messengers IP3 and Ca^{2+}, which regulate Al-inducible genes via calcium-binding proteins and transcription factors (TFs). The TF ANAC071 regulates a phloem-localized transporter *ALS3*, which is involved in the re-distribution of toxic Al^{3+} away from sensitive tissues. STOP1 and CAMTA2 regulate *ALMT1*, encoding a membrane-localized transporter extruding malate in the rhizosphere to chelate Al^{3+}. Another STOP1-regulated gene, *MATE*, encodes a citrate transporter. STOP1 regulates *PGIP1* by binding to the promoters of *PGIP1* and another *TRX* superfamily gene. *PGIP1* is also regulated by a STOP1-independent pathway involving nitric oxide (NO) and the TFs R-R MYB and NAC27. PGIP1 is involved in the stabilization of the cell wall pectin. Other genes *TRM28* and *SFR6* are also involved in cell wall biosynthesis and maintenance under Al. TRX1 is involved in reactive oxygen species (ROS) signaling for cellular protection under Al. The Al tolerance genes identified by GWAS and expression GWAS are red, while those involved in Al signaling are blue. Solid arrows indicate direct regulations, and dashed arrows indicate indirect regulations. ALMT1, Aluminum-activated malate transporter 1; CAMTA2, Calmodulin binding transcription activator 2; CDPK32, Calcium-dependent protein kinase 32; IP3, inositol-1,4,5-triphosphate; MATE, Multidrug and toxic compounds extrusion protein; PGIP1, Polygalacturonase inhibitor protein; PI4K, Phosphatidylinositol 4-kinase; PLC, Phosphatidylinositol-specific phospholipase C; SFR6, Sensitive to freezing 6; STOP1, Sensitive to proton rhizotoxicity 1; TRM28, TON1 recruiting motif 28; TRX1, Thioredoxin H-type 1. *Source:* Figure was created using the BioRender web application (https://biorender.com/).

efflux carrier component, *PIN1*; *Auxin response factor*, *ARF*; *Ethylene-responsive element-binding factor*, *ERF014*), and activation of redox enzymes (*Glutathione peroxidase 2*, *Glutathione S-transferases*, and L-*ascorbate peroxidase 3*) (Rosa-Santos et al. 2020) as crucial determinants of Al tolerance. In peanut, MAPK-promoted induction of *EIL* (*EIN3-like*) gene involved in ethylene signaling transduction was responsible for early activation of Al tolerance genes involved in transcriptional regulation, organic acid, and metal cation transport, and programmed cell death in the root tips of the resistant cultivar, resulting in higher Al tolerance (Xiao et al. 2021).

Low pH and nitric oxide (NO) are coordinating players in Al response. Transcriptome changes in the root meristem of barley seedlings exposed to composite stress of Al and low pH showed that a short-term (24 hours) treatment exerted prominent pH stress, which was aggravated by Al toxicity. However, the long-term (seven days) Al treatment carried out a massive transcriptome change that induced genes responsive to stress, cell wall organization, and iron ion binding, indicating the sole effect of Al in case of persistent combined stress (Szurman-Zubrzycka et al. 2021). In contrast, a study in watermelon showed that exogenous NO could trigger antioxidant enzyme defense system, nitrogen, and phenylpropane metabolism in leaves to alleviate Al toxicity. The study showed participation of *peroxidase (POD)*, *catalase (CAT)*, *nitrate reductase* (*NR*), *phenylalanine ammonia lyase* (*PAL*), and *4-coumarate:CoA ligase* (*4CL*), and *cinnamoyl-CoA reductase* (*CCR*) in withstanding Al stress (Zheng et al. 2021).

Besides stress duration, stress intensity also determines differential tuning of effector genes under Al stress. Kusunoki et al. (2018) studied comparative transcriptome responses under mild and severe Al stress in Arabidopsis. Mild stress induced several calcium transporters such as *auto-regulated* Ca^{2+}*-ATPase* (*ACA7/12/13*) apart from the Al tolerance genes like *ALS3*, *ALMT1*, and *MATE*. In contrast, severe stress-induced genes involved in sulfate transportation (*SULTR 3;1*, *SULTR3;4*, and *SULTR 2;1*), lateral root formation (*LOD16* and *PILS5*), and suberin biosynthesis (*GPAT55* and *FAR4*). However, the induction of malic enzymes (*ME1/2*), glutathione metabolism genes (*GSTF2/3*, *GSTU2/3/5/12*), and SA-responsive genes (*KTI1*, *NHL3*, *GH3.12*, *WRKY38*, *MY112*) indicated that Al tolerance is governed by biochemical pH-stat pathway, ROS detoxification, and SA signaling, regardless of stress intensity.

In addition, the same research group reported five novel determinants of Al-tolerance by combining the comparative transcriptomics and reverse genetic approaches in six Arabidopsis accessions differing in Al tolerance that are highly expressed in the tolerant accessions, suppressed in sensitive ecotypes, and their knock-out conferred shorter roots under Al stress (Kusunoki et al. 2017) (Figure 4.4). *Binding Protein 3* (*BIP3*) encodes an ER-localized chaperone that confers Al tolerance, possibly by aiding protein folding under ER stress. *Arabidopsis Tóxicosen Levadura 30* (*ATL30*) encodes a RING/U-box superfamily protein acting as an E3 ubiquitin ligase which ubiquitinates misfolded proteins, targeting their proteasomal degradation, to combat Al stress. *Target of Avrb Operation1* (*TAO1*) is a membrane-localized Toll-interleukin-1 receptor-nucleotide binding site-leucine rich repeat (TIR-NBS-LRR)-type disease resistance protein that senses a wide range of biotic and abiotic stresses, including Al, and regulatory signaling (Yang et al. 2015). *Carboxylesterase 17* (*CXE17*) is another signal transduction gene identified in the same study regulating Al tolerance (Rui et al. 2022). Moreover, an uncharacterized protein encoded by *AT5G22530*, essential for root growth under Al stress, was also identified in this study that functions by an unknown mechanism (Kusunoki et al. 2017).

4.7 Identification of Al Tolerance Genes from Proteomics

Gene expression studies cannot explain rapid Al stress responses such as post-translational modifications and protein turnover. Proteome analysis by two-dimensional electrophoresis (2-DE) is a well-established whole-system approach that has been widely used to unravel

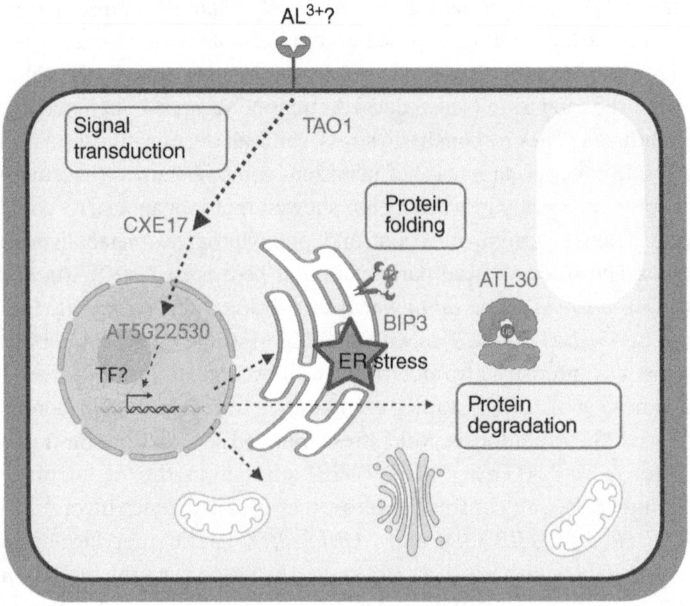

Figure 4.4 New aluminum tolerance genes identified by comparative transcriptome analysis. Aluminum (Al) tolerance genes identified by comparing the root transcriptome of Arabidopsis ecotypes contrasting in Al tolerance are shown in red. TAO1 encodes a membrane-disease-resistance protein possibly acting as an Al receptor. CXE17 is involved in signal transduction. An unknown nucleus-localized factor (AT5G22530) possibly activates other Al tolerance genes. BIP3 acts as a chaperone to guide the proper folding of proteins in the endoplasmic reticulum (ER) under ER stress, an Al-induced condition in which the ER loses the capacity to fold proteins properly. A RING/U-box superfamily protein (ATL30) acts as an E3 ubiquitin ligase targeting damaged proteins for proteasomal degradation. ATL30, Arabidopsis Tóxicosen Levadura 30; BIP3, Binding protein 3; CXE17, Carboxylesterase 17; TAO1, Target of AVRB expression 1.
Source: Figure was created using the BioRender web application (https://biorender.com/).

the general and species-specific changes caused by Al stress at the protein level. However, this approach can identify only ~30% of proteins (Yang et al. 2013). Proteins with low abundance and high membrane-binding affinity are challenging to spot using 2-DE. New approaches such as iTRAQ (Isobaric tags for relative and absolute quantitation) and label-free shot-gun proteomics have enabled high-throughput proteomics to discover complex Al-tolerance mechanisms followed by biochemical and molecular validation (Zhou et al. 2016, 2017).

Chen et al. studied calcium-alleviated Al toxicity in Arabidopsis root tips using 2-DE coupled with MALDI-TOF mass spectroscopy. The study indicated the participation of Ca-regulated proteins involved in organic acid metabolism (AtASL: argininosuccinate lyase), AtCSY3: citrate synthase, AtMDH: malate dehydrogenase, and AtSDH1-1: succinate dehydrogenase), cellular transport (AtATP-α: ATP synthase, AtATP-β, AtCIB [copper ion binding] and AtVHA-A: vacuolar ATP synthase), and signal transduction (AtGST6, AtEF1β-γ: elongation factor 1, and AtEIN3: ethylene insensitive 3), for Al tolerance through the ion-exclusion mechanism in roots (Chen et al. 2019).

A comparative proteome study of root tips in triticale genotypes that differed in Al tolerance identified ten protein functional groups participating in the Al tolerance and recovery, such as cell division, stress-related response, metabolic pathways, lignin synthesis, transport, transcription control, protein folding, protease inhibition, protein synthesis, and degradation (Niedziela et al. 2022). Recently, Zhou et al. applied the iTRAQ approach to study the proteome changes in four-month-long Al-treated tissues of tomato plants (Zhou et al. 2016). Proteins involved in Al^{3+} uptake, transportation, and root development were accumulated in the root. The light reaction centers of photosynthesis in leaf tissue and stress proteins in the embryo and seed coat were the primary targets of the long-period Al treatment (Table 4.2). The plasma-membrane-enriched proteome of Tamba Black soybean under Al stress pointed out the involvement of

Table 4.2 Differentially expressed proteins associated with Al tolerance and signaling.

Crop	Tissue	Genes	Mode of Al tolerance	References
Arabidopsis	Whole root	Copper ion binding (CIB), ATP synthase subunit α/β (ATP α/β), vacuolar ATP synthase subunit A (VHA-A), leucine aminopeptidase (LAP), ethylene-insensitive 3 (EIN3), glutathione *S*-transferase (GST6), citrate synthase (CSY3), aconitase (ACO1), NAD-isocitrate dehydrogenase (DH2), succinate dehydrogenase (SDH1-1), malate dehydrogenase (MDH), argininosuccinate lyase (ASL), etc.	Ca-dependent alleviation of growth inhibition and Al accumulation	Chen et al. (2019)
Triticale	Root tip	protein disulfide-isomerase, oxalate oxidase, 1-Cys peroxiredoxin PER1, flavone *O*-methyltransferase	Recovery of proteome modifications and root growth	Niedziela et al. (2022)
Tomato	Root, leaf, embryo seed coat	Blue copper-binding proteins (BCB), early nodulin-55-1, ATP-BINDING CASSETTE A2, aquaporin, superoxide dismutase, receptor-like kinase (RLK), NTMC2T6.1, heat shock protein 82	Disturbance of mineral homeostasis (Al, P, Fe, S, Zn, and Mn), smaller roots, and plants	Zhou et al. (2016)
Tea	Roots and leaves	Light-harvesting complex II chlorophyll a/b binding protein 2 (LHCB2), photosystem I subunit V (pSAG), pyruvate kinase (PK), pyruvate dehydrogenase (PDH), citrate synthase (CS), ascorbate peroxidase (APX), peroxidase (POD), etc.	Improved photosynthesis, reducing power, and ion homeostasis resulting in growth stimulation	Xu et al. (2017)

(Continued)

Table 4.2 (Continued)

Crop	Tissue	Genes	Mode of Al tolerance	References
Wheat	Root portion	Annexin p35, Ascorbate peroxidase, beta-amylase, phosphoglycerate kinase, intracellular chloride channel, *S*-adenosyl methionine synthetase, oxalate oxidase 2, malate dehydrogenase, cysteine synthase, methionine synthase 2, etc.	Root growth inhibition, decrease in K^+, Mg^{2+} and Ca^{2+}, whereas an increase in $P^{2+}O^{5-}$	Oh et al. (2014)
Tamba Black Soybean	Plasma membrane proteins in root tips	Cellulose synthases (CESAs), Casparian strip proteins (CASPs), aquaporins, NRT1, GmMATE13, GmMATE75, GmMATE87, H^+-ATPase, etc.	Al-induced citrate secretion in the plasma membrane and internal detoxification	Wei et al. (2020)
Sorghum	Root tip	Ammonia lyase 1(AL1), UDP-glucosyl transferase 73C7 (UGT73C7), cinnamyl alcohol dehydrogenase 4 (CAD4), etc.	Phenylpropanoid synthesis	Zhou et al. (2017)
Rice	Leaves	*S*-nitrosoglutathione reductase (GSNOR), Ferredoxin–nitrite reductase (NR), etc.	NO-dependent alleviation of growth inhibition and photosynthetic decline	Yang et al. (2013)
Switchgrass	Root apex and root maturation zone	Rotamase FKBP 1, CDC48 protein, Al-activated malate transporter (ALMT1), xyloglucan endotransglucosylase-hydrolase (XTH), aluminum sensitive 3 (ALS3), etc.	Cell-cycle regulation to secure growth attributes	Rangu et al. (2018)

GmMATE13, GmMATE75, GmMATE87, H^+-ATPase, ABC transporters, receptor-like protein kinases (RLKs), and Ca^{2+} signaling proteins in the internal Al detoxification and signaling (Wei et al. 2020).

The proteome analysis of the tea plant (*Camellia sinensis*), an Al-hyper accumulator species, showed that glycolytic enzymes (PK and PDH), antioxidant enzymes (APX, CAT, and GST), and citrate synthase (CS) were predominantly induced in roots, indicating an Al-activated detoxification response. In contrast, the accumulation of subunits of photosynthetic machinery (LHCB, LHCA, pSAG, etc.) in leaf tissue explained the stimulatory effect of Al on plant growth (Xu et al. 2017). Similarly, a proteome study in an Al-hyper-tolerant crop, i.e. switchgrass (*Panicum virgatum*), which can survive up to 800 μM Al^{3+}, at which species such as tomato are not viable, showed that cell-cycle proteins are markers for distinguishing Al-tolerant and Al-sensitive genotypes (Rangu et al. 2018).

4.8 Conclusion and Future Perspectives

This chapter discusses recent findings of the molecular mechanisms involved in complex Al-tolerance responses of plants from omics approaches. High-throughput forward genetic techniques such as GWAS and reverse genetic techniques such as transcriptomics have enabled a faster understanding of Al tolerance in a genome-wide context. The knowledge from these system approaches is crucial for developing crops using molecular breeding to achieve food and nutritional security in acidic-soil regions covering a significant cultivable area.

Apart from the classical Al tolerance genes such as *STOP1*, *ALMT1*, and *ALS3* that regulate internal and external detoxification of Al^{3+}, several cell-wall biogenesis genes in the roots and photosynthetic genes in the shoots have been recently discovered as novel players of Al tolerance. Functional validation of such genes in the future will throw more light on their significance under Al stress. Secondly, although sufficient information on functional genes and QTLs is available for Arabidopsis and rice, the knowledge of Al-tolerance mechanisms is still in infancy in legume and vegetable crops of commercial interest. On the other hand, plants such as switchgrass and tea should be focused on for a detailed understanding of Al hyper tolerance, hyper accumulation, and growth stimulation effects for future crop improvement under acid soil.

References

Agrahari, R.K., Enomoto, T., Ito, H. et al. (2021). Expression GWAS of *PGIP1* identifies STOP1-dependent and STOP1-independent regulation of PGIP1 in aluminum stress signaling in Arabidopsis. *Front. Plant Sci.* 12: 774687.

Arbelaez, J.D., Maron, L.G., Jobe, T.O. et al. (2017). ALUMINUM RESISTANCE TRANSCRIPTION FACTOR 1 (ART1) contributes to natural variation in aluminum resistance in diverse genetic backgrounds of rice (*O. sativa*). *Plant Direct* 1 (4): e00014.

Bianchi-Hall, C., Carter, T.E., Bailey, M. et al. (2000). Aluminum tolerance associated with quantitative trait loci derived from soybean PI 416937 in hydroponics. *Crop Sci.* 40: 538–545.

Chen, J., Duan, R.-X., Hu, W.-J. et al. (2019). Unravelling calcium-alleviated aluminium toxicity in *Arabidopsis thaliana*: insights into regulatory mechanisms using proteomics. *J. Proteomics* 199: 15–30.

Coelho, C.d.J., Gardingo, J.R., Almeida, M.C.d., and Matiello, R.R. (2019). Mapping of QTL for aluminum tolerance in tropical maize. *Crop Breed. Appl. Biotechnol.* 19: 86–94.

Famoso, A.N., Zhao, K., Clark, R.T. et al. (2011). Genetic architecture of aluminum tolerance in rice (*Oryza sativa*) determined through genome-wide association analysis and QTL Mapping. *PLoS Genet.* 7: 1002221.

Fan, W., Lou, H.Q., Yang, J.L., and Zheng, S.J. (2016). The roles of STOP1-like transcription factors in aluminum and proton tolerance. *Plant Signal. Behav.* 11 (2): e1131371.

Fang, Q., Zhou, F., Zhang, Y. et al. (2021). Degradation of STOP1 mediated by the F-box proteins RAH1 and RAE1 balances aluminum resistance and plant growth in *Arabidopsis thaliana*. *Plant J.* 106: 493–506.

Fujii, M., Yokosho, K., Yamaji, N. et al. (2012). Acquisition of aluminium tolerance by modification of a single gene in barley. *Nat. Commun.* 3: 713.

Godon, C., Mercier, C., Wang, X. et al. (2019). Under phosphate starvation conditions, Fe and Al trigger accumulation of the transcription factor STOP1 in the nucleus of Arabidopsis root cells. *Plant J.* 99: 937–949.

Guimaraes, C.T., Simoes, C.C., Pastina, M.M. et al. (2014). Genetic dissection of Al tolerance QTLs in the maize genome by high density SNP scan. *BMC Genomics* 15: 1–14.

Hoekenga, O.A., Maron, L.G., Piñeros, M.A. et al. (2006). *AtALMT1*, which encodes a malate transporter, is identified as one of several genes critical for aluminum tolerance in Arabidopsis. *Proc. Natl. Acad. Sci. USA* 103: 9738–9743.

Höfte, H. and Voxeur, A. (2017). Plant cell walls. *Curr. Biol.* 27: R865–R870.

Huang, C.F., Yamaji, N., Mitani, N. et al. (2009). A bacterial-type ABC transporter is involved in aluminum tolerance in rice. *Plant Cell* 21: 655–667.

Huang, C.F., Yamaji, N., Chen, Z., and Ma, J.F. (2012). A tonoplast-localized half-size ABC transporter is required for internal detoxification of aluminum in rice. *Plant J.* 69 (5): 857–867.

Ikka, T., Kobayashi, Y., Iuchi, S. et al. (2007). Natural variation of *Arabidopsis thaliana* reveals that aluminum resistance and proton resistance are controlled by different genetic factors. *Theor. Appl. Genet.* 115: 709–719.

Ito, H., Kobayashi, Y., Yamamoto, Y.Y., and Koyama, H. (2019). Characterization of NtSTOP1-regulating genes in tobacco under aluminum stress. *Soil Sci. Plant Nutr.* 65: 251–258.

Iuchi, S., Koyama, H., Iuchi, A. et al. (2007). Zinc finger protein STOP1 is critical for proton tolerance in Arabidopsis and coregulates a key gene in aluminum tolerance. *Proc. Natl. Acad. Sci. USA* 104 (23): 9900–9905.

Kobayashi, Y. and Koyama, H. (2002). QTL analysis of Al tolerance in recombinant inbred lines of *Arabidopsis thaliana*. *Plant Cell Physiol.* 43: 1526–1533.

Kobayashi, Y., Hoekenga, O.A., Itoh, H. et al. (2007). Characterization of AtALMT1 expression in aluminum-inducible malate release and its role for rhizotoxic stress tolerance in Arabidopsis. *Plant Physiol.* 145: 843–852.

Kobayashi, Y., Ohyama, Y., Kobayashi, Y. et al. (2014). STOP2 activates transcription of several genes for Al- and low pH-tolerance that are regulated by STOP1 in Arabidopsis. *Mol. Plant* 7: 311–322.

Kochian, L.V., Piñeros, M.A., Liu, J., and Magalhaes, J.V. (2015). Plant adaptation to acid soils: the molecular basis for crop aluminum resistance. *Annu. Rev. Plant Biol.* 66: 571–598.

Kusunoki, K., Nakano, Y., Tanaka, K. et al. (2017). Transcriptomic variation among six *Arabidopsis thaliana* accessions identified several novel genes controlling aluminium tolerance. *Plant Cell Environ.* 40 (2): 249–263.

Kusunoki, K., Kobayashi, Y., Kobayashi, Y., and Koyama, H. (2018). Comparative characterization of aluminum responsive transcriptome in Arabidopsis roots: comparison with other rhizotoxic ions at different stress intensities. *Soil Sci. Plant Nutr.* 64: 469–481.

Larsen, P.B., Geisler, M.J.B., Jones, C.A. et al. (2004). *ALS3* encodes a phloem-localized ABC transporter-like protein that is required for aluminum tolerance in Arabidopsis. *Plant J.* 41: 353–363.

Li, J.-Y., Liu, J., Dong, D. et al. (2014). Natural variation underlies alterations in Nramp aluminum transporter (*NRAT1*) expression and function that play a key role in rice aluminum tolerance. *Proc. Natl. Acad. Sci. USA* 111: 6503–6508.

Li, W., Sun, Y., Wang, B. et al. (2020). Transcriptome analysis of two soybean cultivars identifies an aluminum responsive antioxidant enzyme GmCAT1. *Biosci. Biotechnol. Biochem.* 84: 1394–1400.

Liu, J., Magalhaes, J.V., Shaff, J., and Kochian, L.V. (2009). Aluminum-activated citrate and malate transporters from the MATE and ALMT families function independently to confer Arabidopsis aluminum tolerance. *Plant J.* 57: 389–399.

Magalhaes, J.V., Liu, J., Guimarães, C.T. et al. (2007). A gene in the multidrug and toxic compound extrusion (MATE) family confers aluminum tolerance in sorghum. *Nat. Genet.* 39 (9): 1156–1161.

Meng, L., Wang, B., Zhao, X. et al. (2017). Association mapping of ferrous, zinc, and aluminum tolerance at the seedling stage in Indica rice using MAGIC populations. *Front. Plant Sci.* 8: 1822.

Nakano, Y., Kusunoki, K., Hoekenga, O.A. et al. (2020a). Genome-wide association study and genomic prediction elucidate the distinct genetic architecture of aluminum and proton tolerance in *Arabidopsis thaliana*. *Front. Plant Sci.* 11: 405.

Nakano, Y., Kusunoki, K., Maruyama, H. et al. (2020b). A single-population GWAS identified *AtMATE* expression level polymorphism caused by promoter variants is associated with variation in aluminum tolerance in a local Arabidopsis population. *Plant Direct* 4 (8): e00250.

Navrot, N., Rouhier, N., Gelhaye, E., and Jacquot, J.P. (2007). Reactive oxygen species generation and antioxidant systems in plant mitochondria. *Physiol. Plant* 129: 185–195.

Nezames, C.D., Sjogren, C.A., Barajas, J.F., and Larsen, P.B. (2012). The Arabidopsis cell cycle checkpoint regulators TANMEI/ALT2 and ATR mediate the active process of aluminum-dependent root growth inhibition. *Plant Cell* 24: 608–621.

Niedziela, A., Domżalska, L., Dynkowska, W.M. et al. (2022). Aluminum stress induces irreversible proteomic changes in the roots of the sensitive but not the tolerant genotype of Triticale seedlings. *Plants* 11: 165.

Oh, M.W., Roy, S.K., Kamal, A.H.M. et al. (2014). Proteome analysis of roots of wheat seedlings under aluminum stress. *Mol. Biol. Rep.* 41: 671–681.

Ohyama, Y., Ito, H., Kobayashi, Y. et al. (2013). Characterization of AtSTOP1 orthologous genes in tobacco and other plant species. *Plant Physiol.* 162: 1937–1946.

Qi, B., Korir, P., Zhao, T. et al. (2008). Mapping quantitative trait loci associated with aluminum toxin tolerance in NJRIKY recombinant inbred line population of soybean (*Glycine max*). *J. Integ. Plant Biol.* 50: 1089–1095.

Rangu, M., Ye, Z., Bhatti, S. et al. (2018). Association of proteomics changes with Al-sensitive root zones in switchgrass. *Proteomes* 6: 15.

Rosa-Santos, T.M., Silva, R.G.d., Kumar, P. et al. (2020). Molecular mechanisms underlying sugarcane response to aluminum stress by RNA-Seq. *Int. J. Mol. Sci.* 21: 7934.

Rui, C., Peng, F., Fan, Y. et al. (2022). Genome-wide expression analysis of carboxylesterase (CXE) gene family implies GBCXE49 functional responding to alkaline stress in cotton. *BMC Plant Biol.* 22: 1–16.

Sadhukhan, A., Agrahari, R.K., Wu, L. et al. (2021). Expression genome-wide association study identifies that phosphatidylinositol-derived signalling regulates *ALUMINIUM SENSITIVE3* expression under aluminium stress in the shoots of *Arabidopsis thaliana*. *Plant Sci.* 302: 110711.

Sasaki, T., Yamamoto, Y., Ezaki, B. et al. (2004). A wheat gene encoding an aluminum-activated malate transporter. *Plant J.* 37 (5): 645–653.

Sawaki, Y., Iuchi, S., Kobayashi, Y. et al. (2009). STOP1 regulates multiple genes that protect Arabidopsis from proton and aluminum toxicities. *Plant Physiol.* 150: 281–294.

Sawaki, K., Sawaki, Y., Zhao, C.R. et al. (2016). Specific transcriptomic response in the shoots of *Arabidopsis thaliana* after exposure to Al rhizotoxicity: – potential gene expression biomarkers for evaluating Al toxicity in soils. *Plant Soil* 409: 131–142.

Sjogren, C.A. and Larsen, P.B. (2017). *SUV2*, which encodes an ATR-related cell cycle checkpoint and putative plant ATRIP, is required for aluminium-dependent root growth inhibition in Arabidopsis. *Plant Cell Environ.* 40: 1849–1860.

Sjogren, C.A., Bolaris, S.C., and Larsen, P.B. (2015). Aluminum-dependent terminal differentiation of the Arabidopsis root tip is mediated through an ATR-, ALT2-, and SOG1-regulated transcriptional response. *Plant Cell* 27: 2501–2515.

Sorek, N., Szemenyei, H., Sorek, H. et al. (2015). Identification of MEDIATOR16 as the *Arabidopsis* COBRA suppressor MONGOOSE1. *Proc. Natl. Acad. Sci. USA* 112: 16048–16053.

Szurman-Zubrzycka, M., Chwiałkowska, K., Niemira, M. et al. (2021). Aluminum or low pH – which is the bigger enemy of barley? Transcriptome analysis of barley root meristem under Al and low pH stress. *Front. Genet.* 12: 675260.

Tokizawa, M., Kobayashi, Y., Saito, T. et al. (2015). SENSITIVE TO PROTON RHIZOTOXICITY1, CALMODULIN BINDING TRANSCRIPTION ACTIVATOR2, and other transcription factors are involved in ALUMINUM-ACTIVATED MALATE TRANSPORTER1 expression. *Plant Physiol.* 167: 991–1003.

Tokizawa, M., Enomoto, T., Ito, H. et al. (2021). High affinity promoter binding of STOP1 is essential for early expression of novel aluminum-induced resistance genes *GDH1* and *GDH2* in Arabidopsis. *J. Exp. Bot.* 72: 2769–2789.

Tovkach, A., Ryan, P.R., Richardson, A.E. et al. (2013). Transposon-mediated alteration of *TaMATE1B* expression in wheat confers constitutive citrate efflux from root apices. *Plant Physiol.* 161: 880–892.

Tsutsui, T., Yamaji, N., and Feng Ma, J. (2011). Identification of a cis-acting element of ART1, a C2H2-type zinc-finger transcription factor for aluminum tolerance in rice. *Plant Physiol.* 156: 925–931.

Tyagi, W., Yumnam, J.S., Sen, D., and Rai, M. (2020). Root transcriptome reveals efficient cell signaling and energy conservation key to aluminum toxicity tolerance in acidic soil adapted rice genotype. *Sci. Rep.* 10: 4580.

Wang, X., Cheng, Y., Yang, C. et al. (2019). QTL mapping for aluminum tolerance in RIL population of soybean (*Glycine max* L.) by RAD sequencing. *PLoS One* 14: e0223674.

Wei, Y., Jiang, C., Han, R. et al. (2020). Plasma membrane proteomic analysis by TMT-PRM provides insight into mechanisms of aluminum resistance in tamba black soybean roots tips. *Peer J.* 8: e9312.

Wu, L., Sadhukhan, A., Kobayashi, Y. et al. (2019). Involvement of phosphatidylinositol metabolism in aluminum-induced malate secretion in Arabidopsis. *J. Exp. Bot.* 70: 3329–3342.

Xia, J., Yamaji, N., Che, J. et al. (2014). Differential expression of Nrat1 is responsible for Al-tolerance QTL on chromosome 2 in rice. *J. Exp. Bot.* 65 (15): 4297–4304.

Xiao, D., Li, X., Zhou, Y.-Y. et al. (2021). Transcriptome analysis reveals significant difference in gene expression and pathways between two peanut cultivars under Al stress. *Gene* 781: 145535.

Xu, Q., Wang, Y., Ding, Z. et al. (2017). Aluminum induced physiological and proteomic responses in tea (*Camellia sinensis*) roots and leaves. *Plant Physiol. Biochem.* 115: 141–151.

Yamaji, N., Huang, C.F., Nagao, S. et al. (2009). A zinc finger transcription factor ART1 regulates multiple genes implicated in aluminum tolerance in rice. *Plant Cell* 21: 3339–3349.

Yan, L., Riaz, M., Liu, J. et al. (2022). The aluminum tolerance and detoxification mechanisms in plants; recent advances and prospects. *Crit. Rev. Environ. Sci. Tech.* 52: 1491–1527.

Yang, L., Tian, D., Todd, C.D. et al. (2013). Comparative proteome analyses reveal that nitric oxide is an important signal molecule in the response of rice to aluminum toxicity. *J. Proteome Res.* 12: 1316–1330.

Yang, J., Yan, L., Song, Y. et al. (2015). New roles for the Arabidopsis TAO1 gene besides disease resistance. *Russian J. Plant Physiol.* 62: 542–550.

Yokosho, K., Yamaji, N., Fujii-Kashino, M., and Ma, J.F. (2016). Retrotransposon mediated aluminum tolerance through enhanced expression of the citrate transporter OsFRDL4. *Plant Physiol.* 172: 2327–2336.

Zhang, P., Zhong, K., Tong, H. et al. (2016). Association mapping for aluminum tolerance in a core collection of rice landraces. *Front. Plant Sci.* 7: 1415.

Zhao, H., Huang, W., Zhang, Y. et al. (2018). Natural variation of CsSTOP1 in tea plant (*Camellia sinensis*) related to aluminum tolerance. *Plant Soil* 431: 71–87.

Zhao, L., Cui, J., Cai, Y. et al. (2020). Comparative transcriptome analysis of two contrasting soybean varieties in response to aluminum toxicity. *Int. J. Mol. Sci.* 21: 4316.

Zheng, Y., Xiao, J., Zheng, K. et al. (2021). Transcriptome profiling reveals the effects of nitric oxide on the growth and physiological characteristics of watermelon under aluminum stress. *Genes* 12: 1735.

Zhou, L.-L., Bai, G.-H., Ma, H.-X., and Carver, B.F. (2007). Quantitative trait loci for aluminum resistance in wheat. *Mol. Breed.* 19: 153–161.

Zhou, S., Okekeogbu, I., Sangireddy, S. et al. (2016). Proteome modification in tomato plants upon long-term aluminum treatment. *J. Proteome Res.* 15: 1670–1684.

Zhou, D., Yang, Y., Zhang, J. et al. (2017). Quantitative iTRAQ proteomics revealed possible roles for antioxidant proteins in sorghum aluminum tolerance. *Front. Plant Sci.* 7: 2043.

Zhu, F., Ahchige, M.W., Brotman, Y. et al. (2022). Bringing more players into play: leveraging stress in genome wide association studies. *J. Plant Physiol.* 271: 153657.

5

Breeding Approaches for Aluminum Toxicity Tolerance in Rice and Wheat

Buu Chi Bui[1] *and Lang Thi Nguyen*[2]

[1] *Department of Biotechnology, Institute of Agriculture Science for Southern Vietnam (IAS), Ho Chi Minh City, Vietnam*
[2] *Department of Rice Breeding, High Agricultural Technology Research Institute (HATRI), Can Tho City, Vietnam*

5.1 Introduction

Stable production of cereal crops is mostly influenced by environmental stresses, such as heavy metals, resulting in a significant decrease in crop yields with the emphasis on rice and wheat. Crop productivity in acid-soil areas is restricted by multiple abiotic stress factors. Since the forms of aluminum (Al) and their solubilities in soil are high at a pH of 5 or less, Al toxicity becomes one of the major growth-limiting factors in acidic soils (Kochian 1995). Acid soils (defined as pH < 5.5) are present in more than 50% of the cultivated and cultivable land area worldwide (Von Uexküll and Mutert 1995). Aluminum toxicity is generally considered as the primary limiting factor for plant growth in acid soils (Kochian et al. 2005; Ma et al. 2014). When soil pH is lower than 5.0, toxic Al ions are released into the soil solution, where they inhibit root growth and ultimately reduce crop yields (Ma et al. 2002, 2014).

Aluminum is a major component of almost all common inorganic soil particles. The most common primary and secondary minerals are the aluminosilicates, which include the feldspars, micas, and clay minerals (McLean 1965). Roots are the main organs through which plants acquire various mineral nutrients from the soil (Garnett et al. 2009; Li et al. 2014). The initial and most dramatic symptom of Al toxicity is inhibition of root elongation as a consequence of toxicity to the root apex (Kochian 1995). Roots injured by high Al concentrations are usually stubby, thick, dark-colored, brittle, poorly branched, with reduced root length and volume. Al toxicity may inhibit shoot growth by limiting the supply of nutrients and water due to poor subsoil penetration or lower root hydraulic conductivity (Kochian 1995; Nguyen et al. 2003). The symptoms in the roots system like inhibition of root growth, reduced water, and mineral uptake caused by Al^{3+} phytotoxicity start to appear within minutes after exposure (Silva et al. 2012).

Heavy Metal Toxicity and Tolerance in Plants: A Biological, Omics, and Genetic Engineering Approach, First Edition. Edited by Mohammad Anwar Hossain, AKM Zakir Hossain, Sylvain Bourgerie, Masayuki Fujita, Om Parkash Dhankher, and Parvez Haris.

One of these important changes in response to Al stress is imbalance in ROS (reactive oxygen species) production, which induces oxidative stress (Darko et al. 2004; Yin et al. 2010). This leads to the induction of antioxidative defense response cascade, which assists to certain extent in containing the oxidative burst (Yin et al. 2010). Glutathione reductase (GR) forms a part of the ascorbate-glutathione cycle (a part of antioxidative defense mechanism) and thus facilitates the conversion of H_2O_2 to H_2O. An increased GR content in both shoot and root tissues in response to Al stress has been observed to limit the oxidative burst and prevent damage to tissues (Kuo and Kao 2003; Sharma and Dubey 2007). Lipid peroxidation is another deleterious process for cellular membrane lipids for initiation of which ROS molecules are responsible (Panda et al. 2009).

Al-responsive genes have important roles to play toward plant Al-tolerance mechanisms. NRAT1 (NRAMP Aluminum Transporter 1) is a plasma-membrane-localized specific Al transporter that is involved in the uptake of Al to cells for sequestration in vacuoles. The silencing of NRAT1 resulted in decreased Al uptake, increased Al binding to the cell wall, and hence enhanced Al sensitivity. With subsequent detoxification via transport and Al accumulation into cell vacuoles, possibly mediated by OsALS1 (aluminum sensitive 1) (Xia et al. 2010; Simoes et al. 2012). NRAT1 transporter plays a major role in rice Al tolerance mechanisms (Li et al. 2014).

Awasthi et al. (2017) assessed 24 indigenous *indica* rice varieties (including Badshahbhog as tolerant check and Mahsuri as sensitive check). They were screened for Al stress tolerance in hydroponic plant growth system. Results showed significant difference in growth parameters (relative growth rate, root tolerance index, fresh and dry weight of root) of rice seedlings due to Al (100 μM) toxicity. Al^{3+} uptake and lipid peroxidation level also increased concomitantly under Al treatment. Histochemical assay was also performed to elucidate uptake of aluminum, loss of membrane integrity, and lipid peroxidation, which were found to be more in sensitive genotypes at higher Al concentration. This study revealed that Al toxicity is a serious harmful problem for rice crop productivity in acid soil. Based on various parameters studied, it was concluded that Disang is a comparatively tolerant variety, whereas Joymati as a sensitive variety. Western blot hybridization further strengthened the claim, as it demonstrated more accumulation of glutathione reductase (GR) protein in Disang rice variety than in Joymati under stressed condition. This study also observed that the emergence of lethal toxic symptoms occurs only after 48 hours irrespective of the dose used in the study (Awasthi et al. 2017).

The root-apex transition zone (TZ), the major perception site for Al toxicity, is crucial for the Al-induced root-growth inhibition. The role of silicon (Si) in alleviating Al-induced damage in the TZ and root-growth inhibition of rice was investigated. Jiang et al. (2022) found that Si had direct alleviative effect on Al toxicity as revealed by less root growth-inhibition, Al accumulation, and callose formation. Si reversed Al-induced decreases of the cell wall elongation and extensibility, and reduced Al-induced increments of cell wall polysaccharides in the TZ. Therefore, Si-mediated cell wall modification in the TZ alleviates Al-induced root-growth inhibition in rice involving the promotion of cell wall extensibility and the decrease of Al accumulation in the cell wall (Jiang et al. 2022). However, the mechanism underlying silicon-mediated alleviation of Al toxicity in the TZ is largely unknown (Jiang et al. 2022).

5.2 Plant Signaling

Plants have evolved different regulatory mechanisms of adaptation to exogenous environmental challenges, such as Al stress, by altering their growth patterns. In the past decades, several key genes involved in plant response to Al stress and the mechanism of Al detoxification have been revealed (Liu et al. 2022). However, the signaling pathways of plant response to Al stress and the regulatory mechanism of plant Al tolerance remain poorly understood. Liu et al. (2022) summarized the findings of recent studies on the plant Al tolerance mechanism and the molecular regulation mechanism of phytohormones in response to Al stress.

Al toxicity in plants could be observed at the cellular level. Roots are the main plant parts exposed to Al stress. When plants are exposed to Al stress, the most obvious symptom is the inhibition of root elongation, following which the acquisition of water and nutrients becomes limited (Sun et al. 2010; Yang et al. 2014; Liu et al. 2016). Adaptive mechanisms of plant tolerance to Al toxicity includes secreting organic acids (OAs) or phosphoric acid into the apoplastic space to chelate external Al (Kochian et al. 2004; Delhaize et al. 2007; Zhang et al. 2019). The internal tolerance mechanisms involve the chelation of Al^{3+} by OAs in the cytosol, its transport, and the storage of its complexes into vacuoles (Kochian et al. 2004; Zhang et al. 2019).

Excessive Al binding to the cell wall leads to the disruption of cell wall extension, thereby inhibiting cell and root elongation (Zhu et al. 2012). WAK1 (cell-wall-associated receptor kinase 1) co-localizes with pectin, which is critical for Al binding (Sivaguru et al. 2003). GRP3, a glycine-rich protein (GRP), is involved in plant response to Al stress by interacting with AtWAK1 (Mangeon et al. 2016).

In plant roots, Al toxicity induces the secretion of chelating agents such as OAs and phosphoric acid (Liu et al. 2009; Zhang et al. 2019). Hormone signaling in plant Al stress response is also important where ethylene and auxin are the key factors (Li et al. 2021; Ranjan et al. 2021; Liu et al. 2022). Al-induced ethylene production is involved in auxin signaling to control root elongation under Al stress. Exogenous application of auxin can increase the expression of ALMT1 (aluminum-activated malate transporter). Malate exudation was not affected by its application (Kobayashi et al. 2013). The cumulative evidence indicates that auxin and OAs exudation independently regulate the Al-induced inhibition of root growth (Yang et al. 2014).

5.3 Rice Genetic Mapping

5.3.1 Linkage Mapping

The natural genetic variation for Al tolerance in other domesticated members of the Poaceae in different tribes, such as rice (Wu et al. 2000; Nguyen et al. 2001, 2002, 2003) and maize (Magnavaca et al. 1987; Ninamango-Cardenas et al. 2003) appears to be quantitative nature. Magalhaes et al. (2004) identified that a single locus, Alt_{SB}, was found to control Al tolerance in two highly Al-tolerant sorghum cultivars. Significant macrosynteny between sorghum and the Triticeae was observed for molecular markers closely linked to putatively orthologous Al tolerance loci present in the group 4 chromosomes of wheat, barley, and rye. Additionally, Alt_{SB} was not located within the homologous region of sorghum but rather mapped near the end of sorghum chromosome 3 (Magalhaes et al. 2004).

Nguyen et al. (2003) studied a population of 171 F_6 recombinant inbred lines (RILs), which were derived from the cross of *Oryza sativa* (IR64), the Al susceptible parent, and *Oryza rufipogon*, the Al-tolerant parent, were evaluated for Al tolerance using a nutrient solution with and without 40 ppm of active Al^{+3}. A genetic map, consisting of 151 molecular markers covering 1755 cM with an average distance of 11.6 cM between loci was constructed. Nine QTLs were identified including one for root length under non-stress conditions, three for root length under Al stress, and five for relative root length (RRL). *Oryza rufipogon* contributed favorable alleles for each of the five QTLs for RRL, which is a primary parameter for Al tolerance, and individually they explained 9.0–24.9% of the phenotypic variation. Comparative genetic analysis showed that QTLs for RRL, which mapped on chromosomes 1 and 9. Interestingly, a major QTL for RRL explained 24.9% of the phenotypic variation (Nguyen et al. 2003).

Famoso et al. (2011) screened 383 diverse rice accessions. They conducted a genome-wide association study (GWAS), and QTL mapping in two bi-parental populations using three estimates of Al tolerance based on root growth. The mean Al tolerance in *japonica* was twice that of *indica*. Forty-eight regions associated with Al tolerance were identified by GWAS, most of which were subpopulation-specific. Four of these regions co-localized with a priori candidate genes and two highly significant regions co-localized with previously identified QTLs (Famoso et al. 2011). Haplotype analysis around the *NRAT1* gene identified susceptible and tolerant haplotypes explaining 40% of the Al tolerance variation within the *aus* subpopulation, and sequence analysis of *Nrat1* identified a trio of non-synonymous mutations predictive of Al sensitivity in our diversity panel (Famoso et al. 2011).

Two immortalized QTL mapping populations were analyzed for Al tolerance. One consisted of 134 recombinant inbred lines (RIL) derived from the cross IR64/Azucena (Ahmadi et al. 2005), and the other was comprised of 78 backcross inbred lines (BIL) derived from the cross Nipponbare/Kasalath//Nipponbare (Lin et al. 1998).

In the RIL population, three Al tolerance (*Alt*) QTL were detected using total root growth (the TRG-RRG index), *AltTRG1.1* on chromosome 1, *AltTRG2.1* on chromosome 2, and *AltTRG12.1* on chromosome 12. The Azucena allele conferred increased tolerance at the loci on chromosomes 1 and 12 and reduced tolerance at the locus on chromosome 2. QTLs were detected in the same positions on chromosomes 1 and 12 using RRG (rate of root growth) based on primary root growth (the PRG-RRG index), although with lower LOD scores (Famoso et al. 2011). Using longest root growth (the LRGRRG index), a single QTL was detected on chromosome 9, *AltLRG9.1*. This QTL was not detected when the other root indices were used. The major QTL on chromosome 12 (*AltTRG12.1*), which explained 19% of the variation in Al tolerance based on TRG-RRG, is located between 2.69 and 5.10 Mb. It encompasses the Al sensitive rice mutant *art1*, which is located at 3.59 Mb (Yamaji et al. 2009).

5.3.2 Association Mapping

Tao et al. (2018) performed an association mapping of Al toxicity tolerance using a core collection of 211 *indica* rice accessions with 700 K high-quality SNP data. A total of

21 putative QTL affecting shoot height (SH), root length (RL), shoot fresh weight (SFW), shoot dry weight (SDW), root dry weight (RDW), and shoot water content (SWC) were identified at seedling stage, including three QTLs detected only under control condition, eight detected only under Al stress condition, ten simultaneously detected in both control and Al stress conditions, and seven were identified by stress tolerance index of their corresponding traits (Tao et al. 2018). Table 5.1 identified the QTLs, which were associated with Al toxicity tolerance-related traits as shoot height (SH), root length (RL), shoot fresh weight (SFW), shoot dry weight (SDW), root dry weight (RDW), and shoot water content (SWC).

A total of 21 candidate genes for 7 important QTL regions associated with Al toxicity tolerance were identified based on combined haplotype analysis and functional annotation. Figure 5.1 expressed the Al toxicity tolerance with 1 mM Al^{3+} in hydroponic nutrient solution. The association panel included 211 *indica* rice accessions. The major QTL for Al toxicity tolerance by linkage analysis was detected by 11KASP SNP markers.

A candidate gene *NRAT1* on chromosome 2 was further fine-mapped using BSA-seq and linkage analysis in the F_2 population, which was derived from the cross of Al-tolerant accession CC105 and super susceptible accession CC180 (Tao et al. 2018). Haplotype analysis of *NRAT1* using 327 3K RGP accessions indicated that minor allele variations in *aus* and *indica* subpopulations decreased Al toxicity tolerance in rice. The minor alleles are important for QTL mapping and its application in rice breeding when natural gene resources are used (Tao et al. 2018).

CC105 (PR) and CC180 (PS), of which the first was identified as Al-tolerant accession with the tolerant allele at the major QTL in the range of 4.61–4.86 Mb on chromosome 2 based on GWAS analysis. The latter as Al super susceptible accession was used to develop F_2 population for BSA-seq and linkage analyses. Moreover, 327 rice accessions consisted of many ecotypes were selected from 3K RGP (GigaScience 2014) and used for haplotype analysis of candidate gene *NRAT1*.

5.4 Root Transcriptome

Aluminum activated malate transporter (*ALMT*) was the first Al toxicity tolerance gene to be identified in wheat (Delhaize et al. 2004, 2012). ALMT protein is involved in secreting/exuding malate from roots (Kochian et al. 2015). A MATE (Multidrug and Toxic Compound Extrusion) ortholog of SbMATE (*Sorghum bicolor* MATE) in rice is FRDL4 (ferric reductase defective-like 4) (Yokosho et al. 2009), which transports citrate transporter and can form complexes with Al. The NRAT1 is responsible for transporting Al across plasma membrane (PM) (Xia et al. 2010). A vacuolar ABC (ATP Binding Cassette) transporter, ALS1 (amyotrophic lateral sclerosis), sequesters Al into vacuoles (Huang et al. 2012), and in this sequestration process NRAT1 works together with ALS1 (Kochian et al. 2015). The Al responsive genes such as STAR1 and STAR2 (sensitive to Al rhizotoxicity) were isolated by mutant screening and these together work as an efflux ABC transporter to transport UDP-glucose into cell wall (Huang et al. 2009; Ryan et al. 2011). This modification of cell wall by altering carbohydrate composition results in reduced Al binding, and thereby reduced accumulation. A zinc finger transcription factor, ART1 (Aluminum rhizotoxicity 1), was characterized as an ortholog of Arabidopsis transcription factor, AtSTOP1 (Iuchi et al. 2007; Sawaki

Table 5.1 QTL identified with significant association to Al toxicity tolerance-related traits (Tao et al. 2018).

Trait	QTL	Chr.	Position (Mb)	Control			Al stress			Al stress/CK ratio			Previously reported QTL or gene
				P value	Effect	R^2 (%)	*P* value	Effect	R^2 (%)	*P* value	Effect	R^2 (%)	
SH	*qSh1*	1	37.93–38.73	8.07E−08	−4.5	6.2	5.76E−07	−4.9	5.7	1.7E−05	−0.06	7.8	*OsMGT1*(43)
	qSh7	7	6.15–6.38	2.66E−05	3.1	3.7	1.28E−06	3.2	5.3				*QalRr7*(19)
RL	*qRl1*	1	28.15–28.16				5.02E−06	−1.1	9.6			11.0	
	qRl8	8	1.28–1.67	3.81E−06	2.3	9.0				1.43E−06			
SFW	*qRl8*	2	4.74–4.86				5.93E−07	−53.2	7.6			10.8	
	qSfw3a	3	15.99–15.77				1.65E−06	81.6	5.5				*PSM377*(7)
	qSfw3b	3	26.62–26.79	7.39E−06	104.1	7.0	2.26E−07	92.1	8.2				
	qSfw4	4	0.38–0.61	3.40E−06	96.5	7.6	8.82E−08	88.9	8.7				
SDW	*qSdw1*	1	37.02–37.60				1.17E−08	13.8	8.0				
	qSdw2	2	4.61–4.86	2.46E−05	−7.3	6.7	1.61E−07	−7.3	9.4				
	qSdw3a	3	26.52–27.07	3.16E−06	13.6	8.3	4.49E−08	12.7	10.3	7.09E−05	−0.04	7.7	
	qSdw3b	3	35.01–35.79	2.61E−05	11.2	6.7	2.78E−07	12.3	9.0				
	qSdw4	4	0.38–0.61	4.65E−07	13.7	9.8	5.44E−08	11.8	10.2				
	qSdw5	5	6.46–7.54	3.36E−05	10.7	6.5	1.95E−06	11.3	7.7				*QAlSDW5*(8)
	qSdw7	7	4.30–4.75				2.96E−06	−11.9	7.4				
RDW	*qRdw2*	2	4.74–4.76	2.12E−05	−1.2	8.4							*QAlSr2.1*(8)
	qRdw4	4	16.39–16.67	3.61E−06	2.0	10.0							
	qRdw5	5	7.01–7.32	1.90E−05	2.2	8.5	2.89E−08	2.9	15.3				
SWC	*qSwc3*	3	15.62–15.74				2.77E−05	1.1	6.1	1.48E−05	0.01	7.6	*PSM377*(7)
	qSwc6	6	19.60–19.76				2.52E−06	1.6	7.8	3.09E−05	0.02	7.0	
	qSwc8	8	1.30–1.58				1.41E−05	2.5	6.6	2.61E−06	0.03	9.2	*qALRR-8*(43)

Trait
RDW, root dry weight; RL, root length; SDW, shoot dry weight; SFW, shoot fresh weight; SH, shoot height; SWC, shoot water content.
Previously reported QTL or gene.
(7): Zhang et al. (2016); (8): Meng et al. (2017); (19): Nguyen et al. (2003); (43): Zhi and Jian (2012); (45): Chen et al. (2013).

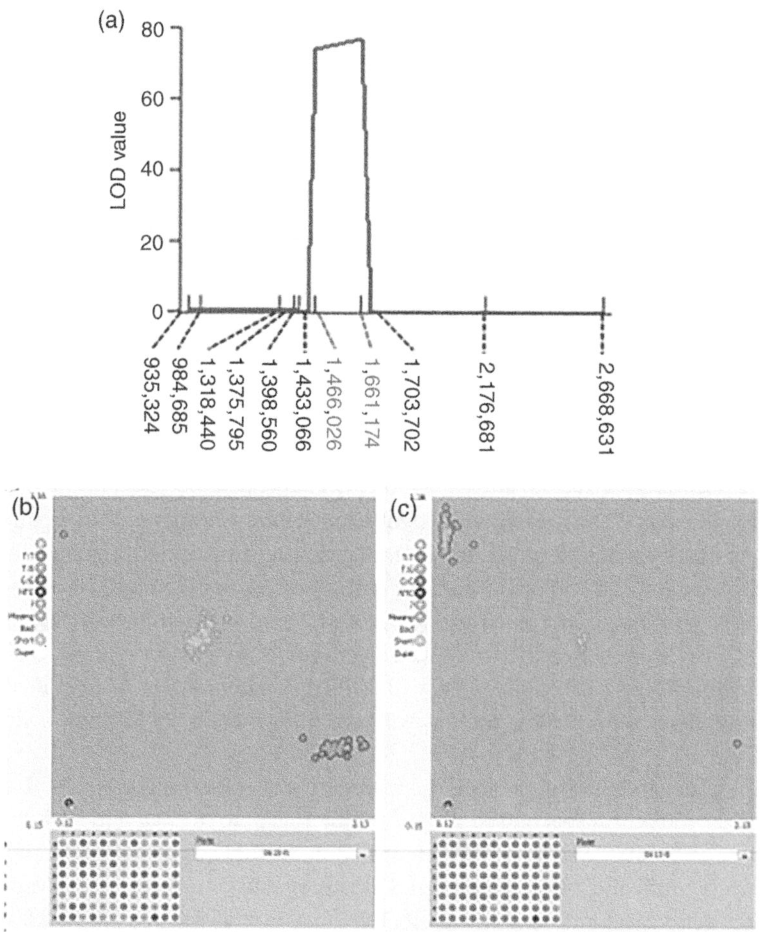

Figure 5.1 Validation of the major QTL for Al toxicity tolerance by linkage analysis with 11 KASP SNP markers (a), genotype of 93 extremely Al toxicity tolerance (b), and sensitive (c) individuals by SNP1661173, the last three sites were NTC, PR, and PS (Tao et al. 2018).

et al. 2009; Yamaji et al. 2009). OsART1 regulates at least 30 downstream genes including STAR1, STAR2, NRAT1, FRDL4, and ALS mostly via a specific *cis*-element (Tsutsui et al. 2011). Apart from the aforementioned genes, two rice genes located on PM, an Mg uptake transporter (*OsMGT1*; Magnesium Transporter 1) and a small peptide (*OsCDT3*; Cadmium Tolerance 3), stop entry of Al inside roots by binding to it (Chen et al. 2012; Xia et al. 2013).

Several genes are reported as ***OsSTAR1***:sensitive to Al rhizotoxicity 1, ***OsSTAR2***:sensitive to Al rhizotoxicity 2, transcription factors **OsART1** (Aluminum rhizotoxicity 1) and transporters **ALMT, OsNRAT1** (NRAMP: natural resistance-associated macrophage protein) (aluminum transporter 1), **OsFRDL4** (Ferric Reductase Defective Like 1), and **OsALS1** (Al-Sensitive 1). Those associated with Al toxicity resistance have been identified up to now (Kochian et al. 2015; Liu et al. 2016).

Tyagi et al. (2020) conducted their study on hydroponics-based screening of three rice genotypes, a tolerant (ARR09), a susceptible (IR1552), and an acid-soil-adapted landrace (Theruvii). They reported that ARR09 accumulates less Al and shows minimum decrease in shoot and root biomass under Al toxicity conditions when compared with IR1552. Transcriptome data generated on roots (grown in presence or absence of Al) led to identification of ~1500 transcripts per genotype with percentage annotation ranging from 21.94% (ARR09) to 29.94% (Theruvii). A total of 511, 804, and 912 DEGs were identified in genotypes ARR09, IR1552, and Theruvii, respectively. IR1552 showed upregulation of transcripts involved in exergonic processes. ARR09 appears to conserve energy by downregulating key genes of glycolysis pathway and maintaining transcript levels of key exergonic step enzymes under Al stress. Figure 5.2 indicated that a total of 8506, 24,725, and 25,120 transcripts passed the cut-off for Theruvii, IR1552, and ARR09, respectively. The percentage annotation ranged from 21.94% (ARR09), 23.54% (IR1552) to 29.94% (Theruvii).

The tolerance in ARR09 appears to be because of novel mechanism as none of the reported Al toxicity genes or QTLs overlap with significant DEGs. Components of signal transduction and regulatory machinery like transcripts encoding zinc finger protein, calcineurin binding protein, and cell-wall-associated transcripts are among the highly upregulated DEGs in ARR09, suggesting increased and better signal transduction in response to Al stress in tolerant rice. Sequencing of NRAT1 and glycine-rich protein A3 revealed distinct haplotype for *indica* type ARR09. The newly identified components of Al tolerance will help in designing molecular breeding tools to enhance rice productivity in acidic soils (Tyagi et al. 2020).

The transcriptome data generated on the three genotypes was checked for quality. A total of 8506; 24,725, and 25,120 transcripts passed the cut-off for Theruvii, IR1552, and ARR09, respectively (Tyagi et al. 2020). 511 DEGs were identified in ARR09 out of which 99 and 100 transcripts were significantly upregulated and downregulated, respectively (Tyagi et al. 2020). This is a novel mechanism of Al toxicity tolerance in *indica* type rice, ARR09. The Al detoxification system was reported through activating transporters and efflux systems. The formation of lateral roots by stimulating auxin players was also recorded. The maintenance of the balance of redox levels was noticed by inducing the antioxidant enzyme system. All are the main reasons for the differences in Al toxicity of two sugarcane species CTC-2 and RB855453, as determined by RNA-seq (Rosa-Santos et al. 2020).

Zhu et al. (2022) demonstrated that hydrogen sulfide (H_2S) improves aluminum (Al) resistance in rice. They conducted the treatment with 30-μM Al significantly inhibited rice root growth and increased the total Al content, apoplastic and cytoplasm Al concentration in the rice roots. However, pretreatment with NaHS (H_2S donor) reversed these negative effects. Pretreatment with NaHS significantly increased energy production under Al toxicity conditions, such as by increasing the content of ATP and nonstructural carbohydrates. In addition, NaHS stimulated the AsA-GSH cycle to decrease the peroxidation damage induced by Al toxicity. Pretreatment with NaHS significantly inhibited ethylene emissions in the rice and then inhibited pectin synthesis and increased the pectin methylation to reduce cell wall Al deposition. The phytohormones "indole-3-acetic" and "brassinolide" were also involved in the alleviation of Al toxicity by H_2S. The transcriptome results further confirmed that H_2S alleviates Al toxicity by

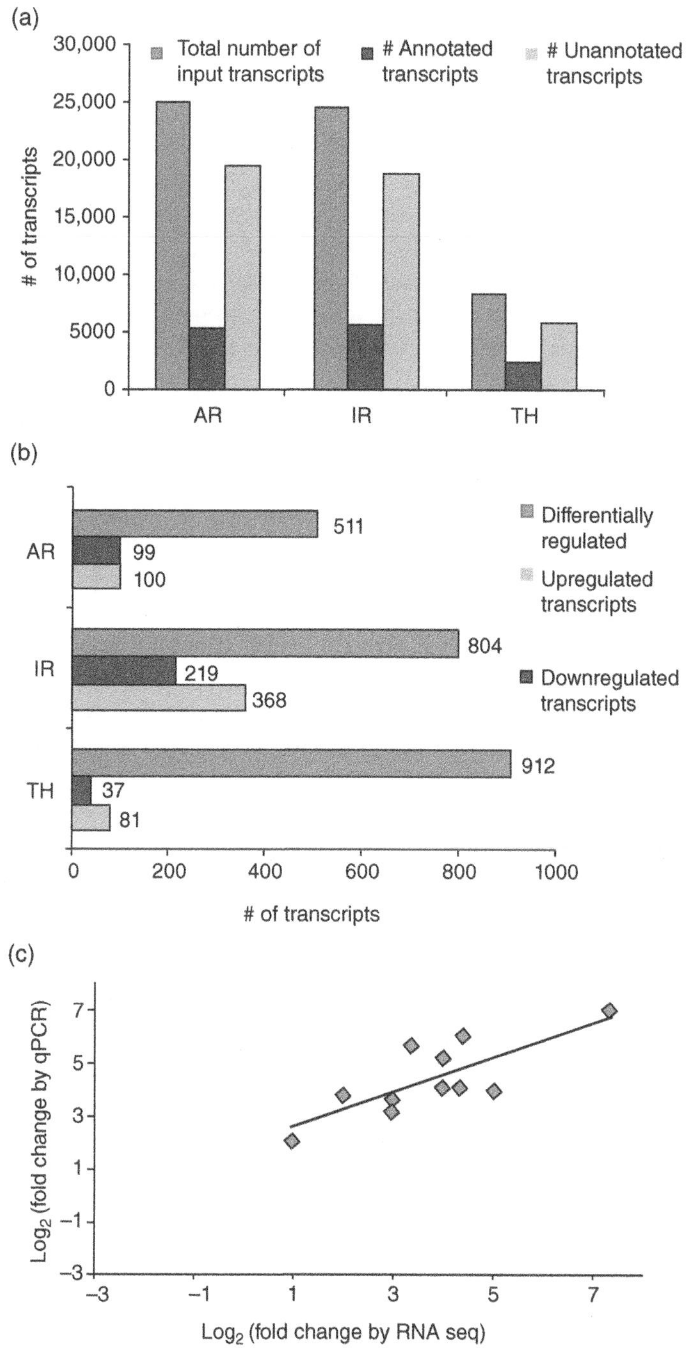

Figure 5.2 Differential gene expression under Al stress condition. (a) The total, unannotated and annotated transcripts identified for the three rice genotypes ARR09 (AR), IR1552 (IR), and Theruvii (TH) is shown in the bar graph. (b) Bar diagram showing the number of up- and downregulated genes in response to Al toxicity in the roots of AR, IR, and TH. The total number of genes differentially expressed under each condition is given on the top of each bar (c). The correlation of gene expression results obtained from qPCR analysis and RNA-seq for fifteen selected genes (Tyagi et al. 2020).

increasing the pathways relating to material and energy metabolism, redox reactions, cell wall components, and signal transduction (Zhu et al. 2022).

5.5 Wheat Genetic Mapping

More than 50% of the world's arable land used for wheat production in developing countries are prone to Al toxicity (Jaskowiak et al. 2018). Some of their adaptation strategies include reduction of stomata opening, which in turn has a negative effect on crop growth and grain yield (Vitorello et al. 2005). On the other hand, the expansion of root systems enhances plants' accessibility to water during the periods of stress while also presenting water potential and contributing to opening of the stomata (Zhang et al. 2018).

Aluminum tolerance is a complex quantitative trait with genetic control, which is significantly influenced by the environment (Dai et al. 2013).

The resulting information is vital for breeders as it can help them to develop Al stress tolerant cultivars (Langridge and Reynolds 2015). Molecular markers may provide new resources for identifying Al-tolerance genes in breeding populations. Marker-assisted selection (MAS) will decrease or can even eliminate the need for phenotypic assays (Ma et al. 2005).

QTLs have been reported for grain yield (Shukla et al. 2015), flag leaf length and width (Xue et al. 2013; Wu et al. 2015), flag leaf area (Fan et al. 2015), relative water content and canopy temperature (Gupta et al. 2017), water soluble carbohydrates (McIntyre et al. 2010), cell membrane stability (Talukder et al. 2014), and aluminum concentration in wheat (Sasaki et al. 2004).

The evidences has reiterated the important genetic components for quantitative traits constitute both epistasis and QTL × environment (QE) interactions (Yu et al. 2002; Liu et al. 2006).

Numerous mapping studies have been reported for Al tolerance in wheat (Sasaki et al. 2004; Raman et al. 2005) under controlled environmental conditions (e.g. hydroponics and greenhouse).

Farokhzadeh et al. (2019) conducted the study to understand the genetic mechanisms of aluminum tolerance underlying wheat (*Triticum aestivum* L.) flag leaf morphological and physiological traits. A wheat RIL population consisting of 167 lines produced from a cross between two semi-dwarf spring wheat varieties, SeriM82 and Babax, with high yield potential was used in this study. An RIL population derived from SeriM82 and Babax was used for mapping quantitative trait loci (QTL) in wheat for tolerance to Al toxicity through 477 DNA markers. Based on a single locus analysis, 48 QTLs including 16 putative and 32 suggestive QTLs were identified for all studied traits (Farokhzadeh et al. 2019). Using single-locus QTL analysis, the colocation of QTLs for several traits as investigated in this study were clearly observed in some chromosomal intervals (Figure 5.3). Figure 5.3 expressed the single-locus QTL and the co-location of QTLs for several traits as investigated that were clearly observed in some chromosomal intervals.

Based on two-locus analysis, 54 additive QTLs and 6 pairs of epistatic effects were detected, among which 29 additive and 5 pairs of epistatic QTLs showed significant QTL 9 environment interactions (Farokhzadeh et al. 2019).

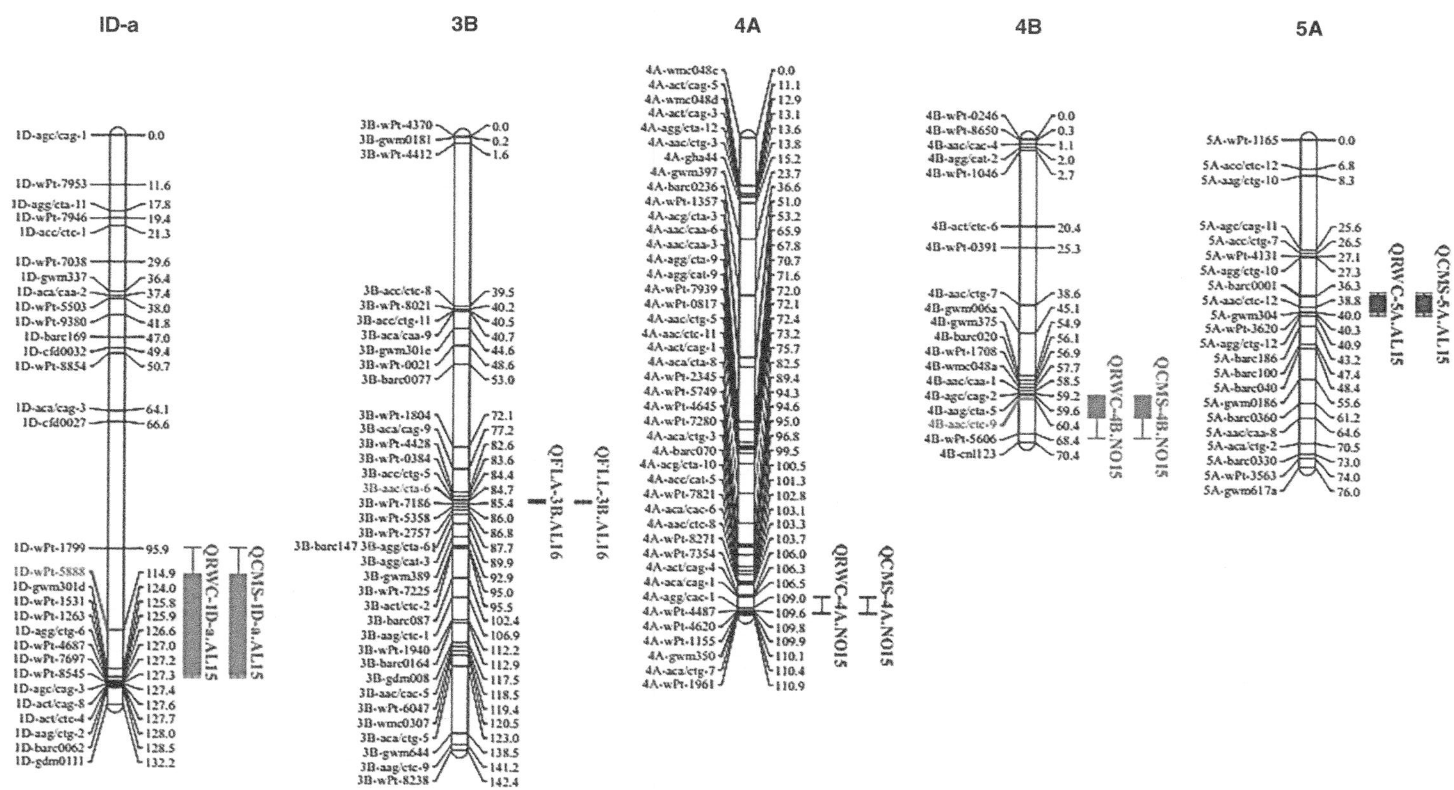

Figure 5.3 The map of co-located QTLs in SeriM82/Babax wheat population. For each QTL, the inner interval (e.g. 1-LOD) is shown as a rectangle and the outer interval (2-LOD) is shown as line segments. Co-located QTL bars and related linked markers are shown in same color in each chromosome (Farokhzadeh et al. 2019).

Han et al. (2014) conducted the introgression of a 4D chromosomal fragment into durum wheat (*Triticum turgidum*) confers aluminum tolerance production on acid soils. The introgression of a chromosomal fragment from hexaploid wheat (*Triticum aestivum*) possesses an Al^{3+} tolerance gene. The 4D(4B) substitution line of durum wheat 'Langdon' was backcrossed to 'Jandaroi', a current semidwarf Australian durum. In the second backcross, using 'Jandaroi' as the recurrent parent, a seedling was identified where TaALMT1 on chromosome 4D was recombined with the Rht-B1b locus on chromosome 4B to yield an Al^{3+} tolerant seedling with a semi-dwarf habit. This seedling was used in a third backcross to generate homozygous sister lines with contrasting Al^{3+} tolerances (Han et al. 2014). The ability to exclude Na^{+} from leaves was also enhanced, indicating that the chromosomal fragment possessed the *Kna1* salt tolerance locus (*Kalium and natrium uptake*). It is possible to increase substantially the Al^{3+} tolerance of an elite durum wheat cultivar by introgression of a 4D chromosomal fragment (Han et al. 2014).

5.5.1 Wheat MATE Gene Family

It is known that the strategies of plants for resisting Al toxicity include external and internal Al detoxification (Ma 2007). Among these mechanisms, organic acid anions (OA) with low molecular weight, such as citrate, have important functions in the external and internal detoxification of Al in different plant species (Ryan et al. 2001, Kochian et al. 2004). To date, numerous studies have revealed the role of two major gene families ALMT (aluminum activated malate transporter) and MATE (multidrug and toxic compound extrusion) in Al tolerance in several plant species that encode membrane proteins and facilitate malate and citrate efflux, respectively (Ryan et al. 2011).

MATE transporters have been found in both prokaryotic and eukaryotic organisms, which exhibit a unique topology (Kuroda and Tsuchiya 2009). These proteins are present in plants in the form of a transporter gene family containing a large number of genes. The "multidrug and toxin efflux" (MATE) gene family plays crucial roles in plant growth and development and response to adverse stresses. Duan et al. (2022) investigated the structural and evolutionary characteristics, expression profiling, and potential functions involved in aluminum (Al) tolerance from a genome-wide level. In total, 211 wheat *MATE* genes were identified, which were classified into four subfamilies and unevenly distributed on chromosomes. Duplication analysis showed that fragments and tandem repeats played the main roles in the amplification of TaMATEs, and Type II functional disproportionation had a leading role in the differentiation of TaMATEs (Duan et al. 2022). Through blast search against the *Triticum aestivum* genome database from WheatOmics 2.1, a total of 211 wheat MATE genes were obtained and named TaMATE1–211. The results showed that the length of the MATE proteins was 197–642 amino acids with the molecular weight from 21.66 to 66.29 kDa and isoelectric point from 4.98 to 9.78 (Duan et al. 2022).

TaMATEs had abundant Al resistance and environmental stress-related elements, and generally had a high expression level in roots and leaves and in response to Al stress. The 3D structure prediction by AlphaFold and molecular docking showed that six TaMATE proteins localized in the plasmalemma could combine with citrate via amino acids in the citrate exuding motif and other sites, and then transport citrate to soil to form citrate aluminum (Duan et al. 2022).

5.6 Wheat Proteomics

In fact, most of the tolerant crop plants are Al-exuders, avoiding the stress by prevention of ion intake into the symplast (Kochian et al. 2005). The biochemical consequences of aluminum intake are an increase in reactive oxygen species, increased fatty acid peroxidation, and inhibition of proton adenosine triphosphatase H^+-ATPases. The displacement of Ca^{2+} in cell membranes contributes to Al^{3+} accumulation in the apoplast, stimulates callose synthesis, and finally inhibits intercellular transport. The decreasing concentration of Ca^{2+} in the cytosol alters the pH balance, which in turn interferes with sugar phosphorylases and the deposition of cell wall polysaccharides (Matsumoto 2000; Kochian 1995; Kochian et al. 2005).

The genetic bases of Al tolerance concern ion chelation and ion transport. The gene families responsible for Al-ion exudation are ALMT (aluminum-activated malate transporter), responsible for malate, and MATE (multidrug and toxin efflux) for citrate exudation (Fontecha et al. 2007; Ryan et al. 2009; Silva-Navas et al. 2012). The transcriptional expression of ALMT and MATE is controlled by a master zinc-finger transcription factor, STOP1 (sensitive to proton rhizotoxicity 1) (Silva-Navas et al. 2012; Garcia-Oliveira et al. 2013)

The NRAT1 (NRAMP aluminum transporter 1) transporter has been identified as a putative Al transporter involved in rice's internal resistance mechanism, which lowers Al-ion concentrations in the root cell wall, transporting the ions inside the root cells for sequestration in vacuoles (Li et al. 2014). The beneficial influence of the rye genome was confirmed by QTL localization on the 7R chromosome, explaining up to 36% of the phenotypic variance, including the malate transporter gene (Niedziela et al. 2014). The other loci were found on chromosomes 3R (Budzianowski and Wos 2004; Niedziela et al. 2012), 4R, and 6R (Niedziela et al. 2012), without a recognized function in the triticale genome.

The two-dimensional electrophoresis (2-DE) with immobilized pH gradients (IPGs), combined with protein identification by mass spectrometry (MS), has been successfully used to identify proteins involved in various stress responses in plants (Kosová et al. 2014, 2015). The upregulation of the enzymes involved in cysteine and methionine metabolism, such as cysteine synthase, *S*-adenosylmethionine synthase, and *O*-methyltransferase, is a common response despite Al-ion tolerance (Fukuda et al. 2007; Zhou et al. 2009; Duressa et al. 2011; Dai et al. 2013; Oh et al. 2014).

Maintaining methyl cycling and glutathione metabolism are important mechanisms that lead to Al detoxification (Ezaki et al. 2001). Niedziela et al. (2022) investigated comparative proteomic studies to use the seedling root tips of triticale plants differing in Al tolerance subjected to recovery after the stress treatment. The triticale, which is a wheat–rye hybrid, exhibited higher abiotic stress tolerance than wheat and is better adapted for cultivation in light-type soils, where aluminum ions are present as Al-complexes that are harmful to plants. The roots are the first plant organs to contact these ions and the inhibition of root growth is one of the first plant reactions. The proteomes of the root apices in Al-tolerant and -sensitive plants were investigated to compare their regeneration effects following stress. The materials used in this study consisted of seedlings of three triticale lines differing in Al^{3+} tolerance, first subjected to aluminum ion stress and then recovered. Two-dimensional electrophoresis (2-DE) was used for seedling root protein separation followed by differential spot analysis using liquid chromatography coupled to tandem mass spectrometry (LC-MS-MS/MS) (Niedziela et al. 2022). The highest, nearly eightfold, difference in upregulated protein spot intensity was detected for spot #6, and a 5.2-fold decline for #24 were detected in the L17 proteome. The plants' tolerance to the stress was evaluated based on biometric screening of seedling root regrowth upon regeneration. The Al-tolerant genotype can recover, without differentiation of proteome profiles, after stress relief, contrary to Al-sensitive genotypes that maintain the proteome modifications caused by unfavorable environments (Niedziela et al. 2022).

5.7 Conclusion

Overall, this chapter provides the basic research activities for molecular breeding of aluminum tolerant rice and wheat varieties, as they are the staple food in the world. The timing, endurance, and strength of aluminum toxicity under natural conditions are dynamic and highly unpredictable so that the stress evaluation has become complicate. The initial and most dramatic symptom of Al toxicity is inhibition of root elongation as a consequence of toxicity to the root apex. Al toxicity may inhibit shoot growth by limiting the supply of nutrients and water. The hormone signaling in plant Al stress response relates to ethylene and auxin as key factors. The successful QTLs' isolating for Al toxicity tolerance will facilitate a better understanding of genetic and molecular mechanisms of aluminum tolerance and marker-assisted selection. The chapter will enable future functional analysis of candidate genes and will contribute to the strategies for improvement of aluminum tolerance in wheat and rice.

References

Ahmadi, N., Dubreuil-Tranchant, C., Courtois, B. et al. (2005). New resources and integrated maps for IR64 x Azucena: a reference population in rice. *Proceedings of the 5th International Rice Genetics Symposium (A platform for exploring developments in rice genetics and their applications) and 3rd International Symposium of Rice Functional Genomics*, Manila, Philippines (19–23 November 2005). International Rice Research Institute.

Awasthi, J.P., Saha, B., Regon, P. et al. (2017). Morpho-physiological analysis of tolerance to aluminum toxicity in rice varieties of North East India. *PLoS One* 12 (4): e0176357.

Budzianowski, G. and Wos, H. (2004). The effect of single D-genome chromosomes on aluminum tolerance of triticale. *Euphytica* 137: 165–172.

Chen, Z.C., Yamaji, N., Motoyama, R. et al. (2012). Up-regulation of a magnesium transporter gene OsMGT1 is required for conferring aluminum tolerance in rice. *Plant Physiol.* 159: 624–1633.

Chen, L.J., Wuriyanghan, H., Zhang, Y.Q. et al. (2013). An S-domain receptor-like kinase, OsSIK2, confers abiotic stress tolerance and delays dark-induced leaf senescence in rice. *Plant Physiol.* 163: 1752–1765.

Dai, J., Bai, G., Zhang, D., and Hong, D. (2013). Validation of quantitative trait loci for aluminum tolerance in Chinese wheat landrace FSW. *Euphytica* 192: 171–179.

Darko, E., Ambrus, H., Stefanovits-Banyai, E. et al. (2004). Aluminum toxicity, al tolerance and oxidative stress in an Al-sensitive wheat genotype and in Al-tolerant lines developed by in vitro microspore selection. *Plant Sci.* 166 (3): 583–591.

Delhaize, E., Ryan, P.R., Hebb, D.M. et al. (2004). Engineering high-level aluminum tolerance in barley with the ALMT1 gene. *Proc. Natl. Acad. Sci. USA* 101: 15249–15254.

Delhaize, E., Gruber, B.D., and Ryan, P.R. (2007). The roles of organic anion permeases in aluminum resistance and mineral nutrition. *FEBS Lett.* 581 (12): 2255–2262.

Delhaize, E., Ma, J.F., and Ryan, P.R. (2012). Transcriptional regulation of aluminum tolerance genes. *Trends Plant Sci.* 17: 341–348.

Duan, W.J., Lu, F.K., Cui, Y. et al. (2022). Genome-wide identification and characterisation of wheat MATE genes reveals their roles in aluminum tolerance. *Mol. Sci.* 23: 4418.

Duressa, D., Soliman, K., Taylor, R., and Senwo, Z. (2011). Proteomic analysis of soybean roots under aluminum stress. *Int. J. Plant Genom.* 2011: 282531.

Ezaki, B., Katsuhara, M., Kawamura, M., and Matsumoto, H. (2001). Different mechanisms of four aluminum (Al)-resistant transgenes for Al toxicity in *Arabidopsis*. *Plant Physiol.* 127: 918–927.

Famoso, A.N., Zhao, K., Clark, R.T. et al. (2011). Genetic architecture of aluminum tolerance in rice (*Oryza sativa*) determined through genome-wide association analysis and QTL mapping. *PloS Genet.* 7 (8): e1002221.

Fan, X., Cui, F., Zhao, C. et al. (2015). QTLs for flag leaf size and their influence on yield related traits in wheat (*Triticum aestivum* L.). *Mol. Breed.* 35: 24.

Farokhzadeh, S., Fakheri, A.B., Nezhad, N.M. et al. (2019). Mapping QTLs of flag leaf morphological and physiological traits related to aluminum tolerance in wheat (*Triticum aestivum* L.). *Physiol. Mol. Biol. Plants* 25 (4): 975–990.

Fontecha, G., Silva-Navas, J., Benito, C. et al. (2007). Candidate gene identification of an aluminum-activated organic acid transporter gene at the *Alt4* locus for aluminum tolerance in rye (*Secale cereale* L.). *Theor. Appl. Genet.* 114: 249–260.

Fukuda, T., Saito, A., Wasaki, J. et al. (2007). Metabolic alterations proposed by proteome in rice roots grown under low P and high Al concentration under low pH. *Plant Sci.* 172: 1157–1165.

Garcia-Oliveira, A.L., Benito, C., Prieto, P. et al. (2013). Molecular characterization of TaSTOP1 homoeologues and their response to aluminium and proton (H^+) toxicity in bread wheat (*Triticum aestivum* L.). *BMC Plant Biol.* 13: 134.

Garnett, T., Conn, V., and Kaiser, B.N. (2009). Root based approaches to improving nitrogen use efficiency in plants. *Plant Cell Environ.* 32: 1272–1283.

GigaScience (2014). The 3,000 rice genomes project. *3K RGP* 3 (1): 2047-217X-3-7.

Gupta, P.K., Balyan, H.S., and Gahlaut, V. (2017). QTL analysis for drought tolerance in wheat: present status and future possibilities. *Agronomy* 7 (1): 5.

Han, C., Ryan, P.R., Yan, Z.H., and Delhaize, E. (2014). Introgression of a 4D chromosomal fragment into durum wheat confers aluminium tolerance. *Ann. Bot.* 114: 135–144.

Huang, C.F., Yamaji, N., Mitani, N. et al. (2009). A bacterial-type ABC transporter is involved in aluminum tolerance in rice. *Plant Cell* 21: 655–667.

Huang, C.F., Yamaji, N., Chen, Z., and Ma, J.F. (2012). A tonoplast-localized half-size ABC transporter is required for internal detoxification of aluminum in rice. *Plant J.* 69: 857–867.

Iuchi, S., Koyama, H., Iuchi, A. et al. (2007). Zinc finger protein STOP1 is critical for proton tolerance in *Arabidopsis* and coregulates a key gene in aluminum tolerance. *Proc. Natl. Acad. Sci. USA* 104: 9900–9905.

Jaskowiak, J., Tkaczyk, O., Slota, M. et al. (2018). Analysis of aluminum toxicity in *Hordeum vulgare* roots with an emphasis on DNA integrity and cell cycle. *PLoS One* 13 (2): e0193156.

Jiang, D., Wu, H., Cai, H., and Chen, G. (2022). Silicon confers aluminum tolerance in rice via cell wall modification in the root transition zone. *Plant Cell Environ.* 45 (6): 1765–1778.

Kobayashi, Y., Kobayashi, Y., Sugimoto, M. et al. (2013). Characterization of the complex regulation of AtALMT1 expression in response to phytohormones and other inducers. *Plant Physiol.* 162 (2): 732–740.

Kochian, L.V. (1995). Cellular mechanisms of aluminum toxicity and resistance in plants. *Annu. Rev. Plant Physiol. Plant Mol. Biol.* 46: 237–260.

Kochian, L.V., Hoekenga, O.A., and Pineros, M.A. (2004). How do crop plants tolerate acid soils? Mechanisms of aluminum tolerance and phosphorous efficiency. *Annu. Rev. Plant Biol.* 55 (1): 459–493.

Kochian, L.V., Piñeros, M.A., and Hoekenga, O.A. (2005). The physiology, genetics and molecular biology of plant aluminum resistance and toxicity. *Plant Soil* 274: 175–195.

Kochian, L.V., Pineros-Miguel, A., Liu, J., and Magalhaes, J.V. (2015). Plant adaptation to acid soils: The molecular basis for crop aluminum resistance. *Annu. Rev. Plant Biol.* 66: 571–598.

Kosová, K., Vítámvás, P., and Prášil, I.T. (2014). Proteomics of stress responses in wheat and barley. Search for potential protein markers of stress tolerance. *Front. Plant Sci.* 5: 711.

Kosová, K., Vítámvás, P., Urban, M.O. et al. (2015). Biological networks underlying abiotic stress tolerance in temperate crops-a proteomic perspective. *Int. J. Mol. Sci.* 16: 20913–20942.

Kuo, M.C. and Kao, C.H. (2003). Aluminum effects on lipid peroxidation and antioxidative enzyme activities in rice leaves. *Biol. Plant.* 46 (1): 149–152.

Kuroda, T. and Tsuchiya, T. (2009). Multidrug efflux transporters in the MATE family. *Biochim. Biophys. Acta* 1794: 763–768.

Langridge, P. and Reynolds, M.P. (2015). Genomic tools to assist breeding for drought tolerance. *Curr. Opin. Biotechnol.* 32: 130–135.

Li, J.Y., Liu, J., Dong, D. et al. (2014). Natural variation underlies alterations in nramp aluminum transporter (NRAT1) expression and function that play a key role in rice aluminum tolerance. *Proc. Natl. Acad. Sci. USA* 111: 6503–6508.

Li, C., Liu, G., Geng, X. et al. (2021). Local regulation of auxin transport in root-apex transition zone mediates aluminium-induced *Arabidopsis* root-growth inhibition. *Plant J.* 108 (1): 55–66.

Lin, S.Y., Sasaki, T., and Yano, M. (1998). Mapping quantitative trait loci controlling seed dormancy and heading date in rice, *Oryza sativa* L., using backcross inbred lines. *Theor. Appl. Genet.* 96: 997–1003.

Liu, P.Y., Zhu, J., and Lu, Y. (2006). Impacts of QTL 9 environment interactions on genetic response to marker-assisted selection. *Acta Genet. Sin.* 33 (1): 63–71.

Liu, J., Magalhaes, J.V., Shaff, J., and Kochian, L.V. (2009). Aluminum-activated citrate and malate transporters from the MATE and ALMT families function independently to confer *Arabidopsis* aluminum tolerance. *Plant J.* 57 (3): 389–399.

Liu, G., Gao, S., Tian, H. et al. (2016). Local transcriptional control of YUCCA regulates auxin promoted root-growth inhibition in response to aluminium stress in Arabidopsis. *PLoS Genet.* 12: e1006360.

Liu, H., Zhu, R., Shu, K. et al. (2022). Aluminum stress signaling, response, and adaptive mechanisms in plants. *Plant Signal. Behav.* 17 (1): e2057060.

Ma, J.F. (2007). Syndrome of aluminum toxicity and diversity of aluminum resistance in higher plants. *Int. Rev. Cytol.* 264: 225–252.

Ma, J.F., Shen, R., Zhao, Z. et al. (2002). Response of rice to Al stress and identification of quantitative trait loci for Al tolerance. *Plant Cell Physiol.* 43: 652–659.

Ma, H.X., Bai, G.H., Carver, B.F., and Zhou, L.L. (2005). Molecular mapping of a quantitative trait locus for aluminum tolerance in wheat cultivar Atlas 66. *Theor. Appl. Genet.* 112 (1): 51–57.

Ma, J.F., Chen, Z.C., and Shen, R.F. (2014). Molecular mechanisms of Al tolerance in gramineous plants. *Plant Soil* 381: 1–12.

Magalhaes, J.V., Garvin, D.F., Wang, Y. et al. (2004). Comparative mapping of a major aluminum tolerance gene in sorghum and other species in the Poaceae. *Genetics* 167: 1905–1914.

Magnavaca, R., Gardner, C.O., and Clark, R.B. (1987). Inheritanceof aluminum tolerance in maize. In: *Genetic Aspectsof Plant Mineral Nutrition* (ed. H.W. Gabelman and B.C. Loughman), 201–212. Dordrecht, The Netherlands: Martinus Nijhoff.

Mangeon, A., Pardal, R., Menezes-Salgueiro, A.D. et al. (2016). AtGRP3 is implicated in root size and aluminum response pathways in Arabidopsis. *PLoS One* 11 (3): e0150583.

Matsumoto, H. (2000). Cell biology of aluminum toxicity and tolerance in higher plants. *Int. Rev. Cytol.* 200: 1–46.

McIntyre, C.L., Mathews, K.L., Rattey, A. et al. (2010). Molecular detection of genomic regions associated with grain yield and yield-related components in an elite bread wheat cross evaluated under irrigated and rainfed conditions. *Theor. Appl. Genet.* 120 (3): 527–541.

McLean, E.O. (1965). Aluminum. In: *Agronomy Journal*, Chapter 67, Agronomy Monographs (ed. A.G. Norman). https://doi.org/10.2134/agronmonogr9.2.c16.

Meng, L., Wang, B., Zhao, X. et al. (2017). Association mapping of ferrous, zinc, and aluminum tolerance at the seedling stage in indica rice using MAGIC populations. *Front. Plant Sci.* 8: 1822.

Nguyen, V.T., Burow, M.D., Nguyen, H.T. et al. (2001). Molecular mapping of genes conferring aluminumtolerance in rice (*Oryza sativa* L.). *Theor. Appl. Genet.* 102: 1002–1010.

Nguyen, V.T., Nguyen, D.B., Sarkarung, S. et al. (2002). Mapping of genes controlling aluminum tolerance in rice: comparison of different genetic backgrounds. *Mol. Genet. Genomics* 267: 772–780.

Nguyen, B.D., Brar, D.S., Bui, C.B. et al. (2003). Identification and mapping of the QTL for aluminumtolerance introgressed from new source, *Oryza rufipogon* Griff., into indica rice (*Oryza sativa* L.). *Theor. Appl. Genet.* 106: 583–593.

Niedziela, A., Bednarek, P.T., Cichy, H. et al. (2012). Aluminum tolerance association mapping in triticale. *BMC Genom.* 13: 67.

Niedziela, A., Bednarek, P.T., Labudda, M. et al. (2014). Genetic mapping of a 7R Al tolerance QTL in triticale (× *Triticosecale* Wittmack). *J. Appl. Genet.* 55: 1–14.

Niedziela, A., Domzalska, L., Dynkowska, W.M. et al. (2022). Aluminum stress induces irreversible proteomic changes in the roots of the sensitive but not the tolerant genotype of triticale seedlings. *Plants* 11: 165.

Ninamango-Cardenas, F.E., Guimaraes, C.T., Martins, P.R. et al. (2003). Mapping QTLs foraluminum tolerance in maize. *Euphytica* 130: 223–232.

Oh, M.W., Roy, S.K., Kamal, A.H.M. et al. (2014). Proteome analysis of roots of wheat seedlings under aluminum stress. *Mol. Biol. Rep.* 41: 671–681.

Panda, S.K., Baluska, F., and Matsumoto, H. (2009). Aluminum stress signaling in plants. *Plant Signal. Behav.* 4: 592–597.

Raman, H., Zhang, K., Cakir, M. et al. (2005). Molecular characterization and mapping of ALMT1, the aluminum-tolerance gene of bread wheat (*Triticum aestivum* L.). *Genome* 48 (5): 781–791.

Ranjan, A., Sinha, R., Lal, S.K. et al. (2021). Phytohormone signaling and cross-talk to alleviate aluminum toxicity in plants. *Plant Cell Rep.* 40 (8): 1331–1343.

Rosa-Santos, T.M., Gonçalves da Silva, R., Kumar, P. et al. (2020). Molecular mechanisms underlying sugarcane response to aluminum stress by RNA-seq. *Int. J. Mol. Sci.* 21: 7934.

Ryan, P.R., Delhaize, E., and Jones, D.L. (2001). Function and mechanism of organic anion exudation from plant roots. *Annu. Rev. Plant Physiol. Plant Mol. Biol.* 52: 527–560.

Ryan, P.R., Raman, H., Gupta, S. et al. (2009). A second mechanism for aluminum resistance in wheat relies on the constitutive efflux of citrate from roots. *Plant Physiol.* 149: 340–351.

Ryan, P.R., Tyerman, S.D., Sasaki, T. et al. (2011). The identification of aluminiumresistance genes provides opportunities for enhancing crop production on acid soils. *J. Exp. Bot.* 62: 9–20.

Sasaki, T., Yamamoto, Y., Ezaki, B. et al. (2004). A wheat gene encoding an aluminum-activated malate transporter. *Plant J.* 37 (5): 645–653.

Sawaki, Y., Iuchi, S., Kobayashi, Y. et al. (2009). STOP1 regulates multiple genes that protect Arabidopsis from proton and aluminum toxicities. *Plant Physiol.* 150: 281–294.

Sharma, P. and Dubey, R.S. (2007). Involvement of oxidative stress and role of antioxidative defense system in growing rice seedlings exposed to toxic concentrations of aluminum. *Plant Cell Rep.* 26 (11): 2027–2038.

Shukla, S., Singh, K., Patil, R.V. et al. (2015). Genomic regions associated with grain yield under drought stress in wheat (*Triticum aestivum* L.). *Euphytica* 203: 449–467.

Silva, S., Pinto, G., Dias, M.C. et al. (2012). Aluminium long-term stress differently affects photosynthesis in rye genotypes. *Plant Physiol. Biochem.* 54: 105–112.

Silva-Navas, J., Benito, C., Téllez-Robledo, B. et al. (2012). The ScAACT1 gene at the *Qalt5*locus as a candidate for increased aluminum tolerance in rye (*Secale cereale* L.). *Mol. Breed.* 30:845–856.

Simoes, C.C., Melo, J.O., Magalhaes, J.V., and Guimaraes, C.T. (2012). Genetic and molecular mechanisms of aluminum tolerance in plants. *Genet. Mol. Res.* 11: 1949–1957.

Sivaguru, M., Ezaki, B., He, Z.H. et al. (2003). Aluminum-induced gene expression and protein localization of a cell wall-associated receptor kinase in Arabidopsis. *Plant Physiol.* 132 (4): 2256–2266.

Sun, P., Tian, Q.Y., Chen, J., and Zhang, W.H. (2010). Aluminum-induced inhibition of root elongation in Arabidopsis is mediated by ethylene and auxin. *J. Exp. Bot.* 61 (2): 347–356.

Talukder, S.K., Babar, M.A., Vijayalakshmi, K. et al. (2014). Mapping QTL for the traits associated with heat tolerance in wheat (*Triticum aestivum* L.). *BMC Genet.* 15: 97.

Tao, Y.H., Niu, Y., Wang, Y. et al. (2018). Genome-wide association mapping of aluminum toxicity tolerance and fine mapping of a candidate gene for *Nrat1* in rice. *PLoS One* 13 (6): e0198589.

Tsutsui, T., Yamaji, N., and Ma, J.F. (2011). Identification of a cis-acting element of ART1, a C2H2-type zinc-finger transcription factor for aluminum tolerance in rice. *Plant Physiol.* 156: 925–931.

Tyagi, W., Yumnam, J.S., Sen, D., and Rai, M. (2020). Root transcriptome reveals efficient cell signaling and energy conservation key to aluminum toxicity tolerance in acidic soil adapted rice genotype. *Sci. Rep.* 10: 4580.

Vitorello, V.A., Capaldi, F.R.C., and Stefanuto, V.A. (2005). Recent advances in aluminum toxicity and resistance in higher plants. *Braz. J. Plant Physiol.* 17 (1): 129–143.

Von Uexküll, H.R. and Mutert, E. (1995). Global extent, development and economic impact of acid soils. *Plant Soil* 171: 1–15.

Wu, P., Liao, C.Y., Hu, B. et al. (2000). QTLs and epistasis for aluminum tolerance in rice (*Oryza sativa* L.) atdifferent seedling stages. *Theor. Appl. Genet.* 100: 1295–1303.

Wu, Q., Chen, Y., Fu, L. et al. (2015). QTL mapping of flag leaf traits in common wheat using an integrated high-density SSR and SNP genetic linkage map. *Euphytica* 208: 337–351.

Xia, J.X., Yamaji, N., Kasai, T., and Ma, J.F. (2010). Plasma membrane-localized transporter for aluminum in rice. *Proc. Natl. Sci. USA* 107: 18381–18385.

Xia, J., Yamaji, N., and Ma, J.F. (2013). A plasma membrane-localized small peptide is involved in rice Al tolerance. *Plant J.* 76: 345–355.

Xue, S., Xu, F., Li, G. et al. (2013). Fine mapping *TaFLW1*, a major QTL controlling flag leaf width in bread wheat (*Triticum aestivum* L.). *Theor. Appl. Genet.* 126: 1941–1949.

Yamaji, N., Huang, C., Nagao, S. et al. (2009). A zinc finger transcription factor ART1 regulates multiple genes implicated in aluminum tolerance in rice. *Plant Cell* 21: 3339–3349.

Yang, Z.B., Geng, X., He, C. et al. (2014). TAA1-regulated local auxin biosynthesis in the root-apex transition zone mediates the aluminum-induced inhibition of root growth in Arabidopsis. *Plant Cell.* 26 (7): 2889–2904.

Yin, L., Mano, J.I., Wang, S. et al. (2010). The involvement of lipid peroxide-derived aldehydes in aluminum toxicity of tobacco roots. *Plant Physiol.* 152 (3): 1406–1417.

Yokosho, K., Yamaji, N., Ueno, D. et al. (2009). OsFRDL1 is a citrate transporter required for efficient translocation of iron in rice. *Plant Physiol.* 149: 297–305.

Yu, S.B., Li, J.X., Xu, C.G. et al. (2002). Identification of quantitative trait loci and epistatic interactions for plant height and heading date in rice. *Theor. Appl. Genet.* 104 (4): 619–625.

Zhang, P., Zhong, K., Tong, H. et al. (2016). Association mapping for aluminum tolerance in a core collection of rice landraces. *Front. Plant Sci.* 7: 1415.

Zhang, J., Gizaw, S.A., Bossolini, E. et al. (2018). Identification and validation of QTL for grain yield and plant water status under contrasting water treatments in fall-sown spring wheats. *Theor. Appl. Genet.* 131: 1741–1759.

Zhang, F., Yan, X., Han, X. et al. (2019). A defective vacuolar proton pump enhances aluminum tolerance by reducing vacuole sequestration of organic acids. *Plant Physiol.* 181 (2): 743–761.

Zhi, C.C. and Jian, F.M. (2012, 2012). Up-regulation of a magnesium transporter gene OsMGT1 is required for conferring aluminum tolerance in rice. *Plant Physiol.* 159: 1624–1633.

Zhou, S., Sauvé, R., and Thannhauser, T.W. (2009). Proteome changes induced by aluminium stress in tomato roots. *J. Exp. Bot.* 60: 1849–1857.

Zhu, X.F., Shi, Y.Z., Lei, G.J. et al. (2012). XTH31, encoding an in vitro XEH/XET-active enzyme, regulates aluminum sensitivity by modulating in vivo XET action, cell wall xyloglucan content, and aluminum binding capacity in Arabidopsis. *Plant Cell.* 24 (11): 4731–4747.

Zhu, C.Q., Wei, Q.Q., Hu, W.J. et al. (2022). Unearthing the alleviatory mechanisms of hydrogen sulfide in aluminum toxicity in rice. *Plant Physiol. Biochem.* 1 (182): 133–144.

6

Chromium Toxicity and Tolerance in Plants

Insights from Omics Studies

Sonali Dubey[1], *Manju Shri*[2], *and Debasis Chakrabarty*[3]

[1] *School of Biosciences, IMS Ghaziabad University Courses Campus, Ghaziabad, Uttar Pradesh, India*
[2] *School of Applied Sciences and Technology, Gujarat Technological University, Ahmedabad, Gujarat, India*
[3] *Molecular Biology & Biotechnology Division, Council of Scientific and Industrial Research – National Botanical Research Institute (CSIR-NBRI), Lucknow, Uttar Pradesh, India*

6.1 Introduction

Chromium (Cr) is the 24th constituent with a molecular weight of 51.1 a.m.u. and a density of 7.19 g cm^{-3}, which commonly exist in rocky soils and volcanic dust (EFSA 2014). Cr has been listed the 17th among the extreme harmful materials by the ATSDR (Agency for Toxic Substances and Disease Registry 2019). CrVI has been categorized as a Group one occupational carcinogen by the IARC (International Agency for Research on Cancer) (Deng et al. 2019). Thus, an understanding of CrVI uptake, translocation, and its accumulation in different parts of plants would allow the practitioners to identify its detoxification mechanism.

Cr contamination in soil has been enhanced due to various human-induced activities with ecological processes. Cr is nondegradable and persistent in soil for an extremely long time, disturbing soil property, as well as plant growth and development (Oliveira 2012). Cr compounds are commercially used in numerous industries, for example tanning leather, fabric and dye production, electroplating, nuclear reactor vessels, alloy products development finally causes its release in the sludge and waste-waters which affects marine as well as land ecosystems. However, Cr is considered as a nonessential component for plants, which is likely to be taken up by the carriers of other essential elements such as sulfate and possibly can induce numerous biochemical variations in plant. Interruption of photosynthesis and respiration activities, oxidative damage, enzyme activities inhibition, and even plant death can be caused by Cr toxicity (Sharma et al. 2022). Among the two most stable and common forms, trivalent CrIII exists commonly in soil, which is considered as nutrient of organisms for the growth and development, while various anthropogenic activities as well as natural oxidation of CrIII are responsible for the production of extremely reactive and hazardous hexavalent CrVI (Monga et al. 2022). It was observed that Cr concentration

Heavy Metal Toxicity and Tolerance in Plants: A Biological, Omics, and Genetic Engineering Approach, First Edition. Edited by Mohammad Anwar Hossain, AKM Zakir Hossain, Sylvain Bourgerie, Masayuki Fujita, Om Parkash Dhankher, and Parvez Haris.

above 75–100 ppm adversely affects the plant growth, and it is toxic as well as inhibitory. Cr Concentration has been exceeded 1000 ppm in the soils of industrial and mining regions (Sehrawat et al. 2021). Thus, it turns into predominantly essential to clean up soils, which possess Cr measure more than 100 ppm. For decontamination of heavy metals, different conventional methods applied at wide range are not only costly but also alters soil composition and productivity. It is essential that soil decontamination methods do not lead to polluting the other natural resources, for example land water and air. Phytoremediation has emerged as an eco-friendly approach to remediate heavy metal contamination of soils in less expensive manner (Yan et al. 2020a). Different metabolic detoxification pathways have been elucidated in response to Cr toxicity. Cr stress has been indicated to induce ROS (reactive oxygen species) accumulation, responses to antioxidant, metal-binding proteins like PC (phytochelatin), MT (metallothionein), and GST (glutathione *S*-transferase) succeeded by vacuole sequestration by this means enhancing the resistance against metal in plants (Dubey et al. 2010; Shanker et al. 2004a, 2004b). Development of transgenic by overexpressing the upregulated genes related to Cr uptake, transport, and sequestration could be one strategy to improve the accumulation as well as tolerance potential of the plant. Considering all, this chapter focuses on Cr uptake, translocation, and sub-cellular distribution of Cr in plants. We will also discuss the various omics studies done on the plants exposed to chromium stress, mainly focused on Cr detoxification mechanisms of the plants.

6.2 Chromium Sources and Bioavailability

Cr is highly reactive due to its electronic configuration (3d5) in the *d* orbital. Its high application in industries made it very hazardous for environment and human health. Human intervention in the environment had made it more complicated and resulted in increased Cr pollution. Cr is released into the environment in larger amounts as a result of human activities, which account for 60–70% of the total emissions of atmospheric chromium (Cheng et al. 2014). Cr has a wide range of industrial application. It is used in wood preservation, electroplating, pulp and paper production, pigment, metallurgy, production of paints and pigments and so on. Along with these industries, Cr is also present in Cr steel, fertilizers, oil well drilling, petroleum, and coal. Utilization of Cr-containing fertilizers in agriculture field is responsible for exposing crop plants to Cr toxicity. Disposal of waste from the tanning, leather industries, as well as sewage to the running water, especially river contaminated or loaded extra amount of Cr to the holy rivers. Tanning and leather industries along with other industries and disposal of sewage as well as application of Cr containing fertilizers are the major culprit of Cr pollution of the environment. Among all the industries utilizing and increasing Cr pollution, the tanning industry ranks first. $Cr_2(SO_4)_3$ is being used in tanning industries and the waste/effluent generated is discharged to the rivers without any proper treatments.

Cr exists in several oxidation states ranging from 0 to +6 due to its electronic configuration and thus making it highly reactive. Among all these oxidation states, +3 [CrIII trivalent chromium] and +6 [hexavalent chromium CrVI] are the most stable and common form. CrIII is very stable in the environment due to its inert complex formation with water, ammonia and urea, organic acids, and halides (Homa et al. 2016). In comparison to CrIII,

CrVI is highly reactive and in terms of toxicity, it is most toxic form of Cr. CrVI is highly soluble in water and therefore more mobile in comparison to CrIII, which is insoluble in water. CrVI is stable in the oxidizing environment with pH almost or above 6. Under reducing environment, hexavalent Cr is reduced to $Cr(OH)_3$, which is more thermodynamically stable in comparison to CrVI. CrVI is a strong oxidizing agent and considered as strong carcinogen compound and is very toxic to animals including human (Homa et al. 2016). The WHO-recommended safe limits for Cr VI in soils and wastewater used for agriculture are 0.05 ppm and 0.1 ppm, respectively (Kinuthia et al. 2020).

Chromium is primarily removed from the atmosphere by fallout and precipitation. The residence time of chromium in atmosphere is about <10 days. The mean concentration of chromium in the air of United States, suburban, and rural areas during 1977–1984 was found to be in the range of 5–525 $ng\,m^{-3}$ (Barceloux 1999). Total chromium concentrations in US soils range from 1 to 2000 $mg\,kg^{-1}$, with a mean of 37.0 $mg\,kg^{-1}$ CrIII in soil is mostly present as insoluble carbonate and oxide of chromium (III); therefore, it will not be mobile in soil (Owhonda et al. 2016). The solubility of CrIII in soil and its mobility may increase due to the formation of soluble complexes with organic matter in soil, with a lower soil pH potentially facilitating complexation (Kamaludeen et al. 2003). Cr has a low mobility for translocation from roots to the aboveground parts of plants (Budak et al. 2011).

A common source of Cr exposure is from food. Total Cr levels in most foods typically range from less than 10 to 1300 $\mu g\,kg^{-1}$. Highest concentration of Cr is found in meat, fish, vegetables, and fruits. Human population is exposed to Cr through breathing air, drinking water, and by the Cr contaminated food they are taking in. However, drinking water may also results in toxicity to humans when the contamination levels in drinking water are >25 $\mu g\,l^{-1}$ (Owhonda et al. 2016).

6.3 Chromium Uptake, Translocation, and Sub-cellular Distribution in Plants

There are very few reports available for the uptake and transport of Cr from soil to the plant. Plant roots play very important role in the entire life of a plant. Roots are accountable for providing the strength, the uptake of nutrients, and water needed for growth of the plant. Roots cannot discriminate between nutrients and toxic elements such as heavy metals (HMs). Therefore, along with nutrients, these HMs are also taken up by the plant roots. Roots are also responsible for the secretion of roots exudates (such as malate and citrate), which can form complexes with the elements present in the soil and may modulates their solubility. The change in the solubility of the root exudate–HM complex may result in the increased bioavailability of the HMs. There was a very interesting study done on the tomato plants to demonstrate the modulation of solubility and bioavailability of inorganic Cr to the plants. The group demonstrated the release of citrate and aspartate from the roots of the tomato plants as root exudates. Root exudates were found to modulate the bioavailability of HMs, which results in modulated accumulation of HMs. Organic form of Cr is more bioavailable in comparison to inorganic Cr. Roots exudates might convert the less bioavailable form of Cr (inorganic) to more bioavailable form (organic Cr) leading to increased accumulation of Cr in the plant (Srivastava et al. 2019).

Uptake of Cr depends on the various factors such as its chemical form (organic or inorganic), temperature, pH of soil, aerobic or anaerobic condition of soil, presence of other elements (competition), and soil and plant moisture content. Cr is not required for the plants in any means so, there is no specific transporter is deployed by the plants for the transport of the Cr. CrVI and CrIII both can be taken up by the plants. However, no specific mechanism for the transporter of the Cr is known (Oliveira 2012; Singh et al. 2013). Cr is transported to the plants via either carrier proteins or the transporters that transport either sulfate or iron (Gajalakshmi et al. 2012; Singh et al. 2013). The mechanism of transport of CrVI and CrIII is different. CrVI is transported by active transport. CrVI is absorbed via active process in comparison to CrIII, which is taken up by the plants via osmosis or diffusion (Barros et al. 2006). Liu et al. (2011) had shown the possibility of involvement of ion channels or transporters of essential ions in the transport of Cr. However, the transport of CrIII is passive transport. CrVI shows structural similarity with sulfate and phosphate and thus share some common transporters for its transportation. The distribution and translocation of Cr within plants depend upon the plant species, the oxidation state of the Cr ions, and also its concentration in the growth medium (Sharma et al. 2012.)

Roots are the first organs of the plant that encounter any type of stress either water or HM. Most of the time heavy metals are absorbed and stored inside the root cells. Most HMs including Cr are sequestered inside the vacuole as a defense mechanism. The sequestration of HMs inside vacuole is helpful in providing the avoidance of interactions of HMs with the cytosolic components that may further lead to toxicity. In literature it has been sited that Cr is stored mainly inside the vacuole and very less amount remains free for further transportation to the above ground parts of plants (Nematshahi et al. 2012; Oliveira 2012). Sequestration of Cr inside the root vacuole results in the accumulation of Cr higher in roots, which is responsible for restriction in movements of Cr from roots to the above ground parts, saving the photosynthetic units (chloroplast of leaves) from the toxic effects of the Cr.

Once the Cr is absorbed, the next step is either its vacuolar compartmentalization inside the vacuole or the transport of Cr to the above ground part. Cr is transported to the above ground parts of the plants via translocation. Translocation occurs at a very slow rate resulting in higher accumulation of Cr inside the roots (Singh et al. 2013; Sundaramoorthy et al. 2010). Cr toxicity may result in the reduction of root growth as Cr interferes with the cell division. Aberrations of the mitotic division were correlated with and attributed to the Cr-induced derangement of the microtubule cytoskeleton, which normally underlies cell division (Eleftheriou et al. 2012). Roots are essential organs as they are responsible for the uptake of nutrients, essential elements, and water. Reduction in root growth has negative impact on all these process, which all together may result in reduced growth of the plants and overall yield of the crops (Shanker et al. 2004b). Transportation of Cr to the above ground parts occur through xylem (Hayat et al. 2012). Symplastic movement is responsible for the translocation of CrVI to the above ground parts, and during its transportation it may be reduced to CrIII. CrIII may be retained by the cortical cells of the roots further leading to lesser transport of Cr to the above ground parts. Like other HMs such as, reduction of CrVI to CrIII is a major defense mechanism of plants because reduced form is less toxic in comparison to oxidized form (CrVI). CrVI is also reduced to CrIII in the leaves of the plants (da Conceicao Gomes et al. 2017)

6.4 Detoxification Mechanisms for Cr

All the HMs including Cr results in toxicity via production of ROS as a result of oxidative stress. Plants are equipped with various antioxidants to combat this oxidative stresses. Antioxidant enzymes such as superoxide dismutase (SODs), peroxidases (PODs), glutathione reductases (GRs), catalases (CATs), and ascorbate peroxidase (APX) have been shown to play important role in Cr detoxification in plants via scavenging ROS resulted as a result of oxidative stress (Srivastava et al. 2019).

CrVI is thought to be more toxic in comparison to CrIII. As soon as CrVI enters the cells, plants try to reduce CrVI to the less toxic form, i.e. CrIII. There are various metabolites that assist in the reduction of CrVI to CrIII such as glutathione, cysteine, sulfite, and thio-sulfates (Whitacre 2010). Rhizospheric bacteria such as *Arlettae, Ochrobacterium intermedium, Pseudomonas* sp., *Bacillus* spp. play very important role in the reduction of CrVI to CrIII (Sinha et al. 2018). Once the CrVI is reduced to CrIII, it may either form complexes with metallothioneins (MT) as MTs can bind only with CrIII not with CrVI (Dubey et al. 2010). These complexes result in reduced toxicity of Cr by avoidance mechanisms. MTs are low-molecular-weight (LMW) cysteine-rich proteins, playing important role in HM detoxification. MTs are localized in Golgi apparatus of the cells. Very recently in 2019, Yu et al. (2019) had demonstrated role of OsMTs in Cr detoxification as scavenging agent for ROS produced as a result of Cr toxicity, and they might stimulate the chelation of the Cr. Along with MTs, there are few other proteins that play important role in the sequestration of Cr. Phytochelatins (PCs) have been well documented for their HM detoxification role in plants. PCs are LMW and HMW cys-rich polypeptides produced by the enzyme phytochelatin synthase. PCs form complexes with the reduced CrIII and these complexes are sequestered inside the vacuole with the help of ABC transporters. Studies done on various plants such as *Vigna radiata, Brassica juncea, Helianthus annuus*, and *Oryza sativa* demonstrated very important role of PCs in the detoxification of Cr (Shahid et al. 2017). Glutathione (GSH) is a LMW peptide with general structure (Glu-Cys-Gly) having predominant role in various HMs including Cr detoxification. GSH may also form complexes with CrIII, and these complexes may be stored inside the vacuole as a detoxification mechanism; GSHs are enzymatically produced by the enzyme glutathione S transferase (GSTs). Along with GSHs, GSTs (mainly Tau class) have been shown to play important role in Cr detoxification. In 2019, Srivastava et al. (2019) over-expressed OsGSTU30 in *Arabidopsis thaliana* and confirmed its role in Cr detoxification. This group demonstrated that OsGSTU30 provides tolerance to Cr via two enzymatic role: one as GST and one as glutathione peroxidase (GPX) under CrVI stress. In summary, Figure 6.1 shows various transporters involved in the transport of Cr and various detoxification mechanisms inside the plants.

In recent years, many efforts have been done to understand the molecular pathways responsible for transport, detoxification, and metabolism of Cr inside the plants. These studies focused on omics approaches that dealt with transcriptomic and metabolomic changes in plants when exposed to Cr stress. MicroRNAs also play important role in the development of plants as well as detoxification of various HMs. In the coming sections, we will discuss the omics-based studies that have been performed on various plants exposed to Cr stress as well as the phytoremediation technologies used to reduce Cr accumulation in plants.

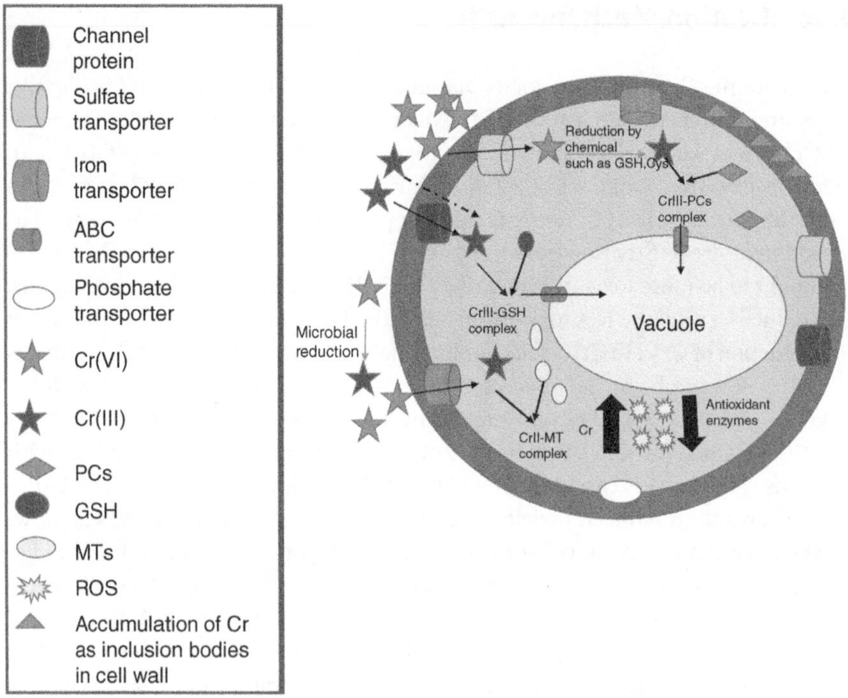

Figure 6.1 Schematic representation of various transporters involved in the transport of chromium from soil to plant roots and detoxification mechanism inside the plants.

6.5 Omics Approaches Used by Plants to Combat Cr Toxicity

6.5.1 Transcriptomics

Toxicity of Cr is deleterious to plant growth, development, as well as crop productivity (Singh et al. 2021). Several changes at the transcriptome, cellular, and physiological levels have been reported in plants in response to different stresses. Omics studies are used to advance the understanding about the physiological and molecular impacts of environmental stress, which can be used to improve stress tolerance by the identification of gene expression, miRNAs, and metabolites (Jamla et al. 2021). Using high-throughput genomic approach, expression of large number of genes can be analyzed with single experiment. Comprehensive genome-wide transcriptome analysis was carried out by Dubey et al. (2010) in rice root under Cr stress. These types of studies are required to identify genes involved in transport and detoxification mechanism in response to Cr stress.

Dubey et al. (2010) have reported that 1138 genes were up and 1610 genes were downregulated during CrVI stress in rice using Affymetrix GeneChip microarrays gene expression analysis. Upregulation of genes represented the involvement of genes related to glutathione *S*-transferase (GST), xenobiotics biodegradation, metabolism of tyrosine, amino acid as well as lipids, MAPK signaling pathway, secondary metabolites synthesis, ABC transporters, signal transduction, carbohydrate, and energy metabolism. Downregulation of genes were associated with growth and death of the cells, peroxidases, biosynthesis of secondary metabolites, carbohydrate, energy, lipid as well as amino acid metabolism, xenobiotics degradation, membrane transporters, pathway related to MAPK signaling, signal transduction, cellular processes, vitamins metabolism, folding, and degradation. Upregulation of one root-specific ABC transporter (Os08g30770) was reported upon CrVI exposure that possibly be involved in Cr transportation within the cell. CrVI is actively taken up by carriers of essential ions involved in plant metabolism. Downregulation of huge numbers of MAPK signaling pathway-associated genes was found under CrVI exposure. It has been reported that cellular processes such as cell division, cell differentiation, and several stress responses can be induced by MAP kinase cascade. Induced expression of cytochrome P450, several NAC domain-containing proteins, WRKY transcription factors, and bZIP (basic region/Leu Zipper) during CrVI stress was observed.

Investigation of the molecular mechanisms involved in signaling pathways as well as transcriptome profiling during CrVI stress in rice roots was performed by Trinh et al. (2019) using microarray analysis. It was observed that genes related to transcription factor, protein kinase, vesicle trafficking, phytohormone, as well as protein phosphatase might perform significant roles during CrVI exposure in rice roots Phytohormones control the growth and environmental stress responses in plants (Waadt et al. 2022). Trinh et al. (2019) demonstrated the involvement of gibberellic acid, ethylene, and jasmonic acid hormones during Cr stress in rice. Various genes involved in the biosynthesis, metabolism, and signaling of these phytohormones were significantly found to be modulated in this study. OsNCED2 and OsNCED3 (ABA biosynthesis encoding genes) transcript levels were significantly enhanced by CrVI treatment. Additionally, upregulation of three ABA signaling genes in rice roots was reported upon CrVI exposure. Inhibition of root growth by Cr in *Arabidopsis* has been indicated by Beaudoin et al. (2000) and Ghassemian et al. (2000) through ETR1 (ethylene response 1) signaling pathway. Further, strong induction of two genes encoding ethylene biosynthesis enzymes, i.e. OsACS and two OsACO genes [OsACS1, OsACS2, OsACO4 and OsACO5, 1-aminocyclopropane-1-carboxylate (ACC) synthase (ACS), and ACC oxidase (ACO)] was studied in response to CrVI treatment. In addition, one ethylene signaling gene (OsEIN3;4) was upregulated in CrVI stress response. 21 AP2/ERF genes were found to be upregulated under CrVI exposure, suggesting their probable involvement in combating the metal toxicity. Trinh et al. (2019) demonstrated the enhanced expression of OsGA2ox genes, i.e. OsGA2ox1, OsGA2ox3, and OsGA2ox9 upon CrVI exposure, which involved in growth inhibition by reducing GA levels. Differential expression of eight genes related to JA (jasmonic acid) biosynthesis as well as signaling was reported (Ghassemian et al. 2000).

DUF26, RLCK, and LRK10-L were differentially expressed among CrVI-responsive RLK genes. Upregulation of CRK (cysteine-rich kinase) and LRK10-L (leaf rust kinase) was identified by microarray analysis during CrVI stress rice roots. Six LRK10L-2 genes were found to be upregulated after CrVI treatment in this study. Therefore, differential expression of transmembrane receptor kinases genes suggested its involvement in the regulation of the plant responses to Cr stress. Ca^{+2} accumulation and proteins involved in its signaling such as calcium sensor proteins like CDPK (calcium dependent protein kinases), CaM (calmodulin), and CBL (calcineurin B-like protein) have also been shown to be modulated during abiotic stress (Das and Pandey 2010). Additionally, CrVI stress elevated the expression of OsCPK4 and OsCPK13 (two CDPK genes) and four CaM genes along with increased accumulation of Ca^{+2}. Modulation in expression of various kinases during Cr stress demonstrates the regulation at transcriptional as well as posttranslational (protein phosphorylation) level.

The vesicle trafficking process, involving exo and endocytosis, plays a significant role during various environmental stresses. The vacuoles play an important role in plant salt tolerance. Accumulation of ROS is generally build up during abiotic stress. In plants ROS are generally produced at chloroplast and mitochondria. However, enzymatic production of ROS has also been observed as is the case of NADPH oxidase-dependent ROS production. ROS that is produced has a significant role in intracellular signaling. Its destination to the cellular compartments is controlled by vesicle trafficking under salinity stress. Accumulation of ROS was significantly decreased in vesicle trafficking inhibitor (BFA)-treated roots (Leshem et al. 2007), therefore ROS accumulation can be reduced by membrane trafficking repression during salt stress. Expression level of Exo70 was found to be induced during CrVI stress in (Trinh et al. 2019) which performs an essential part in vesicle trafficking. ROS inhibition by BFA was recognized under Cr stress. NbExo70 gene silencing repressed the ROS generation in the leaves as well as roots and exhibited an enhanced CrVI tolerance. Thus, the Exo70 may possibly be involved in ROS homeostasis regulation that can activate downstream signal transduction during metal stress.

6.5.2 Chromium-Induced miRNAs in Plants

Several published studies focused on Cr toxicity and tolerance in plants; but still, very limited findings are achieved related to the molecular mechanisms involved in Cr bioaccumulation, uptake, translocation, as well as detoxification in plants. To understand the miRNA-induced regulatory networks involved in Cr stress, posttranscriptional analysis was conducted by Liu et al. (2015) to identify Cr VI responsive miRNAs and target genes in radish; 52 known miRNAs belongs to 24 conserved families from Cr-200 and CK libraries were identified in this study. It has been reported that conserved miRNAs such as miR393, miR397 and miR408, miR159, miR161, miR162, miR171, miR164, miR167 and miR319 were less ample. In contrast, miR169 and miR156/157 along with miR165/167 possessed four and five members, respectively. Enhanced expression level of miR156/157, miR158, miR165/166, miR168, and miR408 were identified in both of the libraries where miR158 family showed the highest expression level. In both the libraries, 29 unique sequences indicating 22 non-conserved miRNA families have been also detected. In two libraries, miR535, miR824, miR845, miR858 and miR7767, miR400, miR825, miR827, miR854, miR860, miR1885 along with a great number of non-conserved miRNAs had low abundance. Higher expression and abundance have been displayed by conserved miRNAs family when compared to non-conserved family in this study.

In recent past, Dubey et al. (2010) predicted 512 known miRNAs from the rice root under CrVI stress. Differential expression of 13 conserved miRNAs such as miR156, miR159, miR160, miR166, miR169, miR171, miR396, miR397, miR408, miR444, miR1883, miR2877, and miR5072 was identified during miRNA analysis. miRNAs regulatory function of various antioxidant responses and defense mechanisms has been confirmed by target prediction of these miRNAs in response to Cr stress. Target genes of these miRNAs were associated with ATP binding cassette transporter, transcription factors, heat shock proteins, as well as auxin response.

6.5.3 Metabolomics

Dubey et al. (2010) studied the metabolic shift in 100 μM Cr (VI) rice roots exposed for 24 hours by GC-MS and NMR analysis. Increase in numerous metabolites such as alanine, uracil, lactate, and fructose content was observed in response to CrVI stress. Accumulation of proline was threefold when compared to control. Proline is an amino acid, which plays an essential role as osmolyte in adaptation to osmotic stress in plants (Meena et al. 2019). Increased content of ornithine was reported in this study, which could be employed as substrate for proline biosynthesis. Signification of urea cycle enriching proline accumulation was suggested by the high content of ornithine in CrVI-treated rice roots. Numerous fatty acids exhibited high accumulation in roots of rice after CrVI exposure (Dubey et al. 2010). In this study, highest content of linoleic acid (C18:2) was observed in CrVI-treated rice root. The outcomes concur to the documented familiar consequence of stress by Cr on the rise in fatty acid unsaturation level in plants. Recently, it has been reported that raising the unsaturated fatty acid volume in membrane lipids may improve the tolerance against environmental stress (Tian et al. 2022). Highly prevailing phytosterols such as beta-sitosterol, stigma-sterol, and campesterol showed their content modulation under stress situations. Enhanced level of linolenic (18 : 3), linoleic (18 : 2) acids and stigmasterol along with reduced level of oleic (18 : 1), palmitoleic (16 : 1) acids, and sitosterol has been identified during salinity stress in broccoli roots (López-Pérez et al. 2009). In rice, reduced concentration of stigmasterol and sitosterol was observed in CrVI-treated roots.

To study the metabolite profiling of *H. annuus* roots during Cr stress, liquid chromatography coupled to high-resolution mass spectrometry were applied by Gonzalez Ibarra et al. (2017). This study was conducted to identify the groups of metabolic compounds predominantly affected by CrVI stress in hydroponic systems. Increased synthesis of the several secondary metabolites such as isocoumarins, benzofurans, sesquiterpenes and their lactones, and glycosides of phenolic compounds has been reported in this study.

6.5.4 Proteomics

Along with transcriptomic and metabolic studies, proteomic analysis is very useful in deciphering the defense mechanisms that organisms employ to combat various environmental stress. Proteomic analysis deals with the change in protein expression and their accumulation in any organism at a particular time. Proteomic analysis is also very useful in comparison of the differential expression and accumulation of proteins in response to changing environmental conditions within the cells (Martin et al. 2001, 2003; Tyers and Mann 2003).

Protein profiles analyzed in *Miscanthus sinensis* roots indicated the involvement of the proteins related to carbohydrate and nitrogen metabolism, molecular chaperone proteins, and novel chromium-responsive proteins such as inositol monophosphatase (E.C. 3.1.3.25), nitrate reductase (E.C. 1.7.99.4), adenine phosphoribosyl transferase (E.C. 2.4.2.7), formate dehydrogenase (E.C. 1.2.1.2.), and a putative dihydrolipoamide dehydrogenase (E.C. 1.8.1.4). The analysis was done by two-dimensional gel electrophoresis (2D) and 36 differential protein spots were detected that were further analyzed by MALDI-TOFF. The modulated expression of these proteins is suggesting that Cr toxicity is linked to heavy metal tolerance and senescence pathways, and associated with altered vacuole sequestration, nitrogen metabolism, and lipid peroxidation (Sharmin et al. 2012).

There are very few studies available on the proteomic analysis of plants exposed to Cr stress. Wang et al. (2020) performed proteomic analysis on the maize leaves in time-dependent manner (1 hour, 6 hours, and 24 hours) of Cr exposure. In the study performed by Wang et al. (2020), they had got 1200 proteins on two-dimensional gel electrophoresis (2D). When they compared the protein bands in control and treated leaves, they got 60 proteins were differentially expressed in treated and non-treated samples. Most of the proteins that were differentially expressed were mainly involved in reactive oxygen species detoxification pathways, chloroplast organization and photosynthesis-related proteins, protein synthesis and folding, DNA damage response, cytoskeleton functions, and the proteins involved in posttranscriptional processing of mRNA and tRNA. Along with these proteins, some novel proteins were also found to be expressed in Cr-treated leaf samples. However, the novel proteins could not been characterized but thought to be important in Cr detoxification.

Another group working on rice (Zeng et al. 2010) performed proteomic analysis in two contrasting rice cultivars differing with each other in terms of Cr tolerance and accumulation. Cr accumulation and proteomic response of rice against Cr stress was genotype and dose-dependent and was found to be tissue specific. In their study, they identified the proteins that were involved in cell wall synthesis, energy production, proteins related to primary metabolism, electron transport system, and the proteins that were involved in detoxification of Cr-stress. There were 64 proteins differentially expressed during Cr stress in rice. However, as stated previously, plant proteomic modifications in response to Cr are still relatively unknown and studies are still scarce, indicating the potential for this field of knowledge in this regard.

6.6 Phytoremediation of Cr Metal by Plants

6.6.1 Phytoremediation Approach for Cr Detoxification

To understand the phytoremediation process for metal detoxification, several hyper-accumulator and tolerant plants have been identified in past couple of decades. Almost 500 plant species possessed by 45 families have been recognized up till now. Numerous hyper-accumulator plants detoxify metals into lesser toxic and immobilized phases (Mahajan and Kaushal 2018). Cr hyper-accumulators can accumulate and tolerate high content of Cr by their sequestration into vacuole through the metal binding ligands, for example peptides, amino acids, and organic acids. The Cr hyper-accumulation includes organic-acid-associated rhizospheric mobilization of metal, uptake by roots using different transporters, afterward root to shoot translocation of metal through xylem loading, and at last vacuolar sequestration to detoxify metal via chelation (Lu et al. 2013). Application of hyper-accumulator plants for contaminated soil and wastewater is an eco-friendly and cost-effective process to resist the toxicity against Cr stress. *Lonicera japonica* has been discovered as a Cr hyper-accumulator plant where high production of anthocyanin, oxalic acid, and carotene can enhance the Cr tolerance (Meng et al. 2019). Kyzioł-Komosińska et al. (2018) studied the efficient bio-absorption potential of macrophyte *Callitriche cophocarpa* in removal of Cr, especially from industrial wastewater. Eze et al. (2018) revealed the high accumulation of Cr only in roots of cowpea when compared to groundnut suggesting it as superior choice for remediation of Cr-polluted soils. Similarly, *Colocasia esculenta* exhibited their potential in removing CrVI from wetlands wastewater by higher accumulation of metal in the plant roots (Men and Ghazi 2018). *Spirodela polyrrhiza* (L.) Schleid was discovered to be an efficient Cr hyper-accumulator as *S. polyrrhiza* showed higher accumulation of Cr in root after seven days of exposure of tannery effluent (Singh and Malaviya 2019). Increased level of malondialdehyde (MDA), antioxidant enzymes, and photosynthetic pigments upon Cr exposure demonstrates the Cr tolerance in vetiver (*Chrysopogon zizanioides* (L.) Roberty) (Rajendran et al. 2019). Efficient Cr accumulation was found in the aerial part as well as roots of *Origanum vulgare* during outdoor pot experiment using soil contaminated with Cr (Levizou et al. 2019). Ram et al. (2019) revealed the phtyoremediation ability of the hybrid Napier grass (*Pennisetum americanus* L. × *Pennisetum purpureum* Schumach) cultivated in Cr-polluted soils. Enhanced antioxidant enzyme activities such as CAT, SOD, and POD exhibit the strong tolerance of hybrid Napier grass against at high CrVI exposure. Due to its high bioaccumulation, tolerance, and transportation index, *Cassia tora* from Sukinda chromite mining areas was suggested to be an efficient phytoremediator of Cr and could be utilized in the phytostabilization program to minimize the Cr toxicity (Patra et al. 2021). Mohanty and Patra (2020) analyzed the phytoremediation potency of profuse native weeds emerging in the South Kaliapani chromite mining region of Odisha, India. The roots of *Diectomis fastigiata* and shoots of *Vernonia cinerea* displayed the maximum Cr concentrations with 2371 $mg\,kg^{-1}$ dry matter and 5500 $mg\,kg^{-1}$ dry matter, respectively, suggesting their Cr phytoremediation potential. Augustynowicz et al. (2020) evaluated the ability of *C. cophocarpa* Sendtn as Cr hyperaccumulator where they added plants in watershed with slits highly contaminated by tannery. Phytoremediation potential of *Cirsium vulgare* medicinal plant has been reported for the remediation of soils contaminated with Cr (Dökmeci and Adiloğlu 2020).

6.6.2 Other Strategies Involved in Cr Remediation

A Cr remediation method was discovered by Wang et al. (2020) using iron-biochar nanocomplex to immobilize the bioavailable metal fraction and uptake by *Leersia hexandra* (Cr hyperaccumulator) in accordance with the help of metal-tolerant microbial consortium (MC). Roots and aerial part of *L. hexandra* could accumulate 485.1–785.0 $mg\,kg^{-1}$ and 147.5–297.2 $mg\,kg^{-1}$ Cr, respectively. Microbial consortium suggested enhancing the Cr translocation to the aerial part that could advance its phytoremediation capability. *Eichhornia* sp. and *Pistia* sp. (two free-floating macrophytes) was reported to be applied for CrVI remediation of water bodies because of their tolerance by efficient bioaccumulation and antioxidant enzyme activity (Mondal and Nayek 2020). Elevated antioxidant activities as well as proline content could improve the Cr bioaccumulation ability of *Gomphrena celosioides* (Adejumo et al. 2019). Root tissues of *Melia azedarach* L. has shown elevated accumulation intensity for Cr and Cd due to amplified concentration of GSH, proline, and soluble protein upon metal stress (Yan et al. 2020b). Enhanced antioxidant enzyme (SOD, CAT,and GR) activities along with higher Cr bioaccumulation property of *Calotropis procera*, make this plant useful for bioremediation of Cr contaminated soil (Usman et al. 2020).

6.6.3 Phytostabilization/Phytoextraction for Cr Decontamination

Numerous Plants have been studied for their phytoextraction and phytostabilization potential for detoxification of metals. Great Cr phytostabilization ability of *Conocarpus erectus* has been identified in metal polluted soil (Tauqeer et al. 2019). Patra et al. (2021) demonstrated the *Sesbania sesban* as efficient phytostabilizer for Cr decontamination in chromite overloaded mining areas. Results exhibited the high antioxidant enzyme activity along with high accumulation ability of Cr as well as high tolerance and transportation index in *S. sesban* in comparison to *Brachiaria mutica* upon Cr exposure. It has been reported that *Mesembryanthemum crystallinum* L. shows significant potential of phytoremediation resulting from distinctive growth capacity in polluted substrates as well as suggested to be appropriate for degenerated soils bio-reclamation. This report suggested that *M. crystallinum* is particularly useful in proficient phytoextraction of chromate-contaminated soils (Śliwa-Cebula et al. 2020).

6.7 Conclusion

This review illustrates that omics approaches including transcriptomics, metabolomics, and next-generation sequencing platform-guided miRNAomics are promising platforms for unraveling the understanding of the Cr-induced toxicity responses in plants. Omics-driven strategies are reviewed to demonstrate the Cr-responsive genes, metabolites, and Cr-associated metabolic pathways as well as miRNAs, which may possibly assist plants in tolerance and detoxification of Cr. Phytoremediation of Cr-polluted soils is efficient, economical, and exponentially emerging area of study. Awareness regarding appropriate native plants for Cr phytoremediation is especially inadequate, which has to be studied further. Thus, detailed information of genes involved in uptake, transport, and sequestration processes is essential for developing transgenic plants, which could accumulate Cr to advance the process of decontamination and remediation of soil.

References

Adejumo, S.A., Tiwari, S., Thul, S., and Sarangi, B.K. (2019). Evaluation of lead and chromium tolerance and accumulation level in *Gomphrena celosoides*: a novel metal accumulator from lead acid battery waste contaminated site in Nigeria. *Int. J. Phytoremediation* 21: 1341–1355.

Agency for Toxic Substances and Disease Registry. (2019). Substance priority list. https://www.atsdr.cdc.gov/spl/index.html.

Augustynowicz, J., Sitek, E., Bryniarski, T. et al. (2020). The use of *Callitriche cophocarpa* Sendtn. for the reclamation of Cr-contaminated freshwater habitat: benefits and limitations. *Environ. Sci. Pollut. Res.* 27: 25510–25522.

Barros, M.A.S.D., Zola, A.S., Arroyo, P.A. et al. (2006). Chromium uptake from tricomponent solution in zeolite fixed bed. *Adsorption* 12 (4): 239–248.

Barceloux, D.G. (1999). Chromium. *Clin Toxicol.* 37 (2): 173–194.

Beaudoin, N., Serizet, C., Gosti, F., and Giraudat, J. (2000). Interactions between abscisic acid and ethylene signaling cascades. *Plant Cell* 12: 1103–1115.

Budak, F., Zaimoglu, Z., and Basci, N. (2011). Uptake and translocation of hexavalent chromium by selected species of ornamental plants. *Polish J. Environ. Stud.* 20: 857–862.

Cheng, H., Zhou, T., Li, Q. et al. (2014). Anthropogenic chromium emissions in China from 1990 to 2009. *PloS One* 9 (2): e87753.

da Conceicao Gomes, M.A., Hauser-Davis, R.A., Suzuki, M.S., and Vitoria, A.P. (2017). Plant chromium uptake and transport, physiological effects and recent advances in molecular investigations. *Ecotoxicol. Environ. Saf.* 140: 55–64.

Das, R. and Pandey, G.K. (2010). Expressional analysis and role of calcium regulated kinases in abiotic stress signaling. *Curr. Genomics* 11: 2–13.

Deng, Y., Wang, M., Tian, T. et al. (2019). The effect of hexavalent chromium on the incidence and mortality of human cancers: a meta-analysis based on published epidemiological cohort studies. *Front. Oncol.* 9: 24.

Dökmeci, A.H. and Adiloğlu, S. (2020). The phytoremediation of chromium from soil using *Cirsium Vulgare* and the health effects. *Biosci. Biotechnol. Res. Asia* 17: 535–541.

Dubey, S., Misra, P., Dwivedi, S. et al. (2010). Transcriptomic and metabolomic shifts in rice roots in response to Cr (VI) stress. *BMC Genom.* 11: 1–9.

EFSA (2014). Panel on Contaminants in the Food Chain (CONTAM). Scientific opinion on the risks to public health related to the presence of chromium in food and drinking water. *EFSA J.* 12: 3595.

Eleftheriou, E.P., Adamakis, I.-D.S., and Melissa, P. (2012). Effects of hexavalent chromium on microtubule organization, ER distribution and callose deposition in root tip cells of*Allium cepa* L. *Protoplasma* 249: 401–416.

Eze, C.N., Odoh, C.K., Eze, E.A. et al. (2018). Chromium (III) and its effects on soil microbial activities and phytoremediation potentials of*Arachis hypogea* and *Vigna unguiculata*. *Afr. J. Biotechnol.* 17: 1207–1214.

Gajalakshmi, S., Iswarya, V., Ashwini, R. et al. (2012). Evaluation of heavy metals in medicinal plants growing in Vellore District. *Eur. J. Exp. Biol.* 2 (5): 1457–1461.

Ghassemian, M., Nambara, E., Cutler, S. et al. (2000). Regulation of abscisic acid signaling by the ethylene response pathway in Arabidopsis. *Plant Cell* 12: 1117–1126.

Hayat, S., Khalique, G., Irfan, M. et al. (2012). Physiological changes induced by chromium stress in plants: an overview. *Protoplasma* 249 (3): 599–611.

Homa, D., Haile, E., and Washe, A.P. (2016). Determination of spatial chromium contamination of the environment around industrial zones. *Int. J. Anal. Chem.* 2016: 7214932.

Gonzalez Ibarra, A.A., Wrobel, K., Yanez Barrientos, E. et al. (2017). Changes of metabolomic profile in *Helianthus annuus* under exposure to chromium (VI) studied by capHPLC-ESI-QTOF-MS and MS/MS. *J. Analy. Methods in Chem.* 2017: 3568621.

Jamla, M., Khare, T., Joshi, S. et al. (2021). Omics approaches for understanding heavy metal responses and tolerance in plants. *Curr. Plant Biol.* 27: 100213.

Kamaludeen, S.P.B., Arunkumar, K., Avudainayagam, S., and Ramasamy, K. (2003). Bioremediation of chromium contaminated environments. *Indian J. Exp. Biol.* 41 (9): 972e985.

Kinuthia, G.K., Ngure, V., Beti, D. et al. (2020). Levels of heavy metals in wastewater and soil samples from open drainage channels in Nairobi, Kenya: community health implication. *Sci. Rep.* 10 (1): 1–13.

Kyzioł-Komosińska, J., Augustynowicz, J., Lasek, W. et al. (2018). *Callitriche cophocarpa* biomass as a potential low-cost biosorbent for trivalent chromium. *J. Environ. Manage.* 214: 295–304.

Leshem, Y., Seri, L., and Levine, A. (2007). Induction of phosphatidylinositol 3-kinase-mediated endocytosis by salt stress leads to intracellular production of reactive oxygen species and salt tolerance. *Plant J.* 51 (2): 185–197.

Levizou, E., Zanni, A.A., and Antoniadis, V. (2019). Varying concentrations of soil chromium (VI) for the exploration of tolerance thresholds and phytoremediation potential of the oregano (*Origanum vulgare*). *Environ. Sci. Pollut. Res.* 26: 14–23.

Liu, J., Duan, C.Q., Zhang, X.H. et al. (2011). Characteristics of chromium (III) uptake in hyperaccumulator *Leersia hexandra* Swartz. *Environ. Exp. Bot.* 74: 122–126.

Liu, W., Xu, L., Wang, Y. et al. (2015). Transcriptome-wide analysis of chromium-stress responsive microRNAs to explore miRNA-mediated regulatory networks in radish (*Raphanus sativus* L.). *Sci. Rep.* 5: 1–7.

López-Pérez, L., del Carmen Martínez-Ballesta, M., Maurel, C., and Carvajal, M. (2009). Changes in plasma membrane lipids, aquaporins and proton pump of broccoli roots, as an adaptation mechanism to salinity. *Phytochemistry* 70: 492–500.

Lu, L., Tian, S., Zhang, J., Yang, X., Labavitch, J.M., Webb, S.M, Latimer, M., Brown, P.H. (2013). Efficient xylem transport and phloem remobilization of Zn in the hyperaccumulator plant species *Sedum alfredii*. *New Phytol.* 198: 721–31https://doi.org/10.1111/nph.12168

Mahajan, P. and Kaushal, J. (2018). Role of phytoremediation in reducing cadmium toxicity in soil and water. *J. Toxicol.* 23: 2018. https://doi.org/10.1155/2018/4864365.

Martin, S.A.M., Cash, P., Blaney, S., and Houlihan, D.F. (2001). Proteome analysis of rainbow trout (*Oncorhynchus mykiss*) liver proteins during short term starvation. *Fish. Physiol. Biochem.* 24: 259–270.

Martin, S.A.M., O Vilhelmsson, F Médale et al., 2003. Proteomic sensitivity to dietary manipulations in rainbow trout. *Biochim. Biophys. Acta, Proteins Proteomics* 1651, 17–29.

Meena, M., Divyanshu, K., Kumar, S. et al. (2019). Regulation of L-proline biosynthesis, signal transduction, transport, accumulation and its vital role in plants during variable environmental conditions. *Heliyon* 5: e02952.

Men, C.K. and Ghazi, R.M. (2018). Phytoremediation of chromium (VI) using *Colocasia esculenta* in laboratory scale constructed wetlands. *J. Trop. For. Sci.* 6: 45–49. https://doi.org/10.47253/jtrss.v6i1.727.

Meng, F., Gao, Y., and Feng, Q. (2019). Discovery and mechanism study of a novel chromium-accumulating plant, *Lonicera japonica* Thunb. *Environ. Sci. Pollut. Res.* 26: 13812–13817.

Mohanty, M. and Patra, H.K. (2020). Phytoassessment of in situ weed diversity for their chromium distribution pattern and accumulation indices of abundant weeds at South Kaliapani chromite mining area with their phytoremediation prospective. *Ecotoxicol. Environ. Saf.* 194: 110399. https://doi.org/10.1016/j.ecoenv.2020.110399.

Mondal, N.K. and Nayek, P. (2020). Hexavalent chromium accumulation kinetics and physiological responses exhibited by *Eichhornia* sp. and *Pistia* sp. *Int. J. Environ. Sci. Technol.* 17: 1397–1410. https://doi.org/10.1007/s13762-019-02418-z.

Monga, A., Fulke, A.B., and Dasgupta, D. (2022). Recent developments in essentiality of trivalent chromium and toxicity of hexavalent chromium: Implications on human health and remediation strategies. *J. Hazard. Mater.* 17: 100113. https://doi.org/10.1016/j.hazadv.2022.100113.

Nematshahi, N., Lahouti, M., and Ganjeali, A. (2012). Accumulation of chromium and its effect on growth of (*Allium cepa* cv. Hybrid). European. *J. Exp. Biol.* 2 (4): 969–974.

Oliveira, H. (2012). Chromium as an environmental pollutant: insights on induced plant toxicity. *J. Bot.* 2012: 1–8. https://doi.org/10.1155/2012/375843.

Owhonda, N.K., Ogali, R.E., and Ofodile, S.E. (2016). Assessment of chromium, nickel, cobalt and zinc in edible flesh of two tilapia fish species found in Bodo River, Rivers State, Nigeria. *J. Appl. Sci. Environ. Manag.* 20 (3): 560–564.

Patra, D.K., Grahacharya, A., Pradhan, C., and Patra, H.K. (2021). Phytoremediation potential of coffee pod (*Cassia tora*): an in situ approach for attenuation of chromium from overburden soil of Sukinda Chromite Mine, India. *Environ. Prog. Sustain. Energy* 40: e13510. https://doi.org/10.1002/ep.13510.

Rajendran, M., An, W.H., Li, W.C. et al. (2019). Chromium detoxification mechanism induced growth and antioxidant responses in vetiver (*Chrysopogon zizanioides* (L.) Roberty). *J. Cent. South Univ.* 26: 489–500.

Ram, B.K., Han, Y., Yang, G. et al. (2019). Effect of hexavalent chromium [Cr (VI)] on phytoremediation potential and biochemical response of hybrid Napier grass with and without EDTA application. *Plants* 8: 515.

Sehrawat, G., Singh, R., and Kaushik, A. (2021). Tolerance of three ornamental plant species to chromium contamination in soil and their potential for phytoextraction and phytostabilization of the toxic metal. *Curr. World Environ.* 16: 386.

Shahid, M., Shamshad, S., Rafiq, M. et al. (2017). Chromium speciation, bioavailability, uptake, toxicity and detoxification in soil-plant system: a review. *Chemosphere* 178: 513–533.

Shanker, A.K., Djanaguiraman, M., Sudhagar, R. et al. (2004a). Differential antioxidative response of ascorbate glutathione pathway enzymes and metabolites to chromium speciation stress in green gram (*Vigna radiata* (L.) R. Wilczek. cv CO 4) roots. *Plant Sci.* 166: 1035–1043.

Shanker, A.K., Djanaguiraman, M., Sudhagar, R. et al. (2004b). Expression of metallothionein 3-like protein mRNA in sorghum cultivars under chromium (VI) stress. *Curr. Sci.* 86 (7): 901–901.

Sharma, P., Jha, A.B., Dubey, R.S., and Pessarakli, M. (2012). Reactive oxygen species, oxidative damage, and antioxidative defense mechanism in plants under stressful conditions. *J. Bot.* 2012: https://doi.org/10.1155/2012/217037.

Sharma, P., Singh, S.P., Parakh, S.K., and Tong, Y.W. (2022). Health hazards of hexavalent chromium (Cr (VI)) and its microbial reduction. *Bioengineered* 13: 4923–4938.

Sharmin, S.A., Alam, I., Kim, K.H. et al. (2012). Chromium-induced physiological and proteomic alterations in roots of *Miscanthus sinensis*. *Plant Sci. 187*: 113–126.

Singh, A. and Malaviya, P. (2019). Chromium phytoaccumulation and its impact on growth and photosynthetic pigments of *Spirodela polyrrhiza* (L.) Schleid. on exposure to tannery effluent. *J. Environ. Sustain.* 2: 157–166.

Singh, H.P., Mahajan, P., Kaur, S. et al. (2013). Chromium toxicity and tolerance in plants. *Environ. Chem. Lett.* 11 (3): 229–254.

Singh, D., Sharma, N.L., Singh, C.K. et al. (2021). Chromium (VI)-induced alterations in physio-chemical parameters, yield, and yield characteristics in two cultivars of mungbean (*Vigna radiata* L.). *Front. Plant Sci.* 2021: 2059.

Sinha, V., Pakshirajan, K., and Chaturvedi, R. (2018). Chromium tolerance, bioaccumulation and localization in plants: an overview. *J. Environ. Manage.* 206: 715–730.

Śliwa-Cebula, M., Kaszycki, P., Kaczmarczyk, A., Nosek, M., Lis-Krzyścin, A., and Miszalski, Z. (2020). The common ice plant (*Mesembryanthemum crystallinum* L.) – phytoremediation potential for cadmium and chromate-contaminated soils. *Plants* 9:1230. https://doi.org/10.3390/plants9091230

Srivastava, D., Verma, G., Chauhan, A.S. et al. (2019). Rice (*Oryza sativa* L.) tau class glutathione *S*-transferase (OsGSTU30) overexpression in *Arabidopsis thaliana* modulates a regulatory network leading to heavy metal and drought stress tolerance. *Metallomics* 11 (2): 375–389.

Sundaramoorthy, P., Chidambaram, A., Ganesh, K.S. et al. (2010). Chromium stress in paddy: (i) nutrient status of paddy under chromium stress; (ii) phytoremediation of chromium by aquatic and terrestrial weeds. *C. R. Biol.* 333 (8): 597–607.

Tauqeer, H.M., Hussain, S., Abbas, F., and Iqbal, M. (2019). The potential of an energy crop "*Conocarpus erectus*" for lead phytoex-traction and phytostabilization of chromium, nickel, and cadmium: an excellent option for the management of multi-metal contaminated soils. *Ecotoxicol. Environ. Saf.* 173: 273–284.

Tian, J., Tian, L., Chen, M. et al. (2022). Low temperature affects fatty acids profiling and key synthesis genes expression patterns in *Zanthoxylum bungeanum* Maxim. *Int. J. Mol. Sci.* 23: 2319.

Trinh, N.N., Le, H.T., and Nguyen, T.P. (2019). Overexpression of the dominant negative nbexo70d1 mutantion confers tolerance to salt stress in transgenic tobacco. *Biol. Plant.* 63: 484–495.

Tyers, M. and Mann, M. (2003). From genomics to proteomics. *Nature* 422 (6928): 193–197.

Usman, K., Al Jabri, H., Abu-Dieyeh, M.H., and Alsafran, M.H. (2020). Comparative assessment of toxic metals bioaccumulation and the mechanisms of chromium (Cr) tolerance and uptake in *Calotropis procera*. *Front. Plant Sci.* 11: 883.

Waadt, R., Seller, C.A., Hsu, P.K. et al. (2022). Plant hormone regulation of abiotic stress responses. *Nat. Rev. Mol. Cell Biol.* 5: 1–5.

Wang, C., Tan, H., Li, H. et al. (2020). Mechanism study of chromium influenced soil remediated by anuptake-detoxification system using hyperaccumulator, resistant microbe consortium, and nano iron complex. *Environ. Pollut.* 257: 113558.

Whitacre, D.M. (2010). Preface. In: *Reviews of Environmental Contamination and Toxicology*, vol. 207 (ed. J.B. Metzler), vii–viii. New York: Springer.

Yan, A., Wang, Y., Tan, S.N. et al. (2020a). Phytoremediation: a promising approach for revegetation of heavy metal-polluted land. *Front. Plant Sci.* 11: 359.

Yan, X., Wang, J., Song, H. et al. (2020b). Evaluation of the phytoremediation potential of dominant plant species growing in a chromium salt–producing factory wasteland, China. *Environ. Sci. Pollut. Res.* 27: 7657–7671.

Yu, X.Z., Lin, Y.J., and Zhang, Q. (2019). Metallothioneins enhance chromium detoxification through scavenging ROS and stimulating metal chelation in *Oryza sativa*. *Chemosphere* 220: 300–313.

Zeng, F., Ali, S., Qiu, B. et al. (2010). Effects of chromium stress on the subcellular distribution and chemical form of Ca, Mg, Fe, and Zn in two rice genotypes. *J. Plant Nutr. Soil Sci.* 173: 135–148.

7

Manganese Toxicity and Tolerance in Photosynthetic Organisms and Breeding Strategy for Improving Manganese Tolerance in Crop Plants

Physiological and Omics Approach Perspectives

Daisuke Takagi

Faculty of Agriculture, Setsunan University, Hirakata, Osaka, Japan

7.1 Introduction

Manganese is an essential mineral nutrient for photosynthetic organisms and animals (Shao et al. 2017). Mn was first discovered in the soil by Scheele in 1774 (McHargue 1923). Mn is the 10th most abundant mineral element in the Earth's crust following Si, Al, Fe, Ca, Na, Mg, K, Ti, and P, although the order changes depending on sampling depth (Turekian and Wedepohl 1961; Wedepohl 1995; Haxel et al. 2002). The average Mn concentration is approximately 700 ppm in continental crust, and Mn exists largely as manganese oxide (MnO), accounting for 0.09–0.11% of the continental crust or tonalites and graywackes (Wedepohl 1995). In contrast to the Earth's crust, the surficial Mn concentration ranges from 1 to 7000 ppm in the conterminous United States of America, and the average Mn concentration is assumed to be 560–850 ppm (Shacklette et al. 1971). Mn exists naturally as Mn(II), Mn(III), and Mn(IV) in the soil, although it has various oxidation states [Mn(I), Mn(II), Mn(III), Mn(IV), Mn (VI), and Mn (VII)] in our planet (Wedepohl 1995; Broadley et al. 2012; Shao et al. 2017).

In terrestrial plants, Mn was first discovered in the ash of *Pinaceae* leaves between 1860 and 1880 (Kelley 1914); subsequently, in 1922, Mn was proven to be essential for plant growth (McHargue 1922). Terrestrial plants can only utilize the Mn(II) form in the soil and absorb it through their roots. The Mn transportation system has been actively studied but has not been completely clarified (Pittman 2005; Socha and Guerinot 2014; Shao et al. 2017). In *Oryza sativa* and *Arabidopsis thaliana,* Mn is transported from the soil to the root by natural resistance-associated macrophage protein (NRAMP), zinc-regulated transporter/iron-regulated transporter (ZRT/IRT1)-related protein (ZIP) family, and Ca^{2+}-permeable channels (Korshunova et al. 1999; Véry and Davies 2000; Pedas et al. 2008; Socha and Guerinot 2014; Shao et al. 2017). After Mn is absorbed through the root, it is transported from the epidermis to the vasculature depending on the Mn transporter family, such as cation exchanger (CAX), calcium cation exchanger (CCX), cation diffusion

Heavy Metal Toxicity and Tolerance in Plants: A Biological, Omics, and Genetic Engineering Approach,
First Edition. Edited by Mohammad Anwar Hossain, AKM Zakir Hossain, Sylvain Bourgerie, Masayuki Fujita, Om Parkash Dhankher, and Parvez Haris.

facilitator/metal tolerance protein (CDF/MTP), and P-type ATPase family proteins (Socha and Guerinot 2014; Shao et al. 2017). When Mn reaches the xylem in the root, it is transferred to the upper part of plant tissues (Broadley et al. 2012).

Manganese is a critical component of oxygenic photosynthesis in photosynthetic organisms. After Mn is absorbed through the roots, it is transported to various organelles in leaf cells, including the Golgi, vacuole, endoplasmic reticulum (ER), mitochondria, and chloroplasts *via* various transporters (Pittman 2005; Socha and Guerinot 2014; Shao et al. 2017). The physiological functions of Mn vary widely among plant cells. For example, Mn activates approximately 35 different enzymes as cofactors, such as RNA polymerase, glycosyltransferase, the tricarboxylic acid (TCA) cycle enzymes (such as malic enzyme and isocitrate dehydrogenase), shikimic acid pathway enzymes, and phosphoenolpyruvate carboxylase in C4-type photosynthesis (Burnell 1988; Broadley et al. 2012). Manganese is also directly involved in the physiological function of Mn-superoxide dismutase (SOD) in the mitochondria to detoxify reactive oxygen species (ROS) and oxalate oxidase (Broadley et al. 2012). Among the various physiological functions of Mn in plant cells, the Mn cluster in the oxygen-evolving complex (OEC) within photosystem II (PSII) is strikingly important for initiating photosynthesis to maintain autotrophic growth (Broadley et al. 2012). PSII contains four atoms of Mn per one OEC and forms a "distorted chair" structure with one atom of calcium and four atoms of oxygen that oxidize H_2O to produce O_2, H^+, and electrons (e^-) by the S-state cycle (Ferreira et al. 2004; Umena et al. 2011; Shen 2015; Suga et al. 2015; Barber 2016). To operate these Mn-related physiological reactions in plant cells, terrestrial plants require Mn concentrations of $1.0\,\mu mol\,g^{-1}$ (dry weight: D.W.) or $50\,mg\,kg^{-1}$·(D.W.) in their shoots (Kirkby 2012). Owing to the importance of Mn in physiological reactions, especially photosynthesis, the effects of Mn deficiency on plant growth and photosynthetic activities have been actively studied in various plants. According to intensive studies focused on Mn deficiency, foliar Mn concentrations of 10–20 $mg\,Mn\,kg^{-1}$ (D.W.) are known as critical thresholds that cause Mn-deficient phenotypes in terrestrial plants (Broadley et al. 2012). In addition to the quantification of leaf Mn concentrations, Mn deficiency can be easily detected in a non-invasive manner by the decrease in the maximum quantum yield of PSII (Fv/Fm) in leaves due to the hyperlocalization of Mn to thylakoid membranes (Anderson et al. 1964). These numerical data are useful for diagnosing Mn deficiency during crop production.

Although Mn is essential for plant growth, excess Mn fertilization attenuates plant growth. This phenomenon is known as Mn toxicity. Although Mn-deficient states can be estimated from certain numerical bases such as foliar Mn concentration or photosynthetic parameters, Mn-toxic phenotypes are more complicated. Indeed, the foliar Mn concentration emerging Mn toxicity cannot be standardized between terrestrial plants. This is because the critical concentration for emerging Mn-toxic symptoms varies among cultivars and species (Table 7.1; Foy et al. 1988). For example, terrestrial plants sensitive to Mn toxicity can exhibit negative physiological responses at 200 $mg\,Mn\,kg^{-1}$ (D.W.). Conversely, Mn hyperaccumulator plants, such as the *Myrtaceae* family, do not show any Mn toxicity symptoms at foliar Mn concentrations of $>10{,}000\,mg\,Mn\,kg^{-1}$ (D.W.) (Fernando et al. 2010).

In this chapter, the author briefly introduces Mn dynamics in soils and the significance of Mn toxicity in crop plants. Thereafter, the author summarizes the physiological effects of excess Mn accumulation in terrestrial plants. Based on recent advances in understanding the mechanisms of Mn toxicity, molecular targets for breeding Mn-tolerant crops are discussed for the future threats of emerging Mn toxicity in agricultural fields.

Table 7.1 The variety of threshold for emerging Mn toxic symptoms in crop plants.

Species	Broadley et al. (2012) and Edwards and Asher (1982)	Foy et al. (1988)	Löhnis (1960)	Johan and Amin (1967)	Blamey et al. (1986)	Führs et al. (2010)
Maize	200					
Lettuce		200–400				
Lespedeza		200–300				
Cassava		250				
Snapbean			300			
Pigeon pea	300					
Barley			452			165
Wheat		380–1296 (depending on cultivars)				
Soybean	600	145–1500 (depending on cultivars)				
Brassica napus			524			
Peacan		557				
Alfalfa			691			
Cucumber		600–800				
Carnation		600–800				
Cotton	750					
Muskmelon		900				
Flax			901			
Mangold			1063			
Rye			>1000			
Oat			>1200			
Coffee		1200–2000				
Sweet potato	1380					
Tomato		1500–2000				
Cotton				4000		
Rice						5500
Sunflower	5300				2205	

7.2 The Change in Mn Availability Within the Soil

Manganese exhibits interesting redox cycles in soil. Mn can exist as Mn(II), which can be utilized by terrestrial plants, and as Mn(III) or Mn(IV), which cannot be absorbed by them. Mn(II) is oxidized to Mn(III)/(IV) oxides in a microbial-dependent or microbial-independent manner (Ross and Bartlett 1981; Wada et al. 1978a, b). Microbial Mn oxidation seems to be related to their metabolism because Mn oxidation is suppressed by antibiotics, such as NaN_3 or streptomycin (Wada et al. 1978a). Several microbial Mn(II) oxidation pathways have been suggested, but the physiological importance of Mn(II) oxidation remains unclear (Tebo et al. 2004; Learman et al. 2011; Li et al. 2021). In microbial-independent reactions, Mn(II) oxidation is favorable under alkaline pH or a higher redox potential conditions (Tebo et al. 2004). In agricultural fields, Mn oxide can be observed in heavily limed soils aimed at raising soil pH or in calcareous soils where Mn deficiency is often problematic for crop production (Alloway 2008). Similar to other abiotic factors, Mn interacts with organic compounds in the soil and forms complexes with them (Tam et al. 1991). The formation of the organic-Mn complex also suppresses Mn absorption by terrestrial plants and causes Mn deficiency (Farley and Draycott 1973).

In contrast, Mn oxide reduction is also driven in a microbial-dependent and microbial-independent manner. Mn-reductive microbes require anaerobic conditions for their growth and reduce MnO_2 by using fermentation products such as acetate as a carbon source (Myers and Nealson 1988; Lovely and Phillips 1988). In a microbial-independent manner, Mn(III)/(IV) can be reduced under acidic conditions or lower redox potential conditions. That is, acidic soils, such as oxisol or waterlogged soils as in paddy fields, show higher Mn availability for terrestrial plants. Under such conditions, terrestrial plants can experience Mn toxicity (Hue et al. 2001). However, heavy rain or good drainage in acidic soils causes Mn deficiency because Mn(II) is leached from the soil (Alloway 2008). Such situations can be observed in tropical or acidic sandy soils. In addition to the pH and redox potentials, saccharides or organics, such as lignin, also contribute to the reduction of Mn(III)/Mn(IV). After severe soil dehydration, Mn reduction is accelerated, although soil drying increases the redox potential (Savant and Kibe 1969; Siman et al. 1974; Bartlett and James 1980). Makino (2001) reported that severe dehydration annihilates microorganisms within the soils and, among the exudates from microorganisms, reducing sugars such as glucose can reduce Mn oxide. This mechanism may also be applicable to an increase in soluble Mn concentration in steam-sterilized soil (Sonneveld and Voogt 1975; Blatt and van Diest 1981). Furthermore, Mn(III) can be regenerated into Mn(II) by functioning as an oxidant for lignin or aromatic litter compounds (Keiluweit et al. 2015; Jones et al. 2020; Li et al. 2021). This mechanism is an important Mn-cycling system in forests.

7.3 Why Should We Consider the Occurrence of Mn Toxicity in Plants? Possible Threats of Mn Toxicity in Agricultural Land

The effects of Mn deficiency are often considered for agricultural croplands. Mn deficiency is mainly caused by leaching from the soil or by intensive crop production using high-yielding cultivars with a large amount of chemical fertilization and by the conversion of available Mn forms into insoluble or unavailable forms depending on the soil environment (Alloway 2008). However, this problem may not apply to all agricultural situations. According to Kabata-Pendias and Wiacek (1985), the total Mn concentration considered phytotoxic is $>1500\,mg\,kg^{-1}$ in the soil. Combined with a report by Shacklette et al. (1971), excessive Mn absorption is likely to occur during agricultural crop production. A representative cause of the increase in the soil Mn concentration is Mn mining. For example, Liu et al. (2018) reported that agricultural soils around Mn mining areas contain approximately 2706–6440 mg Mn kg^{-1}, which means that Mn concentrations are 15.4–36.6 times higher than those in other agricultural fields, and the increase in soil Mn concentration is reflected in crop Mn concentrations (Liu et al. 2018).

Even without Mn fertilization or Mn mining activity, the soil Mn concentration can be increased by natural and anthropogenic emissions (Pacyna and Pacyna 2001). Natural emissions represent wind-borne-rock/soil erosion, wild forest fires, and sea salt spray (Nriagu 1989), and median Mn emissions had been estimated to reach 317×10^3 t $year^{-1}$ as of the mid-1990s. Conversely, Mn emissions from anthropogenic activities include coal and gasoline combustion, metal production, and the steel industry (Pacyna and Pacyna 2001). The emitted Mn is returned to the soil by rainfall. Although anthropogenic Mn emissions are estimated to be much lower (11×10^3 t $year^{-1}$ as of the mid-1990s) than that by natural emissions (Pacyna and Pacyna 2001), they are capable of enriching soil Mn concentrations to a greater extent. Herndon et al. (2011) showed that the soil Mn concentration increases close to the steel plant source, reaching from 900 to 14,400 mg kg^{-1} in the uppermost soil. Owing to the environmental regulation of industrial activities, atmospheric Mn deposits are gradually declining (Herndon and Brantley 2011). However, new anthropogenic problems related to Mn have also been highlighted. McMahon et al. (2019) reported that, in the United States, the Mn(II) concentration in groundwater is high in shallow water tables, where the land surface–soil–aquifer connection is highly effective. The increasing groundwater Mn concentration is due to the higher soil-derived dissolved organic carbon and the lower O_2 and NO_3 concentrations, where microbial Mn reduction activity is increased (McMahon et al. 2019). Interestingly, the dissolved organic carbon and Mn concentrations tended to increase with decreasing distance from rivers (McMahon et al. 2019). In addition, a significant negative relationship between groundwater pH and Mn concentration has also been detected (McMahon et al. 2019). An increase in groundwater Mn concentrations has been observed in other countries (Hamer et al. 2020; Hou et al. 2020).

In addition to these direct human activities, global climate change may also increase Mn availability (Fernando and Lynch 2015; Fernando et al. 2016). For example, with an increase in Earth's surface temperature, global precipitation tends to increase with an increase in the evaporation rate (Blunden and Arndt 2020). The recent precipitation pH is more acidic (pH = 5.5) because of the emissions of sulfate oxide (SO_2), nitrogen oxide (NO_χ), or other acids and alkaline dust from industrial activities (Shah et al. 2020). Acid rain provides H^+ to decrease soil pH, but the degree of acidification depends on soil nutrient conditions. In nitrogen-limited ecosystems, NO_3^- or NH_4^+-contaminated raindrops are consumed by microorganisms or terrestrial plants, whereas in nitrogen-rich ecosystems, they remain in the soil, and microbes further stimulate H^+ input through nitrification (Shah et al. 2020). Extreme precipitation also increases the risk of flooding (Tabari 2020). Prolonged waterlogging in agricultural lands can accelerate the reduction of Mn oxide within soils. However, the increase in precipitation does not occur annually, and seasonal precipitation patterns are different throughout the world (Konapala et al. 2020). Owing to these fluctuations, some parts experience severe drought depending on the season (Konapala et al. 2020) and severe soil dehydration may cause microbial annihilation, which triggers an increase in Mn availability.

From these recent studies on climate change and the circumstances of both natural and anthropogenic activities, the risk of Mn contamination in agricultural fields cannot be disregarded (Fernando and Lynch 2015). To prepare for extreme Mn contamination in agricultural fields, both the physiological effects and molecular mechanisms of Mn toxicity must be investigated. Phenotypic information is informative for the isolation of Mn-tolerant cultivars or mutants in a forward genetic manner for breeding. On the other hand, the molecular mechanisms provide breeding targets for various crops to generate new cultivars in a reverse genetic manner.

7.4 The History of Mn Toxicity

The toxic effects of Mn on crops have been recognized since the 1890s, and a cultivar variety of the Mn toxic sensitiveness has also been identified in the 1900s (Kelley 1914). Mn toxicity symptoms have been reported under acidic soil conditions in Alabama (Funchess 1919). Funchess (1919) found that nitrogen fertilizer stimulates soil acidification, and the increase in soluble Mn could cause injury, especially in leaves, which showed bleaching and abnormal leaf shape. Subsequently, McHargue (1923) suggested that Mn causes injury to leaves under acidic soil conditions and that the injury effects can be reversed by calcium carbonate fertilization. Later, important studies on Mn toxicity in cotton plants were reported in Louisiana (Neal 1937; Adams and Wear 1957). However, as deduced in the title of the Mn toxicity paper, the crinkle leaf phenotype appears to be recognized as a new disease (Neal 1937; Adams and Wear 1957). Subsequently, Neal and Lovett (1938) reported that crinkle leaves can be observed in highly acidic soils and that liming prevents the emergence of crinkle leaves (Adams and Wear 1957). They suggested that excess soluble Mn causes the leaf phenotype in cotton plants. In the 1960s, excessive water-soluble Mn was found to have toxic effects on cotton leaves (Adams and Wear 1957). Based on these studies on cotton plants, the effects of Mn toxicity could be widely investigated in various plants, focusing on phenotypic characteristics such as leaf chlorosis, growth retardation, and leaf morphological changes (Löhnis 1960; Foy 1964). In addition to these Mn-toxic phenotypes, the emergence of brownish spots is one of the major criteria for diagnosing excess Mn fertilization in agriculture (Figure 7.1). As described latter,

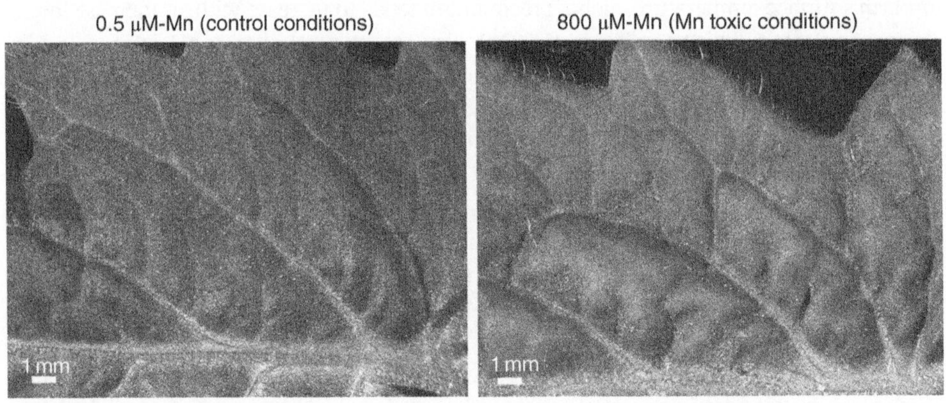

Figure 7.1 The phenotype of the Mn toxicity in tomato (*Lycopersicon esculentum* cv. Micro-Tom). The plants were hydroponically grown with the Hoagland solution being modulated Mn concentration. The leaves were harvested 30 days after their germination. Excess Mn decreased leaf chlorophyll concentration and caused to emerge brownish spots as shown in the pictures. White bars indicate a length of 1 mm. Note: The colored figure can be available in the online version.

Williams and Vlamis (1957a) used radioactive ^{54}Mn to determine that Mn accumulates patchily in leaves, and higher Mn accumulation spots become brownish necrotic spots in oat leaves under excess Mn fertilization conditions. Although growth retardation and leaf chlorosis can be found under other nutrient disorders, the emergence of brownish patchy spots is a characteristic phenotype, especially for Mn toxicity.

7.5 The Features of Mn Toxicity in Terrestrial Plants and Possible Molecular Mechanisms

Before discussing the breeding strategies for overcoming the future threat of Mn toxicity in agriculture, the detailed molecular mechanisms underlying Mn-toxic phenotypes are summarized. Significantly, the chlorosis and brownish patchy spot phenotypes could appear independently depending on the plant species. Thus, although brownish spots are an indicator of Mn toxicity in crop plants, growth retardation may not always be accompanied by the expression of brown spots in leaves under excessive Mn fertilization (Blamey et al. 1986, 2015; Horiguchi 1987; Wissemeier and Horst 1992; González et al. 1998; Chen et al. 2013; Tsunemitsu et al. 2018a). Based on this fact, Fernando et al. (2016) expressed these types of Mn toxicity as apoplastic and symplastic, respectively.

7.5.1 The Mechanisms of Emergence of Brownish Patchy Spots in Leaves: The Apoplastic Mn Toxicity

The Mn-toxic brownish patchy phenotype was reported in the 1910s (Kelley 1914), and the key molecular target responsible for this phenotype was discovered in the 1940s. Kenten and Mann (1949) found that Mn is oxidized in solutions containing H_2O_2 and extracts from horseradish. They hypothesized that Mn can be oxidized by oxidized ether-soluble peroxidase substrates, and the oxidation–reduction cycles of the substrates proceeded with Mn oxidation *in vitro*. Subsequently, they found that monohydric phenolic compounds, such as *p*-cresol, are efficient substrates for Mn oxidation in the presence of H_2O_2 and peroxidase (Kenten and Mann 1950), and at this time, intermediate phenolic oxidized compounds (phenoxyl radical; $PhO^{\cdot}$), but not the final phenolic oxidized products, are suggested to be responsible for Mn oxidation. When the oxidation rate of the intermediate phenolic oxidized compounds exceeds the reduction rate of Mn(II), oxidized phenolic compounds accumulate in *in vitro* experiments. Based on the *in vitro* results and by referring to Kelley (1914), who showed an increase in Mn_3O_4 content in leaf ash in manganiferous soil when compared to that in the control soil, Kenten and Mann (1957) hypothesized that Mn oxidation is stimulated in leaves by the above reactions, where Mn toxicity occurs. They found that Mn showed a higher valency than Mn(II) accumulated in the brownish spots of pea leaves grown under Mn-toxic conditions. Based on these results, they suggested that Mn is oxidized depending on the peroxidase system when excess Mn accumulates in the leaves (Kenten and Mann 1957). Owing to technical advances, radiographic analysis of ^{54}Mn, energy dispersive spectroscopic X-ray scanning, and micro-X-ray fluorescence techniques have revealed that Mn(II)/Mn(IV) are patchily accumulated in the apoplastic region of leaf epidermal cells or trichomes under Mn-toxic conditions, and brownish spots match the Mn distribution in the leaf epidermis (Williams and Vlamis 1957a; Horst and Marschner 1978; Memon et al. 1981; Horst 1983; Blamey et al. 1986, 2015, 2018; Fernando et al. 2016). However, Kenten and Mann (1957) also found that although tomatoes, sprouts, and barley showed severe growth inhibition and brownish spots in leaves where excess Mn was fertilized, a higher valency of Mn was barely detected in leaves. This indicates that the brownish spots might be related to Mn oxidation; however, the brown color is not directly due to the higher valency of Mn or Mn oxide. The same conclusion was drawn for cowpea leaves (Wissemeier and Horst 1992).

Before the brownish spots emerged, the Mn concentration in the apoplast increased. Sasaki et al. (2011) revealed that impairment of OsYSL6 (a transporter localized in the plasma membrane to transport Mn from the apoplast to the symplast) stimulates the emergence of brownish spots in rice leaves. In many crop plants, the apoplastic Mn concentration increases almost linearly with whole-leaf Mn concentration (Horst et al. 1999; Fecht-Christoffers et al. 2003a, b, 2006; Führs et al. 2009, 2010). Subsequently, an increase in Mn concentration stimulates the expression of several class III peroxidases that exhibit NADH-dependent H_2O_2 generation and guaiacol-dependent H_2O_2 consumption activities (Horiguchi 1988; Horst et al. 1999; Iwasaki et al. 2002; Fecht-Christoffers et al. 2003a, b, 2006; Führs et al. 2009, 2010). Simultaneously, the total apoplastic phenolic concentration tends to increase in this situation, which is related to the change in its synthesis activity (Burnell 1988; Fecht-Christoffers et al. 2006). Interestingly, class III peroxidase activity was altered by specific phenolic compounds in a concentration-dependent manner. Führs et al. (2008, 2009) found that the concentration of benzoic acid (which stimulates NADH-dependent H_2O_2 generation in class III peroxidases) is increased, whereas ferulic acid (which suppresses peroxidase activity) is decreased in the apoplast of cowpea leaves under Mn toxicity. In concert with the increase in peroxidase protein content and the change in phenolic composition, the generation of both H_2O_2 and $PhO^\cdot$ is stimulated (Fecht-Christoffers et al. 2006). $PhO^\cdot$ can be reduced by ascorbate or monodehydroascorbate in addition to Mn (Takahama 2004). As a result, the redox state of the apoplast shifts to a more oxidizing environment. In fact, ascorbate concentration in the apoplastic region decreases owing to its oxidation to dehydroascorbate with increasing Mn concentration in the leaf tissue (Takahama and Oniki 1992; Fecht-Christoffers et al. 2003b). As long as Mn(II), ascorbate, or other antioxidants remain, H_2O_2 and $PhO^\cdot$ are re-reduced; however, the more oxidized form of Mn accumulates (Kenten and Mann 1949). The lack of a reduction substrate for $PhO^\cdot$ initiates the accumulation of oxidized phenolics.

Several possibilities for increasing Mn concentration in the apoplast can be considered. For example, blockage or positive external pumping of Mn from the cytoplasm is known as heavy-metal detoxification (Hall 2002). External transport from the cytosol to the apoplast is achieved by ER-localized ER-type calcium ATPase (ECA1), Golgi-apparatus-localized ECA3, and metal-tolerant protein 11 (MTP11) in *A. thaliana*. Rice plants secrete excess Mn into the apoplast through MTP11 localized in the trans-Golgi network (Wu et al. 2002; Peiter et al. 2007; Mills et al. 2008; Farthing et al. 2017; Shao et al. 2017; Tsunemitsu et al. 2018a; Li et al. 2019). In addition to these protective mechanisms for Mn secretion, the increase in apoplast Mn concentration might be related to Mn demand in the apoplast.

Kukavica et al. (2009) reported the presence of Mn-type superoxide dismutase (Mn-SOD) isoform in root cell walls. This result indicates that the apoplast requires soluble Mn for isoform synthesis. Although cell-wall-localized Mn-SOD detoxifies O_2^- produced by PhO$^{\cdot}$ in the apoplastic region (Kukavica et al. 2009), the contribution of cell wall Mn-SOD to alleviating oxidative stress under Mn toxicity is unknown.

Before the formation of brownish spots, callose synthesis and accumulation are locally activated (Wissemeier and Horst 1987, 1992). Callose is a linear polysaccharide formed from glucose through β-1,3-glycoside bonds (Nedukha 2015). The molecular mechanisms that induce callose formation with excess Mn remain unknown. As well as Mn toxicity, aluminum toxicity stimulates callose formation. Al triggers callose formation by accessing the negatively charged membrane surfaces. Al causes membrane rigidity and lipid peroxidation, and simultaneously Al stimulates ROS production by disrupting Ca^{2+} homeostasis (Cakmak and Horst 1991; Ma 2007). An increase in Ca^{2+} activates 1,3-β-glucansynthase and callose is synthesized (Ma 2007; Nedukha 2015). Although Mn^{3+} is a powerful oxidant that stimulates lipid peroxidation, the relationship between Mn^{3+} and lipid peroxidation remains elusive under Mn toxic conditions (Horst 1988; Millaleo et al. 2010). However, ROS concentration is increased in the apoplast region under Mn-toxic conditions (Fecht-Christoffers et al. 2006); therefore, indirect oxidative stress could affect callose synthesis. Alternatively, Mn^{2+}-induced plasma membrane depolarization or Mn^{2+}-dependent 1,3-β-glucansynthase activation instead of Ca^{2+} is also considered (Horst 1988).

The mechanism of callose synthesis under Mn toxicity may resemble a protective response against pathogenic infections. When a pathogen or herbivore physically attacks leaves, cell wall degradation occurs. To withstand biotic infections, terrestrial plants recognize cell wall degradation products and reinforce cell wall robustness (Bellincampi et al. 2014; Lorrai and Ferrari 2021). Callose synthesis is activated by increased ROS concentrations, phenolic concentrations, and class III peroxidases (Bellincampi et al. 2014). From this perspective, brownish spot formation mechanisms might resemble papillae formation mechanisms depending on the damage-associated molecular pattern (DAMP). Several studies have found that the quantities of resistance-related proteins against pathogenic attacks, such as pathogenesis-related (PR) proteins, increase with excess accumulation of Mn in the leaves of tobacco, sunflower, *Stylosanthes* (Edreva 1990; Jung et al. 1995; Liu et al. 2019), and roots of soybean and tomato (Chen et al. 2016; Ceballos-Laita et al. 2018). Moreover, Fecht-Christoffers et al. (2003a) identified that the PR protein content is increased in the apoplast region due to excess Mn accumulation in the leaves of cowpea. Proteomics analyses using roots of tomatoes and *Stylosanthes* also revealed significant changes in β-1,3-glucanase required for antifungal activity and other proteins related to cell wall modulation (Ceballos-Laita et al. 2018; Liu et al. 2019). Interestingly, Eskandari et al. (2020) found that excess Mn-dependent callose formation successfully increased the resistance to *Colletotrichum lagenarium* in cucumber leaves. It is unclear whether these systems are protective mechanisms and whether they are adequate to counteract the toxic effects of Mn. One hypothesis is that blockage or containment of excess Mn might avoid excess Mn accumulation in the symplastic region of other cells. From these considerations, not only studies of physiological responses but also investigation of pathogenic tolerance related to excess Mn might be helpful in elucidating the mechanisms of apoplastic Mn toxicity in leaves.

The accumulation of oxidized phenolics would be the final step in apoplastic Mn toxicity. The ordinal time course for the emergence of visible symptoms of Mn toxicity is as follows: the emergence of brownish spots in leaves, the occurrence of foliar chlorosis, and growth retardation or withering (Horst 1983, 1988; Blamey et al. 1986). As mentioned above, brownish spots can appear on leaf epidermis without growth retardation (Horst 1983; Blamey et al. 1986; Wissemeier and Horst 1992; Fecht-Christoffers et al. 2006). In contrast, some plant species, such as alfalfa and maple, skip the formation of brownish spots and proceed with chlorosis symptoms as Mn toxicity symptoms (Foy 1964; Fernando and Lynch 2015). Thus, although the emergence of the brownish spots is termed apoplastic Mn toxicity, this mechanism might be an important protective mechanism to inactivate Mn toxicity and be the consequence of anchoring excess Mn to reduce the Mn concentration in the symplastic region, where symplastic Mn toxicity occurs.

7.5.2 The Mechanisms of Foliar Chlorosis Under Excess Mn: Symplastic Mn Toxicity

Compared to the mechanisms of apoplastic Mn toxicity, the molecular mechanisms of symplastic Mn toxicity remain elusive. The visible symplastic Mn toxicity phenotype is a decrease in the chlorophyll content of leaves (Löhnis 1960; Foy 1964; Horst 1983; Csatorday et al. 1984; Blamey et al. 1986; Clairmont et al. 1986; Panda et al. 1987; González et al. 1998; Houtz et al. 1988; Nable et al. 1988; Wilkinson and Ohki 1988; Macfie and Taylor 1992; González and Lynch 1997; Kitao et al. 1997; Lidon and Teixeira 2000a, b; Shenker et al. 2004; Clair et al. 2005; Lei et al. 2007; Millaleo et al. 2013; Fernando and Lynch 2015; Takagi et al. 2021). Chlorosis is triggered by an increase in the Mn concentration in the cytosol. In fact, when tonoplast-localized Mn transporters (MTP8.1, MTP 8.2, CAX2), which contribute to the detoxification of Mn, are impaired, the leaf chlorophyll content is lowered (Hirschi et al. 2000; Chen et al. 2013; Eroglu et al. 2016, 2017; Takemoto et al. 2017). Furthermore, inhibition of Mn sequestration from the cytosol to the apoplast region in an ER- or Golgi-dependent manner also lowers the leaf chlorophyll content (Wu et al. 2002; Peiter et al. 2007; Mills et al. 2008; Farthing et al. 2017; Tsunemitsu et al. 2018a). However, a decrease in chlorophyll content can be observed under other stress conditions such as N deficiency, P toxicity, or high light intensity (Makino et al. 1994; Takagi et al. 2019, 2020). Therefore, compared to apoplastic Mn toxicity, the characteristics of symplastic Mn toxicity are less evident.

A decrease in the chlorophyll concentration has been observed in both terrestrial plants and cyanobacteria under excess Mn incorporation. The cyanobacterium *Synechocystis* sp. Strain PCC 6803 is ordinally cultured in a BG11 medium containing 9 μM Mn. However, when the Mn concentration reaches >90 μM, growth is inhibited, and chlorophyll concentration decreases (Zorina et al. 2016; Brandenburg et al. 2017; Gandini et al. 2017). This phenomenon is known as Mn toxicity in cyanobacteria. To avoid excess Mn uptake, *Synechocystis* PCC 6803 possesses regulatory mechanisms to maintain Mn homeostasis within the cell, depending on a histidine kinase self-phosphorylating by capturing Mn (ManS), and a negative transcriptional regulator of the *mntCAB* operon encoding an ABC-type Mn transporter (ManR) (Yamaguchi et al. 2002). This system was activated under sufficient Mn conditions to suppress the Mn uptake. In *ΔmanS* and *ΔmanR* mutants, Mn uptake was not suppressed, and Mn concentration increased within the cell, resulting in growth inhibition and a decrease in chlorophyll content (Zorina et al. 2016). In addition, Gandini et al. (2017) and Brandenburg et al. (2017) reported that thylakoid and plasma membrane-localized protein involved in Mn export to the periplasm or thylakoid lumen [Mn export protein (Mnx), photosynthesis affected mutant 71 (SynPAM71)] plays a critical role in avoiding Mn toxicity in *Synechocystis* PCC 6803, and these defective mutants show high sensitivity to Mn toxicity. However, the mechanisms underlying Mn toxicity in cyanobacteria remain unclear. Csatorday et al. (1984) reported that excess Mn concentration increased Mg-protoporphyrin concentration in the cyanobacterium *Anacystis nidulans*. Based on these results, many studies related to Mn toxicity have hypothesized that excess Mn inhibits the chlorophyll biosynthesis pathway, which could be due to the mismetallation of Mg-protoporphyrin IX monomethyl ester cyclase (Eisenhut 2020). However, molecular evidence to confirm this mismetallation is lacking. In addition to the decrease in chlorophyll concentration, *Synechocystis* PCC 6803 cultured under excess Mn concentrations showed lower electron transport activity in PSII than those cultured under control Mn concentrations (Brandenburg et al. 2017). This is also the case for Mn-hyperaccumulating mutants (*ΔmanS, ΔmanR, Δmnx,* and *Δsynpam71*) (Zorina et al. 2016; Brandenburg et al. 2017; Gandini et al. 2017). Furthermore, hyperMn-accumulating mutants show higher sensitivity to photoinhibition and slower recovery of PSII activity from its photoinhibition (Zorina et al. 2016; Gandini et al. 2017). Future research is needed to investigate whether the delayed recovery of PSII activity is caused by inhibition of chlorophyll biosynthesis or other inhibitory effects of excess Mn accumulation. The effects of excess Mn on chlorophyll biosynthesis have also been reported in terrestrial plants. Clairmont et al. (1986) reported that tobacco callus showed lower chlorophyll concentrations when grown under excess Mn, and the accumulation of protoporphyrin IX was detected. This situation differs from that observed in *A. nidulans*. Wilkinson and Ohki (1988) reported that excess Mn inhibits isoprenoid synthesis in wheat leaves. However, the exact inhibitory mechanisms remain unclear. Importantly, chloroplast Mn concentration in rice plants did not increase linearly with an increase in cellular Mn concentrations (Führs et al. 2010). Therefore, the same situation for cyanobacterial Mn toxicity could appear as a later step in Mn toxicity in terrestrial plants.

Alternative mechanisms for the decrease in leaf chlorophyll concentrations have been widely discussed. One possibility is that the suppression of photosynthesis can cause a decrease in the chlorophyll concentration because of photoinhibition by ROS (Fernando and Lynch 2015). Nable et al. (1988) and Houtz et al. (1988) reported that leaf photosynthetic activity decreases prior to the occurrence of chlorosis in tobacco leaves. These studies found that in the early phase of Mn toxicity, the CO_2 assimilation reaction by Rubisco or other Calvin–Benson cycle enzymes, rather than photosynthetic electron transport reactions on the thylakoid membrane, are targets of Mn toxicity (Nable et al. 1988; Houtz et al. 1988; Kitao et al. 1997). In detail, Nable et al. (1988) suggested that Mn toxicity causes oxidative damage to the Calvin–Benson cycle enzymes, and Houtz et al. (1988) suggested that Mn toxicity decreases Rubisco activity by mismetallation instead of Mg^{2+}. On the other hand, stomata dysfunction and the suppression of stomatal conductance for CO_2 absorption have also been proposed as causes of the decrease in photosynthesis (Ohki 1985; Suresh et al. 1987; Santos et al. 2017). In contrast to the suppression of CO_2 assimilation reactions, direct effects on thylakoid membrane proteins have also been suggested. Millaleo et al. (2013) reported that Mn toxicity decreased the PSI-core subunit content (PsaA and PsaB) in *A. thaliana*. They suggested that because PSI is a high-Fe-demand protein complex, the change in the Fe/Mn ratio disturbs PSI synthesis through Fe deficiency. Panda et al. (1987) suggested that free Mn^{2+} stimulates $^1O_2$ production at the donor side of PSII in chloroplasts isolated from wheat leaves and that $^1O_2$ stimulates thylakoid membrane lipid peroxidation. In this *in vitro* study, the chlorophyll fluorescence at 685 nm decreased in the presence of Mn^{2+}. A recent proteomics analysis also found that PSI subunits (PsaA and PsaC), PSII subunit (PsbC), and some light-harvesting complex proteins were decreased in leaves by excess Mn (Liu et al. 2019). That is, chlorosis would be related to the change in chlorophyll-containing protein content. An increase in oxidative stress certainly occurs due to Mn toxicity. Although significant lipid peroxidation products were not detected, antioxidant enzymes, especially cytosolic SOD activities in the common bean (González et al. 1998) or Cu/Zn-type SOD localized both in the cytosol and chloroplasts of tomato (Shenker et al. 2004), were upregulated under excess Mn conditions. In contrast, Lei et al. (2007) and Liu et al. (2019) detected an increase in both lipid peroxidation products and H_2O_2 in leaves owing to excess Mn. However, the sites of ROS production and ROS production mechanisms have not been determined.

The complicated situation of the symplastic Mn toxicity phenotype is caused by a wide variety of insights into elucidating the mechanisms. The above phenotypes have been separately reported both *in vivo* and *in vitro*. Because of this circumstance, it is difficult to understand whether each phenotype is related to one another. Recently, Takagi et al. (2021) investigated the effects of symplastic Mn toxicity in rice plants by focusing on photosynthesis. They found that excess Mn decreased the CO_2 assimilation rate with a decrease in stomatal conductance (g_s) under ambient air conditions (21 kPa O_2, 40 Pa CO_2) (Takagi et al. 2021). They also examined the relationship between CO_2 assimilation rate and g_s under Mn-toxic conditions. Subsequently, a similar relationship was found between them under the control of Mn and Mn-toxic conditions. This suggests that the differences in the CO_2 assimilation rate between the control and Mn-toxic conditions are determined by g_s. Under the same conditions, the quantum yields of PSII [Y(II)] and PSI [Y(I)] were similar among the plants grown under control and Mn-toxic conditions, even though Mn-toxic conditions caused lower g_s. These results indicated that photosynthetic electron transport activity can be maintained despite stomatal closure. Moreover, Fv/Fm was only decreased by 2% under Mn-toxic conditions compared with that under the control conditions (Takagi et al. 2021). These results indicate that excess Mn is less harmful to the photosynthetic electron transport chain *in vivo*. This interpretation is supported by previous studies that isolated thylakoid membranes from tobacco and rice, which did not show a significant decrease in photosynthetic activity depending on artificial electron donors and acceptors (Nable et al. 1988; Lidon and Teixeira 2000a, b). However, the redox state of PSI shifts to a more oxidized state under Mn-toxic conditions than that under control conditions. The more oxidized state of PSI can be achieved by electron transport limitation from Cyt b_6f, depending on lumenal acidification (Kramer et al. 1999; Tikhonov 2013). Even if the stomata are closed, O_2 can diffuse into the leaves and photorespiration reactions can be driven by Rubisco (Zivcak et al. 2013). When photorespiratory reactions are dominant in photosynthesis, the H^+ conductance of chloroplastic ATP synthase is suppressed because of the lower ATP consumption rate in the photorespiration reaction than in the carboxylation reaction in the Calvin–Benson cycle (Takagi et al. 2016a, 2017; Wada et al. 2018). Consequently, the lumenal pH is more acidic and PSI becomes more oxidized (Takagi et al. 2016a, 2017; Wada et al. 2018). To confirm whether mismetallation in Rubisco changes the photosynthetic activity *in vivo*, the CO_2 assimilation rate between Mg^{2+}- and Mn^{2+}-binding Rubisco was compared using a photosynthetic biochemical model (von Caemmerer and Farquhar 1981). According to their calculation, Mn^{2+}-binding Rubisco hardly operates carboxylation reactions, and the total photosynthetic activity should be decreased by approximately 90% at 40 Pa CO_2 in chloroplasts (Takagi et al. 2021). Based on these observations, Takagi et al. (2021) proposed that the limitation of CO_2 assimilation is mainly caused by stomatal closure, and that Mn toxicity stimulates photorespiration in rice plants. In that study, they referred to two possibilities to avoid mismetallation of chloroplast enzymes within chloroplasts. The first mechanism involves binding Mn to thylakoid membranes as a storage system for Mn homeostasis (Lidon and Teixeira 2000a, b). The second mechanism is the sequestration or limiting of Mn transport within the chloroplasts. The second hypothesis is based on the fact that Mn distribution to the chloroplasts decreases under Mn-toxic conditions in rice (Führs et al. 2010).

When the photorespiration reaction is prolonged instead of carboxylation by Rubisco, a decrease in the total P700 chlorophyll concentration can occur (Wada et al. 2019). This indicates photoinhibition of PSI (Terashima et al. 1994; Takagi et al. 2016c). In other words, the decrease in PSI subunit content under excess Mn toxicity may be related to the suppression of photosynthesis. This would be the case if Mn/Fe homeostasis was maintained under Mn-toxic conditions. At this time, the PSI synthesis activity and PSI metal content were not determined in that study, although the leaf Fe concentration was confirmed to be maintained under Mn-toxic conditions (Takagi et al. 2021). Biochemical analyses would provide a deeper understanding of the molecular mechanism for the change in PSI content under the Mn-toxic conditions.

When the carbon hydrate concentration was quantified at the end of both the day and the night, the leaf sucrose concentration decreased at the end of the day under Mn-toxic conditions compared with that under the control conditions (Takagi et al. 2021). Therefore, a lower CO_2 assimilation rate affects the carbon gain during growth. Diurnal carbon gain is important for catabolism at night (Takagi et al. 2016b). In some studies, leaf respiration was suppressed under Mn-toxic conditions (Sirkar and Amin 1974; Macfie and Taylor 1992). Takagi et al. (2021) also detected a decrease in both O_2 consumption rate and CO_2 emission rate. However, they could not identify a significant molecular target of excess Mn on mitochondrial respiratory enzymes, except for the NAD-malic enzyme, to suppress the respiration rate (Takagi et al. 2021). Therefore, the decrease in leaf respiration rate was caused by lower diurnal carbon gain for catabolism. This could be the cause of the growth inhibition.

The exact cause of stomatal closure remains unclear. However, Takagi et al. (2021) suggested that stomatal function might be altered by disturbance of indole acetic acid (IAA) homeostasis. Previous studies have reported that IAA destruction activity by peroxidase is enhanced in the presence of Mn in etiolated pea internodes (Hillman and Galston 1956). Among the various metals, Ca^{2+}, Ce^{2+}, Co^{2+}, Cu^+, Fe^{2+}, Mg^{2+}, Ni^{2+}, Zn^{2+}, and Mn^{2+}, Mn^{2+} is the exclusive activator of IAA oxidase activity in etiolated pea plumules (Furuya and

Galston 1961). Morgan (1964) reported an increase in IAA oxidase activity during senescence in cotton leaves. Subsequently, Morgan et al. (1966) found that excess Mn accumulation in leaves stimulates IAA oxidase activity and suggested that Mn toxicity causes auxin shortages in cotton leaves. Thereafter, several studies suggested that IAA oxidase could be a peroxidase but Mn^{2+} might inactivate the inhibitor of IAA oxidase, resulting in accelerated oxidative IAA destruction (Stonier et al. 1968; Fowler and Morgan 1972; Sirkar and Amin 1974). Recent studies have attempted to quantify IAA concentrations in plant tissues under Mn-toxic conditions. Zhao et al. (2017) found that auxin concentrations in the root tips decreased in *Arabidopsis* under Mn-toxic conditions. They also found that the expression of genes related to auxin biosynthesis and transporters was suppressed under Mn toxicity. Takagi et al. (2021) reported that leaf IAA concentrations decreased in rice under Mn toxicity. In addition, auxin-responsive gene expressions, such as *ARF* and *AUX/IAA*, were also altered. Jiang et al. (2021) reported that even though *Celosia argentea* L. is a Mn-hyperaccumulator C4 plant, excess Mn fertilization decreased growth and leaf IAA concentration (Jiang et al. 2021). Therefore, Mn toxicity affects the auxin-signaling cascade in terrestrial plants. In fact, stomatal density increases under Mn toxicity in rice and blueberry leaves (Takagi et al. 2021; González-Villagra et al. 2021). Auxin regulates stomatal development and lower auxin concentrations result in high stomatal density by altering ARF-dependent signal transduction (Balcerowicz and Hoecker 2014; Le et al. 2014; Zhang et al. 2014). In addition to the developmental effects of auxin on stomata, auxin stimulates stomatal opening (Klein et al. 2003; Tanaka et al. 2006; Xiao-Ping and Xi-Gui 2006; Jezek and Blatt 2017). Based on these studies, Takagi et al. (2021) suggested a scheme for photosynthetic inhibition related to IAA deficiency under Mn toxicity. In brief, excess Mn accumulation stimulates IAA oxidation, and IAA deficiency alters leaf development and suppresses stomatal opening. As a result, the CO_2 assimilation reaction was limited, and prolonged CO_2 limitation could lead to oxidative stress in both chloroplasts and cytosol. However, molecular evidence for IAA oxidation under Mn toxicity is lacking. Future research is required to determine how IAA concentrations decrease *in vivo* under excess Mn conditions, focusing on both IAA synthesis and degradation.

7.6 Breeding Strategy for Overcoming the Future Threat of Excess Mn Conditions

As proposed by Fernando and Lynch (2015), the threat of Mn toxicity to agricultural land is realistic. Because of the variety of sensitivities toward Mn toxicity among plant species and cultivars, there are various genetic resources to overcome the effects of Mn toxicity. However, as mentioned above, the exact molecular mechanisms that respond to excess Mn concentration are largely unknown. Applying molecular breeding techniques such as genome editing requires genetic information that contributes to increasing Mn tolerance. Owing to the variety of cultivars, breeding techniques such as using recombinant inbred lines can be used to isolate Mn-tolerant loci from various crop plants. Combining breeding with Mn-tolerant cultivars and recent genetic analysis techniques such as mRNA-seq, genome-wide association analysis (GWAS), or genotyping by random amplicon sequencing-direct (GRAS-Di) systems would accelerate the isolation of beneficial molecular targets

(Alqudah et al. 2020; Miki et al. 2020). Omics analyses related to Mn toxicity have also been conducted in both leaves and roots by targeting proteins (Fecht-Christoffers et al. 2003a, b; Führs et al. 2008, 2010; You et al. 2014; Chen et al. 2016; Ceballos-Laita et al. 2018; Liu et al. 2019), mRNAs (Führs et al. 2008), or miRNAs (Valdés-López et al. 2010). These omics analyses indicate that Mn toxicity seems to change protein content, especially in relation to protein metabolism, cell wall, or stress responses. However, from these datasets, the cause-and-effect should be considered separately by referring to the physiological data. Therefore, the critical molecular components being a factor for emerging Mn-toxic symptoms are still awaiting clarification. Data on physiological traits to improve Mn tolerance and phenotypic characteristics are available. Therefore, in the next analytical stage, detailed molecular mapping under Mn toxicity should be performed by utilizing these omics datasets to generate new cultivars that can overcome Mn toxicity for practical applications. Here, the author suggests four strategies to focus on to improve Mn tolerance in crop plants. In addition, the author recommends that other fruitful reviews should also be referred to when considering future breeding strategies related to Mn tolerance (e.g., Millaleo et al. 2010; Shao et al. 2017; Li et al. 2019).

7.6.1 Limiting Mn Absorption from Soil to Root

In various crop plants, Mn concentration depends almost linearly on Mn concentration in the rhizosphere (Horiguchi 1987; González and Lynch 1997; Fecht-Christoffers et al. 2003a; Führs et al.2010). In rice roots, the expression levels of the major Mn transporters (NRAMP5 and MTP9) were not suppressed in response to an increase in Mn concentration (Sasaki et al. 2012; Yang et al. 2014; Ueno et al. 2015; Wang et al. 2020). These results indicate that Mn uptake was less regulated at the mRNA level. However, shoot and leaf Mn concentrations vary among cultivars under excessive Mn concentrations (Blatt and van Diest 1981; Führs et al. 2009; Tsunemitsu et al. 2018b). Furthermore, Wang et al. (2020) found that although *NRAMP5* and *MTP9* expression was not altered, shoot Mn concentration could be reduced in rice under flooded conditions compared to that in upland conditions. That is, it seems that an unknown post-transcriptional Mn uptake regulatory system functions *in vivo* by sensing Mn concentration in the rhizosphere. These Mn-sensing and uptake regulatory networks can be used to modulate Mn uptake, depending on the growth environment, by molecular breeding.

Suppression of the major Mn transporter gene is effective in decreasing Mn uptake activity under excess Mn conditions. The *nramp5* mutant in rice or the *nramp1* mutant in *Arabidopsis* caused lower Mn uptake activity and these plants showed growth inhibition under Mn-deficient conditions (Sasaki et al. 2012; Yang et al. 2014; Bian et al. 2018). However, as the Mn fertilizer concentration increased, the shoot Mn concentration gradually increased and growth inhibition was alleviated (Yang et al. 2014; Bian et al. 2018). That is, inhibition of the major root Mn transporter does not completely inhibit Mn uptake and maintains a relatively normal shoot Mn concentration because of the lower Mn uptake activity under excess Mn conditions.

In addition to controlling Mn uptake by root transporters, the changes in Mn dynamics by organic acid secretion into the soil are also effective (Li et al. 2019). Chen et al. (2015) found that *Stylosanthes guianensis* increases the malate dehydrogenase (SgMDH) content in their roots in response to an increase in Mn concentration in the rhizosphere, and MDH contributes to malate secretion from the roots, which lowers the toxic effects of Mn by decreasing Mn concentration in plants. Furthermore, the overexpression of *SgMDH* improved Mn tolerance in *Arabidopsis*. Therefore, chelation of soil Mn by increasing organic acid secretion from the roots is a practical way to reduce Mn uptake in crop plants.

Even after Mn absorption, the shoot Mn concentration can be lowered by Mn deposition in the roots. Horiguchi (1987) found that crop plants oxidize Mn in roots in response to the high Mn concentration in the rhizosphere and suggested that the deposition of oxidized Mn in roots lowers Mn translocation from root to shoot. This may reduce the toxic effects on leaf photosynthesis.

7.6.2 Sequestration of Mn from Cytosol to the Vacuole or Apoplast

When excessive amounts of Mn flow into cells, terrestrial plants compartment them within the cells to avoid the toxic effects of Mn. The vacuole loading of Mn is a major strategy for detoxifying Mn in terrestrial plants. Indeed, tonoplast-localized influx Mn transporter mutants (*mtp* and *cax*) are highly sensitive to Mn toxicity in both rice and *Arabidopsis* (Hirschi et al. 2000; Chen et al. 2013; Takemoto et al. 2017; Eroglu et al. 2016, 2017). Delhaize et al. (2003) isolated a membrane-bound cation diffusion family protein (ShMTP1) from *Stylosanthes hamata*, an Mn-tolerant legume species. When *ShMTP1* is expressed in *A. thaliana*, ShMTP1 is localized in the tonoplast and improves Mn tolerance despite the increase in Mn concentration in the shoot tissue (Delhaize et al. 2003). This could be due to the sequestration of Mn in the vacuoles. Therefore, the acceleration of Mn compartmentation in vacuoles would be beneficial for attenuating Mn toxicity in crop plants. Vacuole capacity is an important trait in improving Mn tolerance. Fernando et al. (2006) analyzed foliar Mn localization in the Mn-hyperaccumulator plant, *Gossia bidwillii*. Subsequently, they found that the vacuolar Mn concentration in the upper epidermis and first/second palisade cells reached approximately 100–500 mM Mn (Fernando et al. 2006). This concentration was drastically higher than that in the crop plants. For example, Horst et al. (1999) reported that cowpeas reached vacuolar Mn concentrations of up to 6 mM. Later, their analysis revealed that Mn-hyperaccumulator plants deposit Mn(II) in vacuoles with malate and citric acid (Fernando et al. 2010), but it is not clear whether the organic acid concentration in the vacuole is important for determining the vacuolar Mn capacity (Horst et al. 1999).

In addition to vacuoles, the Golgi apparatus and ER also contribute to Mn sequestration from the cytosol. This is evidenced by Mn-sensitive mutant phenotypes such as *mtp11* and *eca1* (Wu et al. 2002; Tsunemitsu et al. 2018a; Farthing et al. 2017). In contrast, overexpression of *AtMTP11* increases Mn tolerance in *Arabidopsis* (Peiter et al. 2007). However, the author wondered why apoplastic Mn sequestration contributed to Mn tolerance. Therefore, apoplastic Mn toxicity should be stimulated in such situations. In fact, the OsYSL6 mutant, which impairs Mn transfer from the apoplast to the symplast, attenuates Mn tolerance in rice (Sasaki et al. 2011). Therefore, Golgi- or ER-dependent Mn sequestration in the apoplast should be connected to another system that transports Mn to other parts of the tissue. Based on the Mn distribution in leaf cross-sections, some directional transport systems may exist, and Mn preferentially moves to the leaf epidermal cells or trichomes (Memon et al. 1981; Blamey et al. 1986, 2018; Fernando et al. 2006, 2010).

7.6.3 Maintenance of Auxin Homeostasis

Auxin deficiency may be one of the causes of Mn toxicity and growth defects in terrestrial plants (Zhao et al. 2017; Jiang et al. 2021; Takagi et al. 2021). Owing to the lower auxin concentration, the auxin-dependent signal cascade is altered under Mn toxicity (Takagi et al. 2021). Interestingly, Gangwar et al. (2011) reported that exogenous IAA application can alleviate the inhibitory effects of excess Mn on the growth of pea seedlings, although higher IAA application causes inverse effects. Therefore, optimization of the auxin signaling cascade would confer tolerance to excess Mn. Both auxin synthesis and auxin degradation are candidate for molecular targets of Mn toxicity (Morgan et al. 1966; Zhao et al. 2017). Detailed research focusing on auxin synthesis and catabolism could provide new insights into the manipulation of auxin concentrations under excess Mn. In particular, the molecular mechanisms underlying IAA oxidation remain unclear. Peroxidases should be considered according to previous studies (Stonier et al. 1968; Fowler and Morgan 1972; Sirkar and Amin 1974). Proteomics analysis detects an increase in the concentration of several peroxidases in both leaves and roots of various terrestrial plants under excess Mn conditions with high probability (Fecht-Christoffers et al. 2003a, b; Führs et al. 2009; Chen et al. 2016; Ceballos-Laita et al. 2018; Liu et al. 2019). First, it should be determined whether certain or nonspecific peroxidases oxidatively catabolize IAA. Genetic modification of IAA oxidizing enzymes may contribute to increased Mn tolerance.

7.6.4 The Reinforcement of Silicon Uptake and Its Distribution

Silicon fertilization successfully ameliorates Mn toxicity in various crop plants, including barley (Williams and Vlamis 1957a, b), rice (Horiguchi 1988), cowpea (Iwasaki et al. 2002), pumpkin (Iwasaki and Matsumura 1999), beans (Horst and Marschner 1978), cucumber (Rogalla and Römheld 2002), and sunflower (Blamey et al. 2018). The visible effect of Si on excess Mn is common. Si fertilization suppresses the emergence of brownish patchy spots because Si contributes to the uniform distribution of Mn in leaves (Williams and Vlamis 1957a; Iwasaki and Matsumura 1999; Blamey et al. 2018). In addition to Si-dependent visible amelioration, there are various molecular possibilities to suppress the toxic effects of Mn by Si. In rice plants, shoot Mn concentrations are lower in the presence of Si rather than that in its absence under excess Mn conditions (Horiguchi 1988). Che et al. (2016) showed that Si suppresses the total Mn concentration in whole plants as a sequel to suppressing Mn translocation from roots to shoots through the xylem stream due to Mn–Si complex formation, which cannot translocate the MTP9 transporter in rice plants (Ueno et al. 2015). Furthermore, an increase in Si concentration suppressed *NRAMP5* expression *via* unknown mechanisms (Che et al. 2016). In contrast, suppression of Mn uptake and Mn translocation was less pronounced in other plant species in the presence of Si. In bean plants, Si accelerates Mn loading into vacuoles under conditions of excess Mn and ameliorates Mn toxicity (Horst and Marschner 1978). Furthermore, Si stimulates Mn sequestration in the apoplastic regions (Iwasaki et al. 2002; Rogalla and Römheld 2002). Blamey et al. (2018) demonstrated that Si and Mn are colocalized in the apoplast of epidermal cells and trichome basal cells and that Si-colocalized Mn maintains a soluble form. In the apoplast, Mn has been suggested to bind to the cell wall *via* Si and is attenuated by its toxic effects (Rogalla and Römheld 2002). These results indicated that the co-existence of both Si and Mn are important for confining the toxic effects of Mn in plant tissues. Therefore, both Mn and Si transport and distribution systems are targets for molecular breeding to enforce Mn tolerance.

7.7 Conclusion and Future Prospects

This chapter introduced the significance of considering the physiological phenomenon of Mn toxicity in agriculture. Mn toxicity phenotypes are often studied using leaves because of the appearance of brownish spots and chlorosis. However, excess Mn disturbs the auxin signaling pathway. Thus, other parts of the plant tissues should be analyzed. In fact, fruit development can be altered by excess Mn, even though the leaves are tolerant of its toxic effects (Aoba 1986). Many issues remain to be investigated regarding the effects of Mn on plants and the threat of Mn toxicity caused by future changes in the global environment. The author hopes that the chapter will inspire future research ideas in readers and become an opportunity to start a wide variety of analyses related to Mn toxicity and its tolerance in plants.

Acknowledgments

The authors thank Editage (https://www.editage.jp) for the English language editing of the manuscript. Part of this work was supported by the Japan Society for the Promotion of Science (JSPS) Research Fellowship (Grant No. 18J00852), Japan, and by Japan Science and Technology Agency (JST), ACT-X (Grant No. JPMJAX21BA) Japan.

References

Adams, F. and Wear, J.I. (1957). Manganese toxicity and soil acidity in relation to crinkle leaf of cotton. *Soil Sci. Soc. Am. J.* 21: 305–308.

Alloway, B.J. (ed.) (2008). *Micronutrient Deficiencies in Global Crop Production*. Heidelberg, Germany: Springer.

Alqudah, A.M., Sallam, A., Baenziger, P.S., and Börner, A. (2020). GWAS: fast-forwarding gene identification and characterization in temperate cereals: lessons from barley–a review. *J. Adv. Res. 22*: 119–135.

Anderson, J.M., Boardman, N.K., and David, D.J. (1964). Trace metal composition of fractions obtained by digitonin fragmentation of spinach chloroplasts. *Biochem. Biophys. Res. Commun.* 17: 685–689.

Aoba, K. (1986). Excess manganese disorder in fruit trees. *Jpn. Agric. Res. Q.* 20: 45–47.

Balcerowicz, M. and Hoecker, U. (2014). Auxin–a novel regulator of stomata differentiation. *Trends Plant Sci.* 19: 747–749.

Barber, J. (2016). Mn_4Ca cluster of photosynthetic oxygen-evolving center: structure, function and evolution. *Biochemistry* 55: 5901–5906.

Bartlett, R. and James, B. (1980). Studying dried, stored soil samples—some pitfalls. *Soil Sci. Soc. Am. J.* 44: 721–724.

Bellincampi, D., Cervone, F., and Lionetti, V. (2014). Plant cell wall dynamics and wall-related susceptibility in plant–pathogen interactions. *Front. Plant Sci. 5:228.* https://doi.org/10.3389/fpls.2014.00228.

Bian, B., Kageshima, S., Yano, K. et al. (2018). Screening *Arabidopsis thaliana* mutants for low sensitivity to manganese identifies novel alleles of NRAMP1 and PGSIP6. *J. Exp. Bot.* 69: 1795–1803.

Blamey, F.P.C., Joyce, D.C., Edwards, D.G., and Asher, C.J. (1986). Role of trichomes in sunflower tolerance to manganese toxicity. *Plant Soil* 91: 171–180.

Blamey, F.P.C., Hernandez-Soriano, M.C., Cheng, M. et al. (2015). Synchrotron-based techniques shed light on mechanisms of plant sensitivity and tolerance to high manganese in the root environment. *Plant Physiol.* 169: 2006–2020.

Blamey, F.P.C., McKenna, B.A., Li, C. et al. (2018). Manganese distribution and speciation help to explain the effects of silicate and phosphate on manganese toxicity in four crop species. *New Phytol.* 217: 1146–1160.

Blatt, C.R. and van Diest, A. (1981). Evaluation of a screening technique for manganese toxicity in relation to leaf manganese distribution and interaction with silicon. *Neth. J. Agric. Sci.* 29: 297–304.

Blunden, J. and Arndt, D.S. (ed.) (2020). State of the climate in 2019. *Bull. Am. Meteorol. Soc.* 101: S1–S429.

Brandenburg, F., Schoffman, H., Kurz, S. et al. (2017). The *Synechocystis* manganese exporter Mnx is essential for manganese homeostasis in cyanobacteria. *Plant Physiol.* 173: 1798–1810.

Broadley, M., Brown, P., Cakmak, I. et al. (2012). Function of nutrients: micronutrients. In: *Marschner's Mineral Nutrition of Higher Plants*, 3e (ed. P. Marschner), 191–249. Sandiego: Elsevier.

Burnell, J.N. (1988). The biochemistry of manganese in plants. In: *Manganese in Soils and Plants* (ed. R.D. Graham, R.J. Hannam and N.C. Uren), 125–137. Dordrecht: Springer.

von Caemmerer, S. and Farquhar, G.D. (1981). Some relationships between the biochemistry of photosynthesis and the gas exchange of leaves. *Planta* 153: 376–387.

Cakmak, I. and Horst, W.J. (1991). Effect of aluminium on lipid peroxidation, superoxide dismutase, catalase, and peroxidase activities in root tips of soybean (*Glycine max*). *Physiol. Plant.* 83: 463–468.

Ceballos-Laita, L., Gutierrez-Carbonell, E., Takahashi, D. et al. (2018). Effects of Fe and Mn deficiencies on the protein profiles of tomato (*Solanum lycopersicum*) xylem sap as revealed by shotgun analyses. *J. Proteom.* 170: 117–129.

Che, J., Yamaji, N., Shao, J.F. et al. (2016). Silicon decreases both uptake and root-to-shoot translocation of manganese in rice. *J. Exp. Bot.* 67: 1535–1544.

Chen, Z., Fujii, Y., Yamaji, N. et al. (2013). Mn tolerance in rice is mediated by MTP8. 1, a member of the cation diffusion facilitator family. *J. Exp. Bot.* 64: 4375–4387.

Chen, Z., Sun, L., Liu, P. et al. (2015). Malate synthesis and secretion mediated by a manganese-enhanced malate dehydrogenase confers superior manganese tolerance in *Stylosanthes guianensis. Plant Physio.* 167: 176–188.

Chen, Z., Yan, W., Sun, L. et al. (2016). Proteomic analysis reveals growth inhibition of soybean roots by manganese toxicity is associated with alteration of cell wall structure and lignification. *J. Proteom.* 143: 151–160.

Clair, S.B.S., Carlson, J.E., and Lynch, J.P. (2005). Evidence for oxidative stress in sugar maple stands growing on acidic, nutrient imbalanced forest soils. *Oecologia* 145: 257–268.

Clairmont, K.B., Hagar, W.G., and Davis, E.A. (1986). Manganese toxicity to chlorophyll synthesis in tobacco callus. *Plant Physiol.* 80: 291–293.

Csatorday, K., Gombos, Z., and Szalontai, B. (1984). Mn^{2+} and Co^{2+} toxicity in chlorophyll biosynthesis. *Proc. Natl. Acad. Sci. U.S.A.* 81: 476–478.

Delhaize, E., Kataoka, T., Hebb, D.M. et al. (2003). Genes encoding proteins of the cation diffusion facilitator family that confer manganese tolerance. *Plant Cell* 15: 1131–1142.

Edreva, A.M. (1990). Induction of 'pathogenesis-related' proteins in tobacco leaves by physiological (non-pathogenic) disorders. *J. Exp. Bot.* 41: 701–703.

Edwards, D.G. and Asher, C.J. (1982). Tolerance of crop and pasture species to manganese toxicity. In: *Proc. 9th Int. Plant Nutr. Coll., Warwick, UK. Ed. A Scaife*, 145–151. UK: Warwick Uni.

Eisenhut, M. (2020). Manganese homeostasis in cyanobacteria. *Plants* 9: 18. https://doi.org/10.3390/plants9010018.

Eroglu, S., Meier, B., von Wirén, N., and Peiter, E. (2016). The vacuolar manganese transporter MTP8 determines tolerance to iron deficiency-induced chlorosis in Arabidopsis. *Plant Physiol.* 170: 1030–1045.

Eroglu, S., Giehl, R.F., Meier, B. et al. (2017). Metal tolerance protein 8 mediates manganese homeostasis and iron reallocation during seed development and germination. *Plant Physiol.* 174: 1633–1647.

Eskandari, S., Höfte, H., and Zhang, T. (2020). Foliar manganese spray induces the resistance of cucumber to *Colletotrichum lagenarium*. *J. Plant Physiol.* 246–247: 153129.

Farley, R.F. and Draycott, A.P. (1973). Manganese deficiency of sugar beet in organic soils. *Plant Soil* 38: 235–244.

Farthing, E.C., Menguer, P.K., Fett, J.P., and Williams, L.E. (2017). OsMTP11 is localised at the Golgi and contributes to Mn tolerance. *Sci. Rep. 7, 15258*. https://doi.org/10.1038/s41598-017-15324-6.

Fecht-Christoffers, M.M., Braun, H.P., Lemaitre-Guillier, C. et al. (2003a). Effect of manganese toxicity on the proteome of the leaf apoplast in cowpea. *Plant Physiol.* 133: 1935–1946.

Fecht-Christoffers, M.M., Maier, P., and Horst, W.J. (2003b). Apoplastic peroxidases and ascorbate are involved in manganese toxicity and tolerance of *Vigna unguiculata*. *Physiol. Plant.* 117: 237–244.

Fecht-Christoffers, M.M., Fuhrs, H., Braun, H.P., and Horst, W.J. (2006). The role of hydrogen peroxide-producing and hydrogen peroxide-consuming peroxidases in the leaf apoplast of cowpea in manganese tolerance. *Plant Physiol.* 140: 1451–1463.

Fernando, D.R. and Lynch, J.P. (2015). Manganese phytotoxicity: new light on an old problem. *Ann. Bot.* 116: 313–319.

Fernando, D.R., Batianoff, G.N., Baker, A.J., and Woodrow, I.E. (2006). *In vivo* localization of manganese in the hyperaccumulator *Gossia bidwillii* (Benth.) N. Snow & Guymer (Myrtaceae) by cryo-SEM/EDAX. *Plant Cell Environ.* 29: 1012–1020.

Fernando, D.R., Mizuno, T., Woodrow, I.E. et al. (2010). Characterization of foliar manganese (Mn) in Mn (hyper) accumulators using X-ray absorption spectroscopy. *New Phytol.* 188: 1014–1027.

Fernando, D.R., Moroni, S.J., Scott, B.J. et al. (2016). Temperature and light drive manganese accumulation and stress in crops across three major plant families. *Environ. Exp. Bot.* 132: 66–79.

Ferreira, K.N., Iverson, T.M., Maghlaoui, K. et al. (2004). Architecture of the photosynthetic oxygen-evolving center. *Science* 303: 1831–1838.

Fowler, J.L. and Morgan, P.W. (1972). The relationship of the peroxidative indoleacetic acid oxidase system to in vivo ethylene synthesis in cotton. *Plant Physiol.* 49: 555–559.

Foy, C.D. (1964). *Toxic Factors in Acid Soils of the Southeastern United States as Related to the Response of Alfalfa to Lime*, USD A Prod. Res. Report 80. Washington, DC: U.S. Govt. Printing Office.

Foy, C.D., Scott, B.J., and Fisher, J.A. (1988). Genetic differences in plant tolerance to manganese toxicity. In: *Manganese in Soil and Plants* (ed. R.D. Graham, R.J. Hannam and N.J. Uren), 293–307. Dordrecht, The Netherlands: Kluwer Academic Publishers.

Führs, H., Hartwig, M., Molina, L.E.B. et al. (2008). Early manganese-toxicity response in *Vigna unguiculata* L.–a proteomic and transcriptomic study. *Proteomics* 8: 149–159.

Führs, H., Götze, S., Specht, A. et al. (2009). Characterization of leaf apoplastic peroxidases and metabolites in *Vigna unguiculata* in response to toxic manganese supply and silicon. *J. Exp. Bot.* 60: 1663–1678.

Führs, H., Behrens, C., Gallien, S. et al. (2010). Physiological and proteomic characterization of manganese sensitivity and tolerance in rice (*Oryza sativa*) in comparison with barley (*Hordeum vulgare*). *Ann. Bot.* 105: 1129–1140.

Funchess, M.J. (1919). Acid soils and the toxicity of manganese. *Soil Sci.* 8: 69.

Furuya, M. and Galston, A.W. (1961). Effect of *in vitro* preincubation with cofactors on the activity of the indoleacetic acid oxidase of peas. *Physiol. Plant.* 14: 750–766.

Gandini, C., Schmidt, S.B., Husted, S. et al. (2017). The transporter Syn PAM 71 is located in the plasma membrane and thylakoids, and mediates manganese tolerance in *Synechocystis* PCC 6803. *New Phytol.* 215: 256–268.

Gangwar, S., Singh, V.P., Prasad, S.M., and Maurya, J.N. (2011). Differential responses of pea seedlings to indole acetic acid under manganese toxicity. *Acta Physiol. Plant.* 33: 451–462.

González, A. and Lynch, J.P. (1997). Effects of manganese toxicity on leaf CO_2 assimilation of contrasting common bean genotypes. *Physiol. Plant.* 101: 872–880.

González, A., Steffen, K.L., and Lynch, J.P. (1998). Light and excess manganese: implications for oxidative stress in common bean. *Plant Physiol.* 118: 493–504.

González-Villagra, J., Escobar, A.L., Ribera-Fonseca, A. et al. (2021). Differential mechanisms between traditionally established and new highbush blueberry (*Vaccinium corymbosum* L.) cultivars reveal new insights into manganese toxicity resistance. *Plant Physiol. Biochem.* 158: 454–465.

Hall, J.Á. (2002). Cellular mechanisms for heavy metal detoxification and tolerance. *J. Exp. Bot.* 53: 1–11.

Hamer, K., Gudenschwager, I., and Pichler, T. (2020). Manganese (Mn) concentrations and the Mn-Fe relationship in shallow groundwater: implications for groundwater monitoring. *Soil Syst.* 4: 49. https://doi.org/10.3390/soilsystems4030049.

Haxel, G.B., Hedrick, J.B., Orris, G.J. et al. (2002). *Rare Earth Elements: Critical Resources for High Technology*. US Geological Survey.

Herndon, E.M. and Brantley, S.L. (2011). Movement of manganese contamination through the critical zone. *Appl. Geochem.* 26: S40–S43.

Herndon, E.M., Jin, L., and Brantley, S.L. (2011). Soils reveal widespread manganese enrichment from industrial inputs. *Environ. Sci. Technol.* 45: 241–247.

Hillman, W.S. and Galston, A.W. (1956). Interaction of manganese and 2, 4-dichlorophenol in the enzymatic destruction of indoleacetic acid. *Physiol. Plant.* 9: 230–235.

Hirschi, K.D., Korenkov, V.D., Wilganowski, N.L., and Wagner, G.J. (2000). Expression of Arabidopsis CAX2 in tobacco. Altered metal accumulation and increased manganese tolerance. *Plant Physiol.* 124: 125–134.

Horiguchi, T. (1987). Mechanism of manganese toxicity and tolerance of plants: II. Deposition of oxidized manganese in plant tissues. *Soil Sci. Plant Nutr.* 33: 595–606.

Horiguchi, T. (1988). Mechanism of manganese toxicity and tolerance of plants: IV. Effects of silicon on alleviation of manganese toxicity of rice plants. *Soil Sci. Plant Nutr.* 34: 65–73.

Horst, W.J. (1983). Factors responsible for genotypic manganese tolerance in cowpea (*Vigna unguiculata*). *Plant Soil* 72: 213–218.

Horst, W.J. (1988). The physiology of manganese toxicity in soils and plants. In: *Manganese in Soils and Plants* (ed. R.D. Graham, R.J. Hannam and N.C. Uven), 175–188. Dordrecht, The Netherlands: Kluwer Academic Publishers.

Horst, W.J. and Marschner, H. (1978). Effect of silicon on manganese tolerance of bean plants (*Phaseolus vulgaris* L.). *Plant Soil.* 50: 287–303.

Horst, W.J., Fecht, M., Naumann, A. et al. (1999). Physiology of manganese toxicity and tolerance in *Vigna unguiculata* (L.) Walp. *J. Plant Nutr. Soil Sci.* 162: 263–274.

Hou, Q., Zhang, Q., Huang, G. et al. (2020). Elevated manganese concentrations in shallow groundwater of various aquifers in a rapidly urbanized delta, south China. *Sci. Total Environ.* 701: 134777.

Houtz, R.L., Nable, R.O., and Cheniae, G.M. (1988). Evidence for effects on the in vivo activity of ribulose-bisphosphate carboxylase/oxygenase during development of Mn toxicity in tobacco. *Plant Physiol.* 86: 1143–1149.

Hue, N.V., Vega, S., and Silva, J.A. (2001). Manganese toxicity in a Hawaiian Oxisol affected by soil pH and organic amendments. *Soil Sci. Soc. Am. J.* 65: 153–160.

Iwasaki, K. and Matsumura, A. (1999). Effect of silicon on alleviation of manganese toxicity in pumpkin (*Cucurbita moschata* Duch cv. Shintosa). *Soil Sci. Plant Nutr.* 45: 909–920.

Iwasaki, K., Maier, P., Fecht, M., and Horst, W.J. (2002). Leaf apoplastic silicon enhances manganese tolerance of cowpea (*Vigna unguiculata*). *J. Plant Physiol.* 159: 167–173.

Jezek, M. and Blatt, M.R. (2017). The membrane transport system of the guard cell and its integration for stomatal dynamics. *Plant Physiol.* 174: 487–519.

Jiang, P., Liu, J., Yu, G. et al. (2021). Moderate Mn accumulation enhances growth and alters leaf hormone contents in the hyperaccumulator *Celosia argentea* Linn. *Environ. Exp. Bot.* 191: 104603.

Johan, H.E. and Amin, J.V. (1967). The influence of foliar and substrate application of manganese on cotton. *Plant Soil* 26: 369–379.

Jones, M.E., LaCroix, R.E., Zeigler, J. et al. (2020). Enzymes, manganese, or iron? Drivers of oxidative organic matter decomposition in soils. *Environ. Sci. Technol.* 54: 14114–14123.

Jung, J.L., Maurel, S., Fritig, B., and Hahne, G. (1995). Different pathogenesis-related-proteins are expressed in sunflower (*Helianthus annuus* L.) in response to physical, chemical and stress factors. *J. Plant Physiol.* 145: 153–160.

Kabata-Pendias, A. and Wiacek, K. (1985). Excessive uptake of heavy metals by plants from contaminated soil. *Soil Sci. Ann. 36*: 33–43.

Keiluweit, M., Nico, P., Harmon, M.E. et al. (2015). Long-term litter decomposition controlled by manganese redox cycling. *Proc. Natl. Acad. Sci. U.S.A.* 112: E5253–E5260.

Kelley, W.P. (1914). The function of manganese in plants. *Bot. Gaz.* 57: 213–227.

Kenten, R.H. and Mann, P.J.G. (1949). The oxidation of manganese by plant extracts in the presence of hydrogen peroxide. *Biochem. J.* 45: 255–263.

Kenten, R.H. and Mann, P.J.G. (1950). The oxidation of manganese by peroxidase systems. *Biochem. J.* 46: 67–73.

Kenten, R.H. and Mann, P.J.G. (1957). Manganese oxidation in the pea plant (*Pisum sativum* L.) grown under conditions of manganese toxicity. *Biochem. J.* 65: 179–185.

Kirkby, E. (2012). Introduction, definition and classification of nutrients. In: *Marschner's Mineral Nutrition of Higher Plants*, 3e (ed. P. Marschner), 3–5. London: Academic.

Kitao, M., Lei, T.T., and Koike, T. (1997). Effects of manganese toxicity on photosynthesis of white birch (*Betula platyphylla* var. *japonica*) seedlings. *Physiol. Plant.* 101: 249–256.

Klein, M., Perfus-Barbeoch, L., Frelet, A. et al. (2003). The plant multidrug resistance ABC transporter AtMRP5 is involved in guard cell hormonal signalling and water use. *Plant J.* 33: 119–129.

Konapala, G., Mishra, A.K., Wada, Y., and Mann, M.E. (2020). Climate change will affect global water availability through compounding changes in seasonal precipitation and evaporation. *Nat. Commun. 11*: 1–10.

Korshunova, Y.O., Eide, D., Clark, W.G. et al. (1999). The IRT1 protein from *Arabidopsis thaliana* is a metal transporter with a broad substrate range. *Plant Mol. Biol.* 40: 37–44.

Kramer, D.M., Sacksteder, C.A., and Cruz, J.A. (1999). How acidic is the lumen? *Photosynth. Res.* 60: 151–163.

Kukavica, B., Mojović, M., Vucčinić, Ž. et al. (2009). Generation of hydroxyl radical in isolated pea root cell wall, and the role of cell wall-bound peroxidase, Mn-SOD and phenolics in their production. *Plant Cell Physiol.* 50: 304–317.

Le, J., Liu, X.G., Yang, K.Z. et al. (2014). Auxin transport and activity regulate stomatal patterning and development. *Nat. Commun.* 5: 1–8.

Learman, D.R., Voelker, B.M., Vazquez-Rodriguez, A.I., and Hansel, C.M. (2011). Formation of manganese oxides by bacterially generated superoxide. *Nat. Geosci.* 4: 95–98.

Lei, Y., Korpelainen, H., and Li, C. (2007). Physiological and biochemical responses to high Mn concentrations in two contrasting *Populus cathayana* populations. *Chemosphere* 68: 686–694.

Li, J., Jia, Y., Dong, R. et al. (2019). Advances in the mechanisms of plant tolerance to manganese toxicity. *Int. J. Mol. Sci.* 20: 5096. https://doi.org/10.3390/ijms20205096.

Li, H., Santos, F., Butler, K., and Herndon, E. (2021). A critical review on the multiple roles of manganese in stabilizing and destabilizing soil organic matter. *Environ. Sci. Technol.* 55: 12136–12152.

Lidon, F.C. and Teixeira, M.G. (2000a). Oxy radicals production and control in the chloroplast of Mn-treated rice. *Plant Sci. 152*: 7–15.

Lidon, F.C. and Teixeira, M.G. (2000b). Rice tolerance to excess Mn: implications in the chloroplast lamellae and synthesis of a novel Mn protein. *Plant Physiol. Biochem.* 38: 969–978.

Liu, K., Fan, L., Li, Y. et al. (2018). Concentrations and health risks of heavy metals in soils and crops around the Pingle manganese (Mn) mine area in Guangxi Province, China. *Environ. Sci. Pollut. Res.* 25: 30180–30190.

Liu, P., Huang, R., Hu, X., Jia, Y., Li, J., Luo, J., Liu, Q., Luo, L., Liu, G. and Chen, Z. (2019). Physiological responses and proteomic changes reveal insights into *Stylosanthes* response to manganese toxicity. *BMC Plant Biol. 19, 212.* https://doi.org/10.1186/s12870-019-1822-y

Löhnis, M.P. (1960). Effect of magnesium and calcium supply on the uptake of manganese by various crop plants. *Plant Soil* 12: 339–376.

Lorrai, R. and Ferrari, S. (2021). Host cell wall damage during pathogen infection: mechanisms of perception and role in plant-pathogen interactions. *Plants 10*: 399. https://doi.org/10.3390/plants10020399.

Lovley, D.R. and Phillips, E.J. (1988). Novel mode of microbial energy metabolism: organic carbon oxidation coupled to dissimilatory reduction of iron or manganese. *Appl. Environ. Microbiol.* 54: 1472–1480.

Ma, J.F. (2007). Syndrome of aluminum toxicity and diversity of aluminum resistance in higher plants. *Int. Rev. Cytol.* 264: 225–252.

Macfie, S.M. and Taylor, G.J. (1992). The effects of excess manganese on photosynthetic rate and concentration of chlorophyll in *Triticum aestivum* grown in solution culture. *Physiol. Plant.* 85: 467–475.

Makino, T. (2001). Redox function and dynamics of manganese in soils. *Bull Natl. Inst. Agro-Environmental Sci. (Japan)* 20: 107–161.

Makino, A., Nakano, H., and Mae, T. (1994). Effects of growth temperature on the responses of ribulose-1, 5-biphosphate carboxylase, electron transport components, and sucrose synthesis enzymes to leaf nitrogen in rice, and their relationships to photosynthesis. *Plant Physiol.* 105: 1231–1238.

McHargue, J.S. (1922). The role of manganese in plants. *J. Am. Chem. Soc.* 44: 1592–1598.

McHargue, J.S. (1923). Effect of different concentrations of manganese sulphate on the growth of plants in acid and neutral soils and the necessity of manganese as a plant nutrient. *J. Agric. Res.* 24: 781–794.

McMahon, P.B., Belitz, K., Reddy, J.E., and Johnson, T.D. (2019). Elevated manganese concentrations in United States groundwater, role of land surface–soil–aquifer connections. *Environ. Sci. Technol.* 53: 29–38.

Memon, A.R., Chino, M., Hidaka, H. et al. (1981). Manganese toxicity in field grown tea plants and the microdistribution of manganese in the leaf tissues as revealed by electron probe X-ray micrography. *Soil Sci. Plant Nutr.* 27: 317–328.

Miki, Y., Yoshida, K., Enoki, H. et al. (2020). GRAS-Di system facilitates high-density genetic map construction and QTL identification in recombinant inbred lines of the wheat progenitor *Aegilops tauschii*. *Sci. Rep.* 10: 21455. https://doi.org/10.1038/s41598-020-78589-4.

Millaleo, R., Reyes-Díaz, M., Ivanov, A.G. et al. (2010). Manganese as essential and toxic element for plants: transport, accumulation and resistance mechanisms. *J. Soil Sci. Plant Nutr.* 10: 470–481.

Millaleo, R., Reyes-Díaz, M., Alberdi, M. et al. (2013). Excess manganese differentially inhibits photosystem I versus II in *Arabidopsis thaliana*. *J. Exp. Bot.* 64: 343–354.

Mills, R.F., Doherty, M.L., López-Marqués, R.L. et al. (2008). ECA3, a Golgi-localized P_{2A}-type ATPase, plays a crucial role in manganese nutrition in Arabidopsis. *Plant Physiol.* 146: 116–128.

Morgan, P.W. (1964). Distribution of indoleacetic acid oxidase and inhibitors in light-grown cotton. *Plant Physiol.* 39: 741–746.

Morgan, P.W., Joham, H.E., and Amin, J.V. (1966). Effect of manganese toxicity on the indoleacetic acid oxidase system of cotton. *Plant Physiol.* 41: 718–724.

Myers, C.R. and Nealson, K.H. (1988). Bacterial manganese reduction and growth with manganese oxide as the sole electron acceptor. *Science* 240: 1319–1321.

Nable, R.O., Houtz, R.L., and Cheniae, G.M. (1988). Early inhibition of photosynthesis during development of Mn toxicity in tobacco. *Plant Physiol.* 86: 1136–1142.

Neal, D.C. (1937). Crinkle leaf, a new disease of cotton in Louisiana. *Phytopathology* 27: 1171–1175.

Neal, D.C. and Lovett, H.C. (1938). Further studies of crinkle leaf, a disorder of cotton plants prevalent in Sintonia and Oliver silt loam soils of Louisiana. *Phytopathology* 28: 582–587.

Nedukha, O.M. (2015). Callose: localization, functions, and synthesis in plant cells. *Cytol. Genet.* 49: 49–57.

Nriagu, J.O. (1989). A global assessment of natural sources of atmospheric trace metals. *Nature* 338: 47–49.

Ohki, K. (1985). Manganese deficiency and toxicity effects on photosynthesis, chlorophyll, and transpiration in Wheat. *Crop Sci.* 25: 187–191.

Pacyna, J.M. and Pacyna, E.G. (2001). An assessment of global and regional emissions of trace metals to the atmosphere from anthropogenic sources worldwide. *Environ. Rev.* 9: 269–298.

Panda, S., Mishra, A.K., and Biswal, U.C. (1987). Manganese induced peroxidation of thylakoid lipids and changes in chlorophyll-α fluorescence during aging of cell free chloroplasts in light. *Phytochemistry* 26: 3217–3219.

Pedas, P., Ytting, C.K., Fuglsang, A.T. et al. (2008). Manganese efficiency in barley: identification and characterization of the metal ion transporter HvIRT1. *Plant Physiol.* 148: 455–466.

Peiter, E., Montanini, B., Gobert, A. et al. (2007). A secretory pathway-localized cation diffusion facilitator confers plant manganese tolerance. *Proc. Natl. Acad. Sci. U.S.A.* 104: 8532–8537.

Pittman, J.K. (2005). Managing the manganese: molecular mechanisms of manganese transport and homeostasis. *New Phytol.* 167: 733–742.

Rogalla, H. and Römheld, V. (2002). Role of leaf apoplast in silicon-mediated manganese tolerance of *Cucumis sativus* L. *Plant Cell Environ.* 25: 549–555.

Ross, D.S. and Bartlett, R.J. (1981). Evidence for nonmicrobial oxidation of manganese in soil. *Soil Sci.* 132: 153–160.

Santos, E.F., Santini, J.M.K., Paixão, A.P. et al. (2017). Physiological highlights of manganese toxicity symptoms in soybean plants: Mn toxicity responses. *Plant Physio. Biochem.* 113: 6–19.

Sasaki, A., Yamaji, N., Xia, J., and Ma, J.F. (2011). OsYSL6 is involved in the detoxification of excess manganese in rice. *Plant Physiol.* 15: 1832–1840.

Sasaki, A., Yamaji, N., Yokosho, K., and Ma, J.F. (2012). Nramp5 is a major transporter responsible for manganese and cadmium uptake in rice. *Plant Cell* 24: 2155–2167.

Savant, N.K. and Kibe, M.M. (1969). Effect of alternate submergence and drying of soil prior to resubmergence on extractable aluminium, iron and manganese in rice soils of the Maharashtra State, India. *Soil Sci.* 108: 440–444.

Shacklette, H.T., Hamilton, J.C., Boerngen, J.G., and Bowles, J.M. (1971). *Elemental composition of surface materials in the conterminous United States*. Washington: United States Government Printing Office.

Shah, V., Jacob, D.J., Moch, J.M. et al. (2020). Global modeling of cloud water acidity, precipitation acidity, and acid inputs to ecosystems. *Atmos. Chem. Phys.* 20: 12223–12245.

Shao, J.F., Yamaji, N., Shen, R.F., and Ma, J.F. (2017). The key to Mn homeostasis in plants: regulation of Mn transporters. *Trends Plant Sci.* 22: 215–224.

Shen, J.R. (2015). The structure of photosystem II and the mechanism of water oxidation in photosynthesis. *Annu. Rev. Plant Biol.* 66: 23–48.

Shenker, M., Plessner, O.E., and Tel-Or, E. (2004). Manganese nutrition effects on tomato growth, chlorophyll concentration, and superoxide dismutase activity. *J. Plant Physiol.* 161: 197–202.

Siman, A., Cradock, F.W., and Hudson, A.W. (1974). The development of manganese toxicity in pasture legumes under extreme climatic conditions. *Plant Soil* 41: 129–140.

Sirkar, S. and Amin, J.V. (1974). The manganese toxicity of cotton. *Plant Physiol.* 54: 539–543.

Socha, A.L. and Guerinot, M.L. (2014). Mn-euvering manganese: the role of transporter gene family members in manganese uptake and mobilization in plants. *Front. Plant Sci.* 5: 106. https://doi.org/10.3389/fpls.2014.00106.

Sonneveld, C. and Voogt, S.J. (1975). Studies on the manganese uptake of lettuce on steam-sterilised glasshouse soils. *Plant Soil.* 42: 49–64.

Stonier, T., Rodriguez-Tormes, F., and Yoneda, Y. (1968). Studies on auxin protectors. IV. The effect of manganese on auxin protector-I of the Japanese morning glory. *Plant Physiol.* 43: 69–72.

Suga, M., Akita, F., Hirata, K. et al. (2015). Native structure of photosystem II at 1.95 Å resolution viewed by femtosecond X-ray pulses. *Nature* 517: 99–103.

Suresh, R., Foy, C.D., and Weidner, J.R. (1987). Effects of excess soil manganese on stomatal function in two soybean cultivars. *J. Plant Nutr.* 10: 749–760.

Tabari, H. (2020). Climate change impact on flood and extreme precipitation increases with water availability. *Sci. Rep.* 10: 13768. https://doi.org/10.1038/s41598-020-70816-2.

Takagi, D., Hashiguchi, M., Sejima, T. et al. (2016a). Photorespiration provides the chance of cyclic electron flow to operate for the redox-regulation of P700 in photosynthetic electron transport system of sunflower leaves. *Photosynth. Res.* 129: 279–290.

Takagi, D., Ifuku, K., Ikeda, K.I. et al. (2016b). Suppression of chloroplastic alkenal/one oxidoreductase represses the carbon catabolic pathway in Arabidopsis leaves during night. *Plant Physiol.* 170: 2024–2039.

Takagi, D., Takumi, S., Hashiguchi, M. et al. (2016c). Superoxide and singlet oxygen produced within the thylakoid membranes both cause photosystem I photoinhibition. *Plant Physiol.* 171: 1626–1634.

Takagi, D., Amako, K., Hashiguchi, M. et al. (2017). Chloroplastic ATP synthase builds up a proton motive force preventing production of reactive oxygen species in photosystem I. *Plant J.* 91: 306–324.

Takagi, D., Ihara, H., Takumi, S., and Miyake, C. (2019). Growth light environment changes the sensitivity of photosystem I photoinhibition depending on common wheat cultivars. *Front. Plant Sci.* 10: 686. https://doi.org/10.3389/fpls.2019.00686.

Takagi, D., Miyagi, A., Tazoe, Y. et al. (2020). Phosphorus toxicity disrupts Rubisco activation and reactive oxygen species defence systems by phytic acid accumulation in leaves. *Plant Cell Environ.* 43: 2033–2053.

Takagi, D., Ishiyama, K., Suganami, M. et al. (2021). Manganese toxicity disrupts indole acetic acid homeostasis and suppresses the CO_2 assimilation reaction in rice leaves. *Sci. Rep.* 11: 20922. https://doi.org/10.1038/s41598-021-00370-y.

Takahama, U. (2004). Oxidation of vacuolar and apoplastic phenolic substrates by peroxidase: physiological significance of the oxidation reactions. *Phytochem. Rev.* 3: 207–219.

Takahama, U. and Oniki, T. (1992). Regulation of peroxidase-dependent oxidation of phenolics in the apoplast of spinach leaves by ascorbate. *Plant Cell Physiol.* 33: 379–387.

Takemoto, Y., Tsunemitsu, Y., Fujii-Kashino, M. et al. (2017). The tonoplast-localized transporter MTP8.2 contributes to manganese detoxification in the shoots and roots of *Oryza sativa* L. *Plant Cell Physiol.* 58: 1573–1582.

Tam, S.C., Sposito, G., and Senesi, N. (1991). Spectroscopic and chemical evidence of variability across a pine litter layer. *Soil Sci. Soc. Am. J.* 55: 1320–1325.

Tanaka, Y., Sano, T., Tamaoki, M. et al. (2006). Cytokinin and auxin inhibit abscisic acid-induced stomatal closure by enhancing ethylene production in Arabidopsis. *J. Exp. Bot.* 57: 2259–2266.

Tebo, B.M., Clement, B.G., Dick, G.J. et al. (2004). Biogenic manganese oxides: properties and mechanisms of formation. *Annu. Rev. Earth Planet. Sci.* 32: 287–328.

Terashima, I., Funayama, S., and Sonoike, K. (1994). The site of photoinhibition in leaves of *Cucumis sativus* L. at low temperatures is photosystem I, not photosystem II. *Planta* 193: 300–306.

Tikhonov, A.N. (2013). pH-dependent regulation of electron transport and ATP synthesis in chloroplasts. *Photosynth. Res.* 116: 511–534.

Tsunemitsu, Y., Genga, M., Okada, T. et al. (2018a). A member of cation diffusion facilitator family, MTP11, is required for manganese tolerance and high fertility in rice. *Planta* 248: 231–241.

Tsunemitsu, Y., Yamaji, N., Ma, J.F. et al. (2018b). Rice reduces Mn uptake in response to Mn stress. *Plant Signal. Behav.* 13: e1422466.

Turekian, K.K. and Wedepohl, K.H. (1961). Distribution of the elements in some major units of the earth's crust. *Geol. Soc. Am. Bull.* 72: 175–192.

Ueno, D., Sasaki, A., Yamaji, N. et al. (2015). A polarly localized transporter for efficient manganese uptake in rice. *Nat. Plants* 1: 1–8.

Umena, Y., Kawakami, K., Shen, J.R., and Kamiya, N. (2011). Crystal structure of oxygen-evolving photosystem II at a resolution of 1.9 Å. *Nature* 473: 55–60.

Valdés-López, O., Yang, S.S., Aparicio-Fabre, R. et al. (2010). MicroRNA expression profile in common bean (*Phaseolus vulgaris*) under nutrient deficiency stresses and manganese toxicity. *New Phytol.* 187: 805–818.

Véry, A.A. and Davies, J.M. (2000). Hyperpolarization-activated calcium channels at the tip of Arabidopsis root hairs. *Proc. Natl. Acad. Sci. U.S.A.* 97: 9801–9806.

Wada, H., Seirayosakol, A., Kimura, M., and Takai, Y. (1978a). The process of manganese deposition in paddy soils: I. A hypothesis and its verification. *Soil Sci. Plant Nutr.* 24: 55–62.

Wada, H., Seirayosakol, A., Kimura, M., and Takai, Y. (1978b). The process of manganese deposition in paddy soils: II. The microorganisms responsible for manganese deposition. *Soil Sci. Plant Nutr.* 24: 319–325.

Wada, S., Suzuki, Y., Takagi, D. et al. (2018). Effects of genetic manipulation of the activity of photorespiration on the redox state of photosystem I and its robustness against excess light stress under CO_2-limited conditions in rice. *Photosynth. Res.* 137: 431–441.

Wada, S., Takagi, D., Miyake, C. et al. (2019). Responses of the photosynthetic electron transport reactions stimulate the oxidation of the reaction center chlorophyll of photosystem I, P700, under drought and high temperatures in rice. *Int. J. Mol. Sci.* 20: 2068. https://doi.org/10.3390/ijms20092068.

Wang, P., Yamaji, N., Inoue, K. et al. (2020). Plastic transport systems of rice for mineral elements in response to diverse soil environmental changes. *New Phytol.* 226: 156–169.

Wedepohl, K.H. (1995). The composition of the continental crust. *Geochim. Cosmochim. Acta* 59: 1217–1232.

Wilkinson, R.E. and Ohki, K. (1988). Influence of manganese deficiency and toxicity on isoprenoid syntheses. *Plant Physiol.* 87: 841–846.

Williams, D.E. and Vlamis, J. (1957a). The effect of silicon on yield and manganese-54 uptake and distribution in the leaves of barley plants grown in culture solutions. *Plant Physiol.* 32: 404–409.

Williams, D.E. and Vlamis, J. (1957b). Manganese and boron toxicities in standard culture solutions. *Soil Sci. Soc. Am. J.* 21: 205–209.

Wissemeier, A.H. and Horst, W.J. (1987). Callose deposition in leaves of cowpea (*Vigna unguiculata* [L.] Walp.) as a sensitive response to high Mn supply. *Plant Soil* 102: 283–286.

Wissemeier, A.H. and Horst, W.J. (1992). Effect of light intensity on manganese toxicity symptoms and callose formation in cowpea (*Vigna unguiculata* (L.) Walp.). *Plant Soil* 143: 299–309.

Wu, Z., Liang, F., Hong, B. et al. (2002). An endoplasmic reticulum-bound Ca^{2+}/Mn^{2+} pump, ECA1, supports plant growth and confers tolerance to Mn^{2+} stress. *Plant Physiol.* 130: 128–137.

Xiao-Ping, S. and Xi-Gui, S. (2006). Cytokinin-and auxin-induced stomatal opening is related to the change of nitric oxide levels in guard cells in broad bean. *Physio. Plant.* 128: 569–579.

Yamaguchi, K., Suzuki, I., Yamamoto, H. et al. (2002). A two-component Mn^{2+}-sensing system negatively regulates expression of the mntCAB operon in *Synechocystis*. *Plant Cell* 14: 2901–2913.

Yang, M., Zhang, Y., Zhang, L. et al. (2014). OsNRAMP5 contributes to manganese translocation and distribution in rice shoots. *J. Exp. Bot.* 65: 4849–4861.

You, X., Yang, L.T., Lu, Y.B. et al. (2014). Proteomic changes of Citrus roots in response to long-term manganese toxicity. *Trees* 28: 1383–1399.

Zhang, J.Y., He, S.B., Li, L., and Yang, H.Q. (2014). Auxin inhibits stomatal development through MONOPTEROS repression of a mobile peptide gene STOMAGEN in mesophyll. *Proc. Natl. Acad. Sci. U.S.A.* 111: E3015–E3023.

Zhao, J., Wang, W., Zhou, H. et al. (2017). Manganese toxicity inhibited root growth by disrupting auxin biosynthesis and transport in Arabidopsis. *Front. Plant Sci.* 8: 272. https://doi.org/10.3389/fpls.2017.00272.

Zivcak, M., Brestic, M., Balatova, Z. et al. (2013). Photosynthetic electron transport and specific photoprotective responses in wheat leaves under drought stress. *Photosynth. Res.* 117: 529–546.

Zorina, A., Sinetova, M.A., Kupriyanova, E.V. et al. (2016). Synechocystis mutants defective in manganese uptake regulatory system, ManSR, are hypersensitive to strong light. *Photosynth. Res.* 130: 11–17.

8

Iron Excess Toxicity and Tolerance in Crop Plants

Insights from Omics Studies

May Sann Aung and Hiroshi Masuda*

Department of Biological Production, Faculty of Bioresource Sciences, Akita Prefectural University, Akita City, Japan

8.1 Iron Uptake and Translocation Mechanism in Plants

8.1.1 Importance of Iron in Living Organisms

Iron (Fe) is an essential element in almost all living organisms, including plants and animals. Fe is a transition element that can take one of multiple oxidation numbers because the d orbitals of its electrons are unsaturated, and the energy level of the next *s* orbital is close to that of the d orbital. The electron configuration of an Fe atom is $1s^2\ 2s^2\ 2p^6\ 3s^2\ 3p^6\ 3d^6\ 4s^2$, where the two electrons in the outermost s orbital and the electrons in the inner d orbital are easily displaced (Demazeau et al. 1982). When the two electrons in the outermost 4s orbital are dislodged, they produce an Fe(II) or Fe^{2+} ion ($1s^2\ 2s^2\ 2p^6\ 3s^2\ 3p^6\ 3d^6$). In addition, the 3d orbital becomes stable in a half-closed shell when it has five electrons, such that one electron tends to be dislodged to become an Fe(III) or Fe^{3+} ion ($1s^2\ 2s^2\ 2p^6\ 3s^2\ 3p^6\ 3d^5$). Thus, an Fe ion can take one of two different oxidation numbers, Fe(II) or Fe(III), and can become an active center for enzymes involved in redox reactions *in vivo*.

In living organisms, elemental Fe becomes functional in either non-heme or heme form. Non-heme Fe exists primarily as Fe–S clusters, which can assume various redox potentials depending on the type of active center (da Silva and Williams 2001); typical Fe-S proteins include rubredoxin and ferredoxin. By contrast, heme Fe is chelated to a porphyrin ring. Non-heme Fe has existed since life began on Earth, whereas heme Fe developed as organisms evolved to utilize O_2 as O_2 levels increased in the environment. Heme proteins include molecules that bind to oxygen molecules, such as hemoglobin, for transport and storage; cytochromes, which constitute the respiration and photosynthesis electron transfer systems; catalases and peroxidases, which are involved in hydrogen peroxide decomposition; and oxidase, which introduces oxygen atoms into organic compounds. Thus, many heme proteins have functions related to oxygen molecules and oxidation reactions.

*Corresponding Author.
Both authors contributed equally

Heavy Metal Toxicity and Tolerance in Plants: A Biological, Omics, and Genetic Engineering Approach, First Edition. Edited by Mohammad Anwar Hossain, AKM Zakir Hossain, Sylvain Bourgerie, Masayuki Fujita, Om Parkash Dhankher, and Parvez Haris.

In plants, Fe takes part in several metabolic processes, including mitochondrial respiration, photosynthesis, and nutrient transport, due to its high redox activity that allows it to readily accept and donate electrons (Marschner 1995). Fe also participates in the regulation of protein stability and many chemical reactions, including hydration, dehydration, redox reactions, and photo redox catalysis, which contribute to Fe detoxification (Zhang et al. 2012; Jain and Connolly 2013; Dufey et al. 2015). Therefore, Fe is an essential mineral element for plant growth and development.

8.1.2 Fe Acquisition Systems in Plants

Higher plants use two major Fe uptake mechanisms to subsist in Fe-limited environments: Fe reduction (Strategy I) and Fe chelation (Strategy II) (Römheld and Marschner 1986). In Strategy I, Fe is readily oxidized in aerobic or alkaline soils, and the oxidized Fe is insoluble and less available to plants. Protons are released from plant roots to acidify the rhizosphere, and Fe^{3+} is reduced to the more soluble Fe^{2+} form through the induction of membrane-localized Fe(III) chelate reductase, followed by the uptake of Fe by plant roots via high-affinity Fe^{2+} transporters. In *Arabidopsis thaliana*, plant roots release H^+-ATPase (*Arabidopsis* H^+-ATPase 2, *AtAHA2*) to acidify the rhizosphere, thereby enhancing Fe solubilization (Santi and Schmidt 2009). Coumarins play a vital role in the adaptation of Arabidopsis to alkaline soils (Rodríguez-Celma et al. 2013; Schmid et al. 2014). Arabidopsis roots secrete coumarins to the rhizosphere as fraxetin in neutral to alkaline pH soils (Tsai and Schmidt 2020), or as sideretin in acidic soils (Gautam et al. 2021) via AtABCG37/AtPDR9 transporters to chelate and mobilize Fe^{3+} (Fourcroy et al. 2014; Sisó-Terraza et al. 2016). Fe^{3+} reduction to Fe^{2+} is followed by root induction of Fe (III)-chelate reductase (*AtFRO2*, ferric reduction oxidase 2) (Robinson et al. 1999), which is likely the rate-limiting step of Fe acquisition from soil (Connolly et al. 2003). Thus, plants depending on Strategy I Fe uptake may suffer from Fe deficiency when AtFRO2 exhibits low efficiency for Fe^{3+} reduction in alkaline soils (Liang 2022), where the bioavailability of Fe^{3+} decreases 1000-fold for each pH unit reduction between pH 9 and 4 (Lindsay 1997). The resulting Fe^{2+} is taken up directly by plant roots via a major iron-regulated transporter (IRT) (Eide et al. 1996).

By contrast, Strategy II involves a Fe chelation-based mechanism exploited by all graminaceous plants, including rice, to acquire Fe from soil (Römheld and Marschner 1986). In Strategy II, plants synthesize and secrete mugineic acid-family phytosiderophores (MAs) into the soil to acquire Fe (Takagi 1976; Marschner et al. 1986). MAs have a high capacity to solubilize and chelate Fe^{3+} in the rhizosphere, particularly in high-pH soils (von Wirén et al. 2000). Plants absorb the Fe(III)-MAs complex via the yellow stripe 1 (YS1) and yellow stripe-like (YSL) transporters (Curie et al. 2001). In the MA biosynthetic pathway, methionine is synthesized into *S*-adenosyl-L-methionines (SAMs) by SAM synthase (SAMS) (Mori and Nishizawa 1987; Shojima et al. 1990). Three molecules of SAM are combined to form nicotianamine (NA), which is synthesized by NA synthase (NAS) (Higuchi et al. 1999). NA is synthesized in both graminaceous and non-graminaceous plants. In graminaceous plants, NA is further converted to a 3″-keto intermediate by NA aminotransferase (NAAT) (Takahashi et al. 1999; Inoue et al. 2008) and then to 2′-deoxymugineic acid (DMA) by DMA synthase (DMAS) (Bashir et al. 2006).

Rice also synthesizes and secretes DMA from the roots that reaches the rhizosphere through the membrane-localized MA transporter (TOM1) (Nozoye et al. 2011). The secreted DMAs chelate insoluble Fe(III) and solubilize it in the rhizosphere to form

Fe(III)-DMA complexes absorbed by root cells via the YSL transporter, OsYSL15 (Inoue et al. 2009; Lee et al. 2009). Unlike other graminaceous plants, rice grows well in anaerobic paddy field conditions, where Fe^{3+} is easily reduced to Fe^{2+} (which is readily available to plants). Thus, rice also uses a partial Strategy I Fe uptake mechanism, in which it takes up Fe^{2+} directly via the Fe^{2+} transporter OsIRT1 (Bughio et al. 2002; Ishimaru et al. 2006). However, rice shows lower tolerance to Fe limitation than other graminaceous plants due to its lack of a complete Strategy II system and low Fe(III)-chelate reductase activity (Ishimaru et al. 2006). Bashir et al. (2011) reported that rice in Fe-limited soil also secretes other compounds, such as protocatechuic acid (PCA) and caffeic acid (CA), into the rhizosphere via the phenolics efflux zero 2 (OsPEZ2) transporter. However, OsPEZ2 was not expressed under Fe deficiency (Kobayashi et al. 2013). Thus, chelation and reduction of Fe^{3+} may be limited in such cases.

8.1.3 Fe Translocation Mechanisms in Plants

NA is as a precursor of DMA that acts as a ubiquitous plant-derived chelator of various divalent cations, including Fe^{2+} and Zn^{2+}, within plants; it is utilized for internal transport of metals including Fe (Hell and Stephan 2003; Takahashi et al. 2003). The NA-defective tomato mutant *chloronerva* (Rudolph et al. 1985) exhibited an Fe-deficient phenotype (Pich and Scholz 1996; Stephan et al. 1996). In rice, NA is biosynthesized by three NAS enzymes: OsNAS1, OsNAS2, and OsNAS3 (Higuchi et al. 2001). Increased Fe and Zn concentrations in brown rice caused by foliar application of NA–Fe to field-grown rice (Yuan et al. 2012) demonstrated the involvement of NA in the long-distance internal transport of Fe and Zn to grains.

YSL transporters play a vital role in Fe homeostasis in plants. OsYSL2 transports Fe(II)-NA and manganese (Mn)-NA in rice (Koike et al. 2004). Lower Fe concentrations were found in *OsYSL2* knockdown rice compared with non-transgenic rice, with 18% and 39% decreases observed in brown and polished seeds, respectively (Ishimaru et al. 2010), demonstrating the role of OsYSL2 in Fe translocation to seeds (Ishimaru et al. 2010). Fe is transported mainly as Fe(III)-citrate within the xylem (Yoneyama et al. 2015). Expression of the Fe(III)-DMA transporter OsYSL15 is induced in the root epidermis and stele and participates in both Fe uptake and internal Fe translocation (Inoue et al. 2009; Lee et al. 2009). Disruption of OsYSL15 leads to less Fe accumulation in seeds (Lee et al. 2009), indicating that OsYSL15 may function in Fe loading to the phloem to transport Fe from roots to leaves and seeds. Kakei et al. (2012) demonstrated that OsYSL16 transports Fe(III)-DMA in rice xylem sap via xylem-to-phloem nodes. Another Fe(III)-DMA transporter, OsYSL18, is expressed in the reproductive organs and phloem of lamina joints (Aoyama et al. 2009). Thus, OsYSL18 may function in unloading Fe(III)-DMA from the phloem in rice panicles. These transporters are all involved in Fe translocation within plants, particularly under Fe-limited conditions.

8.2 Fe Excess Toxicity in Plants

8.2.1 Fe Excess Toxicity in Global Agriculture

Fe excess stress is among the major yield-limiting factors in agriculture worldwide, usually occurring in acid soils and waterlogged environments. Acid soils cover approximately 30% (3950 million ha) of the global land area and are believed to represent >50% of potentially arable land worldwide (Von Uexküll and Mutert 1995; Sumner and Noble 2003). Approximately 60% of lowland paddy fields with poor drainage are at risk of Fe excess toxicity (WARDA Annual Report 2001–2002). Approximately 67% of acidic soils worldwide are Ultisols, Entisols, and Oxisols (Rengel 2015). Among soil types, acid sulfate, acid clay, and peat soils are the main causes of Fe excess toxicity (Becker and Asch 2005). Acid sulfate soil is mainly distributed in Southeast and Pacific Asia, with 4.109 million ha in Indonesia, 2.14 million ha in Vietnam, 0.657 million ha in Thailand, and 0.50 million ha in Malaysia (Dash et al. 2021). Rice is a staple food for half the global population, and demand therefore increases as the size of the global population increases. However, rice is typically grown in flooded conditions such that Fe excess toxicity is a major constraint on its production, especially in Southeast Asia, India, and China (Fageria and Rabelo 1987; Sahrawat 2004; Mahender et al. 2019), western and central Africa (Gridley et al. 2006), and Brazil (Crestani et al. 2009). South Asian countries experience approximately 55% yield loss in wetland rice cultivation (Sikirou et al. 2016). In sub-Saharan Africa, 20–60% of the rice-growing area was affected by Fe excess toxicity, leading to 10–90% yield losses (Rodenburg et al. 2014; Sikirou et al. 2015).

8.2.2 Causes of Fe Excess Toxicity in Soils and Its Interaction with Plants

8.2.2.1 State of Fe in Soils and Soil pH Effects on Fe Excess Toxicity

Fe is the fourth most abundant element in the Earth's crust (Frey and Reed 2012). Most Fe in soil exists in the form of Ferric ions (Fe^{3+}, trivalent iron), which are stable and insoluble oxides or hydroxides. Less Fe exists as ferrous ions (Fe^{2+}, bivalent iron), which are unstable in oxygen-containing environments and readily oxidize to Fe^{3+} (Guerinot and Yi 1994). Thus, Fe availability to plants varies because of highly heterogeneous soil conditions. Under oxidative conditions with sufficient oxygen, such as upland cultivation, Fe exists mainly as Fe^{3+}, which is less soluble in water and less available to plants. Under oxygen-deficient conditions, such as lowland cultivation with waterlogging, Fe exists mainly as Fe^{2+}.

Rice plants are typically cultivated in submerged, anaerobic paddy fields, where Fe^{2+} ions are abundant. Low-pH acidic soils favor the reduction of Fe^{3+} to Fe^{2+}, which is toxic at elevated levels and under high ionic activity (Rengel 2002; Khabaz-Saberi and Rengel 2010). An estimated 10–2000 mg kg^{-1} of Fe^{2+} ions are present in acid sulfate soils with pH < 5 (da Silveira et al. 2007). Ferric ions (Fe^{3+}) are less soluble in water ($Ksp = [Fe^{3+}][OH^-]^3 = 1 \times 10^{-36}$) (Stumm and Lee 1961); thus, Fe^{3+} ions merely dissolve in water ($\sim 1 \times 10^{-9}$ mol l^{-1} at pH 5) (Aung and Masuda 2020). Compared to Fe^{3+} ions, Fe^{2+} are highly soluble in water ($Ksp = [Fe^{2+}][OH^-]^2 = 8 \times 10^{-16}$) at 25 °C (Stumm and Lee 1961), which allows them to dissolve easily (~800 mol l^{-1} at pH 5) (Aung and Masuda 2020). Thus, all Fe^{2+} ions present in soil are soluble in acidic soils (pH ≤ 5), suggesting that excess Fe in acidic soils is caused by Fe^{2+} ions, rather than Fe^{3+} ions. Greater increases in acidity are associated with greater solubility and availability of cation Fe^{2+} in the soil. Therefore, Fe excess toxicity can occur in acidic soils with submerged anaerobic conditions (e.g. paddy fields), or in Fe soils contaminated by factory waste (e.g. in Brazil) (Fageria et al. 2008; Khabaz-Saberi and Rengel 2010). Soil pH is therefore a major cause of nutrient imbalance; plants cannot absorb soil Fe and Fe deficiency occurs at high pH levels (alkaline conditions), whereas Fe toxicity occurs at low pH levels (acidic conditions).

8.2.2.2 Soil Improvement Methods to Ameliorate Fe Excess Toxicity

Low-pH (acidic) soil causes Fe excess toxicity in plants. Excess Fe absorbed by plant roots is transported to shoots via xylem flow, leading to severe Fe excess toxicity, nutrient imbalance, and poor growth and development. To cope with Fe excess stress, lime [calcium oxide (CaO), calcium carbonate ($CaCO_3$), or calcium hydroxide [$Ca(OH)_2$] can be applied to effectively neutralize soil acidity. Fe excess toxicity in flooded rice fields is dramatically decreased through lime application (Sadiq and Babagana 2012). In anaerobic waterlogged soils, lower soil pH is associated with higher Fe^{2+} concentrations and greater toxicity to plants. The application of phosphorus nutrients reduced Fe^{2+} concentrations in anaerobic soils (Alam 1983) and prevented Fe mobility in embryonic shoots and roots (Ramírez et al. 2002). The application of potassium and zinc nutrient fertilizers also reduced Fe excess toxicity in anaerobic soil (Ismunadji and Ardjasa 1989; Ramírez et al. 2002). Other nutrients have also been applied over long periods for avoiding Fe excess toxicity. However, these treatments are costly, and new approaches such as the development of Fe-tolerant genotypes are more desirable for sustainable agriculture.

8.2.2.3 Soil Water and Drainage Effects on Fe Excess Toxicity

Upland and lowland rice cultivars may have different Fe uptake strategies due to their aerobic and anaerobic growth environments, respectively. Fe states also differ between upland and lowland regions (Figure 8.1). For example, in Brazil, i.e. the second largest rice producer

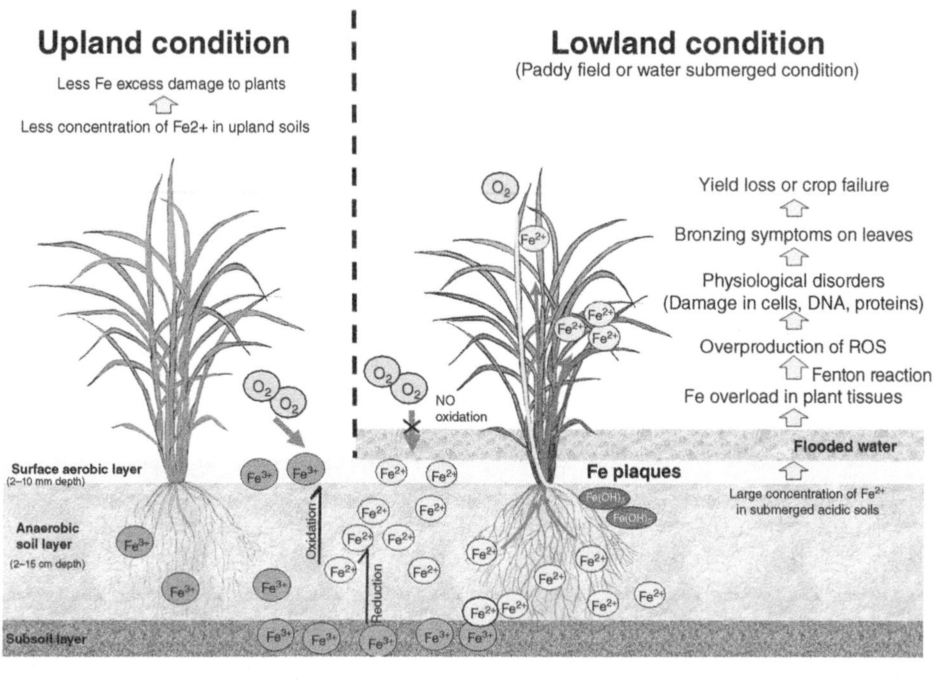

Figure 8.1 Iron (Fe) reduction in submerged soils under lowland conditions and the relative effects of Fe excess toxicity, compared to the rice plants in upland soils. In submerged soils under anaerobic, low-pH conditions, Fe^{3+} is reduced to the more soluble Fe^{2+}. Red arrows indicate Fe uptake and transport by rice. Yellow arrows indicate rhizospheric oxidation of Fe^{2+} to Fe^{3+} through oxygen transport from shoot to root, leading to the formation of Fe plaques on the root surface. ROS, reactive oxygen species. *Source:* Adapted from Aung and Masuda (2020).

worldwide after Asia, Fe excess in soil is a constraint on rice production. When two upland and two lowland rice cultivars were grown under excess Fe conditions in Brazil, and Fe accumulation in the lowland cultivars was higher in roots, but not significantly different in shoots, compared with upland cultivars (Müller et al. 2017). These results suggest that lowland rice varieties have higher Fe retention capacity than upland varieties, probably because lowland varieties have developed Fe uptake and tolerance mechanisms for anaerobic soils with high levels of available Fe^{2+}.

Excess Fe occurs mainly in waterlogged and flooded areas. Soil aeration by drainage can oxidize and precipitate excess Fe^{2+} into Fe^{3+}. Therefore, agronomic crop management practices, including water drainage, may mitigate the negative impacts of Fe excess toxicity. A recent study reported that agronomic water irrigation management successfully alleviated Fe excess stress in rice plants (de Campos Carmona et al. 2021); in a control field with continuous irrigation, a resistant cultivar (IRGA 425) had higher grain yield than a sensitive cultivar (BR-IRGA 409). However, two cycles of water suppression resulted in similarly high grain yields in both cultivars. Thus, field water drainage can significantly improve plant growth and yield, lowering oxidative stress and Fe accumulation in leaves. This field drainage practice is a technically feasible approach for farmers to increase rice production in regions affected by Fe excess toxicity.

8.2.3 Effects of Fe Excess Toxicity on Plant Growth

Fe excess stress causes tissue damage and cellular homeostasis disorders in plants. The sequential effects of Fe-mediated damage to rice plants have been discussed (Figure 8.1). Fe(II) toxicity inhibits cell division and elongation of primary roots, and consequently lateral root growth (Li et al. 2015). Transpiration induces xylem transport of massive amounts of Fe from roots to shoots, where cellular Fe overload occurs in plant tissues (Briat and Lobréaux 1997). Cellular compartments within the plant body vary in pH, causing differences in Fe solubility (Table 8.1). The Fe^{2+} solubility is significantly higher in low-pH cellular compartments, such as the xylem and apoplasts.

Once Fe^{2+} ions are oxidized to Fe^{3+} ions, the Fe^{3+} ions are immobilized, less reactive, and cause less cellular damage because they are readily precipitated in most plant tissues. However, Fe^{2+} ions are highly soluble and reactive in plant tissues. In the absence of chelation or isolation, excessive Fe^{2+} ions can exist freely in the cytosol. Such free Fe^{2+} may cause toxic reactions, such as the Fenton reaction. Reactive oxygen species (ROS) are highly reactive chemicals formed from O_2, including peroxides, superoxide, and hydroxyl radicals. The

Table 8.1 Solubility of ferric ions (Fe^{3+}) and ferrous ions (Fe^{2+}) in various cellular compartments.

Cellular compartments	pH levels	Fe^{3+} solubility (mol l^{-1})	Fe^{2+} solubility (mol l^{-1})
Typical xylem sap	5.5 (Ariga et al. 2014)	~3×10^{-10}	~80
Apoplast	6.0 (Geilfus 2017)	~1×10^{-12}	~8
Cytosol	8.0 (Martinière et al. 2013)	~1×10^{-18}	~8×10^{-4}

Fenton reaction causes ROS overproduction, especially of the cytotoxic hydroxyl radical (•OH) (Becana et al. 1998; Thongbai and Goodman 2000), which causes 60–70% cellular DNA damage. Notably, Fe overload in chloroplasts leads to critical ROS-driven oxidative damage (Balk and Schaedler 2014). This cellular damage interrupts or disrupts photosystem II, and enhances cytochrome b6/f content in thylakoids, leading to lower photosynthetic rates and higher oxygen production (Suh et al. 2002). In turn, this can cause permanent injury and dysfunction of cellular structures, membranes, lipids, proteins, and DNA (Briat et al. 1995; Stein et al. 2009a). Fe-overloaded cells impair biological processes, leading to substantial leaf bronzing and cell death. Cellular oxidative damage caused by Fe excess leads to several changes in morpho-physiological and yield traits in rice (Hell and Stephan 2003; Sikirou et al. 2015), and inhibits the uptake of major nutrients essential for plant growth (Audebert and Sahrawat 2000; Nugraha et al. 2016); this leads to complete crop failure under severe toxicity.

8.3 Crop Defense Mechanisms Against Excess Fe and Genes Regulating Fe Excess

Plants may employ different defense mechanisms to avoid Fe excess. The Fe resistance of a cultivar may partly depend on its Fe excess stress mitigation strategy. Recent studies on Fe excess in plants have applied various analysis techniques including transcriptomics, genome-wide association mapping, RNA sequencing, quantitative trait locus mapping, and bioinformatics approaches. Rice is a special crop for Fe excess tolerance because it is cultivated in paddy fields where Fe^{2+} is abundant. Studies on physiological and transcriptomic analyses of Fe excess responses in rice were reported (Quinet et al. 2012; Finatto et al. 2015; Aung et al. 2018b). For example, comprehensive transcriptomic analysis by Aung et al. (2018b) showed that rice physiological and molecular responses vary among rice tissues [roots, junction node/discrimination center (DC), stems, and old leaves and newest leaves] and Fe excess exposure levels (×1, ×10, ×20, ×50, ×70 Fe). The results of these studies show that expression patterns in various tissues are correlated in several upregulated and downregulated genes. In rice, four defense mechanisms (I–IV) have been hypothesized based on previous findings on physiological and molecular responses and regulation mechanisms in plants (Aung and Masuda 2020). The roles and functions of Fe excess-regulated genes and candidate genes involved in these four defense mechanisms, as well as a hypothesized model, are shown in Figure 8.2.

8.3.1 Defense I: Fe Exclusion from Roots

In Defense I, rice plants inhibit root Fe uptake in response to excess Fe. Shoot-to-root oxygen transport in plants causes rhizospheric oxidation of Fe^{2+} to Fe^{3+}, leading to Fe^{3+} precipitation on the root surface and the formation of Fe plaques. These Fe plaques act as a barrier to the uptake of excess Fe^{2+} into root tissues (Becker and Asch 2005). Substantial Fe plaque formation on root surfaces has been observed in tolerant genotypes with high Fe exclusion ability (Mahender et al. 2019). Wu et al. (2014) reported that wider pith cavity diameters in shoots, increased root aerenchyma volume, and greater numbers of lateral

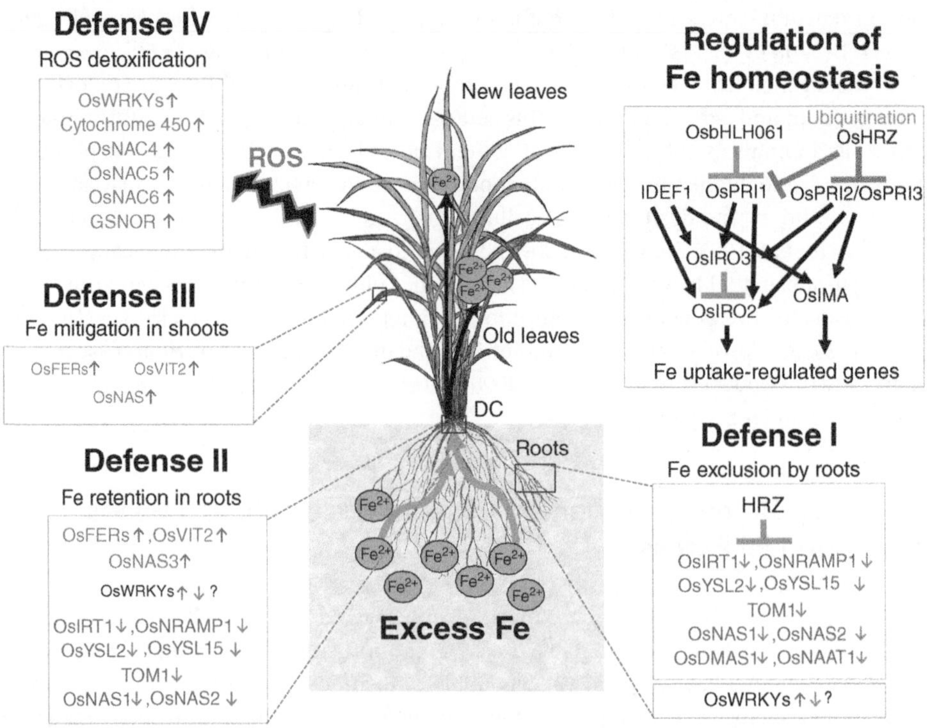

Figure 8.2 Hypothetical model of the four defense mechanisms and the regulation of Fe homeostasis in rice exposed to excess Fe. Defense I refers to Fe exclusion in roots. Defense II refers to Fe retention in roots and avoidance of Fe translocation to shoots. Defense III refers to Fe compartmentalization in shoots. Defense IV refers to ROS detoxification throughout the plant. Red and blue letters indicate highly induced and suppressed genes, respectively. DC, discrimination center. *Source:* Adapted from Aung and Masuda (2020).

roots favored internal oxygen movement in plants, and consequently high root oxidation in an Fe-exclusion genotype. Silicon application can also reduce Fe uptake in rice, thereby alleviating Fe excess toxicity (Ma and Takahashi 1990). Silicon supply can strengthen Casparian bands in the root epidermis, which function as a diffusion barrier and inhibit Fe flux into roots under excess Fe conditions (Hinrichs et al. 2017; Becker et al. 2020). Thus, rice genotypes that have high root oxidation power, large aerenchyma, several lateral roots, and/or strong Casparian bands may exhibit high Fe exclusion by roots, thereby enhancing Fe excess tolerance.

8.3.1.1 Genes Involved in Defense I

Microarray analyses have revealed that, in plants grown under Fe excess stress with mild to high Fe excess (Aung et al. 2018b), high Fe excess (Quinet et al. 2012), or very high Fe excess (Finatto et al. 2015), the expression of Fe uptake- and transport-related genes such as *OsIRT1*, *OsIRT2*, *OsYSL2*, *OsYSL15*, and *OsNRAMP1* is highly suppressed in roots. These findings indicate that rice prevents Fe uptake and transport in response to excess Fe. Under such conditions, the expression of MA biosynthesis genes such as *OsNAS1*, *OsNAS2*,

OsNAAT1, and *OsDMAS1* is highly suppressed in roots, indicating that rice plants restrict DMA release into the rhizosphere under excess Fe (Aung et al. 2018b).

Two Fe-binding ubiquitin ligases, OsHRZ1 and OsHRZ2, negatively regulate Fe deficiency response genes in rice (Kobayashi et al. 2013). When HRZ-knockdown rice lines were subjected to mild to high Fe excess stress, they exhibited hypersensitivity to excess Fe (severe growth defects and leaf bronzing) and 5- to 10-fold hyperaccumulation of Fe in leaves compared with control plants (Aung et al. 2018a). The expression of Fe uptake- and transport-related genes, including *OsIRT1*, *OsYSL2*, *OsYSL15*, *TOM1*, *OsNAS1*, *OsNAS2*, and *OsIRO2*, was highly elevated in the roots of knockdown plants (Aung et al. 2018a). Therefore, HRZ is a crucial protein for protecting plant cells from Fe excess toxicity and maintaining Fe homeostasis in plants through the repression of genes participating in Fe uptake and translocation (Aung et al. 2018a; Aung and Masuda 2020). Together, these physiological and molecular responses demonstrate that rice grown under excess Fe conditions use Defense I Fe exclusion to prevent Fe uptake by roots (Figure 8.2).

8.3.2 Defense II: Fe Retention in Roots and Suppression of Fe Translocation to Shoots

In Defense II, rice plants retain excess Fe in roots to prevent its translocation to shoots (Tadano 1975), thereby avoiding Fe accumulation and maintaining homeostasis in leaves (Becker and Asch 2005; da Silveira et al. 2007). In a field experiment with high-Fe soil, the tolerant genotype EPAGRI 108 had higher Fe concentrations in root symplast cells (7 $mg\,g^{-1}$ dry weight) than the sensitive genotype BR IRGA 409 (1.93 $mg\,g^{-1}$ dry weight) (Stein et al. 2014). Tolerant genotypes allocate approximately 94% of the Fe in their root apoplasts under excess Fe. Thus, some tolerant genotypes may use Defense II to maintain elevated intracellular and apoplastic Fe concentrations in roots.

Protecting aboveground parts or shoots, especially the newest leaves, from Fe overload is crucial for plant survival. Graminaceous plants absorb various metals through the roots and accumulate these in the DC or basal part of the shoot. The DC regulates mineral distribution to other parts of the plant (Mori 1998). Thus, the DC may play a critical role in Fe partitioning to the aerial parts of plants and act as a barrier for root-to-shoot transport of excessive Fe.

8.3.3 Defense III: Fe Compartmentalization in Shoots

Plants may transport Fe from roots to old or mature leaves via the DC to avoid Fe excess through its compartmentalization, disposal, or storage in shoots, when the Defense I and II systems are insufficient (Aung and Masuda 2020). Some tolerant rice genotypes maintain shoot growth through greater Fe accumulation in aerial shoots (Onaga et al. 2013). For example, Indonesian rice genotypes including Siam Saba, Mahsuri, Margasari, and Pokkali have demonstrated shoot-based tolerance, with high shoot Fe concentrations, lower leaf bronzing levels, and higher grain yields (Nugraha et al. 2016). For example, a ninefold increase in root Fe concentrations was observed under 20-fold Fe excess treatment, eventually reaching a constant concentration at higher Fe treatment levels, while shoot Fe concentrations increased in proportion to Fe excess stress levels (Aung et al. 2018b). Thus, the capacity of a rice plant to retain excess Fe in root tissues is often limited. Once Fe levels are enhanced, they cannot be reduced through Defense II, such that rice plants translocate excess Fe from roots to shoots. Notably, higher Fe excess stress levels are associated with greater translocation of Fe to old leaves, which show bronzing earlier than new leaves. Thus, through the function of the DC, Fe is preferentially transported and allocated to old leaves rather than new leaves to avoid damage caused by excess Fe in new leaves (Aung et al. 2018b).

8.3.3.1 Genes Involved in Defense II and III

The transcriptomic analyses have revealed that in Fe-overloaded roots, Fe uptake and transport-related genes such as *OsIRT1*, *OsYSL2*, *OsTOM1*, *OsNRAMP1*, and *OsYSL15* are remarkably suppressed in the DC (Aung et al. 2018b). These genes may help prevent Fe loading in shoots, as well as Fe retention in roots and the DC (Defense II in Figure 8.2). The citrate efflux transporter OsFRDL1 is important for root-to-shoot Fe translocation; its expression in nodes is responsible for Fe distribution to rice grains (Yokosho et al. 2016). Microarray analyses have shown that the expression of *OsFRDL1* and *OsFRDL2* is highly induced in roots and the DC at high Fe excess stress levels (Aung et al. 2018b). Thus, OsFRDL1 and OsFRDL2 may preferentially distribute Fe to old leaves to prevent damage to new ones.

Three genes, *Ferritin*, *OsVIT2*, and *OsNAS3*, are thought to play important roles in Defense II and III (Figure 8.2). Once in the cells, Fe^{2+} ions can be assimilated into proteins or stored in cellular compartments such as ferritins and vacuoles. Ferritin is a ubiquitous Fe storage protein that stores up to 4000 Fe atoms within a complex (Theil 2003). This protein forms non-toxic complexes in plant tissues and protects them against oxidative stress (Ravet et al. 2009; Stein et al. 2009b). The genes encoding rice ferritin genes are *OsFER1* and *OsFER2*. In both monocot and dicot plants, these ferritin genes are transcriptionally induced in response to excess Fe (Stein et al. 2009b; Briat et al. 2010). Fe cannot react with oxygen when it is stored in ferritin molecules; therefore, ferritins are critical for plant defense against the oxidative stress induced by Fe (Ravet et al. 2009; Briat et al. 2010). Microarray analyses of rice tissues have shown that OsFER1 and OsFER2 are strongly upregulated in roots and shoots under excess Fe(II) treatment (Quinet et al. 2012; Finatto et al. 2015; Aung et al. 2018b) (Figure 8.2). Therefore, plants sequester excess Fe into ferritin as a safe and bioavailable form of Fe through OsFER1 and OsFER2 (Briat and Lobréaux 1997; da Silveira et al. 2007). Müller et al. (2017) reported that *OsFER1* expression tends to be higher in lowland cultivars, implying that lowland varieties likely have higher Fe retention capacity (Defense II) than upland ones. This phenomenon may be due to lowland varieties having Fe uptake mechanisms for anaerobic soils, where high Fe^{2+} availability in rice growing areas can result in excessive Fe uptake.

Vacuoles are important cellular organelle for Fe isolation, which protects the plant from cellular toxicity. Vacuoles of various cell types are among the main storage locations for Fe. The vacuolar Fe transporters OsVIT1 and OsVIT2 are responsible for the transport of Fe from the cytosol to vacuoles via the tonoplast (Zhang et al. 2012). *OsVIT1* is less responsive to Fe abundance, whereas *OsVIT2* is highly responsive to Fe abundance (Zhang et al. 2012). Microarray results have also shown that *OsVIT1* expression is suppressed in all tissues, but mainly in roots, whereas *OsVIT2* is significantly upregulated in all tissues (Aung et al. 2018b). Thus, OsVIT2 may play a crucial role in the sequestration of excess Fe into vacuoles in a non-toxic form (Figure 8.2).

In all higher plants, NA is a major chelator of metal cations; it is involved in Fe uptake through the MA synthesis pathway, as well as Fe transport within the plant body. Among the three NAS genes in rice, *OsNAS1* and *OsNAS2* are involved in the Fe deficiency response (Inoue et al. 2003), whereas *OsNAS3* expression is strongly induced in response to excess Fe in various tissues, including roots, DC, stems, and old and new leaves (Aung et al. 2018b). Similarly, methionine cycle-related genes are not entirely suppressed in roots or other tissues under excess Fe. Thus, the methionine cycle appears to produce NA along with OsNAS3 under excess Fe conditions (Aung and Masuda 2020). Under excess Fe, *OsNAS3* expression was dominant in the vascular bundle, epidermis, and exodermis of roots, stems, and old leaves according to promoter-β-glucuronidase (GUS) analyses (Aung et al. 2019). In addition, *OsNAS3* T-DNA knockout revealed high sensitivity to excess Fe, reduced growth, severe leaf bronzing, and Fe accumulation in leaves. These findings suggest that OsNAS3 plays an important role in mitigating the effects of excess Fe. Because Fe chelated with NA does not cause the Fenton reaction, excess Fe chelated by OsNAS3-synthesized NA can be transported into vacuoles in a safe form. NA also participates in Fe buffering to maintain Fe homeostasis (Pich et al. 2001), Fe scavenging, and cell protection from oxidative damage (von Wirén et al. 1999). Thus, NA synthesized by OsNAS3 functions in the chelation of excess Fe and contributes to efficient Fe translocation and sequestration, which demonstrates its important role in the mitigation of excess Fe in the context of the Defense II and III Fe tolerance mechanisms (Figure 8.2). Thus, Defense II and III import Fe to the organelles as a novel strategy for avoiding Fe-driven toxicity.

8.3.3.2 Role of YSL4 and YSL6 Transporters in Preventing Fe Excess in Early Plant Development

YSL4 and YSL6 are further candidate transporter genes contributing to Defense III. Most of the Fe in plants is contained in chloroplasts (Terry and Abadía 1986). YSL4 and YSL6 function in the Fe transport response from the chloroplast to the cytoplasm (Divol et al. 2013); in *Arabidopsis*, knockdown of these transporters significantly reduced the plant's ability to tolerate excess Fe. In the absence of both genes (double-knockout mutants), high amounts of Fe were trapped/accumulated in the chloroplasts of old leaves, and more ferritin accumulated in *ysl4* and *ysl6* mutant embryos than in wild-type ones. The expression of these transporters is also upregulated in response to excess Fe, suggesting that these proteins play a role in Fe detoxification by regulating chloroplast Fe concentrations, by removing Fe from the chloroplast during early plant development stages such as embryo maturation and germination.

8.3.4 Defense IV: ROS Detoxification

Fe inclusion as a mechanism for tolerance to cellular Fe overload is facilitated by enzymatic detoxification in the symplast (Becker and Asch 2005; Stein et al. 2009a), ROS scavenging by antioxidants such as ascorbic acid and reduced glutathione (Fang et al. 2001; Gallie 2013), and the prevention of ROS damage through the activity of antioxidant enzymes such as superoxide dismutase, peroxidase, and catalase (Bode et al. 1995; Fang and Kao 2000). Collectively, these defense mechanisms are categorized as Defense IV. Cultivation of rice with different excess Fe levels revealed mild-to-moderate excess Fe did not seriously affect rice growth where the genes involved in Defense I–III function prominently (Aung et al. 2018b). Genes contributing to Defense IV tend to function under conditions of high to severe Fe excess stress that cause a significant reduction in plant growth and severe leaf bronzing.

8.3.4.1 Genes Involved in Defense IV

Plants induce the expression of genes related to oxidative stress and oxygen and electron transfer, as well as those encoding cytochrome P450 family proteins in response to severe excess Fe (Quinet et al. 2012; Aung et al. 2018b). Plants also induce the expression of transcription factors related to cell death and stress responses (*OsNAC4*, *OsNAC5*, and *OsNAC6*) (Finatto et al. 2015; Aung et al. 2018b) to mitigate ROS-induced excess Fe. *S*-nitrosoglutathione-reductase (GSNOR) enhances root tolerance to excess Fe by inhibiting Fe-dependent nitrosative and oxidative cytotoxicity in plants via the nitric oxide pathway (Li et al. 2019). Arabidopsis GSNOR knockout mutants exhibited decreased root growth under high Fe excess stress (250 μM), suggesting that GSNOR is also important for root tolerance to Fe. WRKY transcription factors are key regulators of several processes associated with abiotic and biotic stress tolerance in plants. In addition to the regulation of responses to various stressors, WRKY genes are involved in ROS-related defense mechanism against Fe excess stress (Figure 8.2).

8.3.4.2 GLY1 as a Detoxifying Agent

Glyoxalase I (GLYI) is an important enzyme in the glyoxalase pathway and a major detoxifying agent of methylglyoxal, which is produced by plants under stress (Singla-Pareek et al. 2020). Recently, *MeGLYI-13* was reported as a dominant gene expressed in young tissues of cassava seedlings that was upregulated in early stress after two hours of exposure to excess Fe (Tang et al. 2022). GLYIs have been identified in various plant species, including 11 *AtGLYI* genes in *A. thaliana* and 11 *OsGLYI* genes in rice (Mustafiz et al. 2011), as well as *GmGLYI* genes in soybean (Ghosh and Islam 2016). Thus, *GLYI* may be candidate genes that involve in response to Fe excess stress in rice and play a role tolerance mechanisms.

8.4 Research Outlook on Fe Excess Response of Plants

Progress has been made in our understanding of Fe excess stress defense mechanisms in plants. Transcription factors may regulate genes induced by Fe excess stress. Some upstream regulatory genes may regulate downstream genes to maintain Fe homeostasis. Direct ROS-mediated signals may stimulate genes involved in Fe excess responses. However, several potent regulators, such as signaling factors, transcription factors, and regulatory genes conserved in the molecular mechanism of the Fe excess response, remain poorly studied.

8.4.1 Regulation of Fe homeostasis in Plants in Response to Fe Excess Stress

The Fe-binding ubiquitin ligases OsHRZ1 and OsHRZ2 negatively regulate Fe deficiency responses in rice (Kobayashi et al. 2013). HRZ, a hemerythrin motif-containing Really Interesting New Gene and zinc finger proteins, is a key Fe homeostasis regulator. Aung et al. (2018a) showed that HRZ-knockdown rice lines are hypersensitive to excess Fe. The expression of Fe uptake- and transport-related genes, including *OsIRT1*, *OsYSL2*, *OsYSL15*, *TOM1*, *OsNAS1*, *OsNAS2*, and *OsIRO2*, is highly elevated in the roots of HRZ-knockdown rice subjected to Fe excess stress (Aung et al. 2018a). Therefore, HRZ is an important protein that protects plant cells from excess Fe by repressing genes involved in Fe uptake and translocation (Figure 8.2). Interestingly, *HRZ* expression is highly induced in response to Fe deficiency, under which several Fe uptake genes are strongly induced in rice (Kobayashi et al. 2013); this suggests that plant Fe uptake must be strictly regulated, and that HRZ prepares to regulate/to repress any time Fe uptake genes since under Fe deficiency in rice (Figure 8.2; Aung and Masuda 2020).

Similar to its homolog BRUTUS in Arabidopsis (Selote et al. 2015; Hindt et al. 2017), the hemerythrin domains of HRZs are predominantly bound by Fe and zinc (Kobayashi et al. 2013). The direct binding of Fe to HRZ or BRUTUS may be responsible for intracellular Fe sensing and signaling events (Kobayashi 2019). Thus, HRZ may play the key role of an Fe-binding sensor in the earliest stage of the response to excess Fe in Defense I (Aung and Masuda 2020).

IRON MAN (IMA) is an Fe-uptake-inducing peptide, and it controls the Fe uptake-regulated genes (Grillet et al. 2018). POSITIVE REGULATOR OF IRON HOMEOSTASIS 2 and 3 (PRI2 and PRI3) enhance the expression of IMA (Kobayashi et al. 2021). Also, HRZ1 suppresses the PRI1/2/3 (Zhang et al. 2017; Zhang et al. 2020). This might be one regulatory pathway by which HRZ suppresses Fe uptake under Fe excess conditions (Figure 8.2).

8.4.2 Transcription Factors

Plant transcription factors induced by excess Fe are important for genetic modifications because they regulate many downstream genes. WRKY transcription factors form one of the most important superfamilies and are associated with several types of stress tolerance (Banerjee and Roychoudhury 2015). Ricachenevsky et al. (2010) reported that OsWRKY80, a senescence- and drought-induced transcription factor, acts as an excess Fe-induced transcription factor in rice plants; after nine days of Fe excess treatment, *OsWRKY80* transcripts increased threefold in both roots and shoots. In another study, 19 WRKYs were upregulated in rice under Fe excess stress according to microarray analysis (Finatto et al. 2015). *OsWRKY55-like*, *OsWRKY46*, *OsWRKY64*, and *OsWRKY113* were upregulated in a Fe-sensitive genotype early in the stress response (Viana et al. 2017). These findings illustrate the important roles of WRKY transcription factors in Fe excess stress tolerance.

WRKY transcription factors play important roles in plant senescence, disease resistance, and salicylic acid signaling. In addition to abiotic stress, OsWRKY80 plays a role in biotic stresses such as pathogen attack (Guo et al. 2004) through salicylic acid signaling. Salicylic acid and its antagonist jasmonic acid play important roles in plant wound healing and systemic acquired resistance, which involve the accumulation of pathogen-related proteins. When an insect attacks a plant, salicylic or jasmonic acid hormones are induced to activate Fe translocation to apoplasts. Fe deficiency in the cytoplasm induces pathogenesis-related genes, and excess Fe mimics insect attack, thereby triggering the same response. Thus, excess Fe induces *OsWRKY80*, leading to increased apoplastic Fe concentrations and signaling of pathogen-related genes (Ricachenevsky et al. 2010).

bHLH transcription factors play crucial roles in Fe homeostasis in plants. Liang (2022) reported the model of Fe uptake regulators and regulation stream. In rice, POSITIVE REGULATOR OF IRON HOMEOSTASIS1 (OsPRI1) works upstream of Iron-related transcription factor 2 (OsIRO2) and Iron-related bHLH transcription factor 3 (OsIRO3) to positively regulate root-to-shoot Fe translocation. OsIRO3 (OsbHLH063) encodes a negative regulator of Fe uptake genes in rice (Zheng et al. 2010). OsIRO3 is also involved in shoot-to-root signal transmission, which is vital to prevent excess Fe in plants (Wang et al. 2020). A recent study demonstrated that OsIRO3 negatively regulates Fe homeostasis by repressing the expression of *OsIRO2* (Li et al. 2022). IDEF1 (IDE-binding factor 1) from the ABI3/VP1 family regulates the gene expression of OsIRO2, OsIRO3 and IMA (Kobayashi et al. 2007; Kobayashi et al. 2021; Wang et al. 2020) (Figure 8.2). Another bHLH transcription factor, OsbHLH61, participates in excess Fe regulation in rice; *OsbHLH061* expression was upregulated by excess Fe and interacted with OsPRI1, followed by recruitment of TOPLESS/TOPLESS-RELATED (OsTPL/TPR) co-repressors by the OsbHLH061–OsPRI1 complex to repress *OsIRO2* and *OsIRO3* expression (Wang et al. 2022). These findings suggest that OsIRO3/OsbHLH063 and OsbHLH61 are important bHLH transcription factors in Fe regulatory mechanisms (Figure 8.2).

8.4.3 Cis-Regulatory Elements

Cis-acting or cis-regulatory elements (CREs) are essential genomic sequences that function cooperatively in promoter regions controlling gene expression. IDE1 and IDE2 are involved in responses to Fe deficiency in rice (Kobayashi et al. 2003). Petit et al. (2001) reported that an Fe-dependent regulatory sequence (IDRS) is responsible for Fe-deficient repression of *ZmFer1* expression in *Zea mays* and *AtFer1* expression in Arabidopsis. However, the transcription factors and CREs that regulate responses to excess Fe in plants remain elusive. CREs involved in abiotic stress, such as light responses and the salicylic acid pathway, may also function in molecular signaling in the context of Fe excess responses (Viana et al. 2017).

However, the functional cis-acting elements directly responsible for Fe excess stress responses have not been well identified in rice. Recent bioinformatics studies using microarray-associated motif analyzer (MAMA) (Kakei et al. 2012) analyses on microarray data (Aung et al. 2018a) identified 560 CRE candidates, and 42 important conserved sequences of CREs directly related to Fe excess stress responses in various rice tissues (Kakei et al. 2021). This study explored several novel candidate CREs for excess Fe, including GCWGCWGC, CGACACGC, and Myb binding-like motifs. Novel CRE candidates were found in the region −500 to +150 bp relative to the transcription start sites of Fe excess response genes in specific tissues. These novel CREs may contain binding sequences of transcription factors that regulate Fe homeostasis in rice under excess Fe. This study also reported known CRE sequences of Fe excess response genes including DCEp2 motif, IDEF1-, Zinc Finger-, WRKY-, Myb-, AP2/ERF-, MADS box-, bZIP-, and bHLH-binding sequences-containing motifs. The known CREs found in Fe excess regulated genes may reflect their new roles in Fe excess responses. The study also built molecular models based on the CRE candidates of promoter structures regulating Fe excess response genes, which could lead to the development of artificial promoters that can be highly expressed under stress conditions.

8.5 Conclusion and Future Prospects

Fe is an essential element for all living organisms. In plants, Fe plays an important role in several metabolic processes including photosynthesis, respiration, and defense mechanisms. However, deficient or excess Fe can severely affect plant growth. Recent studies have discovered candidate genes that may be involved in the physiological and molecular responses of plants to Fe overload, and shed light on their roles in Fe detoxification and tolerance based on four defense mechanisms. Moreover, recent discoveries on new genes involved in the regulation of Fe homeostasis and the identification of Fe excess-responsive CREs reveal the Fe excess regulation pathways in plants. These recent findings may help clarify the Fe excess response mechanism in rice and promote the development of Fe excess-tolerant genotypes, which is essential for increasing crop yield and developing sustainable agricultural practices.

Acknowledgments

We thank Dr. Hiroyuki Hattori for invaluable discussions on the biological chemistry of elemental Fe. This study was funded by a Japan Society for the Promotion of Science (JSPS) KAKENHI grant (no. 21K14759 to M.S.A.).

Author Contributions

M.S.A. and H.M. designed, and wrote the manuscript.

Disclosures

The authors have no conflicts of interest to declare.

References

Alam, S.M. (1983). Influence of phosphorus on the growth and nutrient contents of the rice plant. *J. Plant Nutr.* 6: 291–299.

Aoyama, T., Kobayashi, T., Takahashi, M. et al. (2009). OsYSL18 is a rice iron(III)–deoxymugineic acid transporter specifically expressed in reproductive organs and phloem of lamina joints. *Plant Mol. Biol.* 70: 681–692.

Ariga, T., Hazama, K., Yanagisawa, S., and Yoneyama, T. (2014). Chemical forms of iron in xylem sap from graminaceous and non-graminaceous plants. *Soil Sci. Plant Nutr.* 60: 460–469.

Audebert, A. and Sahrawat, K.L. (2000). Mechanisms for iron toxicity tolerance in lowland rice. *J. Plant Nutr.* 23: 1877–1885.

Aung, M.S., Kobayashi, T., Masuda, H., and Nishizawa, N.K. (2018a). Rice HRZ ubiquitin ligases are crucial for response to excess iron. *Physiol. Plant* 163: 282–296.

Aung, M.S., Masuda, H., Kobayashi, T., and Nishizawa, N.K. (2018b). Physiological and transcriptomic analysis of responses to different levels of iron excess stress in various rice tissues. *Soil Sci. Plant Nutr.* 64: 370–385.

Aung, M.S., Masuda, H., Nozoye, T. et al. (2019). Nicotianamine synthesis by *OsNAS3* is important for mitigating iron excess stress in rice. *Front. Plant Sci.* 10: 660.

Aung, M.S. and Masuda, H. (2020). How does rice defend against excess iron? *Front. Plant Sci.* 11: 1102.

Balk, J. and Schaedler, T.A. (2014). Iron cofactor assembly in plants. *Annu. Rev. Plant Biol.* 65: 125–153.

Banerjee, A. and Roychoudhury, A. (2015). WRKY proteins: signaling and regulation of expression during abiotic stress responses. *Sci. World J.* 2015: 807560.

Bashir, K., Inoue, H., Nagasaka, S. et al. (2006). Cloning and characterization of deoxymugineic acid synthase genes from graminaceous plants. *J. Biol. Chem.* 281: 32395–32402.

Bashir, K., Ishimaru, Y., Shimo, H. et al. (2011). Rice phenolics efflux transporter 2 (PEZ2) plays an important role in solubilizing apoplasmic iron. *Soil Sci. Plant Nutr.* 57: 803–812.

Becana, M., Moran, J.F., and Iturbe-Ormaetxe, I. (1998). Iron dependent oxygen free radical generation in plants subjected to environmental stress: toxicity and antioxidant protection. *Plant Soil* 201: 137–147.

Becker, M. and Asch, F. (2005). Iron toxicity in rice – conditions and management concepts. *J. Plant Nutr. Soil Sci.* 168: 558–573.

Becker, M., Ngo, N.S., and Schenk, M.K.A. (2020). Silicon reduces the iron uptake in rice and induces iron homeostasis related genes. *Sci. Rep.* 10: 5079.

Bode, K., Döring, O., Lüthje, S. et al. (1995). The role of active oxygen in iron tolerance of rice (*Oryza sativa* L.). *Protoplasma* 184: 249–255.

Briat, J.F. and Lobréaux, S. (1997). Iron transport and storage in plants. *Trends Plant Sci.* 2: 187–193.

Briat, J.F., Fobis-Loisy, I., Grignon, N. et al. (1995). Cellular and molecular aspects of iron metabolism in plants. *Biol. Cell.* 84: 69–81.

Briat, J.F., Ravet, K., Arnaud, N. et al. (2010). New insights into ferritin synthesis and function highlight a link between iron homeostasis and oxidative stress in plants. *Ann. Bot.* 105: 811–822.

Bughio, N., Yamaguchi, H., Nishizawa, N.K. et al. (2002). Cloning an iron-regulated metal transporter from rice. *J. Exp. Bot.* 53: 1677–1682.

Connolly, E.L., Campbell, N.H., Grotz, N. et al. (2003). Overexpression of the FRO2 ferric chelate reductase confers tolerance to growth on low iron and uncovers posttranscriptional control. *Plant Physiol.* 133: 1102–1110.

Crestani, M., Silva, J.A.G., Souza, V.Q. et al. (2009). Irrigated rice genotype performance under excess iron stress in hydroponic culture. *Crop Breed. Appl. Biotechnol.* 9: 87–95.

Curie, C., Panaviene, Z., Loulergue, C. et al. (2001). Maize yellow stripe1 encodes a membrane protein directly involved in Fe(III) uptake. *Nature* 409: 346–349.

Dash, A.K., Singh, C., and Varma, S. (2021). Chemistry and recent advances in management of acid sulphate soils. In: *Latest Trends in Soil Science*, 2e (ed. S.S. Porte). New Delhi: Integrated Publication.

da Silva, J.J.R.F. and Williams, R.J.P. (2001). *The Biological Chemistry of the Elements*. Oxford, UK: Clarendon Press.

da Silveira, V.C., Oliveira, A.P., Sperotto, R.A. et al. (2007). Influence of iron on mineral status of two rice (*Oryza sativa* L.) cultivars. *Braz. J. Plant Physiol.* 19: 127–139.

de Campos Carmona, F., Adamski, J.M., Wairich, A. et al. (2021). Tolerance mechanisms and irrigation management to reduce iron stress in irrigated rice. *Plant Soil* 469: 173–191.

Demazeau, G., Buffat, B., Pouchard, M., and Hagenmuller, P. (1982). Recent developments in the field of high oxidation states of transition elements in oxides stabilization of six-coordinated Iron(V). *Z. Anorg. Allg. Chem.* 491: 60–66.

Divol, F., Couch, D., Conéjéro, G. et al. (2013). The Arabidopsis YELLOW STRIPE LIKE4 and 6 transporters control iron release from the chloroplast. *Plant Cell* 25: 1040–1055.

Dufey, I., Draye, X., Lutts, S. et al. (2015). Novel QTLs in an interspecific backcross *Oryza sativa* × *Oryza glaberrima* for resistance to iron toxicity in rice. *Euphytica* 204: 609–625.

Eide, D., Broderius, M., Fett, J., and Guerinot, M.L. (1996). A novel iron-regulated metal transporter from plants identified by functional expression in yeast. *Proc. Natl. Acad. Sci. U.S.A.* 93: 5624–5628. FAOSTAT http://www.fao.org/faostat/en/.

Fageria, N.K. and Rabelo, N.A. (1987). Tolerance of rice cultivars to iron toxicity. *J. Plant Nutr.* 10: 653–661.

Fageria, N.K., Santos, A.B., Filho, M.P.B., and Guimaraes, C.M. (2008). Iron toxicity in lowland rice. *J. Plant Nutr. 31*: 1676–1697.

Fang, W.C. and Kao, C.H. (2000). Enhanced peroxidase activity in rice leaves in response to excess iron, copper and zinc. *Plant Sci.* 158: 71–76.

Fang, W.C., Wang, J.W., Lin, C.C., and Kao, C.H. (2001). Iron induction of lipid peroxidation and effects on antioxidative enzyme activities in rice leaves. *Plant Growth Regul.* 35: 75–80.

Finatto, T., de Oliveira, A.C., Chaparro, C. et al. (2015). Abiotic stress and genome dynamics: specific genes and transposable elements response to iron in rice. *Rice* 8: 13.

Frey, P.A. and Reed, G.H. (2012). The ubiquity of iron. *ACS Chem. Biol.* 7: 1477–1481.

Fourcroy, P., Sisó-Terraza, P., Sudre, D. et al. (2014). Involvement of the ABCG37 transporter in secretion of scopoletin and derivatives by Arabidopsis roots in response to iron deficiency. *New Phytol.* 201: 155–167.

Gallie, D.R. (2013). The role of l-ascorbic acid recycling in responding to environmental stress and in promoting plant growth. *J. Exp. Bot.* 64: 433–443.

Gautam, C.K., Tsai, H.-H., and Schmidt, W. (2021). IRONMAN tunes responses to iron deficiency in concert with environmental pH. *Plant Physiol.* 187: 1728–1745.

Geilfus, C.M. (2017). The pH of the apoplast: dynamic factor with functional impact under stress. *Mol. Plant.* 10: 1371–1386.

Ghosh, A. and Islam, T. (2016). Genome-wide analysis and expression profiling of glyoxalase gene families in soybean (*Glycine max*) indicate their development and abiotic stress specific response. *BMC Plant Biol.* 16: 87.

Gridley, H.E., Efisue, A., Tolou, B., and Bakayako, T. (2006). *Breeding for Tolerance to Iron Toxicity at WARDAl*, 96–111. Cotonou, Benin: Africa Rice Center (WARDA).

Grillet, L., Lan, P., Li, W., Mokkapati, G., and Schmidt, W. (2018). IRON MAN is a ubiquitous family of peptides that control iron transport in plants. *Nat. Plants.* 4: 953–963.

Guerinot, M.L. and Yi, Y. (1994). Iron: nutritious, noxious, and not readily available. *Plant Physiol.* 104: 815e820.

Guo, Z.J., Kan, Y.C., Chen, X.J. et al. (2004). Characterization of a rice WRKY gene whose expression is induced upon pathogen attack and mechanical wounding. *Acta Bot. Sin.* 46: 955–964.

Hell, R. and Stephan, U.W. (2003). Iron uptake, trafficking and homeostasis in plants. *Planta* 216: 541–551.

Higuchi, K., Suzuki, K., Nakanishi, H. et al. (1999). Cloning of nicotianamine synthase genes, novel genes involved in the biosynthesis of phytosiderophores. *Plant Physiol.* 119: 471–479.

Higuchi, K., Watanabe, S., Takahashi, M. et al. (2001). Nicotianamine synthase gene expression differs in barley and rice under Fe-deficient conditions. *Plant J.* 25: 159–167.

Hindt, M.N., Akmakjian, G.Z., Pivarski, K.L. et al. (2017). BRUTUS and its paralogs, BTS LIKE1 and BTS LIKE2, encode important negative regulators of the iron deficiency response in *Arabidopsis thaliana*. *Metallomics* 9: 876–890.

Hinrichs, M., Fleck, A.T., Biedermann, E. et al. (2017). An ABC transporter is involved in the silicon-induced formation of Casparian bands in the exodermis of rice. *Front. Plant Sci.* 8: 671.

Inoue, H., Higuchi, K., Takahashi, M. et al. (2003). Three rice nicotianamine synthase genes, *OsNAS1*, *OsNAS2*, and *OsNAS3* are expressed in cells involved in long-distance transport of iron and differentially regulated by iron. *Plant J.* 36: 366–381.

Inoue, H., Kobayashi, T., Nozoye, T. et al. (2009). Rice *OsYSL15* is an iron-regulated iron (III)-deoxymugineic acid transporter expressed in the roots and is essential for iron uptake in early growth of the seedlings. *J. Biol. Chem.* 284: 3470–3479.

Inoue, H., Takahashi, M., Kobayashi, T. et al. (2008). Identification and localization of the rice nicotianamine aminotransferase gene *OsNAAT1*expression suggests the site of phytosiderophore synthesis in rice. *Plant Mol. Bio.* 166: 193–203.

Ishimaru, Y., Masuda, H., Bashir, K. et al. (2010). Rice metal nicotianamine transporter, OsYSL2, is required for the long-distance transport of iron and manganese. *Plant J.* 62: 379–390.

Ishimaru, Y., Suzuki, M., Tsukamoto, T. et al. (2006). Rice plants take up iron as an Fe^{3+}-phytosiderophore and as Fe^{2+}. *Plant J.* 45: 335–346.

Ismunadji, M. and Ardjasa, W.S. (1989). Potash fertilization for lowland rice can prevent iron toxicity losses. *Better Crop* 3: 12–14.

Jain, A. and Connolly, E.L. (2013). Mitochondrial iron transport and homeostasis in plants. *Front. Plant Sci.* 4: 348.

Kakei, Y., Ishimaru, Y., Kobayashi, T. et al. (2012). OsYSL16 plays a role in the allocation of iron. *Plant Mol. Biol.* 79: 583–594.

Kakei, Y., Masuda, H., Nishizawa, N.K. et al. (2021). Elucidation of novel *cis*-acting elements and promoters involved in iron excess response mechanisms in rice by bioinformatics approach. *Front. Plant Sci.* 12: 660303.

Khabaz-Saberi, H. and Rengel, Z. (2010). Aluminum, manganese, and iron tolerance improves performance of wheat genotypes in waterlogged acidic soils. *J. Plant Nutr. Soil Sci.* 173: 461–468.

Kobayashi, T. (2019). Understanding the complexity of iron sensing and signaling cascades in plants. *Plant Cell Physiol.* 60: 1440–1446.

Kobayashi, T., Nagano, A.J., and Nishizawa, N.K. (2021). Iron deficiency-inducible peptide-coding genes OsIMA1 and OsIMA2 positively regulate a major pathway of iron uptake and translocation in rice. *J. Exp. Bot.* 72: 2196–2211.

Kobayashi, T., Nagasaka, S., Senoura, T. et al. (2013). Iron-binding haemerythrin RING ubiquitin ligases regulate plant iron responses and accumulation. *Nat. Commun.* 4: 2792.

Kobayashi, T., Ogo, Y., Itai, R.N., Nakanishi, H., Takahashi, M., Mori, S., and Nishizawa, N.K. (2007). The transcription factor IDEF1 regulates the response to and tolerance of iron deficiency in plants. *Proc. Natl. Acad. Sci.* USA 104: 19150–19155.

Kobayashi, T., Nakayama, Y., Itai, R.N. et al. (2003). Identification of novel cisacting elements, IDE1 and IDE2, of the barley IDS2 gene promoter conferring iron-deficiency-inducible, root-specific expression in heterogeneous tobacco plants. *Plant J.* 36: 780–793.

Koike, S., Inoue, H., Mizuno, D. et al. (2004). OsYSL2 is a rice metalnicotianamine transporter that is regulated by iron and expressed in the phloem. *Plant J.* 39: 415–424.

Lee, S., Chiecko, J.C., Kim, S.A. et al. (2009). Disruption of *OsYSL15* leads to iron inefficiency in rice plants. *Plant Physiol.* 150: 786–800.

Li, C., Li, Y., Xu, P., and Liang, G. (2022). OsIRO3 negatively regulates Fe homeostasis by repressing the expression of OsIRO2. *Plant J.* 111: 966–978.

Li, G., Song, H., Li, B. et al. (2015). Auxin resistant1 and PIN-FORMED2 protect lateral root formation in *Arabidopsis* under iron stress. *Plant Physiol.* 169: 2608–2623.

Li, B., Sun, L., Huang, J. et al. (2019). GSNOR provides plant tolerance to iron toxicity via preventing iron-dependent nitrosative and oxidative cytotoxicity. *Nat. Commun.* 10: 3896.

Liang, G. (2022). Iron uptake, signaling, and sensing in plants. *Plant Comm.* 3: 100345.

Lindsay, W.L. (1997). *Chemical Equilibria in Soils*, 78–104. Wiley.

Ma, J.F. and Takahashi, E. (1990). Effect of silicon on the growth and phosphorus uptake of rice. *Plant Soil* 126: 115–119.

Mahender, A., Swamy, B.P.M., Anandan, A., and Ali, J. (2019). Tolerance of iron-deficient and -toxic soil conditions in rice. *Plants* 8: 31.

Marschner, H., Romheld, V., and Kissel, M. (1986). Different strategies in higher plants in mobilization and uptake of iron. *J. Plant Nutr.* 9: 695–713.

Marschner, H. (1995). *Mineral Nutrition of Higher Plants*, 2e. London: Academic.

Martinière, A., Desbrosses, G., Sentenac, H., and Paris, N. (2013). Development and properties of genetically encoded pH sensors in plants. *Front. Plant Sci.* 4: 523.

Mori, S. (1998). Iron transport in graminaceous plants. In: *Iron Transport and Storage in Microorganisms, Plants and Animals. Vol. 35, Metal Ions in Biological Systems* (ed. A. Sigel and H. Sigel), 215–237. New York: Marcel Dekker.

Mori, S. and Nishizawa, N. (1987). Methionine as a dominant precursor of Phytosiderophores in Graminaceae plants. *Plant Cell Physiol.* 28: 1081–1092.

Müller, C., Silveira, S.F., Daloso, D.M. et al. (2017). Ecophysiological responses to excess iron in lowland and upland rice cultivars. *Chemosphere* 189: 123.

Mustafiz, A., Singh, A.K., Pareek, A. et al. (2011). Genome-wide analysis of rice and Arabidopsis identifies two glyoxalase genes that are highly expressed in abiotic stresses. *Funct. Integr. Genom.* 11: 293–305.

Nozoye, T., Nagasaka, S., Kobayashi, T. et al. (2011). Phytosiderophore efflux transporters are crucial for iron acquisition in graminaceous plants. *J. Biol. Chem.* 286: 5446–5454.

Nugraha, Y., Utami, D.W., Rosdianti, I.D.A. et al. (2016). Markers-traits association for iron toxicity tolerance in selected Indonesian rice varieties. *Biodivers* 17: 753–763.

Onaga, G., Edema, R., and Asea, G. (2013). Tolerance of rice germplasm to iron toxicity stress and the relationship between tolerance, Fe^{2+}, P and K content in the leaves and roots. *Arch. Agron. Soil Sci.* 59: 213–229.

Petit, J.M., van Wuytswinkel, O., Briat, J.F., and Lobréaux, S. (2001). Characterization of an iron-dependent regulatory sequence involved in the transcriptional control of AtFer1 and ZmFer1 plant ferritin genes by iron. *J. Biol. Chem.* 276: 55845590.

Pich, A., Manteuffel, R., Hillmer, S. et al. (2001). Fe homeostasis in plant cells: does nicotianamine play multiple roles in the regulation of cytoplasmic Fe concentration? *Planta* 213: 967–976.

Pich, A. and Scholz, G. (1996). Translocation of copper and other micronutrients in tomato plants (*Lycopersicon esculentum* Mill.): nicotianamine-stimulated copper transport in the xylem. *J. Exp. Bot.* 294: 41–47.

Quinet, M., Vromman, D., Clippe, A. et al. (2012). Combined transcriptomic and physiological approaches reveal strong differences between short- and long-term response of rice (*Oryza sativa*) to iron toxicity. *Plant Cell Env.* 35: 1837–1859.

Ramírez, L.M., Claassen, N., Ubiera, A.A. et al. (2002). Effect of phosphorus, potassium and zinc fertilizers on iron toxicity in wetland rice (*Oryza sativa* L.). *Plant Soil.* 239: 197–206.

Ravet, K., Touraine, B., Boucherez, J. et al. (2009). Ferritins control interaction between iron homeostasis and oxidative stress in Arabidopsis. *Plant J.* 57: 400–412.

Rengel, Z. (2002). Role of pH in availability of ions in soil. In: *Handbook of Plant Growth. pH as a Master Variable in Plant Growth* (ed. Z. Rengel), 323–350. New York: *Marcel Dekker*.

Rengel, Z. (2015, 2015). Availability of Mn, Zn and Fe in the rhizosphere. *J. Soil Sci. Plant Nutr.* 15 (2): 397–409.

Ricachenevsky, F.K., Sperotto, R.A., Menguer, P.K., and Fett, J.P. (2010). Identification of Fe-excess-induced genes in rice shoots reveals a WRKY transcription factor responsive to Fe, drought and senescence. *Mol. Biol. Rep.* 37: 3735–3745.

Robinson, N.J., Procter, C.M., Connolly, E.L., and Guerinot, M.L. (1999). A ferric-chelate reductase for iron uptake from soils. *Nature* 397: 694–697.

Rodenburg, J., Zwart, S.J., Kiepe, P. et al. (2014). Sustainable rice production in African inland valleys: seizing regional potentials through local approaches. *Agr. Syst.* 123: 1–11.

Rodríguez-Celma, J., Lin, W.-D., Fu, G.-M. et al. (2013). Mutually exclusive alterations in secondary metabolism are critical for the uptake of insoluble iron compounds by Arabidopsis and *Medicago truncatula*. *Plant Physiol.* 162: 1473–1485.

Römheld, V. and Marschner, H. (1986). Evidence for a specific uptake system for iron phytosiderophore in roots of grasses. *Plant Physiol.* 80: 175–180.

Rudolph, A., Becker, R., Scholz, G. et al. (1985). The occurrence of the amino acid nicotianamine in plants and microorganisms: a reinvestigation. *Biochem. Physiol. Pflanz.* 180: 557–563.

Sadiq, A.A. and Babagana, U. (2012). Influence of lime materials to ameliorate acidity on irrigated paddy fields: A review. *Acad. Res. Int.* 3: 17–25.

Sahrawat, K.L. (2004). Iron toxicity in wetland rice and its role of other nutrients. *J. Plant Nutr.* 27: 1471–1504.

Santi, S. and Schmidt, W. (2009). Dissecting iron deficiency-induced proton extrusion in Arabidopsis roots. *New Phytol.* 183: 1072–1084.

Schmid, N.B., Giehl, R.F.H., Döll, S. et al. (2014). Feruloyl-CoA 6′-hydroxylase1-dependent coumarins mediate iron acquisition from alkaline substrates in Arabidopsis. *Plant Physiol.* 164: 160–172.

Selote, D., Samira, R., Matthiadis, A. et al. (2015). Iron-binding E3 ligase mediates iron response in plants by targeting basic helix-loop-helix transcription factors. *Plant Physiol.* 167: 273–286.

Sikirou, M., Saito, K., Achigan-Dako, E.G. et al. (2015). Genetic improvement of iron toxicity tolerance in rice: progress, challenges and prospects in West Africa. *Plant Prod. Sci.* 18: 423–434.

Sikirou, M., Saito, K., Dramé, K.N. et al. (2016). Soil-based screening for iron toxicity tolerance in rice using pots. *Plant Prod. Sci.* 19: 489–496.

Singla-Pareek, S.L., Kaur, C., Kumar, B. et al. (2020). Reassessing plant glyoxalases: large family and expanding functions. *New Phytol.* 227: 714–721.

Sisó-Terraza, P., Luis-Villarroya, A., Fourcroy, P. et al. (2016). Accumulation and secretion of coumarinolignans and other coumarins in *Arabidopsis thaliana* roots in response to iron deficiency at high pH. *Front. Plant Sci.* 7: 1711.

Shojima, S., Nishizawa, N.K., Fushiya, S. et al. (1990). Biosynthesis of phytosiderophores –in vitro biosynthesis of 2′-deoxymugineic acid from l-methionine and nicotianamine. *Plant Physiol.* 93: 1497–1503.

Stein, R.J., Duarte, G.L., Spohr, M.G. et al. (2009a). Distinct physiological responses of two rice cultivars subjected to iron toxicity under field conditions. *Ann. Appl. Biol.* 154: 269–277.

Stein, R.J., Ricachenevsky, F.K., and Fett, J.P. (2009b). Differential regulation of the two rice ferritin genes (OsFER1 and OsFER2). *Plant Sci.* 177: 563–569.

Stein, R.J., Lopes, S., and II, Fett, J.P. (2014). Iron toxicity in field-cultivated rice: contrasting tolerance mechanisms in distinct cultivars. *Theor. Exp. Plant Physiol.* 26: 135–146.

Stephan, U.W., Schmidke, I., Stephan, V.W., and Scholz, G. (1996). The nicotianamine molecule is made-to-measure for complexation of metal micronutrients in plants. *Biometals* 9: 84–90.

Stumm, W. and Lee, G.F. (1961). Oxygenation of ferrous iron. *Ind. Eng. Chem.* 53: 143–146.

Suh, H.J., Kim, C.S., Lee, J.Y., and Jung, J. (2002). Photodynamic effect of iron excess on photosystem II function in pea plants. *Photochem. Photobiol.* 75: 513–518.

Sumner, M.E. and Noble, A.D. (2003). Soil acidification: the world story. In: *Handbook of Soil Acidity* (ed. Z. Rengel), 1–28. New York: Marcel Dekker.

Tadano, T. (1975). Devices of rice roots to tolerant high iron concentrations in growth media. *Japan Agric. Res. Q.* 9: 34–39.

Takahashi, M., Terada, Y., Nakai, I. et al. (2003). Role of nicotianamine in the intracellular delivery of metals and plant reproductive development. *Plant Cell* 15: 1263–1280.

Takahashi, M., Yamaguchi, H., Nakanishi, H. et al. (1999). Cloning two genes for nicotianamine aminotransferase, a critical enzyme in iron acquisition (Strategy II) in graminaceous plants. *Plant Physiol.* 121: 947–956.

Takagi, S. (1976). Naturally occurring iron-chelating compounds in oat- and rice root washing. I: Activity measurement and preliminary characterization. *Soil Sci. Plant Nutr.* 22: 423–433.

Tang, F., Li, R., Zhou, Y. et al. (2022). Genome-wide identification of cassava glyoxalase I genes and the potential function of MeGLYI-13 in iron toxicity tolerance. *Int. J. Mol. Sci.* 23: 5212.

Terry, N. and Abadía, J. (1986). Function of iron in chloroplasts. *J. Plant Nutr.* 9: 609–646.

Theil, E.C. (2003). Ferritin: at the crossroads of iron and oxygen metabolism. *J. Nutr.* 133: 1549–1553.

Thongbai, P. and Goodman, B.A. (2000). Free radical generation and post anoxic injury in an iron toxic soil. *J. Plant Physiol.* 23: 1887–1990.

Tsai, H.-H. and Schmidt, W. (2020). pH-dependent transcriptional profile changes in iron-deficient *Arabidopsis* roots. *BMC Genom.* 21: 694.

Viana, V.E., Marini, N., Finatto, T. et al. (2017). Iron excess in rice: from phenotypic changes to functional genomics of WRKY transcription factors. *Genet. Mol. Res.* 16 (3): gmr16039694.

von Uexküll, H.R. and Mutert, E. (1995). Global extent, development and economic impact of acid soils. In: *Plant–Soil Interactions at Low pH: Principles and Management* (ed. R.A. Date, N.J. Grundon, G.E. Raymet, and M.E. Probert), 5–19. Dordrecht, The Netherlands: Kluwer Academic.

von Wirén, N., Khodr, H., and Hider, R.C. (2000). Hydroxylated phytosiderophore species from rye and barley possess an enhanced chelating efficiency and affinity for iron(III). *Plant Physiol.* 124: 1149–1157.

von Wirén, N., Klair, S., Bansal, S. et al. (1999). Nicotianamine chelates both Fe-III and Fe-II. Implications for metal transport in plants. *Plant Physiol.* 119: 1107–1114.

Wang, F., Itai, R.N., Nozoye, T. et al. (2020). The bHLH protein OsIRO3 is critical for plant survival and iron (Fe) homeostasis in rice (*Oryza sativa* L.) under Fe-deficient conditions. *Soil Sci. Plant Nutr.* 579–592. https://doi.org/10.1080/00380768.2020.1783966.

Wang, W., Ye, J., Xu, H. et al. (2022). OsbHLH061 links TOPLESS/TOPLESS-RELATED repressor proteins with POSITIVE REGULATOR OF IRON HOMEOSTASIS 1 to maintain iron homeostasis in rice. *New Phytol.* 234 (5): 1753–1769.

Wu, L., Shhadi, M.Y., Gregorio, G. et al. (2014). Genetic and physiological analysis of tolerance to acute iron toxicity in rice. *Rice* 7: 8.

Yokosho, K., Yamaji, N., and Ma, J.F. (2016). *OsFRDL1* expressed in nodes is required for distribution of iron to grains in rice. *J. Exp. Bot.* 67: 5485–5494.

Yoneyama, T., Ishikawa, S., and Fujimaki, S. (2015). Route and regulation of zinc, cadmium, and iron transport in rice plants (*Oryza sativa* L.) during vegetative growth and grain filling: metal transporters, metal speciation, grain Cd reduction and Zn and Fe biofortification. *Int. J. Mol. Sci.* 16: 19111–19129.

Yuan, L., Wu, L., Yang, C., and Lv, Q. (2012). Effects of iron and zinc foliar applications on rice plants and their grain accumulation and grain quality. *J. Sci. Food Agric.* 93: 254–261.

Zhang, H., Li, Y., Pu, M., Xu, P., Liang, G., and Yu, D. (2020). Oryza sativa POSITIVE REGULATOR OF IRON DEFICIENCY RESPONSE 2 (OsPRI2) and OsPRI3 are involved in the maintenance of Fe homeostasis. *Plant Cell Environ.* 43: 261–274.

Zhang, H., Li, Y., Yao, X., Liang, G., and Yu, D. (2017). Positive regulator of iron homeostasis1, OsPRI1, facilitates iron homeostasis. *Plant Physiol.* 175: 543–554.

Zhang, Y., Xu, Y.H., Yi, H.Y., and Gong, J.M. (2012). Vacuolar membrane transporters *OsVIT1* and *OsVIT2* modulate iron translocation between flag leaves and seeds in rice. *Plant J.* 72: 400–410.

Zheng, L., Ying, Y., Wang, L. et al. (2010). Identification of a novel iron regulated basic helix-loop-helix protein involved in Fe homeostasis in *Oryza sativa*. *BMC Plant Biol.* 10: 66.

9

Molecular Breeding for Iron Toxicity Tolerance in Rice (*Oryza sativa* L.)

Dorothy A. Onyango[*1], *Mathew M. Dida*[2], *Khady N. Drame*[3], *Benson O. Nyongesa*[4], *and Kayode A. Sanni*[1]

[1] *Department of Product Development and Commercialization, African Agricultural Technology Foundation (AATF), Nairobi, Kenya*
[2] *Department of Applied Plant Sciences, School of Agriculture and Food Security, Maseno University, Kisumu, Kenya*
[3] *Centre d'Etude Régional pour l'Amélioration de l'Adaptation à la Sécheresse (CERAAS)/Institut Sénégalais de Recherches Agricoles (ISRA), Thiès, Sénégal*
[4] *Department of Biological Sciences, School of Science, University of Eldoret, Eldoret, Kenya*

9.1 Introduction

Rice is an important staple crop grown in about 40 out of 54 countries in Africa and its cultivation is the principal activity and source of income for more than 35 million smallholder rice farmers in Africa (https://www.africarice.org/why-rice-matters-for-africa). The demand for rice is growing at more than 6% per annum faster than any other food staple in sub-Saharan Africa (SSA), which has been fueled by population growth, urbanization, and changes in eating habits. This demand has led to most African countries being net importers of rice to close the deficit. To reduce the imports, there is a need to expand the production of rice in areas such as wetlands. SSA estimated 130 million ha of wetland cover, with 23% of these localized in West Africa alone (GRiSP 2013). About 1.066 million hectares representing 12% of soil are affected by iron (Fe) toxicity in SSA (van Oort 2018). Iron toxicity is a nutrient disorder associated with the accumulation of a high amount of Fe^{2+} in the soil solution. Most Fe toxic soils are characterized by a high amount of reducible Fe, low pH, and low CEC and exchangeable K content, and these have been found to be associated with P and Zn deficiency and hydrogen sulfide (H_2S) toxicity (Koike et al. 2004). Iron toxicity in the wetland rice ecosystem is linked to waterlogging under anaerobic soil conditions where microorganisms and plant roots deplete oxygen through respiration leading to the accumulation of Fe^{2+} in the soil (Krohling et al. 2016). Although wetlands offer an additional ecosystem for the expansion of rice cultivation, the presence of iron (Fe) toxicity has contributed to the utilization of 5% of wetlands in SSA (Balasubramanian et al. 2007).

*Corresponding Author

Heavy Metal Toxicity and Tolerance in Plants: A Biological, Omics, and Genetic Engineering Approach, First Edition. Edited by Mohammad Anwar Hossain, AKM Zakir Hossain, Sylvain Bourgerie, Masayuki Fujita, Om Parkash Dhankher, and Parvez Haris.

Therefore, for effective breeding of rice cultivars with enhanced tolerance to Fe toxicity common in wetlands, there is a need to exploit Fe tolerance variations using modern molecular biology tools. This chapter will cover the molecular breeding of rice to expand its cultivation in Fe toxic wetlands in SSA.

9.2 Role of Iron in Plants and Rice

Iron plays numerous roles in plants, including nutrient transport, mitochondrial respiration, photosynthesis, enzyme regulation, and nitrogen assimilation (Bashir et al. 2010; Wu et al. 2014; Brumbarova et al. 2015). Iron also regulates protein stability and participates in many chemical reactions such as hydration, dehydration, redox-dependent catalysis, and photo redox catalysis, all of which detoxify excess Fe^{2+} (Jain and Connolly 2013; Dufey et al. 2015a). The antioxidant defense mechanisms of many such as superoxide dismutase, catalase, and polyphenol oxidase are accelerated by Fe, and aid in the selection of Fe-tolerant rice cultivars (Saikia and Baruah 2012; Pennock and McKenzie 2015). The plastids contain 90% of Fe that is involved in maintaining the structural integrity of the thylakoid membrane contributing to the well functioning of photosynthesis in plants (Rout and Sahoo 2015; Mahender et al. 2019).

9.3 Iron Toxicity and Its Effects on Rice

The Fe^{2+} are abundant in paddy fields and absorption by rice roots causes severe Fe toxicity. The rhizosphere of acid sulfate soils with pH less than 5 contains a massive amount of Fe^{2+} about 10–2000 $mg\,kg^{-1}$ (da Silveira et al. 2007). The Fe^{2+} are highly soluble in water ($Ksp = [Fe^{2+}]\,[OH^{-}]\,2 = 8\times10^{-16}$) compared to ferric ions (Fe^{3+}) ($Ksp = [Fe^{3+}]\,[OH^{-}]\,3 = 1\times10^{-36}$) at 25 °C (Stumm and Lee 1961). This shows that the dissolving capacity of Fe^{3+} and Fe^{2+} in water is approximately 1×10^{-9} $mol\,l^{-1}$ and 800 $mol\,l^{-1}$ at pH 5, respectively. Therefore, these findings demonstrate that Fe^{2+} in the soil can be soluble at pH 5 or below. Among soil types, mainly acid sulfate soil, acid clay soil, and peat soil cause Fe toxicity (Becker and Asch 2005). Other factors that contribute to Fe toxicity include soil types such as alkaline soils, which cause a rise in pH, an increase in calcium carbonate and nitrate, a change in temperature, and poor aeration (Kobayashi et al. 2014; Mongon et al. 2017). Poor drainage, soil organic matter content, more hydrogen sulfides, low soil fertility, and genotypes all contribute to an increase in available forms of Fe in soil (Chandel et al. 2010; Mahender et al. 2019). The distribution of Fe in many types of the soil indicates that it is a major determinant of rice yields and quality. As a result, one of the practical options for overcoming this constraint is to breed and deploy varieties that are resistant to Fe toxicity.

The Fe^{2+} overload in plants causes tissue damage and disrupts cellular homeostasis. Ferrous toxicity inhibits cell division and elongation of the primary roots and the growth of lateral roots (Li et al. 2015). High Fe absorption by the roots and its transport into the leaves by xylem via the transpiration stream result in cellular Fe overload in plant tissues (Briat and Lobréaux 1997). Free Fe^{2+} have a high affinity to bind with oxygen, leading to high oxygen tension and excess Fe accumulation in plant tissues. The Fe^{2+} exist as a highly soluble and reactive form in the plant tissues with Fe overload and even excess Fe^{2+} exist as a free form in the cytosol causing toxic reactions such as the Fenton reaction. The Fenton reaction causes the overproduction of reactive oxygen species (ROS), particularly the cytotoxic hydroxyl radical (OH·) (Becana et al. 1998; Thongbai and Goodman 2000). Plant Fe toxicity disturbs the photosystem and elevates the cytochrome b6/f content of thylakoids, lowering the photosynthetic rate and accelerating oxygen production (Suh et al. 2002). Iron overload in chloroplasts results in oxidative damage caused by ROS (Balk and Schaedler 2014). This may lead to irreversible damage to cellular structures, membranes, DNA, and proteins (Briat et al. 1995; Stein et al. 2009). The excess of Fe in cells impairs biological processes and leads to bronzing symptoms on leaves as a result of cell death. Leaf bronzing induced by Fe toxicity is closely associated with yield loss (Wu et al. 2014). According to Audebert and Fofana (2009), for every one-grain increase in LB, yield is reduced by $500\,kg\,ha^{-1}$. However, a number of studies have found a negative relationship between LB score and yield under Fe toxicity stress (Sahrawat et al. 1996; Nozoe et al. 2008; Onaga et al. 2013). High levels of free Fe^{2+} in soil solutions can directly damage the root system, resulting in reduced nutrient uptake (Wu 2016). Iron toxicity causes a nutrient blockage, which is critical for rice growth (Audebert and Sahrawat 2000; Nugraha et al. 2016), and crop failure can occur under severe conditions (Audebert 2006; Li et al. 2016). Under Fe toxic conditions, rice yield production eventually declines. The yield loss can range from 15 to 100% depending on the timing of the occurrence and the density of the stress (Audebert and Fofana 2009).

9.4 Iron Toxicity Tolerance Mechanisms in Rice Plants

Rice plants use a number of mechanisms to counteract the negative effects of Fe toxicity in the soil. This crop uses four defense mechanisms, namely Fe exclusion from the roots, Fe retention in roots and suppression of Fe translocation to shoots, Fe compartmentalization in shoots, and ROS detoxification to respond to Fe toxicity.

9.4.1 Fe Exclusion from Roots

In paddy fields with excess Fe, rice uses Fe exclusion from the roots strategy also known as defense mechanism 1 to defend itself. This is a root-based tolerant mechanism that prevents the uptake of Fe by the roots. The rhizospheric oxidation of Fe^{2+} to insoluble Fe^{3+} by oxygen transport from shoots to roots results in the formation of Fe plaques on the root surface due to Fe^{3+} precipitation, which acts as a barrier to the uptake of excess Fe^{2+} into the root tissues (Becker and Asch 2005; Wu et al. 2014; Stein et al. 2019). A tolerant rice genotype that uses the Fe exclusion strategy is known to display the high formation of Fe plaque on the root surface (Mahender et al. 2019). Together with a larger diameter of the pith cavity of the shoots, the genotype with Fe excluder has high root oxidation power by an increase in aerenchyma volume and the number of lateral roots that favors the internal oxygen movement (Wu et al. 2014; Onyango et al. 2018, 2020). The application of silicon is known to reduce the uptake of Fe and strengthen the Casparian bands in the root epidermis of the rice plants, which serves as a diffusion barrier and inhibits Fe flux into the roots under excess Fe (Becker et al. 2020; dos Santos et al. 2020). These studies demonstrate that a combination of silicon application and use of rice genotypes with increased ability to exclude Fe such as high root oxidation potential, large aerenchyma, several lateral roots, and strong Casparian bands maybe superior in tolerating excess Fe in the paddy fields.

9.4.2 Fe Retention in Roots and Suppression of Fe Translocation to Shoots

Rice plants have the capacity to retain Fe in the root tissues and decrease Fe translocation from roots to shoots to prevent the accumulation of excessive Fe in the leaves and retain homeostasis (Becker and Asch 2005; da Silveira et al. 2007). Stein et al. (2014) observed that a tolerant genotype EPAGRI 108 accumulated a higher Fe concentration of $7\,mg\,g^{-1}$ dry weight in root symplast than sensitive genotype BR IRGA 409 with $1.93\,mg\,g^{-1}$ dry weight. In this study, it was also noted that the tolerant genotype precipitated approximately 94% of the Fe in its root apoplast during Fe excess. This demonstrates that tolerant genotypes are able to cope with elevated intracellular and apoplastic Fe concentrations in roots. Hence, protecting above ground parts from excessive Fe overload is crucial in determining plants' tolerance levels.

The different metals absorbed by the roots accumulate in the discrimination center (DC) in the basal part of the shoot and the DC is known to regulate mineral distribution to other plant parts in poacea family (Mori 1998). According to Aung et al. (2018), DC preferentially transports Fe to older rather than new leaves in the process protecting the newest leaf and ensuring the survival of rice plants from the toxicity of Fe. These studies indicate that DC plays an important role in partitioning Fe to aerial parts of plants and also acts as a barrier to root-to-shoot transport of excessive Fe. It is well documented that Fe transport-related genes such as *OsIRT1*, *OsYSL2*, *OsTOM1*, *OsNRAMP1* and *OsYSL15* are highly suppressed in the DC, as in roots under Fe excess, suggesting that these genes likely distribute less Fe to shoots, instead of retaining Fe in roots and the DC (Takahashi et al. 2011; Ishimaru et al. 2012).

9.4.3 Fe Compartmentalization in Shoots

When Fe exclusion in the roots and Fe retention in roots and suppression of the Fe translocation to shoots are inadequate, plants may allow transport of Fe from the roots through DC for the subsequent avoidance of Fe toxicity by Fe compartmentalization, disposal, or storage inside shoots. Onaga et al. (2013) observed that some tolerant rice genotypes accumulated more Fe in aerial shoots, and this supported healthy shoot growth. Rice genotypes such as Siam Saba, Mahsuri, Margasari, and Pokkali collected from Indonesia exhibited shoot-based tolerance as evidenced by a high shoot Fe concentration, reduced leaf bronzing, and a high grain yield (Nugraha et al. 2016).

The concentration of Fe in rice roots increased ninefold in the presence of 20-fold Fe excess treatment and plateaued at higher Fe treatments (Aung et al. 2018). On the other hand, the concentration in the shoots of rice plants increased at higher levels of Fe treatments. This shows that the rice plant has the ability to retain a limited amount of Fe in root tissues. At the Fe levels greater than that can be coped with by Fe retention in roots, rice plants are able to translocate excess Fe from roots to shoots. Aung et al. (2018) were able to demonstrate that at the higher Fe excess levels, the more Fe was preferentially translocated to leaves, which showed bronzing earlier than new leaves. In rice, the citrate efflux transporter, *OsFRDL1,* is required for root-to-shoot Fe translocation, and its expression in nodes is required for Fe distribution to rice grains (Yokosho et al. 2016). Microarray analyses showed that the expression of *OsFRDL1* and its homolog *OsFRDL2* is upregulated in roots and the DC at high Fe levels (Aung et al. 2018). These findings demonstrate that *OsFRDL1* and *OsFRDL2* may be involved in dividing Fe in DC under Fe excess, presumably for distributing Fe to old leaves than new leaves. They also demonstrate that rice plants harbor a mechanism that partitions on excess Fe to old leaves to prevent damage to the newest leaf.

Plants also partition Fe at the organ and subcellular levels. For example, it has been proposed that storing Fe in leaf sheaths is less damaging than storing Fe in photosynthetically more active leaf blades of rice (Engel et al. 2012). The vacuole is an important subcellular compartment for the storage of excess metal ions (Moore et al. 2014). Much of the excess Fe in plants is stored in the form of ferritin, a ubiquitous protein found in almost all living species and capable of storing up to 4000 atoms of Fe in a safe and bioavailable form (Stein et al. 2009). This shows that there is a need to search for rice genotypes with the ability to store excess Fe in other organs and subcellular compartments like vacuoles and decipher the mechanisms behind it to be able to exploit it for the development of tolerant materials.

9.4.4 ROS Detoxification

The fourth defense mechanism is Fe inclusion with tolerance to cellular Fe overload and damage in leaves, which is mediated by enzymatic detoxification in the symplast (Becker and Asch 2005; Stein et al. 2009), scavenging of reactive oxygen species (ROS) by antioxidants (e.g., ascorbic acid and reduced glutathione) (Gallie 2013), and protection from ROS damage by the action of antioxidant enzymes (e.g. superoxide dismutase, peroxidase, and catalase) (Fang and Kao 2000). Cultivation with various Fe excess intensities and microarray analyses revealed that the molecular defense mechanism described above works since mild to moderate Fe excess conditions that do not seriously affect rice growth yet. However, the genes that contribute to ROS detoxification work at a molecular level in much severe Fe excess conditions, which causes severe leaf bronzing and remarkably reduces the plant growth (Aung et al. 2018).

Plants switch on genes associated with oxidative stress, oxygen and electron transfer, and those encoding cytochrome P450 family proteins (Quinet et al. 2012; Aung et al. 2018), or transcription factors related to cell death and stress responses such as *OsNAC4*, *OsNAC5* and *OsNAC6* (Finatto et al. 2015; Aung et al. 2018) to alleviate the damage caused by excess Fe-induced ROS. Recent findings indicate that *S*-nitrosoglutathione-reductase (GSNOR) promotes root tolerance to Fe toxicity by inhibiting Fe-dependent nitrosative and oxidative cytotoxicity in plants via the nitric oxide pathway (Li et al. 2019). The knockout mutants in *Arabidopsis* displayed a root growth defect in the presence of 250 μM of Fe, demonstrating the role of GSNOR for root tolerance to Fe toxicity. Furthermore, WRKY transcription factors are key regulators of several processes associated with numerous abiotic and biotic stress tolerance in plants. Most WRKY genes are involved in ROS defense mechanism to various stresses including Fe excess stress (Figure 9.1).

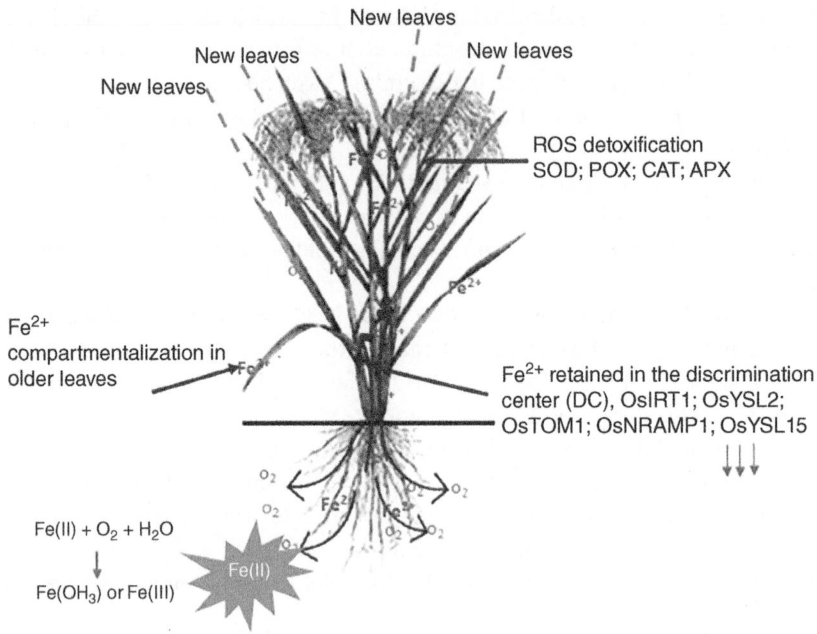

Figure 9.1 Illustration of mechanisms used by rice for iron toxicity tolerance.

9.4.5 Candidate Genes Involved in the Mechanisms of Fe Toxicity

Several studies have cloned and characterized genes involved in the Fe uptake and transport pathways, including some regulatory transcription factors (Quinet et al. 2012; Kobayashi et al. 2014; Wu 2016). Based on Strategy I, Fe^{2+} uptake from the rhizosphere is primarily mediated by the *OsIRT1, OsNRAMP1,* and *OsNRAMP5* genes in rice (Takahashi et al. 2011; Ishimaru et al. 2006, 2012). Rice, like maize and wheat, produces and secretes deoxymugineic acid (DMA) according to Strategy II (Bashir et al. 2006).

The *TOM1,* which is a phytosiderophore (PS) efflux transporter, secretes DMA into the rhizosphere where it binds Fe^{3+}directing the defensive strategy II (Nozoye et al. 2011). The Fe^{3+}-DMA complexes are then absorbed into the root, primarily through the plasma membrane of *OsYSL15,* one of 18 YS1-like proteins in rice (Inoue et al. 2009; Lee et al. 2009). Once inside the rice root epidermal cells, Fe in the form of Fe^{2+}-NA or Fe^{3+}-DMA follows the radial symplastic pathway to the central cylinder. To enter the xylem vessels, Fe is extruded from the symplast into the apoplastic space via an Fe efflux transporter identified in *Arabidopsis* as ferroportin 1/iron-regulated 1 protein (AtFPN1/AtIREG1) located on the pericycle cell plasma membrane (Kim and Guerinot 2007; Morrissey et al. 2009; Onaga et al. 2016). Alternatively, iron movement from the apoplast requires Fe chelators, nicotianamine (NA), DMA, and citrate, which are secreted into the apoplast via the NA efflux transporters *ENA1* and *ENA2,* the PS efflux transporter *TOM1* and *OsFRDL1,* a transporter belonging to the multidrug and toxin efflux MATE family (Yokosho et al. 2009; Nozoye et al. 2011). Citrate, DMA, or NA could function alternatively or jointly to transport Fe in the xylem depending on the Fe form and concentration absorbed by rice cells. In rice, xylem Fe unloading into the phloem can occur directly via Fe^{2+} transporters, *OsIRT1*/*OsIRT2* and *OsNRAMP1* or indirectly via *OsYSL15*/*OsYSL16*/*OsYSL18* and *OsYSL2* (Aoyama et al. 2009; Kobayashi et al. 2014). Further evidence suggests that Fe-citrate could enter xylem-associated cells directly via an unknown transport system or a ZRT-IRT-like transporter after being reduced from Fe^{3+} to Fe^{2+} by a ferric-chelate reductase (FRO) (Roschzttardtz et al. 2011). The *OsYSL2* is a transporter expressed in the sieve elements of the phloem that transports Fe^{2+}-NA (Koike et al. 2004), whereas *OsYSL15* transports Fe^{3+}-DMA (Inoue et al. 2009). As a result, *OsYSL2* and *OsYSL15* provide rice with a one-of-a-kind opportunity to transport both Fe^{2+} and Fe^{3+}. These findings demonstrate that the existence of both strategies I and II in rice could be linked to its adaptation to lowland and upland ecologies.

9.4.6 Genetic Variants for Iron Toxicity Tolerance in Rice Germplasm

Successful improvement of tolerance to Fe toxicity depends heavily on the existence of genetic variations within the rice germplasms. Genetic variations for Fe stress tolerance exist among the rice germplasm (Onaga et al. 2013; Nugraha et al. 2016; Diop et al. 2020; Pawar et al. 2021). Overall, majority of the functional alleles for tolerance to Fe excess toxicity are mainly derived from the *indica* varieties (Melandri et al. 2021). This shows the need for continued search for novel alleles in this gene pool to increase the efficiency of selection in breeding programs designed to enhance tolerance to Fe toxicity in rice. Two genotypes, IR61612-313-16-2-2-1 and Suakoko 8 showed significant high tolerance to Fe toxicity (Onaga et al. 2013). Siam Saba, Mahsuri, Margasari, and Pokkali Indonesian rice genotypes exhibited high shoot-based tolerance under excess Fe in the soil medium (Nugraha et al. 2016). A total of 323 rice accessions and breeding lines cultivated in West Africa spanning two subpopulations of *japonica* and *indica* showed divergent variation for Fe toxicity tolerance under hydroponic and field conditions (Diop et al. 2020). The panel of 119 out of 352 rice germplasms consisting of landraces and released varieties showed morphological and genetic variations for Fe toxicity stress under hydroponic and filed conditions (Pawar et al. 2021). A number of landraces and released varieties with moderate to high tolerance to Fe toxicity have been identified and developed for cultivation in Fe toxic soils worldwide (Table 9.1). Bierschenk et al. (2020) identified four wild and one rice accessions that were tolerant to acute and chronic Fe stress. Onyango et al. (2019) identified two tolerant rice genotypes (CK801 and Suakoko 8) under hydroponic culture system. Kar et al. (2021) identified novel candidate genes for adaptive Fe excess tolerance in Lachit rice variety derived from across between CRM13-3241 and Kalinga 2. These findings point to the existence of diverse rice genotypes with ability to tolerate Fe excess that could be exploited for the improvement of sensitive materials.

9.5 Molecular Breeding for Fe Toxicity Tolerance in Rice

Researchers have used a variety of molecular breeding techniques to improve Fe tolerance in rice. Breeders have identified molecular breeding as a more reliable technique for identifying and transferring genes for improving heavy metal tolerance. To improve Fe

Table 9.1 List of landraces and released varieties showing tolerance to Fe toxicity and their country of origin or release.

Fe tolerant germplasm	Country of origin
Tolerant: Dhusura; moderately tolerant: Punjabniswarna, Dhabalabhuta, Sankaribako, Gurumukhi, Bayabhanda, Ahirman, Veleri, Khandasagar, Kalamara, and Pipalbasa	India (Pawar et al. 2021)
IR61612-313-16-2-2-1 and Suakoko 8	Philippines and West Africa (Onaga et al. 2013; Diop et al. 2020)
Oryza rufipogon 105293 (Australia), *Oryza meridionalis* 105290 (Australia), *Oryza nivara* 100593 (Taiwan), *O. nivara* 103837 (Bangladesh), *Oryza barthii* 106238 (Mali), *O. rufipogon* 105491 (Malaysia)	Australia, Taiwan, Bangladesh, Mali, and Malaysia (Bierschenk et al. 2020)
CK801 and Suakoko 8	West Africa (Onyango et al. 2018)

tolerance in rice, marker-assisted selection (MAS) is being used, which includes the use of markers for the construction of a linkage map to identify putative genes and cloning the genes to speed up molecular breeding. In recent years, molecular breeding approaches have proven to be more beneficial, efficient, and timesaving than traditional breeding approaches, particularly the use of backcross recombinant inbred lines (BRILs) populations, which are ideal populations for targeted gene cloning used to accelerate molecular breeding for improving Fe tolerance in rice (Rasheed et al. 2020). Identification of quantitative trait loci (QTL) has enhanced the understanding of the genetic basis of Fe tolerance in rice (Dramé et al. 2011).

Most of the QTL studies have used a number of populations including double haploid, recombinant inbred lines (RIL), backcrosses, chromosome segment substitution lines (CSSLs), and MAGIC population. Several QTL associated with traits involved in Fe toxicity have been mapped on the rice genome under various environmental conditions using different segregating populations (Wu et al. 1997, 1998, 2014; Wan et al. 2003a, b, 2004, 2005; Shimizu et al. 2005; Dufey et al. 2009, 2012; Shimizu 2009; Fukuda et al. 2012) or interspecific (Dufey et al. 2012, 2015a) crosses or using diversity panels (Matthus et al. 2015). Wu et al. (1997) identified three significant QTL on chromosome 1 and 8 with large phenotypic effects for tolerance to Fe toxicity linked to leaf bronzing score and relative decrease in shoot dry weight. Another QTL associated with the enzymatic activity of ascorbate peroxidase and glutathione reductase, as well as the concentrations of dehydroascorbate and ascorbate in rice leaves were detected in the same region on chromosome 1 (Wu et al. 1998). In other studies, Wan et al. (2003a) identified four QTL for leaf bronzing, root and shoot biomass, and tiller number with relatively strong effects ranging from 20 to 24% using a backcross population developed from Nipponbare and Kasalath. Wan et al. (2003b) also discovered 14 QTL in chromosome segment substitution lines (CSSLs) derived from Asominori and IR24 conditioning variation in leaf bronzing score, stem dry weight, plant height, root length, and root dry weight. In this study, the QTL for leaf bronzing score located on chromosome 3 in the region C515-XNpb279, also was behind the variation observed for chlorophyll content, indicating that the two traits might be under the influence of the same QTL. Another QTL for leaf bronzing score was found near RM221 on chromosome 2 (Shimizu 2009). Yet in another study, Ouyang et al. (2007) using 244 inbred lines derived from Zhenshan97B/Miyang46 identified seven QTL conditioning the elongation of the coleoptile. Under excess Fe, QTL affecting rice shoot Fe concentration on chromosomes 3 and 4 were reported (Shimizu et al. 2005), and it was co-located with a QTL responsible for the high Fe content of shoot in Kasalath (Fukuda et al. 2012).

Two putative QTL for leaf bronzing index (LBI) qFETOX-1-1 and qFETOX-1-2 were observed in recombinant inbred lines (RILs) derived from IR29/Pokkali (Wu et al. 2014). The LBI revealed significant genetic variation in tolerance to Fe toxicity among the parental genotypes. Meng et al. (2017) reported two QTLs, qSL-1 and qRL-8 with 18% phenotypic variation associated with shoot length (SL) and root length (RL), respectively. These traits indicated that there was a lot of genetic variation in Fe tolerance and many genes have been identified as being involved in these QTLs controlling Fe toxicity tolerance in rice. Dufey et al. (2009) evaluated 164 RIL populations in multiple environments and identified one putative QTL qNSP-3 for the number of spikelets in rice, which showed significant variation for iron toxicity tolerance. Similarly, Dufey et al. (2012) identified two more QTLs for 1000 grain qGW-1 weight and chlorophyll content qCC1-7 with varying ratios of variance using the same population.

In other research, Dufey et al. (2015b) evaluated 220 back-cross double-haploid (BC3DH) plants in a hydroponic experiment and identified QTL for shoot water content (SWC) qSWC-3, shoot dry weight (SDW) qSDW-3, and leaf bronzing index (LBI) qLBI-1, which strongly demonstrated that all of these QTL played a significant role in rice response to Fe toxicity tolerance. Dufey et al. (2012) investigated QTLs associated with yield components under Fe toxic field conditions, focusing on West Africa, and found out that despite having small effects, some of these QTLs co-located with previous QTLs, such as the leaf bronzing QTL on chromosome 1, and were consistent across different experimental conditions, environments, and even independent mapping studies. These stable QTLs were found in six regions of the rice genome and could be major QTLs containing genes involved in Fe toxicity tolerance (Dufey et al. 2012). From the previous studies, Dufey et al. (2015a) constructed an integrated map with the previous QTLs identified using the annotated physical map of the rice reference variety Nipponbare and highlighted four candidate regions (CR), which were genomic regions with a high QTL density: CR1 on chromosome 1 between markers RM246 and RM443; CR2 on chromosome 2 between markers RM526 and R758; CR3 on chromosome 3 between markers C515 and C25; and CR4 on chromosome 7 between markers R1245 and RM429. Dufey et al. (2015b) used an interspecific backcross population derived from Caiapo/MG12/Caiapo to identify Fe toxicity tolerance QTLs in *Oryza glaberrima*. Out of the 28 significant putative QTLs identified (Dufey et al. 2015b), 11 QTLs were associated with morphological traits such as leaf bronzing index, shoot dry weight, and root dry weight on chromosomes 1, 2, 3, 5, and 12, with contributions ranging from 5 to 18%, while 17 QTLs were associated with physiological traits such as chlorophyll content index, stomatal conductance, shoot water content, photosystem II efficiency, non-photochemical quenching, blade Fe concentration, sheath Fe concentration, and root-plaque Fe concentration on chromosome 1, 2, 3, 5, 7, and 10 with contributions varying from 17 to 40%.

Qiu et al. (2017) identified two QTLs using IL (Introgression lines) populations derived from *japonica* (02428) and *indica* (Minghui63) and reported two QTLs, qRSDW-11, and qRRDW-2, independent of genetic background. In the past, a QTL qSDW-5 was identified that expressed under both experimental conditions with similar additive effects, implying that there is a genetic overlap between Fe toxicity tolerance and zinc toxicity tolerance in rice (Zhang et al. 2013). Zhang et al. (2017) discovered a putative QTL qSFW-2 with a positive additive effect, indicating that the genes underlying this QTL were inherited from the donor parent. Progresses in molecular breeding for iron toxicity tolerance in rice are summarized in Table 9.2.

QTL mapping has gradually proven to be an effective method for dissecting genetic factors underlying phenotypic traits such as Fe toxicity tolerance (Dufey et al. 2012; Fukuda et al. 2012; Wan et al. 2005). However, a number of screening and mapping experiments for Fe toxicity tolerance in rice have yielded somewhat contradictory tolerance rankings. For example, Wu et al. (1997) and Dufey et al. (2009) classified Azucena as tolerant after screening it in 250 $mg\,l^{-1}$ Fe^{2+} for four weeks in hydroponics. However, in another experiment conducted by Engel et al. (2012), the same *japonica* variety Azucena demonstrated high susceptibility to a pulse stress of 1500 $mg\,l^{-1}$ Fe^{2+}. Audebert and Fofana (2009) found that two varieties, WITA 1 and Matkandu from Africa Rice Center and Malaysia, respectively, were moderately tolerant in Fe-toxic soils in Korhogo, Ivory Coast, but responded sensitively to Fe stress in Kilissi, Guinea. Pokkali, another indica genotype, was found to be sensitive to an acid sulfate soil with chronic toxicity in the Philippines (Gregorio et al. 2000), but it demonstrated marked tolerance when treated with Fe pulse stress of 1500 $mg\,l^{-1}$ Fe^{2+} in hydroponics (Engel et al. 2012; Wu et al. 2014). These findings demonstrate that tolerance attributes of a given rice variety might be dependent on the concentration of Fe and how it interacts with different ions in the medium.

9.6 Conclusion

Despite the advances in molecular breeding for iron tolerance in rice, the use of marker-assisted selection in breeding for tolerance to Fe toxicity is currently severely limited by the fact that most QTLs reported are for minor effects, and for the few major ones, large confidence intervals and/or a lack of validation in other genetic backgrounds and environments are a deterrent to their use. The discovery of numerous QTL with minor effects suggests that tolerance to Fe toxicity may be due to additive genetic effects. This means that in order to have a significant impact on the phenotype, several QTL/genes must be pyramided into the genetic background of sensitive rice lines. Identification of large-effect QTL associated with Fe toxicity tolerance that could be used in MAS can significantly accelerate rice breeding for Fe toxicity tolerance.

Table 9.2 Summary progress in molecular breeding for iron toxicity tolerance in rice.

Mapping population	Traits	Chromosome	References
IR64 × Azucena	SDW, LBI	1, 8	Wu et al. (1997)
IR64 × Azucena	LBI,TFeC	1, 2, 7	Wu et al. (1998)
IR24 × Asominori	LBI, StDW, PH, RL, RDW	3, 6, 7, 9, 11, 12	Wan et al. (2003a)
Nipponbare × Kasalath	LBI, StDW, TN, RDW	1, 3	Wan et al. (2003b)
Kinmaze × DV85	LBI, CC, PH	3	Wan et al. (2004)
Gimbozu × Kasalath	SDW	1, 3, 4, 9	Shimizu et al. (2005)
Longza 8503 XIR64	PH, RI, LBI, CC	1, 2, 3, 4, 5, 6, 8, 9, 11, 12	Wan et al. (2005)
Zhenshan97B × Miyang 46	CER	1, 4, 5, 7	Ouyang et al. (2007)
Gimbozu × Kasalath	LBI, RDWS, SIC	1, 2, 7, 8	Shimizu (2009)
IR64X Azucena	LBI, SWC, SDW, RDW, SIC, CCI	1, 2, 3, 4, 7, 11	Dufey et al. (2009)
IR64X Azucena	NSP	3	Dufey et al. (2009)
Koshihikari × Kasalath	SIC, SDW, RDW	1, 2, 3, 4, 7, 11, 12	Fukuda et al. (2012)
IR64X Azucena	LBI, SWC, SDW, RDW, SIC, CCI, SR, TPB, PDW, SPP, FR, GCL, 1000 GW	1, 2, 3, 7	Dufey et al. (2012)
Lemont × Teqing	SDW	5	Zhang et al. (2013)
Bao Thai × Suakoko 8	LBI, PH, LL, SDW, FCL	3, 4, 5, 8, 10	Elec et al. (2013)
IR29 × Pokkali	Leaf bronzing	1, 2, 4, 7, 12	Wu et al. (2014)
Nipponbare × Kasalath	Leaf bronzing	1, 3, 8	Wu et al. (2014)
Various MP	SC, 100 GW, RDW, LBS, DHA/AS, DHA, GRA, APA, RSVSDW, SK, Fv/Fm, PDW, NSP, FR, GCL, Lfe, ShFe, TPB, RL	1, 2, 3, 7	Dufey et al. (2015a)
Caiapo × MG12	logLBI, LBI, SWC, SDW, RDW, CCI, SC, Fv/Fm, NPQ, Bfe, Sfe, RFe	1, 2, 3, 5, 7, 10, 12	Dufey et al. (2015b)
02428 × Minghui63	RSDW, RRDW	2, 11	Qiu et al. (2017)
MP	SL, RL	1, 8	Meng et al. (2017)
Indica	SFW	2	Zhang et al. (2017)
Japonica/Indica BRILs	RSL, RSFW, RRFW	1, 4, 6, 10, 12	Rasheed et al. (2020)

References

Aoyama, T., Kobayashi, T., Takahashi, M. et al. (2009). OsYSL18 is a rice iron(III)–deoxymugineic acid transporter specifically expressed in reproductive organs and phloem of lamina joints. *Plant Mol. Biol.* 70: 681–692.

Audebert, A. (2006). Iron toxicity in rice – environmental conditions and symptoms. In: *Iron Toxicity in Rice-Based System in West Africa* (ed. A. Audebert, L.T. Narteh, P. Kiepe, et al.), 18–33. Cotonou: WARDA [Africa Rice Center].

Audebert, A. and Fofana, M. (2009). Rice yield gap due to iron toxicity in West Africa. *J. Agron. Crop Sci.* 195: 66–76.

Audebert, A. and Sahrawat, K.L. (2000). Mechanisms for iron toxicity tolerance in lowland rice. *J. Plant Nutr.* 23: 1877–1885.

Aung, M.S., Masuda, H., Kobayashi, T., and Nishizawa, N.K. (2018). Physiological and transcriptomic analysis of responses to different levels of iron excess stress in various rice tissues. *Soil Sci. Plant Nutr.* 64: 370–385.

Balk, J. and Schaedler, T.A. (2014). Iron cofactor assembly in plants. *Annu. Rev. Plant Biol.* 65: 125–153.

Balasubramanian, V., Sie, M., Hijmans, R.J., and Otsuka, K. (2007). Increasing rice production in sub-Saharan Africa: challenges and opportunities. *Adv. Agron.* 94: 55–133.

Bashir, K., Inoue, H., Nagasak, S. et al. (2006). Cloning and characterization of deoxymugineic acid synthase genes in graminaceous plants. *J. Biol. Chem.* 281: 32395–32402.

Bashir, K., Ishimaru, Y., and Nishizawa, N.K. (2010). Iron uptake and loading into rice grains. *Rice* 3: 122–130.

Becana, M., Moran, J.F., and Iturbe-Ormaetxe, I. (1998). Iron-dependent oxygen free radical generation in plants subjected to environmental stress: toxicity and antioxidant protection. *Plant Soil* 201: 137–147.

Becker, M. and Asch, F. (2005). Iron toxicity in rice – condition and management concepts. *J. Plant Nutr. Soil Sci.* 168: 558–573.

Becker, M., Ngo, N.S., and Schenk, M.K.A. (2020). Silicon reduces the iron uptake in rice and induces iron homeostasis related genes. *Sci. Rep.* 10: 5079.

Bierschenk, B., Tagele, M.T., Ali, B. et al. (2020). Evaluation of rice wild relatives as a source of traits for adaptation to iron toxicity and enhanced grain quality. *PLoS One* 15 (1): e0223086.

Briat, J.F. and Lobréaux, S. (1997). Iron transport and storage in plants. *Trends Plant Sci.* 2: 187–193.

Briat, J.F., Fobis-Loisy, I., Grignon, N. et al. (1995). Cellular and molecular aspects of iron metabolism in plants. *Biol. Cell* 84: 69–81.

Brumbarova, T., Bauer, P., and Ivanov, R. (2015). Molecular mechanisms governing *Arabidopsis* iron uptake. *Trends Plant Sci.* 20: 124–133.

Chandel, G., Banerjee, S., and Verulkar, S. (2010). Expression profiling of metal homeostasis related candidate genes in rice (*Oryza* spp.) using semi quantitative RT-PCR analysis. *Rice Genet. Newsl.* 25: 44–47.

Diop, B., Wang, D.R., Drame, K.N. et al. (2020). Bridging old and new: diversity and evaluation of high iron-associated stress response of rice cultivated in West Africa. *J. Expert Bot.* 71: 4188–4200.

Dramé, K.N., Saito, K., Kone, B. et al. (2011). Coping with iron toxicity in the lowlands of sub-Saharan Africa: experience from Africa Rice Center. *Proceedings of the second Africa Rice Congress*, Bamako, Mali 22–26 March 2010: Innovation and Partnerships to realize Africa's Rice Potential.

Dufey, I., Hakizimana, P., Draye, X. et al. (2009). QTL mapping for biomass and physiological parameters linked to resistance mechanisms to ferrous iron toxicity in rice. *Euphytica* 167: 143–160.

Dufey, I., Hiel, M.P., Hakizimana, P. et al. (2012). Multienvironment quantitative trait loci mapping and consistency across environments of resistance mechanisms to ferrous iron toxicity in rice. *Crop Sci.* 52: 539–550.

Dufey, I., Mathieu, A.-S., Draye, X. et al. (2015a). Construction of an integrated map through comparative studies allows the identification of candidate regions for resistance to ferrous iron toxicity in rice. *Euphytica* 203: 59–69.

Dufey, I., Draye, X., Lutts, S. et al. (2015b). Novel QTLs in an interspecific backcross *Oryza sativa* × *Oryza glaberrima* for resistance to iron toxicity in rice. *Euphytica* 14: 1342–1347.

Elec, V., Quimio, C., Mendoza, R. et al. (2013). Maintaining elevated Fe^{2+} concentration in solution culture for the development of a rapid and repeatable screening technique for iron toxicity tolerance in rice (*Oryza sativa* L.). *Plant Soil* 372: 253–264.

Engel, K., Asch, F., and Becker, M. (2012). Classification of rice genotypes based on their mechanisms of adaptation to iron toxicity. *J. Plant Nutr. Soil Sci.* 175: 871–881.

Fang, W.C. and Kao, C.H. (2000). Enhanced peroxidase activity in rice leaves in response to excess iron, copper and zinc. *Plant Sci.* 158: 71–76.

Finatto, T., Oliveira, A.C., Chaparro, C. et al. (2015). Abiotic stress and genome dynamics: specific genes and transposable elements response to iron in rice. *Rice* 8: 12.

Fukuda, A., Shiratsuchi, H., Fukushima, A. et al. (2012). Detection of chromosomal regions affecting iron concentration in rice shoots subjected to excess ferrous iron using chromosomal segment substitution lines between Japonica and Indica. *Plant Prod. Sci.* 15: 183–191.

Gallie, D.R. (2013). The role of l-ascorbic acid recycling in responding to environmental stress and in promoting plant growth. *J. Exp. Bot.* 64: 433–443.

Gregorio, G.B., Senadhira, D., Htut, H., and Graham, R.D. (2000). Breeding for trace mineral density in rice. *Food Nutr. Bull.* 1 (4): 382–386.

GRiSP (2013). *Rice Almanac*, 4e, 1–283. Los Baños: International Rice Research Institute.

Inoue, H., Kobayashi, T., Nozoye, T. et al. (2009). Rice OsYSL15 is an iron-regulated iron(III)-deoxymugineic acid transporter expressed in the roots and is essential for iron uptake in the early growth of the seedlings. *J. Biol. Chem.* 284: 3470–3479.

Ishimaru, Y., Suzuki, M., Tsukamoto, T. et al. (2006). Rice plants take up iron as an Fe^{3+}-phytosiderophore and as Fe^{2+}. *Plant J.* 45: 335–346.

Ishimaru, Y., Takahashi, R., Bashir, K. et al. (2012). Characterizing the role of rice NRAMP5 in manganese, iron and cadmium transport. *Sci. Rep.* 2: 1–8.

Jain, A. and Connolly, E.L. (2013). Mitochondrial iron transport and homeostasis in plants. *Front. Plant Sci.* 4: 1–7.

Kar, S., Mai, H.-J., Khalouf, H. et al. (2021). Comparative transcriptomics of lowland rice varieties uncovers novel candidate genes for adaptive iron excess tolerance. *Plant Cell Physiol.* 62: 624–640.

Kim, S.A. and Guerinot, M.L. (2007). Mining iron: iron uptake and transport in plants. *FEBS Lett.* 581: 2273–2280.

Kobayashi, T., Nakanishi Itai, R., and Nishizawa, N.K. (2014). Iron deficiency responses in rice roots. *Rice* 7: 27.

Koike, S., Inoue, H., Mizuno, D. et al. (2004). OsYSL2 is a rice metal-nicotianamine transporter that is regulated by iron and expressed in the phloem. *Plant J.* 39: 415–424.

Krohling, C.A., Eutrópio, F.J., Bertolazi, A.A. et al. (2016). Ecophysiology of iron homeostasis in plants. *Soil Sci. Plant Nutr.* 62: 39–47.

Lee, S., Chiecko, J.C., Kim, S.A. et al. (2009). Disruption of OsYSL15 leads to iron inefficiency in rice plants. *Plant Physiol.* 150: 786–800.

Li, G., Song, H., Li, B. et al. (2015). Auxin resistant1 and PIN-FORMED2 protect lateral root formation in *Arabidopsis* under iron stress. *Plant Physiol.* 169: 2608–2623.

Li, G., Kronzucker, H.J., and Shi, W. (2016). Root developmental adaptation to Fe toxicity: mechanisms and management. *Plant Signal. Behav.* 11: e1117722.

Li, B., Sun, L., Huang, J. et al. (2019). GSNOR provides plant tolerance to iron toxicity via preventing iron-dependent nitrosative and oxidative cytotoxicity. *Nat. Commun.* 10: 3896.

Mahender, A., Swamy, B.P.M., Anandan, A., and Ali, J. (2019). Tolerance of irondeficient and -toxic soil conditions in rice. *Plants* 8: 31.

Matthus, E., Wu, L.B., Ueda, Y. et al. (2015). Loci, genes, and mechanisms associated with tolerance to ferrous iron toxicity in rice (*Oryza sativa* L.). *Theor. Appl. Gen.* 128: 2085–2098.

Melandri, G., Sikirou, M., Arbelaez, J.D. et al. (2021). Multiple small-effect alleles of *Indica* origin enhance high iron-associated stress tolerance in rice under field conditions in West Africa. *Front. Plant Sci.* 11: 604938.

Meng, L., Wang, B., Zhao, X. et al. (2017). Association mapping of ferrous, zinc, and aluminum tolerance at the seedling stage in indica rice using MAGIC populations. *Front. Plant Sci.* 8: 1–15.

Mongon, J., Chaiwong, N., Bouain, N. et al. (2017). Phosphorus and iron deficiencies influences rice shoot growth in an oxygen dependent manner: insight from upland and lowland rice. *Int. J. Mol. Sci.* 18: 1–7.

Moore, K.L., Chen, Y., van de Meene, A.M.L. et al. (2014). Combined NanoSIMS and synchrotron X-ray fluorescence reveal distinct cellular and subcellular distribution patterns of trace elements in rice tissues. *New Phytol.* 201: 104–115.

Mori, S. (1998). Iron transport in graminaceous plants. In: *Iron Transport and Storage in Microorganisms, Plants and Animals. Vol. 35, Metal Ions in Biological Systems* (ed. A. Sigel and H. Sigel), 215–237. New York: Marcel Dekker.

Morrissey, J., Baxter, I.R., Lee, J. et al. (2009). The ferroportin metal efflux proteins function in iron and cobalt regulation in *Arabidopsis*. *Plant Cell.* 21: 3326–3338.

Nozoe, T., Agbisit, R., Fukuta, Y. et al. (2008). Characteristics of iron tolerant rice lines developed at IRRI under field conditions. *Jpn. Agric. Res. Q.* 42: 187–192.

Nozoye, T., Nagasaka, S., Kobayashi, T. et al. (2011). Phytosiderophore efflux transporters are crucial for iron acquisition in graminaceous plants. *J. Biol. Chem.* 286: 5446–5454.

Nugraha, Y., Ardie, S.W., Ghulamahdi, M. et al. (2016). Generation mean analysis of leaf bronzing associated with iron toxicity in rice seedlings using digital imaging methods. *J. Breed. Gen.* 48: 453–464.

Onaga, G., Edema, R., and Asea, G. (2013). Tolerance of rice germplasm to iron toxicity stress and the relationship between tolerance, Fe^{2+}, P, and K content in the leaves and roots. *Arch. Agron. Soil Sci.* 59: 213–229.

Onaga, G., Drame, K.N., and Ismail, A.M. (2016). Understanding the regulation of iron nutrition: can it contribute to improving iron toxicity tolerance in rice. *J. Funct. Plant Biol.* 43: 709–726.

Onyango, D.A., Entila, F., Dida, M. et al. (2018). Mechanistic understanding of iron toxicity tolerance in contrasting rice varieties from Africa: 1. Morpho-physiological and biochemical responses. *Funct. Plant Biol.* 46: 93–105.

Onyango, D.A., Entila, F., Dida, M. et al. (2019). Mechanistic understanding of iron toxicity tolerance in contrasting rice varieties from Africa: 1. Morpho-physiological and biochemical responses. *Funct. Plant Biol.* 46: 93–105.

Onyango, D.A., Entila, F., Egdane, J. et al. (2020). Mechanistic understanding of iron toxicity tolerance in contrasting rice varieties from Africa: 2. Root oxidation ability and oxidative stress control. *Funct. Plant Biol.* 47: 145–155.

van Oort, P.A.J. (2018). Mapping abiotic stresses for rice in Africa: drought, cold, iron toxicity, salinity and sodicity. *Field Crop. Res.* 219: 55–75.

Ouyang, Y., Zeng, F., Zhuang, J. et al. (2007). Genetic analysis of genotype × iron nutrition interaction on coleoptile elongation rate in rice (*Oryza sativa* L.). *Euphytica* 156: 311–318.

Pawar, S., Pandit, E., Mohanty, I.C. et al. (2021). Population genetic structure and association mapping for iron toxicity tolerance in rice. *PLoS One* 16 (3): e0246232.

Pennock, D. and McKenzie, N. (2015). *Montanarella L. Status of the World's Soil Resources.* Rome, Italy: FAO.

Quinet, M., Vromman, D., Clippe, A. et al. (2012). Combined transcriptomic and physiological approaches reveal strong differences between short- and long-term response of rice (*Oryza sativa*) to iron toxicity. *Plant Cell Env.* 35: 1837–1859.

Qiu, X., Chen, K., Lv, W. et al. (2017). Examining two sets of introgression lines reveals background-independent and stably expressed QTL that improve grain appearance quality in rice (*Oryza sativa* L.). *Theor. Appl. Genet.* 130 (5): 951–967.

Rasheed, A., Fahad, S., Hassan, M.U. et al. (2020). A review on aluminum toxicity and quantitative trait loci mapping in rice (*Oryza sativa* L). *Appl.Ecol. Environ. Res.* 18 (3): 3951–3961.

Roschzttardtz, H., Seguela-Arnaud, M., Briat, J.F. et al. (2011). The FRD3 citrate efflux promotes iron nutrition between symplastically disconnected tissues throughout *Arabidopsis* development. *Plant Cell.* 23: 2725–2737.

Rout, G.R. and Sahoo, S. (2015). Role of iron in plant growth and metabolism. *Agric. Sci. Rev.* 3: 1–24.

Sahrawat, K.L., Mulbah, C.K., Diatta, S. et al. (1996). The role of tolerant genotype and plant nutrients in the management of iron toxicity in lowland rice. *J. Agric. Sci.* 126: 143–149.

Saikia, T. and Baruah, K.K. (2012). Iron toxicity tolerance in rice (*Oryza sativa* L.) and its association with anti-oxidative enzyme activity. *J. Crop Sci.* 3: 90.

dos Santos, M.S., Sanglard, L.M.V.P., Barbosa, M.L. et al. (2020). Silicon nutrition mitigates the negative impacts of iron toxicity on rice photosynthesis and grain yield. *Ecotoxicol. Environ. Saf.* 189: 110008.

Shimizu, A. (2009). QTL analysis of genetic tolerance to iron toxicity in rice (*Oryza Sativa* L.) by quantification of the bronzing score. *J. New Seeds* 10: 171–179.

Shimizu, A., Guerta, C.Q., Gregorio, G.B. et al. (2005). QTLs for nutritional contents of rice seedlings (*Oryza sativa* L.) in solution cultures and its implication to tolerance to iron toxicity. *Plant Soil.* 275: 57–66.

da Silveira, V.C., de Oliveira, A.P., Sperotto, R.A. et al. (2007). Influence of iron on mineral status of two rice (*Oryza sativa* L.) cultivars. *Braz. J. Plant Physiol.* 19: 127–139.
Stein, R.J., Ricachenevsky, F.K., and Fett, J.P. (2009). Differential regulation of the two rice ferritin genes (*OsFER1* and *OsFER2*). *Plant Sci.* 177: 563–569.
Stein, R.J., Lopes, S.I.G., and F. J. P. (2014). Iron toxicity in field-cultivated rice: contrasting tolerance mechanisms in distinct cultivars. *Theor. Expt. Plant Physiol.* 26: 135–146.
Stein, R.J., Duarte, G.L., Scheunemann, L. et al. (2019). Genotype variation in rice (*Oryza sativa* L.) tolerance to Fe toxicity might be linked to root cell wall lignification. *Front. Plant Sci.* 10: 746.
Stumm, W. and Lee, G.F. (1961). Oxygenation of ferrous iron. *Ind. Eng. Chem.* 53: 143–146.
Suh, H.J., Kim, C.S., Lee, J.Y., and Jung, J. (2002). Photodynamic effect of iron excess on photosystem II function in pea plants. *Photochem. Photobiol.* 75: 513–518.
Takahashi, R., Ishimaru, Y., Senoura, T. et al. (2011). The OsNRAMP1 iron transporter is involved in Cd accumulation in rice. *J. Exp. Bot.* 62: 4843–4850.
Thongbai, P. and Goodman, B.A. (2000). Free radical generation and post anoxic injury in an iron toxic soil. *J. Plant Physiol.* 23: 1887–1990.
Wan, J.L., Zhai, H.Q., Wan, J.M., and Ikehashi, H. (2003a). Detection and analysis of QTLs for ferrous iron toxicity tolerance in rice, *Oryza sativa* L. *Euphytica* 131: 201–206.
Wan, J.L., Zhai, H.Q., Wan, J.M. et al. (2003b). Mapping QTL for traits associated with resistance to ferrous iron toxicity in rice (*Oryza sativa* L.), using japonica chromosome segment substitution lines. *Acta Genet. Sin.* 30: 893–898.
Wan, J.L., Zhai, H.Q., Wan, J.M. et al. (2004). Detection and analysis of QTLs associated with resistance to ferrous iron toxicity in rice (*Oryza sativa* L.), using recombinant inbred lines. *Acta Agron. Sin.* 30: 329–333.
Wan, J.L., Zhai, H.Q., and Wan, J.M. (2005). Mapping of QTLs for ferrous iron toxicity tolerance in rice (*Oryza sativa* L.). *Acta Genet. Sin.* 32: 1156–1166.
Wu, L.B. (2016). Genetic and physiological analyses of the tolerance mechanisms to ferrous iron toxicity in rice (*Oryza sativa* L.). PhD thesis. University of Bonn, Bonn, Germany.
Wu, P., Luo, A., Zhu, J. et al. (1997). Molecular markers linked to genes underlying seedling tolerance for ferrous iron toxicity. In: *Plant Nutrition for Sustainable Food Production and Environment* (ed. T. Ando, K. Fujita, T. Mae, et al.), 789–792. Dordrecht: Kluwer Academic.
Wu, P., Hu, B., Liao, C.Y. et al. (1998). Characterization of tissue tolerance to iron by molecular markers in different lines of rice. *Plant Soil.* 203: 217–226.
Wu, L.B., Shhadi, M.Y., Gregorio, G. et al. (2014). Genetic and physiological analysis of tolerance to acute iron toxicity in rice. *Rice* 7: 8.
Yokosho, K., Yamaji, N., Ueno, D. et al. (2009). OsFRDL1 is a citrate transporter required for efficient translocation of iron in rice. *Plant Physiol.* 149: 297–305.
Yokosho, K., Yamaji, N., and Ma, J.F. (2016). *OsFRDL1* expressed in nodes is required for distribution of iron to grains in rice. *J. Exp. Bot.* 67: 5485–5494.
Zhang, J., Soomro, A.A., Chai, L. et al. (2013). Mapping of QTL for iron and zinc toxicity tolerance at seedling stage using a set of reciprocal introgression lines of rice. *Acta Agron. Sin.* 39: 1754–1765.
Zhang, J., Chen, K., Pang, Y. et al. (2017). QTL mapping and candidate gene analysis of ferrous iron and zinc toxicity tolerance at seedling stage in rice by genome-wide association study. *BMC Genomics* 18: 828.

10

Cobalt Induced Toxicity and Tolerance in Plants

Insights from Omics Approaches

Abdul Salam[1], *Muhammad Siddique Afridi*[2], *Ali Raza Khan*[1], *Wardah Azhar*[1], *Yang Shuaiqi*[1], *Zaid Ulhassan*[1], *Jiaxuan Qi*[1], *Nu Xuo*[1], *Yang Chunyan*[1], *Nana Chen*[1], *and Yinbo Gan*[1]

[1] *Zhejiang Key Lab of Crop Germplasm, Department of Agronomy, College of Agriculture and Biotechnology, Zhejiang University, Hangzhou, China*
[2] *Department of Plant Pathology, Federal University of Lavras (UFLA), Lavras, MG, Brazil*

10.1 Introduction

In 1735, Swedish chemist and mineralogist Georg Brandt was the first one to isolate Co from the copper ore (Davis 1933). Cobalt is regarded as heavy metal with 27 atomic number and 58.93 amu atomic mass. It is ranked 32nd in terms of abundance among all metals and 19th among trace elements (Liu 1998). Cobalt may offer both harmful and beneficial effects. Any element that has a positive effect on the growth of a plant at low concentration but turns toxic at concentration beyond an optimum level is labeled a beneficial element (Pais 1992). At low concentrations, Co promotes plant growth while causing phytotoxicity at high concentrations (Hu et al. 2021). As a beneficial element, Co is an essential nutrient for living organisms such as microorganisms and animals including humans but not plants. There has been no physiological requirement of Co reported to date in higher plants. In higher plants, the essentiality of Co is generally examined in leguminous plants due to its role in symbiotic biological nitrogen fixation (Chatterjee and Chatterjee 2003). Its beneficial or harmful effect, accumulation, and tolerance mechanism in vascular plants are still poorly understood. The main environmental causes of Co include both anthropogenic activities and natural happenings. Forest fire, sea-water spray, and volcanic eruption are the major natural sources while metal smelting, processing of Co alloys, sewage sludge, and fossil fuel combustion are the anthropogenic sources (Barceloux and Barceloux 1999). With the rapid pace of urbanization and subsequently industrialization, the production of Co industries is increasing and their hazardous effects on plants and human health are of serious concern receiving less attention. Co contamination mostly occurs in acidic soil, which comprises most of the arable land of the world (Chatterjee and Chatterjee 2003).

Heavy Metal Toxicity and Tolerance in Plants: A Biological, Omics, and Genetic Engineering Approach, First Edition. Edited by Mohammad Anwar Hossain, AKM Zakir Hossain, Sylvain Bourgerie, Masayuki Fujita, Om Parkash Dhankher, and Parvez Haris.

Naturally, Co concentration range from 8 to 20 $mg\,kg^{-1}$ in normal soils, and the maximum toxic threshold level is 20 $mg\,kg^{-1}$ in arable land (Ali et al. 2018; Chatterjee and Chatterjee 2000). Generally, Co contents vary from plant species to species, for example; leguminous plants have more Co contents than cereal plants. In plants, Co contents less than 0.2 ppm are considered to be critical and more than 0.5 ppm to be toxic while the range between 0.2 and 0.5 ppm is deemed to be sufficient (Liu 1998). However, *Nyssa sylvatica* (black gum) is an exceptional case that can tolerate and accumulate high Co content without showing symptoms of toxicity (Chatterjee and Chatterjee 2000; Liu 1998).

10.2 Plant Response to Cobalt Stress

The root is the first organ of a plant that encounters HM in the soil. Under normal conditions, the plant root system absorbs water and nutrients from the deeper part of the soil and transports them through a vascular system throughout the plant's body. However, when they are exposed to a higher concentration of toxic metallic elements, their normal function changes, which ultimately impairs plant growth and development (Khan et al. 2021; Yang et al. 2021; Zeeshan et al. 2021). In such cases, the root must undergo several structural and functional modifications (Noman et al. 2017). Plant changes membrane properties and metabolism, stiffens cell walls, and brings certain molecular, cellular, and phenotypic modifications with suppressed root growth in response to metal stress (Atkinson and Urwin 2012). Besides plant endogenous modifications, the application of exogenous agents (plant growth regulators and nanoparticles) stimulates the plant defense system to cope with abiotic stress (Afridi et al. 2019; Gul et al. 2019; Naila et al. 2021; Ulhassan et al. 2022a, b). Some plant species are capable of metal accumulation within the vacuoles of root cells, thereby inactivating them. However, other plants transfer and accumulate them in shoot systems, such as in hyperaccumulators (Noman et al. 2017). Hyperaccumulator plants can proficiently concentrate metal elements; for example, nickel hyperaccumulator species limit cobalt uptake, and some species can co-accumulate cobalt when grown in ultramafic soils. *Berkheya coddii* is a perennial herb of the Compositae family from South Africa and is one of the most favorable crops known for agromining globally. It may accumulate over 30,000 $\mu g\,g^{-1}$ of nickel in dry leaves, while co-accumulate cobalt up to 600 $\mu g\,g^{-1}$ (Rue et al. 2020). Researchers have documented reduced plantmitotic activity after exposure to HM toxicity in various plant species. The result of curtailed mitosis is reduced cell division and eventually inhibition of plant growth (Sundaramoorthy et al. 2010; Thounaojam et al. 2012). Phytotoxicity of Co triggers various physiological and biochemical changes in the plant. Stunted growth and chlorosis are the most obvious responses of plants to high Co exposure which eventually leads to reduced growth (Salam et al. 2022; Wa Lwalaba et al. 2017). A high concentration of Co in soil impairs the photosynthetic apparatus, affects the process of carbon assimilation, and disturbs cell membrane (Imtiyaz et al. 2014; Sree et al. 2015). High cobalt contents in the plant cause oxidative stress and generate reactive oxygen species (ROS) that result in ultrastructural damage, reduced seed germination, root nodulation, and damage to nuclear materials (Akeel and Jahan 2020; Jayakumar et al. 2008; Salam et al. 2022) (Figure 10.1). Moreover, Co toxicity limit the uptake of key nutrient contents that have a role in plant signaling and ion homeostasis,

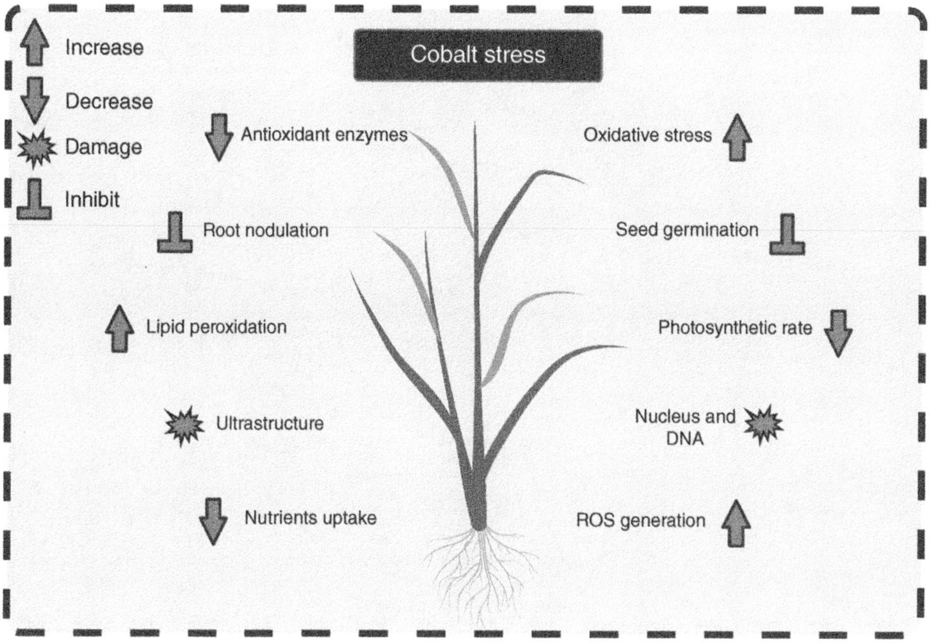

Figure 10.1 Effect of cobalt stress on overall plant growth. Plant exposure to a higher concentration of cobalt induces excessive ROS generation and causes oxidative stress. Excess ROS damages plant ultrastructure and nuclear materials, and inhibits seed germination as well as root nodulation. It also decreases antioxidant enzymatic activity and limits the uptake of nutrients.

which include calcium (Ca), magnesium (Mg), iron (Fe), phosphorus (P), and manganese (Mn) (Salam et al. 2022; Wa Lwalaba et al. 2017) (Figure 10.2).

10.2.1 Uptake and Translocation of Cobalt in Plants

Co uptake from the soil is known to be regulated by various factors including Co bioavailability, soil aeration and texture, soil pH, soil moisture, temperature, and other metals in the soil (Liu 1998). The variable valence of Co accounts for the high value of its oxidation–reduction potential (Co^{+3} to Co^{2+}) in an acid medium (Palit et al. 1994). Co^{2+} is the main bioavailable form of Co in acidic soil and its availability decreases with increasing pH levels (Sanders 1983). The transport of Co^{3+} has not been affirmed so far in plants. The adsorption of Co^{2+} into soil organic matter along with Fe and Mn oxides strongly influences their absorption by the plants (Lange et al. 2014). Soil Mn plays a crucial role in plant Co uptake and these both have an antidoting effect on each other (Bakr Ahmed and Twyman 1953; Kay et al. 2001). Soil with high Mn content regulates the uptake of Co (Li et al. 2004). Plants keep a homeostatic mechanism to maintain a proper concentration of essential metals.

The cobalt transport process is well regulated in plants. One study suggested that Ferroportin-1 (FPN1) may regulate cobalt transport in the xylem to avoid cobalt stress in the roots of wheat (Page and Feller 2005). At the same time, this mode of transport has also been demonstrated in tomatoes (Lyu et al. 2017). A controlled network of metal transport

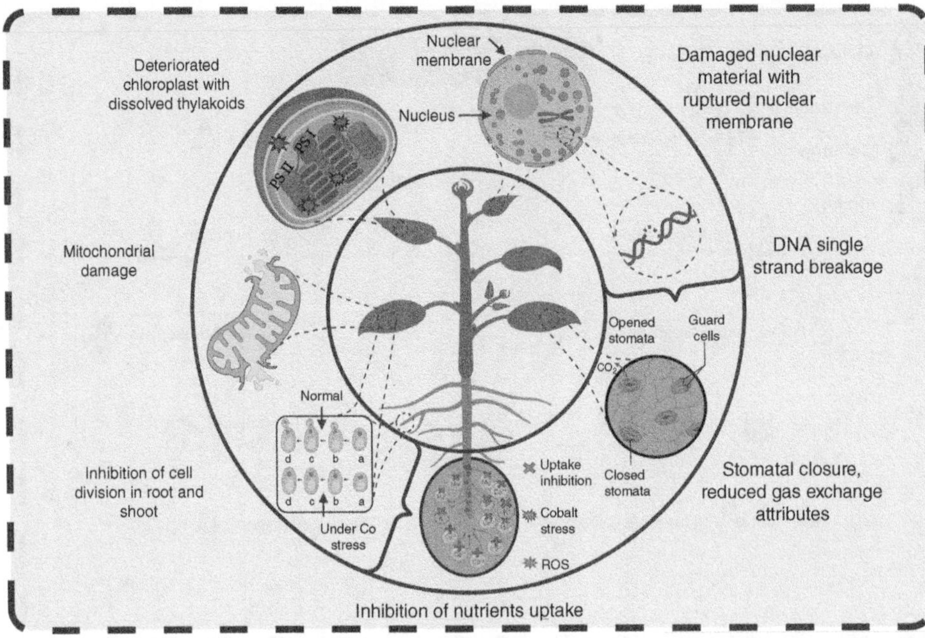

Figure 10.2 Effect of cobalt stress on plant ultrastructure and molecular architecture. Cobalt stress targets plant ultrastructure particularly chloroplast, mitochondria, and nuclear materials, which result in reduced plant growth.

and sequestration regulates the uptake, distribution, and detoxification of excess metal ions. Furthermore, it has been reported that there could be possible competition for the same transporter site between Co and other elements such as Ca; as root to shoot translocation of Co is intensely limited in the presence of Ca (Lwalaba et al. 2017), between Co and Cu, Co and Zn (Lwalaba et al. 2019). The uptake, transport, and distribution of Co are controlled by several mechanisms. Various classes of proteins have been described to be involved in heavy metal transport in plants. Hu et al. (2021) stated that there has been no specific transporter of Co reported to date and the present understanding is that Co is transporting through Fe transporters. For instance, in *Arabidopsis*, Co is transported from the soil into root epidermal cells, regulated by IRON-REGULATED TRANSPORTER-1 (IRT1), which is normally known for Fe uptake (Korshunova et al. 1999). After absorption of Co by the cell, its movement is further regulated by FPN, FPN1, and FPN2. The IREG1/FPN1 is confined to the plasma membrane while expressed in stele, which indicates its involvement in vascular loading. Whereas FPN2 is localized in the vacuole and expressed in response to Fe deficiency in the outer membrane of the root, suggesting its role in buffering cytosolic metal influx. To date, FPN is the only iron efflux transporter reported in animals, and there are two closely linked orthologs in *Arabidopsis thaliana*, i.e. IRON REGULATED1 (IREG1/FPN1) and IREG2/FPN2 (Morrissey et al. 2009). Studies reported that truncation of *FPN2* resulted from an increase in Co level in the shoot that leads to a rise in plant metal sensitivity, whereas loss of *FPN1* stops Co accumulation in the shoot. In *fpn1 fpn2* double mutant, Co cannot translocate to the shoot *via FPN1* and is unable to sequester in the root vacuoles *via FPN2* but instead accumulated in root cytoplasm causing an increase in Co sensitivity (Hu et al. 2021; Morrissey et al. 2009). A cross between cobalt-accumulating *frd3-1* mutant and *fpn1-2* abolished shoot cobalt accumulation. These findings confirm that a constant expression level of FPN1 is the principal route of Co root to shoot translocation (Morrissey et al. 2009). This also supports the idea that Co and Fe follow a similar mechanism for their uptake and subsequent translocation in plants. Furthermore, the ATP-binding cassette (ABC type) transporter is involved in the transport of cobalt and nickel in *Arabidopsis* (Morel et al. 2009), fungi, and bacteria (Eitinger et al. 2005).

10.3 Cobalt-Induced ROS Generation and Their Damaging Effects

Excess cobalt has been reported to be negatively affecting plant growth and development. Co toxicity is a vital growth limiting factor in plants (Salam et al. 2022). The cytotoxicity of Co is due to the overproduction of free radicals including ROS that causes oxidative stress. It acts as a catalyst and generates ROS by several mechanisms including cell metabolism. When oxygen interacts with the cell metabolic pathway, it converts to a highly reactive form such as hydroxyl radicals (•OH), hydrogen peroxide (H_2O_2), superoxide anions ($O_2•^-$), and oxygen singlet ($^1O_2$) (Karuppanapandian et al. 2011; Tewari et al. 2002). ROS are partially reduced or activated forms of oxygen and are a general term for extremely reactive and oxidizing oxygenates. Oxygen is of vital importance for all multicellular organisms including animals and plants, and in particular for certain processes in plants (Stamati et al. 2011). In the presence of oxygen, the cellular processes characterized by rapid electron transfer inevitably lead to electron leakage in the form of molecular oxygen (O_2), resulting in the generation of ROS that is chemically more reactive than ground state oxygen (O). Therefore, ROS is an inevitable by-product of various aerobic metabolism of life on earth depending on their nature of generation (Apel and Hirt 2004). Under normal physiological conditions, these molecules are scavenged by different antioxidant enzymes, but the balance between the production and scavenging of ROS is disturbed under adverse conditions, which results in a rapid rise in intracellular ROS levels (Alscher et al. 1997; Tsugane et al. 1999). Plants have not only acquired a ROS scavenging system to protect themselves from its toxicity over some time, but also ROS play a vital role as signaling molecules to control several processes including program cell death, pathogen defense, and stomatal conductance (Apel and Hirt 2004). However, ROS are highly reactive and high production under adverse conditions resulting in oxidative damage to lipids, proteins, damaged cell nuclei, DNA, and RNA, which may inhibit plant growth and development and in severe cases may result in plant death (Hossain et al. 2015; Jain and Gould 2015) (Figure 10.1).

10.3.1 ROS-Induced Lipid Peroxidation

Co-induced ROS toxicity is usually monitored by measuring lipid peroxidation, which is a typical chain reaction of free radicals and is generally defined as the oxidative denaturation of polyunsaturated fatty acids or lipids (Repetto et al. 2012). ROS preferably targets and attacks polyunsaturated fatty acids within the lipids. The hydroxyl radicals produced as a result of the Fenton reaction cause lipid peroxidation, disrupting cell membrane integrity and fluidity (Grant and Loake 2000). Excess free radicals and ROS adversely modify cellular components and cause damage to macromolecules including protein carbonylation and DNA, which reflects lipid peroxidation and oxidative stress (Erdei 2002; Karuppanapandian and Kim 2013). Lipid peroxidation occurs when membrane phospholipids come in contact with ROS oxidants. The cellular membranes are mostly susceptible to ROS damage because of their high polyunsaturated fatty acid (PUFA) contents, which are termed "lipid peroxidation." Studies have shown that even small amounts of nonessential metals/metalloids including cadmium (Cd), arsenic (As), and Co are toxic to plants and their toxicity can lead to serious damage, such as inhibition of growth and development, reduced production, denaturation of proteins and essential enzymes, overproduction of ROS, lipid peroxidation, and plant death in severe cases (Gashi et al. 2020; Jalmi et al. 2018; Rucinska-Sobkowiak 2016). Lipid peroxidation directly interferes and disrupts the biological function of the membrane, damaging phospholipids that leads to apoptosis (Repetto et al. 2012; Su et al. 2019). Recently, it has been reported that ferroptosis is a new form of apoptosis due to lipid peroxidation, which emphasizes the lipid peroxidation in the physiological course of cell death (Distefano et al. 2021). Hence, it is important to understand how ROS-induced lipid peroxidation leads to cell death and reduces the production of nonessential metal-induced lipid peroxidation.

10.3.2 ROS-Induced Damage to Genetic Material

ROS initially targets molecules with higher molecular mass such as membrane lipids or mitochondrial DNA that form lipids or nucleotide peroxides, particularly at the thymine level. ROS can damage almost all macromolecules of the cell including DNA (Abdul Jaleel et al. 2008; Jaleel et al. 2007; Tuteja and Sopory 2008). The chloroplast and mitochondrial DNA are more vulnerable to oxidative stress than nuclear DNA due to the absence of chromatin organization, their closeness to the ROS production site, absence of protective histones and related proteins, and low DNA repair activities (Das and Roychoudhury 2014; Yakes and Van Houten 1997). ROS compromises the oxidation of nucleic acids, including oxidation of sugar residues, changes of nucleotide bases (insertion or deletion), and extraction of nucleotide breaks in DNA strands, leading to cross-linking of DNA and proteins (Das and Roychoudhury 2014). Among all the ROS, the hydroxyl radical has been shown to induce the most damage to DNA due to its direct reaction with purine/pyrimidine bases and deoxyribose sugar (Soares et al. 2019). The exogenous or endogenous factors mediating ROS radicals can cause permanent damage to DNA, as evidenced by glycosidic shedding and oxidative damage to bases, which are corrected by the base-excision repair (BER) pathway. In DNA, ROS reacts with nitrogenous bases and deoxyribose, which results in significant oxidation reactions. DNA breaks occur forced by the rupture of ribosomes (the basic structure of DNA organization within chromosomes), which leads to problems in the compaction and coiling of DNA within chromatin. Chromatin is very essential in the regulation of gene transcription (Bui et al. 2018). Therefore, changes in its functional properties may lead to errors that can result in mutations. Besides the oxidation of DNA bases, ROS can also lead to DNA strand breaks because of a free radical attack on the DNA glycol-phosphate backbone

(Salehi et al. 2021; Figure 10.2). What's more, oxidation of DNA can cause DNA base changes or double-helix breaks, as well as other mutagenic changes (Pooetsch 2020).

10.4 Cobalt-Induced Plant Antioxidant Defense System

Plants have developed ROS detoxifying antioxidant system in response to stress-induced oxidative damage and increased production of free radicals. Antioxidants play an important role in preventing or quenching free radicals and delaying or impeding cellular damage by catalyzing reduction in ROS production (Chatterjee and Chatterjee 2003). Antioxidant defenses vary from species to species and they include both enzymatic and nonenzymatic antioxidants (Ishtiyaq et al. 2018).

10.4.1 Enzymatic Antioxidants

Half of the antioxidants are composed of enzymatic antioxidant molecules. It takes part in various biochemical and physiological processes of plant growth and development.

10.4.1.1 Superoxide Dismutase (SOD)

SOD is one of the most vital intracellular enzymes that play a key role in the plant defense system against oxidative stress and is responsible for the dismutation of superoxide radical ($O_2^{\bullet -}$) to a partially reduced oxygen species H_2O_2 (Bowler et al. 1992). Hydrogen peroxide is strong enough that oxidizes the thiol group, which is harmful to plants showing thiol-regulated enzyme activity. Other enzymes such as APX, POD, CAT, and GSH further help in the detoxification of H_2O_2. SOD is divided into three isozymes: i.e. iron SOD (Fe-SOD), manganese SOD (Mn-SOD), and zinc and copper (Zn/Cu-SOD) according to the metal ion in their active site (Ahmad et al. 2010; Racchi et al. 2001). Zn/Cu-SOD localizes in cytosol and chloroplast, Fe-SOD in the chloroplast (Jackson et al. 1978), while Mn-SOD lies in mitochondria (del Rio et al. 2006). The upregulation of SOD is related to combating oxidative stress and is of vital importance in plant survival. Increased SOD contents have been reported under Co stress in different plant species and this increase is often linked with plant stress tolerance (Ali et al. 2018; Salam et al. 2022; Wa Lwalaba et al. 2017). Shalata et al. (2001) and Hediye Sekmen et al. (2007) reported higher SOD activities in salt-tolerant species and vice versa in tomatoes and Plantago under salt stress. In the transgenic plants, high levels of Zn/Cu-SOD and APX gene transcripts showed tolerance to plants under metallic stress (Ahmad et al. 2010).

10.4.1.2 Catalases (CAT)

CAT is a ubiquitous heme-containing tetrameric enzyme and is essential for stress defense in plants. It is directly involved in the dismutation of H_2O_2 into H_2O and O_2 (Suman et al. 2021). It is present in all major H_2O_2 production sites in the cellular environment such as mitochondria and peroxisome of higher plants (Sharma and Ahmad 2014). Plant catalases can be divided into three classes. Class-1 catalases participate in the removal of H_2O_2 produced in the course of photorespiration, whereas the exact biological role of class-2 is still unclear. Class-3 is closely bound up with the elimination of excess H_2O_2 produced during fatty acid degradation in the glyoxylate cycle (Willekens et al. 1994). Catalases are essential for ROS detoxification, which can directly decompose H_2O_2 as the principal scavenging enzyme during stress. Wa Lwalaba et al. (2017) reported higher catalase activity in Co-tolerant barley genotype under Co stress. The increase in CAT activity is considered to be an adaptive feature that reduces the toxic levels of H_2O_2 that leads to coping the tissue metabolism damage (Hediye Sekmen et al. 2007). There are three isozymes of CAT, i.e. CAT1, CAT2, CAT3, and three structural genes *cat1, cat2, and cat3* encode them, respectively. CAT deficiency is shown to cause chlorosis and sensitivity to the normal photorespiratory environment in plants (McClung 1997; Sharma and Ahmad 2014).

10.4.1.3 Glutathione Peroxidases (GPX)

GPX is a family of oxidoreductase enzymes present in all almost living organisms. It lies in cell cytosol and mitochondria and has a high affinity to H_2O_2 as compared to CAT. GPX is a selenium-containing enzyme that reduces and breaks down both H_2O_2 and lipid peroxide (LOOH) by catalyzing a redox reaction with an electron donor [reduced glutathione (GSH)] (Higuchi 2014). It reduces hydrogen peroxide and decomposes peroxides into water or alcohol. In addition, GPXs can detoxify lipid peroxidation products produced due to the activity of ROS. It acts together with a high concentration of tripeptide glutathione in cells to eliminate peroxide as a potential substrate of the Fenton reaction. It is the main source that protects against low-level oxidative stress (Ahmad et al. 2010). In plants, GPXs are divided into three categories: that is selenium-independent phospholipid hydroperoxide GPX (PHGPX), selenium-dependent GPX, and glutathione transferase displaying GPX activity (GST-GPX). The two enzymes, GST and GPX, are contrasted in their subunits, the binding properties of selenium at their active site, and their catalytic mechanisms. Overexpression of the two genes (glutathione peroxidase or glutathione *S*-transferase) has been reported to be satisfactory for providing resistance against oxygen singlets in *Escherichia coli* (Ledford et al. 2007).

10.4.1.4 Glutathione Reductase (GR)

GR plays a key defense role in response to overproduction of ROS by restoring the GSH pool with nicotinamide adenine dinucleotide phosphate (NADPH) as an electron donor in the ascorbate–glutathione (ASC-GSH) cycle. It is a homodimeric enzyme that reduces glutathione disulfide (GSSG) to GSH. It is reported to be involved in various enzymatic reactions where it contributes to the oxidation/reduction process (Fagan and Palfey 2010). It is primarily localized in the chloroplast and catalyzes the formation of NADPH-dependent disulfide bonds in GSSG. Nevertheless, small amounts of GR are also present in the cytosol and mitochondria (Gill and Tuteja 2010; Romero-Puertas et al. 2006). Accumulation of GSH leads to increased GR activity in plants and consequently induces stress tolerance. Abiotic stress has been well reported to increase GR activity. Plants have several forms of this enzyme; for example eight in pea (Edwards et al. 1990) and two in wheat (Dalal and Khanna-Chopra 2001). GR is highly conserved and participates in the last step of the ASC-GSH pathway, which is mainly located in the chloroplast matrix. The raised levels of GR activity in plants perhaps could increase the accumulation of GSH. Moreover, with the increasing levels of GR, the ratio of $NADP^+/NADPH$ may grow up ensuring the bioavailability of $NADP^+$ to receive electrons from the photosynthetic electron transport chain.

Under such circumstances, GR maintains a high ratio of GSH/GSSG and ultimately confers tolerance to plants against oxidative stress (Sudhakar et al. 2001). It has been reported that GR activity is increased in Luffa seedlings under Arsenic stress (Singh et al. 2015). Similarly, the activity of GSH was found to be increased in *Helianthus annuus* L. upon exposure to multi-metal stress (Elloumi et al. 2016).

10.4.2 Nonenzymatic Antioxidants

The other half of the antioxidants are nonenzymatic, which takes part in plant growth and development.

10.4.2.1 Ascorbic Acid

Ascorbic acid is one of the widely studied nonenzymatic antioxidants present in chloroplast, cytosol, mitochondria, and apoplast and is considered to be the first wall of defense against ROS (Barnes et al. 2002; Mittler 2002). It is among the most effective antioxidant that donates electrons in various chemical reactions inside plant cells (Blokhina et al. 2003). It is mainly known to regulate the enzymes related to the cell wall and affects many physiological processes in plants. Ascorbic acid is synthesized in mitochondria and exists in reduced form ascorbate, which is transported by proton electrochemical gradient or promoting diffusion (Ahmad et al. 2010). In the ASC-GSH cycle, APX uses two ASA molecules to reduce H_2O_2 to water and produce monodehydroascorbic acid (MDA). MDA is a short-lived free radical, which is out of proportion to dehydroascorbic acid (DHA) and ASA and can also be directly restored to ASA. The electron giver is usually NADPH, which is catalyzed by mono-dehydro-ascorbate reductase (MDAR) or ferredoxin in the water–water cycle of chloroplasts. ASA, as a reducing agent, can remove a variety of free radicals and protect the metabolic process from H_2O_2 in the plant defense system (Asada 1999). It is a plentifully found antioxidant in plants, also called vitamin C, which is a key member of defense against ROS activity (Noman et al. 2014). Under Ni stress, the content of ASC alters in *Solanum nigrum* L. (Soares et al. 2016), and rice seedlings under manganese stress or arsenite stress will also change the ASC level (Mishra et al. 2011; Srivastava and Dubey 2011).

10.4.2.2 Tocopherols

Tocopherols are a group of antioxidants with α-, β-, γ-, and δ-analogs and can be found in almost all parts of plants and algae (Chew and Nyam 2020; Srivalli et al. 2003). Among them, α-tocopherols are the most active antioxidant responsible for the protection against oxidative damage and are located in the chloroplast membrane. The antioxidant scavenges lipid peroxyl free radicals to produce α-tocopheroxyl radical, which can be reused back to the analogous antioxidant by reacting with ascorbic acid or other antioxidants (Igamberdiev and Hill 2004). The tocopherol's most important function is involved in various mechanisms for the protection of polyunsaturated fatty acids from oxidative damage (Ledford and Niyogi 2005). It prevents the increase in lipid peroxidase activity due to oxidative stress by scavenging lipid peroxidative radicals in thylakoid membranes. It is generally believed that higher levels of α-tocopherol help plants cope with stress, while lower level promotes oxidative damage (Munne-Bosch 2005). In plants, α-tocopherols are produced in plastids and work as vitamin E, whose concentration can be increased under adverse conditions. Studies show that α-tocopherols are involved in finishing chain reactions of lipid peroxidation and can scavenge ROS very efficiently (Maeda and DellaPenna 2007; Munne-Bosch 2005).

10.4.2.3 Reduced Glutathione (GSH)

GSH is an important class of metabolites composed of tripeptide and made of glutamate, cysteine, and glycine (Glu–Cys–Gly) molecules. It works as a reducing agent to protect cells from oxidative damage. It is the key constituent of plants' and animals' defense systems in response to adverse environmental conditions. Studies show that GSH is involved in the antioxidant reaction, which makes it one of the key factors to control ROS production (Noctor and Foyer 1998). In the ASC-GSH cycle, GSH is used for enzymatic and nonenzymatic reduction of DHA and oxidizes itself to GSSG (oxidized glutathione). Glutathione reductase catalyzes the regeneration of GSSG to GSH, and NADPH is utilized to reduce force. Residues of the central nucleophilic cysteine are responsible for the high reduction ability of GSH, which scavenges cytotoxic H_2O_2 and engages in nonenzymatic interactions with other ROS (•OH, H_2O_2, $O_2{\bullet}^-$, $^1O_2$). Furthermore, GSH plays a central role in the antioxidant defense system by regenerating another powerful antioxidant: ascorbic acid through the ASC-GSH cycle (Millar et al. 2003).

10.5 Omics Approaches in Cobalt Stress Tolerance

"Omics" is a systematic approach that aims for the collective quantification and characterization of a dynamic biological pool such as transcriptomics, metabolomics, and proteomics. Omics approaches have increased the understanding of plant responses and complex regulatory mechanisms at the molecular level. The transcriptomic studies deal with the RNAs and their expression patterns. Proteomics and metabolomics tools provide insights that how a plant responds in terms of the production of proteins and metabolites under changing environmental conditions. The combination of all these approaches has revealed various well-defined pathways associated with the concepts of plant responses to stress (Cramer et al. 2011).

10.5.1 Transcriptomic

Transcriptomic sequencing is the complete set of transcriptions in a specific type of cell or tissue. This is a method of identifying genes that control desirable traits in crops. The transcriptomic approach not only helps to identify genes through sequence analysis but also helps to characterize the gene through analysis of differentially expressed gene (DEG) with the relevant experimental setup, thus making it one of the fast ways to identify desired genes (Figure 10.3). There are two broad classes of stress-related genes, i.e. functional gene and regulatory gene (Tran et al. 2010). The functional genes encode metabolites, whereas regulatory genes encode various transcription factors. Transcription factor (TF) is a protein-like messenger that binds at a specific site on the promoter and controls the expression of a gene in plants (Huang et al. 2012; Javaid et al. 2022; Wray et al. 2003; Zeng et al. 2014). The transcriptomic sequencing approach can explore functional genomics and pathways that stimulate in response to environmental change. Metals are one of the environmental

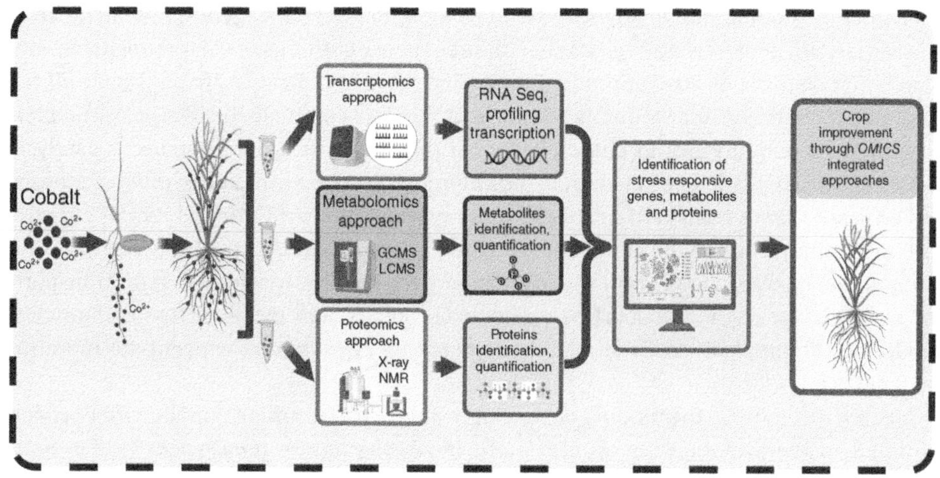

Figure 10.3 Schematic diagram of plant exposure to cobalt stress and subsequent omics approaches to identify desired stress-responsive genes, metabolites, and proteins using emerging tools in molecular crop improvement.

stimuli for plants, and these are natural elements that widely exist in soil, air, and various organisms, and some are necessary trace elements for plants (Kintlova et al. 2017). Plant transcriptomics is widely used to study plant response to several stresses, including HM stress. The activation of stress-responding genes in plants is the first step in response to abiotic stress (Konlechner et al. 2013; Pan et al. 2020). Overexpression of genes during metal stress has exposed a wide range of defense-related pathways. Cell-specific studies and spatial analysis of plant tissues are possible with more advanced microarray technologies (Brady et al. 2007; Spollen et al. 2008).

Lwalaba et al. (2020) commenced a study to analyze the transcriptomic profile of two *Hordeum vulgare* L. genotypes contrasting in Co stress tolerance. They identified 358 genes differentially expressed of which most were stress-responsive genes associated with metal transport and transcription factor in both genotypes under Co-Cu single and combined stress. Furthermore, they found that more stress-responsive genes were expressed in the stress-tolerant genotype in comparison to the sensitive one. Furthermore, in his study, the genes belonging to the NAC family and G2-like family were substantially upregulated under all treatments in both tolerant and sensitive genotypes. The MYB family demonstrated two new transcriptional genes, BG_Inovel_G002824 and BGI_novel_G004877, which were upregulated and downregulated, respectively, under Co stress. However, MYB (BGI_novel_G002824) transcript level was higher in metal tolerant "Yan66" genotype compared to a sensitive "Ea52" genotype (Lwalaba et al. 2021). It is well known that the root is the first organ to interact with metal and usually undergoes lignification, and calluses take the opportunity to form apoplast and plasmodesmata barriers to limit the absorption of heavy metals (Cuypers et al. 2002). Studies show that bacteria *Pseudomonas psychrotolerans* CS51 and *Kocuria rhizophila* 14ASPinoculation induces the synthesis of endogenous indole-3-acetic acid and gibberellin, thus promoting plant growth and enhancing the tolerance to heavy metal stress (cobalt, cadmium, respectively) (Afridi et al. 2021; Kang et al. 2020). Another study shows two isoflavone reductase (IFR) genes, and a chalcone synthase (CHS) gene was up-regulated in different concentrations of Cd treatments in soybean roots (Zhang et al. 2017). Sunlight, water, and essential nutrients are indispensable for plant photosynthesis to carry out metabolic activities and complete the life cycle. The photosynthetic machinery lies in chloroplasts that depend on transition metals as catalysts (Schmidt et al. 2020). During evolution, the chloroplast serves as the site of photosynthesis, and cells are rich in transition metals. A study revealed that ABCI10 and ABCI12 (as the ATPase A-subunit and as the membrane-intrinsic, energy-coupling T-subunit, respectively) promote cobalt absorption (Voith von Voithenberg et al. 2019). In addition, ABC transporters are in charge of the transport of amino acids, sugars, and metals across membranes. ARG1, rice chloroplast-localized ABC transporter, its *arg1* mutant concentrate more Co content (Li et al. 2020).

Cobalt stress inhibits the binding of a metal-response transcriptional modulator encoded by dmeR to the promoter region of dmeRF in *Agrobacterium tumefaciens* (Rubio-Sanz et al. 2013). Natural resistance-associated macrophage protein (NRAMP) transporter family can defend against cellular metal homeostasis perturbations. One study revealed that CrNRAMP1 acts as a Zn and Co transporter in *Chlamydomonas reinhardtii* and *nramp1* mutant increased Co tolerance while CrNRAMP1 overexpression decreased cobalt tolerance (Chang et al. 2020). In *Arabidopsis*, it is also revealed that cobalt concentration decrease in the *nramp* mutant plants (Cailliatte et al. 2010). VirE2-interacting protein1 (VIP1) being a bZIP transcription factor regulates plant–pathogen interaction and responses. Tsugama et al. (2013) stated that the Co influences the interaction between VIP1 and DNA in *A. thaliana*. Moreover, plant leaf senescence is governed by some internal physiological processes during development including leaf age and the onset of the reproductive stage. Plant hormones such as abscisic acid, jasmonic acid, and ethylene mediate the leaf senescence process (Chen et al. 2021). Functional genes play a crucial role in inducing heavy metal stress tolerance in plants. It was reported that Co inhibits ethylene biosynthesis and increases drought resistance to delay leaf senescence (Wang et al. 2020)

10.5.2 Metabolomics

Metabolomics refers to the large-scale scientific study of small molecules (metabolites) present within living tissue, while the term metabolites are the end products of metabolism and refer to the qualitative and quantitative collection of all the low-molecular-weight molecules present in living cell/tissue (Dunn and Ellis 2005). It is involved in plant growth, maintenance, and functioning, and is a powerful tool to investigate plant performance that has shown promising results in breeding and crop improvement programs (Figure 10.3). Until now, several analytical technologies have emerged to measure plant metabolites such as liquid chromatography-mass spectrometry (LCMS), gas chromatography-mass spectrometry (GCMS), matrix-assisted laser desorption ionization (MALDI), nuclear magnetic resonance spectroscopy (NMR), and inductively coupled plasma mass spectrometry (ICP-MS), and these techniques are frequently applied to dissect the state of metabolites in plants. Metabolomics is known as the large-scale study of small molecules; however, this approach is very laborious and complicated but can offer plenty of information about the phenotypes and the specific environmental conditions in which they live. The metabolome not only shows the last downstream product of the genome but also acts as the indicator of many internal and external stressors, and therefore can provide more accurate and direct information than other omics approaches and the interaction between plants and the specific environment (Kumar et al. 2017). Furthermore, the metabolome is believed to be the closest representation of phenotype and shows the huge potential for many applications in environmental research and agriculture (Singh 2020).

The amount of metabolites has a direct relation with plant physiological activities dwelling in different environments (Bundy et al. 2005). It is obvious that plant physiological activities change with change in environment and knowing this change along with specific information of metabolome will help us to understand stress tolerance mechanism. It has been earlier reported that different metabolites play a significant role in tolerating HM. For example, the plant generates a high level of free amino acid to combat the adverse effects of HM stress. These free amino acids can help protect plants by regulating ROS levels, cellular osmotic, protein/enzyme stability, and also protect membrane integrity (Bohnert and Jensen 1996; Raza et al. 2007). Weinstein et al. (1986) reported that Cu and Cd stress significantly increased putrescine amount in bean and oat plants, which help them combat stress factors. Similarly, the amount of putrescine is significantly increased under Cu stress in *Oryza sativa* L. (Lin and Kao 1999). Furthermore, the synthesis of phenolic compounds under heavy metal stress is also enhanced (Jun et al. 2003). Winkel-Shirley (2002) reported that under aluminum exposure, phenolic compounds are induced in maize. Also, Diaz et al. (2001) reported when exposed to Cu stress, phenols are accumulated in leaves of *Phaseolus vulgaris* L.

Lwalaba et al. (2020) reported that phenolic components, particularly cinnamic and benzoic derivatives along with free amino acids are substantially increased under stress conditions, which reveals its involvement in stress tolerance. In duckweed, metabolomics profiling showed that Co stress caused an increase in the production of various amino acids (except aspartic acid and cysteine at 200 μM) and organic acids (except succinic acid) which are believed to be responsible for improving Co tolerance and accumulation in *Lemna minor* L. (Hu et al. 2019). Furthermore, a higher concentration of proline, carbonylated protein, malondialdehyde along with H_2O_2 and $O_2{\bullet}^-$ has been reported under excess Co supply (Karuppanapandian and Kim 2013). Wang et al. (2020) reported that the total glutathione contents are significantly elevated in willow tops under Co stress. Oven et al. (2002) stated that the citric acid, as well as cysteine, is substantially increased under Co exposure in *Crotalaria cobalticola* L. without any change in phytochelatin.

10.5.3 Proteomics

Proteomics is a well-reputed approach in the current post-genomics era (Liu et al. 2013). The proteomic approach deals with the large-scale study of protein expression within an organism's genome (Anderson and Anderson 1998). Proteomics expansively covers the encoded proteins within a living organism in specific situations and plays a crucial role in illustrating all cellular pathways at the molecular level (Mehmood et al. 2021). Recently, proteomics has emerged as an important tool for transmitting plant survival data and adapting toxic/metalloid toxicity in several plants (Figure 10.3).

Plants are constantly exposed to adverse conditions in natural and agricultural environments that threaten their survival and growth. To cope with these changing environments, plants have developed various molecular mechanisms to quickly recognize and acclimatize to the environment. Protein plays an important role in adapting plants to adverse environmental conditions. Therefore, proteomic methods are of crucial importance to elucidate the role of protein accumulation under adverse conditions and its association with stress tolerance (Hossain et al. 2012; Perez-Clemente et al. 2013). Co is considered to participate in plant metabolism as a metal co-factor of some proteins and enzymes (Hu et al. 2021). Among them, carbonic anhydrase works as a metalloenzyme and plays a vital role in plants, especially in C4 and CAM plants. Recent studies reported that the metalloenzymes typically use zinc (Zn^{2+}) as a metal co-factor; however, Zn^{2+} can be substituted by Co^{2+} and Fe^{2+} in some members of Protista, fungi, algae, and Plantae, but it can only be substituted by Co^{2+} in microalgae (Jensen et al. 2020). Methionine aminopeptidase (MetAP) is an enzyme that is Co-dependent and responsible for the splitting of the N-terminal methionine from newly formed polypeptide chains and it is widely present in plants, animals, and bacteria. In addition, there are pieces of evidence that Co^{2+} ions activity plays an important role in *Salmonella typhimurium* and *E. coli* (Permyakov 2021). The desirable substrate needs metal ions such as Mn^{2+}, Zn^{2+}, or Co^{2+}. Studies have shown that Co^{2+} can increase peroxidase activity in horseradish (*Armoracia rusticana* L.) due to their binding to some amino acids in the active site of the enzyme at a concentration below 0.1 mM (Han et al. 2008). Taken together, using the advanced techniques of proteomics will not only help us to unravel the molecular mechanism of the plant under cobalt but will also help us to breed cobalt-tolerant genotypes that will lead us to an improved crop yield (Figure 10.3).

10.6 Conclusion and Future Prospects

As the global population is rapidly increasing, agricultural production needs to be increased with the same pattern to feed the growing population. Global climatic conditions further worsened the problem, imposing extra encumbrance on crop production. On the other side, heavy metal contamination significantly jeopardizes soil health, which leads to compromises on agriculture sustainability and crop production. Co among heavy metals is one major concern. Co is believed to be an essential element for living organisms; however, excess Co is highly harmful to plants. The situation became alarming in recent past years. For instance, new technologies needed to be arrayed in agriculture to achieve crop sustainability and higher yield under heavy metals, particularly Co stress. Omics approaches including transcriptomics, metabolomics, and proteomics are promising tools for elucidating the mechanistic insights of biotic and abiotic stresses and steering up tolerant genotypes that can combat the mentioned adverse environmental conditions. Omics approaches help scientists to get insights into stress factors and the consequent response of plants at transcriptomics, metabolomics, and proteomics levels. Transcriptomic tools have identified various key genes and transporters that can be used in breeding stress-tolerant genotypes. Similarly, proteomics and metabolomics networks have recognized key metabolites and proteins that have a role in combating adverse conditions. However, further advancement in protein and metabolite extraction along with their fractionation is needed to facilitate a deeper study of omics approaches, which will also provide opportunities to compendiously address environmental issues, such as HM pollution of the environment. In addition, the transcriptional reprogramming of genes at large levels in response to HM stimulus also needs to be studied from all perspectives using high-throughput omics approaches. These methods will not only elucidate the molecular mechanisms of plant stress interactions but also a step toward climate-smart agriculture by assisting plant breeding programs aimed at producing HM-tolerant crops.

Acknowledgments

The research was funded by Zhejiang Provincial Natural Science Foundation of China (Grant No. LZ22C130002), the National Key R & D Program of China (2021YFF1000400), and National Natural Science Foundation of China (Grant No. 31570183).

References

Abdul Jaleel, C., Lakshmanan, G.M.A., Gomathinayagam, M. et al. (2008). Triadimefon induced salt stress tolerance in *Withania somnifera* and its relationship to antioxidant defense system. *S. Afr. J. Bot.* 74: 126–132.

Afridi, M.S., Amna, Sumaira et al. (2019). Induction of tolerance to salinity in wheat genotypes by plant growth promoting endophytes: involvement of ACC deaminase and antioxidant enzymes. *Plant Physiol. Biochem.* 139: 569–577.

Afridi, M.S., Van Hamme, J.D., Bundschuh, J. et al. (2021). Biotechnological approaches in agriculture and environmental management-bacterium *Kocuria rhizophila* 14ASP as heavy metal and salt-tolerant plant growth-promoting strain. *Biologia* 76: 3091–3105.

Ahmad, P., Jaleel, C.A., Salem, M.A. et al. (2010). Roles of enzymatic and nonenzymatic antioxidants in plants during abiotic stress. *Crit. Rev. Biotechnol.* 30: 161–175.

Akeel, A. and Jahan, A. (2020). Role of cobalt in plants: its stress and alleviation. In: *Contaminants in Agriculture: Sources, Impacts and Management* (ed. M. Naeem, A.A. Ansari, and S.S. Gill), 339–357. Cham: Springer International.

Ali, S., Gill, R.A., Ulhassan, Z. et al. (2018). Insights on the responses of *Brassica napus* L. cultivars against the cobalt-stress as revealed by carbon assimilation, anatomical changes and secondary metabolites. *Environ. Exp. Bot.* 156: 183–196.

Alscher, R.G., Donahue, J.L., and Cramer, C.L. (1997). Reactive oxygen species and antioxidants: relationships in green cells. *Physiol. Plant.* 100: 224–233.

Anderson, N.L. and Anderson, N.G. (1998). Proteome and proteomics: new technologies, new concepts, and new words. *Electrophoresis* 19: 1853–1861.

Apel, K. and Hirt, H. (2004). Reactive oxygen species: metabolism, oxidative stress, and signal transduction. *Annu. Rev. Plant Biol.* 55: 373–399.

Asada, K. (1999). The water-water cycle in chloroplasts: scavenging of active oxygens and dissipation of excess photons. *Annu. Rev. Plant Physiol. Plant Mol. Biol.* 50: 601–639.

Atkinson, N.J. and Urwin, P.E. (2012). The interaction of plant biotic and abiotic stresses: from genes to the field. *J. Exp. Bot.* 63: 3523–3543.

Bakr Ahmed, M. and Twyman, E.S. (1953). The relative toxicity of manganese and cobalt to the tomato plant. *J. Exp. Bot.* 4: 164–172.

Barceloux, D.G. and Barceloux, D. (1999). Cobalt. *J. Toxicol. Clin. Toxicol.* 37: 201–216.

Barnes, J., Zheng, Y., and Lyons, T. (2002). Plant resistance to ozone: the role of ascorbate. In: *Air Pollution and Plant Biotechnology: Prospects for Phytomonitoring and Phytoremediation* (ed. K. Omasa, H. Saji, S. Youssefian, et al.), 235–252. Tokyo, Japan: Springer.

Blokhina, O., Virolainen, E., and Fagerstedt, K.V. (2003). Antioxidants, oxidative damage and oxygen deprivation stress: a review. *Ann. Bot.* 91: 179–194.

Bohnert, H.J. and Jensen, R.G. (1996). Strategies for engineering water-stress tolerance in plants. *Trends Biotechnol.* 14: 89–97.

Bowler, C., Montagu, M.V., and Inze, D. (1992). Superoxide dismutase and stress tolerance. *Annu. Rev. Plant Biol.* 43: 83–116.

Brady, S.M., Orlando, D.A., Lee, J.-Y. et al. (2007). A high-resolution root spatiotemporal map reveals dominant expression patterns. *Science* 318: 801–806.

Bui, A.D., Sharma, R., Henkel, R. et al. (2018). Reactive oxygen species impact on sperm DNA and its role in male infertility. *Andrologia* 50: e13012.

Bundy, J.G., Willey, T.L., Castell, R.S. et al. (2005). Discrimination of pathogenic clinical isolates and laboratory strains of *Bacillus cereus* by NMR-based metabolomic profiling. *FEMS Microbiol. Lett.* 242: 127–136.

Cailliatte, R., Schikora, A., Briat, J.F. et al. (2010). High-affinity manganese uptake by the metal transporter NRAMP1 is essential for *Arabidopsis* growth in low manganese conditions. *Plant Cell* 22: 904–917.

Chang, P., Yin, H., Imanaka, T. et al. (2020). The metal transporter crnramp1 is involved in zinc and cobalt transports in *Chlamydomonas reinhardtii. Biochem. Biophys. Res. Commun.* 523: 880–886.

Chatterjee, J. and Chatterjee, C. (2000). Phytotoxicity of cobalt, chromium and copper in cauliflower. *Environ. Pollut.* 109: 69–74.

Chatterjee, J. and Chatterjee, C. (2003). Management of phytotoxicity of cobalt in tomato by chemical measures. *Plant Sci.* 164: 793–801.

Chen, C., Galon, Y., Rahmati Ishka, M. et al. (2021). ASCORBATE PEROXIDASE6 delays the onset of age-dependent leaf senescence. *Plant Physiol.* 185: 441–456.

Chew, S.C. and Nyam, K.L. (2020). Refining of edible oils. In: *Lipids and Edible Oils* (ed. C.M. Galanakis), 213–241. Elsevier.

Cramer, G.R., Urano, K., Delrot, S. et al. (2011). Effects of abiotic stress on plants: a systems biology perspective. *BMC Plant Biol.* 11: 1–14.

Cuypers, A.n.n., Vangronsveld, J., and Clijsters, H. (2002). Peroxidases in roots and primary leaves of *Phaseolus vulgaris* copper and zinc phytotoxicity: a comparison. *J. Plant Physiol.* 159: 869–876.

Dalal, M. and Khanna-Chopra, R. (2001). Differential response of antioxidant enzymes in leaves of necrotic wheat hybrids and their parents. *Physiol. Plant.* 111: 297–304.

Das, K. and Roychoudhury, A. (2014). Reactive oxygen species (ROS) and response of antioxidants as ROS-scavengers during environmental stress in plants. *Front. Environ. Sci.* 2: 53.

Davis, T.L. (1933). The discovery of the elements (Weeks, Mary Elvira). *J. Chem. Educ.* 10: 710.

Diaz, J., Bernal, A., Pomar, F. et al. (2001). Induction of shikimate dehydrogenase and peroxidase in pepper (*Capsicum annuum* L.) seedlings in response to copper stress and its relation to lignification. *Plant Sci.* 161: 179–188.

Distefano, A.M., López, G.A., Setzes, N. et al. (2021). Ferroptosis in plants: triggers, proposed mechanisms, and the role of iron in modulating cell death. *J. Exp. Bot.* 72: 2125–2135.

Dunn, W.B. and Ellis, D.I. (2005). Metabolomics: current analytical platforms and methodologies. *TrAC Trends Anal. Chem.* 24: 285–294.

Edwards, E.A., Rawsthorne, S., and Mullineaux, P.M. (1990). Subcellular distribution of multiple forms of glutathione reductase in leaves of pea (*Pisum sativum* L.). *Planta* 180: 278–284.

Eitinger, T., Suhr, J., Moore, L. et al. (2005). Secondary transporters for nickel and cobalt ions: theme and variations. *Biometals* 18: 399–405.

Elloumi, N., Jerbi, B., Zouari, M. et al. (2016). Effects of sewage sludge fertilizer on heavy metal accumulation and consequent responses of sunflower (*Helianthus annuus* L.). *Environ. Sci. Pollut. Res.* 23: 20168–20177.

Erdei, S. (2002). Heavy metal induced physiological changes in the antioxidative response system. *Acta Biol. Szeged.* 46: 89–90.

Fagan, R.L. and Palfey, B.A. (2010). 7.03 – Flavin-dependent enzymes. In: *Comprehensive Natural Products II* (ed. H.-W. Liu and L. Mander), 37–113. Oxford: Elsevier.

Gashi, B., Osmani, M., Aliu, S. et al. (2020). Risk assessment of heavy metal toxicity by sensitive biomarker δ-aminolevulinic acid dehydratase (ALA-d) for onion plants cultivated in polluted areas in Kosovo. *J. Environ. Sci. Health B.* 55: 462–469.

Gill, S.S. and Tuteja, N. (2010). Reactive oxygen species and antioxidant machinery in abiotic stress tolerance in crop plants. *Plant Physiol. Biochem.* 48: 909–930.

Grant, J.J. and Loake, G.J. (2000). Role of reactive oxygen intermediates and cognate redox signaling in disease resistance. *Plant Physiol.* 124: 21–30.

Gul, A., Salam, A., Afridi, M.S. et al. (2019). Effect of urea, bio-fertilizers and their interaction on the growth, yield and yield attributes of *Cyamopsis tetragonoloba*. *Indian J. Agric. Res.* 53: 423–428.

Han, H.Y., Xu, W.A., Lu, Z.R. et al. (2008). Activation and inactivation of horseradish peroxidase by cobalt ions. *J. Biomol. Struct. Dyn.* 26: 83–92.

Hediye Sekmen, A., Türkan, İ., and Takio, S. (2007). Differential responses of antioxidative enzymes and lipid peroxidation to salt stress in salt-tolerant *Plantago maritima* L. and salt-sensitive *Plantago media* L. *Physiol. Plant.* 131: 399–411.

Higuchi, M. (2014). Antioxidant properties of wheat bran against oxidative stress. In: *Wheat and Rice in Disease Prevention and Health* (ed. R.R. Watson, V.R. Preedy, and S. Zibadi), 181–199. Elsevier.

Hossain, Z., Nouri, M.-Z., and Komatsu, S. (2012). Plant cell organelle proteomics in response to abiotic stress. *J. Proteome Res.* 11: 37–48.

Hossain, M.A., Bhattacharjee, S., Armin, S.-M. et al. (2015). Hydrogen peroxide priming modulates abiotic oxidative stress tolerance: insights from ROS detoxification and scavenging. *Front. Plant Sci.* 6: 420.

Hu, D., Cheng, M., Hu, K. et al. (2019). Evaluation of cobalt hyperaccumulation and tolerance potential of the duckweed (*Lemna minor* L.). *Ecotoxicol. Environ. Saf.* 179: 79–87.

Hu, X., Wei, X., Ling, J. et al. (2021). Cobalt: an essential micronutrient for plant growth? *Front. Plant Sci.* 12: 768523.

Huang, T.L., Nguyen, Q.T., Fu, S.F. et al. (2012). Transcriptomic changes and signalling pathways induced by arsenic stress in rice roots. *Plant Mol. Biol.* 80: 587–608.

Igamberdiev, A.U. and Hill, R.D. (2004). Nitrate, no and haemoglobin in plant adaptation to hypoxia: an alternative to classic fermentation pathways. *J. Exp. Bot.* 55: 2473–2482.

Imtiyaz, S., Agnihotri, R., Ganie, S. et al. (2014). Effect of cobalt and lead induced heavy metal stress on some physiological parameters in *Glycine max*. *Int. J. Agric. Crop Sci.* 7: 26–34.

Ishtiyaq, S., Kumar, H., Varun, M. et al. (2018). Heavy metal toxicity and antioxidative response in plants: an overview. In: *Plants Under Metal and Metalloid Stress: Responses, Tolerance and Remediation* (ed. M. Hasanuzzaman, K. Nahar, and M. Fujita), 77–106. Singapore: Springer.

Jackson, C., Dench, J., Moore, A.L. et al. (1978). Subcellular localisation and identification of superoxide dismutase in the leaves of higher plants. *Eur. J. Biochem.* 91: 339–344.

Jain, G. and Gould, K.S. (2015). Are betalain pigments the functional homologues of anthocyanins in plants? *Environ. Exp. Bot.* 119: 48–53.

Jaleel, C.A., Gopi, R., Manivannan, P. et al. (2007). Responses of antioxidant defense system of *Catharanthus roseus* (L.) G. Don. to paclobutrazol treatment under salinity. *Acta Physiol. Plant.* 29: 205–209.

Jalmi, S.K., Bhagat, P.K., Verma, D. et al. (2018). Traversing the links between heavy metal stress and plant signaling. *Front. Plant Sci.* 9: 12.

Javaid, M.H., Khan, A.R., Salam, A. et al. (2022). Exploring the adaptive responses of plants to abiotic stresses using transcriptome data. *Agriculture* 12: 211.

Jayakumar, K., Jaleel, C.A., and Azooz, M.M. (2008). Impact of cobalt on germination and seedling growth of *Eleusine coracana* L. and *Oryza sativa* L. under hydroponic culture. *Glob. J. Mol. Sci.* 3: 18–20.

Jensen, E.L., Maberly, S.C., and Gontero, B. (2020). Insights on the functions and ecophysiological relevance of the diverse carbonic anhydrases in microalgae. *Int. J. Mol. Sci.* 21: 2922.

Jun, M., Fu, H.Y., Hong, J. et al. (2003). Comparison of antioxidant activities of isoflavones from kudzu root (*Pueraria lobata* Ohwi). *J. Food Sci.* 68: 2117–2122.

Kang, S.M., Asaf, S., Khan, A.L. et al. (2020). Complete genome sequence of *Pseudomonas psychrotolerans* CS51, a plant growth-promoting bacterium, under heavy metal stress conditions. *Microorganisms* 8: 382.

Karuppanapandian, T. and Kim, W. (2013). Cobalt-induced oxidative stress causes growth inhibition associated with enhanced lipid peroxidation and activates antioxidant responses in Indian mustard (*Brassica juncea*L.) leaves. *Acta Physiol. Plant.* 35: 2429–2443.

Karuppanapandian, T., Moon, J.-C., Kim, C. et al. (2011). Reactive oxygen species in plants: their generation, signal transduction, and scavenging mechanisms. *Aust. J. Crop Sci.* 5: 709–725.

Kay, J.T., Conklin, M.H., Fuller, C.C. et al. (2001). Processes of nickel and cobalt uptake by a manganese oxide forming sediment in Pinal Creek, Globe mining district, Arizona. *Environ. Sci. Technol.* 35: 4719–4725.

Khan, A.R., Azhar, W., Wu, J. et al. (2021). Ethylene participates in zinc oxide nanoparticles induced biochemical, molecular and ultrastructural changes in rice seedlings. *Ecotoxicol. Environ. Saf.* 226: 112844.

Kintlova, M., Blavet, N., Cegan, R. et al. (2017). Transcriptome of barley under three different heavy metal stress reaction. *Genomics Data* 13: 15–17.

Konlechner, C., Türktaş, M., Langer, I. et al. (2013). Expression of zinc and cadmium responsive genes in leaves of willow (*Salix caprea* L.) genotypes with different accumulation characteristics. *Environ. Pollut.* 178: 121–127.

Korshunova, Y.O., Eide, D., Gregg Clark, W. et al. (1999). The IRT1 protein from *Arabidopsis thaliana* is a metal transporter with a broad substrate range. *Plant Mol. Biol.* 40: 37–44.

Kumar, R., Bohra, A., Pandey, A.K. et al. (2017). Metabolomics for plant improvement: status and prospects. *Front. Plant Sci* 8: 1302.

Lange, B., Faucon, M.-P., Meerts, P. et al. (2014). Prediction of the edaphic factors influence upon the copper and cobalt accumulation in two metallophytes using copper and cobalt speciation in soils. *Plant soil.* 379: 275–287.

Ledford, H.K. and Niyogi, K.K. (2005). Singlet oxygen and photo-oxidative stress management in plants and algae. *Plant Cell Environ.* 28: 1037–1045.

Ledford, H.K., Chin, B.L., and Niyogi, K.K. (2007). Acclimation to singlet oxygen stress in *Chlamydomonas reinhardtii*. *Eukaryot. Cell.* 6: 919–930.

Li, Z., McLaren, R.G., and Metherell, A.K. (2004). The availability of native and applied soil cobalt to ryegrass in relation to soil cobalt and manganese status and other soil properties. *N. Z. J. Agric. Res.* 47: 33–43.

Li, H., Liu, Y., Qin, H. et al. (2020). A rice chloroplast-localized ABC transporter ARG1 modulates cobalt and nickel homeostasis and contributes to photosynthetic capacity. *New Phytol.* 228: 163–178.

Lin, C.C. and Kao, C.H. (1999). Excess copper induces an accumulation of putrescine in rice leaves. *Bot. Bull. Acad. Sin.* 40: 213–221.

Liu, J. (1998). Cobalt: physiological effects and uptake mechanisms in plants. PhD dissertation. University of Adelaide.

Liu, X., Wu, H., Ji, C. et al. (2013). An integrated proteomic and metabolomic study on the chronic effects of mercury in *Suaeda salsa* under an environmentally relevant salinity. *PLoS One* 8: e64041.

Lwalaba, J.L., Zvobgo, G., Fu, L. et al. (2017). Alleviating effects of calcium on cobalt toxicity in two barley genotypes differing in cobalt tolerance. *Ecotoxicol. Environ. Saf.* 139: 488–495.

Lwalaba, J.L.W., Louis, L.T., Zvobgo, G. et al. (2019). Copper alleviates cobalt toxicity in barley by antagonistic interaction of the two metals. *Ecotoxicol. Environ. Saf.* 180: 234–241.

Lwalaba, J.L.W., Zvobgo, G., Mwamba, T.M. et al. (2020). High accumulation of phenolics and amino acids confers tolerance to the combined stress of cobalt and copper in barley (*Hordeum vulagare* L.). *Plant Physiol. Biochem.* 155: 927–937.

Lwalaba, J.L.W., Zvobgo, G., Gai, Y. et al. (2021). Transcriptome analysis reveals the tolerant mechanisms to cobalt and copper in barley. *Ecotoxicol. Environ. Saf.* 209: 111761.

Lyu, S., Wei, X., Chen, J. et al. (2017). Titanium as a beneficial element for crop production. *Front. Plant Sci.* 8: 597.

Maeda, H. and DellaPenna, D. (2007). Tocopherol functions in photosynthetic organisms. *Curr. Opin. Plant Biol.* 10: 260–265.

McClung, C.R. (1997). Regulation of catalases in *Arabidopsis*. *Free Radic. Biol. Med.* 23: 489–496.

Mehmood, S.S., Lu, G., Luo, D. et al. (2021). Integrated analysis of transcriptomics and proteomics provides insights into the molecular regulation of cold response in *Brassica napus*. *Environ. Exp. Bot.* 187: 104480.

Millar, A.H., Mittova, V., Kiddle, G. et al. (2003). Control of ascorbate synthesis by respiration and its implications for stress responses. *Plant Physiol.* 133: 443–447.

Mishra, S., Jha, A.B., and Dubey, R.S. (2011). Arsenite treatment induces oxidative stress, upregulates antioxidant system, and causes phytochelatin synthesis in rice seedlings. *Protoplasma* 248: 565–577.

Mittler, R. (2002). Oxidative stress, antioxidants and stress tolerance. *Trends Plant Sci.* 7: 405–410.

Morel, M., Crouzet, J., Gravot, A. et al. (2009). AtHMA3, a P1B-ATPase allowing Cd/Zn/co/Pb vacuolar storage in *Arabidopsis*. *Plant Physiol.* 149: 894–904.

Morrissey, J., Baxter, I.R., Lee, J. et al. (2009). The ferroportin metal efflux proteins function in iron and cobalt homeostasis in *Arabidopsis*. *Plant Cell.* 21: 3326–3338.

Munne-Bosch, S. (2005). The role of alpha-tocopherol in plant stress tolerance. *J. Plant Physiol.* 162: 743–748.

Naila, S., Haq, Z.U., Abdulllah et al. (2021). Allelopathic effect of *Taraxacum officinale* L. on germination and physiology of wheat. In: *Sustainable Intensification for Agroecosystem Services and Management* (ed. M.K. Jhariya, A. Banerjee, R.S. Meena, et al.), 711–741. Springer: Singapore.

Noctor, G. and Foyer, C.H. (1998). Ascorbate and glutathione: keeping active oxygen under control. *Annu. Rev. Plant Biol.* 49: 249–279.

Noman, A., Ali, Q., Hameed, M. et al. (2014). Comparison of leaf anatomical characteristics of *Hibiscus rosa-sinensis* grown in Faisalabad region. *Pak. J. Bot.* 46: 199–206.

Noman, A., Aqeel, M., Javed, M. et al. (2017). Histological changes in *Hibiscus rosa-sinensis* endorse acclimation and phytoremediation of industrially polluted sites. *J. Anim. Plant Sci.* 27: 1637–1164.

Oven, M., Grill, E., Golan-Goldhirsh, A. et al. (2002). Increase of free cysteine and citric acid in plant cells exposed to cobalt ions. *Phytochemistry* 60: 467–474.

Page, V. and Feller, U. (2005). Selective transport of zinc, manganese, nickel, cobalt and cadmium in the root system and transfer to the leaves in young wheat plants. *Ann. Bot.* 96: 425–434.

Pais, I. (1992). Criteria of essentiality, beneficiality and toxicity of chemical elements. *Acta Aliment.* 21: 145–152.

Palit, S., Sharma, A., and Talukder, G. (1994). Effects of cobalt on plants. *Bot. Rev.* 60: 149–181.

Pan, C., Lu, H., Liu, J. et al. (2020). Sods involved in the hormone mediated regulation of H_2O_2 content in *Kandelia obovata* root tissues under cadmium stress. *Environ. Pollut.* 256: 113272.

Perez-Clemente, R.M., Vives, V., Zandalinas, S.I. et al. (2013). Biotechnological approaches to study plant responses to stress. *BioMed Res. Int.* 2013: 654120.

Permyakov, E.A. (2021). Metal binding proteins. *Encyclopedia* 1: 261–292.

Pooetsch, A.R. (2020). The genomics of oxidative DNA damage, repair, and resulting mutagenesis. *Comput. Struct. Biotechnol. J.* 18: 207–219.

Racchi, M.L., Bagnoli, F., Balla, I. et al. (2001). Differential activity of catalase and superoxide dismutase in seedlings and in vitro micropropagated oak (*Quercus robur*L.). *Plant Cell Rep.* 20: 169–174.

Raza, S.H., Athar, H.R., Ashraf, M. et al. (2007). Glycinebetaine-induced modulation of antioxidant enzymes activities and ion accumulation in two wheat cultivars differing in salt tolerance. *Environ. Exp. Bot.* 60: 368–376.

Repetto, M., Semprine, J., and Boveris, A. (2012). Lipid peroxidation: chemical mechanism, biological implications and analytical determination. In: *Lipid Peroxidation*, vol. 1 (ed. A. Catala), 3–30. London: IntechOpen https://doi.org/10.5772/45943.

del Rio, L.A., Sandalio, L.M., Corpas, F.J. et al. (2006). Reactive oxygen species and reactive nitrogen species in peroxisomes. Production, scavenging, and role in cell signaling. *Plant Physiol.* 141: 330–335.

Romero-Puertas, M.C., Corpas, F.J., Sandalio, L.M. et al. (2006). Glutathione reductase from pea leaves: response to abiotic stress and characterization of the peroxisomal isozyme. *New Phytol.* 170: 43–52.

Rubio-Sanz, L., Prieto, R.I., Imperial, J. et al. (2013). Functional and expression analysis of the metal-inducible dmeRF system from *Rhizobium leguminosarum* bv. Viciae. *Appl. Environ. Microbiol.* 79: 6414–6422.

Rucinska-Sobkowiak, R. (2016). Water relations in plants subjected to heavy metal stresses. *Acta Physiol. Plant.* 38: 1–13.

Rue, M., Paul, A.L.D., Echevarria, G. et al. (2020). Uptake, translocation and accumulation of nickel and cobalt in *Berkheya coddii*, a 'metal crop' from south africa. *Metallomics* 12: 1278–1289.

Salam, A., Khan, A.R., Liu, L. et al. (2022). Seed priming with zinc oxide nanoparticles downplayed ultrastructural damage and improved photosynthetic apparatus in maize under cobalt stress. *J. Hazard. Mater.* 423: 127021.

Salehi, H., Chehregani Rad, A., Raza, A. et al. (2021). Foliar application of CeO_2 nanoparticles alters generative components fitness and seed productivity in bean crop (*Phaseolus vulgaris* L.). *Nanomaterials* 11: 862.

Sanders, J.R. (1983). The effect of pH on the total and free ionic concentrations of mnganese, zinc and cobalt in soil solutions. *Eur. J. Soil Sci.* 34: 315–323.

Schmidt, S.B., Eisenhut, M., and Schneider, A. (2020). Chloroplast transition metal regulation for efficient photosynthesis. *Trends Plant Sci.* 25: 817–828.

Shalata, A., Mittova, V., Volokita, M. et al. (2001). Response of the cultivated tomato and its wild salt-tolerant relative *Lycopersicon pennellii* to salt-dependent oxidative stress: the root antioxidative system. *Physiol. Plant.* 112: 487–494.

Sharma, I. and Ahmad, P. (2014). Catalase: a versatile antioxidant in plants. In: *Oxidative Damage to Plants* (ed. P. Ahmad), 131–148. San Diego: Academic Press.

Singh, A. (2020). Tools for metabolomics. *Nat. Methods* 17: 24–24.

Singh, V.P., Singh, S., Kumar, J. et al. (2015). Investigating the roles of ascorbate-glutathione cycle and thiol metabolism in arsenate tolerance in ridged Luffa seedlings. *Protoplasma* 252: 1217–1229.

Soares, C., de Sousa, A., Pinto, A. et al. (2016). Effect of 24-epibrassinolide on ros content, antioxidant system, lipid peroxidation and ni uptake in *Solanum nigrum* L. under Ni stress. *Environ. Exp. Bot.* 122: 115–125.

Soares, C., Carvalho, M.E.A., Azevedo, R.A. et al. (2019). Plants facing oxidative challenges – a little help from the antioxidant networks. *Environ. Exp. Bot.* 161: 4–25.

Spollen, W.G., Tao, W., Valliyodan, B. et al. (2008). Spatial distribution of transcript changes in the maize primary root elongation zone at low water potential. *BMC Plant Biol.* 8: 32.

Sree, K.S., Keresztes, A., Mueller-Roeber, B. et al. (2015). Phytotoxicity of cobalt ions on the duckweed *Lemna minor* – morphology, ion uptake, and starch accumulation. *Chemosphere* 131: 149–156.

Srivalli, B., Vishanathan, C., and Renu, K.C. (2003). Antioxidant defense in response to abiotic stresses in plants. *J. Plant Biol.* 30: 121–139.

Srivastava, S. and Dubey, R.S. (2011). Manganese-excess induces oxidative stress, lowers the pool of antioxidants and elevates activities of key antioxidative enzymes in rice seedlings. *Plant Growth Regul.* 64: 1–16.

Stamati, K., Mudera, V., and Cheema, U. (2011). Evolution of oxygen utilization in multicellular organisms and implications for cell signalling in tissue engineering. *J. Tissue Eng.* 2: 2041731411432365.

Su, L.J., Zhang, J.H., Gomez, H. et al. (2019). Reactive oxygen species-induced lipid peroxidation in apoptosis, autophagy, and ferroptosis. *Oxid. Med. Cell. Longev.* 2019: 5080843.

Sudhakar, C., Lakshmi, A., and Giridarakumar, S. (2001). Changes in the antioxidant enzyme efficacy in two high yielding genotypes of mulberry (*Morus alba* L.) under NaCl salinity. *Plant Sci.* 161: 613–619.

Suman, S., Bagal, D., Jain, D. et al. (2021). Biotic stresses on plants: reactive oxygen species generation and antioxidant mechanism. In: *Frontiers in Plant-Soil Interaction* (ed. T. Aftab and K.R. Hakeem), 381–411. Academic Press.

Sundaramoorthy, P., Chidambaram, A., Ganesh, K.S. et al. (2010). Chromium stress in paddy: (i) nutrient status of paddy under chromium stress; (ii) phytoremediation of chromium by aquatic and terrestrial weeds. *C. R. Biol.* 333: 597–607.

Tewari, R.K., Kumar, P., Sharma, P.N. et al. (2002). Modulation of oxidative stress responsive enzymes by excess cobalt. *Plant Sci.* 162: 381–388.

Thounaojam, T.C., Panda, P., Mazumdar, P. et al. (2012). Excess copper induced oxidative stress and response of antioxidants in rice. *Plant Physiol. Biochem.* 53: 33–39.

Tran, L.S., Nishiyama, R., Yamaguchi-Shinozaki, K. et al. (2010). Potential utilization of nac transcription factors to enhance abiotic stress tolerance in plants by biotechnological approach. *GM Crops* 1: 32–39.

Tsugama, D., Liu, S., and Takano, T. (2013). Metal-binding ability of VIP1: a bZIP protein in *Arabidopsis thaliana*. *Protein J.* 32: 526–532.

Tsugane, K., Kobayashi, K., Niwa, Y. et al. (1999). A recessive *Arabidopsis* mutant that grows photoautotrophically under salt stress shows enhanced active oxygen detoxification. *Plant Cell* 11: 1195–1206.

Tuteja, N. and Sopory, S.K. (2008). Plant signaling in stress: G-protein coupled receptors, heterotrimeric g-proteins and signal coupling via phospholipases. *Plant Signal. Behav.* 3: 79–86.

Ulhassan, Z., Khan, A.R., Hamid, Y. et al. (2022a). Interaction of nanoparticles with soil–plant system and their usage in remediation strategies. In: *Metals Metalloids Soil Plant Water Systems* (ed. T. Aftab and K. Hakeem), 287–308. Academic Press.

Ulhassan, Z., Khan, A.R., Sheteiwy, M.S. et al. (2022b). Selenium-mediated regulation of antioxidant defense system and improved heavy metals tolerance in plants. In: *Antioxidant Defense in Plants: Molecular Basis of Regulation* (ed. T. Aftab and K.R. Hakeem), 369–382. Singapore: Springer Nature.

Voith von Voithenberg, L., Park, J., Stube, R. et al. (2019). A novel prokaryote-type ECF/ABC transporter module in chloroplast metal homeostasis. *Front. Plant Sci.* 10: 1264.

Wa Lwalaba, J.L., Zvogbo, G., Mulembo, M. et al. (2017). The effect of cobalt stress on growth and physiological traits and its association with cobalt accumulation in barley genotypes differing in cobalt tolerance. *J. Plant Nutr.* 40: 2192–2199.

Wang, Y.-M., Yang, Q., Xu, H. et al. (2020). Physiological and transcriptomic analysis provide novel insight into cobalt stress responses in willow. *Sci. Rep.* 10: 2308.

Weinstein, L.H., Kaur-Sawhney, R., Rajam, M.V. et al. (1986). Cadmium-induced accumulation of putrescine in oat and bean leaves 1. *Plant Physiol.* 82: 641–645.

Willekens, H., Villarroel, R., Van Montagu, M. et al. (1994). Molecular identification of catalases from *Nicotiana plumbaginifolia* (L.). *FEBS Lett.* 352: 79–83.

Winkel-Shirley, B. (2002). Biosynthesis of flavonoids and effects of stress. *Curr. Opin. Plant Biol.* 5: 218–223.

Wray, G.A., Hahn, M.W., Abouheif, E. et al. (2003). The evolution of transcriptional regulation in eukaryotes. *Mol. Biol. Evol.* 20: 1377–1419.

Yakes, F.M. and Van Houten, B. (1997). Mitochondrial DNA damage is more extensive and persists longer than nuclear DNA damage in human cells following oxidative stress. *Proc. Natl. Acad. Sci. U.S.A.* 94: 514–519.

Yang, S., Ulhassan, Z., Shah, A.M. et al. (2021). Salicylic acid underpins silicon in ameliorating chromium toxicity in rice by modulating antioxidant defense, ion homeostasis and cellular ultrastructure. *Plant Physiol. Biochem.* 166: 1001–1013.

Zeeshan, M., Hu, Y.X., Iqbal, A. et al. (2021). Amelioration of AsV toxicity by concurrent application of ZnO-NPs and Se-NPs is associated with differential regulation of photosynthetic indexes, antioxidant pool and osmolytes content in soybean seedling. *Ecotoxicol. Environ. Saf.* 225: 112738.

Zeng, J., He, X., Wu, D. et al. (2014). Comparative transcriptome profiling of two Tibetan wild barley genotypes in responses to low potassium. *PLoS One* 9: e100567.

Zhang, X.N., Piao, C.L., Dong, Y.K. et al. (2017). Transcriptome analysis of response to heavy metal cd stress in soybean root. *Ying Yong Sheng Tai Xue Bao = J. Appl. Ecol.* 28: 1633–1641.

11

Nickel Toxicity and Tolerance in Plants

Sondes Helaoui[1], Marouane Mkhinini[1], Iteb Boughattas[1,2], Noureddine Bousserrhine[3], and Mohamed Banni[1,4]

[1] *Department of plant protection, Laboratory of Agrobiodiversity and Ecotoxicology, Higher Institute of Agronomy Chott-Meriem, Sousse University, Sousse, Tunisia*
[2] *Department of agronomy, Regional Field Crops Research Center of Beja, University of Jendouba, Beja, Tunisia*
[3] *Faculty of Sciences and Technology, Laboratory of Water Environnement and Urban systems, University Paris-Est Créteil, France*
[4] *Higher Institute of Biotechnology, Monastir University, Monastir, Tunisia*

11.1 Introduction

In recent years, heavy metals (HMs) have been widely used in industrial activities and agricultural practices (Zhuo et al. 2019). Therefore, HM has become a serious problem for organisms due to its persistence in soil (Sun et al. 2019). HM can be transferred to plants and animals through food chain (Cai et al. 2019). In addition, they could be accumulated in biological matrices and cause serious harm to human health (Jin et al. 2019). Some of these elements are essential metals (Alloway 1995), but at high concentrations, all are toxic to living organisms. Excessive levels of certain metals could decrease plant growth (Li et al. 2020a) and reduce agricultural production (Li et al. 2020b).

Ni is an essential micronutrient because it plays a vital role in biological systems: in plants, it is involved in the production of vigorous seeds as it is important for seed germination and iron absorption (Fargašová 2012). The causes of Ni contamination in the soil are natural such as alteration of the parent rock, erosion and volcanic eruptions (Iyaka 2011), and anthropogenic activities as the vehicle emissions, fertilizer application, and industrial wastes (Chai et al. 2021). However, Zwolsman and Bokhoven (2007) demonstrated that industrial Ni emissions are more than 100 times higher than those from natural sources. In addition, Ni levels can reach 26,000 $mg\,kg^{-1}$ in contaminated soil (Chen et al. 2009) and 0.2 $mg\,L^{-1}$ in contaminated water (Nriagu 1990).

Due to its high mobility in the environment, Ni can rapidly enter the food chain through plants, causing severe damage to organisms (Jin et al. 2019). Several studies have shown that higher Ni concentrations lead to stunted plant growth (Czajka et al. 2019; Moreira

Heavy Metal Toxicity and Tolerance in Plants: A Biological, Omics, and Genetic Engineering Approach, First Edition. Edited by Mohammad Anwar Hossain, AKM Zakir Hossain, Sylvain Bourgerie, Masayuki Fujita, Om Parkash Dhankher, and Parvez Haris.

et al. 2020). Ni toxicity then reduces photosynthesis; assimilates transport and mineral nutrition in plants (Hassan et al. 2019); and induces changes in physiological and biochemical processes (Hasanuzzaman et al. 2013; Helaoui et al. 2020).

In addition, Ni produces oxidative stress in plants by generating the production of reactive oxygen species like hydrogen peroxide (H_2O_2), singlet oxygen ($^1O_2$), superoxide anion ($O_2^•$), and the hydroxyl radical ($OH^•$) (Rizwan et al. 2018). Then, an excessive amount of ROS may induce oxidative damage to macromolecules (proteins, lipids, and DNA) of cells disrupting the plant's antioxidant system (Das and Roychoudhury 2014). ROS activate the production of antioxidant enzymes such as superoxide dismutase (SOD, EC 1.15.1.1), catalase (CAT, EC 1.11.1.6), and ascorbate peroxidase (APX, EC 1.11.1.11) (Abdal et al. 2021), thereby increasing plant defense mechanisms (Hayat et al. 2016).

Different strategies have been developed to reduce metal accumulation in agricultural soils (Dodds et al. 2016; Xiang et al. 2018). Phytoremediation is a biotechnology that uses plants to remediate sites by removing pollutants from soil and water (DeLorenzo 2018; Duarte et al. 2018; Bell et al. 2019). Plants can degrade organic pollutants and stabilize metal pollutants through metal uptake, rhizome translocation, intracellular chelation, and chemical modification (Hillel 2005; Hanus-Fajerska and Kozminska 2016; Arya et al. 2017).

Due to its invasive appearance, the soil is constantly exposed to high levels of Ni. Therefore, Ni pollution will continue to exist and become a serious environmental hazard for future generations. Therefore, it is urgent to regularly monitor the whereabouts of Ni in the environment to better understand its impact on plants and other organisms in ecosystems. To this end, this review aims to gain a comprehensive understanding of the origin, importance, and toxicological effects of Ni in order to obtain a comprehensive idea of Ni remediation from the environment.

11.2 Sources of Ni

11.2.1 Natural Sources of Ni

Ni is a naturally occurring metal that is widely found in the Earth's crust. Natural sources of Ni in the environment include vegetation fires, volcanic emissions, and aeolian dust, but the highest concentrations are found in basic igneous rocks (Kachenko et al. 2007). The Ni content in the soil is highly dependent on the Ni content of the parent rock. The lowest values were detected in sedimentary rocks containing clay, limestone, sandstone, and shale (Mondal et al. 2019). Ni exists in soils in various forms, including inorganic minerals adsorbed on organic cation surfaces or complexed on inorganic cation exchange surfaces (Gardner et al. 2012). It has been suggested that humic and fulvic acids are significantly more mobile in soil solutions than hydrated divalent cations (Kubicka et al. 2015). These minerals include Ni chromite, garnet, and Ni magnesium silicate. In soil, nickel concentrations on the surface have been estimated to be between 3 and 1000 $mg\,kg^{-1}$ (Iyaka 2011). Although, the concentration in surface soils reflects the additional effect of soil-forming processes and anthropogenic activities (Kabata and Pendias 2001; Straif 2010).

11.2.2 Anthropogenic Sources of Ni

Ni has many industrial uses, such as coin fabrication, electroplating, fabrication of nickel alloys, spark plugs, and other ignition devices, as well as resistance heaters and batteries (Kabata-Pendias and Mukherjee 2007; Prueitt et al. 2020). The major sources of Ni into soils are mainly industrial waste materials, lime, fertilizers, treated wastewater, and sewage sludges (Hussain et al. 2013; Mkhinini et al. 2020). In general, the distribution of Ni in the soil profile is homogeneous, with typical accumulation on topsoil due to anthropogenic deposition (Cempel and Nikel 2006; Guo et al. 2019). Moreover, Ni can also exist in many forms in soil containing adsorbed or complexed to organic cations surface or inorganic cation exchange surface, inorganic crystalline minerals or sediments, water soluble, free ions, or chelated metal complexes in soil solution (Bennett 1982; Tye et al. 2004; Buxton et al. 2019).

11.3 Role of Ni in Plants

Ni is in general recognized as an essential ultra-micronutrient as its exigency is the lowest level of all essential metals at $<0.5\,mg\,kg^{-1}$ of dry weight (Marschner 1995). According to Mulrooney and Hausinger (2003), it has a key role in biological systems: in plants, such as ureolysis, hydrogen metabolism, methane biogenesis, and acetogenesis. In fact, it is essential for plant development and growth (Liu 2001; Kumar et al. 2018). The absence of Ni inhibits seed germination and crop yields (Gratão et al. 2008; Yusuf et al. 2011). Indeed, urease, methyl reductase, hydrogenase, dehydrogenase, and Ni-superoxide dismutase are all essential enzymes containing Ni (Min et al. 2015). Low concentrations of Ni reduce urease activity, thereby disrupting N metabolism and leading to high levels of urea accumulation in plants. This production of urea results in leaf necrosis and chlorosis (Ali et al. 2009). In legumes, low concentrations of Ni are essential for nodule formation and hydrogenase activity. In fact, nitrogen fixation in legumes is entirely dependent on the activity of hydrogenases, as the oxidation of hydrogen provides the ATP required to reduce nitrogen to ammonia (Dalton et al. 1985). Ni also acts as a biocontrol agent by promoting the generation of secondary metabolites of plants that enhance resistance in plants against several diseases and toxic elements (Harasim and Filipek 2015). The critical Ni concentration alters cell morphology, induces oxidative metabolism, reduces glucose metabolism, and increases lipid levels (Kumar and Trivedi 2016). In addition, Ni deficiency minimizes the activity of enzymes that interfere with carbohydrate and amino acid metabolism (Guo et al. 2019).

11.4 Ni Uptake and Accumulation in Plants

Understanding Ni uptake and accumulation is key to controlling plant tolerance to elevated metal levels. The mechanism by which plants take up Ni depends on the amount and solubility of the metal available in soils and the species of plants growing on these soils

(Barman et al. 2000; Rosen and Chen 2014; Rong et al. 2020). The important environmental parameters affecting plant uptake of metals are Ni concentration in soil solutions, soil pH, organic matter, temperature, and concentrations of macro and micronutrients (Adrees et al. 2012; Salinitro et al. 2020). Soil pH plays an important role in the uptake of Ni by plants. Acidic soils increase the solubility and mobility of Ni; therefore, Ni is more toxic to phytotoxicity grown on low pH soils (Rautaray et al. 2003; Shen et al. 2020). Temperatures as low as 2 °C significantly reduced the uptake of Ni by roots in the nutrient solution (Líska et al. 2017). The uptake of Ni can compete with other metals in the soil solution. In fact, the inhibitory effect of metal ions on Ni uptake and transport decreases from root to shoot in the following order: Fe^{2+}> Co^{2+}> Ca^{2+}> Mg^{2+}> K^{2+}> Na^{2+} (Cutillas-Barreiro et al. 2016; Almeida et al. 2017).

Ni uptake by plants is mainly through active and passive diffusion through roots (Seregin and Kozhevnikova 2006; Kamran et al. 2016). However, the rate of active to passive diffusion varies with the species and the Ni concentration in the soil solution (Vogel-Mikuš et al. 2005; Khan et al. 2020). In addition, Ni is transferred from roots to shoots and then to leaves via the xylem (Neumann and Chamel 1986; Peralta-Videaa et al. 2002; Amari et al. 2016). Kamran et al. (2016) applied Ni to *Eruca sativa* and found that more than 50% of Ni was transferred to roots and other organs. Alfalfa absorbs readily available Ni; once absorbed, Ni accumulates mainly in roots (Helaoui et al. 2020). Ni is then highly mobile and can be transported to other parts of the plant, such as seeds, fruits, and flowers (Page et al. 2006; Sachan and Lal 2017). Indeed, Ni translocation in plants is closely associated with Ni–ligand complexes such as organic acids and histidine (Haydon and Cobbett 2007) and proteins that facilitate Ni attachment and transport (Colpas and Hausinger 2000; Chen et al. 2009). These complexes are metal chelators that facilitate Ni translocation in the xylem. But without chelation, Ni transport from roots to leaves is severely affected because xylem cells have a high cation-exchange capacity that inhibits Ni migration (Briat and Lebrun 1999; Chow et al. 2019).

11.5 Ni Toxicity in Plants

Phyto-availability of Ni correlates with free Ni ion activity in soil solution; therefore, plant uptake depends on soil pH, organic matter content, and iron-manganese oxide content (Rooney et al. 2007; Wang et al. 2020). Plants absorb more nickel than they can from natural sources, and plant species vary in their tolerance and ability to absorb Ni from soil. Ni toxicity levels vary widely between 25 and 50 mg kg^{-1} as reported by Mishra and Kar (1974). However, Gregson and Hope (1994) showed that phytotoxic concentrations of Ni occur at leaf levels of 10–100 mg kg^{-1}, depending on the crop species. In addition, Rather et al. (2020) found that, in addition to inhibiting plant growth, other symptoms of Ni toxicity included chlorosis, stunted root growth, and browning including intervening necrosis (Figure 11.1).

11.5.1 Growth Inhibition

Several studies have demonstrated beneficial effects of Ni in plants at low levels, which cannot be replaced by other elements (Fabiano et al. 2015). Ni excess in plant cells leads to

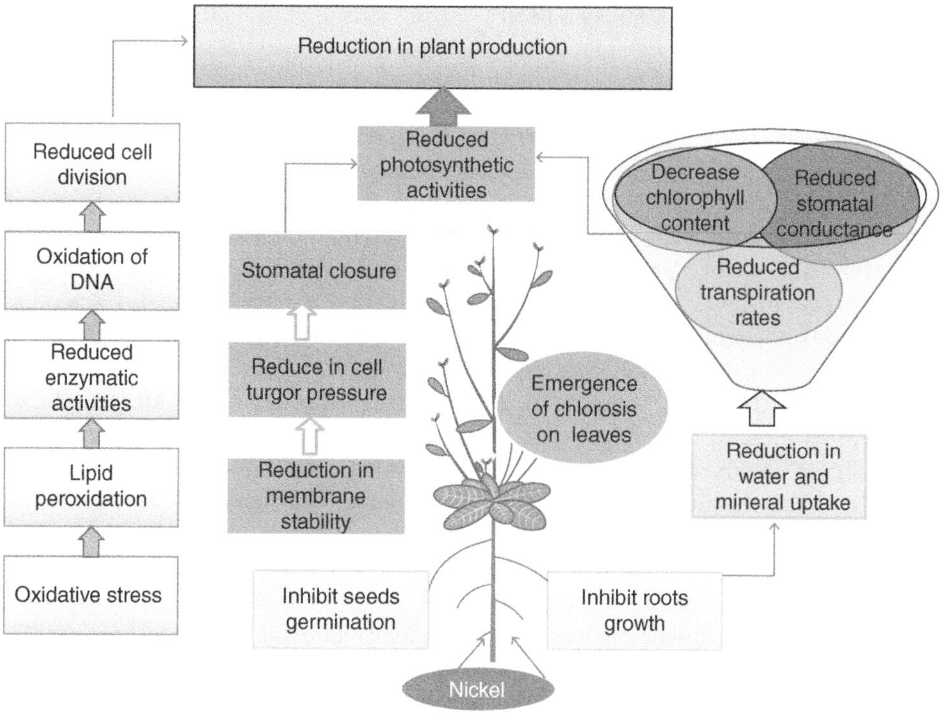

Figure 11.1 High Ni concentration inhibits seed germination and root growth, and thus decreases water and mineral uptake and finally reduces photosynthesis activities. Similarly, Ni toxicity in plant leaves changes the concentration of photosynthetic pigments, and these effects produce leaf chlorosis. Then, Ni toxicity modifies membrane stability and causes cell turgor pressure, leading to the closure of the stoma. Moreover, Ni produces oxidative stress and the production of ROS inhibits protein, lipid synthesis, and enzymatic activities and consequently induces lipid peroxidation, DNA damage, and altered cell division.

dramatic changes at the physiological, biochemical, and cellular levels, resulting in severe cell damage and plant death (Singh et al. 2016). Excessive accumulation of Ni in plants leads to changes in several physiological processes of cell origin, such as enzymatic inactivation, blockage of metabolic molecules, alteration of important molecular structures, disruption of membrane integrity, and substitution of essential elements (Gajewska et al. 2006; Wakeel et al. 2020). In fact, high Ni concentrations disrupt Fe uptake and metabolism, leading to chlorosis and necrosis of plants (Aschmann and Zasoski 1987; Hassan et al. 2019). Multiple studies have shown three events that generate toxicity during the transition of Ni from roots. Thus, Ni can (i) alter the basic composition of biomolecules, (ii) inhibit the function of basic biogroups in the molecule, (iii) alter the activity of enzymes/proteins and the structure of membrane transporters (Rizvi et al. 2020). Ni also reduces root and shoot growth, grain quality, fruit yield, and reduces leaf area by inhibiting mitotic cell activity (Madhava Rao and Sresty 2000; Kopittke et al. 2019). Furthermore, it affects plant anatomy and morphology by inhibiting sugar transport, reducing vascular bundle size, mesophyll size, membrane plasticity, inducing chlorosis, and altering other metabolic processes (Duarte et al. 2007; Gonzalez et al. 2015).

11.5.2 Photosynthesis Inhibition of Ni

The toxic effects of heavy metals vary, and the order of toxicity to plants is: $As^{3+} < Cr^{6+} < Co^{2+} < Ni^{2+} < Cu^{2+} < Pb^{2+} < Hg^{2+} < Cd^{2+} < Ag^{+}$. The main symptoms of Ni^{2+} toxicity are disturbances in physiological mechanisms such as photosynthesis, nutrient uptake, and membrane permeability (Chen et al. 2009; Ameen et al. 2019). Ni accumulation in leaves affects photosynthetic pigments such as chlorophyll a, b, and carotenoids (Lin and Kao 2007; Amjad et al. 2020), which lead to plant necrosis and chlorosis (Hassan et al. 2019). Other consequences of high nickel concentrations are deterioration of chloroplast composition, changes in lipid composition of cells (Hermle et al. 2007; Riaz et al. 2017), deterioration of thylakoid membranes (Molas 2002; Leskova et al. 2017), and reduced particle size (Molas 1997; Kotapati et al. 2017), leading to significant reductions of photosynthetic pigments (Ahmad et al. 2007; Aqeel et al. 2021). In fact, Wheeler et al. (2001) found that 0.025 mM Ni reduced chlorophyll content by 47%. Helaoui et al. (2022a) reported that 500 mg kg^{-1} Ni diminished chlorophyll a and b levels by 30% and 60%, respectively. Ni stress then alters the number of gas exchanges and reduces transpiration rates in different plant species (Pandey and Sharma 2002; Batool 2018). In fact, Ni accumulation disrupts the electron transport chain (Tripathy et al. 1981; Hassan et al. 2019), reduces the enzymes of the Calvin cycle, and promotes stomatal closure, resulting in CO_2 deficiency (Sheoran et al. 1990; Baker and Rosenqvist 2004). Ni toxicity then reduces stomatal opening by inhibiting the flow of K^+ through the cell membrane (Ahmad and Ashraf 2012). Furthermore, high Ni concentrations disrupt electron transport during photoreactions (Malkin et al. 2000; Ekmekçi et al. 2020) in Photosystem II, thereby reducing the rate of photosynthesis (Krupa and Baszynski 1995; Khaliq et al. 2016).

In addition, Ni strongly reduced aldolase, Rubisco dehydrogenase, and F-1,6-bisphosphonate. Depletion of these enzymatic activities results in the accumulation of ATP and NDPH, which create a higher pH gradient around the thylakoid membrane, which blocks the activity of PSII (Bhalerao et al. 2015). Therefore, the reduction in photosynthesis due to Ni toxicity cannot be attributed to a single factor. In fact, this reduction is the effect of various factors, such as the toxic effect of Ni on the electron transport chain, the reduction of enzymatic activity, and the reduction of chlorophyll content (Baran and Ekmekçi 2022).

11.5.3 Induction of Oxidative Stress

Ni accumulation contribute to a decrease in growth of plants, increase in the malondialdehyde (MDA) content and the generation of oxidative stress in plants (Dourado et al. 2015; Ahmad 2016). It has been shown that Ni produce reactive oxygen species such as singlet oxygen ($^1O_2$), superoxide (O_2^-), hydroxyl radicals (OH^-), and hydrogen peroxide (H_2O_2) in plants (Gajewska and Sklodowska 2007), which generate oxidative damage to membrane lipids, proteins, and DNA in plant cells (Verma and Dubey 2003; Riaz et al. 2017; Begum et al. 2022). In the other hand, plants employ complex antioxidant defense systems, including antioxidant enzymes and nonenzymatic antioxidants, to reduce oxidative stress (Gill and Tuteja 2010; Zaid et al. 2019). The important antioxidant enzymes are superoxide dismutase (SOD, EC 1.15.1.1) that catalyzes the dismutation of O_2^- to H_2O_2 and O_2 and presents the first line of antioxidant defense; therefore, H_2O_2 may be recuperated by catalase (CAT, EC 1.11.1.6) (Gajewska and Sklodowska 2008; Amjad et al. 2014). Then nonenzymatic antioxidants are ascorbic acid, glutathione, phenols, and various nitrogen-containing metabolites such as amino acids, especially proline, which have potent antioxidant activity required for protection against reactive oxygen species (Abdal et al. 2021). Although excessive Ni accumulation can significantly reduce the activity of antioxidant enzymes and the ability of plants to eliminate ROS (Amjad et al. 2020).

Thus, several studies have shown that metal accumulation in plants induces cytotoxic effects related to oxidative stress as well as the ability of metal homeostasis (Flora 2011; Khalid et al. 2017; Jan et al. 2019). In fact, homeostasis is achieved by excreting various types of organic molecules that form complexes with heavy metals and transfer them to the vacuole to reduce their mobility and bioavailability (Shukla et al. 2013). The reduced mobility limits their accumulation in plants, thus decreasing their phytotoxicity in different parts of the plant (Yang et al. 1997; Arora et al. 2012; Hattab et al. 2016; Shukla et al. 2022). Several papers have shown that phytochelins (PCs) are the major proteins that form metal-binding ligands and inactivate metals in the cytoplasm (Verbruggen et al. 2009). Metal–PCs complexes are then transferred into the vacuole (Pinter and Stillman 2014; Talebi et al. 2019), thereby reducing the toxicity of heavy metals (Singh et al. 2016). PCs are peptides of 3,-Glu-Cys produced from the S-rich glutathione (GSH) by phytochelatin synthase (Pochodylo and Aristilde 2017). Freeman et al. (2004) demonstrated that the GSH concentration in Arabidopsis plants was significantly increased under Ni stress, thereby increasing the phytochelin content, resulting in a significant reduction in the toxic effects of Ni. According to Emamverdian et al. (2015), used PCs as a biomarker, mainly produced under Cd, Cu, Pb, Hg, and Ni toxicity (Kagi 1991). Plants produce PCs when high Ni concentrations accumulate in plant organs (Helaoui et al. 2020). Indeed, Ni toxicity leads to PC production in different plant species, such as *Nicotiana tabacum* L. (Nakazawa et al. 2001), *Lotus japonicas* L. (Blum et al. 2007), *Thlaspi japonicum* L. (Mizuno et al. 2003), and *Medicago sativa* L. (Helaoui et al. 2020). However, Cobbett and Goldsbrough (2002) showed that Ni is a relatively inefficient PC activator during *in vitro* culture.

More alternative detoxification mechanisms may exist for Ni toxicity in plants. Metallothioneins (MTs) are a superfamily of small molecular weight (6–20 kDa) cysteines (Cys) that bind to heavy metals such as Cu, Zn, Ni, Cd through clusters of thiol bonds (Zhang et al. 2017). Cysteines in MTs have sulfhydryl ligands that provide high-affinity coordination of various metal ions (Blindauer and Leszczyszyn 2010; Leszczyszyn et al. 2013; Ilyas and Rehman 2018) or act as reactive oxygen species chelators (Xue et al. 2009; Hassinen et al. 2011). MTs play important roles in various biological mechanisms, as these proteins are highly conserved in living organisms and protect them from many toxic elements (Mulik et al. 2018).

11.6 Tolerance Mechanisms

Ni-contaminated soil is an important problem faced by organisms and should be addressed by phytoremediation mechanisms (Mahieu et al. 2013; Karimi et al. 2017). Recently, heavy metal phytoremediation is a method produced by plants to significantly reduce the toxicity of heavy metals (Mitra et al. 2018). Plant species are characterized by metal tolerance and accumulation. They are specific through metal uptake, rhizome translocation, cellular sequestration, and chemical accumulation (Mhatre et al. 2019). Several papers have shown that plants have a strong ability to accumulate heavy metals in their tissues without exhibiting phytotoxicity (Helaoui et al. 2022b).There are more than 310 species of Ni hyperaccumulators, such as Compositae, Cruciferae, Staghorne, Caryophyllaceae, Leguminosae, *Rowanaceae, Meliaceae, Myristaceae, Corianderaceae, Rubiaceae, Saponaceae,* and *Stackhousiaceae* (Reeves et al. 1996; Chaney et al. 2005).

Plants can degrade organometals and stabilize pollutants by acting as traps. In fact, some microorganisms (yeasts, fungi, or bacteria) are associated with roots, which break down organic matter for energy (DeLorenzo 2018). Certain microorganisms can be genetically modified to produce enzymes that alter xenobiotic chemicals. A major limitation of bacterial bioremediation is the bioavailability of enzymes that can bioconvert many substances into harmless products and by-products (Kang 2014). To break down pollutants, most bioremediation microorganisms perform metabolic reactions that involve aerobic metabolic pathways that use oxygen as an electron acceptor (Hatzikioseyian 2010).

11.7 Omics Approaches in Ni Stress Tolerance

Heavy metal stress mainly focuses on different regulatory systems and molecular mechanisms. In this context, "omics" approaches aim to study plant responses to metal stress at the molecular level (Frew et al. 2018). These methods include gene regulator identification, differential expression analysis, and molecular target identification. Well-established technologies such as transcriptomics, proteomics, sequencing, metabolomics, chromatography, mass spectrometry, and miRNA omics biology support omics approaches to plant responses to heavy metal stress (Coskun et al. 2019) (Figure 11.2).

11.7.1 Transcriptomics

Many types of studies have shown the effects of Ni on plants at the morphological level, but few studies have examined the effects of Ni on gene expression in plants. Today, transcriptome analysis is used as a novel approach to molecular nickel responses (Helaoui et al. 2020).

In fact, transcriptome analysis uses many of the techniques used to analyze RNA transcripts in organisms. Various techniques are commonly used to quantify the abundance of transcripts of interest (Khalid et al. 2019). Studies have shown that heavy metal stress-induced modification of gene expression can lead to the synthesis of new proteins, metabolic compounds that mediate stress, or transcription factors encoding genes that help regulate stress responses (Zhou et al. 2021).

In fact, cDNA-AFLP was previously developed to study transcriptome dynamics in plants and samples. However, this method provides poor resolution. Microarrays are the second proven technology for analyzing transcriptomes and conducting comparative transcriptomic studies. However, this approach is not adequately covered and leads to false positives.

Furthermore, sequence information is needed to improve hybridization chips and facilitate the use of microarrays as a consistent platform (Muthamilarasan et al. 2019). Sequence

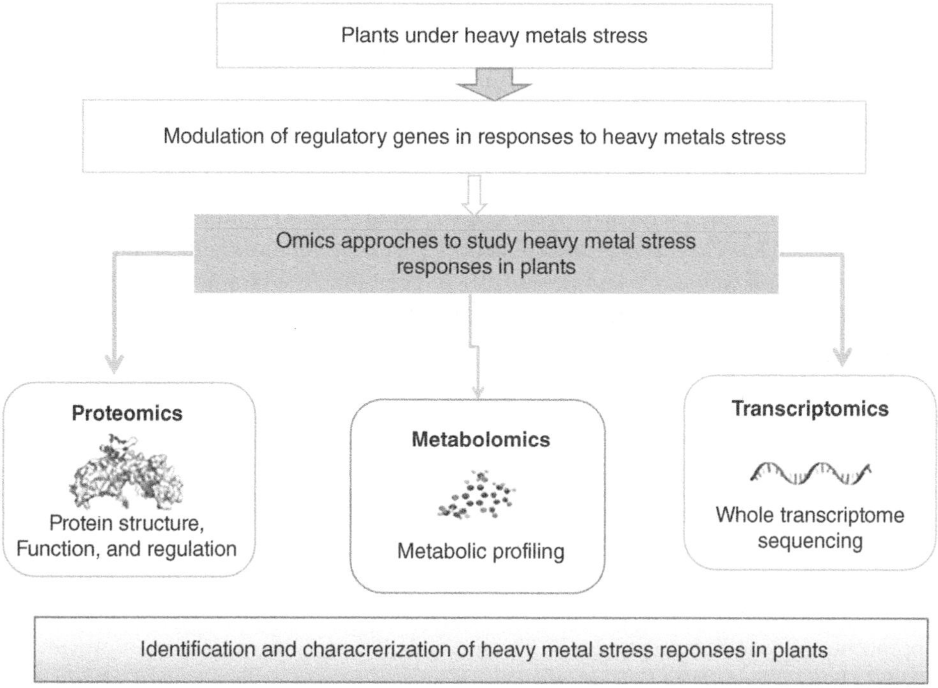

Figure 11.2 Omics technologies for characterizing heavy metal stress responses in plants.

analysis of gene expression (SAGE) is another technique for analyzing the transcriptome. Nonetheless, NGS has also transformed transcriptome research by providing the ability to sequence large amounts of coding and noncoding RNA samples without prior genomic information (Mehta et al. 2019). In addition, processing RNA-seq analyses and matching them (via a reference-based approach) will provide information on the corresponding gene expression levels (Xiao et al. 2019). Phytochelators, metallothioneins, heat shock proteins, metal-induced transporters, especially NRAMP1, sulfate transporters, MATE transporters, glutathione-conjugated transporters in metal stress (such as, Ni, Hg, Cd, and Al) (Fu et al. 2014). DEGs identified using high-throughput techniques revealed DEGs primarily corresponding to cellular structure and metabolism, photosynthesis, transpiration, and redox balance (Paul and Roychoudhury. 2018).

11.7.2 Proteomics

Proteomics plays a crucial role in characterizing the total protein content of cells, tissues, or organisms. Proteomics is the identification and quantification of proteomes to discover reliable biomarkers of different metal stress states by using proteins as biomarkers (Duruflé et al. 2017; Wen et al. 2019). Protein content can be determined by sodium dodecyl sulfate-polyacrylamide gel electrophoresis (SDS-PAGE); however, two-dimensional gel electrophoresis (2DE) can be used to identify proteins (Marouga et al. 2005; Eldakak et al. 2013). In fact, 2DE separates protein content by a pH at which molecules migrate according to their isoelectric pH and SDS-PAGE is published perpendicular to the original orientation (Kosová et al. 2018).

The technique yields important information when applied to genomic or plant samples in response to metal stress. Furthermore, it can identify differentially expressed anti-stress proteins and determine the role of specific proteins in metal stress signaling (Liu et al. 2015). The technique yields important information when applied to genomic or plant samples in response to metal stress. Furthermore, it can identify differentially expressed anti-stress proteins and determine the role of specific proteins in metal stress signaling (Liu et al. 2015). In addition, proteomic methods provide detailed information on protein content and its quantification data. This assay is able to identify post-translational changes (phosphorylation, glycosylation, sulfation (Luan et al. 2018)). Therefore, the phosphorylated proteome has attracted the attention of researchers because the phosphorylated proteins play a key role in the enhanced tolerance of plants (Tamburino et al. 2017).

11.7.3 Metabolomics

Metabolomics is an important technology in functional genomics and systems biology. It plays a crucial role in the functional annotation of genes. Additionally, it contributes to understanding responses to cellular and molecular, biotic and abiotic stresses (Peters et al. 2018). The physiological insights that can be obtained by measuring all small molecules in biological systems have made metabolomics a central pillar of systems biology approaches (Goeddel and Patti 2012). Thus, metabolite profiling patterns can provide an overall characterization of physiological states as well as greater insight into specific biochemical processes (Worley and Powers 2013). Under metal stress, plants behave differently from their natural metabolism, focusing more on responses and survival, and regulating their mechanisms and developmental processes. Thus, these modifications in gene expression patterns and transcriptional expression cause changes in the amount, concentration, and type of metabolites produced. In fact, proline is one of the best-known and most widely used plant metabolites due to its important role in metal chelation and antioxidant enzyme production, and is therefore widely used in the production of transgenic metal-tolerant plants (Villate et al. 2021). Likewise, putrescine has been studied to increase plant tolerance to heavy metal toxicity (Sharma and Dietz 2006).

11.8 Conclusion

Ni is an essential nutritional element based on its role for growth and development. In general, the nickel content of plants is too high due to human pollution. In fact, high nickel accumulation reduces crop yield by affecting seed germination, plant growth, chlorophyll content, and nutrient uptake. In addition, Ni toxicity leads to alterations in membrane permeability, plant water, and mineral uptake. Several papers also highlighted that Ni induces various changes in plant physiological activities, such as changes in antioxidant defense systems, production of reactive oxygen species, and disruption of biological enzymes. Today, Ni pollution is an important environmental problem, causing multiple hazards to plants, animals, and humans. Indeed, the mechanisms of Ni tolerance in contaminated plants are not well understood, so work should be done to understand the mechanisms of Ni tolerance in plants to limit nickel stress.

References

Abdal, N., Abbas, G., Asad, S.A. et al. (2021). Salinity mitigates cadmi-um-induced phytotoxicity in quinoa (*Chenopodium quinoa* Willd) by limiting the Cd uptake and improved responses to oxidative stress: implications for phytoremediation. *Environ. Geochem. Health.* 45: 171–185.

Adrees, M., Ali, S., Rizwan, M. et al. (2015). Mechanisms of silicon-mediated alleviation of heavy metal toxicity in plants: a review. *Ecotoxicol. Environ. Saf.* 119: 186–197.

Ahmad, P. (2016). *Plant Metal Interaction: Emerging Remediation Techniques*. Amsterdam: Elsevier.

Ahmad, M.S.A. and Ashraf, M. (2012). Essential roles and hazardous effects of nickel in plants. In: *Reviews of Environmental Contamination and Toxicology* (ed. D. Whitacre), 125–167. New York: Springer.

Ahmad, M.S.A., Hussain, N.M., Saddiq, R., and Alvi, A.K. (2007). Mungbean: a nickel indicator, accumulator or excluder. *Bull. Environ. Contam. Toxicol.* 78: 319–324.

Ali, M.A., Ashraf, M., and Athar, H.R. (2009). Influence of nickel stress on growth and some important physiological/biochemical attributes in some diverse canola (*Brassica napus* L.) cultivars. *J. Hazard. Mater.* 172: 964–969.

Alloway, B.J. (1995). *Heavy Metals in Soils*, 2e. Blackie Academic and Professional. pp. 11.

Almeida, D.M., Oliveira, M.M., and Saibo, N.J. (2017). Regulation of Na^+ and K^+ homeostasis in plants: towards improved salt stress tolerance in crop plants. *Genet. Mol. Biol.* 40: 326–345.

Amari, T., Lutts, S., Taamali, M. et al. (2016). Implication of citrate, malate and histidine in the accumulation and transport of nickel in *Mesembryanthemum crystallinum* and *Brassica juncea*. *Ecotoxicol. Environ. Saf.* 126: 122–128.

Ameen, N., Amjad, M., Murtaza, B. et al. (2019). Biogeochemical behavior of nickel under different abiotic stresses: toxicity and detoxification mechanisms in plants. *Environ. Sci. Pollut. Res.* 26: 10496–10514.

Amjad, M., Akhtar, J., Haq, M. et al. (2014). Understanding salt tolerance mechanisms in wheat genotypes by exploring antioxidant enzymes. *Pak. Jour. Agric. Sci.* 51: 969–976.

Amjad, M., Ameen, N., Murtaza, B. et al. (2020). Comparative physiological and biochemical evaluation of salt and nickel tolerance mechanisms in two contrasting tomato genotypes. *Physiol. Plant.* 168 (1): 27–37.

Aqeel, M., Khalid, N., Tufail, A. et al. (2021). Elucidating the distinct interactive impact of cadmium and nickel on growth, photosynthesis, metal-homeostasis, and yield responses of mung bean (*Vigna radiata* L.) varieties. *Environ. Sci. Pollut. Res.* 28: 27376–27390.

Arora, P., Bhardwaj, R., and Kanwar, M.K. (2012). Effect of 24-epibrassinolide on growth, protein content and antioxidative defense system of *Brassica juncea* L. subjected to cobalt ion toxicity. *Acta. Physiol. Plant.* 34: 2007–2017.

Arya, S.S., Devi, S., Angrish, R., and Rani, K. (2017). Soil reclamation through phytoextraction and phytovolatilization. In: *Volatiles and Food Security* (ed. D.K. Choudhary, A.K. Sharma, P. Agarwal, et al.), 25–43. Malaysia: Springer Singapore.

Aschmann, S.G. and Zasoski, R.J. (1987). Nickel and rubidium uptake by whole oat plants in solution culture. *Physiol. Plant.* 71: 191–196.

Baker, N.R. and Rosenqvist, E. (2004). Applications of chlorophyll fluorescence can improve crop production strategies: an examination of future possibilities. *J. Exp. Bot.* 55: 1607–1621.

Baran, U. and Ekmekçi, Y. (2022). Physiological, photochemical, and antioxidant responses of wild and cultivated Carthamus species exposed to nickel toxicity and evaluation of their usage potential in phytoremediation. *Environ. Sci. Pollut. Res.* 29: 4446–4460.

Barman, S.C., Sahu, R.K., Bhargava, S.K., and Chaterjee, C. (2000). Distribution of heavy metals in wheat, mustard, and weed grown in field irrigated with industrial effluents. *Bull. Environ. Contam. Toxicol.* 64: 489–496.

Batool, S. (2018). Impact of bioaccumulation of nickel on growth, seed yield and mineral uptake of chickpea (*Cicer arietinum* L.) varieties. *Pak. J. Bot.* 50: 2147–2150.

Begum, W., Rai, S., Banerjee, S. et al. (2022). A comprehensive review on the sources, essentiality and toxicological profile of nickel. *R. Soc. Chem.* 12: 9139.

Bell, C.H., Kalve, E., Ross, I. et al. (2019). *Emerging Contaminants Handbook*. Boca Raton, FL: CRC Press.

Bennett, B.G. (1982). Exposure of man to environmental nickel–an exposure commitment assessment. *Sci. Total. Environ.* 22 (3): 203–212.

Bhalerao, S.A., Sharma, A.S., and Poojari, A.C. (2015). Toxicity of nickel in plants. *Int. J. Pure Appl. Biosci.* 3: 345–355.

Blindauer, C.A. and Leszczyszyn, O.I. (2010). Metallothioneins: unparalleled diversity in structures and functions for metal ion homeostasis and more. *Nat. Prod. Rep.* 27 (5): 720–741.

Blum, R., Beck, A., Korte, A. et al. (2007). Function of phytochelatin synthase in catabolism of glutathione conjugates. *Plant J.* 49: 740–749.

Briat, J.F. and Lebrun, M. (1999). Plant responses tometal toxicity. *C. R. Acad. Sci.* 322: 43–54.

Buxton, S., Garman, E., Heim, K.E. et al. (2019). Concise review of nickel human health toxicology and ecotoxicology. *Inorganics* 7: 89.

Cai, L.M., Wang, Q.S., Wen, H.H. et al. (2019). Heavy metals in agricultural soils from a typical township in Guangdong Province, China: occurrences and spatial distribution. *Ecotoxicol. Environ. Saf.* 168: 184–191.

Cempel, M. and Nikel, G. (2006). Nickel: a review of its sources and environmental toxicology. *Pol. J. Environ. Stud.* 15 (3): 375–382.

Chai, N., Yi, X., Xiao, J. et al. (2021). Spatiotemporal variations, sources, water quality and health risk assessment of trace elements in the Fen River. *Sci. Total Environ.* 757: 143882.

Chaney, R.L., Angle, J.S., McIntosh, M.S. et al. (2005). Using hyperaccumulator plants to phytoextract soil Ni and Cd. *Z. Naturforsc.* 60: 190–198.

Chen, C., Huang, D., and Liu, J. (2009). Functions and toxicity of nickel in plants: recent advances and future prospects. *Clean. Soil. Air. Water.* 37: 304–313.

Chow, Y.N., Lee, L.K., Zakaria, N.A., and Foo, K.Y. (2019). Integrated assessment of nickel electroplating industrial wastewater effluent as a renewable resource of irrigation water using a hydroponic cultivation system. *Front. Plant Sci.* 12: 609396.

Cobbett, C. and Goldsbrough, P. (2002). Phytochelatins and metallothioneins: roles in heavy metal detoxification and homeostasis. *Annu. Rev. Plant. Biol.* 53: 159–182.

Colpas, G.J. and Hausinger, R.P. (2000). In vivo and in vitro kinetics of metal transfer by the *Klebsiella aerogenes* urease nickel metallochaperone, urea. *J. Biol. Chem.* 275: 10731–10737.

Coskun, D., Deshmukh, R., Sonah, H. et al. (2019). The controversies of silicon's role in plant biology. *New. Phytol.* 221: 67–85.

Cutillas-Barreiro, L., Paradelo, R., Igrexas-Soto, A. et al. (2016). Valorization of biosorbent obtained from a forestry waste: competitive adsorption, desorption and transport of Cd, Cu, Ni, Pb and Zn. *Ecotoxicol. Environ. Safe.* 131: 118–126.

Czajka, K.M., Michael, P. and Nkongolo, K. (2019). Differential effects of nickel dosages on in vitro and in vivo seed germination and expression of a high affinity nickel-transport family protein (AT2G16800) in trembling aspen (*Populus tremuloides*). *Ecotoxicology* 28: 92–102.

Dalton, D.A., Evans, H.J., and Hanus, F.J. (1985). Stimulation by nickel of soil microbial urease activity and urease and hydrogenase activities in soybeans grown in a low-nickel soil. *Plant. Soil.* 88: 245–258.

Das, K. and Roychoudhury, A. (2014). Reactive oxygen species (ROS) and response of antioxidants as ROS-scavengers during environmental stress in plants. *Front. Environ. Sci* 2: 1–13.

DeLorenzo, V. (2018). Biodegradation and bioremediation: an introduction. In: *Consequences of Microbial Interactions with Hydrocarbons, Oils, and Lipids: Biodegradation and Bioremediation. Handbook of Hydrocarbon and Lipid Microbiology* (ed. R.J. Steffan), 1–21. Cham, Switzerland: Springer.

Dodds, F., Roesch, J., and Leiva Roesch, J. (2016). *Negotiating the Sustainable Development Goals*. Oxford, UK: Routledge.

Dourado, M.N., Franco, M.R., and Peters, L.P. (2015). Antioxidant enzymes activities of Burkholderia spp. strains-oxidative responses to Ni toxicity. *Environ. Sci. Pollut. Res.* 22: 19922–19932.

Duarte, B., Delgado, M., and Caador, I. (2007). The role of citric acid in cadmium and nickel uptake and translocation, in *Halimione portulacoides*. *Chemosphere* 69: 836–840.

Duarte, A.C., Rocha-Santos, T., and Cachada, A. (2018). *Soil Pollution*. Cambridge, MA: Academic Press.

Duruflé, H., Clemente, H.S., Balliau, T. et al. (2017). Cell wall proteome analysis of *Arabidopsis thaliana* mature stems. *Proteomics* 17 (8): 1–5.

Ekmekçi, Y., Çulha Erdal, Ş., Balkan Nalçaiyi, A.S., and Çiçek, N. (2020). The acquisition of boron tolerance by salt pretreatment in two sunflower cultivars. *Turk. J. Bot.* 44: 153–166.

Eldakak, M., Milad, S.I., Nawar, A.I., and Rohila, J.S. (2013). Proteomics: a biotechnology tool for crop improvement. *Front. Plant Sci.* 4: 35.

Emamverdian, A., Ding, Y., Mokhberdoran, F., and Xie, Y. (2015). Heavy metal stress and some mechanisms of plant defense response. *Sci. World J.* 2015: 756120.

Fabiano, C., Tezotto, T., Favarin, J.L. et al. (2015). Essentiality of nickel in plants: a role in plant stresses. *Front. Plant. Sci.* 6: 754.

Fargašová, A. (2012). Plants as models for chromium and nickel risk assessment. *Ecotox.* 21: 1476–1483.

Flora, S.J. (2011). Arsenic-induced oxidative stress and its reversibility. *Free Radic. Biol. Med.* 51: 257–281.

Freeman, J.L., Persans, M.W., Nieman, K. et al. (2004). Increased glutathione biosynthesis plays a role in nickel tolerance in Thlaspi nickel hyperaccumulators. *Plant Cell.* 16: 2176–2191.

Frew, A., Weston, L.A., Reynolds, O.L., and Gurr, G.M. (2018). The role of silicon in plant biology: a paradigm shift in research approaches. *Ann. Bot.* 121: 1265–1273.

Fu, S.F., Chen, P.Y., Nguyen, Q.T. et al. (2014). Transcriptome profiling of genes and pathways associated with arsenic toxicity and tolerance in *Arabidopsis*. *BMC Plant Biol.* 14: 94.

Gajewska, E. and Skłodowska, M. (2007). Effect of nickel on ROS content and antioxidative enzyme activities in wheat leaves. *Biometals* 20: 27–36.

Gajewska, E. and Sklodowska, M. (2008). Differential biochemical responses of wheat shoots and roots to nickel stress: antioxidative reactions and proline accumulation. *Plant Growth. Regul.* 54: 179–188.

Gajewska, E., Skłodowska, M., Słaba, M., and Mazur, J. (2006). Effect of nickel on antioxidative enzyme activities, proline and chlorophyll contents in wheat shoots. *Biol. Plant.* 50 (4): 653–659.

Gardner, M., Comber, S., Scrimshaw, M. et al. (2012). The significance of hazardous chemicals in wastewater treatment works effluents. *Sci. Total Environ.* 437: 363–372.

Gill, S.S. and Tuteja, N. (2010). Reactive oxygen species and antioxidant machinery in abiotic stress tolerance in crop plants. *Plant Physiol. Biochem.* 48: 909–930.

Goeddel, L.C. and Patti, G.J. (2012). Maximizing the value of metabolomic data. *Bioanalysis* 4: 2199–2201.

González, C.I., Maine, M.A., Cazenave, J. et al. (2015). Ni accumulation and its effects on physiological and biochemical parameters of *Eichhornia crassipes*. *Environ. Exp. Bot.* 117: 20–27.

Gratão, P.L., Pompeu, G.B., Capaldi, F.R. et al. (2008). Antioxidant response of *Nicotiana tabacum* cv. bright yellow 2 cells to cadmium and nickel stress. *Plant Cell. Tiss. Organ Cult.* 94: 73–83.

Gregson, S. and Hope, A. (1994). Review of Phytotoxicity, Uptake and Accumulation of Elements and Organic Chemicals in Terrestrial Higher Plants. AERC Report for Department of the Environment, London.

Guo, H., Liu, H., Wu, H. et al. (2019). Nickel carcinogenesis mechanism: DNA damage. *Int. J. Mol. Sci.* 20: 4690.

Hanus-Fajerska, E.J. and Koźmińska, A. (2016). The possibilities of water purification using phytofiltration methods: a review of recent progress. *Biotechnologia* 97 (4): 315–322.

Harasim, P. and Filipek, T. (2015). Nickel in the environment. *J. Elementol.* 20 (2): 525–534.

Hasanuzzaman, M., Nahar, K., Alam, M.M. et al. (2013). Physiological, biochemical, and molecular mechanisms of heat stress tolerance in plants. *Int. J. Mol. Sci.* 14: 9643–9684.

Hassan, M.U., Chattha, M.U., Khan, I. et al. (2019). Ni toxicity in plants: reasons, toxic effects, tolerance mechanisms, and remediation possibilities. *Environ. Sci. Pollut. Res. Int.* 26 (13): 12673–12688.

Hassinen, V.H., Tervahauta, A.I., Schat, H., and Karenlampi, S.O. (2011). Plant metallothioneins–metal chelators with ROS scavenging activity? *Plant Biol.* 13 (2): 225–232.

Hattab, S., Hattab, S., Flores-Casseres, M.L. et al. (2016). Characterisation of lead-induced stress molecular biomarkers in *Medicago sativa* plants. *Environ. Exp. Bot.* 123: 1–12.

Hatzikioseyian, A. (2010). Principles of bioremediation. In: *Trends in Bioremediation and Phytoremediation* (ed. G. Plaza), 23–54. Kerala: Research Signpost.

Hayat, S., Cheng, Z., Ahmad, H. et al. (2016). Garlic, from remedy to stimulant: evaluation of antifungal potential reveals diversity in phytoalexin allicin content among garlic cultivars; allicin containing aqueous garlic extracts trigger antioxidants in cucumber. *Front. Plant Sci.* 7: 1–15.

Haydon, M.J. and Cobbett, C.S. (2007). Transporters of ligands for essential metal ions in plants. *New Phytol.* 174: 499–506.

Helaoui, S., Boughattas, I., Hattab, S. et al. (2020). Physiological, biochemical and transcriptomic responses of *Medicago sativa* to nickel exposure. *Chemosphere* 249: 126–121.

Helaoui, S., Boughattas, I., Sameh El Kribi-Boukhris, S. et al. (2022a). Assessing the efects of nickel on, e.g., *Medicago sativa* L. nodules using multidisciplinary approach. *Environ. Sci. Pollut. Res.* 29 (51): 77386–77400.

Helaoui, S., Hattab, S., Mkhinini, M. et al. (2022b). The effect of nickel exposure on oxidative stress of *Vicia faba* plants. *Bull. Environ. Contam. Toxicol.* 108: 1074–1080.

Hermle, S., Vollenweider, P., Gunthardt-Goerg, M.S. et al. (2007). Leaf responsiveness of *Populus tremula* and *Salix viminalis* to soil contaminated with heavy metals and acidic rainwater. *Tree Physiol.* 27: 1517–1531.

Hillel, D. (2005). *Encyclopedia of Soils in the Environment*. Cambridge, MA: Academic Press.

Hussain, M.B., Ali, S., Azam, A. et al. (2013). Morphological, physiological and biochemical responses of plants to nickel stress: a review. *Afr. J. Agric. Res.* 8: 1596–1602.

Ilyas, S. and Rehman, A. (2018). Metal resistance and uptake by *Trichosporon asahii* and *Pichia kudriavzevii* isolated from industrial effluents. *Arch. Environ. Prot.* 44: 77–84.

Iyaka, Y.A. (2011). Nickel in soils: a review of its distribution and impacts. *Sci. Res. Essays* 6: 6774–6777.

Jan, R., Khan, M.A., Asaf, S. et al. (2019). Metal resistant endophytic bacteria reduces cadmium, nickel toxicity, and enhances expression of metal stress related genes with improved growth of *Oryza sativa* L., via regulating its antioxidant machinery and endogenous hormones. *Plants* 8: 363.

Jin, Y., O'Connor, D., Ok, Y.S. et al. (2019). Assessment of sources of heavy metals in soil and dust at children's playgrounds in Beijing using GIS and multivariate statistical analysis. *Environ. Int.* 124: 320–328.

Kabata-Pendias, A. and Mukherjee, A.B. (2007). *Trace Elements from Soil to Human*. Berlin: Springer-Verlag.

Kabata-Pendias, A. and Pendias, H. (2001). *Trace Elements in Soils and Plants*, 3e. Boca Raton, FL: CRC Press.

Kachenko, A.G., Singh, B., and Bhatia, N.P. (2007). Heavy metal tolerance in common fern species. *Aust. J. Bot.* 55: 63–73.

Kagi, J.H. (1991). Overview of metallothionein. *Methods Enzymol.* 205: 613–626.

Kamran, M.A., Eqani, S., and Bibi, S. (2016). Bioaccumulation of nickel by *E. sativa* and role of plant growth promoting rhizobacteria PGPRs under nickel stress. *Ecotoxicol Environ. Saf.* 126: 256–263.

Kang, J.W. (2014). Removing environmental organic pollutants with bioremediation and phytoremediation. *Biotechnol. Lett.* 36 (6): 1129–1139.

Karimi, A., Khodaverdiloo, H., and Sadaghiani, M.H.R. (2017). Fungi and bacteria as helping agents for remediation of a Pb-contaminated soil by *Onopordum acanthium*. *Caspian J. Environ. Sci.* 15: 249–262.

Khalid, S., Shahid, M., Niazi, N.K. et al. (2017). Arsenic behaviour in soil-plant system: biogeochemical reactions and chemical speciation influences, enhancing cleanup of environmental pollutants. In: *Enhancing Cleanup of Environmental Pollutants* (ed. N. Anjum, S. Gill, and N. Tuteja), 97–140. Cham: Springer.

Khalid, N., Aqeel, M., and Noman, A. (2019). System biology of metal tolerance in plants: an integrated view of genomics, transcriptomics, metabolomics, and phenomics. In: *Plant Metallomics and Functional Omics* (ed. G. Sablok), 107–144. Cham: Springer.

Khaliq, A., Ali, S., Hameed, A. et al. (2016). Silicon alleviates nickel toxicity in cotton seedlings through enhancing growth, photosynthesis, and suppressing Ni uptake and oxidative stress. *Arch. Agron. Soil. Sci.* 62: 633–647.

Khan, M., Samrana, S., Zhang, Y. et al. (2020). Reduced glutathione protects subcellular compartments from Pb-induced ROS injury in leaves and roots of upland cotton (*Gossypium hirsutum* L.). *Front. Plant Sci.* 11: 412.

Kopittke, P.M., Menzies, N.W., Wang, P. et al. (2019). Soil and the intensification of agriculture for global food security. *Environ. Int.* 132: 105078.

Kosová, K., Vítámvás, P., Urban, M.O. et al. (2018). Plant abiotic stress proteomics: the major factors determining alterations in cellular proteome. *Front. Plant Sci.* 9: 122.

Kotapati, K.V., Palaka, B.K., and Ampasala, D.R. (2017). Alleviation of nickel toxicity in finger millet (*Eleusine coracana* L.) germinating seedlings by exogenous application of salicylic acid and nitric oxide. *Crop J.* 5: 240–250.

Krupa, Z. and Baszynski, T. (1995). Some aspects of heavy metals toxicity towards photosynthetic apparatus direct and indirect effects on light and dark reactions. *Acta Physiol. Plant* 17: 177–190.

Kubicka, K., Samecka-Cymerman, A., Kolon, K. et al. (2015). Chromium and nickel in *Pteridium aquilinum* from environments with various levels of these metals. *Environ. Sci. Pollut. Res.* 22: 527–534.

Kumar, S. and Trivedi, A.V. (2016). A review on role of nickel in the biological system. *Int. J. Curr. Microbiol. Appl. Sci.* 5 (3): 719–727.

Kumar, O., Singh, S.K., Singh, A.P. et al. (2018). Effect of soil application of nickel on growth, micronutrient concentration and uptake in barley (*Hordeum vulgare* L.) grown in Inceptisols of Varanasi. *J. Plant Nutr.* 41: 50–66.

Lešková, A., Giehl, R.F.H., Hartmann, A. et al. (2017). Heavy metals induce iron deficiency responses at different hierarchic and regulatory levels. *Plant Physiol.* 174: 1648–1668.

Leszczyszyn, O.I., Imam, H.T., and Blindauer, C.A. (2013). Diversity and distribution of plant metallothioneins: a review of structure, properties and functions. *Metallomics* 5 (9): 1146–1169.

Li, S.Z., Zhao, B., Jin, M. et al. (2020a). A comprehensive survey on the horizontal and vertical distribution of heavy metals and microorganisms in soils of a Pb/Zn smelter. *J. Hazard. Mater.* 400: 123255.

Li, Y.Z., Chen, H.Y., and Teng, Y.G. (2020b). Source apportionment and source-oriented risk assessment of heavy metals in the sediments of an urban river-lake system. *Sci. Total Environ.* 737: 140310.

Lin, Y.C. and Kao, C.H. (2007). Proline accumulation induced by excess nickel in detached rice leaves. *Biol. Plant.* 51: 351–354.

Líska, D., Soukup, M., Lukačová, Z. et al. (2017). Mechanisms of silicon-mediated alleviation of abiotic stress in plants: recent advances and future perspective. In: *Silicon in Plants* (ed. D.K. Tripathi, V.P. Singh, P. Ahmad, et al.), 1750. CRC Press.

Liu, G.D. (2001). A new essential mineral element–nickel. *Plant Nutr. Fertil. Sci.* 7: 101–103.

Liu, B., Zhang, N., Zhao, S. et al. (2015). Proteomic changes during tuber dormancy release process revealed by iTRAQ quantitative proteomics in potato. *Plant Physiol. Biochem.* 86: 181–190.

Luan, H., Shen, H., Pan, Y. et al. (2018). Elucidating the hypoxic stress response in barley (*Hordeum vulgare* L.) during waterlogging: a proteomics approach. *Sci. Rep.* 8 (1): 9655.

Mahieu, S., Soussou, S., Cleyet-Marel, J.C. et al. (2013). Local adaptation of metallicolous and non-metallicolous *Anthyllis vulneraria* populations: their utilization in soil restoration. *Restor. Ecol.* 21 (5): 551–559.

Malkin, R., Niyogi, K., Buchanan, B.B. et al. (2000). *Biochemistry and Molecular Biology of Plants*, 568–628. Rockville, MD: American Society of Plant Physiologists.

Marouga, R., David, S., and Hawkins, E. (2005). The development of the DIGE system: 2D fluorescence difference gel analysis technology. *Anal. Bioanal. Chem.* 382 (3): 669–678.

Marschner, H. (1995). *Mineral Nutrition of Higher Plants*. London: Academic Press.

Mehta, S., James, D., and Reddy, M.K. (2019). Omics technologies for abiotic stress tolerance in plants: current status and prospects. In: *Omics Technologies for Abiotic Stress Tolerance in Plants: Current Status and Prospects* (ed. S.H. Wani), 1–34. Cham: Springer.

Mhatre, P.H., Karthik, C., Kadirvelu, K. et al. (2019). Plant growth promoting rhizobacteria (PGPR): a potentialalternative tool for nematodes bio-control. *Biocatal. Agric. Biotechnol.* 17: 119–128.

Min, E.Y., Cha, Y.J., and Kang, J.C. (2015). Effects of waterborne nickel on the physiological and immunological parameters of the Pacific abalone *Haliotis discus hannai* during thermal stress. *Environ. Sci. Pollut. Res.* 22: 13546–13555.

Mishra, D. and Kar, M. (1974). Nickel in plant growth and metabolism. *Bot. Rev.* 40: 395–452.

Mitra, S., Pramanik, K., Sarkar, A. et al. (2018). Bioaccumulation of cadmium by *Enterobacter* sp. and enhancement of rice seedling growth under cadmium stress. *Ecotoxicol. Environ. Saf.* 156: 183–196.

Mizuno, T., Sonoda, T., Horie, K. et al. (2003). Cloning and characterization of phytochelatin synthase from a nickel hyperaccumulator *Thlaspi japonicum* and its expression in yeast. *Soil Sci. Plant Nutr.* 49: 285–290.

Mkhinini, M., Helaoui, S., Boughattas, I. et al. (2020). Earthworm *Eisenia andrei* modulate oxidative stress in bean plants *Vicia faba* irrigated with treated waste water. *Ecotoxicology* 29: 1003.

Molas, J. (1997). Changes in morphological and anatomical structure of cabbage (*Brassica oleracea* L.) outer leaves and in ultrastructure of their chloroplasts caused by an in vitro excess of nickel. *Photosynthetica* 34: 513–522.

Molas, J. (2002). Changes of chloroplast ultrastructure and total chlorophyll concentration in cabbage leaves caused by excess of organic Ni II complexes. *Environ. Exp. Bot.* 47: 115–126.

Mondal, M.H., Ali, M.A., Pal, A., and Saha, B. (2019). A review on micellar catalyzed oxidation reactions of organic functional groups in aqueous medium using various transition metals. *Tenside Surfactants Deterg.* 56: 516–525.

Moreira, I.N., Martins, L.L., and Mourato, M.P. (2020). Effect of Cd, Cr, Cu, Mn, Ni, Pb and Zn on seed germination and seedling growth of two lettuce cultivars (*Lactuca sativa* L.). *Plant Physiol. Rep.* 25: 347–358.

Mulik, A., Bhola, J., and Bhadekar, R. (2018). Protein profiling and antioxidant enzyme activity of cadmium and lead tolerant *Kocuria* sp. BRI36. *Int. J. Appl. Eng. Res.* 13: 8357–8363.

Mulrooney, S.B. and Hausinger, R.P. (2003). Nickel uptake and utilization by microorganisms. *FEMS Microbiol. Rev.* 27: 239–261.

Muthamilarasan, M., Singh, N.K., and Prasad, M. (2019). Multi-omics approaches for strategic improvement of stress tolerance in underutilized crop species: a climate change perspective. *Adv. Genet.* 103: 1–38.

Nakazawa, R., Ozawa, T., Naito, T. et al. (2001). Interactions between cadmium and nickel in phytochelatin biosynthesis and the detoxification of the two metals in suspension cultured tobacco cells. *Biol. Plant* 44: 627–630.

Neumann, P.M. and Chamel, A. (1986). Comparative phloem mobility of nickel in nonsenescent plants. *Plant Physiol.* 81: 689–691.

Nriagu, J.O. (1990). Global metal pollution: poisoning the biosphere? *Envir. Sci. Policy Sust. Develop.* 32 (7): 7–33.

Page, V., Weisskof, L., and Feller, U. (2006). Heavy metals in white Lupin: uptake, root-to-shoot transfer and redistribution within the plant. *New Phytol.* 171: 329–341.

Pandey, N. and Sharma, C.P. (2002). Effect of heavy metals Co^{2+}, Ni^{2+} and Cd^{2+} on growth and metabolism of cabbage. *Plant Sci.* 163: 753–758.

Paul, S. and Roychoudhury, A. (2018). Transcriptome profiling of abiotic stress-responsive genes during cadmium chloride-mediated stress in two indica rice varieties. *J. Plant Growth Regul.* 37: 657–667. https://doi.org/10.1007/s00344- 017-9762-y.

Peralta-Videaa, J.R., Gardea-Torresdey, J.L., Gomezc, E.L. et al. (2002). Effect of mixed cadmium, copper, nickel and zinc at different pHs upon alfalfa growth and heavy metal uptake. *Environ. Pollut.* 119: 291–301.

Peters, K., Worrich, A., Weinhold, A. et al. (2018). Current challenges in plant eco-metabolomics. *Int. J. Mol. Sci.* 19: 1385.

Pinter, T.B. and Stillman, M.J. (2014). The zinc balance: competitive zinc metalation of carbonic anhydrase and metallothionein 1A. *Biochemistry* 53: 6276–6285.

Pochodylo, A.L. and Aristilde, L. (2017). Molecular dynamics of stability and structures in phytochelatin complexes with Zn, Cu, Fe, Mg, and Ca: implications for metal detoxification. *Environ. Chem. Lett.* 15 (3): 495–500.

Prueitt, R.L., Li, W., Chang, Y.C. et al. (2020). Systematic review of the potential respiratory carcinogenicity of metallic nickel in humans. *Crit. Rev. Toxicol.* 50: 605–639.

Rao, M.K.V. and Sresty, T.V. (2000). Antioxidative parameters in the seedlings of pigeonpea (*Cajanus cajan* (L.) Millspaugh) in response to Zn and Ni stresses. *Plant Sci.* 157: 113–128.

Rather, B.A., Masood, A., Sehar, Z. et al. (2020). Mechanisms and role of nitric oxide in phytotoxicity-mitigation of copper. *Front. Plant Sci.* 11: 675.

Rautaray, S.K., Ghosh, B.C., and Mittra, B.N. (2003). Effect of fly ash, organic wastes and chemical fertilizers on yield, nutrient uptake, heavy metal content and residual fertility in a rice-mustard cropping sequence under acid lateritic soils. *Bioresour. Technol.* 90 (3): 275–283.

Reeves, R.D., Baker, A.J.M., Borhidi, A., and Berezain, R. (1996). Nickel-accumulating plants from the ancient serpentine soils of Cuba. *New Phytol.* 133: 217–224.

Riaz, L., Mahmood, T., Coyne, M.S. et al. (2017). Physiological and antioxidant response of wheat (*Triticum aestivum*) seedlings to fluoroquinolone antibiotics. *Chemosphere* 177: 250–257.

Rizvi, A., Zaidi, A., Ameen, F. et al. (2020). Heavy metal induced stress on wheat: phytotoxicity and microbiological management. *RSC Adv.* 10: 38379–38403.

Rizwan, M., Mostofa, M.G., Ahmad, M.Z. et al. (2018). Nitric oxide induces rice tolerance to excessive nickel by regulating nickel uptake, reactive oxygen species detoxification and defenserelated gene expression. *Chemosphere* 91: 23–35.

Rong, Q., Zhong, K., Huang, H. et al. (2020). Humic acid reduces the available cadmium, copper, lead, and zinc in soil and their uptake by tobacco. *Appl. Sci.* 10 (3): 1077.

Rooney, C.P., Zhao, F.J., and McGrath, S.P. (2007). Phytotoxicity of nickel in a range of Europian soils: influence of soil properties, Ni solubility and speciation. *Environ. Pollut.* 145: 596–605.

Rosen, V. and Chen, Y. (2014). The influence of compost addition on heavy metal distribution between operationally defined geochemical fractions and on metal accumulation in plant. *J. Soils Sediments* 14: 713–720.

Sachan, P. and Lal, N. (2017). An overview of nickel (Ni^{2+}) essentiality, toxicity and tolerance strategies in plants. *Asian J. Biol.* 2: 1–15.

Salinitro, M., van der Ent, A., Tognacchini, A., and Tassoni, A. (2020). Stress responses and nickel and zinc accumulation in different accessions of *Stellaria media* (L.) Vill. in response to solution pH variation in hydroponic culture. *Plant Physiol. Biochem.* 148: 133–141.

Seregin, I.V. and Kozhevnikova, A.D. (2006). Physiological role of nickel and its toxic effects on higher plants. *Russ. J. Plant Physiol.* 53: 257–277.

Sharma, S.S. and Dietz, K.J. (2006). The significance of amino acids and amino acid-derived molecules in plant responses and adaptation to heavy metal stress. *J. Exp. Bot.* 57: 711–726.

Shen, Z., Zhang, Y., McMillan, O. et al. (2020). The use of biochar for sustainable treatment of contaminated soils. In: *Sustainable Remediation of Contaminated Soil and Groundwater* (ed. D. Hou), 119–167. Butterworth-Heinemann.

Sheoran, I.S., Aggarwal, N., and Singh, R. (1990). Effect of cadmium and nickel on in vivo carbon dioxide exchange rate of pigeon pea (*Cajanus cajan* L.). *Plant Soil* 129: 243–249.

Shukla, D., Tiwari, M., Tripathi, R.D. et al. (2013). Synthetic phytochelatins complement a phytochelatin deficient Arabidopsis mutant and enhance the accumulation of heavy metal(loid)s. *Biochem. Biophys. Res. Commun.* 434: 664–669.

Shukla, D., Sandhu, G., and Prabodh Kumar-Trivedi, P.K. (2022). Role of phytochelatin in cation translocation in plants. In: *Cation Transporters in Plants* (ed. S.K. Upadhyay), 401–415. Academic Press.

Singh, S., Parihar, P., Singh, R. et al. (2016). Heavy metal tolerance inplants: role of transcriptomics, proteomics, metabolomics, and ionomics. *Front. Plant Sci.* 6: 11436.

Straif, K. (2010). The IARC monographs: a review and update on occupational carcinogens, vol. 100C. *Epidemiol. Prev.* 34 (5–6): 38.

Sun, C., Zhang, Z., Cao, H. et al. (2019). Concentrations, speciation, and ecological risk of heavy metals in the sediment of the Songhua River in an urban area with petrochemical industries. *Chemosphere* 219: 538–545.

Talebi, M., Ebrahim, B., Tabatabaei, S., and Akbarzadeh, H. (2019). Hyperaccumulation of Cu, Zn, Ni, and Cd in Azolla species inducing expression of methallothionein and phytochelatin synthase genes. *Chemosphere* 230: 488–497.

Tamburino, R., Vitale, M., Ruggiero, A. et al. (2017). Chloroplast proteome response to drought stress and recovery in tomato (*Solanum lycopersicum* L.). *BMC Plant Biol.* 17: 40.

Tripathy, B.C., Bhatia, B., and Mohanty, P. (1981). Inactivation of chloroplast photosynthetic electron transport activity by Ni^{2+}. *Biochim. Biophys. Acta* 638: 217–224.

Tye, A.M., Young, S., Crout, N.M.J. et al. (2004). Speciation and solubility of Cu, Ni and Pb in contaminated soils. *Eur. J. Soil Sci.* 55: 579–590.

Verbruggen, N., Hermans, C., and Schat, H. (2009). Molecular mechanisms of metal hyperaccumulation in plants. *New Phytol.* 181: 759–776.

Verma, S. and Dubey, R.S. (2003). Lead toxicity induces lipid peroxidation and alters the activities of antioxidant enzymes in growing rice plants. *Plant Sci.* 164: 645–655.

Villate, A., Nicolas, M.S., Gallastegi, M. et al. (2021). Review: metabolomics as a prediction tool for plants performance under environmental stress. *Plant Sci.* 303: 110789.

Vogel-Mikuš, K., Drobne, D., and Regvar, M. (2005). Zn, Cd and Pb accumulation and arbuscular mycorrhizal colonisation of pennycress *Thlaspi praecox* Wulf. (Brassicaceae) from the vicinity of a lead mine and smelter in Slovenia. *Envir. Poll.* 133 (2): 233–242.

Wakeel, A., Xu, M., and Gan, Y. (2020). Chromium-induced reactive oxygen species accumulation by altering the enzymatic antioxidant system and associated cytotoxic, genotoxic, ultrastructural, and photosynthetic changes in plants. *Int. J. Mol. Sci.* 2: 728–747.

Wang, B., Chu, C., Wei, H. et al. (2020). Ameliorative effects of silicon fertilizer on soil bacterial community and pakchoi (*Brassica chinensis* L.) grown on soil contaminated with multiple heavy metals. *Environ. Pollut.* 267: 115411–115421.

Wen, R., Ya, L., Zhou, M.-Y. et al. (2019). Dynamic changes of root proteome reveal diverse responsive proteins in maize subjected to cadmium stress. *J. Integr. Agric.* 18: 2193–2204.

Wheeler, C.T., Hughes, L.T., Oldroyd, J., and Pulford, I.D. (2001). Effects of nickel on Frankia and its symbiosis with *Alnus glutinosa* (L.). *Gaertn. Plant Soil.* 231: 81–90.

Worley, B. and Powers, R. (2013). Multivariate analysis in metabolomics. *Curr. Metabolomics* 1: 92–107.

Xiang, C., Xiahui, C., Bangzhu, Z. et al. (2018). Will developing countries become pollution havens for developed countries? An empirical investigation in the belt and road. *J. Clean. Prod.* 198: 624–632.

Xiao, Y., Du, Z., Qi, X. et al. (2019). RNA-sequencing analysis reveals transcriptional changes in the roots of low-cadmium-accumulating winter wheat under cadmium stress. *Acta Physiol. Plant.* 41: 13.

Xue, T., Li, X., Zhu, W. et al. (2009). Cotton metallothionein GhMT3a, a reactive oxygen species scavenger, increased tolerance against abiotic stress in transgenic tobacco and yeast. *J. Exp. Bot.* 60 (1): 339–349.

Yang, X.E., Baligar, V.C., Foster, J.C., and Martens, D.C. (1997). Accumulation and transport of nickel in relation to organic acids in ryegrass and maize with different nickel levels. *Plant Soil* 196: 271–276.

Yusuf, M., Fariduddin, Q., Hayat, S., and Ahmad, A. (2011). Nickel: an overview of uptake, essentiality and toxicity in plants. *Bull. Environ. Contam. Toxicol.* 86: 1–17.

Zaid, A., Mohammad, F., Wani, S.H., and Siddique, K.M.H. (2019). Salicylic acid enhances nickel stress tolerance by up-regulating antioxidant defense and glyoxalase systems in mustard plants. *Ecotoxicol. Environ. Saf.* 180: 575–587.

Zhang, H., Lv, S., Xu, H. et al. (2017). H_2O_2 is involved in the metallothionein-mediated rice tolerance to copper and cadmium toxicity. *Int. J. Mol. Sci.* 18: 2083.

Zhou, X., Joshi, S., Patil, S. et al. (2021). Reactive oxygen, nitrogen, carbonyl and sulfur species and their roles in plant abiotic stress responses and tolerance. *J. Plant Growth Regul.* 41: 119–142.

Zhuo, H., Wang, X., Liu, H. et al. (2019). Source analysis and risk assessment of heavy metals in development zones: a case study in Rizhao, China. *Environ. Geochem. Health* 42: 135–146.

Zwolsman, J.J.G. and Van Bokhoven, A.J. (2007). Impact of summer droughts on water quality of the Rhine River-a preview of climate change? *Water Sci. Technol.* 56: 45–55.

12

Copper Toxicity and Tolerance in Plants

Insights from Omics Studies

Moreira A[1], Moraes LAC[1], Delfim JJ[2], and Moreti LG[3]

[1] *Department of Soil Science and Ecophysiology, Embrapa Soja, Brazilian Agricultural Research Corporation (Embrapa), Londrina, Paraná State, Brazil*
[2] *Department of Crop Science, UEL-Londrina State University, Londrina, Paraná State, Brazil*
[3] *Department of Crop Science, São Paulo State University, Botucatu, São Paulo State, Brazil*

12.1 Introduction

Copper (Cu) is a transition metal that plays a vital role in plant metabolism. It is required in small amounts and is present at an average concentration of $6.0 \pm 1.0\,mg\,kg^{-1}$ in the shoot based on dry weight; however, its concentration ranges from 2 to $50\,mg\,kg^{-1}$ depending on the crop (Reuter and Robinson 1997; Epstein and Bloom 2005; Moreira et al. 2022a). Copper has an atomic mass and specific gravity above 20 (63.55) and $5.0\,g\,cm^{-3}$ ($8.96\,g\,cm^{-3}$), respectively, which characterizes it as a heavy metal, similar to other micronutrients, such as cobalt (Co), iron (Fe), manganese (Mn), molybdenum (Mo), nickel (Ni), and zinc (Zn), and nonessential elements that play no known role in plant metabolism, such as arsenic (As), cadmium (Cd), lead (Pb), and mercury (Hg) (Kabata-Pendias 2011; Rascioa and Navari-Izzo 2011).

Cu participates in many essential plant metabolic processes such as photosynthesis, respiration, protection against oxidative stress, hormonal signaling, and cell wall metabolism, acting mainly as a structural element in regulatory proteins and cofactors of numerous enzymes, including plastocyanin, cytochrome c oxidase, and superoxide dismutase (SOD) (Burkhead et al. 2009; Yruela 2009; Moreira et al. 2022a). The ability to cyclically switch from the oxidized (Cu^{2+}) to the reduced (Cu^{+}) state enables it to participate in the active processes of electron transport and redox reactions involving the oxygen ion (O_2). However, small increases in its concentration can lead to the formation of reactive oxygen species (ROS) through the Haber-Weiss or Fenton reaction, causing oxidative stress that could damage a wide range of vital cellular macromolecules (Yruela 2009; Liochev and Fridovic 2013).

Even if its concentration is within the tolerable limit in plant tissues, at the cellular level, its presence in the cytoplasm is strictly controlled as only small amounts are tolerated (Kumar et al. 2021; Moreira et al. 2022a). Thus, after crossing the plasma membrane

Heavy Metal Toxicity and Tolerance in Plants: A Biological, Omics, and Genetic Engineering Approach, First Edition. Edited by Mohammad Anwar Hossain, AKM Zakir Hossain, Sylvain Bourgerie, Masayuki Fujita, Om Parkash Dhankher, and Parvez Haris.

barrier, excess Cu ions are removed from the cytosol by an efficient delivery mechanism to the sites of utilization (mitochondria, chloroplasts, and Golgi apparatus) and Cu-dependent proteins. This Cu^{2+} distribution mechanism mainly depends on the activity of specific, soluble, low-molecular-weight cytoplasmic proteins called Cu-chaperones that constitute a fundamental strategy in preventing inappropriate interactions of the element with other cellular components (Migocka and Malas 2018; Shabbir et al. 2020).

Cu deficiency is normally observed when the concentration is below 5.0 mg kg^{-1} in vegetative tissues (Marschner 2012; Moreira et al. 2022a), which causes various symptoms, such as reduced growth rate, curling of leaf margins, and chlorosis, and other symptoms that appear particularly in younger leaves include damage to the apical meristem, inadequacy in cell wall formation and lignification of tissues, including the xylem, thus causing insufficient water transport as a side effect (Epstein and Bloom 2005; Marschner 2012). Conversely, symptoms of toxicity appear in most cultures when the foliar Cu concentration exceeds 20 mg kg^{-1} (Malavolta 2006; Marschner 2012; Moreira et al. 2022a), resulting in a rapid and excessive accumulation of ROS due to interference in the electron transport chain, both in photosynthesis and respiration. Yruela (2009) reported that increased ROS levels expose cells to oxidative stress, leading to lipid peroxidation, deterioration of biological macromolecules and membranes, ion leakage, and DNA strand cleavage (Figure 12.1).

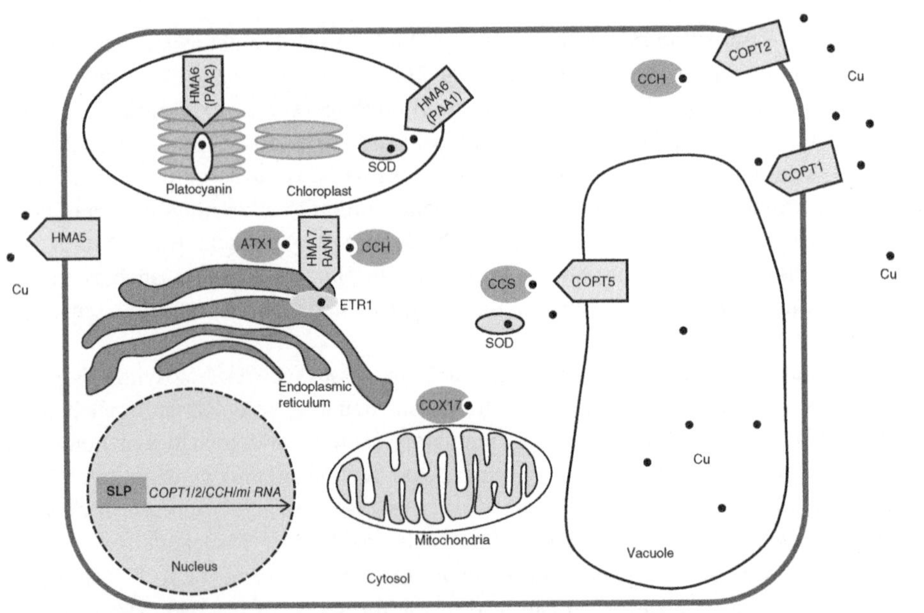

Figure 12.1 Proteins involved in the maintenance of copper (Cu) homeostasis: COPT1 and COPT2 are transporters that import Cu and deliver it to cupro-chaperone proteins (CCH, CCS, COX17, and ATX1); PAA1 and PAA2 are specific chloroplast importers that provide Cu to important proteins such as plastocyanin. RAN1 transports Cu to the endoplasmic reticulum to be loaded onto nascent proteins, such as the ethylene receptor (ETR1); COPT5 and HMA5 act on the Cu efflux through the tonoplast and plasma membrane, respectively. In the nucleus, the Cu-responsive transcriptional regulator, SPL7, activates the expression of CCH, COPT1, and COPT2, among other genes, as well as microRNAs that influence the expression of other Cu-related genes, such as that encoding the Cu/Zn superoxide dismutase (SOD) protein.

The main source of Cu for plants is the soil, where its concentration ranges from 2 to 100 mg kg^{-1} (Willard 1979; Kumar et al. 2021; Moreira et al. 2022a), and the pH, texture and type of soil, organic matter (OM), and cation exchange capacity (CEC) greatly affect the variation in the concentration and distribution of Cu^{2+} in plant tissues (Wu et al. 2010; Moreira and Moraes 2019; Santos et al. 2020). However, an increase in the concentration of Cu in the soil has been observed over the years in areas of anthropogenic activities, such as those where vine (*Vitis labrusca* L.) crops are cultivated (200–500 mg kg^{-1}), due to the excessive application of Cu-based fungicides (Ambrosini et al. 2015), and close to mining areas (700–4000 mg kg^{-1}) (Alloway 2013; Moreira et al. 2022a). Other activities that contribute to the increase in Cu in the environment include refinery, burning of fossil fuels, waste incineration, machines traffic, fertilizers, and soil corrective (limestone) (Kumar et al. 2021). Considering the sessile nature of plants, which makes it impossible to escape from unfavorable environmental conditions, and their role at the primary level of the food chain, the increase in Cu concentration may not only influence the quantity but also the quality of the food and derivatives produced (Liu et al. 2021).

Plants have developed complex defense mechanisms to cancel or mitigate the toxic effects of high Cu concentrations in plant tissues, including compartmentalization and sequestration in cellular organelles, inactivation by the formation of complexes with organic ligands and their exclusion, and through transporters, ion channels, transcription factors, and signaling molecules. Understanding these mechanisms through omics approaches, namely genomics (genes and complete DNA), transcriptomics (coding RNA and its types), proteomics (proteins), metabolomics (metabolite profile), ionomics (micro and macro ion profile), miRNAomics (noncoding RNA), and phenomics (plant phenotype), can help in the development and production of plants that are more tolerant to high concentrations of Cu in the environment (Jamla et al. 2021; Raza et al. 2022).

This chapter presents the current understanding regarding Cu toxicity in plant tissues and the tolerance mechanisms developed by plants using omics approaches, including genomics, transcriptomics, proteomics, metabolomics, miRNAomics, and metallomics.

12.2 Copper in Plants

12.2.1 Functions of Copper

Cu is a vital micronutrient for plant growth, and generally, Cu concentrations between 5 and 30 mg kg^{-1} are considered sufficient in plant tissues of different cultures (Festa and Thiele 2011; Kumar et al. 2021). Copper is absorbed as Cu^{2+} in the assimilation process and is a constituent of several important oxidases, including cytochrome oxidase, lactase, and ascorbic acid oxidase (Fageria 2009; Mydy et al. 2021). Activation of the main Cu-containing enzymes in higher plants (Table 12.1) can occur via electron transport, with or without a valence shift (Malavolta et al. 1997; Marschner 2012). It also affects the permeability of xylem vessels and participates in many physiological processes, such as photosynthesis, respiration, chlorophyll synthesis, carbohydrate distribution, nitrogen (N) reduction and fixation, protein, and cell wall metabolism (Malavolta 2006; Marschner 2012). In the cell, Cu controls DNA and RNA synthesis, and their deficiency inhibits plant

Table 12.1 Main copper-dependent enzymes.

Enzymes	Reaction	MW	Source
Ascorbate oxidase	Enediol oxidation (two electron oxidation)	200,000 4.0 Cu/mol	Zucchini (*Cucurbita pepo*) and fungi
Ribulose diphosphate carboxylase	P Ribulose + $CO_2 \Rightarrow$ 2 P glyceric acid	560,000 1.0 Cu/mol	Spinach (*Spinacia oleracea*)
Cresolase and tyrosinase	$AH_2 + 0.5O_2 \Rightarrow A + H_2O$	119,000 4.0 Cu/mol	Potato (*Solanum tuberosum*) and Mushroom
Diamino oxidase	Putrescine $\Rightarrow$ aminoaldehyde + H_2O_2	73,000 1.0 Cu/mol	Legumes and other plants
Superoxide dismutase (Cu,Zn-SOD)	$O_2^- + O_2^- + 2H^+ \Rightarrow O_2 + H_2O$	32,000 1.0 Cu/mol	Espinafre (*Spinacia oleracea*) and wheat (*Triticum* spp.)
Stelacyanin	$BH_2 + 0.5O_2 \Rightarrow B + H_2O$	27,000 1.0 Cu/mol	Japanese-lacquer (*Rhus vernicefera*)
Laccase	$BH_2 + 0.5O_2 \Rightarrow B + H_2O$	110,000 4.0 Cu/mol	Wax tree (*Toxicodendron succedaneum*) and fungi *Polyporus*
Ascorbic acid oxidase	$C_6H_8O_6 \Rightarrow$ diketogulonic	146,000 8.0 Cu/mol	Curcubitaceae
Platocyanin	Donation of e^- by chlorophyll	11,000–20,000 1.0–2.0 Cu/mol	All higher plants and other photosynthetic
Polyphenol oxidase catecholase	$RH_2 + 0.5O_2 \Rightarrow R + H_2O$	100,000 4.0 Cu/mol	Potato, nanana (*Musa* spp.), tea (*Camellia sinensis*), coffee (*Coffea* spp.), tobacco (*Nicotiana tabacum*)
Umecyanin	Oxidase by O_2 in the presence of laccase	27,000 1.0 Cu/mol	Common turnip (*Brassica rapa*)

MW, molecular weight.
Source: Malavolta (2006), Marschner (2012), and Mydy et al. (2021).

reproduction by sterilizing pollen grains and consequently reducing seed production (Malavolta 2006). Moreover, Cu participates in the electron transport chain of photosystem I and cytochrome oxidase in respiration (Malavolta et al. 1997; Fageria 2009; Moreira et al. 2019a, b); approximately 50% of all Cu is present in the form of protein complexes bound to plastocyanin in the chloroplast and participates directly or indirectly in photosynthesis (Loué 1993; Barker and Pilbean 2007). The proportion of Cu ions in relation to the number of chlorophyll molecules is 3.0–4.0 for 1000 pigment molecules; thus, Cu deficiency results in a decrease in plastocyanin concentration and photosystem I activity (Fageria 2009; Hänsch and Mendel 2009; Marschner 2012).

The physiological effects of Cu on plants have been described in detail by Migocka and Malas (2018), demonstrating the importance of Cu in maintaining the nutritional status of plants. Furthermore, Moreira et al. (2019b) found a linear increase in the photosynthetic rate and chlorophyll content with the increase in Cu rates in soybean (*Glycine max* (L.) Merrill). Dechen et al. (1991) and Fageria (2009) reported that Cu is also involved in resistance mechanisms against fungal diseases and demonstrated an inverse correlation between the degree of infestation and the adequate supply of Cu.

12.2.2 Uptake, Transport, Distribution, and Remobilization Mechanisms

Contact of nutrient with roots occurs predominantly by mass flow (Barber 1995), whereas Cu^{2+} uptake by plants has not yet been fully elucidated (Moreira et al. 2022a). Despite the active participation of Cu in various physiological processes, the amount absorbed is one of the lowest among essential elements (Loué 1993; Fageria 2009). The acquisition of Cu^{2+} by the roots is similar to the uptake mechanism of Fe^{2+}, suggesting that the uptake process involves reductive mechanisms of Cu(I) on the surface of Cu(II) root cells (Printz et al. 2016). Additionally, Cu^{2+} uptake is metabolically mediated and strongly inhibited by other bivalent transition metal ions, particularly Zn^{2+}, which competes for the same carrier site (Graham 1981; Marschner 2012). Cu uptake involves several heavy metal transporters, such as P-type ATPase, COPT-Cu, ZIP family, and NRAMP family, as described in detail by Kumar et al. (2021).

Among the macronutrients, relatively high N and phosphorus (P) applications have induced Cu deficiency in plants grown in soils with low nutrient content (Fageria et al. 2011). There is a tendency for nutrients to accumulate in the roots, and despite Cu mobility is restricted, it can move from the older leaves to the younger ones. In both the raw sap (xylem) and the elaborated sap (phloem), Cu is present in the organic and anionic forms and probably chelated by amino acids (Malavolta 2006; Marschner 2012). Although restricted, the mobility depends on the content of the nutrient present in the plant tissue; in well-nourished plants, the nutrient is easily translocated via the conducting vessels to the grains or reproductive organs, whereas under deficiency, the movement becomes more difficult. Due to the rapid response of plants to the application of Cu-based products, transport via the phloem seems to occur more easily (Malavolta 2006; Shabbir et al. 2020).

12.2.3 Deficient, Sufficient, and Toxic Levels of Copper in Plants

The symptoms of toxicity due to Cu vary greatly between plant species and tissue types (Bergman 1992; Malavolta 2006; Fageria 2009). In general, there is a reduction in the biomass of the shoots and roots because of the preferential accumulation of nutrients in most species, chlorosis of the vegetative tissue, and damage to the thylakoid membranes associated with photosystem II, in addition to reduced Fe uptake (Burkhead et al. 2009; Marschner 2012). Moreover, the darkening of the roots occurs, which leads to loss of efficiency in increasing the concentration of previously absorbed elements, indicating compromised membrane permeability. In the shoot, older leaves show symptoms of toxicity, such as appearance of necrotic spots on the edge of the leaflets, which progress to younger leaves (Malavolta 2006; Moreira et al. 2022a). Cu deficiency causes reduced plant growth; the youngest leaves turn "gray-green" or "bluish-green" in color due to shortening of the internodes as a result of reduced growth. Reduced elongation of the outermost layers of the roots results in the binding of Cu^{2+} to the cell walls, thus altering the biosynthesis of phytohormones involved in the plant development (Kopittke et al. 2009). Excess of limestone due to the presence of Ca^{2+} and Mg^{2+} in the composition decreases Cu^{2+}uptake by plants (Moreira et al. 2000; Santos et al. 2020).

The deficiency, tolerance, and toxicity of a given element "M" depend on the intensity of exposure, duration, and sensitivity of plants to the biotic or abiotic stress, in addition to other climatic factors, such as excess or lack of water and low or high temperatures (Fageria 1998; Kidokoro et al. 2015; Moreira et al. 2022a). The ability of plants to efficiently growing the soil with low or high concentrations of nutrients is mainly due to the selection process, which made them responsive or unresponsive to low, medium, or high concentrations in the soil (Baligar et al. 2001). This occurs when the content of a particular nutrient is below or above the range defined as optimal for use by plants throughout the growth or fruiting cycle (Malavolta 1992; Reuter and Robinson 1997).

12.2.4 Copper Sources: Fertilizers and Fungicides

In agriculture, several fertilizers containing Cu have low, medium, or high solubility in water. The soluble fertilizers are applied by broadcast in the sowing or foliar furrow during the entire vegetative cycle of the plants, whereas the insoluble fertilizers, such as fritted trace elements, must be preferably applied in sowing furrows or broadcasts with restricted use for annual or short-cycle crops (Table 12.2). The amount of Cu applied may vary depending on the type and content of soil organic matter (SOM), species or cultivar, quantity exported (grains, fruits, or forage), and climatic conditions (Fageria et al. 2011; Moreira et al. 2022a). In Europe, the Cu input and output also depend on the type of management adopted. The sources responsible for increasing the concentration of Cu in agricultural soils are fertilizers, biocides, and pesticides. In the case of fertilizers, Cu can be present in manure (mainly poultry, swine, and cattle), mineral fertilizers, sewage sludge and others, as compost, with manure accounting for about 80% of the total Cu input as fertilizers, while sewage sludge and fertilizers minerals account for 18% and 3%, respectively (Heijerick et al. 2006).

In addition, some other factors must be taken into account, such as the presence or absence of negative interactions with other nutrients and the molecular form to be applied, considering that the uptake efficiency in relation to the accompanying ion follows the sequence $Cl^- > NO_3^- >$ chelated $> SO_4^{2-}$ in fertilizers (Malavolta 2006; Moreira et al. 2022a). Moreira et al. (2019b) verified the importance of Cu application in soybean and wheat (*Triticum aestivum* L.) crops in succession by conducting a study in two harvests and found a significant increase in grain yield when Cu application rates were increased up to 4.0 kg ha^{-1}. Moreover, Moreira et al. (2022b) studied the positive relationship between leaf and grains content and grain yield of soybean cultivated in two relevant biomes (Figure 12.2).

Fungicides are classified according to their spectrum of action, which can be site-specific (SS), characterized by having an effect against a specific enzyme or protein necessary for the development of the fungus, or multisite (MS) spectrum, which acts at different metabolic points of the fungus and presents a low risk of resistance, thus

Table 12.2 Main sources of copper used in agriculture.

Source	Formula	Cu (%)[a]	Form
Copper sulfate pentahydrate – soluble	$CuSO_4{\cdot}5H_2O$	25.0	Crystalline
Copper sulfate monohydrate – soluble	$CuSO_4{\cdot}H_2O$	35.0	Crystalline
Bordeaux mixture	$CuSO_4{\cdot}3Cu(OH)_2 + 3CaSO_4$	12.5	Suspension
Cuprous oxide – insoluble	Cu_2O	89.0	Powder
Cupric oxide – insoluble	CuO	75.0	Powder
Copperoxychloride – soluble	$[Cu_2Cl(OH)_3]$	50.0	Powder
Copperchloride – soluble	$CuCl_2$	17.0	Powder
Slag – insoluble	CuO, silicates	2–5	Powder
Fritted trace elements (FTEs) – insoluble	Coppersilicates	45.0	Powder
Chelates (a) – soluble	Na_2Cu, EDTA	13.0	Powder
Chelates (b) – soluble	NaCu, HEDTA	10–13	Powder
Copperlignosulfonate – soluble	—	4.0	Liquid
Copperpoliflavonoid – soluble	—	6.0	Liquid
Cattlemanure – insoluble	—	0.1	Solid
Chicken dung – insoluble	—	0.2–0.3	Solid
Porkmanure – soluble	—	0.2–0.3	Liquid
Sewagesludge – insoluble	—	0.5–4.0	Solid

[a] Approximate value.
Source: Moreira et al. (2022a)/with permission of Elsevier.

being important in the management of the culture in alternation with the SS to prevent the fungus from acquiring resistance to the active principle (McGrath 2004). Cu-based fungicides fit the MS spectrum, and water-soluble fungicides are most commonly used in agriculture, such as copper oxychloride $[Cu_2Cl(OH)_3]$, copper sulfate ($CuSO_4$), and cuprous oxide (Cu_2O), which are applied in the form of a wettable powder (Godoy et al. 2019). These authors verified in soybean crops that, regardless of the solubility type, Cu-based fungicides showed an intermediate spectrum of action in the control of the pathogen (*Phakopsora pachyrhizi*), thus being inferior to SS fungicides and superior to the control.

The most abundant forms of soluble Cu (Cu^{2+} and chelated) can penetrate the apoplasm of the roots, and previous studies have reported the occurrence of specific mechanisms for the movement of the metal through the plasma membrane (Ryan et al. 2013). Two strategies have been devised regarding this movement: strategy I involves reduction of Cu^{2+} to Cu^{+} in the apoplasm by the action of Fe reductases, FRO4 and FRO5 (Bernal et al. 2012), and after reduction, the Cu transporter protein COPT1 facilitates the passage of the metal through the plasma membrane inside the plant cell; strategy II does not depend on redox reactions and involves the complexation of the micronutrient in the oxidized form with phytosiderophores, in which the Cu^{2+}-phytosiderophore complex uptake

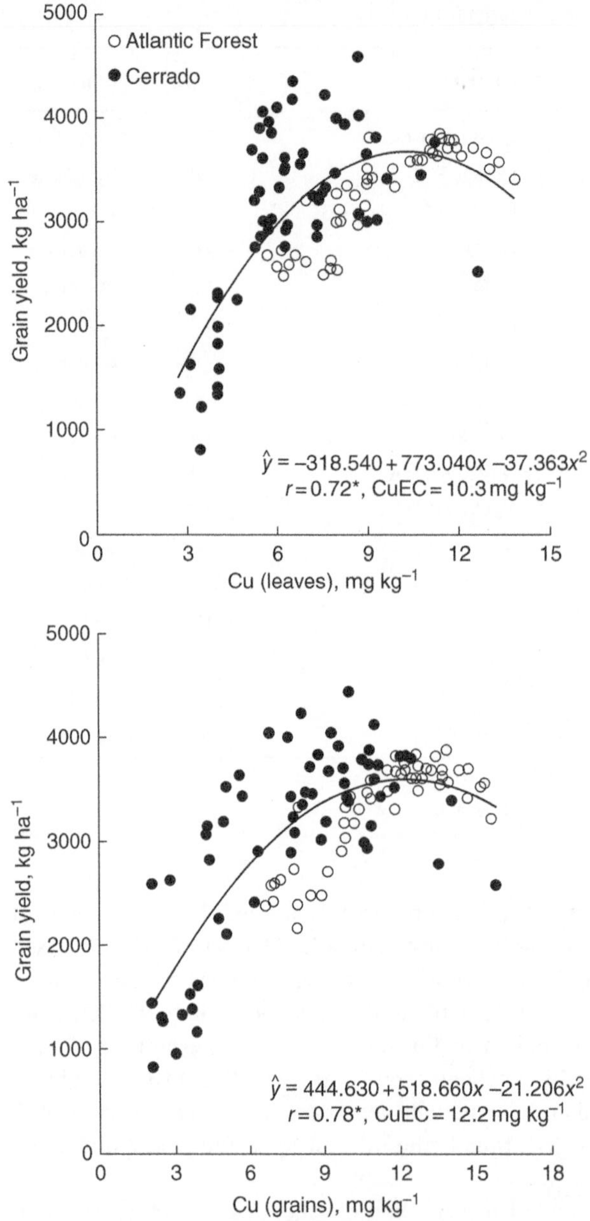

Figure 12.2 Increase in soybean yield grown in two biomes as a function of foliar Cu content. *Source:* From Moreira et al. (2022b)/Taylor & Francis.

occurs through the plasma membrane with the aid of the membrane transporter protein, yellow-stripe-like (YSL), without inducing isotopic fractionation. After crossing the plasma membrane barrier, excess Cu ions are removed from the cytosol, where only small amounts are tolerated, by an efficient delivery mechanism to the sites of use and Cu-dependent proteins (Ryan et al. 2013).

12.3 Omics Approaches for Cu Responses and Tolerance in Plants

12.3.1 Genomics

Genomics involves sequencing of the genomes of species to understand their structure, organization, and function and is divided into structural, functional, and comparative genomics (He et al. 2007; Akpınar et al. 2013). Structural genomics involves the study of the organization and structure of genes, and analysis of transcribed and structural sequences, while comparative genomic seeks to understand the relationships between genomes and the homology between sequences and genes. Functional genomics involves the study of gene function to understand changes in the functioning of the genome at different stages of development and under different conditions (Akpınar et al. 2013).

Functional genomic studies have been performed to identify genes that are directly or indirectly related to the responses of plants exposed to toxic Cu levels. Thus, to identify groups of genes related to excess Cu tolerance, tests were performed with the replacement of chromosomal lines in wheat and positive effects of chromosomes 7A and 1D were observed on shoot and root growth of seedlings cultivated under excess Cu, whereas chromosomes 3A, 3B, 3D, 4A, and 4D mainly affected the roots, and chromosomes 5BL, 7BL, 1B, and 6D affected the shoot; only chromosome 1A had a negative effect (Ganeva et al. 2003). Using the same tests, Bálint et al. (2003) reported a positive effect of chromosomes 3D, 5A, 5B, 5D, 6B, and 7D on Cu^{2+} stress tolerance. These results indicate the polygenic character of Cu tolerance and the possibility of different gene expression patterns in response to different toxic Cu levels in different populations (Bálint et al. 2007).

Quantitative trait loci (QTLs) for Cu tolerance were determined in wheat grown in the substrate containing adequate and excessive amount of Cu under controlled conditions. A major QTL for Cu tolerance was found on chromosome 5DL, while minor effects of chromosomes 1AL, 2DS, 4AL, 5BL, and 7DS were determined. A QTL controlling Cu content in shoots was found on chromosome 1BL, and on chromosome 5AL, a QTL was observed that influences the ability of the plant to accumulate Cu. A significant negative correlation was observed between the tolerance and the ability to accumulate Cu in the shoots of wheat ($r = -0.31$, $p \leq 0.05$), which suggested that the main mechanisms to increase Cu tolerance in plants would be the restriction of Cu uptake in the roots or the reduced translocation from the roots to the shoots (Bálint et al. 2007); this can be achieved by increasing the soil pH (Malavolta 2006) or by increasing the SOM content (Moreira et al. 2022a).

12.3.2 Transcriptomics

While genomics reveals the presence or absence of target genes in an organism, the functional role of these genes can be elucidated through transcriptomic techniques. The transcriptomic profile enhances our understanding of the plant response under a particular physiological state or stress condition, as the transcriptome varies depending on the cultivar, tissue, developmental stage, and environmental condition (Mehta et al. 2019).

Transcriptomic studies have identified and quantified RNA in different tissues under different physiological conditions. The most used techniques in transcriptomic studies are as follows: reverse transcriptase quantitative polymerase chain reaction (RT–qPCR), which evaluates gene expression in a punctual way (gene by gene); qPCR array, which uses RT–qPCR to assess changes in the expression of tens to hundreds of genes; microarrays, which use nucleic acid hybridization to assess gene expression; and RNA-Seq, which employs sequencing to quantify transcripts (Jamla et al. 2021).

All these techniques (RT–qPCR, qPCR arrays, microarrays, and RNA-Seq) present the results in terms of *fold-change*, indicating how many times the RNA concentration increases or decreases. The resulting data can be used in comparative studies; for example, to analyze the changes in gene expression resulting from the increase in Cu concentration in the medium (Mehta et al. 2019). Gene expression in response to Cu stress in rice leaves (*Oryza sativa* L.) was quantified using DNA microarray and real-time PCR analyses. Microarray analysis revealed 305 Cu-responsive genes, and their expression profiles showed that 146 general stress response and defense genes were upregulated under Cu excess conditions, with most of the defense genes being involved in the phenylpropanoid pathway for the biosynthesis of flavonoids, phytoalexins, and lignin, whereas 159 genes related to photosynthesis and transport were downregulated. A study on vine leaves (*Vitis* sp.) showed a similar pattern to that reported for rice leaves by Sudo et al. (2008) in the differential expression of genes under control and Cu stress condition, confirming that the genes related to photosynthesis are negatively regulated under excess Cu, such as those involved in the chlorophyll synthesis. Upregulation of many genes related to stress tolerance was also observed, including antioxidant defense system and secondary metabolites and those associated with anthocyanin and lignin biosynthesis. Furthermore, genes related to the synthesis of hormones involved in plant stress response, such as ethylene and jasmonic acid, were also found to be upregulated (Jiang et al. 2015).

The upregulation of genes that are usually activated by pathogen attack, such as genes encoding pathogenesis-related (PR) proteins (Jiang et al. 2015), is expected, considering that Cu is an abiotic elicitor that induces resistance to pathogen attack (Graham 1980). Thus, defense-related genes may act not only to inhibit pathogens but also to tolerate Cu stress (Sudo et al. 2008).

Genes encoding Cu transport proteins show different patterns of Cu stress regulation. The mRNA level of COPT1, a transporter associated with the uptake of available Cu in the soil, is negatively regulated (Sancenón et al. 2003), as is the transcription of plastid-associated transporters PAA1 and PAA2 (Schiavon et al. 2007). In contrast, increased expression of HMA5 and HMA9, which are P1B-ATPases-type transporters involved in Cu detoxification by increased efflux or accumulation in the vacuole, was observed (Andrés-Colás et al. 2006, 2013; Lee et al. 2007).

The chelation of Cu ions by Cu-chaperone proteins, with a specific delivery function for transport proteins, plays a fundamental role in transport across cell membranes and is necessary not only for Cu uptake but also in detoxification processes and regulates its expression (Yruela 2009), with AtCOX17 expression being upregulated and that of AtATX1 being downregulated in response to excess Cu supply (Schiavon et al. 2007). Thus, these transporters and Cu-chaperones may be involved in Cu tolerance mechanism, contributing to homeostasis, and minimizing damage to plant tissues (Figure 12.3). These results indicate that several classes of genes may have a specific role under Cu stress and help to understand the complex molecular mechanisms involved in cellular responses to stress.

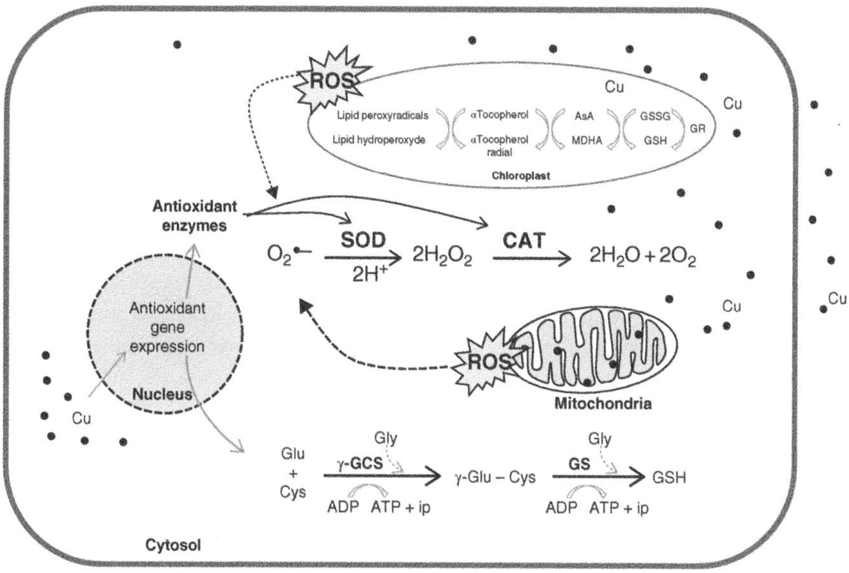

Figure 12.3 Induction of oxidative stress by excess copper (Cu) in the plant cell with generation of reactive oxygen species (ROS) such as superoxide radicals ($O_2^{\bullet-}$) and hydrogen peroxide (H_2O_2) in the main organelles (mitochondria and chloroplast). Detoxification mechanism by the enzymes superoxide dismutase (SOD) and catalase (CAT) and nonenzymatic compounds such as reduced glutathione (GSH), synthesized by the junction of three amino acids – glutamine (Glu), cysteine (Cys), and glycine (Gly) – by the action of the enzymes γ-glutamylcysteine synthetase (γ-GCS) and glutathione synthetase (GS). Detoxification of chloroplast membrane lipid peroxidation by the action of α-tocopherol, ascorbic acid (AsA), monodehydroascorbate (MDHA), oxidized glutathione (GSSG), reduced glutathione (GSH), and glutathione reductase (GR).

12.3.3 Proteomics

Although genomic and transcriptomic analyses help understand plant responses to stress by elucidating changes in gene expression, these changes are not always reflected at the level of protein synthesis (Hossain and Komatsu 2013). Thus, proteomics has emerged as a vital tool for assessing the profile of proteins at the organ, tissue, cell, and organelle levels under various stress conditions, including heavy metal stress, such as Cu (Ahsan et al. 2009; Raza et al. 2022). It contributes to the understanding of the processes that are triggered in these conditions because proteins are functionally versatile macromolecules that act in cell signaling, regulation, catalysis, intra-and intercellular movement of nutrients and other molecules, membrane fusion, structural support, and protection (Amm et al. 2014).

Chen et al. (2015) evaluated the root protein profile of two rice cultivars that were tolerant and sensitive to excess Cu (8.0 μM Cu^{2+} for three days) and found that Cu-responsive proteins included those involved in antioxidant defense processes and detoxification, pathogenesis, regulation of gene transcription, amino acid synthesis, protein synthesis, modification, transport, and degradation, cell wall synthesis, molecular signaling, and salt stress.

The increase in the number of proteins related to antioxidant defense and detoxification was more pronounced in the tolerant cultivar than in the sensitive cultivar, suggesting that these proteins, together with the related physiological processes, play important roles in the detoxification of excess Cu and maintenance of cellular homeostasis (Yruela 2009; Chen et al. 2015).

Compounds involved in antioxidant defense protect the cell from damage caused by highly reactive species, mainly ROS, such as hydrogen peroxide (H_2O_2), singlet oxygen ($^1O_2$), superoxide anion ($O_2^{\bullet-}$), and hydroxyl radicals (•OH), which can interact with macromolecules essential for cell function (DNA, pigments, proteins, lipids, and others) and are abundantly produced in plant tissues under excess Cu^{2+} (Yruela 2009). Enzymes such as SOD, catalase (CAT), monodehydroascorbate reductase (MDHAR), glutathione reductase (GR), glutathione-*S*-transferase (GST), and ascorbate peroxidase (APX) have antioxidant activity and are responsible for removing and neutralizing ROS and increasing their expression under conditions of excess Cu (Nanda and Agrawal 2018; Gong et al. 2019; Nazir et al. 2019; Huang et al. 2020). Additionally, nonenzymatic antioxidants (ascorbate, glutathione, proline, and alpha-tocopherol) form a coordinated network (Figure 12.4), which confers greater tolerance to biotic and abiotic stresses (Hossain et al. 2012; Singh et al. 2016).

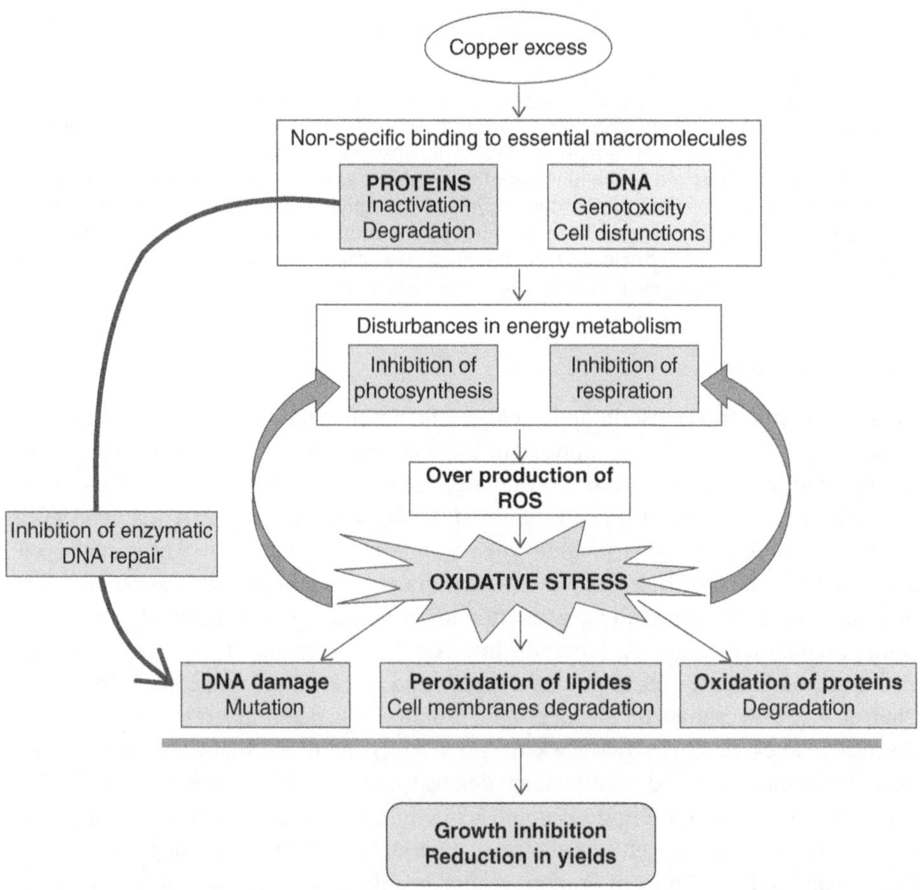

Figure 12.4 Main aspects of the physiological and biochemical mechanisms of copper toxicity in plants.

In wheat (*Triticum durum* L.) seedlings exposed to excess Cu (100 μM $CuSO_4$ for three days), 93 proteins that showed reduced or increased abundance in response to stress were identified in the roots and leaves. An increase was reported for proteins that play a role in signal transduction, defense against stress, energy production, and transport, whereas those associated with carbohydrate metabolism, protein metabolism, and photosynthesis were severely reduced (Li et al. 2013).

Studies on different species have reported a reduction in proteins associated with carbohydrate metabolism (Li et al. 2019a), photosynthetic rate, and particularly, those involved in electron transport chain, CO_2 assimilation, and chlorophyll biosynthesis (Vassilev et al. 2003; Gonzalez-Mendoza et al. 2013; Huang et al. 2020; Ju et al. 2021), protein synthesis, and total amount of proteins (Llorens et al. 2000; Liu et al. 2014) in plants under stress by ascorbic acid. These factors are directly related to the reduction in plant growth observed under phytotoxic conditions (Santos et al. 2020; Moreira et al. 2022a).

In contrast, an increase in the abundance of proteins involved in carbohydrate metabolism has been observed in *Citrus grandis* (Huang et al. 2020) and *Oenothera glazioviana* (Wang et al. 2017). These increases were related with tolerance mechanisms that promote energy availability and maintain metabolic pathways, such as glycolysis, which guarantee the oxidation of substrates that lead to ATP synthesis.

The results of protein profile analysis of plants under Cu stress agreed with the results of the transcriptomic profiles that showed an increase in the expression of proteins involved in decreasing oxidative stress and a decrease in the expression of proteins involved in photosynthesis and chlorophyll biosynthesis (Migocka and Malas 2018; Shabbir et al. 2020).

12.3.4 Metabolomics

Metabolomics involves the simultaneous detection, identification, characterization, and quantification of low-molecular-weight compounds (molecular mass of less than 2000 Da) and the most varied classes of both primary (amino acids, lipids, and sugars) and secondary (flavonoids, terpenoids, and glucosinolates) metabolites, thus providing information on the total active and/or present metabolites in an organism, organ, tissue, or cell at a time under specific conditions (Mehta et al. 2019).

Pumpkin (*Cucumis sativus* L.) leaves treated with increasing concentrations of $CuSO_4$ (0, 10, 100, and 500 $mg\,l^{-1}$) for seven days showed variation in metabolites compared to the control. At the lowest rates (10 and 100 $mg\,l^{-1}$), reductions were observed in the amount of organic acids (citric, malic, and aconitic acid) and components of the tricarboxylic acid (TCA) cycle, which is one of the main pathways of energy metabolism. These acids, in addition to participating in energy generation, are precursors of essential amino acids that are involved in protein synthesis. The amount of the aromatic amino acids, phenylalanine, and tyrosine that are precursors of the shikimate–phenylpropanoid pathway responsible for generating antioxidants were increased (Zhao et al. 2018).

The amounts of unsaturated fatty acids were increased and those of sterols were decreased, indicating regulation to maintain fluidity and adequate functioning of the plasma membrane, which is one of the first structures to be affected by Cu toxicity. These results suggest that doses slightly above the appropriate dose can cause a reduction in energy generation and activate antioxidant defense responses. The highest rate of Cu resulted in a twofold increase in phytol, which is formed due to the breakdown of chlorophyll. Phenolic compounds, such as flavonoids, were downregulated, while the non-flavonoids, 4-hydroxycinnamic acid and trans-2-hydroxycinnamic acid, were upregulated 4- and 26-fold, respectively, compared to the control (Zhao et al. 2018). Jung et al. (2015) reported similar results showing increased levels of phenols and unsaturated fatty acids and reduced levels of primary metabolites (sugars, amino acids, and organic acids) and sterols in copper-treated burdock (*Arctium lappa* L.) roots. These results suggest that Cu stress leads to activation of the phenylpropanoid pathway and growth inhibition.

Few studies have analyzed the total metabolites under excess Cu (Gonzalez-Mendoza et al. 2013); however, in the literature, a series of compounds of primary and secondary metabolism associated with stress can be found, especially those with chelating and/or antioxidant action, which participate in the tolerance mechanism, as shown below.

The synthesis of phytochelatins (PCs) from glutathione (GSH) can be induced by excess Cu, resulting in the formation of a stable compound with the metal that implies collaboration with Cu detoxification and homeostasis (Gonzalez-Mendoza et al. 2013; Navarrete et al. 2019). In addition to acting as a precursor of metal-binding peptides (PCs), GSH acts as an antioxidant in association with ascorbic acid (AsA) in a cycle that eliminates free oxygen radicals formed in large quantities under excess Cu, with the aid of enzymes such as monodehydroascorbate reductase (MDHAR) and glutathione reductase (GR) (Mellidou et al. 2012; Sharma et al. 2019). An increase in the amount of AsA and GSH have been reported in both roots and shoots of common bean (*Phaseolus vulgaris* L.) plants when exposed to high Cu concentration (Younis et al. 2018).

Salicylic acid (SA) production in plants can be induced in response to Cu toxicity to prevent oxidative damage (Li et al. 2019b). In rice seedlings (*O. sativa* L.) cultivated under excess of Cu, SA application attenuated oxidative stress, increased proline content, and increased the activity of enzymes related to defense response (Mostofa and Fujita 2013). Proline is an antioxidant with osmoregulatory action that helps maintain the integrity of cellular structures, such as membranes, which are damaged by oxidative stress induced by Cu toxicity (Nazir et al. 2019).

12.3.5 miRNAomics

Alteration of gene expression at the transcriptional level is a well-documented response mechanism in plants under biotic and abiotic stresses, including excess metals/metalloids, such as Cu (Sudo et al. 2008; Jiang et al. 2015; Xiangpeng et al. 2015; Jamla et al. 2021). Recently, posttranscriptional regulation has been studied with the recognition of the key role of small RNAs, with microRNAs (miRNAs) being the best-characterized subgroup (Rogers and Chen 2013). miRNAs are defined as small sequences (approximately 21 nucleotides long) of noncoding endogenous RNA that participate in the regulation of posttranscriptional gene expression. They are complementary to the target coding RNA to which they bind and act as a translational barrier or can even cause their cleavage. Thus, miRNAs modulate protein expression in plant tissues (Khraiwesh et al. 2012).

In plants, miRNAs are not only involved in stress responses but also participate in fundamental processes such as development, metabolism, and phytohormonal regulation. miRNA precursors, usually transcribed by RNA polymerase II or III, have a stem-loop structure that is excised and methylated by DICER-LIKE1 (DCL1) and HUA ENHANCER1 (HEN1) proteins, respectively. Mature miRNAs are exported from the nucleus to the cytosol and incorporated into an Argonaute protein, forming an RNA-induced silencing complex (RISC) (Voinnet 2009; Rogers and Chen 2013).

Many miRNAs responsive to metal/metalloids, such as cadmium (Cd), mercury (Hg), and aluminum (Al), have been identified in several species, including rice (Huang et al. 2009), corn (*Zea mays* L.) (Gao et al. 2019), and canola (*Brassica napus* L.) (Yu et al. 2019). In general, the targets of miRNAs in plants under heavy metal stress are genes that encode transcription factors and proteins related to metabolic processes or stress responses associated with the uptake and complexation of metals, oxidative stress, the antioxidant system, and signal transduction (Figure 12.5), thus maintaining homeostasis and exhibiting tolerance under excess or deficiency of heavy metals (Ding et al. 2020; Dubey et al. 2021).

The posttranscriptional regulation of the amount of cuproproteins in plants under Cu deficiency directed by miRNAs leads to an increase in the expression of miR398, which promotes the reduction of the expression of the enzyme Cu/ZnSOD, thus releasing limited Cu for plastocyanin and prioritizing photosynthesis (Yamasaki et al. 2009). In contrast, Sunkar et al. (2006) observed the downregulation of miR398 and increased expression of Cu/ZnSOD

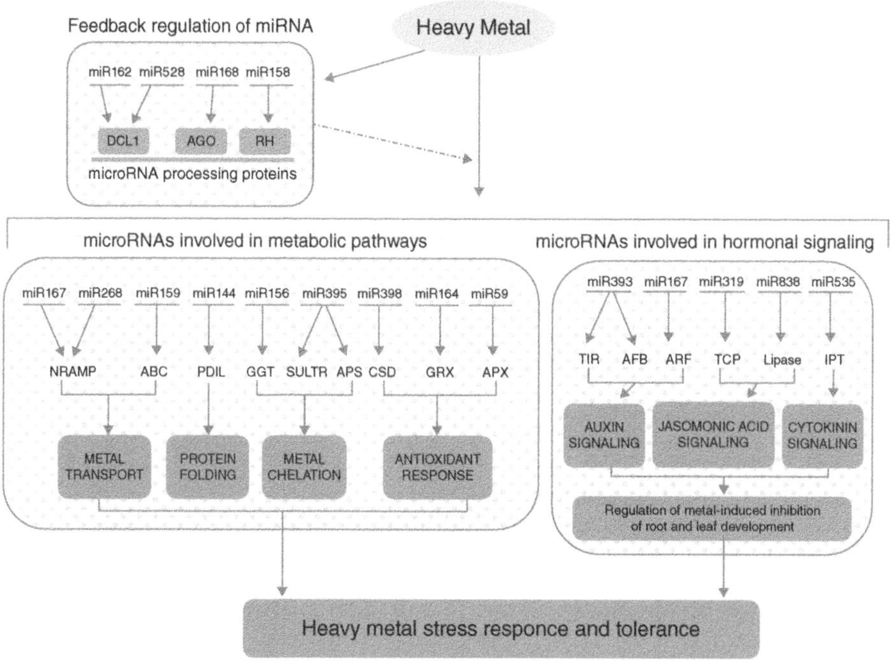

Figure 12.5 MiRNAs and target genes involved in plant responses to excess heavy metal stress. ABC, ATP-binding cassette; AFB, auxin-signaling F-box protein; AGO, ARGONAUTE; APS, ATP sulfurylase; APX, ascorbate peroxidase; ARF, auxin response factor; CSD, Cu/Zn superoxide dismutase; DCL, dicer-like protein; GGT, glutathione-γ-glutamylcysteinyl transferase; GRX, glutaredoxin; IPT, isopentyl transferase; NRAMP, natural resistance associated macrophage protein; PDIL, protein disulfide isomerase-like protein; RH, RNA helicase; SULTR, sulfate transporter; TCP, TEOSINTE BRANCHED 1-CYCLOIDEA-PCF; TIR, transport inhibitor response. *Source:* Adapted from Ding et al. (2020).

under excess Cu. The transcription factor squamosa promoter binding protein-like7 (SPL7) has been reported as a putative sensor of Cu availability in plant cells and an inducer of miR398 expression during Cu deficiency (Gielen et al. 2016). SPL7 recognizes and binds to the GTAC sequence in the miR398 promoter, thereby inducing its expression.

Other Cu-responsive miRNAs, whose expression is also regulated by SPL7, are miR397, miR408, and miR857; their target genes are members of the laccase family that encode enzymes that play a key role in the oxidation of monolignols during lignin biosynthesis (Schuetz et al. 2014). These miRNAs, as well as miR398 and SPL7, are highly conserved and present in several species, such as soybean (Song et al. 2011), peanut (*Arachis hypogea* L.) (Zhao et al. 2018), *Arabidopsis thaliana* (Gielen et al. 2016), and grapevine (*Vitis vinifera* L.) (Jiu et al. 2019).

Recent studies have revealed several new miRNAs associated with plant responses to excess Cu stress. In grapevine, Jiu et al. (2019) observed 158 known and 98 new miRNAs, of which 100 known and 47 new miRNAs were identified as differentially expressed under excess Cu stress and control conditions that regulate transcription factors (AP2, SBP, NAC, MYB, and ARF) involved in the uptake, transport, and chelation of metals; signaling; and development of roots and shoots. The research on the origin, synthesis, and mode of regulatory action of miRNAs involved in excess Cu stress responses indicates that its manipulation can generate effective tools for the development of stress-tolerant plant varieties.

12.4 Concluding Remarks

Due to the increasing industrialization, pollution of the biosphere by heavy metals/metalloids, such as Cu, is becoming a growing environmental problem (Kosakivska et al. 2020). In plant cultivation, the inappropriate use of fertilizers and fungicides with high concentrations of Cu, despite their essentiality (Marschner 2012; Ambrosini et al. 2015), disrupts fundamental physiological and biochemical processes that influence metabolism by altering shoot growth and root system development, and even hampers survival when applied at high concentrations (Shabbir et al. 2020; Kumar et al. 2021).

In this context, omics technology has proven to be a tool with great potential and has helped in the functional characterization of the genome, coding, and noncoding RNAs, and metabolites, which has promoted the elucidation of the possible mechanisms developed by plants to deal with excess Cu stress (Jamla et al. 2021). However, further use of omics technology is required to broaden the knowledge about these mechanisms to guide the search for potential responsive targets, leading to the development of stress-tolerant plants.

Acknowledgments

The National Council for Scientific and Technological Development (CNPq), Brazilian Agricultural Research Corporation (Embrapa), and Coordination of Superior Level Staff Improvement (CAPES) for financial support.

Orcid

Adônis Moreira: 0000-0003-4023-5990

Larissa Alexandra Cardoso Moraes: 0000-0001-7296-2936

Jorge João Delfim: 000-0002-3366-2698

Luiz Gustavo Moreti: 0000-0001-7693-7826

References

Ahsan, N., Renaut, J., and Komatsu, S. (2009). Recent developments in the application of proteomics to the analysis of plant responses to heavy metal. *Proteomics* 9: 2602–2621. https://doi.org/10.1002/pmic.200800935.

Akpınar, B.A., Lucas, S.J., and Budak, H. (2013). Genomics approaches for crop improvement against abiotic stress. *Sci. World J.* 2013: 361921. https://doi.org/10.1155/2013/361921.

Alloway, B.J. (2013). Sources of heavy metals and metalloids in soils. In: *Heavy Metals in Soils Environmental Pollution* (ed. E. Steinnes), 11–50. London: Springer.

Ambrosini, V.G., Rosa, D.J., Prado, J.P.C. et al. (2015). Reduction of copper phytotoxicity by liming: a study of the root anatomy of young vines (*Vitis labrusca* L.). *Plant Physiol. Biochem.* 96: 270–280. https://doi.org/10.1016/j.plaphy.2015.08.012.

Amm, I., Sommer, T., and Wolf, D.H. (2014). Protein quality control and elimination of protein waste: the role of the ubiquitin–proteasome system. *Biochim. Biophys. AcaBBA Molecul. Cell Res.* 1843: 182–196. https://doi.org/10.1016/j.bbamcr.2013.06.031.

Andrés-Colás, N., Perea-García, A., Andrés, S.M. et al. (2013). Comparison of global responses to mild deficiency and excess copper levels in *Arabidopsis* seedlings. *Metallomics* 5: 1234–1246. https://doi.org/10.1039/c3mt00025g.

Andrés-Colás, N., Sancenón, V., Rodríguez-Navarro, S. et al. (2006). The *Arabidopsis* heavy metal P-type ATPasa HMA5 interacts with metallochaperones and functions in copper detoxification of roots. *Plant J.* 45: 225–236. https://doi.org/10.1111/j.1365-313X.2005.02601.x.

Baligar, V.C., Fageria, N.K., and He, Z.L. (2001). Nutrient use efficiency in plants. *Commun. Soil Sci. Plant Anal.* 32 (7–8): 921–950. https://doi.org/10.1081/CSS-100104098.

Bálint, A.F., Kovács, G., Börner, A. et al. (2003). Substitution analysis of seedling stage copper tolerance in wheat. *Acta Agron. Hung.* 51: 397–404. https://doi.org/10.1556/AAgr.51.2003.4.4.

Bálint, A.F., Röder, M.S., Hell, R. et al. (2007). Mapping of QTLs affecting copper tolerance and the Cu, Fe, Mn, and Zn contents in the shoots of wheat seedlings. *Biol. Plant* 51 (1): 129–134. https://doi.org/10.1007/s10535-007-0025-9.

Barber, S.A. (1995). *Soil Nutrient Bioavailability: A Mechanistic Approach*. New York: Wiley.

Barker, A.V. and Pilbean, D.J. (2007). *Handbook of Plant Nutrition*. Boca Raton: CRC Press.

Bergman, W. (1992). *Nutritional Disorders of Plants*. New York: Gustav Fischer Verlag Jena.

Bernal, M., Casero, D., Singh, V. et al. (2012). Transcriptome sequencing identifies *SPL7*-regulated copper acquisition genes *FRO4*/*FRO5* and the copper dependence of iron homeostasis in *Arabidopsis*. *Plant Cell* 24: 738–761. https://doi.org/10.1105/tpc.111.090431.

Burkhead, J.L., Reynolds, K.A.G., Abdel-Ghany, S.E. et al. (2009). Copper homeostasis. *New Phytol.* 182: 799–816. https://doi.org/10.1111/j.1469-8137.2009.02846.x.

Chen, C., Song, Y., Zhuang, K. et al. (2015). Proteomic analysis of copper-binding proteins in excess copper-stressed roots of two rice (*Oryza sativa* L.) varieties with different cu tolerances. *PLoS ONE* 10 (4): e0125367. https://doi.org/10.1371/journal.pone.0125367.

Dechen, A.R., Haag, H.P., and Carmello, Q.A.C. (1991). Functions of micronutrients in plants. In: *Micronutrients in Agriculture* (ed. M.E. Ferreira and M.C.P. Cruz), 65–78. Piracicaba, Brazil: Potafos.

Ding, Y., Ding, L., Xia, Y. et al. (2020). Emerging roles of microRNAs in plant heavy metal tolerance and homeostasis. *J. Agric. Food Chem.* 68: 1958–1965. https://doi.org/10.1021/acs.jafc.9b07468.

Dubey, S., Shri, M., and Chakrabarty, D. (2021). MicroRNA mediated regulation of gene expression in response to heavy metals in plants. *J. Plant Biochem. Biotechnol.* 30: 744–755. https://doi.org/10.1007/s13562-021-00718-5.

Epstein, E. and Bloom, A.J. (2005). *Mineral Nutrition of Plants: Principles and Perspectives.* Sunderland, MA: Sinauer Associates Inc.

Fageria, N.K. (1998). Optimizing nutrient use efficiency in crop production. *Rev. Bras. Eng. Agric. Ambient.* 2 (1): 6–16. https://doi.org/10.1590/1807-1929/agriambi.v02n01p6-16.

Fageria, N.K. (2009). *The Use of Nutrients in Crop Plants.* Boca Raton: CRC Press.

Fageria, N.K., Baligar, V.C., and Jones, C.A. (2011). *Growth and Mineral Nutrition of Field Crops.* Boca Raton: CRC Press.

Festa, R.A. and Thiele, D.J. (2011). Copper: an essential metal in biology. *Curr. Biol.* 21: R877–R883. https://doi.org/10.1016/j.cub.2011.09.040.

Ganeva, G., Landjeva, S., and Merakchijska, S. (2003). Effects of chromosome substitutions on copper toxicity tolerance in wheat seedlings. *Biol. Plant.* 47 (4): 621–623. https://doi.org/10.1023/B:BIOP.0000041075.87242.eb.

Gao, J., Luo, M., Peng, H. et al. (2019). Characterization of cadmium-responsive MicroRNAs and their target genes in maize (*Zea mays*) roots. *BMC Mol. Biol.* 20 (1): 1–9. https://doi.org/10.1186/s12867-019-0131-1.

Gielen, H., Remans, T., Vangronsveld, J., and Cuypers, A. (2016). Toxicity responses of Cu and Cd: the involvement of miRNAs and the transcription factor SPL7. *BMC Plant Biol.* 16: 145. https://doi.org/10.1186/s12870-016-0830-4.

Godoy, C.V., Utiamada, C.M., Meyer, M.C. et al. (2019). *Eficiência de fungicidas multissítios no controle da ferrugem-asiática da soja, Phakopsora pachyrhizi, na safra 2018/19: resultados sumarizados dos experimentos cooperativos.* Londrina, Brazil: Embrapa Soja.

Gong, Q., Wang, L., Dai, T. et al. (2019). Effects of copper on the growth, antioxidant enzymes and photosynthesis of spinach seedlings. *Ecotoxicol. Environ. Saf.* 171: 771e780. https://doi.org/10.1016/j.ecoenv.2019.01.016.

Gonzalez-Mendoza, D., Espadas y Gil, F., Escoboza-Garcia, F. et al. (2013). Copper stress on photosynthesis of black mangle (*Avicennia germinans*). *An Acad. Bras. Ciênc.* 85 (2): 665e670. https://doi.org/10.1590/S0001-37652013000200013.

Graham, R.D. (1980). Susceptibility to powdery mildew of wheat plants deficient in copper. *Plant Soil* 56: 81–185. https://doi.org/10.1007/BF02197968.

Graham, R.D. (1981). *Absorption of Copper by Plants Roots. Copper in Soils and Plants*, 141–163. London: Academic Press.

Hänsch, R. and Mendel, R.R. (2009). Physiological functions of mineral micronutrients (Cu, Zn, Mn, Fe, Ni, Mo, B, Cl). *Curr. Opin. Plant Biol.* 12: 259–266. https://doi.org/10.1016/j.pbi.2009.05.006.

He, D., Chung, M., Chan, E. et al. (2007). Comparative genomics of elastin: sequence analysis of a highly repetitive protein. *Matrix Biol.* 26: 524–540. https://doi.org/10.1016/j.matbio.2007.05.005.

Heijerick, D.G., Van Sprang, P.A., and Van Hyfte, A.D. (2006). Ambient copper concentrations in agricultural and natural European soils: an overview. *Environ. Toxicol. Chem.* 25: 858–864. https://doi.org/10.1897/04-671R.1.

Hossain, M.A., Piyatida, P., Silva, J.A.T., and Fujita, M. (2012). Molecular mechanism of heavy metal toxicity and tolerance in plants: central role of glutathione in detoxification of reactive oxygen species and methylglyoxal and in heavy metal chelation. *J. Bot.* 2012: 872875. https://doi.org/10.1155/2012/872875.

Hossain, Z. and Komatsu, S. (2013). Contribution of proteomic studies towards understanding plant heavy metal stress response. *Front. Plant Sci.* 3: 310. https://doi.org/10.3389/fpls.2012.00310.

Huang, S.Q., Peng, J., Qiu, C.X., and Yang, Z.M. (2009). Heavy metal-regulated new microRNAs from rice. *J. Inorg. Biochem.* 103: 282–287. https://doi.org/10.1016/j.jinorgbio.2008.10.019.

Huang, W.L., Wu, F.L., Huang, H.Y. et al. (2020). Excess copper-induced alterations of protein profiles and related physiological parameters in citrus leaves. *Plants* 9: 291. https://doi.org/10.3390/plants9030291.

Jamla, M., Khare, T., Joshi, S. et al. (2021). Omics approaches for understanding heavy metal responses and tolerance in plants. *Curr. Plant Biol.* 27: 100213. https://doi.org/10.1016/j.cpb.2021.100213.

Jiang, L., Wu, J., Fan, S. et al. (2015). Isolation and characterization of a novel pathogenesis-related protein gene (GMPRP) with induced expression in soybean (*Glycine max*) during infection with *Phytophthora sojae*. *PLoS ONE* 10 (6): e0129932. https://doi.org/10.1371/journal.pone.01299321-19.

Jiu, S., Leng, X., Haider, M.S. et al. (2019). Identification of copper (Cu) stress-responsive grapevine microRNAs and their target genes by high-throughput sequencing. *R. Soc. Open Sci.* 6: 180735. https://doi.org/10.1098/rsos.180735.

Ju, Y.H., Roy, S.K., Roy Choudhury, A. et al. (2021). Proteome changes reveal the protective roles of exogenous citric acid in alleviating cu toxicity in *Brassica napus* L. *Int. J. Mol. Sci.* 22: 5879. https://doi.org/10.3390/ ijms22115879.

Jung, Y., Ha, M., Lee, J. et al. (2015). Metabolite profiling of the response of burdock roots to copper stress. *J. Agric. Food Chem.* 63 (4): 1309–1317. https://doi.org/10.1021/jf503193c.

Kabata-Pendias, A. (2011). *Trace Elements in Soils and Plants*. Boca Raton: CRC Press.

Khraiwesh, B., Zhu, J.K., and Zhu, J. (2012). Role of miRNAs and siRNAs in biotic and abiotic stress responses of plants. *Biochim. Biophys. Acta* 1819: 137–148. https://doi.org/10.1016/j.bbagrm.2011.05.001.

Kidokoro, S., Watanabe, K., Ohori, T. et al. (2015). Soybean DREB1/CBF-type transcription factors function in heat and drought as well as cold stress-responsive gene expression. *Plant J.* 81: 505–518. https://doi.org/10.1111/tpj.12746.

Kopittke, P.M., Asher, C.J., Blamey, F.P.C., and Menzies, N.W. (2009). Toxic effects of Cu^{2+} on growth, nutrition, root morphology, and distribution of Cu in roots of Sabigrass. *Sci. Total Environ.* 407: 4616–4621. https://doi.org/10.1016/j.scitotenv.2009.04.041.

Kosakivska, I.V., Babebnko, L.M., Romanenko, K.O. et al. (2020). Molecular mechanisms of plant adaptive responses to heavy metals stress. *Cell. Biol. Int.* 45 (2): 258–272.

Kumar, V., Pandita, S., Sidhu, G.P.S. et al. (2021). Copper bioavailability, uptake, toxicity, and tolerance in plants: a comprehensive review. *Chemosphere* 262: 127810. https://doi.org/10.1016/j.chemosphere.2020.127810.

Lee, S., Kim, Y.Y., Lee, Y., and An, G. (2007). Rice P1B-ATPase, OsHMA9, is a metal efflux protein. *Plant Physiol.* 145: 831–842. https://doi.org/10.1104/pp.107.102236.

Li, G., Peng, X., Xuan, H. et al. (2013). Proteomic analysis of leaves and roots of common wheat (*Triticum aestivum* L.) under copper-stress conditions. *J. Proteome Res.* 12: 4846–4861. https://doi.org/10.1021/pr4008283.

Li, Q., Chen, H.H., Qi, Y.P. et al. (2019a). Excess copper effects on growth, uptake of water and nutrients, carbohydrates, and PSII photochemistry revealed by OJIP transients in Citrus seedlings. *Environ. Sci. Pollut. Res.* 26: 30188–30205. https://doi.org/10.1007/s11356-019-06170-2.

Li, Q., Wang, G., Wang, Y. et al. (2019b). Foliar application of salicylic acid alleviate the cadmium toxicity by modulation the reactive oxygen species in potato. *Ecotoxicol. Environ. Saf.* 172: 317e325. https://doi.org/10.1016/j.ecoenv.2019.01.078.

Liochev, S.I. and Fridovic, I. (2013). The Haber-Weiss cycle—70 years later: an alternative view. *Redox Rep.* 7: 55–57. https://doi.org/10.1179/135100002125000190.

Liu, H.L., Zhou, J., Li, M. et al. (2021). Chemical speciation of trace metals in atmospheric deposition and impacts on soil geochemistry and vegetable bioaccumulation near a large copper smelter in China. *J. Hazard. Mater.* 413: 125346. https://doi.org/10.1016/j.jhazmat.2021.125346.

Liu, T., Shen, C., Wang, Y. et al. (2014). New insights into regulation of proteome and polysaccharide in cell wall of *Elsholtzia splendens* in response to copper stress. *PLoS ONE* 9 (10): e109573. https://doi.org/10.1371/journal.pone.0109573.

Llorens, N., Arola, L., Bladé, C., and Mas, A. (2000). Effects of copper exposure upon nitrogen metabolism in tissue cultured *Vitis vinifera*. *Plant Sci.* 160: 159–163. https://doi.org/10.1016/s0168-9452(00)00379-4.

Loué, A. (1993). *Micronutrients in Agriculture [Oligoéléments em Agriculture]*. Antibes, FR: SCPA-Nathan.

Malavolta, E. (1992). *ABC of Soil and Leaf Analysis [ABC da Análise de Solos e Folhas]*. São Paulo, Brazil: Agronômica Ceres.

Malavolta, E. (2006). *Manual of Mineral Nutrition of Plants [Manual de Nutrição Mineral de Plantas]*. Piracicaba, Brazil: Agronômica Ceres.

Malavolta, E., Vitti, G.C., and Oliveira, S.A. (1997). *Evaluation of Nutritional Status of Plants; Principles and Application [Avaliação do Estado Nutricional das Plantas; princípios e aplicações]*. Piracicaba, Brazil: Associação Brasileira para Pesquisa da Potassa e do Fosfato.

Marschner, H. (2012). *Marschner's Mineral Nutrition of Higher Plants*. London: Academic Press.

McGrath, M.T. (2004). What are fungicides. In: *The Plant Health Instructor* (ed. E. Cantonwine). Saint Paul, MN: American Phytopathological Society https://doi.org/10.1094/PHI-I-2004-0825-01.

Mehta, S., James, D., and Reddy, M.K. (2019). Omics technologies for abiotic stress tolerance in plants: current status and prospects. In: *Recent Approaches in Omics for Plant Resilience to Climate Change* (ed. S.H. Wani), 1–34. Cham: Springer International Publishing https://doi.org/10.1007/978-3-030-21687-0_1.

Mellidou, I., Keulemans, J., Kanellis, A.K., and Davey, M.W. (2012). Regulation of fruit ascorbic acid concentrations during ripening in high and low vitamin C tomato cultivars. *BMC Plant Biol.* 12: 239. https://doi.org/10.1186/1471-2229-12-239.

Migocka, M. and Malas, K. (2018). Plant responses to copper: molecular and regulatory mechanisms of copper uptake, distribution, and accumulation in plants. In: *Plant Micronutrient Use Efficiency: Molecular and Genomic Perspectives in Crop Plants* (ed. A.H. Mohammad, T. Kamiya, D. Burrit, et al.), 71–86. London: Elsevier.

Moreira, A., Carvalho, J.G., Moraes, L.A.C., and Salvador, J.O. (2000). Effect of limestone calcium and magnesium ratio on micronutrients in alfalfa. *Pesq. Agropec. Bras.* 35: 2051–2056. https://doi.org/10.1590/S0100-204X2000001000017.

Moreira, A. and Moraes, L.A.C. (2019). Soybean response to copper applied to two soils with different levels of organic matter and clay. *J. Plant Nutr.* 42: 2247–2258. https://doi.org/10.1080/01904167.2019.1655039.

Moreira, A., Moraes, L.A.C., Loreto, R.B., and Delfim, J.J. (2022b). Cooper sufficiency concentration in leaves and grains in two representative biomes of soybean crops in Brazil. *J. Plant Nutr.* 45: 2039–2048. https://doi.org/10.1080/01904167.2022.2046082.

Moreira, A., Moraes, L.A.C., Nogueira, T.A.R., and Canizella, B.T. (2019a). Copper use efficiency in soybean cultivars. *Pesq. Agropec. Bras.* 54: e01140. https://doi.org/10.1590/S1678-3921.pab2019.v54.01140.

Moreira, A., Moraes, L.A.C., Melo, T.R. et al. (2022a). Management of copper for crop production. *Adv. Agron.* 173: 257–298. https://doi.org/10.1016/bs.agron.2022.02.005.

Moreira, A., Moraes, L.A.C., and Schroth, G. (2019b). Copper fertilization in soybean-wheat intercropping under no-till management. *Soil Tillage Res.* 193: 133–141. https://doi.org/10.1016/j.still.2019.06.001.

Mostofa, M.G. and Fujita, M. (2013). Salicylic acid alleviates copper toxicity in rice (*Oryza sativa* L.) seedlings by up regulating antioxidative and glyoxalase systems. *Ecotoxicology* 22: 959e973. https://doi.org/10.1007/s10646-013-1073-x.

Mydy, L.S., Chigumba, D.N., and Kersten, R.D. (2021). Plant copper metalloenzymes as prospects for new metabolism involving aromatic compounds. *Front. Plant Sci.* 29: 692108. https://doi.org/10.3389/fpls.2021.692108.

Nanda, R. and Agrawal, V. (2018). *Piriformospora indica*, an excellent system for heavy metal sequestration and amelioration of oxidative stress and DNA damage in *Cassia angustifolia* Vahl under copper stress. *Ecotoxicol. Environ. Saf.* 156: 409e419. https://doi.org/10.1016/j.ecoenv.2018.03.016.

Navarrete, A., González, A., Gómez, M. et al. (2019). Copper excess detoxification is mediated by a coordinated and complementary induction of glutathione, phytochelatins and metallothioneins in the green seaweed *Ulva compressa*. *Plant Physiol. Biochem.* 135 (2): 423–431. https://doi.org/10.1016/j.plaphy.2018.11.019.

Nazir, F., Hussain, A., and Fariduddin, Q. (2019). Hydrogen peroxide modulate photosynthesis and antioxidant systems in tomato (*Solanum lycopersicum* L.) plants under copper stress. *Chemosphere* 230: 544e558. https://doi.org/10.1016/j.chemosphere.2019.05.001.

Printz, B., Lutts, S., Hausman, J.F., and Sergeant, K. (2016). Copper trafficking in plants and its implication on cell wall dynamics. *Front. Plant Sci.* 7: 1–16. https://doi.org/10.3389/fpls.2016.00601.

Rascioa, N. and Navari-Izzo, F.N. (2011). Heavy metal hyperaccumulating plants: how and why do they do it? And what makes them so interesting? *Plant Sci.* 180: 169–181. https://doi.org/10.1016/j.plantsci.2010.08.016.

Raza, A., Tabassum, J., Zahid, Z. et al. (2022). Advances in "Omics" approaches for improving toxic metals/metalloids tolerance in plants. *Front. Plant Sci.* 12: 794373. https://doi.org/10.3389/fpls.2021.794373.

Reuter, D.J. and Robinson, J.B. (1997). *Plant Analysis: An Interpretation Manual.* Collingwood, AU: CSIRO Publishing.

Rogers, K. and Chen, X. (2013). Biogenesis, turnover, and mode of action of plant microRNAs. *Plant Cell* 25: 2383–2399. https://doi.org/10.1105/tpc.113.113159.

Ryan, B.M., Kirby, J.K., Degryse, F. et al. (2013). Copper speciation and isotopic fractionation in plants: uptake and translocation mechanisms. *New Phytol.* 199: 367–378. https://doi.org/10.1111/nph.12276.

Sancenón, V., Puig, S., Mira, H. et al. (2003). Identification of a copper transporter family in *Arabidopsis thaliana*. *Plant Mol. Biol.* 51: 577–587. https://doi.org/10.1023/A:1022345507112.

Santos, L.O., Moraes, L.A.C., Petineli, R. et al. (2020). Yield, yield components, soil fertility, and nutritional status of soybean as influenced by limestone and copper interactions. *J. Plant Nutr.* 43: 2445–2454. https://doi.org/10.1080/01904167.2020.1783308.

Schiavon, M., Zhang, L.H., Abdel-Ghany, S.E. et al. (2007). Variation in copper tolerance in *Arabidopsis thaliana* accessions Columbia, *Landsberg erecta* and Wassilewskija. *Physiol. Plant.* 129: 342–350. https://doi.org/10.1111/j.1399-3054.2006.00809.x.

Schuetz, M., Benske, A., Smith, R.A. et al. (2014). Laccases direct lignification in the discrete secondary cell wall domains of protoxylem. *Plant Physiol.* 166: 798–807. https://doi.org/10.1104/pp.114.245597.

Shabbir, Z., Sardar, A., Shabbir, A. et al. (2020). Copper uptake, essentiality, toxicity, detoxification, and risk assessment in soil-plant environment. *Chemosphere* 259: 127436. https://doi.org/10.1016/j.chemosphere.2020.127436.

Sharma, R., Bhardwaj, R., Thukral, A.K. et al. (2019). Oxidative stress mitigation and initiation of antioxidant and osmoprotectant responses mediated by ascorbic acid in *Brassica juncea* L. subjected to copper (II) stress. *Ecotoxicol. Environ. Saf.* 182: 109436. https://doi.org/10.1016/j.ecoenv.2019.109436.

Singh, S., Parihar, P., Singh, R. et al. (2016). Heavy metal tolerance in plants: role of transcriptomics, proteomics, metabolomics, and ionomics. *Front. Plant Sci.* 6: 1143. https://doi.org/10.3389/fpls.2015.01143.

Song, Q.X., Liu, Y.F., Hu, X.Y. et al. (2011). Identification of miRNAs and their target genes in developing soybean seeds by deep sequencing. *BMC Plant Biol.* 11: 5. https://doi.org/10.1186/1471-2229-11-5.

Sudo, E., Itouga, M., Yoshida-Hatanaka, K. et al. (2008). Gene expression and sensitivity in response to copper stress in rice leaves. *J. Exp. Bot.* 59 (12): 3465–3474. https://doi.org/10.1093/jxb/ern196.

Sunkar, R., Kapoor, A., and Zhu, J. (2006). Posttranscriptional induction of two Cu/Zn superoxide dismutase genes in *Arabidopsis* is mediated by downregulation of miR398 and important for oxidative stress tolerance. *Plant Cell* 18 (8): 2051–2065. https://doi.org/10.1105/tpc.106.041673.

Vassilev, A., Lidon, F., Ramalho, J.C. et al. (2003). Effects of excess Cu on growth and photosynthesis of barley plants. Implication with a screening test for Cu tolerance. *J. Cent. Eur. Agric.* 4 (3): 225e236.

Voinnet, O. (2009). Origin, biogenesis, and activity of plant microRNAs. *Cell* 136: 669–687. https://doi.org/10.1016/j.cell.2009.01.046.

Wang, C., Wang, J., Wang, X. et al. (2017). Proteomic analysis on roots of *Oenothera glazioviana* under copper-stress conditions. *Sci. Rep.* 7: 10589. https://doi.org/10.1038/s41598-017-10370-6.

Willard, L.L. (1979). *Chemical Equilibria in Soils*. Chichester: Wiley.

Wu, C., Luo, Y., and Zhang, L. (2010). Variability of copper availability in paddy fields in relation to selected soil properties in southeast China. *Geoderma* 156: 200–206. https://doi.org/10.1016/j.geoderma.2010.02.018.

Xiangpeng, L., Haifeng, J., Xin, S. et al. (2015). Comparative transcriptome analysis of grapevine in response to copper stress. *Sci. Rep.* 5: 17749. https://doi.org/10.1038/srep17749.

Yamasaki, H., Hayashi, M., Fukazawa, M. et al. (2009). Squamosa promoter binding protein-like7 is a central regulator for copper homeostasis in *Arabidopsis*. *Plant Cell* 21: 347–361. https://doi.org/10.1105/tpc.108.060137.

Younis, M., Tourky, S., and Elsharkawy, S. (2018). Symptomatic parameters of oxidative stress and antioxidant defense system in *Phaseolus vulgaris* L. in response to copper or cadmium stress. *South Afr. J. Bot.* 117: 207e214. https://doi.org/10.1016/j.sajb.2018.05.019.

Yruela, I. (2009). Copper in plants: acquisition, transport, and interactions. *Funct. Plant Biol.* 36: 409–430. https://doi.org/10.1071/FP08288.

Yu, Y., Zhang, Y., Xuemei, C., and Chen, Y. (2019). Plant noncoding RNAs: hidden players in development and stress responses. *Annu. Rev. Cell Dev. Biol.* 35: 407–431. https://doi.org/10.1146/annurev-cellbio-100818-125218.

Zhao, L., Yuxiong, H., Kelly, P. et al. (2018). Metabolomics reveals the molecular mechanisms of copper induced cucumber leaf (*Cucumis sativus*) Senescence. *Environ. Sci. Technol.* 52: 7092–7100. https://doi.org/10.1021/acs.est.8b00742.

13

Zinc Toxicity and Tolerance in Plants

Insights from Omics Studies

Imran Haider Shamsi[1], Qichun Zhang[2], Zhengxin Ma[1], Sibgha Noreen[3], Muhammad Salim Akhter[3], Ummar Iqbal[4], Muhammad Faheem Adil[1], Muhammad Fazal Karim[5], and Najeeb Ullah[6]

[1] *Department of Agronomy, College of Agriculture and Biotechnology, Key Laboratory of Crop Germplasm Resource, Zhejiang University, Hangzhou, China*
[2] *MOE Key Laboratory of Environment Remediation and Ecosystem Health, College of Environmental and Resource Sciences, Zhejiang University, Hangzhou, China*
[3] *Department of Botany, Institute of Pure and Applied Biology, Bahauddin Zakariya University, Multan, Pakistan*
[4] *Department of Botany, The Islamia University of Bahawalpur, Bahawalpur, Pakistan*
[5] *Department of Agronomy, Pir Mehar Ali Shah Arid Agriculture University, Rawalpindi, Pakistan*
[6] *Agricultural Research Station, Office of VP for Research and Graduate Studies, Qatar University, Doha, Qatar*

13.1 Introduction

Soils are enriched with micronutrients (B, Zn, Mn, Fe, Cu, Mo, Ni, Cl) but relatively trace amounts of these elements are required by plants for completion of their life cycle (Imtiaz et al. 2010). They constitute about 1% dry weights of plants and play a crucial role in almost all the physiological and biochemical processes of life including photosynthesis, chlorophyll production, and yield development. In plants the accumulation and distribution may be altered among different plant species and growing conditions (Rahman et al. 2020; Mapodzeke et al. 2021a). Among these micronutrients, zinc ($^{65.37}Zn_{30}$) belongs to transition elements (group IIB) and is ranked among most abundant transition element (Gentzel et al. 2020). It plays a central role in structural and catalytic activities in different processes i.e., cell division and elongation, protein synthesis, gene expression, metabolism of lipids, protein, carbohydrates and nucleic acid (Kaur and Garg 2021; Paradisone et al. 2022), and for synthesis of tryptophan, precursor of auxin.

13.1.1 Zinc Uptake and Translocation Mechanisms in Plants

Zinc mobility is very high in soils rich in organic matter and clay fractions, as they have the ability to hold Zn strongly at neutral or nearly alkaline pH (Azouzi et al. 2015; Mapodzeke et al. 2021a). Uptake of Zn in plants occur as divalent Zn^{2+} or by forming complex with

Heavy Metal Toxicity and Tolerance in Plants: A Biological, Omics, and Genetic Engineering Approach, First Edition. Edited by Mohammad Anwar Hossain, AKM Zakir Hossain, Sylvain Bourgerie, Masayuki Fujita, Om Parkash Dhankher, and Parvez Haris.

organic ligands via xylem of the root cells that later on is translocated to upper aerial parts, which directly depend upon its concentration present in nutrient medium (Naskar et al. 2016). However, other factors like moisture content, pH, temperature, organic matter content of soil, and plant root distribution also directly influence the availability and translocation of Zn in plants (Azouzi et al. 2015). Transport of Zn from external medium to plant root cell is mediated by the process of diffusion in which a potential gradient in root cell is formed and Zn^{2+} ions are transported to the negatively charged cytoplasm of root cell through plasma lemma with the help of protein transporters. The intake of Zn over the plasma membrane is thermodynamically favored, and it is supposed that no transport system is responsible for active transport of Zn^{2+} in the root cells (Hacisalihoglu and Kochian 2003). Its elevated level in the external media results in the uptake of Zn influx to xylem of root cells by apoplastic pathways or through connecting link of cytoplasm of root cells to plasmodesmata by symplastic movement (Figure 13.1). There are particular cations that can be selected by specific transporters for transport through plasma membrane by symplastic movement (Rafie et al. 2022).

In *Thlaspi caerulescens*, the Zn transport to xylem occurs via symplastic pathway by entering the root cells across the plasma membrane. Its translocation from root to shoot tissues of the plants requires its loading in the apoplastic xylem (Kozhevnikova et al. 2014).

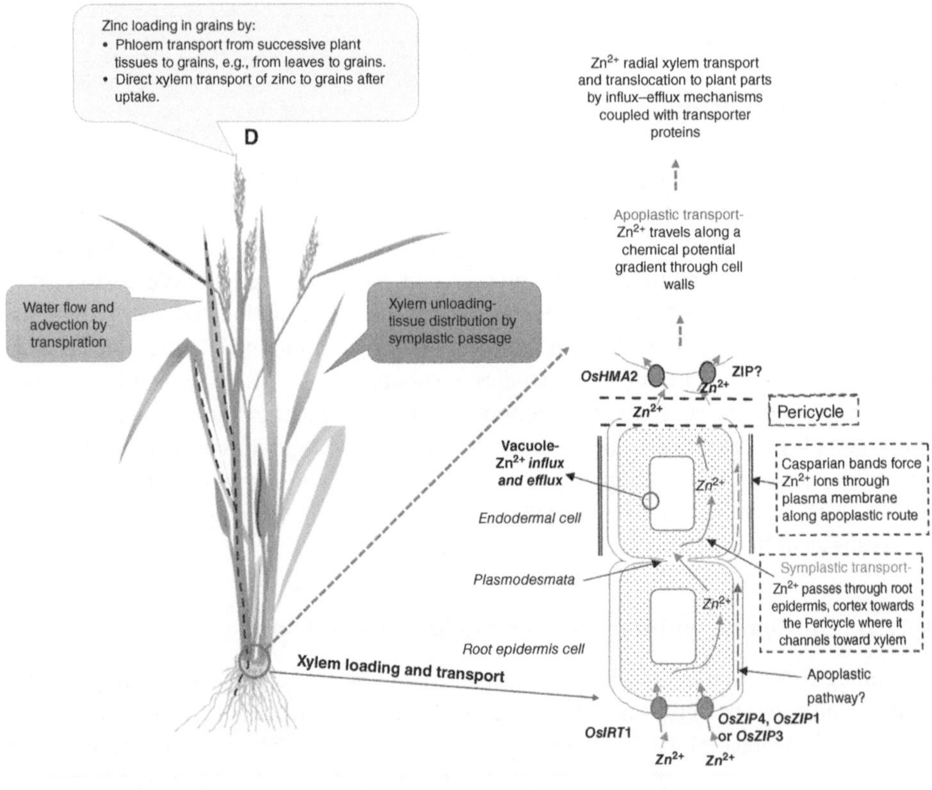

Figure 13.1 Diagram illustrates pathways of root uptake and translocation mechanism. *Source:* Reproduced from Mapodzeke et al. (2021a).

Inside the xylem, Zn form chelation by low-molecular-weight ligands, thus prevent its binding to the neighboring cell walls or uptake into the cells via Zn transporters. According to Sinclair and Krämer (2012), Zn uptake from the xylem is mediated over the plasma membrane of neighboring cells present in the shoots.

13.1.2 Transporters and Metal-Binding Compounds Involved in Zinc Homeostasis

The transport and homeostasis of Zn is accomplished by several protein transporter and proteins (Ricachenevsky et al. 2013; Adil et al. 2022), including Zn-regulated transporters (ZRT)/iron-regulated transporters (IRT), Zn-/iron-related proteins (ZIP), heavy metal ATPase proteins (HMA), and metal tolerance proteins (MTP). Similarly, the transporters such as natural resistance-associated macrophage protein (NRAMP) and vacuolar iron transporter (VIT) are also involved in transport and homeostasis of Zn (Song et al. 2014). Moreover, various number of metal-binding compounds and their corresponding transporters like yellow stripe-like proteins (YSLs) are also involved in Zn homeostasis (Curie et al. 2009). However, the overexpression of some *ZIP* gene members, like *AtIRT3* (Lin et al. 2009), *OsZIP4* (Ishimaru et al. 2007; Yang et al. 2020), *OsZIP8* and *OsIRT1* (Adil et al. 2020; Mapodzeke et al. 2021a), *OsZIP5* (Adil et al. 2022), and *HvZIP7* (Tiong et al. 2014), increases the Zn concentrations in shoots and/or roots, suggesting roles of these transporters in Zn homeostasis control. List of various Zn transporters are provided in Table 13.1.

13.2 Impact of Excess Zinc on Physio-genetics Aspects of Plants

In most parts of the world, Zn is a major pollutant of both aquatic and terrestrial ecosystems, while soils having low pH have generally high Zn availability and in this acidic pH it can easily be taken up by the plants and can cause harm to them at higher levels (Rafie et al. 2022). Zinc toxicity occurs in plants range between 400 and 500 $\mu g\,g^{-1}$ in leaves organ. However, the threshold level of Zn toxicity is based on species types or genotypes, growth media or levels of Zn contamination in growing media, and time of exposure. Cellular homeostasis for Zn concentration is necessary because presence of Zn to toxic levels in plant's cellular system compete with binding site of other elements (Fe^{2+}, Mn^{2+}), which lead to mineral imbalance (Schvartzman et al. 2018). In crop plants, the visible symptoms of oxidative stress caused by Zn toxicity are necrosis, leaf chlorosis, curling of younger leaves, stunted growth and low yield (Kosová et al. 2018; Kaur and Garg 2021). In plants, Zn toxicity diminishes the growth of primary roots and also impedes the growth of secondary roots. Because Zn is accumulated more in roots than in leaves and limits plant uptake of water and nutrients by retarding elongation and root growth (Mapodzeke et al. 2021b). It has been found that addition of 400–1600 mM $ZnSO_4$ inhibits the growth of *Eucalyptus maculata* and *Eucalyptus urophylla* after a treatment of five weeks (Soares et al. 2001), whereas treatment range from 1 to 10 mM $ZnSO_4$ retards the growth of ryegrass with complete inhibition at 50 mM $ZnSO_4$ (Bonnet et al. 2000). According to Doncheva et al. (2001), growth inhibition in *Pisum sativum* became apparent at 1000 μM Zn concentration. Crops growing under polluted and acidic soils suffer Zn toxicity, as some food crops like beet and

Table 13.1 List of transporters involved in zinc homeostasis.

Plant name	Transporters involved in zinc transportation	References
Arabidopsis thaliana	*AtIRT1*	Connolly et al. (2002)
Oryza sativa	*OsZIP1*, *OsZIP3*, *OsZIP4*, *OsZIP5*, *OsZIP7*, *OsZIP8*, and *OsZIP9*	Ishimaru et al. (2007) and Yang et al. (2020)
Zea mays	*ZmZIP1-ZmZIP12*	Mondal et al. (2014)
Wheat	*TaZIP3*, *TaZIP5*, *TaZIP6*, *TaZIP7*, and *TaZIP13*	Evens et al. (2017)
Peanut	*AhNRAMP1*	Xiong et al. (2012)
Noccaea caerulescens *Arabidopsis halleri*	*NcZNT1*, *AtZIP4* and *AtIRT3*,	Milner et al. (2012) and Lin et al. (2009)
Sedum alfredii	*SaZIP4*	Yang et al. (2018)
Tobacco	*NtZIP11*	Kozak et al. (2019)
Poncirus trifoliata	*PtIRT1*, *PtZIP1-PtZIP3* and *PtZIP5-PtZIP14*	Fu et al. (2017)
Setaria italic	*SiZIPs*	Alagarasan et al. (2017)
Medicago sativa	*MsZIP3*	Cardini et al. (2021)
Hordeum vulgare	*HvZIP7*	Tiong et al. (2014)
Nicotiana tabacum var. *Xanthi*	*NtZIP4A/B*	Barabasz et al. (2018) and Maslinska-Gromadka et al. (2021)
Panicum hallii	*ZIP1* (*PhZIP1*) and *SbZIP1*	Ramesh et al. (2003)
Citrus sinensis	*CsZIP1*, *CsZIP2*, *CsZIP3*, and *CsZIP4*	Fei et al. (2016)
Foxtail millet	*SiZIP2*, *SiZIP3*, *SiZIP4*, and *SiZIP5*	Alagarasan et al. (2017)
Common bean	*PvZIP12*, *PvZIP13*, and *PvZIP16*	Astudillo et al. (2013)

spinach can accumulate high levels of Zn in their tissues and exhibit growth retardation. For example, in *Populus deltoides* × *Populus nigra,* Di-Baccio et al. (2003) observed that application of 100 and 1000 μM Zn treatments exerted a significant reduction in total dry mass and foliage content of plant. Khudsar et al. (2004) noted reduction in root and shoot growth parameters in *Artemisia annua* at different stages of development with varying Zn concentration ranges from 100 to 400 $\mu g\, g^{-1}$ (soil dry matter).

13.2.1 Effect of Zinc Toxicity on Seed Germination and Growth of Plants

Excessive presence of Zn in soils either due to natural or external means can pose many structural and functional abnormalities in plants depending on plant species and developmental stage at which plant exposed to elevated Zn levels in soil (Wei et al. 2022). Germination, the first step of plant life cycle, which determines growth and yield at later stages, is significantly affected by Zn toxicity (Zhang et al. 2019). The amylase activity and mobilization of reserve sugars to embryo were reduced when Zn was present in excess, which reduced germination of *Cicer arietinum* seeds (Ullah et al. 2020). Its toxicity inhibits cell division and elongation, which reduces leaf growth and number in different plant species (Glińska et al. 2016). Root length, diameter, surface area, and volume of root system is reduced due to Zn-induced damages on root tips. Transmission electron microscopy revealed that Zn stress significantly affected the radicle elongation compared to that of the plumule extension. It is also reported that zinc treatment at 7.5 mM completely inhibits the elongation of root cortical cells in *Cajanus cajan* cv. ICPL 87 after 24 hours (Rout and Das 2009). Zinc application also caused significant reduction in the height of *A. annua* at 50 and 100 $\mu g\, g^{-1}$ soil dry matter and affects the root length (Khudsar et al. 2004). However, excess Zn imparts an inhibitory effect on plants belonging to family Leguminosae, whereas elongation of shoot and root was retarded by Zn application in *Phaseolus mungo* and *Bacopa monniera*. Zinc toxicity also restricts the leaf growth in two species like *A. annua* and *Phaseolus vulgaris*. Elevated Zn level influence the blockage of xylem, thus lower the water content in leaf. All these factors can disturb the stomatal balance, resulting in decreased rate of transpiration and photosynthetic efficiency (Sagardoy et al. 2010). Higher Zn concentration interferes with the thylakoid electron transport rates (ETRs). Activity of Rubisco responsible for driving the biochemical reactions occurring in the stroma of the chloroplasts. Presence of divalent cations like Co^{2+}, Ni^{2+}, and Zn^{2+} ensures the activity of Rubisco. However, these bivalent cations compete with Mg^{2+} cations present in Rubisco, which affects its activity in plants. Cherif et al. (2011) reported that 150 $\mu mol\, l^{-1}$ Zn concentration severely reduce the level of Chl-*a* and Chl-*b* in *Solarium lycopersicum* compared to the control. In *Zea mays*, treatment with 1000 $mg\, kg^{-1}$ Zn in soil induces significant reduction in chlorophyll content in plants (Cui and Zhao 2011). Rout and Das (2009) reported disintegration of cell organelles by Zn toxicity. Such disintegration justifies the reduction in the rate of photosynthesis in leaves of rye plants exposed to higher concentration of Zn (2.7 $mg\, g^{-1}$) for 15 days (Monnet et al. 2001).

13.2.2 Effect of Zinc Toxicity on Oxidative Metabolism in Plants

The deleterious phytotoxic effect of Zn is the induction of oxidative stress in different plant parts due to generation and also accumulation of reactive oxygen species (ROS) that disturb the cellular redox status of the plants (Pei et al. 2022). Unlike other essential elements, Zn in plants is a redox inactive metal but elevated and long term exposure of Zn may induce cytotoxicity, thus it severely disrupts normal cellular metabolic functioning. Different ROS species which include singlet oxygen ($^1O_2$), superoxide anion ($O^{2\cdot-}$), hydroxyl ion ($OH^\cdot$), and hydrogen peroxide (H_2O_2) are accumulated in plants during normal cellular metabolism under Zn stress (Paradisone et al. 2022). These oxygen free radicals are extremely reactive and led to toxicity or even death of plants by damaging lipid membranes, proteins, and nucleic acids (Ozdener and Kutbay 2009). Zinc at higher levels induces oxidative stress by disturbing antioxidant defense system and photosynthetic electron transport. Zinc induces lipid peroxidation when H_2O_2 and $O^{2\cdot-}$ produced by Fenton reaction are converted into extremely toxic $OH^\cdot$ radical that causes peroxidation of lipids (Ozdener and Kutbay 2009). Biochemical estimations showed that in *P. vulgaris*, treatment of both Cd and Zn with concentrations (5 μM Cd $(NO_3)_2{\cdot}4H_2O$ and 100 μM $ZnSO_4{\cdot}7H_2O$) in nutrient media induced oxidative stress in all organs of plant seedlings, as was evidenced by the enhancement in lipid peroxidation. Elevated levels of lipid peroxides with increasing concentration of

Zn-induced oxidative stress in *Hibiscus esculentus* plants. They further opined that in shoots of *H. esculentus*, the H_2O_2 content decreased at 5 mM Zn^{2+} but incremented later depending upon the increased concentrations of Zn, when compared with the control. In *S. lycopersicum*, Zn exposure to 100 μmol l^{-1} promotes toxicity and induces lipid peroxidation by evaluating the increased amount of TBARS in Zn-treated plants (Cherif et al. 2011). In leaves of wheat plants, significant increase in H_2O_2 content was observed under 1 and 3 mM Zn treatment when compared with the control but no such significant effect was observed on lipid peroxides in the leaves of wheat seedlings. Furthermore, remarkable increase in H_2O_2 level in roots of wheat seedlings was observed when exposed to varied Zn concentrations, and malondialdehyde (MDA) content also elevated with the elevated Zn levels in roots of wheat seedlings. Mapodzeke et al. (2021b) found that exposure to varied Zn concentrations resulted in an increase in MDA level in various organs of rice plants as compared to that of the control.

13.2.3 Effect of Zn Toxicity on Physiology and Biochemistry of Plants

The active/passive transport of Zn from roots is accumulated in leaves, which is cytotoxic leading to leaf chlorosis and necrosis. These damages are attributed to decreased activity of V-ATPase, as observed in *Arabidopsis thaliana* (Fukao et al. 2011). The organization of mitochondria is damaged, which decreased synthesis of nicotinamide (B-group vitamins), thus enhancing NAD^+ breakdown that lowers energy metabolism (Zhang et al. 2019). In coriander, lettuce, and *Arabidopsis*, Zn toxicity caused redness of leaves and shoots probably due to anthocyanin synthesis. The excessive Zn in soil hampers nitrogen metabolism and uptake of nitrate (Blasco et al. 2019) due to dis-organization of chloroplast and decreased nitrate reductase (NR) synthesis. Low NR activity reduces free amino acid (asparagine and glutamate) synthesis required for chlorophyll and ATP biosynthesis. The inhibition in chlorophyll biosynthesis and carbon dioxide (CO_2) fixation under Zn stress might leads to inefficient nitrogen assimilation, consequently reducing NR activity (Khudsar et al. 2004). Similarly, nodulation (number and weight) and nitrogen fixation activity in legumes is also badly affected by excessive Zn in root zone (Souza et al. 2020) due to reduction of population of free living rhizobia in the soil, thus directly reducing nitrogen fixing capability of legumes (Paradisone et al. 2022).

Elevated Zn results render the plants to the condition of physiological drought through alteration in water balance in plants by either blocking or restricting the transport of water. Exposer of plants to excessive Zn in soils lowers water absorption capabilities of root due to blocking of xylem elements that results in lowering leaf relative water contents (LRWC) resulting in dehydration of leaves and plants as a whole. Reduced LRWC leads to loss turgidity of leaves, thus reducing cell expansion ultimately leading to growth inhibition and reduced biomass accumulation (Garg and Singh 2018). The processes of photosynthesis and respiration are reduced under Zn toxicity possibly due to restriction in mesophyll and stomatal conductance, reduced photochemistry, enhanced respiration, and weakened biochemical activities. Higher levels of soil Zn results in decreased leaf area and stomatal conductance, which it further lowers CO_2 flow within the leaves of chloroplasts, thus reducing carbon fixation during photosynthesis. Excess Zn damage disrupted the efficiency of PSII and non-cyclic photophosphorylation by disrupting the structure of PSII core complex and interacting with donor side of PSII or dislocating non-heme group of PSII (Sagardoy et al. 2010). Moreover, the mitochondrial activity is disturbed by excessive Zn presence, which results in abnormal respiration in plants. It is also reported that due to higher Zn levels, the process of respiration was enhanced due to higher production of different organic acids (citric acid, oxalic acid, fumaric acid, malic acid) through TCA cycle (Kaur and Garg 2021).

13.3 Plants Stress Adaptation to Zinc Toxicity

Plants has distinct mechanism to cope with Zn toxicity. The non-accumulators avoid Zn toxicity by exclusion, sequestration to vacuole, retaining or immobilizing in root cell walls through Zn-specific transporters that limit the translocation of Zn to aerial parts of plants resulting in improved resistance to Zn toxicity (Fu et al. 2017). Similarly, Zn accumulator plants increase transport of large amount of Zn from roots to leaves and then squeeze extra Zn to vacuoles of leaves without showing symptoms of Zn toxicity (Sytar et al. 2021). Around 450 plant species have been discovered as hyperaccumulators for heavy metals. Generally heavy metals translocated from root to shoot, mostly in leaves. Zinc hyperaccumulator plants would accumulate 10,000 ppm dry weight under natural environmental conditions. For instance, *Arabidopsis halleri* and *Noccaea caerulescens* have great potential to accumulate the maximum concentration of Zn (13,620 and 43,710 ppm) when growing naturally in metalliferous soil (Peer et al. 2006). Various zinc hyperaccumulator plants are listed in Table 13.2.

13.4 Multi-omics Approaches for Zinc Toxicity and Tolerance in Plants

Recent developments and crop enhancement strategies in toxic heavy metal stress are mainly concentrated on multidimensional regulation linkage at molecular level. Well-established omics technologies including transcriptomics, genomics, and metabolomics, as well as contemporary techniques like RNA omics and multi-omics approaches are playing a critical role in illuminating plant stress responses and agricultural enhancements. In order to identify regulators and their candidate genes, identify molecular targets, analyze differential expression, and create signaling networks, omics methods have become increasingly important, which include microarrays, sequencing, mass spectroscopy, and chromatographic high-throughput method (Pandita et al. 2021).

13.4.1 Genomics and Metabolomics

Genomic analysis is the study of multiple genes and how they interact across a range of physiological conditions. However, stress response is molecularly controlled by a group of genes and shifting mechanisms from transport to metabolic processes (Sahoo et al. 2020). It comprises several methods for identifying the structural and as well as functional properties of genes and their inter-genic areas (Malla et al. 2022). At the genomic level, plant

Table 13.2 List of zinc accumulator plants.

Plant name	Plant organ for accumulation of zinc	References
Cymodocea nodosa	Leaves	Llagostera et al. (2011)
Arabidopsis thaliana	Root, shoot, and leaves	
Phragmites australis	Roots and leaves	
Kosteletzkya virginica	Shoot and flowers	
Verbascum olympicum	Root, stem, and leaves	
Ageratina glechonophylla	Stem and leaves	
Spartina maritima	Roots	
Linum usitatissimum	Stem and leaves	
Lactuca sativa	Leaves	
Brassica oleracea	Stem and leaves	
Sinapis alba	Leaves	
Kosteletzkya virginica	Roots	
Typha angustifolia	Roots	
Phragmites australis	Stem and leaves	
Potamogeton pectinatus	Root and stem	Demırezen et al. (2004)
Rumex alpinus	Rhizome	
Brassica juncea	Roots	
Typha latifolia	Stem and leaves	
Oryza sativa	Roots and leaves	Adil et al. (2020)
Arabidopsis thaliana	Roots	
Euphorbia pithyusa	Roots	
Arabidopsis helleri	Roots	Peer et al. (2006)
Noccaea caerulescens	Root dry wt.	
Thlaspi ochroleucum	Shoot and root ratio	
Brassica napus	Leaf	
Oryza sativa	Seed germination, leaf sheath, and roots epidermis	
Triticum polonicum	Leaf blades, roots, and nodes at the grain filling stage	
Triticum turgidum L.	Roots	
Crotalaria juncea	Stem, leaves, and roots	
Spirodela polyrhiza	Roots	Chen et al. (2021)
Nicotiana tabacum	Conductive tissues of leaves	
Malus hupehensis	Roots	
Malus xiaojinensis	Roots and leaves	
Hordeum vulgare	Roots	
Brassica rapa L.	Roots	

response to heavy metal stress is mostly regulated by structure of chromatin, retarded responsive genes such as *cis*-regulatory DNA sequences, and transcriptional factors; they are primarily responsible for controlling how plants respond to heavy metal stress (Liu and Yan 2019). Database and computational methods both are frequently employed in genomics investigations for gene identification, sequence method, phylogenetic method, orthologous method, gene-function studies, chromosomal placement, *cis*-regulators, introns, and 5′- and 3′-untranslated regions (Joshi et al. 2021). The genomic variation under hyperaccumulation of Cd/Zn was assessed by random amplification polymorphic DNA (RAPD) method. Recently simple sequence repeats (SSRs) and single nucleotide polymorphisms (SNPs) are identified as molecular markers in *S. alfredii*, which are used as important methodology for genome mapping under hyperaccumulation of metals in plants. Schvartzman et al. (2018) studied the intraspecific variation in two populations of *A. halleri*, and major variations were found in both populations to adopt Zn toxicity in soil. Metabolomics can be considered as an efficient determinant of plant responses; the relative protein activity and differences in protein amount within the plant lead to changes in metabolic fluxes and enzymatic activities (Wada et al. 2020). Excess Zinc interferes with normal plant functioning and causes metabolic disturbances, enzyme inhibition, retardation of growth, and ultimately plant death. Zhang et al. (2017) studied the metabolomics profile of tea (*Camellia sinensis* L.) using UPLC-MS-MS and GC-MS analysis and observed that Zn concentrations changed carbohydrate, caffeine, nitrogen, and flavonoid response in tea leaves. The increasing Zn levels reduced carbohydrate metabolites, i.e. glucose-6-phosphate and fructose-6-phosphate while fructose, galactose and cellobiose were enhanced in tea leaves when Zn was present in excess.

13.4.2 Proteomics and Transcriptomics

Proteomics study is the most recent biotechnological approach that is used in plants in order to elaborate the biological functions of proteome and proteins when exposed to stress (Wu et al. 2020). It presents the relationship of changes in mRNA expression related proteins under stressful environmental conditions (Kosová et al. 2018). The role of plant proteins under stress has a direct relationship in adjusting plant's physiological traits to environmental conditions as it gives a clear picture among protein network and metabolic pathway especially in detoxification of ROS produced under heavy metal toxicity (Sagonda et al. 2021). Proteomic analysis strongly indicated that accumulation of antioxidants, *i.e.* superoxide dismutase, catalase, peroxidase, glutathione, and ascorbic acid, that helps the cellular system to mitigate HM-induced oxidative stress. Over the last two decades, the genomic-sequencing technologies has emerged a major tool for understanding plant response to different biotic and abiotic stresses. In current years' advancement in techniques such as RNA sequencing next-generation sequencing (NGS), microarrays, representational difference analysis differential display, reverse transcription polymerase chain reaction, serial analysis of gene expression (SAGE), and massively parallel signature sequencing (MPSS) have made possible in analyzing known/unknown differentially expressed genes (DEGs) for plants under metal stress (Zhou et al. 2022). However, RNA-Seq technologies have appeared as advanced techniques for transcriptomic analysis that take advantage of high-throughput next-generation sequencing. Moreover, these methods

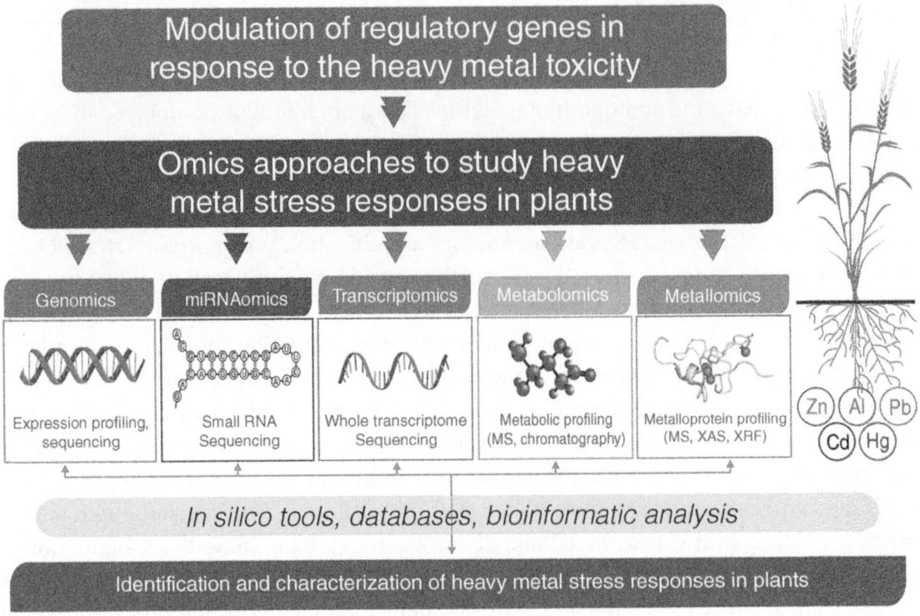

Figure 13.2 Representing gene ontologies of both up- and down-regulated genes under various level of metal stress conditions in plants by transcriptomic analysis.

continue to impose challenge on full-length transcript structure identification, primarily due to inherent limitations of read length. The transcriptomic analyses focus integrated response of different metals and metalloids to various abiotic and biotic stresses in plants (Figure 13.2 and Table 13.3).

13.4.3 miRNA Omics and CRISPR/Cas9 System

The micro-RNA genes (miRNA or MIR), a class of small noncoding RNAs, are usually present in intergenic areas and are also found in antisense or sense orientation in introns of genes (Patil et al. 2021). They are found as clustered in genome and are normally transcribed together as long polycistronic RNAs about 20–24 nucleotides. They are emerged as important regulators of gene expression, usually by cleavage and inhibition in translation of the target mRNAs during or after transcription. miRNA plays a central role in regulating different biological processes in plants, which include maintenance of genome reliability, development, metabolism, and responses toward environmental stresses (Jones-Rhoades 2012). The expression of different miRNAs in *Carthamus tinctorius* regulate the expression of target genes when exposed to Cd stress (Kouhi et al. 2020). However, heavy metal resistant miRNA has already been reported in mays (Gao et al. 2019), wheat (Qiu et al. 2016), canola (Jian et al. 2018), and barley (Yu et al. 2019). Clustered regularly interspaced short palindromic repeats (CRISPR)/CRISPR-associated protein 9 (CRISPR/Cas9) has become an emerging new genome-editing tool, which is extensively useful in various organisms, including plants (Manghwar et al. 2020). The use of CRISPR/Cas9 allows generating transgene-free genome-edited plants ("null segregants") in a very precise time (Wada

Table 13.3 Genes or proteins involved in the accumulation of zinc in plants.

Plant name	Proteins and/or genes names	References
Cymodocea nodosa	Differential abundant proteins (DAPs)	
Arabidopsis halleri	MT2A(At3g09390), MT2B(At5g02380)	
Arabidopsis thaliana	ZINC-INDUCED FACILITATOR1 (ZIF1)	
Triticum aestivum	Inositol Pentakisphosphate 2-Kinase 1 (TaIPK1)	
Oryza sativa	*OsHMA2, AtHMA2* and *AtHMA4*	
Cereals	ZIP (ZRT, IRT-related protein) family, metal tolerance protein (MTP)	
Phaseolus vulgaris	ZIP1-4 genes	Hacisalihoglu and Kochian (2003)
Spartina maritima *Spartina alterniflora*	FtsH genes	
Nicotiana tabacum	ZIP, NRAMP, MTP, and MRP/ABCC	
Thlaspi caerulescens	NRAMP3 and NRAMP4	
Nicotiana glauca	*NgMTP1*	
Thlaspi goesingense	MTP1	
Linum usitatissimum	Heavy-metal-associated (HMA) gene, Pinoresinol–Lariciresinol Reductase 1 (LuPLR1) gene	
Brassica juncea	AtATM3, BjCET3, and BjCET4	
Brassica rapa	Natural resistance-associated macrophage protein (NRAMP), BrRZFP1	Song et al. (2014)
Populus trichocarpa	*PtGSTF4*	
Brassica napus	BnMTP3	
Medicago sativa	MsZIP1-7, MsZIF1, MsMTP1, MsYSL, MsHMA4, and MsNAS1	Cardini et al. (2021)
Hordeum vulgare	Metal tolerance proteins	Cardini et al. (2021)

et al. 2020). Gene editing using CRISPR-associated proteins (CRISPR/Cas) has modernized biotechnology and offers genetic tools for both medicine and life sciences. Nevertheless, the use of this technology to miRNAs, with the function as negative gene regulators, has not been extensively reviewed in plants. This system is considered as pivotal technology for gene editing based on endonucleases, which has revolutionized biotechnology in various disciplines (Zhu et al. 2020). Due to the high efficacy and cost-effectiveness of CRISPR/Cas systems (Chen et al. 2021), they are now widely used in gene editing of Arabidopsis (Ursache et al. 2021) and many important crops, i.e. rice, tomato, maize, wheat, cotton, and potato (Johansen et al. 2019; Zhu et al. 2019; Gentzel et al. 2020; Kwon et al. 2020) through the precise genetic manipulation of traits such as yield, quality, disease resistance, and stress tolerance, leading to improved breeding technologies for crop plants (Chen et al. 2021).

13.4.4 Quantitative Trait Locus Mapping and Genome-Wide Association Study

Different methods to comprehend multiple characters in plants through quantitative trait loci (QTL) mapping and genome-wide association studies (GWAS) are employed nowadays (Huang and Han 2014). It is based on statistical approach that helps in connecting two types of information or data, that is a complex phenotype (characters with extents) with genotypes (particular sections of chromosomes) (Evans et al. 2021). Molecular markers, i.e. single nucleotide polymorphisms (SNPs) and amplified fragment length polymorphism (AFLPs), are utilized for mapping QTLs, and ultimately they are associated with known phenotypic data. Various QTL for tolerance of Zn toxicity has been recognized via germplasm assets or biparental populations (Zhang et al. 2013). Genome-wide association studies (GWAS) also known as whole genome association studies used to investigate a genome-wide set of genetic variants in different varieties to see if any variant is associated with a trait (Manolio 2013). It generally gives emphasis to linking among SNPs and traits, while it involves testing genetic variants across the genomes of many individuals of a population to identify genotype–phenotype association, which is based on designs, genotyping technologies, and statistical concepts for analysis, replication, interpretation, and follow up of association results (Bush and Moore 2012). However, in plants GWAS studies initially focused on single feature polymorphism, recombination, and linkage disequilibrium but has now been comprised by various disciplines (Alseekh et al. 2021).

13.5 Conclusion and Future Prospective

Zinc is an important micronutrient, which is required by plants in various metabolic and physiological functions. But presence of Zn in cellular system effects overall health of plants by posing negative effect on biochemical, physiological, gene expression, etc. Zinc toxicity leads to impairment in functions of various important cellular structures including transporter lead by alteration in gene expression. The modifications, purifications, and expression of membrane proteins potentially help out to comprehend the role of membrane transporters and gene expression involved in trafficking of Zn within plant system. Further understanding of miRNA and editing of genome through CRISPR-Case9 system for inserting genes that are associated with tolerance of Zn toxicity will work an important tool for enhancing metal stress in plants. Similarly, understanding the response of proteomics, genomics, and metabolomics under Zn toxicity will open new avenues for development of Zn-resistant varieties of plants. Therefore, the omics approach may facilitate the plant breeders to develop new resistant varieties of crops, which can tolerate fluctuation in environment more efficiently in order to feed ever-increasing human population.

Acknowledgments

The authors gratefully acknowledge the Sino-Pakistan Project NSFC Grant No. 31961143008, National Natural Science Foundation of China, International (Regional) Cooperation and Exchange Program, Research fund for International young scientists Grant Nos. 31750110462, 32250410280 and Jiangsu Collaborative Innovation Center for Modern Crop Production (JCIC-MCP) China.

References

Adil, M.F., Sehar, S., Han, Z.G. et al. (2020). Zinc alleviates cadmium toxicity by modulating photosynthesis, ROS homeostasis, and cation flux kinetics in rice. *Environ. Pollut.* 265 (Part B): 114979.

Adil, M.F., Sehar, S., Chen, S. et al. (2022). Stress signaling convergence and nutrient crosstalk determine zinc-mediated amelioration against cadmium toxicity in rice. *Ecotoxicol. Environ. Saf.* 230: 113128.

Alagarasan, G., Dubey, M., Aswathy, K.S., and Chandel, G. (2017). Genome wide identification of orthologous ZIP genes associated with zinc and iron translocation in *Setaria italica*. *Front. Plant Sci.* 8: 775.

Alseekh, S., Kostova, D., Bulut, M., and Fernie, A.R. (2021). Genome-wide association studies: assessing trait characteristics in model and crop plants. *Cellular Mol. Life Sci.* 78: 5743–5754.

Astudillo, C., Fernandez, A., Blair, M.W., and Cichy, K.A. (2013). The *Phaseolus vulgaris* ZIP gene family: identification, characterization, mapping, and gene expression. *Front. Plant Sci.* 4: 286.

Azouzi, R., Charef, A., and Hamzaoui, A.H. (2015). Assessment of effect of pH, temperature and organic matter on zinc mobility in a hydromorphic soil. *Environ. Earth Sci.* 74: 2967–2980.

Barabasz, A., Palusinska, M., Papierniak, A. et al. (2018). Functional analysis of *NtZIP4B* and Zn status-dependent expression pattern of tobacco ZIP genes. *Front. Plant Sci.* 9: 1984.

Blasco, B., Navarro-León, E., and Ruiz, J.M. (2019). Study of Zn accumulation and tolerance of HMA4 TILLING mutants of *Brassica rapa* grown under Zn deficiency and Zn toxicity. *Plant Sci.* 287: 110201.

Bonnet, M., Camares, O., and Veisseire, P. (2000). Effects of zinc and influence of *Acremonium lolii* on growth parameters, chlorophyll a fluorescence and antioxidant enzyme activities of ryegrass (*Lolium perenne* L. cv *Apollo*). *J. Exp. Bot.* 51: 945–953.

Bush, W.S. and Moore, J.H. (2012). Genome-wide association studies. *PLoS Comput. Biol.* 8: e1002822.

Cardini, A., Pellegrino, E., White, P.J. et al. (2021). Transcriptional regulation of genes involved in zinc uptake, sequestration and redistribution following foliar zinc application to *Medicago sativa*. *Plants* 10 (3): 476.

Chen, Y., Fu, M., Li, H. et al. (2021). High-oleic acid content, non-transgenic allotetraploid cotton (*Gossypium hirsutum* L.) generated by knockout of GhFAD2 genes with CRISPR/Cas9 system. *Plant Biotechnol. J.* 19: 424.

Cherif, J., Mediouni, C., Ammar, W.B., and Jemal, F. (2011). Interactions of zinc and cadmium toxicity in their effects on growth and in antioxidative systems in tomato plants (*Solarium lycopersicum*). *J. Environ. Sci.* 23: 837–844.

Connolly, E.L., Fett, J.P., and Guerinot, M.L. (2002). Expression of the *IRT1* metal transporter is controlled by metals at the levels of transcript and protein accumulation. *Plant Cell* 14: 1347–1357.

Cui, Y. and Zhao, N. (2011). Oxidative stress and change in plant metabolism of maize (*Zea mays* L.) growing in contaminated soil with elemental sulfur and toxic effect of zinc. *Plant Soil Environ.* 57: 34–39.

Curie, C., Cassin, G., Couch, D. et al. (2009). Metal movement within the plant: contribution of nicotianamine and yellow stripe 1-like transporters. *Ann. Bot.* 103: 1–11.

Di-Baccio, D., Tognetti, R., Sebastiani, L., and Vitagliano, C. (2003). Responses of *Populus deltoides*× *Populus nigra* (Populus× euramericana) clone I-214 to high zinc concentrations. *New Phytol.* 159: 443–452.

Doncheva, S., Stoynova, Z., and Velikova, V. (2001). Influence of succinate on zinc toxicity of pea plants. *J. Plant Nutr.* 24: 789–804.

Evans, K.S., van Wijk, M.H., McGrath, P.T. et al. (2021). From QTL to gene: *C. elegans* facilitates discoveries of the genetic mechanisms underlying natural variation. *Trends Genet.* 37: 933–947.

Evens, N.P., Buchner, P., Williams, L.E., and Hawkesford, M.J. (2017). The role of ZIP transporters and group F bZIP transcription factors in the Zn-deficiency response of wheat (*Triticum aestivum*). *Plant J.* 92: 291–304.

Fei, X., Fu, X.Z., Wang, N.Q. et al. (2016). Physiological changes and expression characteristics of ZIP family genes under zinc deficiency in navel orange (*Citrus sinensis*). *J. Integr. Agric.* 15: 803–811.

Fu, X.Z., Zhou, X., Xing, F. et al. (2017). Genome-wide identification, cloning and functional analysis of the zinc/iron-regulated transporter-like protein (ZIP) gene family in trifoliate orange (*Poncirus trifoliata* L. Raf.). *Front. Plant Sci.* 8 (3): 588.

Fukao, Y., Ferjani, A., Tomioka, R. et al. (2011). iTRAQ analysis reveals mechanisms of growth defects due to excess zinc in Arabidopsis. *Plant Physiol.* 155: 1893–1907.

Gao, J., Luo, M., Peng, H. et al. (2019). Characterization of cadmium-responsive microRNAs and their target genes in maize (*Zea mays*) roots. *BMC Mol. Biol.* 20: 1–9.

Garg, N. and Singh, S. (2018). *Arbuscular mycorrhiza Rhizophagus irregularis* and silicon modulate growth, proline biosynthesis and yield in *Cajanus cajan* L. Millsp. (pigeonpea) genotypes under cadmium and zinc stress. *J. Plant Growth Regul.* 37: 46–63.

Gentzel, I.N., Park, C.H., Bellizzi, M. et al. (2020). A CRISPR/dCas9 toolkit for functional analysis of maize genes. *Plant Methods* 16: 1–9.

Glińska, S., Gapińska, M., Michlewska, S. et al. (2016). Analysis of *Triticum aestivum* seedling response to the excess of zinc. *Protoplasma* 253: 367–377.

Hacisalihoglu, G. and Kochian, L.V. (2003). How do some plants tolerate low levels of soil zinc? Mechanisms of zinc efficiency in crop plants. *New Phytol.* 159 (2): 341–350.

Huang, X. and Han, B. (2014). Natural variations and genome-wide association studies in crop plants. *Annu. Rev. Plant Biol.* 65: 531–551.

Imtiaz, M., Rashid, A., Khan, P. et al. (2010). The role of micronutrients in crop production and human health. *Pak. J. Bot.* 42: 2565–2578.

Ishimaru, Y., Masuda, H., Suzuki, M. et al. (2007). Overexpression of the *OsZIP4* zinc transporter confers disarrangement of zinc distribution in rice plants. *J. Exp. Bot.* 58: 2909–2915.

Jian, H., Yang, B., Zhang, A. et al. (2018). Genome-wide identification of microRNAs in response to cadmium stress in oilseed rape (*Brassica napus* L.) using high-throughput sequencing. *Int. J. Mol. Sci.* 19: 1431.

Johansen, I.E., Liu, Y., Jørgensen, B. et al. (2019). High efficacy full allelic CRISPR/Cas9 gene editing in tetraploid potato. *Sci. Rep.* 9: 1–7.

Jones-Rhoades, M.W. (2012). Conservation and divergence in plant microRNAs. *Plant Mol. Biol.* 80: 3–16.

Joshi, S., Kaur, K., Khare, T. et al. (2021). Genome-wide identification, characterization and transcriptional profiling of NHX-type (Na^+/H^+) antiporters under salinity stress in soybean. *3 Biotech* 11: 1–17.

Kaur, H. and Garg, N. (2021). Zinc toxicity in plants: a review. *Planta* 253: 1–28.

Khudsar, T., Iqbal, M., and Sairam, R. (2004). Zinc-induced changes in morpho-physiological and biochemical parameters in *Artemisia annua*. *Biol. Plant.* 48: 255–260.

Kosová, K., Vítámvás, P., Urban, M.O. et al. (2018). Plant abiotic stress proteomics: the major factors determining alterations in cellular proteome. *Front. Plant Sci.* 9: 122.

Kouhi, F., Sorkheh, K., and Ercisli, S. (2020). MicroRNA expression patterns unveil differential expression of conserved miRNAs and target genes against abiotic stress in safflower. *PLoS One* 15: e0228850.

Kozak, K., Papierniak, A., Barabasz, A. et al. (2019). *NtZIP11*, a new Zn transporter specifically upregulated in tobacco leaves by toxic Zn level. *Environ. Exp. Bot.* 157: 69–78.

Kozhevnikova, A.D., Seregin, I.V., Erlikh, N.T. et al. (2014). Histidine-mediated xylem loading of zinc is a species-wide character in *Noccaea caerulescens*. *New Phytol.* 203: 508–519.

Kwon, C.T., Heo, J., Lemmon, Z.H. et al. (2020). Rapid customization of Solanaceae fruit crops for urban agriculture. *Nat. Biotechnol.* 38: 182–188.

Lin, Y.F., Liang, H.M., Yang, S.Y. et al. (2009). Arabidopsis IRT3 is a zinc regulated and plasma membrane localized zinc/iron transporter. *New Phytol.* 182: 392–404.

Liu, H.J. and Yan, J. (2019). Crop genome-wide association study: a harvest of biological relevance. *Plant J.* 97: 8–18.

Llagostera, I., Pérez, M., and Romero, J. (2011). Trace metal content in the seagrass *Cymodocea nodosa*: differential accumulation in plant organs. *Aquat. Bot.* 95 (2): 124–128.

Malla, M.A., Dubey, A., Raj, A. et al. (2022). Emerging frontiers in microbe mediated pesticide remediation: Unveiling role of omics and in silico approaches in engineered environment. *Environ. Pollut.* 299: 118851.

Manghwar, H., Li, B., Ding, X. et al. (2020). CRISPR/Cas systems in genome editing: methodologies and tools for sgRNA design, off-target evaluation, and strategies to mitigate off-target effects. *Adv. Sci.* 7: 1902312.

Manolio, T.A. (2013). Bringing genome-wide association findings into clinical use. *Nat. Rev. Genet.* 14: 549–558.

Mapodzeke, J.M., Adil, M.F., Sehar, S. et al. (2021a). Myriad of physio-genetic factors determining the fate of plant under zinc nutrient management. *Environ. Exp. Bot.* 189: 104559.

Mapodzeke, J.M., Adil, M.F., Wei, D. et al. (2021b). Modulation of key physio-biochemical and ultrastructural attributes after synergistic application of zinc and silicon on rice under cadmium stress. *Plants* 10 (1): 87.

Maslinska-Gromadka, K., Barabasz, A., Palusinska, M. et al. (2021). Suppression of *NtZIP4*A/B changes Zn and Cd root-to-shoot translocation in a Zn/Cd status-dependent manner. *Int. J. Mol. Sci.* 22: 5355.

Milner, M.J., Craft, E., Yamaji, N. et al. (2012). Characterization of the high affinity Zn transporter from *Noccaea caerulescens*, *NcZNT1*, and dissection of its promoter for its role in Zn uptake and hyperaccumulation. *New Phytol.* 195: 113–123.

Mondal, T.K., Ganie, S.A., Rana, M.K., and Sharma, T.R. (2014). Genome-wide analysis of zinc transporter genes of maize (*Zea mays*). *Plant Mol. Biol. Rep.* 32: 605–616.

Monnet, F., Vaillant, N., Vernay, P. et al. (2001). Relationship between PSII activity, CO_2 fixation, and Zn, Mn and Mg contents of *Lolium perenne* under zinc stress. *J. Plant Physiol.* 158: 1137–1144.

Naskar, B., Modak, R., Sikdar, Y. et al. (2016). A simple Schiff base molecular logic gate for detection of Zn^{2+} in water and its bio-imaging application in plant system. *J. Photochem. Photobiol. A Chem.* 321: 99–109.

Ozdener, Y. and Kutbay, H.G. (2009). Toxicity of copper, cadmium, nickel, lead and zinc on seed germination and seedling growth in *Eruca sativa*. *Fresenius Environ.* 18: 26–31.

Pandita, D., Pandita, A., Wani, S.H. et al. (2021). Crosstalk of multi-omics platforms with plants of therapeutic importance. *Cells* 10: 1296.

Paradisone, V., Navarro-León, E., Albacete, A. et al. (2022). Improvement of the physiological response of barley plants to both zinc deficiency and toxicity by the application of calcium silicate. *Plant Sci.* 319: 111259.

Patil, S., Joshi, S., Jamla, M. et al. (2021). MicroRNA-mediated bioengineering for climate-resilience in crops. *Bioengineered* 12: 10430–10456.

Peer, W.A., Mahmoudian, M., Freeman, J.L. et al. (2006). Assessment of plants from the Brassicaceae family as genetic models for the study of nickel and zinc hyperaccumulation. *New Phytol.* 172: 248–260.

Pei, X., Jiang, H., Xu, G. et al. (2022). Lethality of zinc oxide nanoparticles surpasses conventional zinc oxide via oxidative stress, mitochondrial damage and calcium overload: a comparative hepatotoxicity study. *Int. J. Mol. Sci.* 23: 6724.

Qiu, Z., Hai, B., Guo, J. et al. (2016). Characterization of wheat miRNAs and their target genes responsive to cadmium stress. *Plant Physiol. Biochem.* 101: 60–67.

Rafie, M., Khoshgoftarmanesh, A., Shariatmadari, H., and Darabi, A. (2022). Apoplastic and symplastic zinc concentration of intact leaves of field onion (*Allisum cepa*) as affected by foliar application of $ZnSO_4$ and Zn-amino chelates. *J. Plant Nutr.* 46: 1–12.

Rahman, R., Sofi, J.A., Javeed, I. et al. (2020). Role of micronutrients in crop production. *Int. J. Curr. Microbiol. Appl. Sci.* 11: 2265–2287.

Ramesh, S.A., Shin, R., Eide, D.J., and Schachtman, D.P. (2003). Differential metal selectivity and gene expression of two zinc transporters from rice. *Plant Physiol.* 133: 126–134.

Ricachenevsky, F.K., Menguer, P.K., and Sperotto, R.A. (2013). kNACking on heaven's door: how important are NAC transcription factors for leaf senescence and Fe/Zn remobilization to seeds? *Front. Plant Sci.* 4: 226.

Rout, G.R. and Das, P. (2009). *Effect of Metal Toxicity on Plant Growth and Metabolism: I. Zinc Sustainable Agriculture*, 873–884. Springer.

Sagardoy, R., Vázquez, S., Florez-Sarasa, I. et al. (2010). Stomatal and mesophyll conductances to CO_2 are the main limitations to photosynthesis in sugar beet (*Beta vulgaris*) plants grown with excess zinc. *New Phytol.* 187: 145–158.

Sagonda, T., Adil, M.F., Sehar, S. et al. (2021). Physio-ultrastructural footprints and iTRAQ-based proteomic approach unravel the role of *Piriformospora indica*-colonization in counteracting cadmium toxicity in rice. *Ecotoxicol. Environ. Saf.* 220 (3): 112390.

Sahoo, J.P., Behera, L., Sharma, S.S. et al. (2020). Omics studies and systems biology perspective towards abiotic stress response in plants. *Am. J. Plant Sci.* 11: 2172.

Schvartzman, M.S., Corso, M., Fataftah, N. et al. (2018). Adaptation to high zinc depends on distinct mechanisms in metallicolous populations of *Arabidopsis halleri*. *New Phytol.* 218: 269–282.

Sinclair, S.A. and Krämer, U. (2012). The zinc homeostasis network of land plants. *Biochimica et Biophysica Acta (BBA)-Molecular. Cell Res.* 1823: 1553–1567.

Soares, C.R.F.S., Grazziotti, P.H., Siqueira, J.O. et al. (2001). Zinc toxicity on growth and nutrition of *Eucalyptus maculata* and *Eucalyptus urophylla* in nutrient solution. *Pesq. Agrop. Brasileira* 36: 339–348.

Song, Y., Hudek, L., Freestone, D. et al. (2014). Comparative analyses of cadmium and zinc uptake correlated with changes in natural resistance-associated macrophage protein (NRAMP) expression in *Solanum nigrum* L. and *Brassica rapa*. *Environ. Chem.* 11 (6): 653–660.

Souza, S.C., Souza, L.A., Schiavinato, M.A. et al. (2020). Zinc toxicity in seedlings of three trees from the Fabaceae associated with Arbuscular mycorrhizal fungi. *Ecotoxicol. Environ. Saf.* 195: 110450.

Sytar, O., Ghosh, S., Malinska, H. et al. (2021). Physiological and molecular mechanisms of metal accumulation in hyperaccumulator plants. *Physiol. Plant.* 173: 148–166.

Tiong, J., McDonald, G.K., Genc, Y. et al. (2014). *HvZIP7* mediates zinc accumulation in barley (*Hordeum vulgare*) at moderately high zinc supply. *New Phytol.* 201: 131–143.

Ullah, A., Farooq, M., Nadeem, F. et al. (2020). Zinc seed treatments improve productivity, quality and grain biofortification of desi and kabuli chickpea (*Cicer arietinum*). *Crop Pasture Sci.* 71: 668–678.

Ursache, R., Fujita, S., Tendon, V.D., and Geldner, N. (2021). Combined fluorescent seed selection and multiplex CRISPR/Cas9 assembly for fast generation of multiple Arabidopsis mutants. *Plant Methods* 17: 1–14.

Wada, N., Ueta, R., Osakabe, Y., and Osakabe, K. (2020). Precision genome editing in plants: state-of-the-art in CRISPR/Cas9-based genome engineering. *BMC Plant Biol.* 20: 1–12.

Wei, C., Jiao, Q., Agathokleous, E. et al. (2022). Hormetic effects of zinc on growth and antioxidant defense system of wheat plants. *Sci. Total Environ.* 807: 150992.

Wu, S., Guo, Y., Adil, M.F. et al. (2020). Comparative proteomic analysis by iTRAQ reveals that plastid pigment metabolism contributes to leaf color changes in tobacco (*Nicotiana tabacum* L.) during curing. *Int. J. Mol. Sci.* 21: 2394.

Xiong, H., Kobayashi, T., Kakei, Y. et al. (2012). *AhNRAMP1* iron transporter is involved in iron acquisition in peanut. *J. Exp. Bot.* 63: 4437–4446.

Yang, Q., Ma, X., Luo, S. et al. (2018). *SaZIP4*, an uptake transporter of Zn/Cd hyperaccumulator *Sedum alfredii* Hance. *Environ. Exp. Bot.* 155: 107–117.

Yang, M., Li, Y., Liu, Z. et al. (2020). A high activity zinc transporter *OsZIP9* mediates zinc uptake in rice. *Plant J.* 103: 1695–1709.

Yu, J., Wu, L., Fu, L. et al. (2019). Genotypic difference of cadmium tolerance and the associated microRNAs in wild and cultivated barley. *Plant Growth Regul.* 87: 389–401.

Yu, Q., He, L., Huo, C. et al. (2021). Genome-wide identification and expression analysis of heavy metal stress–responsive metallothionein family genes in *Nicotiana tabacum*. *Plant Mol. Biol. Rep.* 39: 443–454.

Zhang, J., Soomro, A., Chai, L. et al. (2013). Mapping of QTL for iron and zinc toxicity tolerance at seedling stage using a set of reciprocal introgression lines of rice. *Acta Agron. Sin.* 39: 1754–1765.

Zhang, Y., Wang, Y., Ding, Z. et al. (2017). Zinc stress affects ionome and metabolome in tea plants. *Plant Physiol. Biochem.* 111: 318–328.

Zhang, Q., Chen, H., Xu, C. et al. (2019). Heavy metal uptake in rice is regulated by pH-dependent iron plaque formation and the expression of the metal transporter genes. *Environ. Exp. Bot.* 162: 392–398.

Zhou, X., Joshi, S., Patil, S. et al. (2022). Reactive oxygen, nitrogen, carbonyl and sulfur species and their roles in plant abiotic stress responses and tolerance. *J. Plant Growth Regul.* 41: 119–142.

Zhu, Y., Lin, Y., Chen, S. et al. (2019). CRISPR/Cas9-mediated functional recovery of the recessive rc allele to develop red rice. *Plant Biotechnol. J.* 17: 2096–2105.

Zhu, H., Li, C., and Gao, C. (2020). Applications of CRISPR–Cas in agriculture and plant biotechnology. *Nat. Rev. Mol. Cell Biol.* 21: 661–677.

14

Arsenic Toxicity and Tolerance in Plants

Insights from Omics Studies

Barsha Majumder, Palin Sil, and Asok K. Biswas

Department of Botany, Plant Physiology and Biochemistry Laboratory, Centre of Advanced Study, University of Calcutta, Kolkata, India

14.1 Introduction

Anthropogenic intrusions along with soil mineralization and weathering have accentuated heavy metal accumulation in the biosphere perturbing ecosystems worldwide. Heavy metal accumulation in soils adversely affects environmental health of soil organisms, hampers crop growth, and lowers productivity (Gill 2014). Bioaccumulation of these heavy metals in the food chain is an environmental distress and a critical health concern for plants as well as animals (Emamverdian et al. 2015). The nonessential ubiquitous environmental contaminant arsenic (As) has received attention lately owing to the ecotoxicological risks posed by metalloid. Arsenic is the 20th most abundant element in earth crust and constitutes about 5 $mg\,kg^{-1}$ of earth's crust existing in soil at concentrations ranging from 0.1 to 40 $mg\,kg^{-1}$ (Mirza et al. 2014). Arsenic contamination in soil and ground water is one of the worst natural geo environmental disasters reported globally (Dey et al. 2014). However, the crisis is acute in South East Asian countries particularly Bangladesh, India (West Bengal), Taiwan, and China where the concentration of the metalloid exceeds far beyond the permissible threshold of 0.01 $mg\,l^{-1}$ as stated by the World Health Organization (McCarty et al. 2011). The US Environmental Protection Agency (EPA) and the International Agency for Research on Cancer (IARC) have classified arsenic as a Class I human carcinogen. More than 200 million people worldwide, with approximately 38 million residing in the Indo-Bangladesh region are chronically exposed to elevated concentrations of the toxicant through food, water, soil and air are afflicted with long-term health adversities. Wide range of health consequences in humans include arsenicosis, black foot disease, cognitive impairment, neurological defects, diabetes mellitus, chronic kidney disease, various forms of cancer, cardiovascular, and peripheral vascular diseases (Flora 2011; Luis et al. 2019).

In soil, arsenic exists principally in two biologically relevant oxidative states: pentavalent arsenate [As(V)] and the trivalent arsenite [As(III)]. Arsenate is found under aerobic conditions, whereas arsenite predominates the anaerobic environment (Abedin et al. 2002;

Heavy Metal Toxicity and Tolerance in Plants: A Biological, Omics, and Genetic Engineering Approach, First Edition. Edited by Mohammad Anwar Hossain, AKM Zakir Hossain, Sylvain Bourgerie, Masayuki Fujita, Om Parkash Dhankher, and Parvez Haris.

Zhao et al. 2009). Arsenite can also be methylated by soil microorganisms to produce methylated As species (Di et al. 2019). The toxicological concerns of arsenic depend on the concentration of the metalloid in the soil and its oxidative state. Generally, trivalent aromatic arsenicals are considered to be the most toxic followed by trivalent methylated arsenicals, trivalent inorganic As(III), and the pentavalent As(V) species in a sequential manner (Luis et al. 2019). The two inorganic As species are readily inter convertible depending on the environmental conditions. Plants can take up different As species through varied mechanisms. As(V) taken up by the high-affinity phosphate uptake system is loaded into xylem vessels by phosphate transporter (PHT) proteins (Ullrich-Eberius et al. 1989; Meharg and Macnair 1992; Zhao et al. 2010; Wu et al. 2011) while As(III) after its uptake through the aquaporin channels enters the stele through the silicon (Si) transport pathway (Ma et al. 2008; Zhao et al. 2010). Both inorganic arsenic species are highly toxic to plants.

Alterations in cellular mechanisms and aberrant gene expression patterns are the noticeable effects in plants grown in arsenic contaminated areas. Phytotoxic symptoms under As exposure include decline in plant biomass, chlorosis, inhibition of growth, photosynthesis, water and nutrient imbalance ultimately senescence and plant death (Choudhury et al. 2011; Tripathi et al. 2013; Dixit et al. 2015b; Saha et al. 2017b; Sil et al. 2018; Singh et al. 2020). Being a structural analogue of Pi (phosphate), As(V) has the ability to replace Pi in mitochondrial ATP synthase to form unstable ADP-As(V) esters that undergoes rapid hydrolysis and generates futile reaction cycles to uncouple respiratory and photosynthetic electron transport chain disturbing cellular energy production (Finnegan and Chen 2012). As(III) on the other hand, binds with the thiol groups of proteins and cofactors consequently disrupts protein structure and protein-protein interactions that affects crucial metabolic processes including fatty acid metabolism, glucose uptake and glutathione production (Martínez-Castillo et al. 2022). Oxidative stress is one of the most important manifestation of arsenic toxicity that incurs drastic physiological and molecular readjustments, disrupting cellular homeostasis resulting in oxidative injuries (Singh et al. 2016; Jung et al. 2019; Sil and Biswas 2020). Assemblage of reactive oxygen species (ROS) within the plant cell triggers complex defense mechanisms therein to facilitate production of both enzymatic and non-enzymatic antioxidants as well as thiols to safeguard the cellular machinery against the toxic effects of arsenic (Hartley-Whitaker et al. 2001). Arsenic-incited synthesis and production of low molecular weight thiols and phytochelatins (PCs) help to bind As(III) forming As(III)–PC complexes that are compartmentalized and sequestered in vacuoles for detoxification (Dixit et al. 2015b).

Additionally, application of omics approaches have assisted to explore modifications in molecular responses involving biological components, viz., genes, mRNA, proteins, and metabolites under arsenic toxicity (Kumar et al. 2015). Genome-wide gene expression analysis has divulged alterations in expression profiles of a wide array of genes, viz. stress-responsive gene, heat-shock proteins, metallothioneins, sulfate-metabolizing proteins, and assimilation pathway under As stress in rice (Dixit et al. 2015c). Arsenic exposure has been reported to induce genotoxic responses making DNA susceptible to ROS injury. ROS-mediated DNA-protein adducts, strand breaks, chromosomal aberrations, and mutations are common in arsenic-contaminated tissues (Duquesnoy et al. 2010; Ahmad et al. 2012). A decrease in genomic template stability (GTS) along with an increase in random amplified polymorphic DNAs (RAPDs) and DNA methylation has also been elucidated in maize and rice cultivars (Erturk et al. 2015; Majumder et al. 2019a, b). Arsenic imposition brought about modulations in metabolic networks to influence diverse developmental pathways including lipid metabolism, phytohormone signaling, thiol metabolism, as well as sugar metabolism essential for growth and development (Choudhury et al. 2010; Yu et al. 2012; Sil et al. 2019a; Majumder et al. 2022). Proteomic analysis of rice shoots under As(V) stress indicated presence of 38 differentially displayed proteins related to photosynthesis (Liu et al. 2013). Moreover, there was deformation of chloroplast structure and distortion of grana thylakoid structure upon As exposure that reduced the carbon fixation. Results obtained from infrared (FTIR) and near-infrared (FTNIR) spectroscopic techniques have correlated arsenic-incited molecular modifications with interactions of cellular biomolecules including nucleic acids, carbohydrates, lipids, and proteins in broad bean (*Vicia faba*) (Boccia et al. 2013).

Apart from physiological and biochemical alterations, molecular alterations can therefore, emerge as potent determinants of stress responses in plants. With advancements in instrumentation and analytical techniques, arsenic research has evolved, and newer insights involving utilization of *Omics* approaches encompassing genomics, transcriptomics and metabolomics under arsenic stress further affords to decipher the arsenic-induced alterations in growth and metabolism and serves as a platform to develop As-tolerant crops. The present chapter intends to put forth the current trends in arsenic research that incorporates arsenic uptake, transport, sequestration, and chelation along with *omics* approaches that would assist to comprehend in developing mitigation measures to counteract the crisis of food-chain contamination by As and to build up arsenic tolerance.

14.2 Occurrence and Distribution of As in the Environment

The presence of arsenic (As) in food chain and trophic levels in consequence of As pollution in soil and groundwater through natural and anthropogenic activities causes undesirable adversities to ecosystem and food security (Hettick et al. 2015; Vithanage et al. 2017; Pandey et al. 2020). It has been reported that As is a chief constituent of several forms of mineral, viz. realgar (AsS), claudetite (As_2O_3), scorodite ($FeAsO_4{\cdot}2H_2O$), arsenopyrite (FeAsS), orpiment (As_2O_3), diarsenic pentoxide (As_2O_5), and arsenolite (As_4O_6) where arsenopyrite is found to be the most common source (Sarkar and Paul 2016). Different forms of As present in natural water and sediments include arsenate (AsO_4^{3-}), arsenite (AsO_3^{3-}), methyl arsenic acid, dimethylarsinic acid, arsenious acids (H_3AsO_3, $H_3AsO_3^{2-}$), arsenic acids (H_3AsO_4, $H_3AsO_4^{-}$, $H_3AsO_4^{2-}$), and arsine (Nabi et al. 2021). Srivastava and Sharma (2013a, b) proclaimed that seven states of India are censoriously affected by As-polluted groundwater. It has been reported that various anthropogenic activities, viz. mining, copper smelting, irrigation with As-polluted water, paints, excess use of pesticides, herbicides and fertilizers in agriculture, slag, electrode manufacturing, dyes, additives of poultry feed, cigarettes, timber preservatives, and coal combustion contribute to arsenic contamination and its wide distribution in the environment (Basu et al. 2014). Liao et al. (2005) documented in their study that approx. 52,000–112,000 tons of As distributed in the environment annually due to anthropogenic activities. Agency for Toxic Substances and Disease Registry (ATSDR) has registered As in the first position by reason of acute toxicity and health hazards caused by As exposure to human (ATSDR 2019).

14.3 Arsenic Uptake, Accumulation, and Detoxification in Plants

Arsenic in soil mainly exists in inorganic as well as organic forms and the inorganic species are considered to be more toxic in comparison to the organic ones. The major inorganic forms of As in soil include arsenate [As(V)] and arsenite [As(III)], while the methylated As species are the organic species (Jia et al. 2012).

14.3.1 Uptake of Inorganic Arsenic

Arsenate [As(V)] is the predominant As species in aerobic soils, while arsenite [As(III)] dominates under anoxic conditions. Microorganisms in the soil transform inorganic arsenic to organic forms (Heidari 2017). The solubility and speciation of As, as well as its uptake from the soil is governed by various factors. The solubility of arsenate increases with increasing pH, while with decreasing pH solubility of arsenite is improved. Being a phosphate (Pi) analogue, arsenate and phosphate share similar transport pathway where the uptake mechanism involves cotransport of phosphate or arsenate and protons, with $2H^+$ for each phosphate or arsenate (Ullrich-Eberius et al. 1989). Under low phosphate conditions, As(V) subdues the entry of phosphate and intensifies Pi-deprivation symptoms in plants. A number of unidirectional phosphate transporters belonging to the phosphate transporter 1 (*Pht1*) family with considerable role in phosphate acquisition from both low- and high-phosphorus environments have been identified (Wu et al. 2011; DiTusa et al. 2016; Wang et al. 2016). These are strongly expressed in roots and are implicated in phosphate uptake. As(V) uptake in rice utilizes *Pht1* transporters, OsPht1;1 and OsPht1;8. The *Pht1* transporter OsPht1;1 localized in the plasma membrane is involved in As(V) uptake, while OsPht1;8 expressed in both root and shoot has high affinity for Pi and As(V) and its overexpression markedly enhanced As(V) uptake and translocation in rice (Jia et al. 2011; Kamiya et al. 2013). Again, knockout of OsPht1;8 was shown to decrease As(V) uptake, thereby increasing As(V) tolerance (Wang et al. 2016). However, in flooded soils As (V) is not the predominant form of As and so Pi transport pathway may not be considered to be the major route for As uptake (Li et al. 2015).

Arsenite [As(III)] dominates in reducing environments and its uptake is mediated by nodulin-26-like intrinsic proteins (NIPs) belonging to the aquaporin family (Ma et al. 2008; Kamiya et al. 2009; Xu et al. 2015). Owing to the structural analogy between As(III) and Si, As(III) uses silicon transporters (Figure 14.1). The Si influx transporter Lsi1 (Low silicon rice 1; OsNIP2;1) and Si efflux transporter Lsi2 (Low silicon rice 2) mediate As(III) uptake and efflux, respectively, in rice (Ma et al. 2006, 2007, 2008). These transporters are located in the plasma membrane of exodermal and endodermal cells in roots having opposite polarities (Ma et al. 2006, 2007). Concerted actions of these transporters cause an effectual flow of Si as well as As(III) across the endodermis and toward the xylem (Ma et al. 2008). However, Lsi2 has more critical role than Lsi1 as mutation in Lsi2 decreases As concentrations significantly and disrupts Si uptake inhibiting growth and yield in rice (Ma et al. 2007). Other plasma membrane intrinsic proteins (PIPs of the aquaporin family), viz. OsPIP2;4, OsPIP2;6, and OsPIP2;7 are also involved in As(III) transport and tolerance (Mosa et al. 2012).

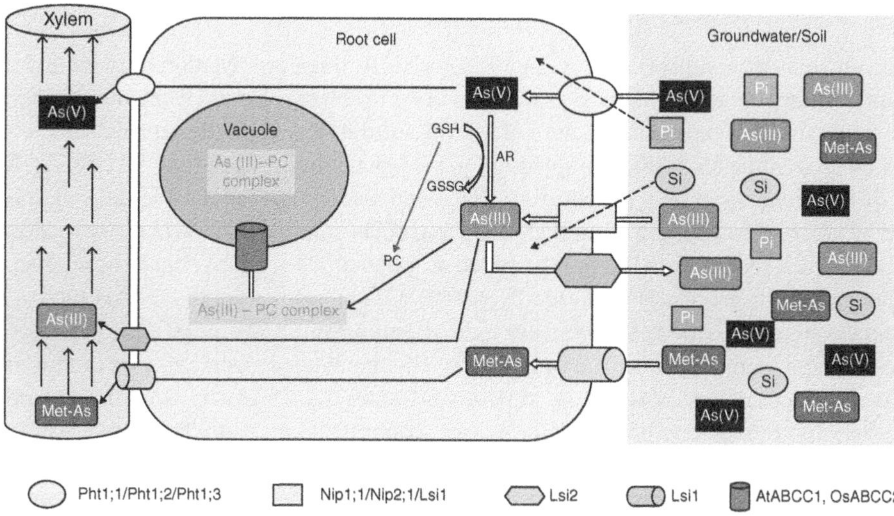

Figure 14.1 Schematic diagram representing transporter assisted uptake, translocation and vacuolar sequestration of arsenic (As). As(V) – arsenate; As(III) – arsenite; PC – phytochelatin; As(III)-PC complex – arsenite–phytochelatin complex; AR – arsenate reductase; GSH – reduced glutathione; GSSG – oxidized glutathione; Pi – phosphate; Si – silicon; Met-As – methylated species of arsenic; Pht – phosphate transporter; Nip – nodulin 26 like intrinsic protein; Lsi – silicon transporter; ABCC – ATP binding cassette transporter subfamily C member.

14.3.2 Uptake of Methylated Arsenic

Uptake of methylated As species (MMA and DMA) by plants also occurs but with lesser efficiency in comparison to As(V) or As(III), partly due to lower abundance in soil and diminishes with the increase in number of methyl group (Abedin et al. 2002; Raab et al. 2007; Zhao et al. 2013; Zheng et al. 2013). MMA uptake is mediated by the silicic acid transporter Lsi1, while the transport pathway of DMA has not been elucidated (Li et al. 2009; Carey et al. 2011). MMA(V) is partially reduced to trivalent MMA(III) in rice roots but only MMA(V) is translocated to shoots (Abedi and Mojiri 2020). On the other hand, DMA(V) moves easily from root to shoot probably due to its poor SH (sulfhydryl) coordination and altered molecular structure, in comparison to As(III) (Raab et al. 2007).

14.3.3 Arsenic Accumulation and Detoxification

The pentavalent As(V) after its uptake is reduced to trivalent As(III) before its extrusion or sequestration. The reduction of As(V) to As(III) may occur both enzymatically and non-enzymatically. As(V) can be directly reduced to As(III) via arsenate reductase or reduction of As(V) to As(III) may occur using glutathione (GSH) as reductant by GSH reductase (Mukhopadhyay et al. 2000; Foyer and Noctor 2011). Cytosolic arsenite, either from direct uptake via aquaglyceroporin or formed after arsenate reduction, is detoxified by binding to phytochelatins (PCs) and are sequestered to vacuoles for detoxification (Brait 2010). Transport of these complexes across tonoplast is mediated by a C-type ATP-binding cassette (ABC) transporter and has crucial role for As resistance in plants (Song et al. 2014).

14.3.4 Arsenic Methylation and Volatilization

Microbe-mediated modification of As may also occur through methylation or demethylation and changes the toxicity as well as mobility of As in soil to substantial extents. Arsenic methylation involves biotransformation of arsenic from inorganic to the organic state followed by its volatilization, thus completing the biogeochemical cycle of the metalloid (Di et al. 2019). Here, As(V) is reduced to As(III) by reduced glutathione inside the cell, followed by progressive addition of methyl group leading to continuous enzymatic reduction of methylated arsenic from MMA to the nontoxic gaseous TMAs(III) as the final product. The detoxification process is catalyzed by As(III) *S*-adenosylmethionine (SAM) methyltransferase enzymes (ArsM) that transfer methyl group from SAM to As(III) to produce mono-, di-, and tri-methyl arsenical compounds. The methylated arsenic species produced in this sequence include MMAs(III), MMAs(V), DMAs(III), DMAs(V), TMAs(V)O, and TMAs(III) and the toxicity of arsenic decreases sequentially. As methylation can lead to biovolatilization of methylarsine gases that decreases As concentration in the soil and can be utilized as a potential technology for As bioremediation. More than 30,000 arsM genes responsible for methylation have been deposited in the National Centre for Biotechnology Information (NCBI) database, with highly diverse sequences belonging to fungi, bacteria, methanoarchaea, and other eukaryotic microorganisms but not in higher plants (Qin et al. 2006; Chen et al. 2017b). Although a gene from rice (Os02g51030) contains similar motif with ArsM/AS3MT but does not contain the core region of ArsM/AS3MT responsible for methyl group transfer (Ye et al. 2012). Transformation of the RpArsM gene to rice by Meng et al. (2011) led to the production of methylated As species and volatile arsenicals to greater extents over control, thus providing a strategy to reduce accumulation of As species in rice grains.

Biotransformation between inorganic arsenicals and methyl arsenicals is a bidirectional process and methylated arsenic species can be demethylated to more toxic and carcinogenic inorganic arsenic (Yoshinaga et al. 2011). It is a two-step process that involves reduction of MMAs(V) to MMAs(III) followed by demethylation of MMAs(III) to inorganic arsenic and is also coordinated by microorganisms.

14.4 Influence of Arsenic on Phytotoxicity

14.4.1 Germination and Growth

Presence of arsenic in the rhizosphere above the threshold level hinders normal functioning of plant metabolic activities. Decrease in germination and early seedling growth has been documented in cereals under arsenic exposure (Chun-Xi et al. 2007; Shri et al. 2009). Germination percentage, germination index, and vitality index displayed a decreasing trend with increasing concentrations of arsenic. Arsenic-induced reduction in root and shoot growth has been reported in rice (*Oryza sativa* L.), black gram (*Vigna mungo* L.), pea (*Pisum sativum* L.), maize (*Zea mays* L.), wheat (*Triticum aestivum* L.), spinach (*Spinacia oleracea* L.) (Choudhury et al. 2011; Srivastava and Sharma 2014; Singh et al. 2015a; Tripathi et al. 2016; Sil et al. 2018; Amna et al. 2020). Reduction in root growth is greater in comparison to shoot as roots are the first to come in contact with arsenic and are damaged most. Arsenic exposure also suppresses the number of leaves; reduces leaf area; and induces chlorosis, necrosis, defoliation, and senescence (Abbas et al. 2018). Arsenic accumulation in the roots and other plant parts decreases growth, fresh and dry weights, as well as relative water content (Hasanuzzaman and Fujita 2013; Srivastava and Sharma 2014; Sil et al. 2018; Majumder et al. 2019a, b). As stress has been explicated to decrease endogenous levels of three auxins, viz. indole acetic acid, naphthoxy acetic acid, and indole-3-butyric acid in mustard (*Brassica juncea*) that reduces growth (Srivastava et al. 2013).

14.4.2 Nutrient Uptake

Alterations in the uptake of essential elements caused shift in elemental homeostasis under arsenic toxic conditions (Mallick et al. 2013; Sil et al. 2019a). Arsenic manipulates uptake of macro- and micronutrients by competing with nutrient ions for binding to uptake carriers (Gusman et al. 2013a). Arsenic-induced disruption in plant water status also impedes nutrient absorption to generate diverse stresses that ultimately reduces growth (Farnese et al. 2014; Sil et al. 2019a). Readjustments in transport and metabolic processes affect plant biomass, causes wilting, curling, chlorosis, as well as necrosis of leaf blades and decreases leaf area (Singh et al. 2006). Micromorphologic and ultrastructural alterations of parenchyma include initiation of circular necrotic patches at specific regions due to intense membrane vesiculation, which triggers plasma membrane rupture and extravasation of cytoplasm culminating in cell death (Farnese et al. 2017).

14.4.3 Oxidative Stress and Antioxidative Defense

Uptake and accumulation of arsenic in the plant tissues induces generation of ROS during conversion from As(V) to As(III) that triggers oxidative stress and upsets the plant metabolism, as well as cellular redox (Finnegan and Chen 2012). As(III) has the ability to react with the thiol (–SH) groups of enzymes and disturbs cellular, molecular, and physiological processes in plant species (Choudhury et al. 2011; Tripathi et al. 2013; Singh et al. 2016; Tripathi et al. 2016; Sil and Biswas 2020; Singh et al. 2020). Again, biomethylation of As produces organic As species that react with oxygen (O_2) to generate ROS (Sharma 2012). ROS including singlet oxygen ($^1O_2$), superoxide (O_2^-), hydrogen peroxide (H_2O_2), and hydroxyl radicals (OH^-) has the ability to damage amino acids, proteins, nucleic acids, and causes peroxidation of membrane lipids. Activation of nonenzymatic antioxidants, viz. ascorbate (ASA), reduced glutathione (GSH), carotenoids, and proline, as well as enzymatic antioxidants, viz. superoxide dismutase (SOD), ascorbate peroxidase (APX), catalase (CAT), glutathione reductase (GR), glutathione *S*-transferase (GST), and glutathione peroxidase (GPX) have been reported to counter arsenic-mediated oxidative stress and protect plant cells against the ROS-incited injuries (Zvobgo et al. 2019; Sil and Biswas 2020; Singh et al. 2020). Arsenic exposure alters the contents of antioxidants to differential extents (Farooq et al. 2016; Jung et al. 2019; Singh et al. 2020). Production of these antioxidants causes metabolic alterations and involves diversion of carbon, nitrogen, sulfur and energy from normal growth and development.

Build up of O_2^- and H_2O_2 under As exposure initiates membrane lipid peroxidation and causes malondialdehyde (MDA) production that promotes electrolyte leakage in cellular membranes (Hasanuzzaman and Fujita 2013; Rahman et al. 2015; Tripathi et al. 2016; Sil et al. 2018; Majumder et al. 2019a, b). Accumulation of H_2O_2 and MDA within the tissues inactivates enzymes by oxidizing their thiol groups that upset antioxidative defense.

14.4.4 Ascorbate–Glutathione Cycle

The ascorbate–glutathione (AsA–GSH) cycle participates in reducing H_2O_2 accumulation within the cells. In this cycle, AsA and GSH are successively oxidized and reduced with cyclic transfer of electrons to scavenge H_2O_2. The oxidized AsA is regenerated to maintain the reduced AsA pool at the expense of GSH as an electron donor by the enzymes of the AsA–GSH cycle that include ascorbate peroxidase (APX), monodehydroascorbate reductase (MDHAR), glutathione-dependent dehydroascorbate reductase (DHAR), and glutathione reductase (GR). Highly reduced GSH and AsA pools are necessary to sustain optimal functioning of the AsA–GSH cycle. GR is one of the vital enzymes of AsA–GSH pathway that sustains the reduced status of GSH and acts as a substrate for glutathione *S*-transferase (GST). GPX catalyzes H_2O_2 degradation and forms oxidized glutathione (GSSG), which is further reduced to GSH by GR. GST mediates the conjugation of various electrophilic xenobiotics with GSH to catalyze their detoxification. Higher levels of GSH and GSH/GSSG ratio are vital to maintain the antioxidative defense. Arsenic imposition has been reported to enhance the activity of APX along with a decline in activity of DHAR that led to exhaustion in the ASA pool in wheat (*T. aestivum* L.) seedlings (Hasanuzzaman and Fujita 2013). The activities of GR and GST were also enhanced along with an increase in GSH and GSSG contents, but the GSH/GSSG ratio decreased. Deviation from the mentioned trend was also evident where APX, DHAR, and GR activities were inhibited by excess As(V) that lowered AsA and GSH pools in maize (*Z. mays* L.) as well (Tripathi et al. 2016). On the other hand, escalations in APX activity accompanied by marked reduction in activities of MDHAR, DHAR, and GR have also been demonstrated in rice (*O. sativa* L.) seedlings (Jung et al. 2019). This enhanced the content of DHA and GSSG and concomitantly decreased the GSH redox ratio (GSH/GSSG) and AsA redox ratio (AsA/DHA) to dismantle the cellular redox homeostasis.

14.4.5 Photosynthesis

Arsenic is reported to reduce the rate of photosynthesis (Gusman et al. 2013a; Tripathi et al. 2016; Farnese et al. 2017; Sil et al. 2019a; Zemanová et al. 2021). Damages elicited in the inner membrane of chloroplast along with vesiculation in the outer membrane decreased levels of photosynthetic pigments consequently suppressing carbon assimilation rate and the activities of photosystems. Decrease in the levels of photosynthetic pigments under arsenic stress has been reported in lettuce (*Lactuca sativa* L.) (Gusman et al. 2013a); spinach (*S. oleracea* L.) (Srivastava and Sharma 2013c); maize (*Z. mays* L.) (Da silva et al. 2015); brinjal (*Solanum melongena* L.) (Singh et al. 2015); wheat (*T. aestivum* L.) (Sil et al. 2019a); rice (*O. sativa* L.) (Majumder et al. 2020). Such reduction in photosynthetic pigments, viz. chlorophylls and carotenoids, resulted in a concomitant decrease in the gas exchange parameters, viz. net photosynthesis, stomatal conductance, internal CO_2 concentration affected carbon assimilation, growth, survival, and reproduction (Sil et al. 2019a; Majumder et al. 2020). Decline in net photosynthesis under As(V) stress may be attributed to the inactivation of enzymes of the carbon fixation pathway (Finnegan and Chen 2012). This reduces the CO_2 demand and subsequently stomatal conductance. Additionally, As(V)-induced stomatal closure also decreased stomatal conductance and transpiration rate, thereby slowing down water and nutrient absorption (Sil et al. 2019a; Zemanová et al. 2021). Decrease in internal CO_2 concentration due to decreased stomatal conductance in turn lowered the net photosynthetic rate. Arsenic stress also affected the light harvesting complexes and reduced electron flow through the thylakoid membranes that decreased the production of ATP and NADPH. Decrease in chlorophyll a/b ratio in oat (*Avena sativa*) and Chinese ginseng (*Panax notoginseng*) has been explicated to cause structural damages in the thylakoid membrane that reduced the efficiency of quantum energy conversion into chemical energy and ultimately lowered photosynthetic rate (Stoeva and Bineva 2003; Zu et al. 2016).

Additionally, altered activities of ribulose-1,5-biphosphate carboxylase/oxygenase (RuBisCO) precursors have been documented to downregulate the RuBisCO-mediated reaction leading to reduced carbon fixation under As(V) stress that lowered photosynthetic rate in rice (Finnegan and Chen 2012; Dixit et al. 2015c). Inactivation of enzymes under As(V) stress was ascribed to reduce CO_2 demand, stomatal conductance, internal CO_2 concentration, and consequently net photosynthesis in wheat seedlings (Sil et al. 2019a). Alterations in fluorescence parameters indicating an increase of minimal fluorescence yield (F_0), decrease of maximal fluorescence yield (F_v), as well as basal quantum yield (F_v:F_0) and maximum quantum efficiency of PSII (F_v:F_m) ratios in As-stressed plants also indicated damage to the PSII in As-stressed maize and mustard (Tripathi et al. 2016; Singh et al. 2018).

14.4.6 Respiration

Arsenic contamination considerably affects the respiratory process in plants (Shao et al. 2011; Saha et al. 2017a; Sil et al. 2018). The respiration rate in leaves and roots of wheat seedlings was enhanced at lower concentrations of As(V) (Shao et al. 2011). Increment in respiratory activity under low concentrations of As(V) has been attributed to the chemical similarity between the phosphate and arsenate, which compete for the same active site of the mitochondrial ATP synthase. However, formation of highly unstable As-ADP adducts generates futile reaction cycles of As-ADP generation and uncouples respiratory and photosynthetic electron transport chains causing a decrease in ATP production (Finnegan and Chen 2012; Shrivastava et al. 2015). Disturbance in ATP production compromises the energy status of the cell by decreasing the production of ATP and NADPH needed to fuel the carbon fixation reactions and enhances ROS production. Excess of ROS disrupts the mitochondrial membrane system and lowers enzyme activities. Higher concentration of As(V) on the contrary lowered the activities of dehydrogenases in rice and wheat (Shao et al. 2011; Saha et al. 2017a; Sil et al. 2018). Decline in reducing power due to down regulation in dehydrogenase activity disrupted cellular redox, lowered respiration and reduced growth. As(III) inhibits the activities of PDH complex and the 2-oxoglutarate dehydrogenase (OGDH) complex by binding to the lipoic acid moiety (Shen et al. 2013). This depletes the mitochondrial NADH pool and leads to oxidative stress. Decreased activity of the dehydrogenase enzymes in the TCA cycle decreased entry of carbon into the cycle and brought about changes in the levels of TCA cycle intermediates as well. As(V)-incited enhancements in the levels of TCA cycle intermediates, viz. pyruvate, citrate, succinate, and malate, in wheat were probably an attempt to chelate the metalloid in rhizosphere and counteract its toxic effects (Sil et al. 2018). However, MMA(III) is considered to be a more potent inhibitor of the PDH complex than As(III). Such inhibition of the enzyme complexes can obstruct the citric acid cycle and eventually reduce ATP production resulting in cellular damage and death (Shen et al. 2013).

14.4.7 Carbohydrate Metabolism

Arsenic toxicity has been reported to perturb starch and sugar metabolism (Jha and Dubey 2005; Chun-Xi et al. 2007; Choudhury et al. 2011; Pandey and Tripathi 2011; Sil et al. 2019a). Escalation in the activity of starch-degrading enzyme, starch phosphorylase, decreased starch contents, while enhancement in activities of sucrose synthesizing enzymes, sucrose phosphate synthase, sucrose synthase, and sucrose hydrolyzing enzyme acid invertase, lead to sugar accumulation in wheat (*T. aestivum* L.) seedlings subjected to As(V) (Sil et al. 2019a). Accumulation of soluble sugar is an adaptive mechanism that helped to retain cellular organization, prevented dehydration of membranes, and stimulated respiration to provide energy under As stress (Chun-Xi et al. 2007). Sucrose and glucose serve as osmolytes to maintain cellular homeostasis, while fructose is associated with the synthesis of secondary metabolites under stressful conditions (Rosa et al. 2009). Marked decline in the activities of starch degrading enzymes, α-amylase and β-amylase, in the endosperm and embryo axes of germinating rice seeds was evident. As(V) suppressed the activity of starch phosphorylase in the endosperm, whereas in embryo axes an enhancement in the enzyme activity was observed in rice seedlings accompanied with an increase in acid invertase activity (Jha and Dubey 2005). Presence of higher levels of starch in the endosperm along with decline in the activity of amylolytic enzymes indicated inhibition in starch mobilization to the embryo axes that decreased reducing and nonreducing sugars in shoots, thereby limiting growth. Similar metabolic impairments caused due to alterations in carbohydrate levels and inhibition in activities of the carbohydrate metabolizing enzymes under the exogenous application of As restrained growth and development of the rice seedlings (Choudhury et al. 2011).

14.4.8 Nitrogen Metabolism

The macronutrient nitrogen is indispensible for plant growth and forms an integral component of coenzymes, photosynthetic pigments, polyamines, as well as secondary metabolites. Arsenic toxicity has been known to affect nitrogen fixation in both legumes and non-legumes (Finnegan and Chen 2012). Arsenic exposure reduced the total number of root nodules in alfalfa (*Medicago sativa* L.) by inducing necrosis and root hair damage in the root region that reduced the number of rhizobial infections (Pajuelo et al. 2008). In non-legumes, As perturbed nitrogen metabolism by hampering the enzyme activities of nitrogen and ammonium assimilatory pathways (Singh et al. 2009, 2018; Shahid et al. 2019; Sil et al. 2019b; Srivastava et al. 2019). As(V)-mediated reduction in nitrate (NO_3^-) uptake decreased the activity of nitrate reductase (NR) in Indian mustard (*B. juncea* L.), potato (*Solanum tuberosum* L.), and wheat (*T. aestivum* L.) (Singh et al. 2018; Shahid et al. 2019; Sil et al. 2019b). Downregulation in the activity of NR decreased nitrite (NO_2^-) contents and subsequently lowered the activity of nitrite reductase (NiR) in the mentioned seedlings. Arsenic has been reported to bind the functional –SH groups present in the active sites of these enzymes causing conformational changes in the enzyme structure, thereby reducing their activities (Singh et al. 2018). Additionally, As-induced accumulation of endogenous H_2O_2 along with restricted availability of NO_3^- decreased the internal concentration of NO_3^- and hence NR activity (Singh et al. 2009). Similar decrease in the concentration of NO_3^- limited the activity of NR and NiR and decreased protein contents significantly in arsenic hyperaccumulating and non-hyperaccumulating ferns (Singh et al. 2009). Suppression in the activities of glutamine synthetase (GS) and glutamate synthase (GOGAT) due to oxidative modifications by excessive ROS disturbed C/N metabolism, lowered amino acid contents, and impairment in NH_4 assimilation causing its accumulation that decreased growth (Singh et al. 2018; Sil et al. 2019b). Arsenic exposure has been reported to alter amino acid production in chickpea (*Cicer arietinum* L.), mustard (*B. juncea* L.), ribbon fern (*Pteris cretica*), and spinach (*S. oleracea*) differentially indicating shifts in the regulation of amino acid biosynthesis (Malik et al. 2011; Pathare et al. 2013; Zemanová et al. 2021). Reduction in levels of lysine, methionine, cysteine, phenylalanine, tyrosine, proline, threonine, tryptophan, and valine in chickpea (*C. arietinum* L.) seedlings reduced the protein content of seeds (Malik et al. 2011). However, enhancement in levels of amino acids was attributed to activation of the defense system in spinach (*S. oleracea*) under As toxicity (Zemanová et al. 2021).

14.5 Modulation in "Omics" Profiling Under As Challenged Environment

Plethora of studies have been documented on the subject of next-generation sequencing, microarray, and gene expression profiling for demonstration of varied gene expression pattern under arsenic (As) exposure (Tripathi et al. 2012; Kumar and Trivedi 2016b). Principally, "omics" studies are focused on the detection of modulation in the function of genes, mRNA, proteins, and metabolites being ascribed by terms genomics, transcriptomics, proteomics, and metabolomics, respectively (Kumar et al. 2015). In As-stressed plants, several stress-responsive and defensive genes of heat shock proteins, metallothioneins, antioxidants regulatory proteins, and small peptides have been found to be overexpressed. Additionally, several groups have analyzed expression of genes at mRNA level to investigate the responses of plants under As stress (Abercrombie et al. 2008; Norton et al. 2008; Chakrabarty et al. 2009). During last decade genes associated with As uptake, translocation, modulation of oxidative stress responses, and mechanism of detoxification have been identified (Tuli et al. 2010). Recently, several reports have explored As metabolism in various plants via transcriptome sequencing and proteomic profiling that provide detailed information of the genes and transcriptional regulation pathways involved in stress-defensive mechanism of As.

14.5.1 Genomic Profiling

The genome profiling has been evaluated in many plant species where genes of a group of stress-responsive transporters upregulated variably under As exposure. A comparison about up and/or downregulation of genes under As(V) and As(III) treatments has been documented (Tripathi et al. 2012). Recently, genetic mapping technique identified As tolerance genes by utilizing the population of two rice varieties – Bala X Azucena and inferred the presence of epistatic interactions among three principal genes, one presents in chromosome 10 and other two in chromosome 6. Under As imposition, As contents in different tissues of maize were estimated by using recombinant inbred line (RIL) population and three quantitative trait locus (QTL) responsible for higher As contents in leaves were located on chromosomes 1, 5, and 8, respectively (Ding et al. 2011). Chakrabarty et al. (2009) studied several genes of rice (*O. sativa*) associated with zinc transporters, ATP binding cassette transporters, glutathione-conjugated transporters, and amino acid transporters along with other genes linked to plant growth and metabolism that were differentially modulated under As(V) and As(III) treatments as defensive responses of rice plants are presented in Table 14.1.

14.5.2 Transcriptomic Profiling

Abercrombie et al. (2008) investigated the transcriptional regulation of thale cress (*Arabidopsis thaliana*) under As(V) challenged condition using oligonucleotide microarrays and recorded strong induction of Cu/Zn superoxide dismutase activity (SOD), proteins of cell wall growth, and tonoplast proteins along with repressed production of FeSOD. Alteration of expression of selective genes associated with glutathione *S*-transferases (GSTs), heat shock proteins, multidrug and toxic compound extrusion transporter (MATE), metallothioneins, sulfate-metabolizing proteins, multidrug resistance proteins, and sulfate transporters have been documented under As stress in different plants (Norton et al. 2008; Chakrabarty et al. 2009; Srivastava et al. 2009; Gautam et al. 2012). In arsenite (AsIII)-exposed rice, expression profiles of genes assisted with fatty acid metabolism, phytohormone transport, and its signaling have been detected (Yu et al. 2012). Similarly, in As-stressed mustard (*B. juncea*), microarray profiling of miRNA detected a total of 69 miRNAs that were linked to stress-responsive genes (Srivastava et al. 2013). Liu and Zhang (2012) reported that under As(III) treatment, 13 osa-miRNA members of 7 miRNA families significantly upregulated, while 54 osa-miRNA members of 19 miRNA families were found to be downregulated in rice plants. A summary of differential expression of miRNA genes under As(V) and As(III) exposures in rice plants and their function associated with regulation of various genes of rice (*O. sativa*) and mustard (*B. juncea*) are presented in Tables 14.2 and 14.3, respectively (Liu and Zhang 2012; Sharma et al. 2015; Vasupalli et al. 2020). Additionally, under both As(V) and As(III) exposure, rice genotypes Triguna, PNR 546, IET-4786, and IR-36 exhibited differential transcriptome expression profiles, which led to reduced sulfate assimilation pathways and antioxidative defense mechanism in sensitive IET-4786 compared to tolerant genotype Triguna (Rai et al. 2011). Differential expression pattern of transcripts of oilseed crop (*Crambe abyssinica*) under As(V) exposure using PCR-select suppression subtraction hybridization (SSH) technique disclosed novel insights about the modulation of gene networks and acted as molecular evidence for the physiochemical responses to As(V) toxicity in plants (Paulose et al. 2010). Dubey et al. (2014) documented genome-wide alteration in transcriptome from hydroponically grown As(V) challenged rice root, where genes of IN2-1 protein, transporters, mannitol dehydrogenase, cytochrome P450, sulfite oxidase and different transcription factors

Table 14.1 List of genes modulated under As(V) and As(III) imposition in *Oryza sativa* L. seedlings.

Genes	As(V) stress	As(III) stress
Aquaporin transporter (*Os10g36924*)		Downregulated
Amino acid transporters (*Os04g39489, Os04g56470*)	Downregulated	
ATP-binding cassette transporter (*Os06g38950*)	Downregulated	
Phosphate:H^+-symporter (*Os03g05640*)	Downregulated	
Proteins of the zinc-finger family (*Os06g04920, Os01g60730, Os07g34120*)	Downregulated	
Zinc transporter (*Os01g74110*)	Downregulated	
MATE-efflux transporters (*Os03g0890, Os10g20350*) (*Os03g62270*)	Upregulated Downregulated	Upregulated Downregulated
Zinc-finger C3HC4-typeprotein (*Os07g43740*)	Upregulated	Upregulated
Glutaredoxin peptides (*Os01g26912*–A, *Os02g40500*–B)	Upregulated (B)	Upregulated (A, B)
Glutathione conjugated transporter (*Os03g13210*)	Upregulated	
Glutathione *S*-transferase peptides (*Os01g27390, Os01g27630*) (*Os09g20220, Os10g34020, Os10g38189, Os10g38350, Os10g38610, Os10g38720, Os01g49710, Os01g49720, Os01g72140, Os01g72150*)	Downregulated } Upregulated	} Upregulated
Glutathione-related transporters (*Os03g54000, Os04g13210*)	Upregulated	
Heat shock proteins (*Os04g01740, Os01g08860, Os03g14180, Os03g16020, Os03g16030, Os03g16920, Os04g36750, Os04g45480, Os02g52150, Os02g54140*)	Upregulated	Upregulated
Helix-loop-helix DNA-binding proteins (*Os03g26210, Os04g31290*)	Upregulated	
Metal transporter, e.g. NRAMP1 (*Os07g15460*)	Upregulated	
Metallothioneins (*Os12g38051, Os12g38064, Os12g38290, Os12g38300*)	Upregulated	Upregulated
Multidrug resistance protein – MRP (*Os01g50100*)	Upregulated	Upregulated

(Continued)

Table 14.1 (Continued)

Genes	As(V) stress	As(III) stress
Protein kinases		
(*Os10g10130, Os04g25060, Os04g25650 Os1g46720, Os06g47530, Os01g53920*)	Upregulated	
(*Os3g28300, Os11g46860*)		Upregulated
Helix-turn-helix protein (*Os06g39240*)	Upregulated	Upregulated
Sulfate transporter (*Os03g09970*)	Upregulated	Upregulated
WRKY proteins (*Os11g29870, g44010*)	Upregulated	
Zinc/iron-transport family protein (*Os03g46470*)	Upregulated	

Table 14.2 List of differential expression of miRNA genes under As(V) and As(III) imposition in *Oryza sativa* L. seedlings.

miRNA genes	As(V) stress	As(III) stress
miR164	Downregulated	Downregulated
miR171	Downregulated	Downregulated
miR395	Downregulated	Downregulated
miR396	Downregulated	Upregulated
miR399	Down/Upregulated	Upregulated
miR408	Upregulated	Upregulated
miR528	Upregulated	Downregulated
miR529	Downregulated	Downregulated
miR820	Downregulated	Downregulated
miR1432	Downregulated	Downregulated
miR1846	Up/Downregulated	Downregulated
miR1861	Upregulated	Upregulated
miR2102	Upregulated	Upregulated
miR2907	Upregulated	Upregulated

were overexpressed. Moreover, upregulation in expression profile of C-type ATP-binding cassette transporter (*OsABCC7*) along with overexpression of phytochelatin synthase in As-exposed rice plant exhibited quenching of As(III)–phytochelatin complex into vacuoles as an innate defensive responses (Tang et al. 2019).

Table 14.3 As(III) responsive miRNAs associated with regulation of various genes of *Oryza sativa* and *Brassica juncea* seedlings.

Plant	miRNA	Gene name
Brassica juncea	*miR159*	GAMYB, ACC synthase
	miR164, miR167, miR390	ARF6, ARF8, NAC1, ARF3/4
	miR395, miR838, miR854	Sulfate transporters, APS, SAT
	miR156, miR162, miR165,miR169, miR172, miR426,miR535, miR1436	TFs and other genes
	miR319, miR838	Lipase, TCP4
Oryza sativa	*miR1318, miR1432*	Ca^{2+} binding protein or Ca^{2+} ATPase
	miR397b	Laccases
	miR169a	NFYA or HAP2 TFs
	miR408	SPX, BCP
	miR172c	AP2 like TFs
	miR528	IAR1, CBP

14.5.3 Proteomic Profiling

Knowledge development about As stress-elicited modulation at translational and posttranslational levels can propose an intense insights into the functional interplay of protein with other stress-responsive attributes. So far little information is known about As-induced alteration in proteome profiling (Tripathi et al. 2012; Kumar et al. 2015). Initially, Requejo and Tena (2005, 2006) evaluated expression pattern of proteins in As-challenged maize root and leaf and demonstrated that the identified proteins are involved in cellular redox homeostasis led to conclude that oxidative stress is a major state of redox imbalance underlying As toxicity in plants. Ahsan et al. (2008) performed physio-biochemical and proteome analyses together and detected 23 differentially expressed proteins in rice roots that are functionally connected to stress signaling, detoxification, oxidative overload defensive mechanisms, cellular development, and protein biosynthesis. Likewise, the first proteome map of As-exposed rice leaves noted an enhanced activity of different proteins associated with energy metabolism of respiratory cycle indicating an adaptive feature of plants to As stress, whereas downregulation of chloroplast 29 kDa ribonucleoproteins and RuBisCO suggests the possible reasons behind reduced net photosynthetic rate under As challenged environment (Ahsan et al. 2010). A comparative proteomic analysis was executed by Liu et al. (2013) in rice shoots under As(V) exposure and revealed differential expression patterns of 38 proteins conferred resistance toward As-induced toxicity. However, an upregulation in protein expression profiles of glyceraldehyde-3-phosphate dehydrogenase, enolase, and phosphoglycerate kinase documented in As treated fronds of Chinese brake fern (*Pteris vittata*), suggesting the pivotal role of glycolytic enzymes in As metabolism (Bona et al. 2010; Vasupalli et al. 2020). It has been reported that in As-stressed bentgrass (*Agrostis tenuis*) and sweet wormwood (*Artemisia annua*), multiple number of differentially expressed proteins linked to enzymes of Calvin cycle as well as several proteins connected to cyclic electron transport, ATP synthase, and multidrug resistant (MDR) ABC transporter were upregulated and led to As detoxification (Duquesnoy et al. 2009; Rai et al. 2014).

14.5.4 Metabolomic Profiling

Metabolomics is a subject that enables identification, quantification, and evaluation of primary and secondary metabolites that are characterized with low molecular weight (<1500 Da), designated as metabolome (Burdisso et al. 2016; Yan and Xu 2018; Villate et al. 2020). In recent years, metabolomics is considered to be the predominant "omic science" where several modern technologies, viz. spectrophotometric measurements, nuclear magnetic resonance, fluorescence detectors coupled to chromatography along with mass spectrometry linked to liquid or gas chromatography are utilized to complement metabolomics analyses (Duman et al. 2010; Leterrier et al. 2012; Zhang et al. 2020). The aim of metabolomics study is to deduce the responses of different metabolic pathways under As stress, which are of key importance in gaining phenotypic fingerprint of a plant under stressed environment (Matich et al. 2019; Martínez-Castillo et al. 2022). The metabolites analyzed in As-stressed plants are found to be linked to antioxidant machinery, phytochelatins, and thiol ligand compounds (Tripathi et al. 2012). Gupta and Ahmad (2014) performed a comparative metabolomics study between two rice genotypes – PB1 and IR-64 and reported that PB1 genotype ameriolated As(V)-induced oxidative stress by induction of antioxidant defense system and production of cysteine and proline. A summary of modulated primary and secondary metabolites along with antioxidants enzymes under As stress has been presented in Tables 14.4 and 14.5, respectively.

14.6 Progress in Molecular Biotechnology to Evolve As-Tolerant Plants

In recent years, advanced technologies have been employed to develop transgenic crops for their sustainable endurance in As-contaminated soil by using the data recorded as alteration in genomic, transcriptomic, proteomic, and metabolomics profiling. Additionally, "genoremediation" is another field of molecular biotechnology where gene-targeted bioremediation programs are utilized to genetically manipulate different plants with target genes that are associated with the regulation of As uptake, transport, homeostasis, defensive responses to oxidative overloads, and its detoxification process (Mani and Kumar 2014). The gene targets being used are the area of studies involved in the restriction of As influx and efflux by roots, biotransformation of toxic As forms to methylated As species, formation of As-cysteine-rich peptide complex, and their sequestration into vacuoles (Zhao et al. 2010; Duan et al. 2012; Shukla et al. 2012). Heterologous expression of *ACR3;1* gene in thale cress (*A. thaliana*) and tobacco (*Nicotiana tabacum*) obtained from Chinese brake fern (*P. vittata)* inhibited the translocation of As from root to shoot compared to wild plants (Chen et al. 2017a). In As-exposed tobacco (*N. tabacum*), overexpression of thale cress (*A. thaliana)* metallothionein gene – (*AtMT2b*) led to

Table 14.4 Alteration of primary metabolites in different plants induced by arsenic.

Plant	Arsenic stress applied	Modulation of primary metabolites under As stress	References
Rice (*Oryza sativa* L.)	Plants grown in sand soil contaminated with 25 and 50 μM As(III) for 20 days	Soluble protein, proline, and carboxypeptidase content increased, free amino acid and leucine amino peptidase content decreased in both root and shoot	Mishra and Dubey (2006)
Aquatic plant (*Ceratophyllum demersum* L.)	Plants exposed to 10, 50, and 250 μM As(V) for seven days	Elevation in non-protein thiol and cysteine contents at 50 μM As(V) and phytochelatin contents at 100 μM As(V) concentrations were observed	Mishra et al. (2008)
Duckweed (*Lemna minor* L.)	Plants exposed to 1, 4, 16, and 64 μM As(III) and As(V) for six days	Increment in protein contents was at 16 μM As(V)/As(III), while at >64 μM As(V)/As(III) concentrations, protein contents reduced	Duman et al. (2010)
Spinach (*Spinacia oleracea* L.)	Plants exposed to 25, 50, and 75 mg As/kg soil for 50 days	Amino acids content increased; asparagine contents decreased at higher As concentrations	Pavlík et al. (2010)
Rice (*Oryza sativa* L.)	Plants exposed to 25 and 50 μM As(III) for 20 days	Phytochelatin and total acid soluble thiol contents increased at 50 μM As(III) exposure	Mishra et al. (2011)
Aquatic plant (*Hydrilla verticillata*)	Plants exposed to 100 and 500 μM As(V) for four days	Phytochelatin contents increased under 500 μM As(V) exposure	Srivastava et al. (2011)
Rice (*Oryza sativa* L.)	Plants exposed to 50, 150, and 300 μM As(V) for four days	Soluble protein contents decreased	Ahmad et al. (2012)
Rice (*Oryza sativa* L.)	Plants exposed to irrigation water containing 17, 27, and 53 μg/l As	Essential and nonessential amino acids content decreased in grains at higher As concentrations	Dwivedi et al. (2012)

significant reduction in As levels in root (Grispen et al. 2009). Similarly, in transgenic rice, expression of *ACR3* gene obtained from baker's yeast (*Saccharomyces cerevisiae*) increased As(III) efflux and led to reduced accumulation of As in grains (Duan et al. 2012). In transgenic rice, expression of *arsM* gene, involved in As(III) methylation by producing *S*-adenosylmethionine methyltransferase enzyme resulted in elevated production of volatile arsenical, less toxic monomethylarsenate (MMA-V), and dimethylarsenate (DMA-V) that eventually led to lower accumulation of As in rice grain (Meng et al. 2011). Several studies exhibited that overexpression of phytochelatin synthase (*PCS*) genes considered as one of the promising mechanisms to gain As tolerance in plants (Wojas et al. 2010; Shukla et al. 2013). In As-exposed Indian mustard plant expression of *AtPCS* gene obtained from thale cress (*A. thaliana*) showed elevated degree of tolerance toward As toxicity (Gasic and Korban 2007). In addition to *PCS* gene, other genes of PC

Table 14.5 Arsenic-induced alteration in secondary metabolites in different plants.

Plant	Arsenic stress applied	Modulation of secondary metabolites under As stress	References
Red clover (*Trifolium pratense* L.)	Plants exposed to As contaminated soil (5, 10, and 50 mg As/kg) for 70 days	Chlorophylls and carotenoids content decreased, putrescine contents increased under 50 mg As/kg soil where spermine and spermidine contents elevated under 10 mg As/kg soil concentrations	Mascher et al. (2002)
Aquatic plant (*Ceratophyllum demersum* L.)	Plants exposed to As(V)-contaminated irrigation water at concentrations of 10, 50, and 250 μM for seven days	Phytochelatin synthase, cysteine synthase, glutamylcysteine synthetase, and arsenate reductase activities increased at higher As(V) concentrations	Mishra et al. (2008)
Spinach (*Spinacia oleracea* L.)	Plants exposed to As contaminated soil (25, 50, and 75 mg As/kg) for 50 days	Glutamate kinase activity decreased	Pavlík et al. (2010)
Rice (*Oryza sativa* L.)	Plants exposed to 25 and 50 μM As(III) for seven days	Levels of thiobarbituric acid reactive substances, reduced glutathione, ascorbate, and dehydroascorbate along with activities of superoxide dismutase, catalase, ascorbate peroxidase, glutathione peroxidase, chloroplast-ascorbate peroxidase, monodehydroascorbate reductase, and dehydroascorbate reductase increased	Mishra et al. (2011)
Aquatic plant (*Hydrilla verticillata*)	Plants exposed to As(V) contaminated irrigation water at a concentrations of 100 and 500 μM for four days	Malondialdehyde, reduced and oxidized glutathione, dehydroascorbate and NADPH content increased while reduced ascorbate contents decreased under As treatment	Srivastava et al. (2011)
Arabidopsis thaliana	Plants exposed to 100, 150, 200, 250, and 500 μM As(V) for seven days	NADPH oxidase and superoxide dismutase activities were unaffected	Leterrier et al. (2012)
Rice (*Oryza sativa* L.)	Plant exposed to 50, 150, and 300 μM As(V) concentrations in culture medium for four days	Chlorophylls content declined at 300 μM As concentration	Gupta and Ahmad (2014)

biosynthesis have been found to contribute tolerance to As (Kumar et al. 2015). Le Blanc et al. (2011) developed a transgenic eastern cottonwood plant by overexpressing bacterial gamma-glutamylcysteine synthetase (*ECS*) gene and increased thiol production led to As tolerance. Likewise, Yamazaki et al. (2018) characterized the influence of PC synthases

(*PCS*) gene in rice under As exposure and showed that transgenic plants developed using *OsPCS* gene produced 20–100 times higher biomass compared to wild type grown under 250 and 300 μM As(V) concentrations. However, in transgenic line of thale cress (*A. thaliana*) overexpressing *TIP4;1 aquaporin* gene obtained from Chinese brake fern (*P. vittata*) resulted in pronounced increments in As(III) transport, and its accumulation showed induced As sensitivity that led to development of idea of mutating the gene for better tolerance toward As toxicity (He et al. 2016).

14.7 Conclusion and Future Perspective

Arsenic (As) interrupts the basic metabolic pathways and signaling networks of plants and is injurious to mankind as well. Extensive studies about occurrence of As, their distribution, speciation, acquisition, translocation, signaling crosstalk with different biomolecules and their impact on physio-chemical and molecular aspects of plants resulting in growth and productivity of plants have been extensively carried out to determine defensive mechanisms involved in As-induced detoxification process. Recent progress on functional genomics, transcriptomics, proteomics, and metabolomics exhibited how the primary and the secondary metabolism and their associated genes of plants are affected under As-challenged environment. Moreover, the resistant genes obtained from different plants and microorganisms have been reported to be efficiently minimizing the As-induced toxicity and transformed them into less toxic As species. The evaluation of "omics" analyses suggesting that phytochelatins (PCs) and their precursor molecule glutathione hold a promising position in As detoxification process and a way to attenuate As-induced oxidative stress. Further exploration of omics approaches using various advanced tools might be needed to format a common hypothesis regarding development of safer crops for human consumption. Studies on genetic manipulation along with altered protein and metabolite profiles identified different regulatory components comprised of miRNA and transcription factors that participate in defensive responses against As stress in plants. Precisely, the genes responsible for translocation of As from root to grain required to be assessed in detail to check the entry of As into food chain. Additionally, the genes involved in transformation of toxic inorganic As species to less harmful methylated As species in plants are another beneficial target and could be utilized for phytoremediation of As. A better understanding of multidisciplinary "omics" approach at field levels and its associated bioinformatics could help in breeding of crops to produce economically important plants with low As contents in edible parts to develop As tolerant transgenic plants by using As-resistant genes and provide a perspective to effectively remediate arsenic-polluted ecosystems.

Acknowledgment

The authors acknowledge the infrastructural assistance provided by Department of Botany, University of Calcutta during preparation of the manuscript.

Author Contributions

AKB conceived the study and designed the manuscript. BM and PS contributed equally toward drafting the manuscript; AKB revised and finalized the manuscript. All authors equally approve the publication.

References

Abbas, G., Murtaza, B., Bibi, I. et al. (2018). Arsenic uptake, toxicity, detoxification, and speciation in plants: physiological, biochemical, and molecular aspects. *Int. J. Environ. Res. Public Health* 15: 59.

Abedi, T. and Mojiri, A. (2020). Arsenic uptake and accumulation mechanisms in rice species. *Plants* 9 (2): 129.

Abedin, M.J., Feldmann, J., and Meharg, A.A. (2002). Uptake kinetics of arsenic species in rice plants. *Plant Physiol.* 128: 1120–1128.

Abercrombie, J.M., Halfhill, M.D., Ranjan, P. et al. (2008). Transcriptional responses of *Arabidopsis thaliana* plants to As(V) stress. *BMC Plant Biol.* 8: 87.

Ahmad, A.M., Gaur, R., and Gupta, M. (2012). Comparative biochemical and RAPD analysis in two varieties of rice (*Oryza sativa*) under arsenic stress by using various biomarkers. *J. Hazard. Mat.* 217–218: 141–148.

Ahsan, N., Lee, D.G., Alam, I. et al. (2008). Comparative proteomic study of arsenic-induced differentially expressed proteins in rice roots reveals glutathione plays a central role during As stress. *Proteomics* 8 (17): 3561–3576.

Ahsan, N., Lee, D.G., Kim, K.H. et al. (2010). Analysis of arsenic stress-induced differentially expressed proteins in rice leaves by two-dimensional gel electrophoresis coupled with mass spectrometry. *Chemosphere* 78 (3): 224–231.

Amna, S., Qamar, S., Naqvi, A.A.T. et al. (2020). Role of sulfur in combating arsenic stress through upregulation of important proteins, and in-silico analysis to study the interaction between phosphate transporter (PHO1), arsenic and phosphate in spinach. *Plant Physiol. Biochem.* 157: 348–358.

ATSDR (2019). Agency for Toxic Substances and Disease Registry. Substance Priority List. CERCLA.

Basu, A., Saha, D., Saha, R. et al. (2014). A review on sources, toxicity and remediation technologies for removing arsenic from drinking water. *Res. Chem. Intermed.* 40 (2): 447–485.

Boccia, P., Meconi, C., Mecozzi, M., and Sturchio, E. (2013). Molecular modifications induced by inorganic arsenic in *Vicia faba* investigated by FTIR, FTNIR spectroscopy and genotoxicity testing. *J. Toxicol. Environ. Health A Curr. Issues* 76: 281–290.

Bona, E., Cattaneo, C., Cesaro, P. et al. (2010). Proteomic analysis of *Pteris vittata* fronds: two arbuscular mycorrhizal fungi differentially modulate protein expression under arsenic contamination. *Proteomics* 10: 3811–3834.

Briat, J.F. (2010). Arsenic tolerance in plants: Pas de deux between phytochelatin synthesis and ABCC vacuolar transporters. *Proc. Natl. Acad. Sci. U.S.A.* 107: 20853–20854.

Burdisso, P., Rasia, R.M., and Vila, A.J. (2016). Metabolomica y medicina de precision. *Rev. Med. Rosario* 82: 75–76.

Carey, A.M., Norton, G.J., Deacon, C. et al. (2011). Phloem transport of arsenic species from flag leaf to grain during grain filling. *New Phytol.* 192: 87–98.

Chakrabarty, D., Trivedi, P.K., Misra, P. et al. (2009). Comparative transcriptome analysis of arsenate and arsenite stresses in rice seedlings. *Chemosphere* 74 (5): 688–702.

Chen, Y., Han, Y.H., Cao, Y. et al. (2017a). Arsenic transport in rice and biological solutions to reduce arsenic risk from rice. *Front. Plant Sci.* 8: 268.

Chen, P., Li, J., Wang, H.Y. et al. (2017b). Evaluation of bioaugmentation and biostimulation on arsenic remediation in soil through biovolatilization. *Environ. Sci. Pollut. Res.* 24 (27): 21739–21749.

Choudhury, B., Mitra, S., and Biswas, A.K. (2010). Regulation of sugar metabolism in rice (*Oryza sativa* L.) seedlings under arsenate toxicity and its improvement by phosphate. *Physiol. Mol. Biol. Plants* 16: 59–68.

Choudhury, B., Chowdhury, S., and Biswas, A.K. (2011). Regulation of growth and metabolism in rice (*Oryza sativa* L.) by arsenic and its possible reversal by phosphate. *J. Plant Interact.* 6 (1): 15–24.

Chun-Xi, L.I., Feng, S.L., Shao, Y. et al. (2007). Effects of arsenic on seed germination and physiological activities of wheat seedling. *J. Environ. Sci.* 19: 725–732.

Da Silva, A.J., Nascimento, C.W., Gouveia-Neto, A.D.S., and Silva, E.A. Jr. (2015). Effects of silicon on alleviating arsenic toxicity in maize plants. *Rev. Bras. Cienc. Solo.* 39: 289–296.

Dey, T.K., Banerjee, P., Bakshi, M. et al. (2014). Groundwater arsenic contamination in West Bengal: current scenario, effects and probable ways of mitigation. *Int. Lett. Nat. Sci.* 13: 45–58.

Di, X., Beesley, L., Zhang, Z. et al. (2019). Microbial arsenic methylation in soil and uptake and metabolism of methylated arsenic in plants: a review. *Int. J. Environ. Res. Public Health* 16: 5012.

Ding, D., Li, W., Song, G. et al. (2011). Identification of QTLs for arsenic accumulation in maize (*Zea mays* L.) using a RIL population. *PloS One* 6: e25646.

DiTusa, S.F., Fontenot, E.B., Wallace, R.W. et al. (2016). A member of the phosphate transporter 1 (Pht1) family from the arsenic-hyperaccumulating fern *Pteris vittata* is a high affinity arsenate transporter. *New Phytol.* 209: 762–772.

Dixit, G., Singh, A.P., Kumar, A. et al. (2015b). Sulfur mediated reduction of arsenic toxicity involves efficient thiol metabolism and the antioxidant defense system in rice. *J. Hazard. Mater.* 298: 241–251.

Dixit, G., Singh, A.P., Kumar, A. et al. (2015c). Sulfur alleviates arsenic toxicity by reducing its accumulation and modulating proteome, amino acids and thiol metabolism in rice leaves. *Sci. Rep.* 5: 16205.

Duan, G., Kamiya, T., Ishikawa, S. et al. (2012). Expressing ScACR3 in rice enhanced arsenite efflux and reduced arsenic accumulation in rice grains. *Plant Cell Physiol.* 53: 154–163.

Dubey, S., Shri, M., Misra, P. et al. (2014). Heavy metals induce oxidative stress and genome-wide modulation in transcriptome of rice root. *Funct. Integr. Genomics.* 14: 401–417.

Duman, F., Ozturk, F., and Aydin, Z. (2010). Biological responses of duckweed (*Lemna minor* L.) exposed to the inorganic arsenic species As(III) and As(V): effects of concentration and duration of exposure. *Ecotoxicology* 19: 983–993.

Duquesnoy, I., Goupil, P., Nadaud, I. et al. (2009). Identification of *Agrostis tenuis* leaf proteins in response to As(V) and As(III) induced stress using a proteomics approach. *Plant Sci.* 176: 206–213.

Duquesnoy, I., Champeau, G.M., Evray, G. et al. (2010). Enzymatic adaptations to arsenic-induced oxidative stress in *Zea mays* and genotoxic effect of arsenic in root tips of *Vicia faba* and *Zea mays*. *CR Biol.* 333: 814–824.

Dwivedi, S., Mishra, A., Tripathi, P. et al. (2012). Arsenic affects essential and non-essential amino acids differentially in rice grains: inadequacy of amino acids in rice based diet. *Environ. Int.* 46: 16–22.

Emamverdian, A., Ding, Y., Mokhberdoran, F., and Xie, Y. (2015). Heavy metal stress and some mechanisms of plant defense response. *Sci. World J.* 2015: 756120.

Erturk, F.A., Aydin, M., Sigmaz, B. et al. (2015). Effects of As_2O_3 on DNA methylation, genomic instability, and LTR retrotransposon polymorphism in *Zea mays*. *Environ. Sci. Pollut. Res.* 22: 18601–18606.

Farnese, F.S., Oliveira, J.A., Farnese, M.S. et al. (2014). Uptake arsenic by plants: effects on mineral nutrition, growth and antioxidant capacity. *Idesia* 32 (1): 99–106.

Farnese, F.S., Oliveira, J.A., Paiva, E.A.S. et al. (2017). The involvement of nitric oxide in integration of plant physiological and ultrastructural adjustments in response to arsenic. *Front. Plant. Sci.* 8: 516.

Farooq, M.A., Gill, R.A., Islam, F. et al. (2016). Methyl jasmonate regulates antioxidant defense and suppresses arsenic uptake in *Brassica napus* L. *Front. Plant Sci.* 7: 468.

Finnegan, P.M. and Chen, W. (2012). Arsenic toxicity: the effects on plant metabolism. *Front. Physiol.* 3 (182): 1–18.

Flora, S.J.S. (2011). Arsenic-induced oxidative stress and its reversibility. *Free Radic. Biol. Med.* 51: 257–281.

Foyer, C.H. and Noctor, G. (2011). Ascorbate and glutathione: the heart of the redox hub. *Plant Physiol.* 155: 2–18.

Gasic, K. and Korban, S.S. (2007). Transgenic Indian mustard (*Brassica juncea*) plants expressing an *Arabidopsis* phytochelatin synthase (AtPCS1) exhibit enhanced As and Cd tolerance. *Plant. Mol. Biol.* 64: 361–369.

Gautam, N., Verma, P.K., Verma, S. et al. (2012). Genome-wide identification of rice Class I metallothionein gene: tissue expression patterns and induction in response to heavy metal stress. *Funct. Integr. Genomics.* 12: 635–647.

Gill, M. (2014). Heavy metal stress in plants: a review. *Int. J. Adv. Res.* 2: 1043–1055.

Grispen, V.M.J., Irtelli, B., Hakvoorta, H.J. et al. (2009). Expression of the *Arabidopsis* metallothionein 2b enhances arsenite sensitivity and root to shoot translocation in tobacco. *Environ. Exp. Bot.* 66: 69–73.

Gupta, M. and Ahmad, M.A. (2014). Arsenate induced differential response in rice genotypes. *Ecotoxicol. Environ. Saf.* 107: 46–54.

Gusman, G.S., Oliveira, J.A., Farnese, F.S., and Cambraia, J. (2013a). Arsenate and arsenite: the toxic effects on photosynthesis and growth of lettuce plants. *Acta Physiol. Plant.* 35: 1201–1209.

Hartley-Whitaker, J., Ainsworth, G., and Meharg, A.A. (2001). Copper and arsenate-induced oxidative stress in *Holcus lanatus* L. clones with differential sensitivity. *Plant Cell Environ.* 24: 713–722.

Hasanuzzaman, M. and Fujita, M. (2013). Exogenous sodium nitroprusside alleviates arsenic-induced oxidative stress in wheat (*Triticum aestivum* L.) seedlings by enhancing antioxidant defense and glyoxalase system. *Ecotoxicology* 22: 584–596.

He, Z., Yan, H., Chen, Y. et al. (2016). An aquaporin PvTIP4;1 from *Pteris vittata* may mediate arsenite uptake. *New Phytol.* 209: 746–761.

Heidari, M. (2017). Plant responses to arsenic: uptake, metabolism and resistance to arsenic toxicity – a review. *Zemdirbyste* 104 (4): 369–376.

Hettick, B.E., Canas-Carrell, J.E., French, A.D., and Klein, D.M. (2015). Arsenic: a review of the element's toxicity, plant interactions, and potential methods of remediation. *J. Agric. Food Chem.* 63: 7097–7107.

Jha, A.B. and Dubey, R.S. (2005). Carbohydrate metabolism in growing rice seedlings under arsenic toxicity. *J. Plant Physiol.* 161: 867–872.

Jia, H.F., Ren, H.Y., Gu, M. et al. (2011). The phosphate transporter gene OsPht1;8 is involved in phosphate homeostasis in rice. *Plant Physiol.* 156: 1164–1175.

Jia, Y., Huang, H., Sun, G.X. et al. (2012). Pathways and relative contributions to arsenic volatilization from rice plants and paddy soil. *Environ. Sci. Technol.* 46: 8090–8096.

Jung, H., Kong, M.S., Lee, B.R. et al. (2019). Exogenous glutathione increases arsenic translocation into shoots and alleviates arsenic-induced oxidative stress by sustaining ascorbate–glutathione homeostasis in rice seedlings. *Front. Plant Sci.* 10: 1089.

Kamiya, T., Tanaka, M., Mitani, N. et al. (2009). NIP1;1, an aquaporin homolog, determines the arsenite sensitivity of *Arabidopsis thaliana*. *J. Biol. Chem.* 284: 2114–2120.

Kamiya, T., Islam, M.R., Duan, G.L. et al. (2013). Phosphate deficiency signaling pathway is a target of arsenate and phosphate transporter OsPT1 is involved in As accumulation in shoots of rice. *Soil Sci. Plant Nutr.* 59: 580–590.

Kumar, S., Dubey, R.S., Tripathi, R.D. et al. (2015). Omics and biotechnology of arsenic stress and detoxification in plants: current updates and prospective. *Environ. Int.* 74: 221–230.

Kumar, S. and Trivedi, P.K. (2016b). Transcriptome modulation in rice under abiotic stress. In: *Plant Environment Interaction: Responses and Approaches to Mitigate Stress* (ed. M.M. Azooz and P. Ahmed), 70–80. New York: Wiley.

Le Blanc, M.S., Lima, A., Montello, P. et al. (2011). Enhanced arsenic tolerance of transgenic eastern cottonwood plants expressing gammaglutamylcysteine synthetase. *Int. J. Phytoremediation* 13: 657–673.

Leterrier, M., Airaki, M., Palma, J.M. et al. (2012). Arsenic triggers the nitric oxide (NO) and S-nitrosoglutathione (GSNO) metabolism in *Arabidopsis*. *Environ. Pollut.* 166: 136–143.

Li, R.Y., Ago, Y., Liu, W.J. et al. (2009). The rice aquaporin Lsi1 mediates uptake of methylated arsenic species. *Plant Physiol.* 150: 2071–2080.

Li, N., Wang, J., and Song, W.Y. (2015). Arsenic uptake and translocation in plants. *Plant Cell Physiol.* 57 (1): 4–13.

Liao, X.Y., Chen, T.B., Xie, H., and Liu, Y.R. (2005). Soil as contamination and its risk assessment in areas near the industrial districts of Chenzhou City, Southern China. *Environ. Int.* 31 (6): 791–798.

Liu, Q. and Zhang, H. (2012). Molecular identification and analysis of arsenite stress-responsive miRNAs in rice. *J. Agric. Food Chem.* 60 (26): 6524–6536.

Liu, Y., Li, M., Han, C. et al. (2013). Comparative proteomic analysis of rice shoots exposed to high arsenate. *J. Integr. Plant Biol.* 55: 965–978.

Luis, D.G., Barry, P.R., and Jian, C. (2019). Pathways of arsenic uptake and efflux. *Environ. Int.* 126: 585–597.

Ma, J.F., Tamai, K., Yamaji, N. et al. (2006). A silicon transporter in rice. *Nature* 440: 688.

Ma, J.F., Yamaji, N., Tamai, K., and Mitani, N. (2007). Genotypic difference in silicon uptake and expression of silicon transporter genes in rice. *Plant Physiol.* 145 (3): 919–924.

Ma, J.F., Yamaji, N., Mitani, N. et al. (2008). Transporters of arsenite in rice and their role in arsenic accumulation in rice grain. *Proc. Nat. Acad. Sci. U. S. A.* 105: 9931–9935.

Majumder, B., Das, S., Mukhopadhyay, S., and Biswas, A.K. (2019a). Identification of arsenic-tolerant and arsenic-sensitive rice (*Oryza sativa* L.) cultivars on the basis of arsenic accumulation assisted stress perception, morpho-biochemical responses, and alteration in genomic template stability. *Protoplasma* 256: 193–211.

Majumder, B., Das, S., Pal, B., and Biswas, A.K. (2019b). Evaluation of arsenic induced toxicity based on arsenic accumulation, translocation and its implications on physio-chemical changes and genomic instability in Indica rice (*Oryza sativa* L.) cultivars. *Ecotoxicology* 29 (1): 13–13.

Majumder, B., Das, S., Biswas, S. et al. (2020). Differential responses of photosynthetic parameters and its influence on carbohydrate metabolism in some contrasting rice (*Oryza sativa* L.) genotypes under arsenate stress. *Ecotoxicology* 29: 912–931.

Majumder, B., Das, S., Pal, B., and Biswas, A.K. (2022). Influence of arsenate imposition on modulation of antioxidative defense network and its implication on thiol metabolism in some contrasting rice (*Oryza sativa* L.) cultivars. *BioMetals* 35 (3): 451–478.

Malik, J.A., Goel, S., Sandhir, R., and Nayyar, H. (2011). Uptake and distribution of arsenic in chickpea: effects on seed yield and seed composition. *Commun. Soil Sci. Plant Anal.* 42: 1728–1738.

Mallick, S., Kumar, N., Singh, A.P. et al. (2013). Role of sulfate in detoxification of arsenate-induced toxicity in *Zea mays* L. (SRHM 445): nutrient status and antioxidants. *J. Plant. Interact.* 8 (2): 140–154.

Mani, D. and Kumar, C. (2014). Biotechnological advances in bioremediation of heavy metals contaminated ecosystems: an overview with special reference to phytoremediation. *Int. J. Environ. Sci. Technol. (Tehran)* 11 (3): 843–872.

Martínez-Castillo, J.I., Saldaña-Robles, A., and Ozuna, C. (2022). Arsenic stress in plants: a metabolomic perspective. *Plant Stress.* 3 (2022): 100055.

Mascher, R., Lippmann, B., Holzinger, S., and Bergmann, H. (2002). Arsenate toxicity: effects on oxidative stress response molecules and enzymes in red clover plants. *Plant Sci.* 163: 961–969.

Matich, E.K., Soria, N.G.C., Aga, D.S., and Atilla-gokcumen, G.E. (2019). Applications of metabolomics in assessing ecological effects of emerging contaminants and pollutants on plants. *J. Hazard. Mater.* 373: 527–535.

McCarty, K.M., Hanh, H.T., and Kim, K.W. (2011). Arsenic geochemistry and human health in South East Asia. *Rev. Environ. Health* 26: 71–78.

Meharg, A.A. and Macnair, M.R. (1992). Suppression of the high affinity phosphate uptake system: a mechanism of arsenate tolerance in *Holcus lanatus* L. *J. Exp. Bot.* 43 (4): 519–524.

Meng, X.Y., Qin, J., Wang, L.H. et al. (2011). Arsenic biotransformation and volatilization in transgenic rice. *New Phytol.* 191: 49–56.

Mirza, N., Mahmood, Q., Shah, M.M. et al. (2014). Plants as useful vectors to reduce environmental toxic arsenic content. *Sci. World J.* 2014: 921581.

Mishra, S. and Dubey, R.S. (2006). Inhibition of ribonuclease and protease activities in arsenic exposed rice seedlings: role of proline as enzyme protectant. *J. Plant Physiol.* 163: 927–936.

Mishra, S., Srivastava, S., Tripathi, R.D., and Trivedi, P.K. (2008). Thiol metabolism and antioxidant systems complement each other during arsenate detoxification in *Ceratophyllum demersum* L. *Aquat. Toxicol.* 86: 205–215.

Mishra, S., Jha, A.B., and Dubey, R.S. (2011). Arsenite treatment induces oxidative stress, upregulates antioxidant system, and causes phytochelatin synthesis in rice seedlings. *Protoplasma* 248: 565–577.

Mosa, K.A., Kumar, K., Chhikara, S. et al. (2012). Members of rice plasma membrane intrinsic proteins subfamily are involved in arsenite permeability and tolerance in plants. *Transgenic Res.* 21: 1265–1277.

Mukhopadhyay, R., Shi, J., and Rosen, B.P. (2000). Purification and characterization of Acr2p, the *Saccharomyces cerevisiae* arsenate reductase. *J. Biol. Chem.* 275: 21149–21157.

Nabi, A., Naeem, M., Aftab, T. et al. (2021). A comprehensive review of adaptations in plants under arsenic toxicity: physiological, metabolic and molecular interventions. *Environ. Pollut.* 1 (290): 118029.

Norton, G.J., Lou-Hing, D.E., Meharg, A.A., and Price, A.H. (2008). Rice–arsenate interactions in hydroponics: whole genome transcriptional analysis. *J. Exp. Bot.* 59: 2267–2276.

Pajuelo, E., Rodríguez-Llorente, I.D., Dary, M., and Palomares, A.J. (2008). Toxic effects of arsenic on Sinorhizobiume-*Medicago sativa* symbiotic interaction. *Environ. Pollut.* 154: 203–211.

Pandey, P. and Tripathi, A.K. (2011). Effect of heavy metals on morphological and biochemical characteristics of *Albizia procera* (Roxb.) Benth. seedlings. *Int. J. Environ. Sci.* 1 (5): 1009–1018.

Pandey, A.K., Gedda, M.R., and Verma, A.K. (2020). Effect of arsenic stress on expression pattern of a rice specific miR156j at various developmental stages and their allied coexpression target networks. *Front. Plant Sci.* 11: 752.

Pathare, V., Srivastava, S., and Suprasanna, P. (2013). Evaluation of effects of arsenic on carbon, nitrogen, and sulfur metabolism in two contrasting varieties of *Brassica juncea*. *Acta Physiol. Plant.* 35: 3377–3389.

Paulose, B., Kandasamy, S., and Dhankher, O.P. (2010). Expression profiling of *Crambe abyssinica* under arsenate stress identifies genes and gene networks involved in arsenic metabolism and detoxification. *BMC Plant Biol.* 10: 108.

Pavlík, D., Staszkova, L., Neuberg, M. et al. (2010). The effect of arsenic contamination on amino acids metabolism in *Spinacia oleracea* L. *Ecotoxicol. Environ. Saf.* 73: 1309–1313.

Qin, J., Rosen, B.P., Zhang, Y. et al. (2006). Arsenic detoxification and evolution of trimethylarsine gas by a microbial arsenite S-adenosylmethionine methyltransferase. *Proc. Natl. Acad. Sci. U.S.A.* 103: 2075–2080.

Raab, A., Williams, P.N., Meharg, A., and Feldmann, J. (2007). Uptake and translocation of inorganic and methylated arsenic species by plants. *Environ. Chem.* 4: 197–203.

Rahman, A., Mostofa, M.G., Alam, M.M. et al. (2015). Calcium mitigates arsenic toxicity in rice seedlings by reducing arsenic uptake and modulating the antioxidant defense and glyoxalase systems and stress markers. *BioMed. Res. Int.* 2015: 340812.

Rai, A., Tripathi, P., Dwivedi, S. et al. (2011). Arsenic tolerances in rice (*Oryza sativa*) have a predominant role in transcriptional regulation of a set of genes including sulphur assimilation pathway and antioxidant system. *Chemosphere* 82: 986–995.

Rai, R., Pandey, S., Shrivastava, A.K., and Pandey Rai, S. (2014). Enhanced photosynthesis and carbon metabolism favor arsenic tolerance in *Artemisia annua*, a medicinal plant as revealed by homology-based proteomics. *Int. J. Proteom.* 1–21.

Requejo, R. and Tena, M. (2005). Proteome analysis of maize roots reveals that oxidative stress is a main contributing factor to plant arsenic toxicity. *Phytochemistry* 66 (13): 1519–1528.

Requejo, R. and Tena, M. (2006). Maize response to acute arsenic toxicity as revealed by proteome analysis of plant shoots. *Proteomics* 6 (Suppl 1): S156–S162.

Rosa, M., Prado, C., Podazza, G. et al. (2009). Soluble sugars-metabolism, sensing and abiotic stress. *Plant Signal. Behav.* 4 (5): 388–393.

Saha, J., Dutta, M., and Biswas, A.K. (2017a). Influence of arsenate and phosphate on the regulation of growth and TCA cycle in the rice (*Oryza sativa* L.) cultivars IR64 and Nayanmani. *Am. J. Plant Sci.* 8: 1868–1887.

Saha, J., Majumder, B., Mumtaz, B., and Biswas, A.K. (2017b). Arsenic induced oxidative stress and thiol metabolism in two cultivars of rice and its possible reversal by phosphate. *Acta. Physiol. Plant.* 39: 263.

Sarkar, A. and Paul, B. (2016). The global menace of arsenic and its conventional remediation-A critical review. *Chemosphere* 158: 37–49.

Shahid, M.A., Balal, R.M., Khan, N. et al. (2019). Selenium impedes cadmium and arsenic toxicity in potato by modulating carbohydrate and nitrogen metabolism. *Ecotoxicol. Environ. Saf.* 180: 588–599.

Shao, Y., Jiang, L., Zhang, D. et al. (2011). Effects of arsenic, cadmium and lead on growth and respiratory enzymes activity in wheat seedlings. *Afr. J. Agric. Res.* 6: 4505–4512.

Sharma, I. (2012). Arsenic induced oxidative stress in plants. *Biologia* 67: 447–453.

Sharma, D., Tiwari, M., Lakhwani, D. et al. (2015). Differential expression of microRNAs by arsenate and arsenite stress in natural accessions of rice. *Metallomics* 7: 174–187.

Shen, S., Li, X.F., Cullen, W.R. et al. (2013). Arsenic binding to proteins. *Chem. Rev.* 113: 7769–7792.

Shri, M., Kumar, S., Chakrabarty, D. et al. (2009). Effect of arsenic on growth, oxidative stress, and antioxidant system in rice seedling. *Ecotoxicol. Environ. Saf.* 72: 1102–1110.

Shrivastava, A., Ghosh, D., Dash, A., and Bose, S. (2015). Arsenic contamination in soil and sediment in India: sources, effects, and remediation. *Curr. Pollut. Rep.* 1: 35–46.

Shukla, D., Kesari, R., Mishra, S. et al. (2012). Expression of phytochelatin synthase from aquatic macrophyte *Ceratophyllum demersum* L. enhances cadmium and arsenic accumulation in tobacco. *Plant Cell Rep.* 31: 1687–1699.

Shukla, D., Tiwari, M., Tripathi, R.D. et al. (2013). Synthetic phytochelatins complement a phytochelatin-deficient *Arabidopsis* mutant and enhance the accumulation of heavy metal(loid)s. *Biochem. Biophys. Res. Commun.* 434 (3): 664–669.

Sil, P., Das, P., and Biswas, A.K. (2018). Silicon induced mitigation of TCA cycle and GABA synthesis in arsenic stressed wheat (*Triticum aestivum* L.) seedlings. *S. Afr. J. Bot.* 119: 340–352.

Sil, P., Das, P., Biswas, S. et al. (2019a). Modulation of photosynthetic parameters, sugar metabolism, polyamine and ion contents by silicon amendments in wheat (*Triticum aestivum* L.) seedlings exposed to arsenic. *Environ. Sci. Pollut. Res.* 26 (13): 13630–13648.

Sil, P., Das, P., and Biswas, A.K. (2019b). Impact of exogenous silicate amendments on nitrogen metabolism in wheat seedlings subjected to arsenate stress. *Silicon* 12: 535–545.

Sil, P. and Biswas, A.K. (2020). Silicon nutrition modulates arsenic-inflicted oxidative overload and thiol metabolism in wheat (*Triticum aestivum* L.) seedlings. *Environ. Sci. Pollut. Res.* 27: 45209–45224.

Singh, N., Ma, L.Q., Srivastava, M., and Rathinasabapathi, B. (2006). Metabolic adaptations to arsenic-induced oxidative stress in *Pteris vittata* L. and *Pteris ensiformis* L. *Plant Sci.* 170: 274–282.

Singh, N., Ma, L.Q., Vu, J.C., and Raj, A. (2009). Effects of arsenic on nitrate metabolism in arsenic hyperaccumulating and non-hyperaccumulating ferns. *Environ. Pollut.* 157: 2300–2305.

Singh, M., Singh, V.P., Dubey, G., and Prasad, S.M. (2015). Exogenous proline application ameliorates toxic effects of arsenate in *Solanum melongena* L. seedlings. *Ecotoxicol. Environ. Saf.* 117: 164–173.

Singh, V.P., Singh, S., Kumar, J., and Prasad, S.M. (2015a). Hydrogen sulphide alleviates toxic effects of arsenate in pea seedlings through up-regulation of the ascorbate-glutathione cycle: possible involvement of nitric oxide. *J. Plant Physiol.* 181: 20–29.

Singh, A.P., Dixit, G., Kumar, A. et al. (2016). Nitric oxide alleviated arsenic toxicity by modulation of antioxidants and thiol metabolism in rice (*Oryza sativa* L.). *Front. Plant Sci.* 6: 1272.

Singh, R., Parihar, P., and Prasad, S.M. (2018). Sulfur and calcium simultaneously regulate photosynthetic performance and nitrogen metabolism status in as-challenged *Brassica juncea* L. seedlings. *Front. Plant Sci.* 9: 772.

Singh, R., Parihar, P., and Prasad, S.M. (2020). Interplay of calcium and nitric oxide in improvement of growth and arsenic-induced toxicity in mustard seedlings. *Sci. Rep.* 10: 6900.

Song, W.Y., Yamaki, T., Yamaji, N. et al. (2014). A rice ABC transporter OsABCC1, reduces arsenic accumulation in the grain. *Proc. Natl. Acad. Sci. U.S.A.* 111: 15699–15704.

Srivastava, S. and Sharma, Y.K. (2013a). Arsenic occurrence and accumulation in soil and water of eastern districts of Uttar Pradesh, India. *Environ. Monit. Assess.* 185: 4995–5002.

Srivastava, S. and Sharma, Y.K. (2013b). Impact of arsenic toxicity on black gram and its amelioration using phosphate. *ISRN Toxicol.* 22 (2013): 340925.

Srivastava, S. and Sharma, Y.K. (2013c). Altered growth, photosynthetic machinery and induced oxidative stress in spinach in response to arsenic stress. *J. Plant. Physiol. Pathol.* 1: 4.

Srivastava, S. and Sharma, Y.K. (2014). Arsenic induced changes in growth and metabolism of black gram seedlings (*Vigna mungo* L.) and the role of phosphate as an ameliorating agent. *Environ. Process.* 1: 431–445.

Srivastava, S., Srivastava, A.K., Suprasanna, P., and D'Souza, S.F. (2009). Comparative biochemical and transcriptional profiling of two contrasting varieties of *Brassica juncea* L. in response to arsenic exposure reveals mechanisms of stress perception and tolerance. *J. Exp. Bot.* 60: 3419–3431.

Srivastava, S., Suprasanna, P., and D′Souza, S.F. (2011). Redox state and energetic equilibrium determine the magnitude of stress in *Hydrilla verticillata* upon exposure to arsenate. *Protoplasma* 248: 805–815.

Srivastava, S., Srivastava, A.K., Suprasanna, P., and D'Souza, S.F. (2013). Identification and profiling of arsenic stress-induced microRNAs in *Brassica juncea*. *J. Exp. Bot.* 64 (1): 303–315.

Srivastava, S., Pathare, V.S., Sounderajan, S., and Suprasanna, P. (2019). Nitrogen supply influences arsenic accumulation and stress responses of rice (*Oryza sativa* L.) seedlings. *J. Hazard. Mater.* 367: 599–606.

Stoeva, N. and Bineva, T. (2003). Oxidative changes and photosynthesis in oat plants grown in As-contaminated soil. *Bulg. J. Plant Physiol.* 29: 87–95.

Tang, Z., Chen, Y., Miller, A.J., and Zhao, F.J. (2019). The C-type ATP-binding cassette transporter OsABCC7 is involved in the root-to-shoot translocation of arsenic in rice. *Plant Cell Physiol.* 60 (7): 1525–1535.

Tripathi, R.D., Tripathi, P., Dwivedi, S. et al. (2012). Arsenomics: omics of arsenic metabolism in plants. *Front. Physiol.* 3: 1–14.

Tripathi, P., Tripathi, R.D., Singh, R.P. et al. (2013). Silicon mediates arsenic tolerance in rice (*Oryza sativa* L.) through lowering of arsenic uptake and improved antioxidant defence system. *Ecol. Eng.* 52: 96–103.

Tripathi, D.K., Singh, S., Singh, V.P. et al. (2016). Silicon nanoparticles more efficiently alleviate arsenate toxicity than silicon in maize cultiver and hybrid differing in arsenate tolerance. *Front. Environ. Sci.* 4: 46.

Tuli, R., Chakrabarty, D., Trivedi, P.K., and Tripathi, R.D. (2010). Recent advances in arsenic accumulation and metabolism in rice. *Mol. Breed.* 26: 307–323.

Ullrich-Eberius, C.I., Sanz, A., and Novacky, A.J. (1989). Evaluation of arsenate- and vanadate associated changes of electrical membrane potential and phosphate transport in *Lemna gibba* G1. *J. Exp. Bot.* 40: 119–128.

Vasupalli, N., Koramutla, M.K., Aminedi, R. et al. (2020). Omics approaches and biotechnological perspectives of arsenic stress and detoxification in plants. *Metall. Plants Adv. Futur. Prospect.* 14: 249–273.

Villate, A., San Nicolas, M., Gallastegi, M. et al. (2020). Review: metabolomics as a prediction tool for plants performance under environmental stress. *Plant Sci.* 303: 110789.

Vithanage, M., Herath, I., Joseph, S. et al. (2017). Interaction of arsenic with biochar in soil and water: a critical review. *Carbon* 113: 219–230.

Wang, P., Zhang, W., Mao, C. et al. (2016). The role of OsPT8 in arsenate uptake and varietal difference in arsenate tolerance in rice. *J. Exp. Bot.* 67: 6051–6059.

Wojas, S., Clemens, S., Sklodowska, A., and Antosiewicz, D. (2010). Arsenic response of AtPCS1- and CePCS-expressing plants – effects of external As(V) concentration on As accumulation pattern and NPT metabolism. *J. Plant Physiol.* 167: 169–175.

Wu, Z., Ren, H., McGrath, S.P. et al. (2011). Investigating the contribution of the phosphate transport pathway to arsenic accumulation in rice. *Plant Physiol.* 157: 498–508.

Xu, W., Dai, W., Yan, H. et al. (2015). *Arabidopsis* NIP3;1 plays an important role in arsenic uptake and root-to-shoot translocation under arsenite stress conditions. *Mol. Plant.* 8: 722–733.

Yamazaki, S., Ueda, Y., Mukai, A. et al. (2018). Rice phytochelatin synthases OsPCS1 and OsPCS2 make different contributions to cadmium and arsenic tolerance. *Plant Direct* 2 (1): e00034.

Yan, M. and Xu, G. (2018). Current and future perspectives of functional metabolomics in disease studies: a review. *Anal. Chim. Acta.* 1037: 41–54.

Ye, J., Rensing, C., Rosen, B.P., and Zhu, Y.G. (2012). Arsenic biomethylation by photosynthetic organisms. *Trends Plant Sci.* 17: 155–162.

Yoshinaga, M., Cai, Y., and Rosen, B.P. (2011). Demethylation of methylarsonic acid by a microbial community. *Environ. Microbiol.* 13 (5): 1205–1215.

Yu, L.J., Luo, Y.F., Liao, B. et al. (2012). Comparative transcriptome analysis of transporters, phytohormone and lipid metabolism pathways in response to arsenic stress in rice (*Oryza sativa*). *New Phytol.* 195: 97–112.

Zemanová, V., Pavlíková, D., Hnili˘cka, F., and Pavlík, M. (2021). Arsenic toxicity-induced physiological and metabolic changes in the shoots of *Pteris cretica* and *Spinacia oleracea*. *Plants* 10: 2009.

Zhang, X.W., Li, Q.H., and Dou, J.J. (2020). Mass spectrometry-based metabolomics in health and medical science: a systematic review. *RSC Adv.* 10 (6): 3092–3104.

Zhao, F.J., Ma, J.F., Meharg, A.A., and McGrath, S.P. (2009). Arsenic uptake and metabolism in plants. *New Phytol.* 181: 777–794.

Zhao, F.J., McGrath, S.P., and Meharg, A.A. (2010). Arsenic as a food chain contaminant: mechanisms of plant uptake and metabolism and mitigation strategies. *Annu. Rev. Plant. Biol.* 61: 535–559.

Zhao, F.J., Zhu, Y.G., and Meharg, A.A. (2013). Methylated arsenic species in rice: geographical variation, origin, and uptake mechanisms. *Environ. Sci. Technol.* 47: 3957–3966.

Zheng, M.Z., Li, G., Sun, G.X. et al. (2013). Differential toxicity and accumulation of inorganic and methylated arsenic in rice. *Plant Soil* 365: 227–238.

Zu, Y.Q., Sun, J.J., He, Y.M. et al. (2016). Effects of arsenic on growth, photosynthesis and some antioxidant parameters of *Panax notoginseng* growing in shaded conditions. *Int. J. Agric. Res.* 4: 78–88.

Zvobgo, G., Lwalaba, J.L.W., Sagonda, T. et al. (2019). Alleviation of arsenic toxicity by phosphate is associated with its regulation of detoxification, defense, and transport gene expression in barley. *J. Integr. Agric.* 18 (2): 381–394.

15

Selenium Toxicity and Tolerance in Plants

Insights from Omics Studies

Ali Kıyak[1], Selman Uluısık[2], Ertugrul Filiz[3], and Firat Kurt[4]

[1] *Research Center for Scientific and Technology Applications, Burdur Mehmet Akif Ersoy University, Burdur, Turkey*
[2] *Burdur Food Agriculture and Livestock Vocational School, Burdur Mehmet Akif Ersoy University, Burdur, Turkey*
[3] *Department of Crop and Animal Production, Cilimli Vocational School, Duzce University, Duzce, Turkey*
[4] *Department of Plant Production and Technologies, Faculty of Applied Sciences, Mus Alparslan University, Mus, Turkey*

15.1 Introduction

Selenium (Se), abundant in nature, is a trace element whose distribution in earth crust shows temporal and seasonal variation in soils; but it is needed at very small amounts by all organisms. Although initially identified as a toxic element for living things, later, it was taken into the essential element category when proteins containing Se discovered. Selenium is a chalcogen element with physicochemical properties similar to sulfur (S), which causes nonspecific binding of Se rather than S in biochemical reactions. These substitutions can damage cell metabolism by disrupting protein structures (Bodnar et al. 2012). The importance of Se for metabolism peaked in 1957 when Schwartz and Foltz showed that the addition of Se to feeds prevents muscular dystrophy and liver cirrhosis in rats (Rayman 2000) and found that it participates in the structure of the enzyme glutathione peroxidase (Rotruck et al. 1980). Since then, the antioxidant and the anticancer (Reid et al. 2008; Hatfield et al. 2014) properties of selenium have been brought to forward with many studies, and it has been accepted as an essential element in human nutrition (Hartikainen 2005). Humans have 25 different seleno-proteins known to contribute to immune enhancement, gastrointestinal and thyroid protection, and normal spermatogenesis development (Rayman 2012; Schmidt and Simonović 2012).

The presence of Se in nature is changing day by day due to natural and anthropogenic activities (Zhu et al. 2009). Therefore, people suffer from both Se deficiency and toxicity (Kumar and Prasad 2021). For example, countries such as China, Siberia, Scandinavia, Sub-Saharan Africa, and New Zealand have Se-poor soils, and therefore, the people in these countries suffer from Se deficiency-related diseases (Combs 2001).

Edible plants are one of the major sources of Se for humans, and Se-rich plants can help compensate for Se deficiency (Schiavon and Pilon-Smits 2017).As mentioned above, the

Heavy Metal Toxicity and Tolerance in Plants: A Biological, Omics, and Genetic Engineering Approach,
First Edition. Edited by Mohammad Anwar Hossain, AKM Zakir Hossain, Sylvain Bourgerie,
Masayuki Fujita, Om Parkash Dhankher, and Parvez Haris.

distribution of selenium on world soils is uneven and its phyto-availability for plants is generally low (Fordyce 2013). However, although the necessity of Se for plants is still controversial for some scientists, it is still considered as a beneficial element for many of them. In particular, when applied at low concentrations, Se is reported to improve plant growth and development, and food quality, and it is suggested to have important roles in bestowing tolerance to the negative effects of abiotic stresses (Hasanuzzaman et al. 2014). On the contrary, it has been reported that the excessive Se accumulation causes both oxidative and nitrosative stresses by disrupting the structures of SeCys and SeMet proteins, thereby impairing cellular metabolism (Brown and Shrift 1982; Dimkovikj et al. 2015; Van Hoewyk 2013). In addition, Se at higher concentrations in the soil decreases root length and biomass in plants while altering the uptake of some essential minerals (Lavu et al. 2012; Garousi et al. 2017; Mostofa et al. 2017). Selenium also interacts with toxic metals/metalloids in soils, and this increases its toxic effect on the plants (Ali et al. 2021). In this chapter, the phytotoxicity mechanisms of Se in plants as well as the tolerance mechanisms against Se toxicity developed by plants are discussed in terms of omics studies.

15.2 Selenium Toxicity in Plants

The two main inorganic forms of Se with the highest bioavailability in soil are selenite (SeO_3^{2-}), common in oxidized soils, and selenate (SeO_4^{2-}), common in anoxic soils. Both inorganic forms can be taken up by plant roots in the form of ions, but the transporters involved in Se transport are not unique to them. Selenate is taken up by sulfate transporters (SULTRs), located in the plasma membrane of plant roots, while the uptake of selenite are performed by phosphate transporters and aquaporins (White 2018; Schiavon et al. 2020). Selenite is rapidly metabolized after being taken into plant body, while selenate can be found in its natural form in all parts of plants (Huang et al. 2017). Nonetheless, during the metabolic processes, higher concentrations of selenite and selenate induce cytotoxicity if ROS generation takes place or Cys replacement with selenocysteine (Sec) occurs in proteins and other biomolecules (Tobe et al. 2017).

15.2.1 Se-Induced Protein Malformation

In Se-sensitive plants like *Vigna radiata*, when Cys is replaced with Sec, the exchange leads to disruption of Cys-containing proteins. This metabolic error in Se-sensitive plants is explained by the fact that cysteinyl-tRNA (tRNAcys) cannot distinguish between Sec and Cys. In other words, tRNAcys' are able to bind to either Sec or Cys randomly (Brown and Shrift 1980). When accumulator and sensitive *Astragalus* species were grown in selenate-containing medium, it was found that the production of seleno-amino acids (Sec or SeMet) does not occur and seleno-amino acids are excluded from protein structures in accumulator plants (Brown and Shrift 1981). On the contrary to seleno-amino acids, non-protein seleno-amino acids, such as Se-methylselenocysteine (MeSec), γ-glutamyl-Se-methylselenocysteine (γ-GluMeSeCys), and seleno cystathionine, are produced in the tolerant plants grown in Se-rich soil sand thereby store higher amounts of Se in their bodies (Brown and Shrift 1981; Tobe et al. 2017). It is also suggested that because non-protein seleno-amino acids are synthesized in accumulators in lieu of seleno-amino acids, it makes impossible tRNAcys to cause the binding error explained above since there would be no substrate to bind (Brown and Shrift 1981). On the other hand, the substitution of SeMet to Met during translation stage of protein synthesis does not induce as severe protein disruptions as the substitution of Cys to Sec (Brown and Shrift 1982). The fact that SeMet constitutes over half of total Se in rice is given as evidence to this suggestion (Van Hoewyk 2013). The fact the replacement of SeMet to Met is less detrimental is attributed to nonspecific role of Met in protein structures as well. Additionally, although it is known that SeMet has more hydrophobic character than Met, thereby taking part in hydrophobic reactions, detrimental effect of substitution of Met to SeMet in terms of enzymatic reactions has not been found so far (Brown and Shrift 1982). However, if Met is in the catalytic site of Met-containing enzymes and replaces with SeMet, the replacement highly likely impairs such enzymes' activities. The replacement of Cys to Sec also affects the formation of iron-sulfur clusters, the strength and the stabilization of chemical bonds and bond lengths between co-factors and proteins, and the functions of electron transport chain in mitochondria and chloroplasts. These ramifications are caused by larger atomic size of Se, formation of di-selenide bridges affecting chemical properties of biomolecules, and formation of Fe–Se clusters instead of Fe–S clusters (Van Hoewyk 2013). Other than that, Sec and SeMet are not just used as direct antioxidants in plants but also employed for the production of Se-dependent antioxidants or selenoenzymes, such as glutathione peroxidases (GPX), thioredoxin reductases (TRs), and methionine sulfoxide reductases (Chauhan et al. 2017). Higher rate of thiol/desulfide exchange reactions are regulated by these enzymes and the higher thiol/desulfide exchanges occur in the presence of Se since Se excels S in terms of nucleophilicity, electrophilicity, and atomic size (Hondal et al. 2012).

15.2.2 ROS-Induced Se Phytotoxicity

As mentioned above, Se toxicity occurs when excessive amounts of selenate and selenite are present in the growth medium of plants. Other than Se dose, exposure time to Se and chemical form of Se-containing compounds are other factors affecting plants to have Se stress (Van Hoewyk 2013). Selenate induces less detrimental effects in plant cells compared to selenite (Mézes and Balogh 2009) since selenate does not react with reduced glutathione, a compound leading to ROS generation when interacted with selenite. ROS formation occurs as a result of disruption of electron transport chain as well (Freeman et al. 2010). As an adaptation strategy to Se toxicity, seleno-amino acids are modified by being added to methyl groups, thereby evading their misincorporations into proteins. Also, selenate, selenite, and other organic forms of Se are sequestered in chloroplasts (Hugouvieux et al. 2009). Moreover, seleno-proteins, known as antioxidants, such as GPX, TR, and methionine-*R*-sulfoxide reductase 1 (MsrB1) are used to keep redox homeostasis under the control by preserving reduced state of Cys-containing-redox-sensitive proteins. With involvement of seleno-proteins, hydrogen peroxide, thioredoxin, and methionine sulfoxide generations are kept under the control by the reduction reactions of GPX, TR, and MsrB1, respectively (Lee and Jeong 2012). The fact that plants combat Se stress is vigorously associated with synthesis, accumulation, and depletion glutathione. As is known, the glutathione is the primary antioxidant alleviating cellular oxidative stress symptoms by decreasing ROS accumulation

(Koprivova et al. 2010). When APR2, involved in sulfate reduction as a catalyzer, was mutated in Arabidopsis, it was found that the tolerance of mutants (apr2-1) to selenate decreased twofold and had less selenoproteins compared to control group (Grant et al. 2011) and the mutant plants showed decreased glutathione concentrations, which will have ramifications on antioxidant mechanism (Trippe and Pilon-Smits 2021).

15.3 Selenium Tolerance in Plants

Due to the variability of the physiological and biochemical nature, the uptake of Se by plants is controlled by physical, chemical, and biological factors, and the metabolism of Se differs significantly in different soil and plant systems. The existence of Se in soils is affected by oxidation/reduction, mineralization, and evaporation processes, while its bioavailability and mobility are influenced by its concentration and speciation (Wang et al. 2013; Etteieb et al. 2020). In the nature, Se exists in either organic or inorganic forms, and plants are able to take up both forms through their roots. The uptake, the transport, and the accumulation of Se by plants are associated with various factors such as plant type and variety, the chemical form of Se, redox potential, mineralogy, mineral fertilizer application, and local rains (Galeas et al. 2007; Gupta and Gupta 2017).

The Se concentrations in the earth crust are about 0.01–2.0 $mg\,kg^{-1}$ (Johnson et al. 2010). It is known that there is a very narrow range in terms of concentration between harmful and beneficial doses of Se in plants (Zhu et al. 2009). While a small amount of Se (1 $\mu mol\,l^{-1}$) can boost plant growth, yield, yield quality, and antioxidant capacity and regulate multiple stress responsive genes (Chauhan et al. 2017), it is generally considered to be toxic when the amount of Se in plant tissues exceeds 5 $mg\,kg^{-1}$ (Reilly 1996). Nonetheless, Se tolerance of plants varies greatly in plant kingdom. For example, while 2 $mg\,Se\,kg^{-1}$ dry weight (DW) is toxic for rice, wheat show toxic symptoms at 4.9 $mg\,Se\,kg^{-1}$ DW (Tripathi and Misra 1974). However, Dutch alfalfa can tolerate around 330 $mg\,Se\,kg^{-1}$ DW in their tissues (Mikkelsen et al. 1989).

Excessive accumulation of Se at whole plant causes symptoms such as chlorosis, wilting, and stunted shoot and root growth (Molnár et al. 2018). Selenium toxicity affects both non-proteomic processes (e.g. lipid peroxidation and altered redox state) and causes protein damage, which is associated with abiotic stress (e.g. oxidized or nitrated proteins). To overcome this problem, Se is either removed or its concentration levels in soils are reduced by phytoremediation, an environmentally friendly approach (Salt et al. 1995). Some members of Fabaceae (*Astralagus* spp. *Astragalus bisulcatus*), and Brassicaceae (*Stanleya pinnata*), two Se hyperaccumulators, are reported to accumulate Se between 0.1% and 1.5% of their DWs in their natural habitats and to evaporate excess Se further into the atmosphere. The identification of Se accumulators or hyperaccumulators are particularly important to improve Se uptake capacity of non-accumulator/non-hyperaccumulator plants for Se phytoremediation or to develop new crop varieties for Se biofortification. Moreover, identification of the genes or of the gene networks, involved in Se accumulation/hyperaccumulation, may pave the way for the development of transgenic plants. In this regard, a transcriptomic study conducted to compare Se hyperaccumulator *Stanleya* species to non-accumulator ones revealed that the genes associated with Se accumulation are either involved in sulfate/selenate transports or in assimilation of Se or oxidative stress resistance (glutathione-related genes and peroxidases) (Wang et al. 2018). In another study, conducted by RNA sequencing (RNA-Seq) and quantitative real-time PCR (RT-qPCR) in *Cardamine hupingshanensis* (a member of *Brassicaceae* family), treated with selenite, it was found that Se storage, oxidation, transamination, and selenation processes are significant metabolic processes having roles in the plant's selenium tolerance mechanism. Particularly, selenite was found significantly to affect 25 genes in four pathways in *C. hupingshanensis* seedlings (Zhou et al. 2018).

Only a few genes involved in the mechanism of Se accumulation and tolerance have been identified so far. It is stated that when *ATP sulfurylase* (APS transgenics) and *cystathionine-γ-synthase* (CGS), involved in Se accumulation and tolerance, were overexpressed in Indian mustard *(Brassica juncea)*, the transgenic plants had 2.5-fold higher Se in their shoots than those in the wild-type. The amount of Se in the shoots of Indian mustard was reported to be similar to those in *S. pinnata*, an hyperaccumulator used for comparison in the study (Huysen et al. 2004). *ERF96* (*Ethylene response factor96*) is involved in the regulation of selenite tolerance in *Arabidopsis*. While *ERF96*-silenced plants showed wild-type (WT) resistance to selenite, overexpressing *ERF96* increased *Arabidopsis* sensitivity to selenite stress. *ERF96* increased Se tolerance through Se absorption, Se allocation, and boosting antioxidant responses. The study shows that *ERF96* plays a positive role in the regulation of Se tolerance in *Arabidopsis* (Jiang et al. 2020). Likewise, *WRKY47* gene was found as another gene taking part in Se homeostasis and tolerance (Wu et al. 2020).

To the best of our knowledge, there are very few Se studies conducted using omics technologies. The study, carried out by Rao et al. (2021), aimed at understanding process of Se accumulation and tolerance using selenate in *Cardamine violifolia* seedlings, which accumulate more Se than 9000 mg Se kg^{-1} DW. Rao et al. (2021) reported that the expression of selenium-binding protein 1 (*SBP1*) and protein sulfur deficiency-induced 2 (*SDI2*) were upregulated, while calcium protein and cysteine-rich kinase genes were downregulated upon the treatment of selenate. The up and downregulations of the mentioned genes are important elements of the Se tolerance mechanism in *C. violifolia* (Rao et al. 2021). Moreover, the transcriptome and proteome analyses of the study revealed that the Se detoxification mechanism in *C. violifolia* was induced by modulations of gene expressions of ribosomal protein, chaperone, and ubiquitin. Also, posttranslational modifications are reported to play important roles in Se tolerance mechanism in *C. violifolia*.

The uptake of Se in plants are achieved in the form of selenite and selenate, as mentioned above, and the uptake of Se is preferred in the form of selenate considerably much more than selenate; probably selenite has more toxic effects in plants. In order for selenate to be stored in chloroplasts, selenate should be reduced to selenide as a final product by glutathione. Selenium accumulation is managed by two processes in non-accumulator and accumulator plants. These two metabolic processes possess common steps; however, differences occur after selenocysteine formation. Selenium non-accumulators have a specific mechanism of reducing and converting selenate into dimethylselenide (DMSE), whereas Se accumulators convert bi-methylates selenocysteine into dimethyl diselenide (DMDSe) (Mehdi et al. 2013). Also, less ROS generation, and higher antioxidant formation occur in hyperaccumulator plants and maybe because of this reason, ROS scavenging capacity of hyperaccumulator plants is similar to sensitive plants. Freeman et al. (2010) stated that when both sensitive and hyperaccumulator plants were grown in media with and without Se, Se-hyperaccumulator plants such as *S. pinnata* have higher level of ascorbic acid, GSH, GSSG, and total glutathione in their young leaves in each growth conditions than *Synallaxis albescens*, a Se-sensitive species. The ROS generation are triggered as a result of reaction of selenite with reduced glutathione. However, hydrogen peroxide and superoxide generation were lower than *S. albescens* when both species were treated with 20 μm selenate for 10 weeks. Also, ROS scavenging capacity of *S. pinnata* was equal to that of *S. albescens* under Se-containing growth conditions. In parallel to this, *S. pinnata* showed significant differences in its gene regulation regarding antioxidant and ROS generation such as glutathione *S*-transferase, glutaredoxin, catalase, and dehydroascorbate reductase in mitochondria. The other differential gene expressions in *S. pinnata* versus *S. albescens* were observed in defense related, chaperone, and S assimilation genes, which were described in the study of Freeman et al. (2010). Kolbert et al. (2018) reported that the response of the hyperaccumulator plant of *A. bisulcatus* to Se toxicity is explained by less formation of superoxide dismutase enzymes and nitric oxide (NO) in root tips and cotyledons compared to those in *Astragalus membranaceus*, a sensitive species. NADPH oxidase enzymes (NOXs), causing formation of superoxide, were one quarter ratio in *A. bisulcatus* versus *A. membranaceus*. Therefore, less NOX formation occurs in *A. bisulcatus* as a response to excess Se. In a similar manner, peroxynitrite formation in cotyledons of *A. bisculatus* was insignificant, whereas it was found significant in the same organs of *A. membranaceus*. Most importantly, besides ROS, generation of reactive nitrogen species (RNS) only occurred in the sensitive species. *S*-nitrosoglutathione (GSNO) showed a decreased activity after Se treatment in roots and cotyledons of *A. bisulcatus*, whereas *S*-nitrosoglutathione reductase (GSNOR) activity was impeded by Se treatment. The decrease observed in GSNO formation is attributed to its buffering capacity of NO by the authors.

15.4 Selenium Biofortification in Plants

The objective of biofortification is to enhance the nutrient concentration of crops by fertilizers, conventional plant breeding, biotechnological approaches, or beneficial microbes and thereby increase food and fodder quality where the essential elements are deficient in soil (Hossain et al. 2021). In many parts of the world, plants are the main source of Se, followed by meat and fish (Broadley et al. 2010). The amount of Se in soils and Se concentrations competing with S compounds in soils affect the plant Se levels in crops (Malagoli et al. 2015). Agronomic Se biofortification has many advantages over direct Se supplementation since inorganic Se, absorbed by plants, is converted into organic Se forms with higher bioavailability (D'Amato et al. 2020). Selenium can be biofortified in plants using a variety of agronomic management techniques, such as soil inorganic fertilization, foliar application, and organic fertilizers. Selenium -biofortification through agronomic management aims at regulating the Se concentrations in crops by taking application timing and application rate of Se fertilizer into consideration. Agronomic Se biofortification is more effective to control Se supplementation in many ways because it converts the inorganic Se, taken by plants, into an organic form and make it more bioavailable (Chasteen and Bentley 2003).

The genetics of Se metabolism has received little attention so far, despite the fact that inter- and intraspecific diversity in Se accumulation in plants are extensively known. When the tolerance mechanism in *Arabidopsis thaliana* to selenite was studied by quantitative trait loci (QTLs), analysis in recombinant inbred lines (RILs), a total of three quantitative trait loci (QTLs) on chromosomes 1, 3, and 5 are reported to account for 32% of the phenotypic variation associated with root length differences between control and selenate-treated plants. However, when the root length ratios between control and selenate-treated plants were taken into account, it was reported that 18–24% phenotypic variation was explained by the same QTLs (Zhang et al. 2006).

It is known that S-related enzymes improve Se accumulation, tolerance, and volatilization of Se. At this juncture, the enzymes have received a lot of attention in genetic engineering studies. Overexpression of *ATP sulfurylase* (*APS*) in *B. juncea*, encoding a key enzyme for selenate-to-selenite transition, leads to enhanced selenate reduction. Put it another way, the transgenic APS plants accumulate organic Se (likely methyl-SeCys) when supplied with selenate, while wild type controls accumulation of only selenite (Pilon-Smits et al. 1999). The APS transgenics stores more than three times Se than the wild-type *B. juncea* when the plants are grown in pots filled with naturally seleniferous soil in a greenhouse experiment (Van Huysen et al. 2004). Similarly, the APS transgenics accumulate four times more Se than wild-type of *B. juncea*; and SMT transgenics have two times higher Se levels in two field experiments, which were carried out on selenate-contaminated soil in central California (Bañuelos et al. 2005). In another study, when *SeCys methyltransferase* (*SMT*) was overexpressed in *A. thaliana* and *B. juncea*, the SMT transgenic plants showed enhanced Se accumulation in the form of methyl-SeCys. Similarly, Freeman et al. (2010) also reported that SMT was expressed seven times more in a Se-hyperaccumulator plant, *S. pinnata*, than Se-sensitive species, *S. albescens*, and Kolbert et al. (2018) stated that higher accumulation of Se in shoots of *A. bisulcatus* is explained by detoxification mechanism of Se, which is probably maintained by expression of SMT. As a result, the mentioned studies show that APS and SMT enzymes are significantly important enzymes to develop biofortified crops (Ellis et al. 2004; LeDuc et al. 2004).

15.5 Conclusion

Selenium is a very important trace element for living things. However, due to variation of its bioavailability in earth crust, approximately one billion people suffer from Se deficiency. For plants, at low doses, it reduces oxidative stress and increases plant growth and development by regulating the uptake and translocation of trace elements and generation of reactive oxygen species. On the other hand, excessive Se in plants changes the structures of proteins and enzymatic reactions, disrupts the metabolism of intracellular metabolites, and causes high oxidative stress by disrupting both the structure and functioning of chloroplast and mitochondria. However, the optimum amount of Se for each plant species and variety is quite different, and it is still controversial whether it is necessary or not for plants. Currently, the recent studies have focused on either phytoremediation or biofortification-related topics. Although Se transporters are identified in few plants, the omics studies are very limited in terms of shedding light on Se uptake, assimilation, and detoxification pathways. The association between Se and S uptake metabolism is known; however, no information has been gained so far about which genes are involved in Se–S uptake regulation. There is almost no proteomics, metabolomics, metagenomics, phenomics, and transcriptomics studies to give insights about Se metabolism. QTL studies are very insufficient to pinpoint the genes involved in Se metabolism as well. Consequently, much more research is needed to enlighten Se metabolism in plants using omics studies.

References

Ali, W., Zhang, H., Junaid, M. et al. (2021). Insights into the mechanisms of arsenic-selenium interactions and the associated toxicity in plants, animals, and humans: a critical review. *Crit. Rev. Environ. Sci. Technol 51* (7): 704–750.

Bañuelos, G., Terry, N., LeDuc, D.L. et al. (2005). Field trial of transgenic Indian mustard plants shows enhanced phytoremediation of selenium-contaminated sediment. *Environ. Sci. Technol.* 39: 1771–1777.

Bodnar, M., Konieczka, P., and Namiesnik, J. (2012). The properties, functions, and use of selenium compounds in living organisms. *J. Environ. Sci. Health C Toxicol. Carcinog. 30* (3): 225–252.

Broadley, M.R., Alcock, J., Alford, J. et al. (2010). Selenium biofortification of high-yielding winter wheat (*Triticum aestivum* L.) by liquid or granular Se fertilisation. *Plant Soil.* 332: 5–18.

Brown, T.A. and Shrift, A. (1980). Identification of selenocysteine in the proteins of selenate-grown *Vigna radiata. Plant Physiol.* 66: 758–761.

Brown, T.A. and Shrift, A. (1981). Exclusion of selenium from proteins of selenium-tolerant astragalus species. *Plant Physiol.* 67: 1051–1053.

Brown, T.A. and Shrift, A. (1982). Selenium: toxicity and tolerance in higher plants. *Biol. Rev. 57* (1): 59–84.

Chasteen, T.G. and Bentley, R. (2003). Biomethylation of selenium and tellurium: microorganisms and plants. *Chem. Rev.* 103: 1–26.

Chauhan, R., Awasthi, S., Tripathi, P. et al. (2017). Selenite modulates the level of phenolics and nutrient element to alleviate the toxicity of arsenite in rice (*Oryza sativa* L.). *Ecotoxicol. Environ. Saf.* 138: 47–55.

Combs, G.F. (2001). Selenium in global food systems. *Br. J. Nutr. 85* (5): 517–547.

D'Amato, R., Regni, L., Falcinelli, B. et al. (2020). Current knowledge on selenium biofortification to improve the nutraceutical profile of food: a comprehensive review. *J. Agric. Food Chem.* 68 (14): 4075–4097.

Dimkovikj, A., Fisher, B., Hutchison, K., and Van Hoewyk, D. (2015). Stuck between a ROS and a hard place: analysis of the ubiquitin proteasome pathway in selenocysteine treated *Brassica napus* reveals different toxicities during selenium assimilation. *J. Plant Physiol.* 181: 50–54.

Ellis, D.R., Sors, T.G., Brunk, D.G. et al. (2004). Production of Se-methylselenocysteine in transgenic plants expressing selenocysteine methyltransferase. *BMC Plant Biol.* 4: 1–11.

Etteieb, S., Magdouli, S., Zolfaghari, M., and Brar, S.K. (2020). Monitoring and analysis of selenium as an emerging contaminant in mining industry: a critical review. *Sci. Total Environ.* 698: 134339.

Fordyce, F.M. (2013). Selenium deficiency and toxicity in the environment. In: *Essentials of Medical Geology: Revised Edition* (ed. O. Selinus), 375–416. Springer Science+Business.

Freeman, J.L., Tamaoki, M., Stushnoff, C. et al. (2010). Molecular mechanisms of selenium tolerance and hyperaccumulation in *Stanleya pinnata. Plant Physiol.* 153: 1630–1652.

Galeas, M.L., Zhang, L.H., Freeman, J.L. et al. (2007). Seasonal fluctuations of selenium and sulfur accumulation in selenium hyperaccumulators and related nonaccumulators. *New Phytol.* 173 (3): 517–525.

Garousi, F., Kovács, B., Domokos-Szabolcsy, E., and Veres, S. (2017). Biological changes of green pea (*Pisum sativum* L.) by selenium enrichment. *Acta Biol. Hung.* 68 (1): 60–72.

Grant, K., Carey, N.M., Mendoza, M. et al. (2011). Adenosine 5-phosphosulfate reductase (APR2) mutation in Arabidopsisimplicates glutathione deficiency in selenate toxicity. *Biochem. J.* 438: 325–335.

Gupta, M. and Gupta, S. (2017). An overview of selenium uptake, metabolism, and toxicity in plants. *Front. Plant Sci. 7*: 1–14.

Hartikainen, H. (2005). Biogeochemistry of selenium and its impact on food chain quality and human health. *J. Trace Elem. Med. Biol. 18* (4): 309–318.

Hasanuzzaman, M., Nahar, K., Alam, M.M., and Fujita, M. (2014). Modulation of antioxidant machinery and the methylglyoxal detoxification system in selenium-supplemented *Brassica napus* seedlings confers tolerance to high temperature stress. *Biol. Trace Elem. Res.* 161 (3): 297–307.

Hatfield, D.L., Tsuji, P.A., Carlson, B.A., and Gladyshev, V.N. (2014). Selenium and selenocysteine: roles in cancer, health, and development. *Trends Biochem. Sci.* 39 (3): 112–120.

Hondal, R.J., Marino, S.M., and Gladyshev, V.N. (2012). Selenocysteine in thiol/disulfide-like exchange reactions. *Antioxid. Redox Signal.* 18 (13): 1675–1689.

Hossain, A., Skalicky, M., Brestic, M. et al. (2021). Selenium biofortification: roles, mechanisms, responses and prospects. *Molecules* 26: 881.

Huang, Q.Q., Wang, Q., Wan, Y.N. et al. (2017). Application of X-ray absorption near edge spectroscopy to the study of the effect of sulphur on selenium uptake and assimilation in wheat seedlings. *Biol. Plant.* 61: 726–732.

Hugouvieux, V., Dutilleul, C., Jourdain, A. et al. (2009). Arabidopsis putative selenium-binding protein1 expression is tightly linked to cellular sulfur demand and can reduce sensitivity to stresses requiring glutathione for tolerance. *Plant Physiol.* 151: 768–781.

Huysen, T., Terry, N., and Pilon-Smits, E.A.H. (2004). Exploring the selenium phytoremediation potential of transgenic indian mustard overexpressing ATP sulfurylase or cystathionine-γ-synthase. *Int. J. Phytoremediat.* 6: 111–118.

Jiang, L., Yang, J., Liu, C. et al. (2020). Overexpression of ethylene response factor *ERF96* gene enhances selenium tolerance in Arabidopsis. *Plant Physiol. Biochem.* 149: 294–300.

Johnson, C.C., Fordyce, F.M., and Rayman, M.P. (2010). Symposium on 'geographical and geological influences on nutrition' factors controlling the distribution of selenium inthe environment and their impact on health and nutrition. *Proc. Nutr. Soc.* 69: 119–132.

Kolbert, Z., Molnar, A., Szöllösi, R. et al. (2018). Nitro-oxidative stress correlates with Se tolerance of *Astragalus* species. *Plant Cell Physiol.* 59 (9): 1827–1843.

Koprivova, A., Mugford, S.T., and Kopriva, S. (2010). Arabidopsis root growth dependence on glutathione is linked to auxin transport. *Plant Cell Rep.* 29: 1157–1167.

Kumar, A. and Prasad, K.S. (2021). Role of nano-selenium in health and environment. *J. Biotechnol.* 325: 152–163.

Lavu, R.V.S., Du Laing, G., Van De Wiele, T. et al. (2012). Fertilizing soil with selenium fertilizers: impact on concentration, speciation, and bioaccessibility of selenium in leek (*Allium ampeloprasum*). *J. Agric. Food Chem. 60* (44): 10930–10935.

LeDuc, D.L., Tarun, A.S., Montes-Bayon, M. et al. (2004). Overexpression of selenocysteine methyltransferase in Arabidopsis and Indian mustard increases selenium tolerance and accumulation. *Plant Physiol.* 135: 377–383.

Lee, K.H. and Jeong, D. (2012). Bimodal actions of selenium essential for antioxidant and toxic pro-oxidant activities: the selenium paradox (review). *Mol. Med. Rep.* 5: 299–304.

Malagoli, M., Schiavon, M., dall'Acqua, S., and Pilon-Smits, E.A.H. (2015). Effects of selenium biofortification on crop nutritional quality. *Front. Plant Sci.* 6: 280.

Mehdi, Y., Hornick, J.L., Istasse, L., and Dufrasne, I. (2013). Selenium in the environment, metabolism and involvement in body functions. *Molecules* 18: 3292–3311.

Mézes, M. and Balogh, K. (2009). Prooxidant mechanisms of selenium toxicity – a review. *Acta Biol. Szeged.* 53: 15–18.

Mikkelsen, R.L., Page, A.L., and Bingham, F.T. (1989). Factors affecting selenium accumulation by agricultural crops. In: *Selenium in Agriculture and the Environment, SSSA Special Publications*, vol. 23 (ed. L.W. Jacobs), 64–94.

Molnár, Á., Kolbert, Z., Kéri, K. et al. (2018). Selenite-induced nitro-oxidative stress processes in *Arabidopsis thaliana* and *Brassicajuncea*. *Ecotoxicol. Environ. Saf.* 148: 664–674.

Mostofa, M.G., Hossain, M.A., Siddiqui, M.N. et al. (2017). Phenotypical, physiological and biochemical analyses provide insight into selenium-induced phytotoxicity in rice plants. *Chemosphere* 178: 212–223.

Pilon-Smits, E.A.H., Hwang, S.B., Lytle, C.M. et al. (1999). Overexpression of ATP sulfurylase in *Brassica juncea* leads to increased selenate uptake, reduction and tolerance. *Plant Physiol.* 119: 123–132.

Rao, S., Yu, T., Cong, X. et al. (2021). Transcriptome, proteome, and metabolome reveal the mechanism of tolerance to selenate toxicity in *Cardamine violifolia*. *J. Hazard. Mater.* 406: 124283.

Rayman, M.P. (2000). The importance of selenium to human health. *Lancet* 356 (9225): 233–241.

Rayman, M.P. (2012). Selenium and human health. *Lancet* 379 (9822): 1256–1268.

Reid, M.E., Duffield-Lillico, A.J., Slate, E. et al. (2008). The nutritional prevention of cancer: 400 Mcg per day selenium treatment. *Nutr. Cancer* 60 (2): 155–163.

Reilly, C. (1996). *Selenium in Food and Health*, 1e. Great Britain: Cornwell.

Rotruck, J.T., Pope, A.L., and Ganther, H.E. (1980). Selenium: biochemical role as a component of glutathione peroxidase. *Nutr. Rev. 38* (8): 280–283.

Salt, D., Blaylock, M., Kumar, N. et al. (1995). Phytoremediation: a novel strategy for the removal of toxic metals from theenvironment using plants. *Nat. Biotechnol.* 13: 468–477.

Schiavon, M. and Pilon-Smits, E.A.H. (2017). Selenium biofortification and phytoremediation phytotechnologies: a review. *J. Environ. Qual.* 46 (1): 10–19.

Schiavon, M., Nardi, S., dalla Vecchia, F., and Ertani, A. (2020). Selenium biofortification in the 21st century: status and challenges for healthy human nutrition. *Plant Soil* 453 (1–2): 245–270.

Schmidt, R.L. and Simonović, M. (2012). Synthesis and decoding of selenocysteine and human health. *Croat. Med. J.* 53 (6): 535–550.

Tobe, R., Koga, H., Tani, Y. et al. (2017). Identification of selenomethionine, selenocysteine, and Se-methylselenocysteine in the selenium-enriched common ice plant. *Biol. Trace Elem. Res.* 28: 135–142.

Tripathi, N. and Misra, S.G. (1974). Uptake of applied selenium by plants. *Indian J. Agric. Sci.* 44: 804–807.

Trippe, R.C. and Pilon-Smits, E.A.H. (2021). Selenium transport and metabolism in plants: phytoremediation and biofortification implications. *J. Hazard. Mater.* 404: 124178.

Van Hoewyk, D. (2013). A tale of two toxicities: malformed selenoproteins and oxidative stress both contribute to selenium stress in plants. *Ann. Bot.* 112 (6): 965–972.

Van Huysen, T., Terry, N., and Pilon-Smits, E.A.H. (2004). Exploring the selenium phytoremediation potential of transgenic *Brassica juncea* overexpressing ATP sulfurylase or cystathionine-g-synthase. *Int. J. Phytoremed.* 6: 111–118.

Wang, J., Li, H., Li, Y. et al. (2013). Speciation, distribution, and bioavailability of soil selenium in the tibetan plateau kashin-beck disease area – a case study in Songpan County, Sichuan Province, China. *Biol. Trace Elem. Res.* 156: 1–3.

Wang, J., Cappa, J.J., Harris, J.P. et al. (2018). Transcriptome-wide comparison of selenium hyperaccumulator and nonaccumulator Stanleya species provides new insight into key processes mediating the hyperaccumulation syndrome. *Plant Biotechnol. J.* 16: 1582–1594.

White, P.J. (2018). Selenium metabolism in plants. *Biochim. Biophys. Acta Gen. Subj.* 1862 (11): 2333–2342.

Wu, X., Tao, M., Meng, Y. et al. (2020). The role of WRKY47 gene in regulating selenium tolerance in *Arabidopsis thaliana*. *Plant Biotechnol. Rep.* 14: 121–129.

Zhang, L., Byrne, P.F., and Pilon-Smits, E.A.H. (2006). Blackwell Publishing Ltd Mapping quantitative trait loci associated with selenatetolerance in *Arabidopsis thaliana*. *New Phytol.* 170: 33–42.

Zhou, Y., Tang, Q., Wu, M. et al. (2018). Comparative transcriptomicsprovides novel insights into themechanisms of selenium tolerance in the hyperaccumulator plant *Cardamine hupingshanensis*. *Sci. Rep.* 8: 2789.

Zhu, Y.G., Pilon-Smits, E.A.H., Zhao, F.J. et al. (2009). Selenium in higher plants: understanding mechanisms for biofortification and phytoremediation. *Trends Plant Sci.* 14 (8): 436–442.

16

Breeding for Rice Cultivars with Low Cadmium Accumulation

Li Tang, Yaokui Li, Yan Peng, Bigang Mao, Ye Shao, Zhongying Ji, and Bingran Zhao

State Key Laboratory of Hybrid Rice, Hunan Hybrid Rice Research Center, Changsha, China

16.1 Introduction

Cadmium (Cd) is a heavy metal with strong biological toxicity, and its half-life in human body can be as long as 30 years (Uraguchi and Fujiwara 2013). Cd contamination in farmland has become a global environmental problem (Gallego et al. 2012). Rice is the staple food for more than half of the world's population. Compared with other crops, rice has a stronger ability to accumulate Cd in shoots and grains (Rizwan et al. 2016; Zhang et al. 2022). The Cd concentration in grains of *indica* rice is generally higher than that of *japonica* rice. For Cd-contaminated areas where rice is the staple food, rice is the most important source of Cd intake (Zhao et al. 2022). About 90% of the world's rice is produced in Asia. Sampling surveys show that the average Cd concentration of rice grain from Asia is higher than that from Europe, the Middle East, and North America (Jallad 2015). Excessive Cd in rice grains has become an urgent problem to ensure food security and rice quality. Breeding rice cultivars with low Cd accumulation is the most direct and economical way to solve this problem.

16.2 Molecular Mechanisms of Cd Accumulation in Rice

The accumulation of Cd in rice involves physiological processes such as root uptake of Cd, Cd translocation from roots to shoots mediated by xylem, and Cd distribution in shoots (Clemens and Ma 2016). These physiological processes are accomplished via the transmembrane transport of Cd transporters. It has been found that natural resistance-related macrophage proteins (NRAMP), heavy metal ATPase (HMA), zinc-regulated/iron-regulated transporter-like protein (ZIP), and other family transporters have Cd transport activities, and participate in the accumulation of Cd in grains by mediating Cd uptake or transport in rice.

Heavy Metal Toxicity and Tolerance in Plants: A Biological, Omics, and Genetic Engineering Approach, First Edition. Edited by Mohammad Anwar Hossain, AKM Zakir Hossain, Sylvain Bourgerie, Masayuki Fujita, Om Parkash Dhankher, and Parvez Haris.

16.2.1 Cd Uptake

It has been found that the metal transporters including OsNRAMP5, OsNRAMP1, OsIRT1, OsIRT2, OsZIP5, OsZIP9, and OsCd1 are involved in Cd uptake of rice. As they are not Cd-specific transporters, Cd entering into rice roots is accompanied by the uptake of Mn, Fe, Zn, and other essential mineral elements (Table 16.1) (Nakanishi et al. 2006; Sasaki et al. 2012; Yan et al. 2019; Chang et al. 2020b; Tan et al. 2020). Unlike most plants, the roots of rice have two Casparian strips at both the exodermis and endodermis. Moreover, there are almost no cortical cells between the exodermis and endodermis of the mature roots of rice, instead a highly developed aerenchyma. Therefore, there must be influx and efflux transporters on the plasma membrane of exodermis and endodermis of rice roots (Sasaki and Ashikari 2018). OsNRAMP5 is the major transporter responsible for Cd and Mn uptake in rice and also has Fe transport activity (Ishikawa et al. 2012; Ishimaru et al. 2012; Sasaki et al. 2012). OsNRAMP5 is polarly localized at the distal side of the plasma membranes in both exodermis and endodermis cells (Table 16.1) (Sasaki et al. 2012). Among the Cd transporters identified so far, OsNRAMP5 contributed the most to Cd accumulation in rice grains. Knockout of *OsNRAMP5* significantly reduced the Cd and Mn uptake capacity of roots, resulting in a decrease of more than 90% of Cd concentration in grains. In different reports, mutation of *OsNRAMP5* either had no effect or different effects on agronomic traits (Ishikawa et al. 2012; Sasaki et al. 2012; Yang et al. 2014; Tang et al. 2017; Wang et al. 2019; Yang et al. 2019). This inconsistence may be due to different growth conditions, genetic background, or mutation sites on *OsNRAMP5*, resulting in the distinct Mn concentration in mutant *osnramp5* plants. However, with increasing environmental Cd concentrations, the effect of *OsNRAMP5* knockout on reducing root Cd uptake weakened, while its effect on improving root-to-shoot Cd translocation was constant. As a result, its mutation dramatically reduced Cd accumulation in shoots under low and moderate Cd stress, but inversely increased that under high Cd conditions (Tang et al. 2022). Interestingly, Cd tolerance of mutant *osnramp5* was persistently enhanced, irrespective of lower or higher Cd concentrations in shoots, compared with that of wild-type plants. Additionally, knockout of *OsNRAMP5* conferred preferential delivery of Mn to young leaves and grains, which contribute to the improvement of the environmental adaptation of *osnramp5* (Tang et al. 2022).

OsNRAMP1 is also a transporter that mediates the uptake of Cd and Mn, which plays a similar but nonredundant function to OsNRAMP5. OsNRAMP1 is localized to the plasma membranes of most root cells except the vascular bundle in the stele (Table 16.1) (Chang et al. 2020b). The mutant *osnramp1* exhibited significantly lower Cd and Mn concentrations in roots, shoots, and grains than the wild-type (WT) plants, but the reduction was much less than that of mutant *osnramp5*, while the straw biomass and grain yields were not affected by the moderate reduction of Mn accumulation in *osnramp1* (Chang et al. 2020b). OsCd1, a member of the major facilitator superfamily, also has Cd and Mn uptake activities. Knockout of the *OsCd1* could significantly reduce the uptake of Cd in roots and Cd accumulation in grains, but it also caused Mn deficiency in plants, resulting in the decreased seed setting rate and yield reduction (Yan et al. 2019).

In the ZIP family, OsZIP5 and OsZIP9 synergistically mediate root uptake of Zn and Cd. The uptake of Cd and Zn by the double mutant *oszip5/oszip9* was significantly lower than that of

Table 16.1 Encoding genes of Cd transporters in rice.

Gene name	Gene ID	Main expression regions	Subcellular localization	Function	References
OsNRAMP1	LOC_Os07g15460	Most root cells except the central vasculature	Plasma membrane	Cd and Mn uptake	Chang et al. (2020b)
OsNRAMP5	LOC_Os07g15370	Exodermis and endodermis cells in roots	Plasma membrane	Cd and Mn uptake	Sasaki et al. (2012)
OsIRT1	LOC_Os03g46470	Roots	Plasma membrane	Cd and Fe uptake	Nakanishi et al. (2006)
OsIRT2	LOC_Os03g46454	Roots	Plasma membrane	Cd and Fe uptake	Nakanishi et al. (2006)
Cd1	LOC_Os03g02380	Root exodermis, parenchyma in cortex, endodermis, and stele cells	Plasma membrane	Cd and Mn uptake	Yan et al. (2019)
OsZIP5	LOC_Os05g39560	Root epidermis	Plasma membrane	Cd and Zn uptake	Tan et al. (2020)
OsZIP9	LOC_Os05g39540	Root epidermis and glumes	Plasma membrane	Cd and Zn uptake	Tan et al. (2020)
OsHMA2	LOC_Os06g48720	Root pericycle and stem nodes phloem	Plasma membrane	Cd and Zn transport	Yamaji et al. (2013)
OsHMA3	LOC_Os07g12900	Roots	Tonoplast	Cd sequestration in vacuole	Ueno et al. (2010)
OsZIP7	LOC_Os05g10940	Parenchyma cells of vascular bundles in roots and nodes	Plasma membrane	Cd and Zn transport	Tan et al. (2019)
OsCAL1	LOC_Os02g41904	Root exodermis and xylem parenchyma cells	Cell wall (secretory protein)	Cd transport	Luo et al. (2018)
OsCCX2	LOC_Os03g45370	Xylem region of vascular tissues at the nodes	Plasma membrane	Cd and Ca transport	Hao et al. (2018)
OsLCT1	LOC_Os06g38120	Leaf blades and nodes	Plasma membrane	Cd transport	Uraguchi et al. (2011)
OsLCT2	GenBank accession number:MW757982	Most root cells	Endoplasmic reticulum	Cd transport	Tang et al. (2021a)
OsABCC9	LOC_Os04g13210	Epidermis, exodermis, and stele of roots	Tonoplast	Cd sequestration in vacuole	Yang et al. (2021)

the WT. However, knockout of *OsZIP5* or *OsZIP9* resulted in the lower Zn concentration and less grain yields (Tan et al. 2020; Yang et al. 2020). *OsIRT1* and *OsIRT2*, two transporters in ZIP family, had induced expression under Fe-deficient conditions and exhibited Cd transport activity in yeast, presumably enhancing Cd uptake in rice plants under Fe-deficient conditions (Nakanishi et al. 2006).

16.2.2 Radial Transport and Xylem Loading

Following Cd uptake by the root epidermis and exodermis from the soil, Cd is radially transferred across the cortex, endodermis, and pericycle cells to the vascular bundle of the stele and subsequently translocated to the shoots through the xylem. The Cd and Zn transporters, OsHMA3 and OsABCC9, is localized at the tonoplast of root cells and could sequester Cd into the root cell vacuoles (Table 16.1), consequently reducing Cd accumulation in the shoots (Ueno et al. 2010; Yang et al. 2021). Rice varieties with loss-of-function alleles of *OsHMA3* had high Cd concentrations in the shoots and grains due to a weakened ability to sequester Cd in the root vacuoles (Ueno et al. 2010; Yan et al. 2016; Sui et al. 2019). The *OsHMA3* promoter has naturally variation between *indica* and *japonica* rice subspecies, which partly contributes to the divergence of Cd accumulation between the two subspecies (Liu et al. 2020). Unlike *OsHMA3*, which is expressed in all root cells, *OsABCC9* is mainly expressed in the epidermis, exodermis, and stele, so OsABCC9 has less effect on root Cd compartmentalization than OsHMA3 (Yang et al. 2021).

Metal transporters OsHMA2, OsZIP7, OsCCX2, and CAL1 can promote translocation of Cd to the aboveground parts (Satoh-Nagasawa et al. 2012; Takahashi et al. 2012; Hao et al. 2018; Luo et al. 2018; Tan et al. 2019). OsHMA2 is a Zn and Cd transporter, which is localized at the pericycle of root (Table 16.1). Its encoding gene is mainly expressed in the mature region of the root during the vegetative growth stage (Yamaji et al. 2013). Compared with the WT, mutant *oshma2* showed the lower ratio of root-to-shoot translocation and decreased concentrations of Zn and Cd in shoots (Satoh-Nagasawa et al. 2012). OsZIP7 is also a transporter of Zn and Cd, which is localized at the stele of the root. Knockout of *OsZIP7* reduced the translocation of Zn and Cd from roots to shoots and grain yields (Tan et al. 2019). *CAL1* encodes a defensins protein and is mainly expressed in the exodermis and xylem parenchyma cells of roots (Table 16.1). CAL1 specifically chelates with Cd, driving Cd secretion from the cytoplasm of xylem companion cells to xylem vessels, thereby promoting Cd translocation from roots to shoots. CAL1 directionally regulates Cd accumulation in vegetative organs such as leaves, but does not affect Cd accumulation in grains (Luo et al. 2018). Calcium/cation exchange protein OsCCX2 is a Ca and Cd transporter (Table 16.1). In *osccx2* lines, the roots-to-shoots translation rate of Cd was significantly lower than that of the WT (Hao et al. 2018).

16.2.3 Distribution of Cd in Shoots

After Cd is translocated to the aboveground parts, it is transported over long distances via the xylem and phloem. The rice stem nodes play a crucial role in the distribution of Cd to grains. Some metal transporters have been demonstrated to participate not only in the translocation of Cd from roots to shoots but also in the distribution of Cd of stem nodes, such as OsHMA2, OsZIP7, and OsCCX2. In the stem nodes, OsHMA2 is localized at the phloem of the vascular bundle, whereas OsZIP7 is localized at the parenchyma cells of the large vascular bundle. The Zn concentrations in the upper stem nodes and flag leaves, and the grain yields were significantly lower in *oshma2* and *oszip7* than in WT plants (Yamaji et al. 2013; Tan et al. 2019). OsCCX2 is mainly expressed in the xylem parenchyma of stem nodes (Table 16.1). The 1000-grain weight is significantly lower in *osccx2* than in the WT, probably due to the impaired Ca transport in *osccx2* (Hao et al. 2018). The low-affinity cation transporter gene *OsLCT1* is highly expressed in the stem nodes and flag leaves at the ripening stage of rice. OsLCT1 participates in the intervascular transfer of Cd in stem nodes and affect Cd accumulation in grains by regulating the phloem transport of Cd (Uraguchi et al. 2011). The Cd concentrations of brown rice in *OsLCT1* RNAi lines were reduced to about half of that of WT plants, while the plant growth and concentrations of minerals in brown rice were not significantly affected in these RNAi lines (Uraguchi et al. 2011).

16.3 Transgenic Approach for Breeding Low-Cd Rice

Cloning and characterization of genes associated with Cd accumulation in rice provide a theoretical basis for low-Cd rice breeding. Genetic modification of genes involved in Cd uptake, transport in rice through genetic engineering can decrease Cd accumulation in rice grains, and consequently breed rice cultivars with low-Cd accumulation in rice grains.

16.3.1 Traditional Transgenic Technology

Traditional transgenic methods include gene overexpression and RNA interference (RNAi). Both overexpression of genes negatively regulated grain Cd accumulation via constitutive or tissue-specific strong promoters and expression reduction of genes positively regulated grain Cd accumulation via RNAi can reduce the Cd level in rice grains. *OsHMA3* negatively regulates grain Cd concentration by sequestering Cd into vacuoles of root cell. *OsHMA3* is an important target for restricting Cd accumulation in rice grains. Ueno et al. (2010) overexpressed *OsHMA3* through the maize ubiquitin promoter in the *japonica* cultivars Nipponbare. They found that when gown in Cd-contaminated soil, *OsHMA3* overexpression line showed significantly lower grain Cd concentration, but similar concentrations of Zn and Fe in grain compared to those of the WT. Lu et al. (2019) overexpressed the *OsHMA3* gene by CaMV 35S promoter in *indica* cultivars Zhongjiazao 17 and found that overexpression of the *OsHMA3* reduced the Cd concentrations in grains by 94–98%, whereas had no effect on yield and Zn, Mn, Fe concentration of grains, when compared with the WT plants. Shao et al. (2018) used the *OsHMA2* promoter to drive *OsHMA3* expression and found that the Cd concentrations of brown rice in transgenic plants was declined to less than 10% of the WT.

OsHMA2 is involved in the transport of Cd from roots to shoots. Takahashi et al. (2012) used the CaMV 35S and sucrose transporter (SUT1) promoters to drive the overexpression of *OsHMA2* in a *japonica* cultivar Tsukinohikari. They found the Cd grain concentrations we decreased by 50%, while the other metal elements did not differ significantly in transgenic lines, compared with the WT. Wang et al. (2016) overexpressed the fragments of

OsO3L2 and *OsO3L3* in the *japonica* cultivar Zhonghua 11, and found the Cd concentrations of brown rice were considerably lower in the transgenic line than in the WT without affecting grain yield or levels of Fe, Mn, Cu and Zn. Recently, Tang et al. (2021a) identified a new member of the rice LCT family, OsLCT2. In the Cd-contaminated field, plants overexpressing *OsLCT2* exhibited 24.1–27.5% lower Cd concentrations and similar concentrations of Zn, Fe, Mn and Cu in brown rice, compared with the WT plants. Chang et al. (2020a) used the promoters of rice *OsActin1* and maize *Ubiquitin* to drive *OsNRAMP5* expression, respectively, and found that grain Cd concentrations were reduced by 49–94% in *OsNRAMP5*-overexpressed plants, probably due to the disruption of OsNRAMP5 polar localization and subsequently decreased Cd translocation from roots to shoots.

OsLCT1 regulates the transport of Cd through the phloem to rice grain. Uraguchi et al. (2011) silenced the gene in Nipponbare using RNAi technology, which resulted in the grain Cd concentrations of the RNAi lines was about half that of the WT plants, without influence on yield and other mineral metal concentrations. PCS is a phytochelatin synthase. Das et al. (2017) suppressed the expression of both OsPCS1 and OsPCS2 in rice seeds by endosperm-specific RNAi, reducing the Cd concentrations in grains by about 50% without significant effects on yield.

16.3.2 Genome-Editing Technology

The genome editing system could targeted knockout genes positively regulated Cd accumulation, causing them to generate frameshift mutations and lose function. At present, genome editing technology represented by the CRISPR/Cas system is widely used, low-cost, and simple. They have played an important role in genetic improvement of crop important traits.

OsNRAMP5 is an important mutation target for breeding of rice cultivars with low Cd accumulation. OsNRAMP5 is the predominant transporter involved in Cd uptake in rice. Mutant *osnramp5* has good utilization potential in Cd-contaminated areas. However, the ecological adaptability of *osnramp5* lines with different genetic backgrounds and *OsNRAMP5* mutation sites is divergent. Tang et al. (2017) used CRISPR/Cas9 technology to edit the *ninth* exon of *OsNRAMP5* in both parental lines of an *indica* rice hybrid, Huazhan and Longke 638S, and obtained *osnramp5* lines in the background of hybrid combination Longliangyouhuazhan by crossing. It was found that the Cd concentrations in brown rice drastically decreased by more than 90%, and main agronomic traits such as yield and rice quality were not significantly affected in these *osnramp5* lines, compared with the WT, when grown in Cd-contaminated paddy fields. CRISPR/Cas9-mediated the mutagenesis of the first exon of *OsNRAMP5* in *japonica* cultivars Nanjing 46 and Huaidao 5 showed that the Cd and Mn accumulation of *osnramp5* lines were greatly reduced, while the plant height, number of grains per panicle and seed setting rate were also decreased, resulting in a yield decline by 15–24% compared with the WT (Yang et al. 2019). Long et al. (2019) generated *osnramp5* lines by CRISPR/Cas9-based mutation of the tenth exon of *OsNRAMP5* in *indica* cultivars, Huang Huazhan, Huazhan, Wufeng B, Wushan Simiao, and Zhongzao 35. They found that the mutation of *OsNRAMP5* significantly diminished the accumulation of Cd in rice, and reduced yields slightly under certain conditions.

Wang et al. (2019) used CRISPR/Cas9 system to edit the second, sixth, and ninth exons of *OsNRAMP5* in *indica* cultivar Huanghuazhan and found that mutation of *OsNRAMP5* not only significantly reduced the Cd concentration in brown rice, but also decreased yield and rice quality. However, with the backward shift of the *OsNRAMP5* mutagenesis position, its effects on Mn accumulation, yield and rice quality were also weakened. In addition, the mutation of *OsNRAMP5* probably resulted in reduced tolerance to low Mn, high temperature, brown spot disease and other stresses, owing to the defective Mn uptake and dramatical decrease in Mn content in *osnramp5* plants (Ishikawa 2020; Dong et al. 2021).

Liu et al. (2019) knocked out *OsLCT1* in the *japonica* variety Xidao 1 and found that the Cd concentrations in grains were reduced by about 50%, and the grain yield was 17.3% lower in *oslct1* than in the WT. Recently, Chen et al. generate *oslcd* mutants in different backgrounds such as the *japonica* cultivar Zhonghua 11, *indica* cultivars Yuejingsimiao and Yuxiangyouzhan. When grown in Cd-contaminated paddy soils, *oslcd* reduced the Cd concentrations in grains by about 20% and had no significant effect on the accumulation of essential elements. In addition, the seed setting rate and grain yields of the *oslcd* mutant in the *japonica* background were severely impaired, but these traits of *oslcd* in the *indica* background was not significantly affected, compared with that of the corresponding WT (Chen et al. 2022).

Transgenic technology can achieve directional improvement of Cd accumulation traits in rice. The ideal low-Cd accumulation rice should has the following characteristics: (i) The Cd concentrations in grains is significantly lower than that of the WT plants, and consistently lower than maximum permissible limit when grown in Cd-contaminated soil; (ii) The concentrations of other harmful elements in grains do not increase, and the concentrations of essential mineral elements in grains do not decrease; (iii) The growth, yield and rice quality are not negatively influenced or have little impact. Therefore, in the rice production areas where Cd level in rice grains is slightly exceeding $0.2\,mg\,kg^{-1}$, overexpression of genes such as *OsHMA2* and *OsLCT2*, RNAi of *OsPCS*, and knockout of genes such as *OsLCT1*, *OsLCD*, could be used to cultivate low-Cd accumulation rice varieties with little effect on the other metal elements and yield traits. Moreover, in rice production areas with extremely high Cd-contaminated soils, overexpression of *OsHMA3* or mutation of *OsNRAMP5* could minimize the Cd concentration of rice grain to below $0.2\,mg\,kg^{-1}$.

16.4 Mutation Breeding for Low-Cd Rice Cultivars

Mutation breeding is an important way for plant germplasm innovation (Prasanna and Jain 2017; Melsen et al. 2021), due to the characteristics of obtaining massive mutations in absence of genetic engineering technology.

At present, some rice germplasms with low-Cd accumulation trait have been created by physical and chemical mutagenesis (Ishikawa et al. 2012; Zhu et al. 2013a, b; Cao et al. 2019; Ishikawa 2020; Lv et al. 2020; Wang et al. 2021b). Ishikawa et al. (2012) used the heavy ion beams to irradiate the seeds of *japonica* cultivar Koshihikari. The irradiated 2592 M_2 plants were grown in Cd-contaminated soil and measured the concentrations of Cd in grains, and consequently three low-Cd mutants were screened out. *OsNRAMP5* in the three low-Cd mutants were loss of function, resulting in the impaired uptake of Cd in root and significantly lower Cd concentrations in grains and straw (Ishikawa et al. 2012). In order to obtain three-line sterile line with early maturity character, Zhu et al. (2013b) successfully bred Luohong 3A, a new Honglian-type cytoplasmic male sterile line, using γ-ray radiation mutagenesis. Luohong 4A, a new Honglian-type cytoplasmic male sterile line with brown planthopper resistance, was bred by hybridization and molecular marker-assisted selection on the basis of Luohong 3A (Zhu et al. 2013a). By re-sequencing analysis of 1143 rice accessions mostly selected from the parents of superior hybrid rice cultivars of China, Lv et al. (2020) found that a 408 kb deletion existed on chromosome 7 of Luohong 3A and Luohong 4A, including *Ghd7*, the main locus controlling of rice heading date, *OsNramp5* and *OsNramp1*, two genes for Cd uptake. At the same time, Wang et al. (2021b) conducted Cd concentration determination of 275 rice resources grown in the heavily Cd-contaminated fields and isolated Luohong 3A and Luohong 4A, which showed low-Cd accumulation trait. Cao et al. (2019) identified a mutant with extremely low Cd accumulation (*lcd1*) in *indica* cultivar 9311 by EMS mutagenesis technology. An amino acid substitution of proline to leucine at position 236 of OsNRAMP5 is likely responsible for the low Cd accumulation in the *lcd1* mutant (Cao et al. 2019).

The mutation sites from physical or chemical mutagenesis are random. The traditional mutagenesis screening is mainly based on the phenotypic identification of M_2 generation or its progeny, which is time consuming and costly. The main reason is that (i) the phenotypic variation of M_1 plants is mainly attributed to the physiological damage caused by mutagenesis, which is unable to inherit; and (ii) the heritable variation in M_1 generation is genotypically chimeric (Jankowicz-Cieslak et al. 2017). The targeting induced local lesions in genomes (TILLING) technology could be used to perform targeted selection for gene mutations, but the detection accuracy is limited, and selection has been performed in M_2 or the later generations (Chen et al. 2014). The M_1 generation contains all the mutant genotypes with a minimum sample size, but there are some technical obstacles how to find the possible chimeric mutation of the targeted gene in the population. Shao et al. (2022) established a technology of a high-throughput screening for chimeric mutations in M_1 generation based on targeted deep sequencing (M_1TDS), inspired by "liquid biopsy" technology for early noninvasive diagnosis of tumors (Przybyl et al. 2018; He et al. 2019). Using the M_1TDS technique, a chimeric mutant of *OsNRAMP5* was screened out from the M_1 plants obtained by the heavy ion mutagenesis (Shao et al. 2022). The *osnramp5* homozygous mutants, male sterile line Lian 1S, and the restorer line R001 were obtained in the M_2 progeny of the chimeric mutant of *OsNRAMP5*. Lianliangyou 1, a hybrid rice combination by crossing Lian 1S and R001, exhibited much lower grain Cd concentration (below 0.03 $mg\,kg^{-1}$) and parallel agronomic traits, compared to that of Zhenliangyou 8612, the control hybrid rice combination when grown in a Cd-contaminated paddy field with the bioavailable Cd concentration of 0.44 $mg\,kg^{-1}$ in soils (Shao et al. 2022).

Physical and chemical mutagenesis is an effective way to create rice germplasms with low-Cd accumulation trait. With the continuous reduction of high-throughput sequencing costs, M_1TDS technology would efficiently achieve improvement of Cd accumulation traits in rice and accelerate mutation breeding of new rice cultivars with low-Cd accumulation trait.

16.5 Molecular Marker-Assisted Breeding for Low-Cd Rice Cultivars

The Cd accumulation of rice grain is a quantitative trait controlled by multiple genes. Some metal transporter genes including *OsNRAMP5, OsHMA3, OsNRAMP1* and *OsCd1* have been reported to significantly affect Cd accumulation in rice grains (Ueno et al. 2010; Takahashi et al. 2011; Sasaki et al. 2012; Yan et al. 2019; Chang et al. 2020b). Some quantitative trait loci (QTLs) and genes associated with Cd accumulation in grains were identified by genome-wide association studies (GWAS) (Zhao et al. 2018; Yan et al. 2019; Pan et al. 2020). The functional allelic variations of these genes play an important role in the Cd accumulation divergence of rice.

The development of molecular markers closely linked to low-Cd accumulation genotypes or QTLs can efficiently and accurately develop low-Cd rice cultivars through molecular marker-assisted selection (MAS) breeding. Tang et al. (2021b) developed the single nucleotide polymorphisms (SNP) marker TT for the excellent allele of *OsNRAMP5*. Sun and Chen (2021) developed the cleaved amplified polymorphic sequence (CAPS) marker that can distinguish strong and weak alleles of *OsHMA3*. Natural variation in OsCd1 with a substitution at 449 from valine (Val) to aspartic (Asp) contributes to the divergence of grain Cd accumulation between *indica* and *japonica* rice. Most *japonica* rice varieties have $OsCd1^{Val449}$ allele, whereas most *indica* rice varieties have $OsCd1^{Asp449}$ allele (Yan et al. 2019). A kompetitive allele-specific PCR (KASP) marker LCd-38 was developed for the natural variation of the missense mutation Val449Asp in *OsCd1* (Xu et al. 2022). Yang et al. (2022) have developed the molecular marker R2ID3116 of the *qGCd2* gene, which associated with Cd concentrations in rice grains. Wang et al. (2021b) developed a molecular marker Tons that is closely associated with the low-Cd accumulation in rice grains based on the insertion sequence on chromosome 7 of Luohong 4A.

Chen and Tang (2018) conducted genotype screening of 5769 hybrid rice parental lines derived from *indica-japonica* cross using MAS with 4 low-Cd molecular markers and obtained 8 male sterile lines and 72 restorer lines. Huang et al. (2019) introduced rice blast resistance genes and the low-Cd allele of *OsHMA3* into the male sterile line Chuang 5S using MAS technology, and created the new materials with high blast resistance and low-Cd accumulation in grains. Ishikawa (2020) successfully introduced the *osnramp5-2* allele in *lcd-kmt2* into more than 100 rice varieties through MAS technology. Wang et al. (2021a) generated near-isogenic line (NIL) by introgression of *OsHMA3-OsNRAMP5-OsNRAMP1* genes cluster from *japonica* landrace IRAT29 into *indica* cultivar 93-11 and found the NIL showed 31.8% lower Cd concentrations of grains and no penalty on yield compared with 93-11. Wang et al. (2021b) used the low-Cd molecular marker Tons to screen the offspring crossed by Luohong 4A and IR28 to obtain a new low-Cd material, Tonys. In the future, high-efficiency breeding of low-Cd rice cultivars can be achieved by the integration of excellent low-Cd alleles and QTLs.

16.6 Future Perspectives

The excessive Cd accumulation in rice grains caused by soil Cd contamination threatens food security and human health. In recent years, a series of progress has been made in understanding the molecular mechanisms of uptake, translocation, and tolerance of Cd in rice, as well as the breeding of low-Cd rice cultivars. Mutagenesis of *OsNRAMP5* or overexpression of *OsHMA3* can reduce the Cd concentration in rice grains by more than 90%, which is the most effective method for breeding low-Cd rice cultivar. At present, commercial cultivation of transgenic rice is not allowed in many countries, which limits the application of *OsHMA3* overexpression. Strikingly, there are mutant *osnramp5* obtained by mutagenesis, so the breeding of low-Cd rice cultivars based on mutant *osnramp5* is expected to achieve safe rice production in most Cd-contaminated paddy fields with sufficient Mn in soil.

In the future, the following aspects need to be studied further. (i) The research on the molecular mechanisms of Cd accumulation in rice is mostly limited to the functional analysis of a single transporter. The research on the relationship and pyramiding effect of different transporters is relatively insufficient, while the regulatory network of the transporter, especially the sensing mechanism of Cd, is still unclear. (ii) The mutations of most Cd transporter genes not only reduce Cd accumulation in rice but also cause the deficiency of Mn, Fe, Zn, or other essential micronutrients in rice. The low-Cd alleles that do not affect the growth, development, agronomic traits, and stress tolerance of rice are still insufficient. Therefore, it is urgent to identify more low-Cd excellent allelic variation in natural rice populations to carry out molecular design breeding of low-Cd rice. (iii) Most of the existing studies on the major gene *OsNRAMP5* for Cd and Mn uptake are limited to loss-of-function mutations, so there may be a risk of reduced stress resistance due to the loss of its Mn uptake activity. It is possible to create an ideal genotype variant of *OsNRAMP5* that specifically uptake Mn rather than Cd by mutation of key amino acid residues of *OsNRAMP5* mediated by genome-editing technology.

References

Cao, Z.Z., Lin, X.Y., Yang, Y.J. et al. (2019). Gene identification and transcriptome analysis of low cadmium accumulation rice mutant (*lcd1*) in response to cadmium stress using MutMap and RNA-seq. *BMC Plant Biol.* 19: 250.

Chang, J.D., Huang, S., Konishi, N. et al. (2020a). Overexpression of the manganese/cadmium transporter OsNRAMP5 reduces cadmium accumulation in rice grain. *J. Exp. Bot.* 71: 5705–5715.

Chang, J.D., Huang, S., Yamaji, N. et al. (2020b). OsNRAMP1 transporter contributes to cadmium and manganese uptake in rice. *Plant Cell Environ.* 43: 2476–2491.

Chen, C.Y. and Tang, W.B. (2018). A perspective on the selection and breeding of low-Cd rice (in Chinese). *Res. Agric. Modernization* 39: 1044–1051.

Chen, H., Ye, R., Liang, Y. et al. (2022). Generation of low-cadmium rice germplasms via knockout of *OsLCD* using CRISPR/Cas9. *J. Environ. Sci.* 126: 138–152.

Chen, L., Hao, L., Parry, M.A. et al. (2014). Progress in TILLING as a tool for functional genomics and improvement of crops. *J. Integr. Plant Biol.* 56: 425–443.

Clemens, S. and Ma, J.F. (2016). Toxic heavy metal and metalloid accumulation in crop plants and foods. *Annu. Rev. Plant Biol.* 67: 489–512.

Das, N., Bhattacharya, S., Bhattacharyya, S., and Maiti, M.K. (2017). Identification of alternatively spliced transcripts of rice phytochelatin synthase 2 gene *OsPCS2* involved in mitigation of cadmium and arsenic stresses. *Plant Mol. Biol.* 94: 167–183.

Dong, J.Y., Wu, T.H., Sun, Y.T. et al. (2021). Effects of *OsNRAMP5* mutation on heat tolerance and main economic traits of rice under the conditions of different manganese concentration (in Chinese). *Hybrid Rice* 36: 79–88.

Gallego, S.M., Pena, L.B., Barcia, R.A. et al. (2012). Unravelling cadmium toxicity and tolerance in plants: insight into regulatory mechanisms. *Environ. Exp. Bot.* 83: 33–46.

Hao, X., Zeng, M., Wang, J. et al. (2018). A node-expressed transporter OsCCX2 is involved in grain cadmium accumulation of rice. *Front. Plant Sci.* 9: 476.

He, G., Chen, Y., Zhu, C. et al. (2019). Application of plasma circulating cell-free DNA detection to the molecular diagnosis of hepatocellular carcinoma. *Am. J. Trans. Res.* 11: 1428–1445.

Huang, X.G., Jiang, H.Y., Zhang, J.P. et al. (2019). Improvement of blast resistance and low cadmium accumulation of dual-purpose genic male sterile rice line chuang 5S by molecular marker-assisted selection (in Chinese). *Hybrid Rice* 34: 51–56.

Ishikawa, S. (2020). Mechanisms of cadmium accumulation in rice grains and molecular breeding for its reduction. *Soil Sci. Plant Nutr.* 66: 28–33.

Ishikawa, S., Ishimaru, Y., Igura, M. et al. (2012). Ion-beam irradiation, gene identification, and marker-assisted breeding in the development of low-cadmium rice. *Proc. Natl. Acad. Sci. U.S.A.* 109: 19166–19171.

Ishimaru, Y., Takahashi, R., Bashir, K. et al. (2012). Characterizing the role of rice NRAMP5 in manganese, iron and cadmium transport. *Sci. Rep.* 2: 286.

Jallad, K.N. (2015). Heavy metal exposure from ingesting rice and its related potential hazardous health risks to humans. *Environ. Sci. Pollut. Res. Int.* 22: 15449–15458.

Jankowicz-Cieslak, J., Tai, T.H., Kumlehn, J. et al. (2017). *Biotechnologies for Plant Mutation Breeding: Protocols.* Switzerland: Springer Nature.

Liu, C.L., Gao, Z.Y., Shang, L.G. et al. (2020). Natural variation in the promoter of OsHMA3 contributes to differential grain cadmium accumulation between *Indica* and *Japonica* rice. *J. Integr. Plant Biol.* 62: 314–329.

Liu, S.M., Jiang, J., Liu, Y. et al. (2019). Characterization and evaluation of *OsLCT1* and *OsNramp5* mutants generated through CRISPR/Cas9-mediated mutagenesis for breeding low Cd rice. *Rice Sci.* 26: 88–97.

Long, Q.Z., Huang, Y.L., Tang, X.Y. et al. (2019). Creation of low-Cd-accumulating *indica* rice by disruption of *OsNramp5* gene via CRISPR/Cas9 (in Chinese). *Chin. J. Rice Sci.* 33: 407–420.

Lu, C., Zhang, L., Tang, Z. et al. (2019). Producing cadmium-free Indica rice by overexpressing *OsHMA3*. *Environ. Int.* 126: 619–626.

Luo, J.S., Huang, J., Zeng, D.L. et al. (2018). A defensin-like protein drives cadmium efflux and allocation in rice. *Nat. Commun.* 9: 645.

Lv, Q., Li, W., Sun, Z. et al. (2020). Resequencing of 1,143 indica rice accessions reveals important genetic variations and different heterosis patterns. *Nat. Commun.* 11: 4778.

Melsen, K., Wouw, M., and Contreras, R. (2021). Mutation breeding in ornamentals. *Hort Sci.* 56: 1–12.

Nakanishi, H., Ogawa, I., Ishimaru, Y. et al. (2006). Iron deficiency enhances cadmium uptake and translocation mediated by the Fe^{2+} transporters OsIRT1 and OsIRT2 in rice. *Soil Sci. Plant Nutr.* 52: 464–469.

Pan, X., Li, Y., Liu, W. et al. (2020). QTL mapping and candidate gene analysis of cadmium accumulation in polished rice by genome-wide association study. *Sci. Rep.* 10: 11791.

Prasanna, S. and Jain, S.M. (2017). Mutant resources and mutagenomics in crop plants. *Emir. J. Food Agric.* 29: 651.

Przybyl, J., Chabon, J.J., Spans, L. et al. (2018). Combination approach for detecting different types of alterations in circulating tumor DNA in leiomyosarcoma. *Clin. Cancer Res.* 24: 2688–2699.

Rizwan, M., Ali, S., Adrees, M. et al. (2016). Cadmium stress in rice: toxic effects, tolerance mechanisms, and management: a critical review. *Environ. Sci. Pollut. Res. Int.* 23: 17859–17879.

Sasaki, A., Yamaji, N., Yokosho, K., and Ma, J.F. (2012). Nramp5 is a major transporter responsible for manganese and cadmium uptake in rice. *Plant Cell* 24: 2155–2167.

Sasaki, T. and Ashikari, M. (2018). *Rice Genomics, Genetics and Breeding*. Springer Book.

Satoh-Nagasawa, N., Mori, M., Nakazawa, N. et al. (2012). Mutations in rice (*Oryza sativa*) heavy metal ATPase 2 (OsHMA2) restrict the translocation of zinc and cadmium. *Plant Cell Physiol.* 53: 213–224.

Shao, J.F., Xia, J., Yamaji, N. et al. (2018). Effective reduction of cadmium accumulation in rice grain by expressing *OsHMA3* under the control of the *OsHMA2* promoter. *J. Exp. Bot.* 69: 2743–2752.

Shao, Y., Peng, Y., Mao, B.G. et al. (2022). M1TDS technology and creation of low-cadmium accumulation parents for hybrid rice breeding (in Chinese). *Hybrid Rice* 37: 1–11.

Sui, F., Zhao, D., Zhu, H. et al. (2019). Map-based cloning of a new total loss-of-function allele of *OsHMA3* causes high cadmium accumulation in rice grain. *J. Exp. Bot.* 70: 2857–2871.

Sun, L. and Chen, C.Y. (2021). An intracellular molecular marker Caps7 for cadmium accumulation related gene OsHMA3 in rice and its application. China Patent. ZL 201810299208.

Takahashi, R., Ishimaru, Y., Nakanishi, H., and Nishizawa, N.K. (2011). Role of the iron transporter OsNRAMP1 in cadmium uptake and accumulation in rice. *Plant Signal. Behav.* 6: 1813–1816.

Takahashi, R., Ishimaru, Y., Shimo, H. et al. (2012). The OsHMA2 transporter is involved in root-to-shoot translocation of Zn and Cd in rice. *Plant Cell Environ.* 35: 1948–1957.

Tan, L., Qu, M., Zhu, Y. et al. (2020). ZINC TRANSPORTER5 and ZINC TRANSPORTER9 function synergistically in zinc/cadmium uptake. *Plant Physiol.* 183: 1235–1249.

Tan, L., Zhu, Y., Fan, T. et al. (2019). OsZIP7 functions in xylem loading in roots and inter-vascular transfer in nodes to deliver Zn/Cd to grain in rice. *Biochem. Biophys. Res. Commun.* 512: 112–118.

Tang, L., Dong, J., Tan, L. et al. (2021a). Overexpression of *OsLCT2*, a low-affinity cation transporter gene, reduces cadmium accumulation in shoots and grains of rice. *Rice (NY)* 14: 89.

Tang, L., Mao, B., Li, Y. et al. (2017). Knockout of *OsNramp5* using the CRISPR/Cas9 system produces low Cd-accumulating *indica* rice without compromising yield. *Sci. Rep.* 7: 14438.

Tang, L., Dong, J., Qu, M. et al. (2022). Knockout of *OsNRAMP5* enhances rice tolerance to cadmium toxicity in response to varying external cadmium concentrations via distinct mechanisms. *Sci. Total Environ.* 832: 155006.

Tang, L., Zhao, B.R., He, H.J. et al. (2021b). SNP marker of *OsNRAMP5* uptaking cadmium in rice and their application. China Patent. ZL 201911296360.X.

Ueno, D., Yamaji, N., Kono, I. et al. (2010). Gene limiting cadmium accumulation in rice. *Proc. Natl. Acad. Sci. U.S.A.* 107: 16500.

Uraguchi, S. and Fujiwara, T. (2013). Rice breaks ground for cadmium-free cereals. *Curr. Opin. Plant Biol.* 16: 328–334.

Uraguchi, S., Kamiya, T., Sakamoto, T. et al. (2011). Low-affinity cation transporter (OsLCT1) regulates cadmium transport into rice grains. *Proc. Natl. Acad. Sci. U.S.A.* 108: 20959–20964.

Wang, C., Guo, W., Ye, S. et al. (2016). Reduction of Cd in rice through expression of OXS3-like gene fragments. *Mol. Plant* 9: 301–304.

Wang, K., Yan, T.Z., Xu, S.L. et al. (2021a). Validating a segment on chromosome 7 of *japonica* for establishing low-cadmium accumulating *indica* rice variety. *Sci. Rep.* 11: 6053.

Wang, T., Li, Y., Fu, Y. et al. (2019). Mutation at different sites of metal transporter gene OsNramp5 affects Cd accumulation and related ggronomic traits in rice (*Oryza sativa* L.). *Front. Plant Sci.* 10: 1081.

Wang, T.K., Li, Y.X., Song, S.F. et al. (2021b). Excavation of rice resources with low cadmium accumulation in grains and development of new materials (in Chinese). *Hybrid Rice* 36: 68–74.

Xu, J., Li, T., Hu, M.J. et al. (2022). Development and utilization of KASP marker LCd-38 for cadmium accumulation in rice grain (in Chinese). *J. Agric. Sci. Technol.* 24: 40–47.

Yamaji, N., Xia, J., Mitani-Ueno, N. et al. (2013). Preferential delivery of zinc to developing tissues in rice is mediated by P-type heavy metal ATPase OsHMA2. *Plant physiol.* 162: 927–939.

Yan, H., Xu, W., Xie, J. et al. (2019). Variation of a major facilitator superfamily gene contributes to differential cadmium accumulation between rice subspecies. *Nat. Commun.* 10: 2562.

Yan, J., Wang, P., Wang, P. et al. (2016). A loss-of-function allele of *OsHMA3* associated with high cadmium accumulation in shoots and grain of Japonica rice cultivars. *Plant Cell Environ.* 39: 1941–1954.

Yang, C.H., Zhang, Y., and Huang, C.F. (2019). Reduction in cadmium accumulation in *japonica* rice grains by CRISPR/Cas9-mediated editing of *OsNRAMP5*. *J. Integr. Agric.* 18: 688–697.

Yang, G., Fu, S., Huang, J. et al. (2021). The tonoplast-localized transporter OsABCC9 is involved in cadmium tolerance and accumulation in rice. *Plant Sci.* 307: 110894.

Yang, M., Li, Y., Liu, Z., et al. (2020). A high activity zinc transporter OsZIP9 mediates zinc uptake in rice. Plant J. 103:1695–1709.

Yang, M., Zhang, Y., Zhang, L. et al. (2014). OsNRAMP5 contributes to manganese translocation and distribution in rice shoots. *J. Exp. Bot.* 65: 4849–4861.

Yang, Y.Z., Li, L.Y., Zhou, Y.B. et al. (2022). Application of molecular markers of *qGCd2* gene related to cadmium content in rice grains. China Patent. ZL 201710374342.

Zhang, J., Zhu, Y., Yu, L. et al. (2022). Research advances in cadmium uptake, transport and resistance in rice (*Oryza sativa* L.). *Cells* 11: 569.

Zhao, F.J., Tang, Z., Song, J.J. et al. (2022). Toxic metals and metalloids: uptake, transport, detoxification, phytoremediation, and crop improvement for safer food. *Mol. Plant* 15: 27–44.

Zhao, J., Yang, W., Zhang, S. et al. (2018). Genome-wide association study and candidate gene analysis of rice cadmium accumulation in grain in a diverse rice collection. *Rice (N Y)* 11: 61.

Zhu, R., Huang, W., Jun, H.U. et al. (2013a). Breeding of the new sterile line Luohong 4A of honglian type hybrid rice (in Chinese). *J. Wuhan University (Nat. Sci. Ed.)* 59: 33–36.

Zhu, R., Liu, W., Shaoqing, L.I., and Zhu, Y. (2013b). Breeding and application of honglian type hybrid rice Luoyou 8 and sterile line Luohong 3A (in Chinese). *Journal of Wuhan University (Natural Science Edition)* 59: 29–32.

17

Mercury Toxicity

Plant Response and Tolerance

Arifin Sandhi[1], *Abu Bakar Siddique*[2], *and Meththika Vithanage*[3]

[1] *Department of Biology and Environmental Science, Faculty of health and life sciences, Linnaeus University, Kalmar, Sweden*
[2] *Department of Plant Physiology, Umeå Plant Science Centre (UPSC), Umeå University, Umeå, Sweden*
[3] *Ecosphere Resilience Research Center, Faculty of Applied Sciences, University of Sri Jayewardenepura, Nugegoda, Sri Lanka*

17.1 Introduction

Mercury (Hg) has been considered a naturally occurring element, and it could directly mobilize into terrestrial and aquatic systems due to anthropogenic activities (e.g. mining, metal extraction, paint and electronic device production, and industrial uses) for thousands of years (Driscoll et al. 2013; Liu et al. 2020). Mercury emissions through natural sources have been considered one of the vital components in global Hg circulation as it is the major part of the global Hg cycle (Gworek et al. 2016). Among heavy metals and metalloid pollution, Hg contamination got attention due to its prolonged environmental persistence, high mobility, and biotoxicity characteristics (Cozzolino et al. 2016). Several studies have already revealed the toxicity of Hg and its neurodegenerative effects on humans (Carocci et al. 2014; Cariccio et al. 2019; Fung et al. 1997). Two significant epidemics happened in this world due to Hg toxicity and its exposure to humans; one is the Minamata Bay disaster by consumption of MeHg (methylmercury)-contaminated fish from industrial wastewater discharge in Japan (1953), and another one is the consumption of bread content with organomercury fungicide (1971–1972) in Iraq (Cariccio et al. 2019). In October 2013, a new global convention named Minamata convention on Hg had signed at Minamata city, Japan, and it included several controls related to Hg mining import–export process, dumping and decreasing mercury pollution (Mackey et al. 2014)

Mercury can occur in organic and inorganic compounds, and its toxicity greatly depends on its form (Liu et al. 2020). Mainly, two forms (such as methylmercury and mercury vapor) could be absorbed in the organisms and cause toxicity (Cariccio et al. 2019). Mercury in the soil mostly depends on elemental mercury, inorganic and organic mercury. In contrast, in the aquatic medium, the dissolved Hg are elemental mercury (Hg^0), which complexes of

Heavy Metal Toxicity and Tolerance in Plants: A Biological, Omics, and Genetic Engineering Approach, First Edition. Edited by Mohammad Anwar Hossain, AKM Zakir Hossain, Sylvain Bourgerie, Masayuki Fujita, Om Parkash Dhankher, and Parvez Haris.

Hg(II) and methylmercury (MeHg) and dimethylmercury (DMHg) (Gworek et al. 2016; Ullrich et al. 2001). Mercury got attention mainly because of its extremely high toxicity and causing health hazards even at low concentrations. According to the United States EPA (Environmental Protection Agency), the maximum inorganic Hg contaminant level in the water is 0.002 $mg\,l^{-1}$ in national primary drinking water regulations (EPA 2022a). On the other hand, the WHO (World Health Organization) has also made a guideline value for Hg content in the drinking water as 0.006 $mg\,l^{-1}$ or 6 ppb and 100 countries (out of 104) also have Hg value (WHO 2018).

Plants can accumulate mercury from either terrestrial soil or aquatic systems. Plants uptake mercury in the forms of elemental mercury (Hg^0), Hg (II) and organic methylmercury (MeHg) and terrestrial plant species have less MeHg in their body compared to the aquatic plant species (Li et al. 2017). The mercury uptake in the plants occurs from the soil, sediment, water, and the atmosphere (Patra and Sharma 2000). Xylem sap in plants plays a vital role in mobilizing total mercury and methylmercury inside the plant tissue (Bishop et al. 1998). A study regarding a meta-analysis of soil Hg accumulation by different vegetables suggests that green pepper, spinach, cabbage, and Chinese cabbage have a higher mercury bioconcentration factor (BCF) and Hg uptake depends on soil pH and soil organic matter (Yu et al. 2018). On the other hand, plants can also accumulate Hg from the atmosphere by using plant stomata or the foliar adsorption process (Millhollen et al. 2006). However, Hg transfer from the soil–plant system greatly depends on several factors, including Hg concentration, its speciation, bioavailability, and the plant species (Fernandez-Martinez et al. 2015). Due to various uptake points available in the plants, it acts as an essential pathway component for Hg cycling in the ecosystem. A recent study has also found that rice consumption can be a significant pathway of methylmercury exposure in humans besides consumption of fish, fish-derived products, and marine animals (Meng et al. 2014).

17.2 Global Mercury Pollution

Mercury is released into the environment by natural/geogenic and anthropogenic processes. Among the natural processes are volcanoes, geothermal activities, water bodies, vegetation surface, waterbodies, wildfire, and even re-emission from the deposited Hg (Feng et al. 2008). Compared to natural processes, the number of anthropogenic sources is relatively more significant in number that comprises coal combustion, water incineration, metal refining-mining, artisanal, and small-scale gold mining, acetaldehyde production, mercuric compound production, mercury-contained devices (e.g. mirrors, paint, batteries), health care, e-waste recycling, dentistry, cosmetics, construction (EPA 2022b; Feng et al. 2008; ILO 2022; Mackey et al. 2014; Meng et al. 2014; Qiu et al. 2008). The Hg emission can also be classified into oceanic and terrestrial parts. The global marine and terrestrial annual Hg emissions are approximately 800–2600 tons and 1000–3200 tons, respectively (Bergan et al. 1999; Li et al. 2009).

The regional attribution for Hg emission shows a similar trend since the beginning of this millennium. The Asian countries contributed 54% of total global emissions, whereas China (600 tons of Hg) secured the top position in the list of Hg-emitting countries in 2000

Table 17.1 Regional global emission of Hg to air from anthropogenic sources (2000–2015).

	Regional total Hg emission (tons)	
Region	**Year 2000 Pacyna et al. (2006)**	**Year 2015 AMAP/UN Environment (2019)**
Asia	1179.3 (excluding Russia)	859 (East and southeast Asia) 225 (South Asia)
Africa	398.4	20.9 (North Africa) 360 (Sub-Saharan Africa)
Europe	175.1 (excluding Russia)	77.2 (EU28) 124 (CIS & other European countries)
Russia	72.6	—
South America	92.1	409
North America	145.8	40.4

by anthropogenic activities (Pacyna et al. 2006). After 15 years, the global Hg emission trend was not changed much, especially in Asia. Most Hg emissions occurred in Asia (49%), followed by South America (18%). In Table 17.1, a comparative trend shows that Hg emission increased in South America on a massive scale within 15 years. In the case of Asia, China (604.7 tons $year^{-1}$) has the top position for Hg emission and is followed by India (149.9 tons $year^{-1}$) (Li et al. 2009). Mainly mining activities have been considered the major source of Hg emission in China. Several studies have been performed regarding Hg emission from mining activities, especially in Guizhou, southwestern China (Meng et al. 2012; Qiu et al. 2008). Besides mining activities, the elemental Hg–Au amalgamation method is also extensively used in gold mines in China and the Southeast Asian region, and that causes severe Hg pollution in both the terrestrial and aquatic systems of those areas (Li et al. 2009).

Several studies have also reported regarding Hg contamination in Peru, Brazil, Columbia, and Suriname, located in South America and in most cases, the contamination occurred due to extensive Hg use for gold mining and production-related activities (Martinez et al. 2018; Olivero et al. 2002; Ouboter et al. 2012; Pestana and Formoso 2003; Torres and De-la-Torre 2022). A significant portion (37% of global Hg emission) of Hg emission also occurs due to ASGM (artisanal and small-scale gold mining) activities used during the gold refining process. It is considered one of the major sources of Hg contamination in South America (Martinez et al. 2018). For example, Peru is one of the top gold producers in the world, and a more significant portion of the produced gold comes from ASGM activities. Their discharged Hg is highly mobile and influenced by soil erosion and surface runoff (Torres and De-la-Torre 2022). Two major gold rush activities occurred in Brazil during the eighteenth and the middle of the nineteenth centuries. At the beginning of the twenty-first century, the annual Hg input from gold mining to the environment in the Brazilian Amazon was approximately 180 tonnes $year^{-1}$ (Pestana and Formoso 2003).

17.3 Mercury Uptake and Toxicity in Plants

There is no physiological function of plants depending on Hg content, but still it could be found in plants in both organic and inorganic forms (Ansari et al. 2021). Besides plant species and soil type, the bioavailability of Hg in soil greatly depends on several factors of soil physicochemical properties, including cation exchange capacity (CEC), organic matter (OM) content, pH, texture, and metal (e.g. Fe or Al) oxides in the soil (Ding et al. 2014; Hussain et al. 2022). The effect of Hg on plants' physiology depends on their concentration. A higher concentration is relatively phytotoxic and causes cell damage, physiological disorders, and in some cases, injuries also noticeable (Tiwari and Lata 2018). A study also found that higher doses of Hg (10 μM) reduce biomass, increase chlorophyll content, malondialdehyde, total thiols and phenols in white Lupin (*Lupinus albus* L.) grown in the nutrient medium (Esteban et al. 2008). Lipophilic and protein binding properties of MeHg could rapidly be uptake in the organisms and later transfer into the food chain through food consumption (Amin et al. 2021). The terrestrial plant species accumulate Hg mainly in the root zone, and its uptake process follows the same channels used for Fe. Cu and Zn are transported in the plant's root part, and a small proportion of accumulated Hg is translocated in the above-ground parts, such as leaves and flowers (Chen and Yang 2012). A study has found that about 80% of accumulated Hg is bound in the cell wall (Wang and Greger 2004). The reasons for Hg bounding in the root (accumulated by the root zone) could be explained by two factors: (i) Hg ion easily creates interaction with anionic compounds and turns insoluble precipitates, (ii) Hg bonds with root cells with high cationic capacity (Chen et al. 2009; Chen and Yang 2012). Among the terrestrial plant species, accumulation of Hg in rice plants and grain has been investigated extensively for several reasons, including (i) rice is ranked as the second primary source of MeHg exposure to human health, (ii) rice has been considered an intensive Hg bioaccumulator based on Hg bioaccumulation factor, (iii) MeHg stored in the rice grain part as methylmercury-L-cysteinate (Me–Hg–Cysteine) and which is identical complex stored in the fish and predicted as responsible for MeHg transfer toward the nervous system, (iv) the main rice growing areas also located in close vicinity of different Hg mining sites located in Guizhou, Human, Chongqing, Yunnan, and Shanxi province in China and those have already identified with high Hg contamination due to mining activities (Amin et al. 2021; Meng et al. 2014; Zhou et al. 2021). Hg content in brown rice shows a positive correlation between MeHg concentration in the root and grain due to the transfer from the root to the shoot parts (Li et al. 2014). Already many studies have found that Hg concentration in rice varied in different countries (e.g. Egypt, Sri Lanka, Thailand, China, India, United States, Canada, and Japan) and that could be a threat to human health (Amin et al. 2021).

However, the terrestrial plant species can also accumulate Hg from the aboveground parts. The annual Hg assimilation in the terrestrial plant's vegetation is approximately $3062 \pm 607\,\text{Mg year}^{-1}$, combined below ground $571 \pm 253\,\text{Mg year}^{-1}$ from root uptake and aboveground $2491 \pm 551\,\text{Mg year}^{-1}$ from the aerial parts of plants (Zhou and Obrist 2021).

The atmospheric Hg accumulation is controlled mainly by the stomatal uptake, especially in trees and agronomic crop species such as wheat and maize (Pleije et al. 2021). The accumulation of Hg by plant stomata depends on several factors, including Hg air concentration, illumination, temperature and catalase activity, tissue age, and CO_2 content in the atmosphere (Stamenkovic and Gustin 2009). Generally, the forest ecosystem plays a vital role as 60–90% of Hg assimilation occurs due to the uptake of atmospheric gaseous Hg(0), and lichen and moss accumulate higher Hg levels than other vascular plant species (Zhou et al. 2021). Another recent study has also found that more than 90% of Hg accumulation occurs in aboveground parts of rice through stomata, and elevated CO_2 could reduce Hg content due to its effect on stomatal conductance (Tang et al. 2021). Plants have the potential to release accumulated Hg from the root and then transfer it to the aboveground parts and final release it in the air by using stomata. Few studies have already found that plant species such as garlic vine (*Pseudocallymma alliaceum*), avocado (*Persea americana*), and haole-koa (*Leucaena glauca*) released Hg as the volatile form in the air (Siegel et al. 1974). However, the foliage content of Hg was reported as depending on the plant species-specific (Wholgemuth et al. 2020). On the other hand, several plant species (garden pea, spring wheat, sugar beet, rape seed, white clover, and *Salix*) also showed low translocation of Hg and their release into the atmosphere as gaseous form (Greger et al. 2005).

Hg toxicity on plants greatly depends on the concentration of Hg as it has been found that mercuric cations (Hg^{2+}) can be uptake by both terrestrial and aquatic plant species, and only a high level is considered phytotoxic to the cells and can induce apparent cell injuries and physiological disorders (Zhou et al. 2008b). For example, higher concentrations of $HgCl_2$ (250 and 500 μM) reduced the root length of cucumber (*Cucumis sativus* L) by 96% and 98%, respectively, than the control treatment (Cargnelutti et al. 2006).

As Hg^{2+} is a transition metal, it can cause oxidative stress in plants, including lipid peroxidation, modification, or change in antioxidant enzymatic activities and is also responsible for thiol comprising compounds (Zhou et al. 2007). The excessive production of reactive oxygen species (ROS) has been considered one of Hg's significant plant exposure outcomes; the effect is discussed in Section 17.4.7. The application of Hg-contained fungicide also has influenced plants, like abnormal germination and hypertrophy of the roots and S–H (sulfhydryl) systems of the embryo (Patra and Sharma 2000).

17.4 Existence of Differential Plant Response to Hg Stress

Mercury has been considered one of the most harmful environmental pollutants that enter the plant biological systems, altering cellular structures and processes and changing the plant morphology and physiology. The responses of plants against Hg are both genotypic and dose-dependent manner. However, plants from different genotypic backgrounds share common responses to mercury stress. This section covers plants' responses (morphological, anatomical, cellular, hormonal, photosynthetic, and oxidative) against Hg stress with related examples.

17.4.1 Plant Morphological Responses

A high Hg level affects the plant's morphological features irrespective of plant developmental stages (Safari et al. 2019). The accumulation of Hg occurs more in the root area than in all other aerial parts because of the higher concentration of Hg in the soil available compared to the aerial environment. The leaf accumulates less Hg in the plants due to the barrier of Hg translocation from root to shoot and leaf. However, the leaf part acts as a site for Hg loss to the environment (Nyman and Lindau 2016; Patra and Sharma 2000; Raj et al. 2020). Among different plant organs, roots are severely affected due to the higher concentration of Hg^{2+} in soil solution. The root growth, including the lateral root formation and root branching, is affected by Hg stress (Mathur et al. 2022; Nibau et al. 2008). A study has reported that the root and stem diameter reduction due to the anatomical alteration occurs in response to Hg stress (Batool et al. 2015). Reduction in germination percentage, stem and root elongation, plant height, and root and shoot fresh weight has been reported to be decreased mercury stress in cotton, maize, rice, alfalfa, and soybean (Cruz et al. 2021; Mei et al. 2021; Ahmed et al. 2022; El-Shehawi et al. 2022; Shao et al. 2022). Cotton seeds grown under Hg stress resulted in depressed shoot elongation and darkened roots surface with the lowest percentage of seed germination (Mei et al. 2021). *Jatropha curcas* and *Gossypium herbaceum* resulted in smaller leaves with a sick phenotype under Hg-contaminated soil (Marrugo-Negrete et al. 2016; Mei et al. 2021). Besides these, another study has also found that Mung bean (*Vigna radiata*) produced fewer leaf and nodules with shoot rot and nodule deformation under Hg exposure (Mondal et al. 2015). Similar to *Vigna radiatia, Phaseolus vulgaris* and *Avena sativa* also show lower shoot growth, higher leaf chlorosis, and necrosis in leaf margins and tips with shoot tip rotting due to Hg stress (Lima et al. 2021).

17.4.2 Plant Anatomical Responses

Plants grown under Hg-contaminated soil experience structural changes in the root, shoot, leaf tissues, and cells. High accumulation of Hg in plants causes misplacement and structural abnormalities in the root cortex, protoxylem, mechanical tissues, epidermis, and vascular bundles in the root (Shao et al. 2022). Rice has less developed Casparian bands and suberin lamellae in root endodermal cells when grown in a hydroponics medium containing Hg (Wang et al. 2015). Due to the translocation ability to the aerial part from the root, Hg also affects the anatomical features of the stem and leaf. Deformed pith cells with thicker cell walls were observed in stem tissue of *Trigonella foenum-graccum* (Askari and Azmat 2016). High accumulation of Hg in the cellular compartment reduces intercellular space due to the severely damaged vascular bundle with breakage of parenchyma tissue in *Chilopsis linearis* (Rodriguez et al. 2009). A number of multicellular and unicellular trichome with larger vascular bundle were observed in the leaf of *Chromolaena odorata* under a high level of Hg (Swapna et al. 2015). A study has also found that ruptured leaf blades with disintegrated leaf aerenchyma were observed in *Salvinia biloba* under Hg concentration (Casagrande et al. 2018).

17.4.3 Cellular Responses

Besides physiological and anatomical, Hg affects the cellular organization of plants with several abnormalities. Cell wall lignification is greatly hampered due to the effect of Hg on enzymes and genes responsible for lignin biosynthesis and accumulation (Cabane et al. 2012). Mercury concentration deteriorates the membrane by reducing photosynthetic capacity and respiration and inducing lipid peroxidation (Zhang et al. 2017). A high level of Hg causes a lower number of grana cascade in chloroplast with smaller and damaged thylakoid, resulting in smaller-sized chloroplast with distorted structure (Upadhyay 2014). A study has also indicated that cellular morphological abnormalities, including the constricted cytoplasm and chromatin shrinkage along with DNA damage, result from Hg stress (Zi et al. 2017). The shrinkage of chromatin under Hg stress has also been reported to reduce cell division. Mercury can also bind with the DNA and causes injury to chromosomes, lowering cell growth, and division with a higher number of cell death (Kapoor et al. 2016; Mohammadi et al. 2021). The posttranslational phase (protein assembly, degradation, and turnover) is severely affected due to the Hg stress, resulting in cellular protein homeostasis and decreases cellular viability (Hasan et al. 2017; Amir et al. 2020). MiRNA-target gene regulatory network is involved in plant response to heavy metals, including Hg stress. Both genes and transcription factors related to metal transport, metal chelation, protein folding, antioxidant responses, and hormonal biosynthesis and signaling (auxin, jasmonic acid, and cytokinin), which are required for root and shoot development, have been reported to be affected due to Hg contamination (Ding et al. 2020).

17.4.4 Plant Photosynthetic Response

Plant photosynthesis has been considered one of the most crucial events for energy production required for plant growth and development. High Hg accumulation in plant leaves cause reduction and ceases of photosynthesis. Water, CO_2, and the light source are the substrates needed for photosynthesis, which is greatly affected due to Hg accumulation in the leaves (Mao et al. 2021). Hg accumulation triggers the stomatal closure and thus affects the availability of CO_2 by reducing the gas exchange capacity (Lima et al. 2021). Hg binds with the water channel and is a physical barrier to water uptake. Besides this, Hg lowers transpiration and nutrient availability, resulting in less photosynthesis (Ahmad et al. 2021; Lima et al. 2021; Mao et al. 2022; Mei et al. 2021; Yadav 2010). Due to the reduction in lignin biosynthesis, photosynthesis responsible cell division and growth are restricted, reducing net photosynthesis (Choudhary et al. 2021).

The chloroplast is the leading site for photoreaction, which is also affected due to the Hg accumulation in the leaf. Hg accumulated in the leaf adversely affects the structure and size of the grana cascade and thylakoid in the chloroplast (Zhang et al. 2014). Smaller and damaged chloroplast with distorted structure due to the Hg results in less photosynthesis (Upadhyay 2014). Hg stress reduces chlorophyll metabolism, which is responsible for light harvesting for photosynthesis (Houri et al. 2020; Li et al. 2019). Hg inhibits the activity of NADPH and protochlorophyllide oxidoreductase (POR), which are required for photosynthesis (Mei et al. 2021). These events are caused by Hg accumulation in the leaves, which initiates the reduction of net photosynthesis in plants.

17.4.5 Enzymatic and Metabolic Responses

High accumulation and absorption of Hg in plant organs lead to the generation of ROS and cellular homeostasis, which alters enzymatic activity and biosynthesis. Lipid peroxidation is one of the most common phenomena due to Hg. Hg^{2+} has an affinity to react with enzymes having thiol groups and active groups of ATP and ADP (Sytar et al. 2013).

Antioxidants, including superoxide dismutase (SOD) and ascorbate peroxidase (APX), have been reported to be affected due to high levels of Hg^{2+} (Mei et al. 2021). A study has found that Tartary buckwheat (*Fagopyrum tataricum*) showed a reduction in SOD, POD, and GR activity grown in Hg-contaminated soil (Pirzadah et al. 2018). However, catalase (CAT) enzymes and glutathione reductase (GR) are increased in response to Hg^{2+} (Amir et al. 2020; Kim et al. 2014). MDA level in plants has been reported to be affected as well in plants grown under Hg stress (Cargnelutti et al. 2006; Mei et al. 2021; Mondal et al. 2015). Besides these, Hg also inhibits the activity of the NADPH enzyme, which is responsible for producing the first product of photosynthesis.

Hg affects most of the metabolism in plant cells, including nitrogen metabolism, chlorophyll biosynthesis, and phenylpropanoid pathways and results in lower metabolic content (Ahmad et al. 2021; Pirzadah et al. 2018). Mitochondrial activity interferes with the high level of Hg^{2+} and causes the disruption in bio-membrane lipids and cellular metabolisms (Israr and Sahi 2006). Anthocyanins, carotenoids, and total chlorophyll contents were decreased in *F. tataricum, V. radiata,* and cotton is grown at higher concentrations of Hg (Mei et al. 2021; Mondal et al. 2015; Pirzadah et al. 2018). It also affects the heat shock proteins, which play crucial roles in plants' early response to abiotic stresses (Cruz et al. 2021).

17.4.6 Plant Hormonal Responses

A number of plant hormones such as auxin, jasmonic acid, ABA, ethylene, and cytokinin play critical roles in the regulation of plant growth and development process. Genes involved in hormone biosynthesis are upregulated due to Hg exposure (Ding et al. 2020). The distribution of auxin in plant organs is altered due to Hg stress, which results in morphological defects, especially in roots (Mathur et al. 2022). Plants exhibit reduced primary root growth and lateral root branching due to the differences in auxin distribution in the meristem (Nibau et al. 2008). Ethylene has also played a critical role in root architecture, which is affected due to Hg. Hg blocks the ethylene perception in the root area and inhibits the rapid root growth in *Arabidopsis thaliana* and *Medicago sativa* (Montero-Palmero et al. 2013). Meanwhile, ABA is another crucial plant hormone that increases under heavy metal and Hg stress (Nambara et al. 2010). In the case of the agronomic crop, species such as rice, wheat, cucumber, and chickpea grown in Hg contamination result in higher ABA levels and higher genes responsible for ABA biosynthesis (Kim et al. 2014; Wang et al. 2015).

17.4.7 Reactive Oxygen Species and Oxidative Responses

The overproduction of ROS is a typical response in plants to abiotic stresses, including the Hg. The generation of ROS due to the Hg leads to oxidative stress in plants (Baier et al. 2005; Dang et al. 2013; Montero-Palmero et al. 2013; Zhang et al. 2017). Another important effect of exposure to Hg in plant species is excessive production of reactive oxygen species (ROS) (Ali et al. 2000; Cho and Park 2000; Zhou et al. 2008b). Lipid peroxidation is the common phenomenon of Hg stress, which is a biochemical marker for the presence of free-radical-mediated injury by ROS (Sytar et al. 2013). A study has also found that the increased ROS in plants influences the activities of different antioxidant enzymes, such as superoxide dismutase (SOD) and peroxidase (POD) in roots (Zhou et al. 2007). Meanwhile, redox homeostasis is altered due to Hg and produces ROS (Ahmad et al. 2018; Safari et al. 2019). Mainly, an increase in MDA level due to the Hg stress causes overproduction of ROS and leads to oxidative damage (Cargnelutti et al. 2006; Mei et al. 2021; Mondal et al. 2015). Besides this, the application of Hg also decreased the chlorophyll and thylakoid content and inhibited the activities of NADPH, which is responsible for the biosynthesis of chlorophyll (Chen and Yang 2012). The overgeneration of ROS due to Hg leads to higher electrolyte leakages, resulting in cell death (El-Shehawi et al. 2022; Mei et al. 2021). ROS triggers the expression of genes responsible for hypersensitive responses (HR), and HR leads to cell death, which is a prerequisite for programmed cell death (PCD) (Tamas et al. 2010; Zi et al. 2017). Additionally, ROS induces cell toxicity leading to cellular abnormalities and cell death. A recent study has found that maize grown in Hg-contaminated soil showed higher levels of ROS, which positively correlated with cell death (İşkil et al. 2022).

17.5 Plant Tolerance Mechanisms

High accumulation of Hg into plant cells leads to ROS generation, which seriously damages the cellular process and restricts plant growth and development. However, plants activate an array of defense mechanisms to tolerate and cope with the changes due to Hg exposure. The most common tolerance mechanisms include the chelation, sequestration, and compartmentation of heavy metals and compounds to prevent them from interfering with cellular metabolisms (Anjitha et al. 2021; Chen and Yang 2012; Natasha et al. 2020). Besides the reduction of heavy metals by the above processes, plants exhibit diverse antioxidant systems, including enzymatic and nonenzymatic antioxidants, which stable the redox state of the cells by scavenging and detoxification of excess ROS (Abbas et al. 2018; Natasha et al. 2020; Shahid et al. 2017; Sousa et al. 2008). These defense mechanisms also allow the plants to repair the cellular damages caused by heavy metal exposure and ensure their growth and development.

17.5.1 Chelation

Chelation is one of the most potential mechanisms which triggers the heavy metal tolerance in plant cells by reducing the concentration at a low level (Anjum et al. 2015). Plants have a larger number of molecules and compounds having the capacity to make bonds with metal ions and restrict them from interacting with the cellular processes. Glutathione, phytochelatin (PCs), ascorbic acid, and proline can chelate heavy metals and Hg in plant biological systems.

Glutathione is a tripeptide belonging to the thiol group and is present in all cell compartments, including cytosol, chloroplast, endoplasmic reticulum, vacuole, and mitochondria. It is the most abundant low molecular weight (Rahman 2020; Yadav 2010). Both the reduced (GSH) and oxidized form (GSSH) of glutathione can be present, and their redox ratio depends on the cellular pH, total GSH concentration, biosynthesis, and catabolism (Mullineaux and Rausch 2005; Yadav 2010). Glutathione has higher water solubility and can react with heavy metals, including Hg, making it an important biochemical for ROS detoxification and heavy metal sequestration (Foyer and Noctor 2005; Kim et al. 2017; Rausch et al. 2007). GSH binds with Hg to form a stable complex to restrict the entry of Hg into the root cells and rescue the cell vitality and plant growth by relieving the Hg-induced oxidative stress in *A. thaliana* (Kim et al. 2017). GSH acts as an antioxidant, which reduces the ROS levels and the stress caused by ROS in plant cells (Shao et al. 2008). Among all the ROS generated during oxidative stress, H_2O_2 causes severe damage to cellular processes, including metabolism. The participation of GSH and GR activity in H_2O_2 scavenging under heavy metal stress was previously reported (Bari et al. 2019; Chou et al. 2012; Hsu and Kao 2007). Changes in the ratio of reduced GSH to oxidized GSH regulate the H_2O_2 degradation, and thus the overproduction of ROS under Hg stress is checked (Miller et al. 2003; Yamazaki et al. 2018). Kim et al. (2017) has shown that exogenous application of GSH inhibits the biosynthesis of H_2O_2, decreases lipid peroxidation, increases the leaf chlorophyll content, and improves the tolerance ability in *A. thaliana*.

Besides the ROS degradation and restriction of Hg translocation in plants, GSH also acts as a precursor for the PC biosynthesis, which also binds to Hg in the root cell wall, resulting in minimizing the Hg accumulation in leaves that stabilize the photosynthesis and cellular functions (Rahman 2020). PCs are the oligomers of GSH, which is non-ribosomal polypeptides and synthesized enzymatically in response to metal stress (Cahoon et al. 2015; Natasha et al. 2020; Yadav 2010). Lack of PCs synthesized gene mutant *Arabidopsis* plant showed higher accumulation and hypersensitive phenotypes under Hg stress (Fischer et al. 2014), which showed the involvement of PCs in the ROS degradation process.

17.5.2 Enzymatic and Antioxidative Tolerance

It has already been discussed in the earlier sections that Hg-induced overproduction of ROS leads to oxidative stress. However, the overproduction of ROS initiates the defense mechanisms, including the enzymatic and antioxidative defense mechanisms. Activating the antioxidative defense mechanism is critical and is considered the most efficient method to protect against Hg-induced oxidative damage in plant cells. Hg-induced oxidative stress triggers the signaling for the biosynthesis of superoxide dismutase (SOD), ascorbate peroxidase (APX), glutathione reductase (GR), peroxidase (POD), catalase (CAT), monodehydroascorbate reductase (MDAR), dehydro-ascorbate reductase (DHAR), and other peroxidases including glutathione peroxidase (GPX), and glutathione-*S*-transferase (GST) (Bela et al. 2015; Hasanuzzaman et al. 2019; Liebthal et al. 2018; Mhamdi et al. 2012; Vaish et al. 2020). Some of the nonenzymatic antioxidative defense mechanisms have also been reported, including glutathione (GSH), carotenoids, vitamins (A, C, E), and phenol-based molecules, which can play roles in preventing oxidative damage in plants (Foyer and Noctor 2005).

Antioxidants are involved in scavenging the ROS and protecting the cellular components from oxidizing (Hasan et al. 2019). They also cover the cellular damages and maintain the redox potential by preventing the unnecessary reactions required for generating the free radicals (Li et al. 2019). SOD responds to Hg-induced oxidative stress and plays a vital role in tolerance to ROS. SOD reacts with the first formed free radicals (O_2^-) and converts them into H_2O_2, catalyzed by APX, CAT, GPX, and POD to result in H_2O (Kim et al. 2022; Mittler et al. 2011). SOD in plant tolerance to Hg-induced oxidative stress has been reported in

Arabidopsis, *Brassica juncea* L., *Syntrichia caninervis*, and *Clitoria ternatea* (Kim et al. 2022; Mao et al. 2022; Priya et al. 2014; Shiyab et al. 2008).

Ascorbic acid, also called vitamin C, is another water-soluble antioxidant, which plays a role in minimizing the detrimental effect of ROS induced by Hg stress. The glutathione conversion is somewhat dependent on the ascorbic acid, and this conversion promotes the chelation process. Ascorbic acid is also involved in regulating the redox state and scavenging the ROS by reacting with the free radicals (Hossain et al. 2010). Proline is an amino acid that functions in plant growth and signaling process, including maintaining cellular homeostasis, osmotic adjustment, and redox balance to mitigate Hg-induced oxidative stress (Ghosh et al. 2021; Iskil et al. 2022). Proline triggers the biosynthesis of glutamate kinase and glutamic acid. Also, it has a role in the biosynthesis of GSH and PCs, which participates in Hg chelation to reduce toxicity (Chandrakar et al. 2016). Proline biosynthesis related genes showed higher expression under the high level of Hg stress, resulting in higher proline accumulation in *Eichhornia crassipes* (Malar et al. 2015) and *Melissa officinalis* (Safari et al. 2019).

17.5.3 Hormonal Regulations

Phytohormones are involved in almost every aspect of plant growth, development, and protection against biotic and abiotic stresses, including heavy metals (Bücker-Neto et al. 2017; Masood et al. 2012). Plant biosynthesis includes a diverse range of plant hormones, including abscisic acid (ABA), salicylic acid (SA), ethylene, cytokinin, and brassinosteroids (BR), which play roles in plant defense either directly or indirectly.

Stomatal closing is one of the mechanisms, which results in changes in other functions, including the reduction in transpiration and water uptake rate. These events result in lower uptake of heavy metals and slower heavy metal translocation from the root to the shoot and leaf area. ABA-induced stomatal closure helps plants to lower the uptake and accumulation of the heavy metal in the shoot and root. Besides, ABA is involved in activating the mitogen-activated protein kinase (MAPK) signaling, which provides the tolerance to heavy metal stress, including Hg (Bücker-Neto et al. 2017). Auxin is involved in plant tolerance and adaptation against heavy metals, including Hg. Hg alleviation mechanism governed by auxin occurs in root and shoot areas. Auxin biosynthesis is triggered due to the plant–microbe interaction at the root level of heavy-metal-contaminated soils and provides a diverse beneficial role. Auxin contributes to root architecture, root branching, lateral root formation, water and nutrient uptake, and secretion of PCs and organic acids to combat heavy metals (Mathur et al. 2022). Auxin is synthesized at the root zone, transferred to the shoot and leaf area, and plays critical roles in antioxidant defense, upregulation of gene expressions related to tolerance, maintenance in redox, and osmotic status under Hg stress (Singh et al. 2020). Auxin is reported to mitigate the damages caused by Hg-induced ROS.

Meanwhile, brassinosteroids trigger the expression of certain genes which are involved in the biosynthesis of various antioxidants such POD (peroxidase), SOD (superoxide dismutase) and CAT (catalase) (Kapoor et al. 2014; Mathur et al. 2022; Xia et al. 2009). Cytokinin also plays a role in overall plant growth and development and the defense against biotic and abiotic stresses (Perilli et al. 2009). Cytokinin triggers the antioxidant enzyme activity, including the proline, free amino acids, and soluble sugars, which fight against the

Hg-induced ROS. Hg stress inhibits photosynthesis by reducing the chlorophyll biosynthesis and chloroplast functions. However, cytokinin is involved in restoring photosynthetic pigment metabolism under heavy metal stress and enabling plants to take photosynthesis (Jalmi et al. 2018). Ethylene is a gaseous plant hormone known for its triple responses (leaf abscission, fruit ripening and senescence). 1-Aminocyclopropane-1-carboxylic acid (ACC) synthase (ACS) and ACC oxidase (ACO) are involved in the biosynthesis of auxin in the plant. It has been reported that Hg increased the accumulation of ethylene by upregulating ACC and ACO-related gene expression (Khan et al. 2015). Ethylene activates the overexpression of ROS, which activates the diverse tolerance mechanisms. Like other phytohormones, salicylic acid (SA) is involved in metabolic and physiological processes, including photosynthesis, growth, development, nitrate metabolism, flowering, and fruit ripening (Hayat et al. 2007; Natasha et al. 2020). Besides SA interacts with the ROS signal pathway resulting in the termination of ROS generation under Hg stress. SA mitigates the Hg toxicity by reducing the H_2O_2 and MDA levels in plants as we all increase the enzymatic activities. The association of SA with POD and APX has been found in *M. sativa* under Hg exposure (Zhou et al. 2009).

17.5.4 miRNA-Mediated Tolerance

Plant tolerance to abiotic stress, including heavy metals, is observed at both transcriptional and posttranscriptional levels (Jamla et al. 2021). Plant microRNA (miRNA) is involved in posttranscriptional regulations of miRNA-mediated mRNA cleavage, translational repression, chromatin remodeling, and epigenetic mechanisms. Along with these functions, miRNA plays a role in regulating heavy metal uptake and transport, chelation, homeostasis, and antioxidant defense. Hg-induced stress affects and alters the metabolism and assimilation process. miRNAs counteract these by targeting the genes involved in these processes. Besides the target of the genes involved in the Hg-induced metabolic changes, miRNAs affect the signaling of phytohormone and antioxidative defense, which are involved in ROS scavenge. Different miRNAs have been identified in *B. juncea, B. napus, M. truncatula, O. sativa,* and *P. vulgaris* under Hg exposure (Chen and Yang 2012). Under the high level of Hg, miR156/157 has been reported to target the flowering genes, including the squamosa-promoter-binding protein (SBP)-like proteins (SPL) family (Yamasaki et al. 2009). Some miRNAs have been identified to function as a regulator of Ca signaling, glutathione and PC biosynthesis, and cellular homeostasis (Çelik et al. 2020). Different miRNAs, including miR319, miR171, and miR393, showed an alteration in gene expression under heavy metal stress (Xie et al. 2012; Zhou et al. 2008a).

17.6 Phytoremediation Prospects

Since the early 1990s, plant-based heavy metal remediation or phytoremediation has received attention due to its environmental friendliness and economic benefits (Salt et al. 1995). Then a number of plant species have been discovered to remove various kinds of heavy metals from the soil, water, and air systems. There are a number of branches within phytoremediation techniques like phytostabilization, phytoextraction, phytofiltration/rhizofiltration, and phytomining (Salt et al. 1998; Sandhi 2017). Generally, hyperaccumulator plant species have been considered to make the phytoremediation process faster. The hyperaccumulator plant species could uptake a certain amount of heavy metal, like 0.01% of dry weight biomass (1000 mg kg^{-1}), in their body (Marrugo-Negrete et al. 2015). Therefore, a hyperaccumulator plant species for Hg could uptake approximately 10 mg kg^{-1} of Hg in their body (Lasat 2002). However, the background concentration of Hg in the plants in the noncontaminated soil did not reach more than 0.01 mg kg^{-1} in their body (Rodriguez et al. 2007). Hg phytoremediation from the contaminated soil depends on factors like soil properties, weather, Hg bioavailability, and implementation duration (Liu et al. 2020; Marrugo-Negrete et al. 2015). The quest to find Hg hyperaccumulator plant species started over two decades ago. Few plant species have been found and termed as "potential Hg hyperaccumulator" but are not recognized as Hg hyperaccumulator, and most of them are under the herb plant category (Liu et al. 2020). Different plant species have investigated both phytoextraction and phytostabilization of Hg. A recent study found that the halophyte plant species *Juncus maritimus* showed strong potential regarding Hg accumulation from the contaminated sediment and stabilizes in their root part (Garcia-Ordiales et al. 2020). The terrestrial plant species are like different clones of *Salix* spp. It could grow in 1 μM Hg solution (200 μg l^{-1}) without showing any growth and water transpiration inhibition symptoms (Wang and Greger 2004). On the other hand, another study has reported that *Salix* spp. accumulated less Hg from contaminated soil compared to the solution, which could be explained as Hg bioavailability being relatively lower in soil (Liu et al. 2020). Meanwhile, assisted phytoremediation (or chelator application) could be an effective tool for different kinds of heavy metal remediation by using plant species. The chelator could transform the non-bioavailable Hg into bioavailable form Hg in the soil by changing soil pH or soluble Hg–ligand complexes, which could trigger the uptake process (Liu et al. 2020). However, the chelator application is not always found sufficiently effective for all cases; for example, the application of iodine could enhance Hg accumulation in *Salix* spp. But the translocation of Hg in the shoot part was low, which reduces its potentiality for the Hg phytoextraction process (Wang and Greger 2006).

17.7 Conclusion

Recently, Hg contamination in Asia got attention as a high percentage of rice cultivation is also practicing relatively close to the contaminated sites. It has already been discovered that rice plants could uptake Hg from the contaminated soil. Besides a short overview of global Hg status, this chapter covers Hg uptake, stress physiology, and brief information about potential plant species for Hg phytoremediation. However, the application of chelators to pace up the Hg phytoremediation process and conduct experiments *in situ* conditions need more research as they could provide new information regarding the sustainable remediation process of Hg, especially Hg mining sites to reduce Hg transfer in the human food chain.

References

Abbas, G., Murtaza, B., Bibi, I. et al. (2018). Arsenic uptake, toxicity, detoxification, and speciation in plants: physiological, biochemical, and molecular aspects. *Int. J. Environ. Res. Public Health.* 15: 59.

Ahmad, P., Ahanger, M.A., Egamberdieva, D. et al. (2018). Modification of osmolytes and antioxidant enzymes by 24-epibrassinolide in chickpea seedlings under mercury (Hg) toxicity. *J. Plant Growth Regul.* 37: 309–322. https://doi.org/10.1007/s00344-017-9730-6.

Ahmad, P., Alyemeni, M.N., Wijaya, L. et al. (2021). Nitric oxide donor, sodium nitroprusside, mitigates mercury toxicity in different soybean cultivars. *J. Hazard. Mater.* 408: https://doi.org/10.1016/j.jhazmat.2020.124852.

Ahmed, S.F., Kumar, P.S., Rozbu, M.R. et al. (2022). Heavy metal toxicity, sources, and remediation techniques for contaminated water and soil. *Environ. Tech. Innov.* 25: 102114.

Ali, M., Vajpayee, P., Tripathi, R. et al. (2000). Mercury bioaccumulation induces oxidative stress and toxicity to submerged macrophyte *Potamogeton crispus* L. *Bull. Environ. Contam. Toxicol.* 65: 573–582.

AMAP/UN Environment (2019). Technical Background Report for the Global Mercury Assessment 2018. Arctic Monitoring and Assessment Programme, Oslo, Norway/UN Environment Programme, Chemicals and Health Branch, Geneva, Switzerland. viii + 426 pp including E-Annexes, ISBN-978-82-7971-108-7.

Amin, S., Khan, S., Sarwar, T., and Nawab, J. (2021). Mercury methylation and its accumulation in rice and paddy soil in degraded lands: a critical review. *Environ. Tech. Innov.* 23: 101638.

Amir, W., Farid, M., Ishaq, H.K. et al. (2020). Accumulation potential and tolerance response of *Typha latifolia* L. under citric acid assisted phytoextraction of lead and mercury. *Chemosphere* 257: https://doi.org/10.1016/j.chemosphere.2020.127247.

Anjitha, K.S., Sameena, P.P., and Puthur, J.T. (2021). Functional aspects of plant secondary metabolites in metal stress tolerance and their importance in pharmacology. *Plant Stress* 2: 100038.

Anjum, N.A., Hasanuzzaman, M. et al. (2015). Jacks of metal/metalloid chelation trade in plants-an overview. *Front. Plant. Sci.* 6: 192.

Ansari, M.K.A., Ahmad, A., Umar, S. et al. (2021). Suitability of Indian mustard genotypes for phytoremediation of mercury-contaminated sites. *S. Afr. J. Bot.* 142: 12–18.

Askari, S. and Azmat, R. (2016). Codium mediated anatomical improvement in *Trigonella foenum-graecum* shoot under mercury stress. *J. Pharmacognosy Phytochem.* 5 (2): 170–174.

Baier, M., Kandlbinder, A., Golldack, D., and Dietz, K.J. (2005). Oxidative stress and ozone: perception, signalling and response. *Plant Cell Environ.* 28: 1012–1020.

Bari, M.A., Akther, M.S., Reza, M.A., and Kabir, A.H. (2019). Cadmium tolerance is associated with the root-driven coordination of cadmium sequestration, iron regulation, and ROS scavenging in rice. *Plant Physiol. Bioch.* 136: 22–33. https://doi.org/10.1016/j.plaphy.2019.01.007.

Batool, R., Hameed, M., Ashraf, M. et al. (2015). Physio-anatomical responses of plants to heavy metals. In: *Phytoremediation for Green Energy* (ed. M. Öztürk, M. Ashraf, A. Aksoy, and M. Ahmad). *Springer* https://doi.org/10.1007/978-94-007-7887-0_5.

Bela, K., Horvath, E., Galle, A. et al. (2015). Plant glutathione peroxidase: emerging role of the antioxidant enzymes in plant development and stress responses. *J. Plant Physiol.* 176: 192–201.

Bergan, T., Gallardo, L., and Rodhe, H. (1999). Mercury in the global troposphere: a three-dimensional model study. *Atmos. Environ.* 33: 1575–1585.

Bishop, K.H., Lee, Y.H., Munthe, J., and Dambrine, E. (1998). Xylem sap as a pathway for total mercury and methylmercury transport from soils to tree canopy in the boreal forest. *Biogeochemistry* 40: 101–113.

Bücker-Neto, L., Paiva, A.L.S., Machado, R.D. et al. (2017). Interactions between plant hormones and heavy metals responses. *Genet. Mol. Biol.* 40: 373–386. https://doi.org/10.1590/1678-4685-GMB-2016-0087.

Cabane, M., Afif, D., and Hawkins, S. (2012). Lignins and abiotic stresses lignins-biosynthesis, biodegradation and bioengineering. *Adv. Bot. Res.* 61: 219–262. https://doi.org/10.1016/B978-0-12-416023-1.00007-0.

Cahoon, R.E., Lutke, W.K., Cameron, J.C. et al. (2015). Adaptive engineering of phytochelatin-based heavy metal tolerance. *J. Biol. Chem.* 290 (28): 17321–17330. https://doi.org/10.1074/jbc.M115.652123.

Cargnelutti, D., Luciane, A., Spanevello, R.M. et al. (2006). Mercury toxicity induces oxidative stress in growing cucumber seedlings. *Chemosphere* 65: 999–1006. https://doi.org/10.1016/j.chemosphere.2006.03.037.

Cariccio, V.L., Samà, A., Bramanti, P. et al. (2019). Mercury involvement in neuronal damage and in neurodegenerative diseases. *Biol. Trace. Elem. Res.* 187: 341–356.

Carocci, A., Rovito, N., Sinicropi, M.S., and Genchi, G. (2014). Mercury toxicity and neurodegenerative effects. *Rev. Environ.Contam. Toxicol.* 229: 1–18.

Casagrande, G.C.R., Reis, C., Arruda, R. et al. (2018). Bioaccumulation and biosorption of mercury by *Salvinia biloba* Raddi (*Salviniaceae*). *Water Air Soil Pollut.* 229: 166.

Çelik, Ö., Ayan, A., Meriç, S., and Atak, C. (2020). Heavy metal stress responsive phyto-miRNAs. In: *Cellular and Molecular Phytotoxicity of Heavy Metals Nanotechnology in the Life Sciences* (ed. M. Faisal, Q. Saquib, A.A. Alatar, and A.A. Al-Khedhairy), 137–155. Springer.

Chandrakar, V., Dubey, A., and Keshavkant, S. (2016). Modulation of antioxidant enzymes by salicylic acid in arsenic exposed *Glycine max* L. *J. Soil Sci. Plant Nut.* 16: 662–676.

Chen, J. and Yang, Z.M. (2012). Mercury toxicity, molecular response, and tolerance in higher plant species. *Biometals* 25: 847–857.

Chen, J., Yang, Z.M., Su, Y. et al. (2009). Phytoremediation of heavy metal/metalloid-contaminated soils. In: *Contaminated Soils: Environmental Impact, Disposal and Treatment* (ed. R.V. Steinberg), 181–260. Nova Science Publishers.

Cho, U. and Park, J. (2000). Mercury-induced oxidative stress in tomato seedlings. *Plant Sci.* 156 (1): 1–9.

Chou, T.S., Chao, Y.Y., and Kao, C.H. (2012). Involvement of hydrogen peroxide in heat shock- and cadmium-induced expression of ascorbate peroxidase and glutathione reductase in leaves of rice seedlings. *J. Plant Physiol.* 169: 478–486.

Choudhary, S., Zehra, A., Mukarram, M. et al. (2021). Salicylic acid-mediated alleviation of soil boron toxicity in *Mentha arvensis* and *Cymbopogon flexuosus*: growth, antioxidant responses, essential oil contents and components. *Chemosphere* 276: https://doi.org/10.1016/j.chemosphere.2021.130153.

Cozzolino, V., De Martino, A., Nebbioso, A. et al. (2016). Plant tolerance to mercury in a contaminated soil is enhanced by the combined effects of humic matter addition and inoculation with arbuscular mycorrhizal fungi. *Environ. Sci. Pollut. Res.* 23: 11312–11322. https://doi.org/10.1007/s11356-016-6337-6.

Cruz, Y., Villar, S., Gutiérrez, K. et al. (2021). Gene expression and morphological responses of *Loliumperenne* L. exposed to cadmium (Cd^{2+}) and mercury (Hg^{2+}). *Sci. Rep.* 11: https://doi.org/10.1038/s41598-021-90826-y.

Dang, Z.H., Zheng, L.I., Wang, J. et al. (2013). Transcriptomic profiling of the salt-stress response in the wild recretohalophyte *Reaumuria trigyna*. *BMC Genomics.* 14: 29.

Ding, C., Zhang, T., Li, X., and Wang, X. (2014). Major controlling factors and prediction models for mercury transfer from soil to carrot. *J. Soils Sed.* 14: 1136–1146.

Ding, Y., Ding, L., Xia, Y. et al. (2020). Emerging roles of microRNAs in plant heavy metal tolerance and homeostasis. *J. Agric. Food Chem.* 68 (7): 1958–1965. https://doi.org/10.1021/acs.jafc.9b07468.

Driscoll, C.T., Mason, R.P., Chan, H.M. et al. (2013). Mercury as global pollutant: sources, pathways and effects. *Environ. Sci. Technol.* 47 (10): 4967–4983. https://doi.org/10.1021/es305071v.

El-Shehawi, A.M., Rahman, M.A., Elseehy, M.M., and Kabir, A.H. (2022). Mercury toxicity causes iron and sulphur deficiencies along with oxidative injuries in alfalfa (*Medicago sativa*). *Plant Biosyst.* 284–291. https://doi.org/10.1080/11263504.2021.1985005.

EPA (2022a). National primary drinking water regulations. https://www.epa.gov/ground-water-and-drinking-water/national-primary-drinking-water-regulations (accessed 2 July 2022).

EPA (2022b). Mercury emissions: the global context. https://www.epa.gov/international-cooperation/mercury-emissions-global-context (accessed 4 July 2022).

Esteban, E., Moreno, E., Penalosa, J. et al. (2008). Short and long-term uptake of Hg in white lupin plants: kinetics and stress indicators. *Environ. Expt. Bot.* 62: 316–322.

Feng, X., Li, P., Qiu, G. et al. (2008). Human exposure to methylmercury through rice intake in mercury mining areas, Guizhou Province, China. *Environ. Sci. Technol.* 42 (1): 326–332.

Fernández-Martínez, R., Larios, R., Hrachovinová, J., and Grmela, J. (2015). Mercury accumulation and speciation in plants and soils from abandoned cinnabar mines. *Geoderma* 253: 30–38.

Fischer, S.K., Kuhnlenz, T., Thieme, M. et al. (2014). Analysis of plant Pb tolerance at realistic sub micromolar concentrations demonstrates the role of phytochelatin synthesis for Pb detoxification. *Environ. Sci. Technol. 48*: 7552–7559.

Foyer, C.H. and Noctor, G. (2005). Oxidant and antioxidant signalling in plants: a re-evaluation of the concept of oxidative stress in a physiological context. *Plant Cell Environ.* 28: 1056–1071. https://doi.org/10.1111/j.1365-3040.2005.01327.x.

Fung, Y.K., Meade, A.G., Rack, E.P., and Blotcky, A.J. (1997). Brain mercury in neuro-degenerative disorders. *J. Toxicol. Clin. Toxicol.* 35 (1): 49–54.

Garcia-Ordiales, E., Roqueñí, N., and Loredo, J. (2020). Mercury bioaccumulation by *Juncus maritimus* grown in a Hg contaminated salt marsh (northern Spain). *Mar. Chem.* 226: https://doi.org/10.1016/j.marchem.2020.103859.

Ghosh, U.K., Islam, M.N., Siddiqui, M.N. et al. (2021). Proline, a multifaceted signaling molecule in plant responses to abiotic stress: understanding the physiological mechanisms. *Plant Biol.* https://doi.org/10.1111/plb.13363.

Greger, M., Wang, Y., and Neuschütz, C. (2005). Absence of Hg transpiration by shot after Hg uptake by roots of six terrestrial plant species. *Environ. Pollut.* 134: 201–208.

Gworek, B., Bemowska-Kałabun, O., Kijeńska, M. et al. (2016). Mercury in marine and oceanic waters – a review. *Water Air Soil Pollut.* 227: https://doi.org/10.1007/s11270-016-3060-3.

Hasan, M.K., Cheng, Y., Kanwar, M.K. et al. (2017). Response of plant proteins to heavy metals stress – a review. *Front. Plant Sci.* 8: 1492.

Hasan, M.M., Uddin, M.N., Ara-Sharmeen, I.F. et al. (2019). Assisting phytoremediation of heavy metals using chemical amendments. *Plants* 8: https://doi.org/10.3390/plants8090295.

Hasanuzzaman, M., Bhuyan, M.H.M.B., Anee, T.I. et al. (2019). Regulation of ascorbate-glutathione pathway in mitigating oxidative damage in plants under abiotic stress. *Antioxidants* 8: https://doi.org/10.3390/antiox8090384.

Hayat, S., Ali, B., and Ahmad, A. (2007). Salicylic acid: biosynthesis, metabolism and physiological role in plants. In: *Salicylic Acid: A Plant Hormone* (ed. S. Hayat and A. Ahmad). Springer Nature https://doi.org/10.1007/1-4020-5184-0_1.

Hossain, M.A., Hasanuzzaman, M., and Fujita, M. (2010). Up-regulation of antioxidant and glyoxalase systems by exogenous glycine betaine and proline in mung bean confer tolerance to cadmium stress. *Physiol. Mol. Biol. Plants.* 16 (3): 25–272. https://doi.org/10.1007/s12298-010-0028-4.

Houri, T., Khairallah, Y., Zahab, A.A. et al. (2020). Heavy metals accumulation effects on the photosynthetic performance of geophytes in Mediterranean reserve. *J. King Saud Univ. Sci.* 32: 874–880. https://doi.org/10.1016/j.jksus.2019.04.005.

Hsu, Y.T. and Kao, C.H. (2007). Cadmium induced oxidative damage in rice leaves is reduced by polyamines. *Plant Soil.* 291: 27–37.

Hussain, S., Jianjun, Y., Hussain, J. et al. (2022). The rhizospheric transformation and bioavailability of mercury in pepper plants are influenced by selected Chinese soil types. *Environ. Geochem. Health* 45: 41–52. https://doi.org/10.1007/s10653-022-01209-9.

ILO (2022). Exposure to Mercury in the World of Work: A Review of the Evidence and Key Priority Actions. Report, ISBN: 9789220358047. https://www.ilo.org/global/topics/safety-and-health-at-work/areasofwork/chemical-safety-and-the-environment/WCMS_834578/lang—en/index.htm (accessed 29 June 2022).

İşkil, R., Surgun-Acar, Y., Çatav, Ş.S. et al. (2022). Mercury toxicity affects oxidative metabolism and induces stress responsive mechanisms in wheat (*Triticum aestivum* L.). *Physiol. Mol. Biol. Plants.* 28: 911–920. https://doi.org/10.1007/s12298-022-01171-x.

Israr, M. and Sahi, S.V. (2006). Antioxidative responses to mercury in the cell cultures of *Sesbania drummondii. Plant Physiol. Biochem.* 44: 590–595.

Jalmi, S.K., Bhagat, P.K., Verma, D. et al. (2018). Traversing the links between heavy metal stress and plant signaling. *Front. Plant Sci.* 5: https://doi.org/10.3389/fpls.2018.00012.

Jamla, M., Khare, T., Joshi, S. et al. (2021). Omics approaches for understanding heavy metal responses and tolerance in plants. *Curr. Plant Biol.* 27: 100213. https://doi.org/10.1016/j.cpb.2021.100213.

Kapoor, D., Rattan, A., Gautam, V. et al. (2014). 24-Epibrassinolide mediated changes in photosynthetic pigments and antioxidative defense system of radish seedlings under cadmium and mercury stress. *Physiol. Biochem.* 10: 110–121.

Kapoor, D., Rattan, A., Gautam, V. et al. (2016). Alleviation of cadmium and mercury stress by supplementation of steroid hormone to *Raphanus sativus* seedlings. *Proc. Natl. Acad. Sci. U.S.A.* 86: 661–666. https://doi.org/10.1007/s40011-015-0501-5.

Khan, M.I.R., Nazir, F., Asgher, M. et al. (2015). Selenium and sulfur influence ethylene formation and alleviate cadmium-induced oxidative stress by improving proline and glutathione production in wheat. *J. Plant Physiol.* 173: 9–18.

Kim, Y.H., Khan, A.L., Kim, D.H. et al. (2014). Silicon mitigates heavy metal stress by regulating P-type heavy metal ATPases, *Oryza sativa* low silicon genes, and endogenous phytohormones. *BMC Plant Biol.* 14: https://doi.org/10.1186/1471-2229-14-13.

Kim, Y., Bae, H., Chao, E., and Kang, H. (2017). Exogenous glutathione enhances mercury tolerance by inhibiting mercury entry into plant cells. *Front. Plant Sci.* https://doi.org/10.3389/fpls.2017.00683.

Kim, Y., Rudasingwa, G., Cho, S.H. et al. (2022). Comparison of the concentrations of heavy metals in pm 2.5 analyzed in three different global research institutions using X-ray fluorescence. *Appl. Sci.* 12: https://doi.org/10.3390/app12094572.

Lasat, M.M. (2002). Phytoextraction of toxic metals – a review of biological mechanisms. *J. Environ. Qual.* 31: 109–120.

Li, P., Feng, F.B., Qiu, G.L. et al. (2009). Mercury pollution in Asia: a review of the contaminated sites. *J. Hazard. Mater.* 168: https://doi.org/10.1016/j.jhazmat.2009.03.031.

Li, W.C., Ouyang, Y., and Ye, Z.H. (2014). Accumulation of mercury and cadmium in rice from paddy soil near a mercury mine. *Environ Toxicol. Chem.* 33 (11): 2438–2447.

Li, R., Wu, H., Ding, J. et al. (2017). Mercury pollution in vegetables, grains and soils from areas surrounding coal-fired power plants. *Sci. Rep.* https://doi.org/10.1038/srep46545.

Li, C., Kehai, Z., Wenqiang, Q. et al. (2019). A review on heavy metals contamination in soil: effects, sources, and remediation techniques. *Soil Sediment Contam. Intl. J.* 28: 380–394. https://doi.org/10.1080/15320383.2019.1592108.

Liebthal, M., Maynard, D., and Dietz, K.J. (2018). Peroxidase and redox signaling in plants. *Antioxid. Redox Signal.* 28 (7): https://doi.org/10.1089/ars.2017.7164.

Lima, F.R.D., Barbosa, A.P., Vasques, I.C.F. et al. (2021). Physiological effects of mercury contaminated Oxisols on common bean and oat. *Environ. Sci. Pollut. Res.* 28: 11279–11128. https://doi.org/10.1007/s11356-020-11286-x.

Liu, Z., Chen, B., Wang, L. et al. (2020). A review on phytoremediation of mercury contaminated soils. *J. Hazard. Mater.* 400: 123–138.

Mackey, T.K., Contretas, J.T., and Liang, B.A. (2014). The minamata convention on mercury: attempting to address the global controversy of dental amalgam use and mercury waste disposal. *Sci. Total Environ.* 472: https://doi.org/10.1016/j.scitotenv.2013.10.115.

Malar, S., Sahi, S.V., Favas, P.J.C., and Venkatachalam, P. (2015). Assessment of mercury heavy metal toxicity-induced physiochemical and molecular changes in *Sesbania grandiflora* L. *Int. J. Environ. Sci. Technol.* 12: https://doi.org/10.1007/s13762-014-0699-4.

Mao, Q., Tang, L., Ji, W. et al. (2021). Elevated CO_2 and soil mercury stress affect photosynthetic characteristics and mercury accumulation of rice. *Ecotoxic. Environ. Saf.* 208: https://doi.org/10.1016/j.ecoenv.2020.111605.

Mao, Y., Liu, W., Yang, X. et al. (2022). *Syntrichia caninervis* adapt to mercury stress by altering sub-microstructure and physiological properties in the Gurbantünggüt Desert. *Sci. Rep.* 12: https://doi.org/10.1038/s41598-022-15822-2.

Marrugo-Negrete, J., Durango-Hernández, J., Pinedo-Hernández, J. et al. (2015). Phytoremediation of mercury-contaminated soils by *Jatropha curcas. Chemosphere* 127: 58–63.

Marrugo-Negrete, J., Durango-Hernández, J., Pinedo-Hernández, J. et al. (2016). Mercury uptake and effects on growth in *Jatropha curcas*. *J. Environ. Sci.* 48: https://doi.org/10.1016/j.jes.2015.10.036.

Martinez, G., McCord, S.A., Driscoll, C.T. et al. (2018). Mercury contamination in riverine sediments and fish associated with artisanal and small-scale gold mining in Madre de Dios, Peru. *Int. J. Environ. Res. Public Health* 15: 1584.

Masood, A., Iqbal, N., and Khan, N.A. (2012). Role of ethylene in alleviation of cadmium-induced photosynthetic capacity inhibition by sulfur in mustard. *Plant Cell Environ.* 35: 524–533.

Mathur, P., Tripathi, D.K., Baluška, F., and Mukherjee, S. (2022). Auxin-mediated molecular mechanisms of heavy metal and metalloid stress regulation in plants. *Environ. Exp. Bot.* 196: https://doi.org/10.1016/j.envexpbot.2022.104796.

Mei, L., Zhu, Y., Zhang, X. et al. (2021). Mercury-induced phytotoxicity and responses in upland cotton (*Gossypium hirsutum* L.) seedlings. *Plants* 10: 1494. https://doi.org/10.3390/plants10081494.

Meng, B., Feng, X., Qiu, G. et al. (2012). Inorganic mercury accumulation in rice (*Oryza sativa* L.). *Environ. Toxicol. Chem.* 31 (9): 2093–2098. https://doi.org/10.1002/etc.1913.

Meng, B., Feng, X., Qiu, G. et al. (2014). Localization and speciation of mercury in brown rice with implications for pan-Asian public health. *Environ. Sci. Technol.* 48 (14): 7974–7981. https://doi.org/10.1021/es502000d.

Mhamdi, A., Noctor, G., and Baker, A. (2012). Plant catalases: peroxisomal redox guardians. *Arch. Biochem. Biophys.* 15: 181–194. https://doi.org/10.1016/j.abb.2012.04.015.

Miller, J.R., Hudson-Edwards, K.A., Lechler, P.J. et al. (2003). Heavy metal contamination of water, soil and produce within riverine communities of the Río Pilcomayo basin, Bolivia. *Sci. Total Environ.* 320: 189–209. https://doi.org/10.1016/j.scitotenv.2003.08.011.

Millhollen, A.G., Gustin, M.S., and Obrist, D. (2006). Foliar mercury accumulation and exchange for three tree species. *Environ. Sci. Technol.* 40 (19): 6001–6006.

Mittler, R., Vanderauwera, S., Suzuki, N. et al. (2011). ROS signalling: the new wave? *Trends Plant Sci.* 16: 300–309.

Mohammadi, S., Pourakbar, L., Moghaddam, S.S., and Popović-Djordjević, J. (2021). The effect of EDTA and citric acid on biochemical processes and changes in phenolic compounds profile of okra (*Abelmoschus esculentus* L.) under mercury stress. *Ecotox. Environ. Saf.* 208: 111607.

Mondal, N.K., Das, C., and Datta, J.K. (2015). Effect of mercury on seedling growth, nodulation and Uultrastructural deformation of *Vigna radiate* (L) Wilczek. *Environ. Monit. Assess.* 187: 241. https://doi.org/10.1007/s10661-015-4484-8.

Montero-Palmero, M.B., Amanda, M.B., Escobar, C., and Hernández, L.E. (2013). Early transcriptional responses to mercury: a role for ethylene in mercury-induced stress. *New Phytol.* 21: 116–130. https://doi.org/10.1111/nph.12486.

Mullineaux, P.M. and Rausch, T. (2005). Glutathione, photosynthesis and the redox regulation of stress-responsive gene expression. *Photosynth. Res.* 86: https://doi.org/10.1007/s11120-005-8811-8.

Nambara, E., Okamoto, M. et al. (2010). Abscisic acid and the control of seed dormancy and germination. *Seed Sci. Res.* 20: 55–67.

Natasha, Shahid, M., Khalid, S. et al. (2020). A critical review of mercury speciation, bioavailability, toxicity and detoxification in soil-plant environment: ecotoxicology and health risk assessment. *Sci. Total Environ.* 711: 134749.

Nibau, C., Gibbs, D., and Coates, J. (2008). Branching out in new directions: the control of root architecture by lateral root formation. *New Phytol.* 179: 595–614.

Nyman, J.A. and Lindau, C.W. (2016). Nutrient availability and flooding stress interact to affect growth and mercury concentration in *Taxodium distichum* (L.). *Environ. Exp. Botany.* 125: 77–86. https://doi.org/10.1016/j.envexpbot.2016.02.004.

Olivero, J., Johnson, B., and Arguello, E. (2002). Human exposure to mercury in San Jorge River Basin, Colombia (South America). *Sci. Total Environ.* 289: 41–47.

Ouboter, P.E., Landburg, G.A., Quik, J.H.M. et al. (2012). Mercury levels in pristine and gold mining impacted aquatic ecosystems of Suriname, South America. *AMBIO* 41: 873–882.

Pacyna, E.G., Pacyna, J.M., Steenhuisen, F., and Wilson, S. (2006). Global anthropogenic mercury emission inventory for 2000. *Atmos. Environ.* 40 (22): 4048–4063.

Patra, M. and Sharma, A. (2000). Mercury toxicity in plants. *Bot. Rev.* 66: 379–422. https://doi.org/10.1007/BF02868923.

Perilli, S., Moubayidin, L., and Sabatini, S. (2009). The molecular basis of cytokinin function. *Curr. Opin. Plan Biol.* 13 (1): 21–26. https://doi.org/10.1016/j.pbi.2009.09.018.

Pestana, M. and Formoso, M. (2003). Mercury contamination in Lavras do Sul, south Brazil: a legacy from past and recent gold mining. *Sci. Total Environ.* 307: 125–140.

Pirzadah, T.B., Bisma, M., Tahir, I. et al. (2018). Characterization of mercury-induced stress biomarkers in *Fagopyrum tataricum* plants. *Int. J. Phytoremed.* 20: 225–236. https://doi.org/10.1080/15226514.2017.1374332.

Pleije, L.H., Klingberg, J., Mastromonaco, M.N. et al. (2021). Mercury accumulation in leaves of different plant types – the significant of tissue age and specific leaf area. *Biogeosciences* 18: 6313–6328.

Priya, M., Gurung, N., Mukherjee, K., and Bose, S. (2014). Microalgae in removal of heavy metal and organic pollutants from soil. In: *Microbial Biodegradation and Bioremediation* (ed. S. Das). Elsevier https://doi.org/10.1016/B978-0-12-800021-2.00023-6.

Qiu, G., Feng, X., Li, P. et al. (2008). Methylmercury accumulation in rice (*Oryza sativa* L) grown in abandoned mines in Guizhou China. *J. Agric. Food. Chem.* 56 (7): 2465–2468. https://doi.org/10.1021/jf073391a.

Rahman, Z. (2020). An overview on heavy metal resistant microorganisms for simultaneous treatment of multiple chemical pollutants at co-contaminated sites, and their multipurpose application. *J. Hazard. Mater.* 396: https://doi.org/10.1016/j.jhazmat.2020.122682.

Raj, D., Kumar, A., and andMaiti, S.K. (2020). *Brassica juncea* (L.) Czern. (Indian mustard): a putative plant species to facilitate the phytoremediation of mercury contaminated soils. *Int. J. Phytoremediation* 22 (7): 733–744.

Rausch, T., Gromes, R., Liedschulte, V. et al. (2007). Novel insight into the regulation of GSH biosynthesis in higher plants. *Plant Biol.* 9: 565–572. https://doi.org/10.1055/s-2007-965580.

Rodriguez, L., Jesusa, R.J., Isaac, A.I., and Rodríguez-Castellanos, L. (2007). Capability of selected crop plants for shoot mercury accumulation from polluted soils: phytoremediation perspectives. *Int. J. Phytoremediation* 9: 1–13.

Rodríguez, E., José, R., Peralta-Videa, M.I. et al. (2009). Effect of mercury and gold on growth, nutrient uptake, and anatomical changes in *Chilopsis linearis*. *Environ. Exp. Bot.* 65: 253–262. https://doi.org/10.1016/j.envexpbot.2008.09.014.

Safari, F., Akramian, M., Arjmad, H.S., and Khadivi, A. (2019). Physiological and molecular mechanisms underlying salicylic acid-mitigated mercury toxicity in lemon balm (*Melissa officinalis* L.). *Ecotox. Environ. Saf.* 183: https://doi.org/10.1016/j.ecoenv.2019.109542.

Salt, D.E., Blaylock, M., Kumar, N.P. et al. (1995). Phytoremediation: a novel strategy for the removal of toxic metals from the environment using plants. *Biotechnology* 13 (5): 468–474. https://doi.org/10.1038/nbt0595-468.

Salt, D.E., Smith, R.D., and Raskin, I. (1998). Phytoremediation. *Annu. Rev. Plant Physiol. Plant Mol. Biol.* 49: 643–668. https://doi.org/10.1146/annurev.arplant.49.1.643.

Sandhi, A. (2017). Arsenic removal by Phyto filtration and silicon treatment-a potential solution for lowering arsenic concentrations in food crops. TRITA-LWR PHD-2017:02.

Shahid, M., Shamshad, S., Rafiq, M. et al. (2017). Chromium speciation, bioavailability, uptake, toxicity and detoxification in soil-plant system-a review. *Chemosphere* 178: https://doi.org/10.1016/j.chemosphere.2017.03.074.

Shao, H., Chu, L., Shao, M. et al. (2008). Higher plant antioxidants and redox signalling under environmental stresses. *Comptes Rendus Biol.* 331: 433–441. https://doi.org/10.1016/j.crvi.2008.03.011.

Shao, R., Zhang, J., Shi, W. et al. (2022). Mercury stress tolerance in wheat and maize is achieved by lignin accumulation controlled by nitric oxide. *Environ. Pollut.* 307: https://doi.org/10.1016/j.envpol.2022.119488.

Shiyab, S., Chen, J., Han, F.X. et al. (2008). Mercury-induced oxidative stress in Indian mustard (*Brassica juncea* L.). *Environ. Toxicol.* 24 (5): 462–471. https://doi.org/10.1002/tox.20450.

Siegel, S.M., Puerner, N.J., and andSpeitel, T.W. (1974). Release of volatile mercury from vascular plants. *Physiol. Plant.* 32: 174–176.

Singh, H., Kumar, D., and Soni, V. (2020). Copper and mercury induced oxidative stresses and antioxidant responses of *Spirodela polyrhiza* (L.) Schleid. *Biochem. Biophys. Rep.* 23: 100781.

Sousa, A.I., Cacador, I., Lillebø, A.I., and Pardal, M.A. (2008). Heavy metal accumulation in *Halimione portulacoides*: intra- and extra-cellular metal binding sites. *Chemosphere* 70: https://doi.org/10.1016/j.chemosphere.2007.07.012.

Stamenkovic, J. and Gustin, M. (2009). Nonstomatal versus stomatal uptake of atmospheric mercury. *Environ. Sci. Technol.* 43 (5): 1367–1372.

Swapna, K.S., Salim, N., Chandra, R. et al. (2015). Structural changes in response to bioaccumulation of iron and mercury in *Chromolaena odorata* (L.). *Environ. Monit. Assess.* 187: https://doi.org/10.1007/s10661-015-4732-y.

Sytar, O., Kumar, A., Latowski, D. et al. (2013). Heavy metal-induced oxidative damage, defense reactions, and detoxification mechanisms in plants. *Acta Physiol. Plant.* 35: 985–999. https://doi.org/10.1007/s11738-012-1169-6.

Tamas, L., Mistrik, I., Huttova, J. et al. (2010). Role of reactive oxygen species-generating enzymes and hydrogen peroxide during cadmium, mercury, and osmotic stresses in barley root tip. *Planta* 231: 221–231.

Tang, B., Chen, J., Wang, Z. et al. (2021). Mercury accumulation response of rice plant (*Oryza sativa* L.) to elevated atmospheric mercury and carbon dioxide. *Ecotoxical. Environ. Saf.* 224: 112628.

Tiwari, S. and Lata, C. (2018). Heavy metal stress, signalling, and tolerance due to plant-associated microbes: an overview. *Front. Plant Sci.* 9: 452.

Torres, F.G. and De-la-Torre, G.E. (2022). Mercury pollution in Peru: geographic distribution, health hazards, and sustainable removal technologies. *Environ. Sci. Pollut. Res.* https://doi.org/10.1007/s11356-022-21152-7.

Ullrich, S.M., Tanton, T.W., and andAbdrashitova, S.A. (2001). Mercury in the aquatic environment: a review of factors affecting methylation. *Critic. Rev. Environ. Sci. Technol.* 31 (3): 241–293. https://doi.org/10.1080/2001649108922.

Upadhyay, R.K. (2014). Metal stress in plants: its detoxification in natural environment. *Braz. J. Bot.* 37: 377–382. https://doi.org/10.1007/s40415-014-0087-9.

Vaish, S., Gupta, D., Mehrotra, R. et al. (2020). Glutathione S-transferase: a versatile protein family. *Biotechnology* 10 (7): https://doi.org/10.1007/s13205-020-02312-3.

Wang, Y. and Greger, M. (2004). Clonal differences in mercury tolerance, accumulation, and distribution in willow. *J. Environ. Qual.* 33: 1779–1785.

Wang, Y. and Greger, M. (2006). Use of iodide to enhance the phytoextraction of mercury-contaminated soil. *Sci. Total Environ.* 368: 30–39.

Wang, X., Tam, N.F.Y., and He, H. (2015). The role of root anatomy, organic acids and iron plaque on mercury accumulation in rice. *Plant Soil* https://doi.org/10.1007/s11104-015-2537-y.

WHO (2018). *A Global Overview of National Regulations and Standards for Drinking-Water Quality*. Geneva: World Health Organization (WHO). https://www.who.int/publications/i/item/9789241513760 (accessed 30 June 2022).

Wholgemuth, L., Osterwalder, S., Joseph, C. et al. (2020). A bottom-up quantification of foliar mercury uptake fluxes across Europe. *Biogeosciences* 17: 6441–6456.

Xia, X.J., Huang, L.F., Zhou, Y.H. et al. (2009). Brassinosteroids promote photosynthesis and growth by enhancing activation of Rubisco and expression of photosynthetic genes in *Cucumis sativus*. *Planta* 230 (6): 1185–1196.

Xie, F., Xiao, P., Chen, D. et al. (2012). miRDeepFinder: a miRNA analysis tool for deep sequencing of plant small RNAs. *Plant Mol. Biol.* 80: 75–84.

Yadav, S.K. (2010). Heavy metals toxicity in plants: an overview on the role of glutathione and phytochelatins in heavy metal stress tolerance of plants. *S. Afr. J. Bot.* 76: 167–179. https://doi.org/10.1016/j.sajb.2009.10.007.

Yamasaki, H., Hayashi, M., Fukazawa, M. et al. (2009). SQUAMOSA promoter binding protein-like7 is a central regulator for copper homeostasis in *Arabidopsis*. *Plant Cell.* 21: 347–461.

Yamazaki, S., Ueda, Y., Mukai, A. et al. (2018). Rice phytochelatin synthases OsPCS1 and OsPCS2 make different contributions to cadmium and arsenic tolerance. *Plant Direct 2: e00034*. https://doi.org/10.1002/pld3.34.

Yu, H., Li, J., and Luan, Y. (2018). Meta-analysis of soil mercury accumulation by vegetables. *Sci. Rep.* 19 (1): 1261. https://doi.org/10.1038/s41598-018-19519-3.

Zhang, X., Zhao, X., Li, B. et al. (2014). SRO_1 regulates heavy metal mercury stress responses in *Arabidopsis thaliana*. *Chin. Sci. Bull.* 59: 3134–3141.

Zhang, T., Lu, Q., Su, C. et al. (2017). Mercury induced oxidative stress, DNA damage, and activation of antioxidative system and Hsp70 induction in duckweed (*Lemna minor*). *Ecotoxical. Environ. Saf.* 143: https://doi.org/10.1016/j.ecoenv.2017.04.058.

Zhou, J. and Obrist, D. (2021). Global mercury assimilation by vegetation. *Environ. Sci. Technol.* 55 (20): 14245–14257.

Zhou, Z.S., Huang, S.Q., Guo, K. et al. (2007). Metabolic adaptations to mercury-induced oxidative stress in roots of *Medicago sativa* L. *J. Inorg. Biochem.* 101 (1): 1–9.

Zhou, Z.S., Zeng, H.Q., Liu, Z.P., and Yang, Z.M. (2008a). Genome-wide identification of *Medicago truncatula* microRNAs and their targets reveals their differential regulation by heavy metal. *Plant Cell Environ.* 35 (1): 86–99. https://doi.org/10.1111/j.1365-3040.2011.02418.x.

Zhou, Z.S., Wang, S.H., and Yang, Z.M. (2008b). Biological detection and analysis of mercury toxicity to alfalfa (*Medicago sativa*) plants. *Chemopshere* 70 (8): 1500–1509.

Zhou, Z.S., Guo, K., Elbaz, A.A., and Yang, Z.M. (2009). Salicylic acid alleviates mercury toxicity by preventing oxidative stress in roots of *Medicago sativa*. *Environ. Exp. Bot.* 65: 27–34. https://doi.org/10.1016/j.envexpbot.2008.06.001.

Zhou, J., Obrist, D., Dastoor, A. et al. (2021). Vegetation uptake of mercury and impacts on global cycling. *Nat. Rev. Earth Environ.* 2: 269–284.

Zi, L., Jiang, L.P., Liu, X.F. et al. (2017). Influence of boron deprivation on Hg stress-induced programmed cell death in suspension-cultured tobacco BY-2 cells. *J. Plant Nutri.* 40: 23362348. https://doi.org/10.1080/01904167.2017.1346122.

18

Lead Toxicity and Tolerance in Plants

Insights from Omics Studies

Sayyeda Hira Hassan[1,2], *Yassine Chafik*[2,3], *Manhattan Lebrun*[2,4], *Gabriella Sferra*[1], *Sylvain Bourgerie*[2], *Gabriella Stefania Scippa*[1], *Domenico Morabito*[2], *and Dalila Trupiano*[1]

[1] *Department of Biosciences and Territory, University of Molise, Pesche, Italy*
[2] *Laboratoire de Biologie des Ligneux et des Grandes Cultures, Université d'Orléans, INRAE USC1328, Orléans, France*
[3] *Département de Biologie, Faculté des Sciences, Université Mohammed Premier, Oujda, Morocco*
[4] *Department of Environmental Geosciences, Faculty of Environmental Sciences, Czech University of Life Sciences Prague, Suchdol, Czech Republic*

18.1 Introduction

Lead (Pb) is ranked as the second most toxic heavy metal (HM) by the given guidelines by the Agency for Toxic Substances and Disease Registry (ATSDR 2022). Apart from the natural weathering like volcanic eruptions, soil erosion, and forest fires, an excessive amount of Pb is continuously released into the environment due to several anthropogenic activities like mining and smelting of Pb ores (Obiora et al. 2016). It is widely used in industries, gasoline additives, storage batteries, paints, pesticide production, ammunition, glasses, cosmetic products, etc. (Eick et al. 1999; Sharma and Dubey 2005). As compared to other HMs, Pb tends to persist longer, and its high concentration ($>400\,mg\,kg^{-1}$) affects human health (Laidlaw and Filippelli 2008).

Regardless of the presence of Pb everywhere in the environment, its soil contamination is contemplated as the most serious threat to life (Arshad et al. 2008; Ma et al. 2016). Pb has restricted bioavailable solubility in soils as it forms stable complexes with organic matter, oxide, and clays and accumulates more in the surface ground layer and decreases with the increase in depth (McBride 1994; De Abreu et al. 1998). Plants have the natural capability to thrive in Pb-contaminated soils, establishing different internal defense mechanisms to cope with the adverse effects of this toxic metal (Chandrasekhar and Ray 2019). Generally, most Pb uptake is confined to roots, with restricted translocation to the shoots (Kumar et al. 1995; Małkowski et al. 2019). However, some hyperaccumulator plants such as *Thlaspi rotundifolium*, *Armeria martima*, *Thlaspi alpestre*, *Polycarpaea synandra*, and

Heavy Metal Toxicity and Tolerance in Plants: A Biological, Omics, and Genetic Engineering Approach, First Edition. Edited by Mohammad Anwar Hossain, AKM Zakir Hossain, Sylvain Bourgerie, Masayuki Fujita, Om Parkash Dhankher, and Parvez Haris.

Alyssum wulfenianum are known to translocate and accumulate Pb at the rate of >1000 mg kg^{-1} in dry biomass in their aboveground tissues without showing considerable toxicity symptoms (Baker et al. 1991; Adhikari et al. 2010; Singh et al. 2010). Thus, the threshold for Pb has been set to 1 mg g^{-1} (Reeves et al. 2018), and it is toxic to most plants with a soil concentration of higher than 30 mg kg^{-1} (Al-Thani and Yasseen 2018).

Nevertheless, these defense mechanisms are not adequate for some other plants to overwhelm the metal toxicity, even at a low proportion of Pb, producing improper root growth, variation in membrane permeability, hormonal and water status, alteration in mineral nutrition, disarray in photosynthesis, disturbed transpiration rate, DNA synthesis, and increased reactive oxygen species (ROS) production (Herbette et al. 2006; Sengar et al. 2008; Pourrut et al. 2011). However, Pb toxicity is related to its bioavailability and is based on the plant species, concentration, temperature, season, physicochemical properties of soil, redox potential, and microbial community (Cullen and McAlister 2017; Kushwaha et al. 2018). The organic form of Pb is found to be more toxic than the inorganic form, and it is absorbed to root hairs passively as Pb^{2+} (divalent ions) (Uzu et al. 2009; Kumar and Prasad 2018). At higher concentrations, Pb competes with the uptake of essential divalent mineral nutrients such as Ca^{2+}, Fe^{2+}, Mg^{2+}, Mn^{2+}, and Zn^{2+} and thus causes significant damage to the plants (Zulfiqar et al. 2019). However, applications of some amendments/fertilizers such as compost, biochar, organic acids, and EDTA are being explored to alter the bioavailability of toxic concentration of Pb (Lebrun et al. 2017; Simiele et al. 2020). Different pathways have been reported in plants for Pb uptake and sequestration involving ABC transporters, P-type pumps, pleiotropic drug resistance (PDR1), inner membrane proteins of mitochondria, ATM1, leucine-rice repeat proteins (LRR), Ca^{2+}-gated channels, cyclic nucleotide ion-gated channels, and K^{+}-gated channels (Lee et al. 2005; Wang et al. 2007; Verbruggen et al. 2009; Zhu et al. 2013). In addition, Das et al. (2011) reported a higher expression level of yellow stripe-like protein (BjYSL 2.8), which belongs to a class of oligopeptide transporter proteins, in *Brassica juncea* plants under Pb exposure or other alternative nonselective pathways such as low-affinity cation transporters (TaLCT1) and calmodulin-binding protein (NtCBP4) in tobacco, transmembrane proteins (HvCBT1) in barley, and cyclic nucleotide-gated ion channels (AtCNGC1-6) in *Arabidopsis thaliana* (Arazi et al. 1999; Köhler et al. 1999; Wojas et al. 2007; Gupta et al. 2013).

There has been a huge interest in understanding the mechanistic insights behind Pb uptake, transport, translocation and sequestration/chelation in cellular compartments of Pb-stressed plants, and in identifying potential Pb hyperaccumulating plant species, for phytoremediation of Pb-contaminated soils. In recent years, "omics" approaches have been widely in use to characterize the molecular mechanism in system biology, giving a clear and wide view of biological networks (Urano et al. 2010; Booth et al. 2011). Omics approaches like genomics, transcriptomics, proteomics, and metabolomics have documented detailed molecular changes in plants under Pb stress, with the purpose to take advantage of adaptation skills to utilize plants in phytoremediation strategies to alleviate the effects of increasing metal and metalloid concentrations in important soils around the globe.

This chapter tends to summarize the information on omics techniques used to uncover the plant response toward Pb stress and provide a mechanistic understanding of Pb toxicity and tolerance in plants based on the omics findings.

18.2 Omics' Contribution in Uncovering Molecular Alterations in Plants Under Pb Exposure

In recent years, "omics" approaches like epigenomics/genomics, transcriptomics, proteomics, and metabolomics have emerged as advanced tools to identify genetic and epigenetic, and transcription factors, target genes, proteins, and metabolites, involved in HM toxicity and tolerance response in plants (Chaudhary et al. 2019). Several omics-based designs, summarized in Table 18.1, revealed the stress-responsive genes, proteins, metabolites, miRNAs, stress-inducible phenotypes, and metabolic pathways that could potentially help plants to cope up with Pb toxicity in the face of fluctuating environmental conditions. Epigenomics is the least considered "omics" approach, which allows gaining insight into the molecular adaptation of plants in response to stress (Perrone and Martinelli 2020). Epigenetic regulations involve the modification of gene expression without any alteration in DNA sequence and it includes chromatin reconstruction, histone posttranslational modifications, DNA methylation, and noncoding RNAs associated with posttranscriptional modifications (Perrone and Martinelli 2020). Very few studies highlighted the association of epigenetic regulations with Pb-associated stress in plants among which the methylated DNA immunoprecipitation-sequencing (MeDIP-seq), methylation sensitive amplified polymorphism (MSAP), and Chop polymerase chain reaction (Chop-PCR) techniques (Table 18.1). Genomics includes MSAP data sequencing, structural organization, genetic alignments, interactions, and functional characterization of a whole organism (Bohnert et al. 2006; Khalid et al. 2019). Moreover, the stress response in plants at the molecular level is governed by an association of genes and regulatory mechanisms from transport to metabolic processes (Sahoo et al. 2020); thus, it is necessary to sequence and annotate these genes by implementing a genomics approach. Specific genomics-based design and tools have been applied to uncover novel molecular markers associated with Pb tolerance and accumulation, such as Pb tolerance and accumulation associated quantitative trait loci (QTLs), single nucleotide polymorphism (SNPs), and some candidate genes have been identified by implementing genomic approaches like GWAS and QTL mapping (Table 18.1).

Where genomics specifies the presence or absence of the targeted genes, it is also important to implement the expression analysis to uncover the function of those genes. Here, transcriptomics plays an important role to understand the variation in gene expression under specific conditions (Bräutigam and Gowik 2010). Advancements in transcriptomics approaches made it possible to identify transcriptional factors (TFs) and known/unknown differentially expressed unigenes such as by using next-generation sequencing (NGS), and RNA sequencing (RNA-seq). Wang et al. (2015a) and Li et al. (2017) identified 16,246 and 25,415 differential expressed genes (DEGs) in Pb-stressed *Platanus acerifolia* and *Festuca arundinacea*, respectively. Most of these DEGs were related to metal transporters, antioxidative enzymes, chelating compounds, carbohydrates, and energy processes. Moreover, Zhang et al. (2017) used RNA-seq to identify 262 differentially expressed transcripts (DETs) in *Zea mays* under Pb exposure.

Proteomics is a well-known technique that deals with the large-scale expression of proteins in an organism encoded by a genome. It is not only an advanced tool to describe whole protein change but also compares alteration in protein profiles in an organism from organ to cellular level under any specific condition (Ahsan et al. 2009; Liu et al. 2013).

Table 18.1 Literature relevant to "omics" techniques to uncover Pb-induced changes in different species of plants.

Plant species	Technique used	Findings in response to Pb exposure	References
Zea mays (Maize)	MeDIP-seq	Enhanced methylation on CpG sites followed by intergenic regions and 140 genes with altered methylation, including stress-responsive transcription factors and proteins	Ding et al. (2014)
Isoetes sinensis (Isoetaceae)	MSAP	Altered methylation profile; more hemi-methylation and less full-methylation level	Ding et al. (2019)
Triticum aestivum (Wheat)	Chop-PCR	Altered DNA methylation and expression of metal detoxification transporters to confer resistance	Shafiq et al. (2019)
Zea mays (Maize)	Chop-PCR	Altered DNA methylation levels at the promoter of zinc-regulated transporters	Shafiq et al. (2020)
Brassica napus (Rapeseed)	GWAS	Four QTLs associated with Pb tolerance	Zhang et al. (2020)
Zea mays (Maize)	QTL mapping	QTLs controlling Pb content on chromosomes 1 and 4	Zhao et al. (2014)
Zea mays (Maize)	GWAS	Pb accumulation correlated with identification of 21 SNPs	Zhao et al. (2018a)
Hirschfeldia incana (Mediterranean mustard)	Microarray, reverse transcriptase PCR (RT–PCR)	Gene expression, global expression using microarrays between a plant susceptible to Pb (*Arabidopsis thaliana*) and a Pb-tolerant plant (*H. incana*)	Auguy et al. (2016)
Lycopersicon esculentum (Tomato)	Microarray, RT–qPCR	78 genes related to 20 biological processes including DNA metabolic process, monosaccharide catabolic process, and cell division were identified as potential biomarkers	Hou et al. (2015)
Arabidopsis thaliana	DNA microarray, qRT–PCR	Expressions of over 1310 genes identified, some of the important genes were activated including genes encoding enzymes or proteins involved in sulfur assimilation, GSH metabolism, indole-3-acedic acid, and jasmonic acid (JA) biosynthesis	Liu et al. (2009)
Festuca arundinacea (Tall fescue)	NGS	3696 differentially expressed unigenes	Li et al. (2017)
Platanus acerifolia (London plane tree)	NGS	16,246 differentially expressed unigenes	Wang et al. (2015a)

Table 18.1 (Continued)

Plant species	Technique used	Findings in response to Pb exposure	References
Trifolium pratense (Red clover)	NGS, UPLC, MS/MS, QTRAP	Regulation of gene expression leading to change in glycine and dicarboxylic acid metabolism, and amino acid biosynthesis pathways	Meng et al. (2022)
Medicago sativa (Alfalfa)	De novo and strand-specific RNA-sequencing	A total of 415,350 unigenes, with 5416 DEGs	Xu et al. (2017)
Zea mays (Maize)	RNA-sequencing, qRT-PCR	36 genes with altered expression levels in four different developmental phases	Gao et al. (2015)
Allium cepa (Onion)	Single-cell gel electrophoresis, 2-DE, MS, qRT–PCR	17 DEPs involved in various cellular functions	Lyu et al. (2020)
Raphanus sativus (Radish)	iTRAQ, Strong cation exchange (SCX) chromatography, LC-MS/MS	721 DEPs, related to translation, processing, and degradation, ROS scavenging, photosynthesis, and carbon metabolism were successfully identified	Wang et al. (2016)
Zea mays (Maize)	RNA-sequencing, qRT–PCR	262 DETs involved in plant hormone signal transduction, basal transcription factors, and base excision repair	Zhang et al. (2017)
Medicago sativa (Alfalfa)	SDS-PAGE, HPLC, non-denaturing electrophoresis, UV spectroscopy, RT–qPCR	Increase in oxidative stress-related transcripts, higher GSH, and heat shock proteins	Hattab et al. (2016)
Gossypium hirsutum (Cotton)	qRT–PCR	Significant changes in the expression pattern of miRNA and their targets	He et al. (2014)
Zea mays (Maize)	iTRAQ	Identification of a total of 252, 72, and 116 DEPs after 12, 24, and 48 hr of Pb exposure	Li et al. (2016)
Populus trichocarpa (Populus)	SWATH-MS	4388 unique proteins, among which 542 proteins showed significant abundance changes	Shen et al.(2021)
Glycine max (Soybean)	2-DE (SDS-PAGE), MALDI-TOF MS/MS, qRT–PCR	Identification of 55 proteins, functionally associated with numerous cellular functions	Baig et al. (2018)

(Continued)

Table 18.1 (Continued)

Plant species	Technique used	Findings in response to Pb exposure	References
Talinum triangulare (Ceylon spinach)	2-DE, MALDI-TOF MS	Increase in relative abundance of 20 major proteins, involved in energy metabolism, ROS detoxification, root architecture, cell signaling, primary and secondary metabolisms, and molecular transport systems	Kumar and Majeti (2014)
Arabidopsis thaliana [ecotype Columbia (Col-0)]	SWATH-MS, HPLC–MS/MS	231 proteins with significant abundance changes, involved in photosynthesis and biosynthesis of isoprenoids and tetrapyrroles in chloroplasts	Zhu et al. (2016)
Cannabis sativa (Hemp)	RPLC, LC-ESI-MS/MS	A total of 63 and 372 DEPs in Pb tolerant and Pb-sensitive variety	Xia et al. (2019)
Rhoeo discolor (L. Her.) Hance	2D-DIGE, LC–MS/MS	A predominant decrease in protein levels and a significant effect on the nuclear proteome, as well as photosynthesis, carbon fixation, and metabolism	Enot et al. (2021)
Theobroma cacao	2-DE SDS-PAGE, MS, qPCR	Induction of oxidative stress-related proteins	Reis et al. (2015)
Helianthus annuus (Dwarf Sunflower)	2-DE, ESI-MS/MS	Upregulated proteins related to photosynthesis and antioxidation	Walliwalagedara et al. (2010)
Cichorium intybus (Chicory)	2-DE, SDS-PAGE	168 protein spots were detected, and 81 spots were differentially expressed, having roles in various cell functions including imparting of defense	Malik and Pirzadah (2020)
Suaeda salsa (Halophyte)	NMR	Increase of isocaproate, glucose and fructose, and decrease of malate, citrate and sucrose	Wu et al. (2012)
Chrysopogon zizanioides (vetiver), *Zea mays* (Maize)	LC-MS/MS	Differences in metabolomics of vetiver and maize under lead stress, metabolites showing large increase include amino acids, organic acids, and coenzymes	Pidatala et al. (2018)
Zea mays (Maize)	LC–MS/MS	226 metabolites in maize roots were accurately quantified mainly involved in glycolysis, purine, pyrimidine, and phospholipids pathways	Li et al. (2014)
Typha angustifolia (Narrow-leaved cattail)	2-DE, MS	Upregulated carbohydrate metabolic pathway, enzymes of fructokinase, and improved RuBisCO activase and large subunit, Mg-protoporphyrin IX chelatase	Bah et al. (2010)

Table 18.1 (Continued)

Plant species	Technique used	Findings in response to Pb exposure	References
Raphanus sativus (Radish)	GC-MS	Altered metabolites, mainly involved in carbohydrate, energy, and glutathione metabolism	Wang et al. (2015c)
Arabidopsis thaliana	HPLC–ESI–MS, tandem MS/MS, RT–PCR	Relationship between Pb stress and flavanol	Zhang et al. (2020)

Moreover, despite the role of genomics and transcriptomics in the enhanced understanding of plants' response to HM stress, sometimes transcriptomic changes do not reflect at the protein level due to posttranscriptional/translational modifications, protein folding, stability, and interactions with other proteins (DalCorso et al. 2013). For instance, Liu et al. (2016) reported a contrary expression level of RUBISCO at gene and protein expression levels under a combined treatment of Pb and Zn in halophyte *Suaeda salsa*. Also, protein immunodetection coupled with gene expression analysis in *Medicago sativa* under Pb stress showed some conflicting results, which suggest that Pb stress can lead to posttranslational modifications (Hattab et al. 2016). Thus, a detailed analysis of the proteome of an organism is necessary to study complex biological processes that involve a network of proteins (Ahsan et al. 2009). In the past decades, conventional two-dimensional electrophoresis (2-DE) technology has been often used for the identification of proteins, but with some limitations such as low sensitivity, poor reproducibility, and low automation (Sazuka et al. 2004). However, 2-DE combined with mass spectrometry (MS) has been a commonly used technique to investigate HM- and Pb-induced proteome alterations (Romeo et al. 2014; Simiele et al. 2021). As compared to this traditional approach, fluorescence two-dimensional difference gel electrophoresis (2-D DIGE) approach is more advanced, which allows comparison of differentially expressed proteins (DEPs) of HM-stress and control tissue on one single gel (Kieffer et al. 2008; Alvarez et al. 2009). Using 2D-DIGE coupled with liquid chromatography–tandem mass spectrometry (LC-MS/MS) and quadrupole time-of-flight mass spectrometry (QTOF-MS), a total of 181 differentially abundant protein spots were detected and subsequent analysis identified a prominent decrease in protein content with a significant effect on nuclear proteome, photosynthesis, and carbon fixation, giving insight into the response to Pb stress in *Rhoeo discolor* (Enot et al. 2021). Electrospray ionization (ESI) is the most effective interface of LC-MS, and this technique was used to reveal several DEPs in Pb tolerance and Pb sensitive variety of *Cannabis sativa* (Xia et al. 2019). Where DIGE is a gel-based method, isobaric tag for relative and absolute quantification (iTRAQ) is a gel-free proteomic approach used to find DEPs in *Raphanus sativus* (Radish) and *Z. mays* (Maize) in response to Pb stress (Li et al. 2016; Wang et al. 2016). Sequential windowed acquisition of all theoretical mass spectra (SWATH-MS) is emerging from recent years, which is a novel label-free quantitative proteomic platform. It was applied to detect proteome changes in poplar seedling roots following Pb treatment and proved to be a high reproducible, quantitatively consistent, and accurate proteomic technique (Shen et al. 2021). Both iTRAQ and SWATH-MS match peptide with a database to recognize the most probable proteins, isoelectric points, and molecular weight of the protein. In recent years, matrix-assisted laser desorption ionization–time-of-flight mass spectrometry (MALDI-TOF-MS) has also emerged as a potential tool for proteomics studies and has been applied to study proteomic response in Pb-stressed plant species (Table 18.1).

As compared to other approaches, the metabolomics approach is wider and more advanced, giving more extensive data on the internal working and response under stress conditions. Metabolomics deals with metabolic profiles of plants, like detection, identification, characterization, and quantification, and the metabolome of an organism is related to the total active metabolite pool present in an organism under specified conditions (Peters et al. 2018). It is reported that various metabolites are involved under Pb tolerance strategies like amino acids, organic acids, glutathione (GSH), ascorbate, and phenolic compounds by implementing various mechanisms. At present, some bioanalytical techniques such as high-performance liquid chromatography (HPLC), nuclear magnetic resonance (NMR) spectroscopy, and other combinational tools including gas chromatography mass spectrometry (GC-MS) and liquid chromatography-MS (LC-MS), ultra-performance liquid chromatography coupled with quadrupole time-of-flight mass spectrometry (UPLC-QTOF-MS/MS) have been used to identify and analyze metabolic alterations under Pb exposure (Table 18.1). Among these techniques, UPLC-QTOF-MS/MS is more advanced that allows high sensitivity and accurate molecular weight detection (Yan et al. 2021). By using this technique, Meng et al. (2022) reported a total of 905 metabolites in *Trifolium pratense* against Pb toxicity and showed that the content of total metabolites decreased in general. A combination of chromatography-mass spectrometry techniques also gives the high resolution and throughput aspects of chromatography with the high specificity and sensitivity of mass spectrometry (Putri et al. 2013). Whereas HPLC and NMR are more common analytical techniques, these are not as sensitive as the above-mentioned techniques (Emwas et al. 2019). Despite advancements in omics techniques, omics studies about Pb toxicity and tolerance in plants are very limited at present. The following sections intend to describe findings in detail and understandings of plants' response toward Pb stress and tolerance at the molecular level by integrating omics-based results.

18.3 Genetics and Epigenetics Regulations of Pb Toxicity and Tolerance

Information on the genetic basis of Pb toxicity and tolerance is very scarce. Among different methods, genome-wide association study (GWAS) offers advance resolution to relate definite genetic variations with certain traits in different individuals (Busoms et al. 2021). By using this tool, Zhang et al. (2020) analyzed a large collection of *Brassica napus* and identified Pb-tolerance-associated four QTLs in six Pb-tolerant genotypes, selected by quantifying radicle length (RL > 80%) under Pb stress ($100\,mg\,l^{-1}$). Similarly, GWAS along with Q + K model conducted on a population of 269 maize accessions identified 21 single nucleotide polymorphisms (SNPs) significantly associated with the Pb accumulation in maize kernel and leaves (Zhao et al. 2018a). Further, candidate genes were predicted based on the SNPs and were related to DNA binding, electron carrier activity, catalytic activity, and transporter activity (Zhao et al. 2018a). Five QTLs were detected on chromosomes 1, 4,

6, and 12 associated with Pb accumulation in grains and leaves of F6 RIL population derived from the cross between *japonica* and *indica* cultivars (Norton et al. 2010) and 20 QTLs for Pb accumulation in maize kernel, root, stem, and leaves by combining composite interval mapping using the program QTL Cartographer (Zhao et al. 2018b).

Moreover, the availability of sequencing data from various plant cultivars made it possible to use and combine computational analysis in order to characterize and annotate the data. Computational tools and databases are often used in combination with genomics such as the Darmor-Bzh (reference genome of the *B. napus*; http://www.genoscope.cns.fr/brassicanapus/), nucleic acid basic local alignment search tool in *Arabidopsis* genomic database (BLASTN; https://www.arabidopsis.org/) with Kyoto Encyclopedia of Genes and Genomes (KEGG; http://www.genome.jp/) databases were used to annotate all 115 genes located in the QTL regions of Pb tolerant genotypes identified by GWAS and revealed nine pivotal genes including ubiquitin-specific protease 13 (UBP13), three glutathione *S*-transferases (GSTs) related, trichome birefringence (TBR), two branched-chain-amino-acid aminotransferases (BCATs) related, and one trichome birefringence (HIPP01) involved in Pb tolerance by regulating cell wall development, GSH, ubiquitination, and amino acid metabolism (Zhang et al. 2020). These genetic studies and provided information can be a valuable resource to breed Pb-tolerant plant species and understand the genetic basis behind Pb-tolerance and accumulation and valuable resource for breeding Pb-tolerant plant species.

Apart from genetic recombination, DNA methylation and related mechanisms can also modify plant genomic DNA, making it adapt to the complex environment (Goll and Bestor 2005; Lukens and Zhan 2007; Boyko and Kovalchuk 2008). Epigenetic modification in genes occurs via DNA hypomethylation or hypermethylation at specific loci (Dowen et al. 2012). In plants, DNA methylation can occur at both asymmetric (CpNpN) and symmetric (CpG and CpNpG) sites by methyltransferases (Henderson and Jacobsen 2007). Moreover, excessive methylation is believed to protect the DNA against multi-copy transposition and endonuclease cleavage (Bender 1998), which favors tolerance to HMs. This point has been supported by a genome-wide study of DNA methylation in Pb-stressed maize root by methylated DNA immunoprecipitation-sequencing (MeDIP-seq) revealed enhanced methylation on CpG sites followed by intergenic regions and 140 genes with altered methylation. Among these genes, there were stress-responsive transcription factors and proteins like bZIP, serine-threonine/tyrosine proteins, AP2/ERF, RING zinc finger proteins, leucine-rich proteins, tetratricopeptide repeat proteins, and F-box proteins (Ding et al. 2014). Further, in another study, total methylation level was increased with the increase in Pb concentration (He et al. 2014). Whereas a study stated that Pb-treated wheat showed increased CG DNA hypomethylation in the promoter region of metal detoxification transporters such as ATP-binding cassette (ABC) and heavy metal ATPase 2 (HMA2), which also resulted in enhanced transcription of these transporters (Shafiq et al. 2019). However, in the case of *Isoetes sinensis*, an endangered plant, where methylation level was the same for control and Pb-treated plants, but different methylation profiles. In general, there was more hemi-methylation and less full-methylation level in Pb-stressed groups and nearly equal of methylation and demethylation (Ding et al. 2019). This indicates that some specific loci may be altered through demethylation to regulate gene expression and cope with Pb toxicity as also stated by

Yang et al. (2007) that increased methylation in promotor and coding region can reduce the gene transcription and cause gene silencing, but demethylation favors gene expression. These studies highlight the association of DNA methylation with Pb stress tolerance in plants. But there is very little information available about the methylation modification at specific genomic loci under Pb stress.

18.4 The Role of Plant Cell Wall, Cell Signaling, and Transduction

The plant cell wall is a barrier limiting metal uptake and penetration into the protoplast and thus has an important role in HM defense and detoxification (Pourrut et al. 2011). In particular, the root cell wall is directly in contact with the metals in the soil solution and Pb enters the plant mainly by root uptake. Wang et al. (2016) identified the "cell wall" gene ontology terms (GO: 0005618) as the most enriched among all 135 common differential abundant protein species involved in Pb stress response by proteomic analysis of *R. sativus*. Moreover, proteomic study from poplar plants revealed 13 DEPs with increased abundance and mainly related to pectin and hemicellulose biosynthesis enzymes. Both pectin and hemicellulose are important component of cell wall (Shen et al. 2021). Pb stress reportedly caused the sharp increase in the transcript level of laccases (Shen et al. 2013), which are copper-containing oxidase enzymes and are vital for plant lignin production. Lignin is an integral part of plant cell walls and a well-known adsorbent material for abstracting HMs from water (Guo et al. 2008; Quintana et al. 2008). Caffeic acid 3-*O*-methyltransferase 2 (COMT) is also a vital enzyme for lignin synthesis, and its protein content was increased under 5.0 μM Pb treatment, but increased Pb concentration (15.0 μM) leads to the reduced expression (Lyu et al. 2020). These factors can be the base of morphological alterations in roots implied by different concentration of Pb. Moreover, two inhibiting proteins polygalacturonase inhibitor and pectinesterase inhibitor modulate the activities of two proteins involved in cell wall remodeling named as polygalacturonases (PG) and pectinesterases (PE) in plants and were down-accumulated in exposure to Pb stress in radish root (Wang et al. 2016). Some enzymes including cinnamyl alcohol dehydrogenase, involved in lignin biosynthesis, were abundantly increased in poplar seedlings under Pb treatment (Shen et al. 2021). These studies indicate that regulating cell wall compositional change could be an effective strategy against Pb stress in plants.

The cell wall is also considered an active metabolic site and could initiate some specific stress-responsive signaling mechanism to protect the cell when exposed to external stress stimulus (DalCorso et al. 2010; Pourrut et al. 2011). Among these specific processes are calcium-dependent protein kinases (CDPK) and mitogen-activated protein kinases (MAPKs) (Jonak et al. 2002). MAPK pathway has a profound effect on tolerating stress and MAPK cascade involves three activated protein kinases: MAPK kinase (MAPKKs), MAPK kinase kinase (MAPKKKs), and MAPKs mediating phosphorylation reactions from upstream receptor to downstream target. MAPKs are known to get activated by metal ligands or ROS produced as a result of metal stress and regulates a vast variety of cellular processes by mediating stress signal transmissions (Cristina et al. 2010; Smeets et al. 2013;

Jalmi and Sinha 2015). Transcriptomics analysis of radish root by RNA-seq technology revealed that Pb stress causes the upregulation of four MAPKs like MAPK6, MAPK18, MAPK20, and MAPKKK7 in *R. sativus* root (Wang et al. 2013), which revealed Pb-ions ability to bind with MAPKs. Huang and Huang (2008) showed that Pb induces ROS production and activated MAPK (40- and 42-kDa) in *Oryza sativa* (rice) root, specifically in the apical region. Treatment with GSH decreased the Pb-induced root cell death as well as MAPK activation.

In general, MAPK cascade is connected to other signal transduction pathways to create a molecular interaction network and derive multiple cellular functions including growth, development, and response toward biotic or abiotic stresses by targeting and regulating various transcription factors (TFs) (Andreasson and Ellis 2010). However, knowledge about the downstream signaling targets of MAPKs is scarce under Pb stress. Transcription factors (TFs) are a group of proteins that regulate target gene expression and are vital for normal development and cellular functions in an organism (Yusuf et al. 2012). Several studies indicated the involvement of MAPK-regulated transcription factors in regulating plant response toward Pb stress, such as Wang et al. (2013) found 58 DEGs related to basic leucine zipper (bZIP), WRKY, myeloblastosis protein (MYB), dehydration-responsive element-binding protein (DREB), and ethylene-responsive factor (ERF) have been proved to contribute to regulating the expression level of stress-related genes in radish under Pb stress. Pb-tolerant maize inbred line expressed 262 differentially expressed TFs, specifically bZIP being the major class associated with tolerance under Pb treatment. Also, two maize bZIP TF genes ZmbZIP54 and ZmbZIP107 were highly expressed in a Pb tolerance maize line as compared to Pb sensitive line under Pb treatment (Zhang et al. 2017). WRK and MYB protein TFs are essential for regulating development, metabolism and response to biotic and abiotic stresses (Li et al. 2022). Genes (WRKY-binding protein, WRKY17, Myb39/45/63/93/94) encoding MYB and WRKY transcription factors were also induced by Pb treatment in *Arabidopsis* (Liu et al. 2009). Karanja et al. (2017) identified 20 WRKY transcripts induced by Pb treatment in *R. sativus*. DEAD-box AT-dependent RNA helicase is involved in every aspect of RNA metabolism and utilizes energy from ATP hydrolysis to help correct the folding of RNA by actively destroying misfolded RNA structures (Jones et al. 1996; Mohr et al. 2002; Macovei et al. 2012). However, it was found less abundant in Pb-stress plants (Szuba et al. 2020). Apart from these, heat shock transcription factors (HSF) are well-known responder to heat stress in plants and the member of class A has been reported to be involved in Pb-stress response. For instance, heat shock transcription factor A4a (HSF-08/16/22) were found highly expressed under Pb stress in *R. sativus* (Tang et al. 2019).

CDPKs are a class of calcium-binding related proteins and possess a main role in regulating stress response by transducing Ca^{2+} signals into downstream events (Shi et al. 2018). The functions of these CDPKs are entirely based on Ca^{2+} signatures (Atif et al. 2019) and the increased cytosolic Ca^{2+} levels is the prime response toward most abiotic stresses and it has been confirmed that MAPK and CDPK work together for transmitting the signals against any change in the environment (Takahashi et al. 2011; Opdenakker et al. 2012). A study suggested the involvement of Ca^{2+} in Pb-mediated cell death and enhancing the MAPK activity via CDPK pathway (Huang and Huang 2008). Moreover, CDPKs are important regulators of various signaling pathways including abscisic acid (ABA) signaling

(Sheen 1996) and a transcriptomic analysis of *Hirschfeldia incana*, which is a Pb-tolerant plant, revealed upregulated genes involved in ABA biosynthesis in root and shoot subjected to Pb (Auguy et al. 2016). Besides CDPKs, calmodulin-like proteins (CML), having Ca^{2+}-binding domains, are well-known Ca^{2+} sensors (Zeng et al. 2015) and a study showed an increase in CML 36/43 in soybean (*Glycine max* L.) root nodules under Pb stress (Baig et al. 2018). Wang et al. (2013) also discovered DEGs similar to calcium-binding-related protein genes including EICBP5, CDPK9, and CML45/22, and the majority of them were highly expressed under Pb stress. In addition to this, Annexin D8 (which belongs to Annexins, a family of calcium- and membrane-binding proteins) were upregulated by Pb in soybean (*G. max* L.) (Baig et al. 2018). Another Ca^{2+}-binding protein, calreticulin (CRT), was found to be induced by Pb stress in radish root (Wang et al. 2016). Taken together, plants regulate MAPK and Pb induced Ca^{2+} signaling by modulating CDPK to tackle Pb stress. However, the information regarding these cascades in relation to Pb toxicity and tolerance is very scarce.

Apart from the contribution of signaling pathways in Pb tolerance, other regulators like microRNA and RNA processing have a profound effect on Pb-stress response. MicroRNAs are 20–24 nucleotide long noncoding RNAs that control gene expression at posttranscription levels by translation repression by targeting mRNA degradation (Zhang et al. 2022). HTS and degradome sequencing was used to identify miRNAs responsive to Pb stress in *P. acerifolia* (street tree) and major targets of most miRNAs were TFs involved in mediating stress under Pb toxicity. For instance, Pla-miR159a, with highest fold change, was downregulated under Pb stress and predicted to target MYB TF as well as its targeted APX genes were upregulated (Wang et al. 2015b). Other miRNAs including pla-miR3, pla-miR45-3p, and pla-miR108-3p were also downregulated under Pb stress in *P. acerifolia*, which pointed toward ROS detoxification and PC formation (Wang et al. 2015b). Pb toxicity in cotton (*Gossypium hirsutum* L.) leaves showed downregulated miR156, miR399, miR398, miR414, miR833a, and miR5658 and upregulated miR162, miR167, miR159, miR395, and miR396 (He et al. 2014). These downregulated miRNAs such as miR399 could be related to phosphate homeostasis with less mineral uptake and improper root growth (He et al. 2014; Bari et al. 2006), and miR398 controls CSD transcripts encoding SOD in metal stress (Sunkar et al. 2006). miR414, miR833a, and miR5658 target RAX1 (Rac Family Small GTPase 1), RAX2, and BIT1 (peptidyl-tRNA hydrolase 2), respectively, and involved in root development (He et al. 2014). While upregulated miRNAs such as miR159 and miR396 target TCP and GRF (growth regulating factor), respectively, which are related to plant development (He et al. 2014). This indicated the importance of investigating miRNAs and their targets in plants in response to Pb stress.

18.5 Pb-Binding Proteins/Transporters and Their Involvement in Tolerance

Membrane-associated protein transporters are necessary to facilitate metal ions across the lipid bilayer of the plasma membrane of the root, an initial step in metal transportation and trafficking (Pilon-Smits 2005; Clemens 2006; Krämer et al. 2007). However, being a nonessential and toxic element for plants, they are not likely to have specific transporters for Pb (Arazi et al. 1999). As previously mentioned, studies showed that Pb ions are transported via plasma membrane by some specific transporters, and for the first time, Arazi et al. (1999) discovered a specific membrane-associated protein channels, NtCBP4 (calmodulin-binding protein-4), in *Nicotiana tabacum* and revealed their ability to carry Pb ions through the plasma membrane into the plant cell. But overexpression of NtCBP4 showed enhanced sensitivity to Pb accumulation in transgenic tobacco lines leading to chlorosis and retarded growth (Arazi et al. 1999). On the other hand, several studies suggested the possible role of these metal transporters in increased Pb tolerance and accumulation like an enhanced expression of Acyl-CoA-binding proteins (AtACBP1) showed increased Pb tolerance and accumulation in shoots, and its knockout caused enhanced Pb sensitivity in *Arabidopsis* (Xiao et al. 2008).

Active transport of ions through the plasma membrane also requires the enhanced activity of the proton pump to generate a proton gradient (Janicka-Russak 2011). The proton-pumps H^+-ATPases are located at the plasma membrane and generates proton gradient and difference in electric potential necessary for most of the ion and metabolite transport (Falhof et al. 2016). Here, membrane-associated transporter proteins belonging to the P-type ATPase family such as heavy metal-ATPase 4 (AtHMA4) showed enhanced tolerance and accumulation of Pb (Verret et al. 2004; Xiao et al. 2008), and AtHMA3 involved in Pb detoxification by taking part in vacuolar sequestration (Morel et al. 2009). Three members of ATPase-binding cassette (ABC) transporters were linked with Pb tolerance in *Arabidopsis*. The induced mRNA expression of ABC transporter of the mitochondria (AtATM3) and pleiotropic drug resistance (AtPDR12) under Pb exposure was associated with Pb resistance and lower Pb content (Lee et al. 2005; Kim et al. 2006; Bhuiyan et al. 2011a) and AtPDR12 knockout plants showed less growth and higher Pb content (Lee et al. 2005). Similarly, Pb-inducible PDR8 conferred Pb tolerance but did not show any alteration in Pb content (Kim et al. 2007). Yeast cadmium factor 1 (YCF1) isolated from *Saccharomyces cerevisiae*, is also characterized as ABC transporter genes (Li et al. 1997; Tong et al. 2004), was found to be associated with Pb resistance. Transgenic *Arabidopsis* plants with overexpressed YCF1 showed enhanced Pb content in *Arabidopsis* shoot and improved Pb resistance (Song et al. 2003). Also, YCF1-transgenic *B. juncea* plants accumulated approximately twofold higher Pb than that of wild-type plants (Bhuiyan et al. 2011b). However, it is not clear how these transporters contribute toward Pb tolerance in plants. After entering the symplast, Pb ions bind to various chelators and are subsequently sequestrated into the vacuoles, the final site of Pb sequestration in the protoplast.

18.6 Pb-Induced Oxidative Stress and Antioxidative Mechanisms

Although Pb is not a redox metal and cannot produce ROS, Pb-induced ROS has been observed in terrestrial and aquatic plants (Reddy et al. 2005; Pourrut et al. 2008; Singh et al. 2010). Indeed, in the case of inefficient Pb sequestration processes, Pb ions in the cytosol stimulate the production of ROS (Jalmi et al. 2018), including hydroxyl radical (OH·), superoxide anion (O_2^-), and hydrogen peroxide (H_2O_2), which are hazardous to plants as they rapidly target nucleic acids, proteins, and lipids (Apel and Hirt 2004;

Gopal and Rizvi 2008). Brunet et al. (2009) found over-accumulated phospholipase D α (PLDα) transcripts in *Lathyrus sativus* (grass pea) roots under Pb treatment, and it has been already defined that HMs stimulate PLDα signaling pathways, which leads to the production of ROS and cause cell death (Navari-Izzo et al. 2006; Yakimova et al. 2007). An antioxidant cysteine-rich protein glutaredoxin (GRX), which are GSH-dependent oxidoreductases and involved in redox signaling and ROS scavenging, was downregulated under 15.0 μM Pb treatment in *Allium cepa* (Lyu et al. 2020). Other proteins including Annexin D1 (AnxD1), which can detoxify H_2O_2 by using GSH as a substrate and nucleoside diphosphate kinase (NDPK), which can be induced by H_2O_2, were reduced and increased, respectively (Lyu et al. 2020). Moreover, oxidative stress leads to the lipid peroxidation, which causes internal damage to the membranes, as resulted ROS during oxidative stress causes hydrogen ion removal from unsaturated fatty acids leading to production of lipid radicals (Chen et al. 2009). Malondialdehyde (MDA) is one of the products of lipids peroxidation, which reflects membrane damage due to oxidative stress. MDA content was found to increase in *Ceratophyllum demersum, Vallisneria spinulosa, Vallisneria natans,* and *Cichorium intybus* (chicory) under Pb stress (Mishra et al. 2006; Yan et al. 2006; Wang et al. 2012; Malik and Pirzadah 2020). These studies revealed the reduction of antioxidation ability under Pb stress. However, there are not many "omics"-based studies to pen the molecular reason behind Pb-induced ROS and associated damage.

H_2O_2 is further converted into H_2O and O_2 by some enzymes such as superoxide dismutase (SOD), catalase (CAT), and peroxidase (POD/PER) (Mahdavian et al. 2016). SOD is an important metalloenzyme and a ROS scavenger that cause the disproportionation of O_2^-. It also protects the cells against the delirious effect of very reactive superoxide radicals by reacting with it and catalyzing them into H_2O_2 in various cellular compartments (Del Rio et al. 2002; Fredj et al. 2010). Whereas CAT and POD catalyze the oxidation of H_2O_2-dependent substrates to H_2O. (Podazza et al. 2012), thus the accumulation of O_2^- and H_2O_2 is effectively prevented. Transcriptional profiling of Pb-stressed *Arabidopsis* revealed induced expression level of four genes encoding superoxidase (SOD) enzymes and 20 genes encoding (POD) enzymes (Liu et al. 2009). Also, six POD protein species were identified as differential abundance protein species (DAPS) during the proteomic analysis of radish root under Pb stress (Wang et al. 2016).

Ascorbate peroxidase (APX) is also an important enzyme that detoxifies H_2O_2 by utilizing ascorbate as substrate in the GSH-ascorbate cycle (Apel and Hirt 2004; Sharma and Dubey 2005). A study showed that the mutant *Arabidopsis* with knockout APX1 protein were more resistant to Pb stress, with significantly enhanced CAT (33%) and glutathione peroxidase (GPX) (170%) activity (Jiang et al. 2017). However, there were mild increase in RNA messenger content related to cytosolic and peroxisomal ascorbate peroxidases in (APX) in *L. sativus* (Brunet et al. 2009). *Medicago sativa* plants also showed increased expression of APX and GPX genes under Pb stress (Hattab et al. 2016), which is related to Pb tolerance.

In addition to enzymatic antioxidants, there is another mechanism to detoxify Pb stress is the synthesis of metal-chelating compounds (nonenzymatic), which can sequester and detoxify the excess metal ions (Lin and Aarts 2012). In general, the first step of Pb sequestration/detoxification relies on chelation by thiol-rich protein (metallothioneins, phytochelatins) or nonprotein thiols (GSH, cysteine). Metallothionein (MT) is a cysteine-rich protein with low molecular weight. The sulfhydryl group of cysteine can bind with HM ions to create a partially or nontoxic complex that increases plant resistance toward HM toxicity and promote their accumulation (Zimeri et al. 2005; Turchi et al. 2012). Expression of genes (At4g25100, At1g08330, and At2g28190) encoding MT proteins were induced under Pb stress in *Arabidopsis* (Liu et al. 2009). Also, Xu et al. (2007) reported the induction of MTs expression by Pb in rice (Xu et al. 2007) and three DEGs, homologous with MT, were found in metallothioneins under Pb stress (Wang et al. 2013).

Phytochelatins (PCs) belong to small thiol (SH)-rich peptides that bind HMs via thiolate coordination. Due to this characteristic, PCs are considered important chelating agents (Gupta et al. 1999; Nishikawa et al. 2006). Previously, it was reported that PCs do not form complex with Pb (Leopold et al. 1999), but some studies reported the strong induction of PCs in response to Pb stress, such as Zhang et al. (2008) reported the PC induction in stem and root of *Sedum alfredii* exposed to 700 μM Pb. Other studies also confirmed its induction in *C. demersum* and legume plants (Piechalak et al. 2002; Mishra et al. 2006). Piechalak et al. (2002) reported that PCs sequestrate cytoplasmic soluble Pb before transporting it into the vacuoles, but the mechanism involved in the transportation of Pb–PC complex via tonoplast is not known yet. However, a study reported the noncorrelation between Pb hyperaccumulation and PCs in *S. alfredii* (Sun et al. 2005), which is a substantial argument on the role of PCs in Pb detoxification and tolerance in plants. Gupta et al. (2010) suggested that mainly GSH involves in Pb detoxification rather than PCs, by making a Pb–GSH complex catalyzed by GSH *S*-transferase (GST) (Chasseaud 1979). This can be understandable from the fact that PCs are not formed as a result of gene expression but rather as a product of the biosynthetic pathway (Cobbett 2000). Various genetic, biochemical, and physiological studies have pointed out that GSH is used as a substrate for PC synthesis (Anjum et al. 2012). GSH is the main antioxidant in plant cells. The GSH conversion into PCs can be catalyzed by a special γ-glutamyl cysteine dipeptidyl transpeptidase/phytochelatin synthase (PCS) (Yadav 2010). According to a model, proposed by Hirata et al. (2005), HMs are first chelated by GSH, getting a higher affinity to activate PC synthesis and transfer HMs to these peptides. Pb-stressed grass pea plants showed unaltered transcript levels for PCS and enhanced GST transcripts, which clearly correlated the sequestration of Pb ions by the formation of GSH–Pb complex (Brunet et al. 2009). A transcriptome profiling of Pb-treated *R. sativus* L.'s root showed only one DEG-encoding PCS, and there were 163 transcripts were found, encoding 18 enzymes involved in GSH metabolism (Wang et al. 2013). *Salvinia minima* and *M. sativa* plants also responded to Pb by increasing the expression level of glutathione synthase (GS) genes (Estrella-Gómez et al. 2012; Hattab et al. 2016). GSH can be regenerated from GSSG (oxidized glutathione) by the activity of glutathione reductase (GR). Interestingly, Pb-stressed *V. natans* showed decreased GSH content with enhanced GR activity (Wang et al. 2012), which is possibly due to the GSH consumption to synthesize PCs. Transformed shrub tobacco (*Nicotiana glauca* R. Graham), overexpressing genes encoding phytochelatin synthase (PCS1: enzyme to synthesize from GSH) from *Triticum aestivum* (shrub tobacco) plants, showed better growth and enhanced Pb tolerance than wild-type plants. Moreover, seedling of these transgenic plant grown in Pb-contaminated soil (1572 mg kg^{-1} Pb) accumulated twice the concentration of Pb than the wild type (Gisbert et al. 2003). The overexpression of PCS1 in transgenic lines of aspen (*Populus tremula* × *tremuloides* cv. Etrepole) resulted in higher Pb accumulation and total biomass than control plants (Couselo et al. 2010). In addition, genes involved in the GSH metabolism pathway, encoding GS, glutathione peroxidase, GR, and γ-glutamyl cysteine synthetase were induced by Pb stress in *Arabidopsis* (Liu et al. 2009). All of these studies support the idea that Pb mainly enhances GSH level by regulating at transcript level, which further activates PC synthesis.

Once metals are transported to extra-cytosolic compartments, they eventually bind to oxygen-based chelators such as organic acids like oxalate, malate, or citrate in the vacuole, and pectin in the cell wall (Hall 2002). Pb tolerance has been correlated with the increased oxalate influx in Pb-tolerant rice varieties (Yang et al. 2000). There was a positive correlation of Pb with citrate content in xylem sap of *B. juncea* and *Sesuvium portulacastrum* and a negative correlation with malate content in the roots of radish (Ghnaya et al. 2013; Wang et al. 2015c). Moreover, other organic acids such as α-linolenic acid, linoleic acid, hexadecanoic and octadecanoid acids were found accumulated under Pb stress (Wang et al. 2015c).

Amino acids and derivatives are known to chelate metal ions, thus participating in metal tolerance in plants (Rai 2002; Sharma and Dietz 2006). Pidatala et al. (2018) conducted a comparative metabolic study on vetiver (a hyperaccumulator) and maize (a hypersensitive plant), which showed an overall increase in amino acid content in vetiver than maize. In case of Pb stress, the content of all altered amino acids was reduced, which revealed the possible inhibition in their synthesis (Wang et al. 2015c). Proline is an eminent stress-related amino acid and its accumulation contributed to various function including osmolytic role, and act as ROS scavenger and molecular chaeperone. Roots of vetiver plants showed enhanced concentration of proline under Pb stress, which decreased after 10 days and an inverse relation was seen for glutamate (source of proline synthesis) with the increased Pb amount (Pidatala et al. 2016). In addition, microarray results from Pb-stressed *Arabidopsis* revealed high expressed genes encoding a metabolic enzyme Δ-pyrroline-5-carboxylate synthetase (P5CS) (Liu et al. 2009), which is attributed to catalyze the first step in proline synthesis (Hu et al. 1992). A non-protein amino acid γ-amino butyric acid (GABA) and other amino acids such as serine, alanine, histidine, betaine, spermidine, glutamate, glutamine, and threonine also showed the same trend in vetiver plants but decreased with high concentration Pb in 20 days of root (Pidatala et al. 2016).

18.7 Metabolic Pathways Associated with Pb Tolerance

18.7.1 Sugar/Carbohydrate and Energy Metabolic Pathway

Carbohydrate and energy metabolic pathways were among the most overrepresented pathways in Pb tolerant and sensitive cultivar of *F. arundinacea* under Pb stress (Li et al. 2017). In plants, light-harvesting chlorophyll protein complexes (LHC), antenna proteins, are vital for efficient light absorption in photosynthesis (Nield et al. 2000). Shen et al. (2013)

reported that pathways associated with photosynthesis including photosynthesis-antenna proteins were affected by Pb stress in maize root. Also, DEGs involved in photosynthesis-antenna proteins were suppressed in tall fescue after Pb treatment (Li et al. 2017). Metabolic analysis of Pb-stressed radish root also highlighted altered metabolites involved in energy metabolism and carbohydrate metabolic pathway (Wang et al. 2015c). This suggests the disturbance in light energy absorption, subsequent photosynthesis, and carbohydrate metabolism due to Pb stress.

Also, increased energy demand linked with the stress requires the activation of multiple proteins, such as enhanced abundance of photosynthesis-related proteins was reported in *Helianthus annuus* under Pb stress (Walliwalagedara et al. 2010). This increased energy demand requires the activation of carbohydrate metabolic pathway, which is the most common pathway affected by abiotic stress as soluble sugars serve as signals in the form of osmoprotectants to maintain turgor pressure, stabilize and protect cell membrane, and act as ROS scavengers (Rosa et al. 2009; Van de Ende and Valluru 2009; Muller et al. 2011; Keunen et al. 2013). Pb-stressed plants showed a temporarily increased concentration of photoassimilates such as hexoses, which decreased with passage of time (Pidatala et al. 2016). Besides this, Pb stress also affects carbon assimilation process as its exposure negatively impacted the Calvin cycle intermediates such as glyceraldehyde-3-phosphate and 3-phosphoglycerate in vetiver root and shoot. Metabolites involved in TCA were also decreased in Pb treatment groups in vetiver plants after 20 days of treatment, but opposite trend was seen after 10 days of treatment (Pidatala et al. 2016). Also, the organic acids involved in TCA cycle like citrate and malate were decreased under Pb stress, which shows disturbance in energy process. Other carbohydrates such as sucrose, turanose, maltose, and α/β-D-glucopyranose were also decreased, while the contents of glucose, fructose, and galactose were increased under Pb exposure, suggesting that photoassimilates are stored as hexoses under Pb treatment (Wang et al. 2015c; Hu et al. 2016).

As plants activate the carbohydrate and energy metabolic pathways to maintain the normal growth and development and to protect the cells against abiotic/biotic stress, some vital metabolic enzymes involved in glycolysis, TCA cycle including β-glucosidase, ATP-citrate lyase (ACL), and pyruvate dehydrogenase were upregulated under Pb stress, revealing their ability to compensate high energy demand during Pb stress (Wang et al. 2016). Also, genes encoding for two vital enzymes, trehalose-6-phosphate (TPP) and trehalose-6-phosphate synthase (TPS), involved in trehalose synthesis were strongly induced under Pb treatment in *Arabidopsis* (Liu et al. 2009). The increased level of these genes has shown enhanced abiotic tolerance in many plants (Li et al. 2011); thus, the activation of energy metabolic pathways might be beneficial for plant to combat the toxicity.

18.7.2 Phenylpropanoid Pathway

Studies have highlighted the potential role of phenylpropanoid pathway and its metabolites in abiotic stress tolerance (Hernandez et al. 2009; Mishra et al. 2014). Phenylalanine ammonia-lyase (PAL) is a rate-limiting enzyme in phenylpropanoid synthesis and catalyze the first step in the phenylpropanoid pathway. Pb stress induced its mRNA coding in *G. max* (Pawlak-Sprada et al. 2011a), and proteomic analysis of poplar plant also showed PAL with elevated abundance under Pb stress (Shen et al. 2021). An enhanced phenylpropanoid content was also seen to be correlated with Pb tolerance in *Lupinus luteus* root and *Prosopis* shoots in response to Pb stress (Izbiańska et al. 2014; Zafari et al. 2016). It has been shown that phenylpropanoids are involved in lignin production and its deposition in cell walls, which make it more potential mechanical barrier against HM entrance into the cell (Maestri et al. 2010; Feng et al. 2018). Moreover, flavonoids interact with phospholipids in the membrane and this interaction leads to decrease membrane's fluidity and stabilize its integrity to prevent the access of ROS and diffusion to the bilayer (Ortega et al. 2017). Flavonoids are secondary metabolites produced through phenylpropanoid pathway and flavanone 3-hydroxylase (F3H) has a crucial role in regulating flavonoid biosynthesis (Pelletier and Shirley 1996). F3H was found in increased abundance in poplar under Pb exposure (Shen et al. 2021). Pb also enhanced the total flavonoid content by 46% in lupine roots, with higher fold increase in glycosylated and hydroxylated derivatives of genistein (Pawlak-Sprada et al. 2011b).

18.7.3 Sulfur-Related Pathway and Phytohormones

Sulfur metabolism involves sulfur uptake, assimilation, biosynthesis, and degradation of sulfur-containing compounds (Capaldi et al. 2015). Pb-stressed *Arabidopsis* has shown highly activated two important enzymes, serine acetyltransferase (SAT) and *O*-acetylserine sulfhydrylase/*O*-acetylserine (thiol) lyase (OAA/OASTL), involved in sulfur assimilation, which is further linked to GSH and PCs synthesis, the main products of sulfur metabolism (Liu et al. 2009). During the sulfur assimilation pathway, SAT uses serine and acetyl-coenzyme A and forms *O*-acetylserine (Saito et al. 1995). In the final step, OAA catalyzes the combining of sulfide and *O*-acetylserine to produce cysteine (Bonner et al. 2005), which then used for the synthesis of GSH and phytochelatin peptides under HM stress (Cobbett and Goldsbrough 2002; Maruyama-Nakashita et al. 2003). Cystein level was found lower in metallicolous population of *Peganum harmala* L. with a higher rate of GSH activity under Pb stress (Mahdavian et al. 2016), which suggests the activation of this pathway under Pb exposure.

Moreover, Liu et al. (2009) showed that sulfur assimilation and auxin biosynthesis pathways have a crosstalk in response to Pb stress. Genes encoding proteins, involved in the auxin biosynthetic pathway, including myrosinase-associated proteins (At3g14210, At1g54010, At5g25980), nitrilase 3 (At3g44320), and indoleacetic acid (IAA)-induced related proteins (At1g60710, At5g65670, and At1g51950) were significantly activated in *Arabidopsis* under Pb stress (Liu et al. 2009). It has been known that during stress condition, the stored glucosinolates are hydrolyzed into sulfates and indole-acetonitrile (a direct IAA precursor) by myrosinases (Visvalingam et al. 1998), thus providing additional sulfate. Glucosinolate metabolism is catalyzed by cytochrome P450 (Hogge et al. 1988), and there were 28 genes related to cytochrome P450 family highly expressed under Pb treatment in *Arabidopsis* (Liu et al. 2009).

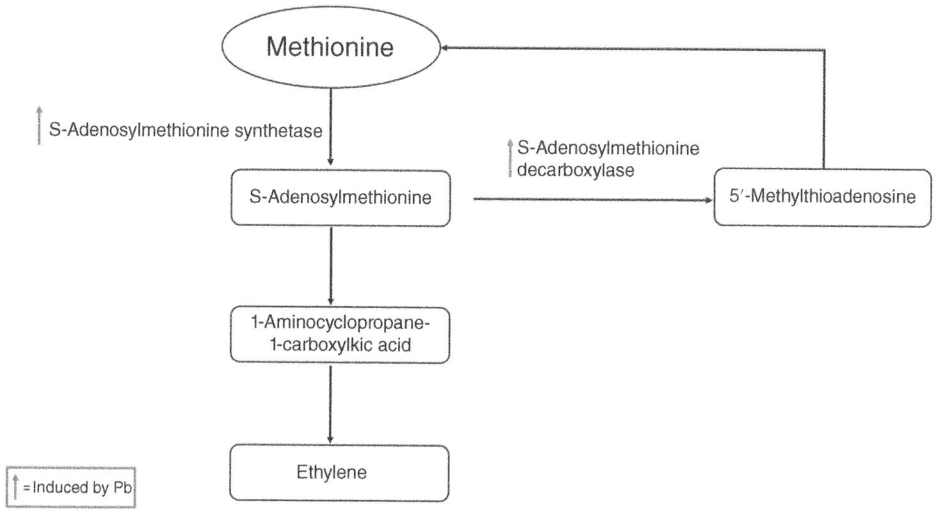

Figure 18.1 Involvement of Pb in methionine-recycling pathway.

Pb exposure can cause alteration of another sulfur-related pathway, the methionine-recycling pathway, enhancing the expression level of genes encoding *S*-adenosylmethionine synthetase-like protein (At3g17390), *S*-adenosylmethionine decarboxylase (At3g02470), and *S*-adenosylmethionine synthetase 2 (At4g01850) in *Arabidopsis* Pb-stressed plant (Liu et al. 2009) (Figure 18.1). In methionine-recycling pathway, methionine is converted to *S*-adenosylmethionine, which is used as a substrate to produce 1-aminocyclopropane-1-carboxylic acid (ACC) and release 5′-methylthioadenosine, which is recycled back to methionine. Further ACC is oxidized to produce ethylene (Sharma and Dubey 2005). Also, Ethylene-Insensitive 2 (EIN2) is a transducer of ethylene, and its transcript expression level was induced under Pb treatment in *Arabidopsis* and ein2-1 mutant showed enhanced sensitivity to Pb with reduced GSH content, leading to the impaired GSH-dependent tolerance against Pb (Cao et al. 2009). It shows that ethylene production can also be enhanced in plants in response to Pb stress by modulating methionine-recycling pathway. However, future research should be focused on investigating the detailed role of methionine-recycling pathway in tolerance against Pb.

ROS production by oxidative stress also initiates hormonal signaling including the above-mentioned IAA together with jasmonic acid (JA), salicylic acid (SA), and abscisic acid (ABA) (Xia et al. 2015). Phytohormones contribute to defense against stress by stimulating antioxidant system (Singh et al. 2010). Microarray data on Pb-stressed *Arabidopsis* revealed the induction of several JA biosynthesis genes including allene oxide synthase (AOS) and 12-oxo-phytodienoate reductase (OPR1) (Liu et al. 2009). In another study, expression level of 12-oxo-phytodienoic acid reductase (OPR), a key enzyme of JA biosynthesis pathway, revealed differential abundance levels during the Pb stress in *R. sativus* (Wang et al. 2016). Principal component analysis on data from Pb-stressed leaves of *Pisum sativum* revealed that JA is positively correlated with APX and GR (Dias et al. 2019), which suggests the protective role of JA in Pb. ABA is usually linked with stomatal opening (Rodriguez et al. 2015) and multivariate analysis from *P. sativum* showed the correlation of ABA with Pb exposure in leaves instead of roots (Dias et al. 2019). Moreover, Dias et al. (2019) suggested the correlation of SA with oxidative stress as well as antioxidant enzymes in leaves and roots of *P. sativum*. This correlation is even proposed by Singh et al. (2010) stating that high levels of SA can alleviate metal toxicity by stimulating antioxidant enzymes, also promoting cell death by oxidative stress.

18.8 Conclusion and Future Perspective

Omics approaches have accelerated the understanding of plant responses to Pb toxicity and tolerance-related genetic factors, TFs, genes, proteins, and metabolites in plants, which can further be exploited to engineer plants with improved tolerance against Pb toxicity. Several studies reported the nonselective entrance of Pb by transporter proteins into the roots and showed their involvement into Pb tolerance, but it is not much clear that how these transporter proteins contribute toward the Pb resistance in plants. ROS production and oxidative stress are the most obvious and well-known effects of Pb and plants cope with this problem by regulating the content, gene-transcript level of enzymatic and nonenzymatic antioxidants. Further, ROS production can activate MAPK and CDPK cascades, which in turn lead to activation of various transcription factors including bZIP, WRKY, MYB, DREB, and ERF to mediate Pb stress. Very few Pb-responsive miRNAs are reported so far, including pla-miR159a, pla-miR3, pla-miR45-3p, pla-miR108-3p in *P. acerifolia* and miR156, miR399, miR398, miR414, miR833a, miR5658, miR162, miR167, miR159, miR395, and miR396 in *G. hirsutum*. The identification of miRNAs and associated genes in response to Pb stress could assist in characterizing the role of miRNAs in plants under heavy metal stress. Moreover, the identification of molecular basis of alteration in metabolic pathways including carbohydrate, energy, phenylpropanoid, and sulfur-related pathways adapted by plants under Pb stress can be the key area of future research. Figure 18.2 presents a global overview of Pb-tolerance-related mechanisms in plants.

Despite acceleration in the use of omics, there are still gaps in uncovering the molecular basis of Pb toxicity and tolerance in plants, and they are essential to figure out in order to improve phytoremediation of Pb-contaminated sites. Moreover, a multidisciplinary approach with well-integrated approach, combining genomics, transcriptomics, metabolomics, and proteomics datasets with bioinformatics-assisted analyses, can be used for comprehensive understanding of the processes involved in plant Pb toxicity and tolerance, identification of potential candidates to improve knowledge about hyperaccumulator/stabilizing plants and to encourage phytoremediation technologies.

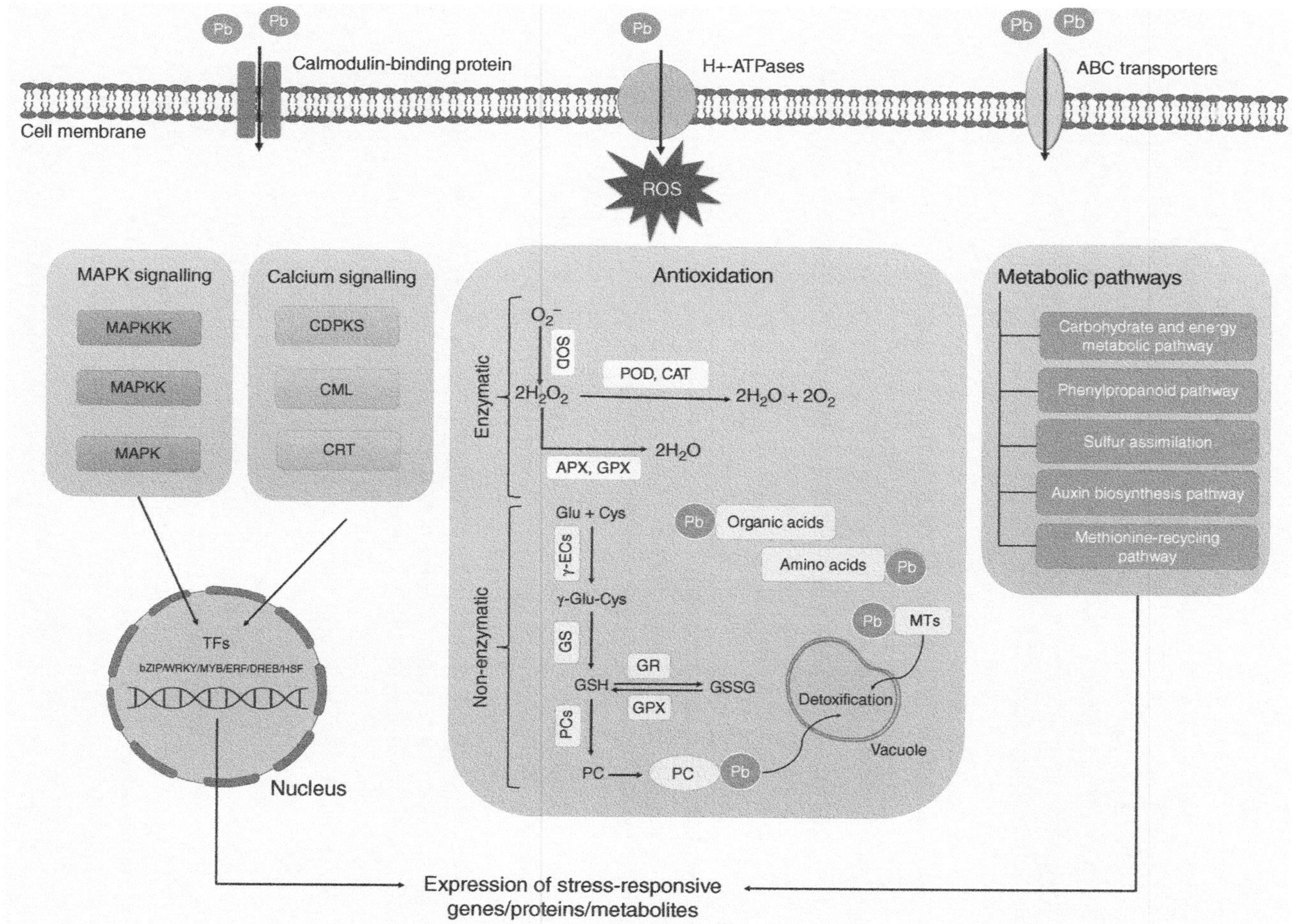

Figure 18.2 Schematic representation of major mechanisms involved in Pb tolerance. *Source:* Based on "omics" studies. These mechanisms involved MAPK and calcium signaling, antioxidative mechanism including enzymatic and nonenzymatic, and metabolic pathways resulting in expression of stress-responsive genes, proteins, and metabolites.

References

Adhikari, T., Kumar, A., Singh, M.V. et al. (2010). Phytoaccumulation of lead by selected wetland plant species. *Commun. Soil Sci. Plant Anal.* 41: 2623–2632.

Ahsan, N., Renaut, J., and Komatsu, S. (2009). Recent developments in the application of proteomics to the analysis of plant responses to heavy metals. *Proteomics* 9: 2602–2621.

Al-Thani, R.F. and Yasseen, B.T. (2018). Biological soil crusts and extremophiles adjacent to native plants at Sabkhas and Rawdahs, Qatar: the possible roles. *Front. Environ. Microbiol.* 4: 55–70.

Alvarez, S., Berla, B.M., Sheffield, J. et al. (2009). Comprehensive analysis of the *Brassica juncea* root proteome in response to cadmium exposure by complementary proteomic approaches. *Proteomics* 9: 2419–2431.

Andreasson, E. and Ellis, B. (2010). Convergence and specificity in the *Arabidopsis* MAPK nexus. *Trends Plant Sci.* 15: 106–113.

Anjum, N.A., Ahmad, I., Mohmood, I. et al. (2012). Modulation of glutathione and its related enzymes in plants' responses to toxic metals and metalloids—a review. *Environ. Exp. Bot.* 75: 307–324.

Apel, K. and Hirt, H. (2004). Reactive oxygen species: metabolism, oxidative stress, and signal transduction. *Annu. Rev. Plant Biol.* 55: 373–399.

Arazi, T., Sunkar, R., Kaplan, B., and Fromm, H. (1999). A tobacco plasma membrane calmodulin-binding transporter confers Ni^{2+} tolerance and Pb^{2+} hypersensitivity in transgenic plants. *Plant J.* 20: 171–182.

Arshad, M., Silvestre, J., Pinelli, E. et al. (2008). A field study of lead phytoextraction by various scented *Pelargonium* cultivars. *Chemosphere* 71: 2187–2192.

Atif, R.M., Shahid, L., Waqas, M. et al. (2019). Insights on calcium-dependent protein kinases (CPKs) signaling for abiotic stress tolerance in plants. *Int. J. Mol. Sci.* 20: 5298.

ATSDR (Agency for Toxic Substances and Disease Registry) (2022). ATSDR's Substance Priority List. https://www.atsdr.cdc.gov/spl/index.html.

Auguy, F., Fahr, M., Moulin, P. et al. (2016). Transcriptome changes in *Hirschfeldia incana* in response to lead exposure. *Front. Plant Sci.* 6: 1231.

Bah, A.M., Sun, H., Chen, F. et al. (2010). Comparative proteomic analysis of *Typha angustifolia* leaf under chromium, cadmium and lead stress. *J. Hazard. Mater.* 184: 191–203.

Baig, M.A., Ahmad, J., Bagheri, R. et al. (2018). Proteomic and ecophysiological responses of soybean (*Glycine max* L.) root nodules to Pb and hg stress. *BMC Plant Biol.* 18: 1–21.

Baker, A.J.M., Reeves, R.D., and McGrath, S.P. (1991). In situ decontamination of heavy metal polluted soils using crops of metal-accumulating plants—a feasibility study. In: *In Situ Bioreclamation*, 600–605. Butterworth-Heinemann.

Bari, R., Datt Pant, B., Stitt, M., and Scheible, W.R. (2006). PHO_2, microRNA399, and PHR1 define a phosphate-signaling pathway in plants. *Plant Physiol.* 141: 988–999.

Bender, J. (1998). Cytosine methylation of repeated sequences in eukaryotes: the role of DNA pairing. *Trends Biochem. Sci.* 23: 252–256.

Bhuiyan, M.S.U., Min, S.R., Jeong, W.J. et al. (2011a). Overexpression of AtATM3 in *Brassica juncea* confers enhanced heavy metal tolerance and accumulation. *Plant Cell Tissue Organ Cult.* 107: 69–77.

Bhuiyan, M.S.U., Min, S.R., Jeong, W.J. et al. (2011b). Overexpression of a yeast cadmium factor 1 (YCF1) enhances heavy metal tolerance and accumulation in *Brassica juncea*. *Plant Cell Tissue Organ Cult.* 105: 85–91.

Bohnert, H.J., Gong, Q., Li, P., and Ma, S. (2006). Unraveling abiotic stress tolerance mechanisms–getting genomics going. *Curr. Opin. Plant Biol.* 9: 180–188.

Bonner, E.R., Cahoon, R.E., Knapke, S.M., and Jez, J.M. (2005). Molecular basis of cysteine biosynthesis in plants: structural and functional analysis of *O*-acetylserinesulfhydrylase from *Arabidopsis thaliana*. *J. Biol. Chem.* 280: 38803–38813.

Booth, S.C., Workentine, M.L., Weljie, A.M., and Turner, R.J. (2011). Metabolomics and its application to studying metal toxicity. *Metallomics* 3: 1142–1152.

Boyko, A. and Kovalchuk, I. (2008). Epigenetic control of plant stress response. *Environ. Mol. Mutagen.* 49: 61–72.

Bräutigam, A. and Gowik, U. (2010). What can next generation sequencing do for you? Next generation sequencing as a valuable tool in plant research. *Plant Biol.* 12: 831–841.

Brunet, J., Varrault, G., Zuily-Fodil, Y., and Repellin, A. (2009). Accumulation of lead in the roots of grass pea (*Lathyrus sativus* L.) plants triggers systemic variation in gene expression in the shoots. *Chemosphere* 77: 1113–1120.

Busoms, S., Pérez-Martín, L., Llimós, M. et al. (2021). Genome-wide association study reveals key genes for differential lead accumulation and tolerance in natural *Arabidopsis thaliana* accessions. *Front. Plant. Sci.* 12: 689316.

Cao, S., Chen, Z., Liu, G. et al. (2009). The *Arabidopsis* ethylene-insensitive 2 gene is required for lead resistance. *Plant Physiol. Biochem.* 47: 308–312.

Capaldi, F.R., Gratão, P.L., Reis, A.R. et al. (2015). Sulfur metabolism and stress defense responses in plants. *Trop. Plant Biol.* 8: 60–73.

Chandrasekhar, C. and Ray, J.G. (2019). Lead accumulation, growth responses and biochemical changes of three plant species exposed to soil amended with different concentrations of lead nitrate. *Ecotoxicol. Environ. Saf.* 171: 26–36.

Chasseaud, L.F. (1979). The role of glutathione and glutathione *S*-transferases in the metabolism of chemical carcinogens and other electrophilic agents. In: *Advances in Cancer Research*, vol. 29, 175–274. Academic Press.

Chaudhary, J., Khatri, P., Singla, P. et al. (2019). Advances in omics approach for abiotic stress tolerance in tomato. *Biology* 8: 90.

Chen, J., Shiyab, S., Han, F.X. et al. (2009). Bioaccumulation and physiological effects of mercury in *Pteris vittata* and *Nephrolepis exaltata*. *Ecotoxicology* 18: 110–121.

Clemens, S. (2006). Toxic metal accumulation, responses to exposure and mechanisms of tolerance in plants. *Biochimie* 88: 1707–1719.

Cobbett, C.S. (2000). Phytochelatin biosynthesis and function in heavy-metal detoxification. *Curr.Opin. Plant Biol.* 3: 211–216.

Cobbett, C. and Goldsbrough, P. (2002). Phytochelatins and metallothioneins: roles in heavy metal detoxification and homeostasis. *Annu. Rev. Plant Biol.* 53: 159–182.

Couselo, J.L., Navarro-Avino, J., and Ballester, A. (2010). Expression of the phytochelatin synthase TaPCS1 in transgenic aspen, insight into the problems and qualities in phytoremediation of Pb. *Int. J. Phytoremediation* 12: 358–370.

Cristina, M.S., Petersen, M., and Mundy, J. (2010). Mitogen-activated protein kinase signaling in plants. *Annu. Rev. Plant Biol.* 61: 621–649.

Cullen, J.T. and McAlister, J. (2017). Biogeochemistry of lead. Its release to the environment and chemical speciation. *Met. Ions. Life Sci.* 17: 21–48.

DalCorso, G., Farinati, S., and Furini, A. (2010). Regulatory networks of cadmium stress in plants. *Plant Signal. Behav.* 5: 663–667.

DalCorso, G., Fasani, E., and Furini, A. (2013). Recent advances in the analysis of metal hyperaccumulation and hypertolerance in plants using proteomics. *Front. Plant. Sci.* 4: 280.

Das, S., Sen, M., Saha, C. et al. (2011). Isolation and expression analysis of partial sequences of heavy metal transporters from *Brassica juncea* by coupling high throughput cloning with a molecular fingerprinting technique. *Planta* 234: 139–156.

De Abreu, C.A., De Abreu, M.F., and De Andrade, J.C. (1998). Distribution of lead in the soil profile evaluated by DTPA and Mehlich-3 solutions. *Bragantia* 57: 185–882.

Del Río, L.A., Corpas, F.J., Sandalio, L.M. et al. (2002). Reactive oxygen species, antioxidant systems and nitric oxide in peroxisomes. *J. Exp. Bot.* 53: 1255–1272.

Dias, M.C., Mariz-Ponte, N., and Santos, C. (2019). Lead induces oxidative stress in *Pisum sativum* plants and changes the levels of phytohormones with antioxidant role. *Plant Physiol. Biochem.* 137: 121–129.

Ding, H., Gao, J., Qin, C. et al. (2014). The dynamics of DNA methylation in maize roots under Pb stress. *Int. J. Mol. Sci.* 15: 23537–23554.

Ding, G.H., Guo, D.D., Guan, Y. et al. (2019). Changes of DNA methylation of *Isoetes sinensis* under Pb and Cd stress. *Environ. Sci. Pollut. Res.* 26: 3428–3435.

Dowen, R.H., Pelizzola, M., Schmitz, R.J. et al. (2012). Widespread dynamic DNA methylation in response to biotic stress. *Proc. Natl. Acad. Sci. U.S.A.* 109: E2183–E2191.

Eick, M.J., Peak, J.D., Brady, P.V., and Pesek, J.D. (1999). Kinetics of lead adsorption/desorption on goethite: residence time effect. *Soil Sci.* 164: 28–39.

Emwas, A.H., Roy, R., McKay, R.T. et al. (2019). NMR spectroscopy for metabolomics research. *Metabolites* 9: 123.

Enot, M.M., Weiland, F., Mittal, P. et al. (2021). Differential proteome analysis of the leaves of lead hyperaccumulator, *Rhoeo discolor* (L. Her.) Hance. *J. Mass Spectrom.* 56: e4689.

Estrella-Gómez, N.E., Sauri-Duch, E., Zapata-Pérez, O., and Santamaría, J.M. (2012). Glutathione plays a role in protecting leaves of *Salvinia minima* from Pb^{2+} damage associated with changes in the expression of SmGS genes and increased activity of GS. *Environ. Exp. Bot.* 75: 188–194.

Falhof, J., Pedersen, J.T., Fuglsang, A.T., and Palmgren, M. (2016). Plasma membrane H^+-ATPase regulation in the center of plant physiology. *Mol. Plant.* 9: 323–337.

Feng, J., Jia, W., Lv, S. et al. (2018). Comparative transcriptome combined with morpho-physiological analyses revealed key factors for differential cadmium accumulation in two contrasting sweet sorghum genotypes. *Plant Biotechnol. J.* 16: 558–571.

Fredj, F.B., Irie, M., Han, J. et al. (2010). Stress response of heavy metal mixture present in wastewater and leachate on heat-shock protein 47-transfected cells. *Environ. Toxicol. Chem.* 29: 1637–1647.

Gao, J., Zhang, Y., Lu, C. et al. (2015). The development dynamics of the maize root transcriptome responsive to heavy metal Pb pollution. *Biochem. Biophys. Res. Commun.* 458: 287–293.

Ghnaya, T., Zaier, H., Baioui, R. et al. (2013). Implication of organic acids in the long-distance transport and the accumulation of lead in *Sesuvium portulacastrum* and *Brassica juncea. Chemosphere* 90: 1449–1454.

Gisbert, C., Ros, R., De Haro, A. et al. (2003). A plant genetically modified that accumulates Pb is especially promising for phytoremediation. *Biochem. Biophys. Res. Commun.* 303: 440–445.

Goll, M.G. and Bestor, T.H. (2005). Eukaryotic cytosine methyltransferases. *Annu. Rev. Biochem.* 74: 481–514.

Gopal, R. and Rizvi, A.H. (2008). Excess lead alters growth, metabolism and translocation of certain nutrients in radish. *Chemosphere* 70: 1539–1544.

Guo, X., Zhang, S., and Shan, X.Q. (2008). Adsorption of metal ions on lignin. *J. Hazard. Mater.* 151: 134–142.

Gupta, D.K., Huang, H.G., Yang, X.E. et al. (2010). The detoxification of lead in *Sedum alfredii* H. is not related to phytochelatins but the glutathione. *J. Hazard. Mater.* 177: 437–444.

Gupta, D.K., Huang, H.G., and Corpas, F.J. (2013). Lead tolerance in plants: strategies for phytoremediation. *Environ. Sci. Pollut. Res.* 20: 2150–2161.

Gupta, M., Tripathi, R.D., Rai, U.N., and Haq, W. (1999). Lead induced synthesis of metal binding peptides (Phytochelatins) in submerged macrophyte *Vallisneria spiralis* L. *Physiol. Mol. Biol. Plants* 5: 173–180.

Hall, J.Á. (2002). Cellular mechanisms for heavy metal detoxification and tolerance. *J. Exp. Bot.* 53: 1–11.

Hattab, S., Hattab, S., Flores-Casseres, M.L. et al. (2016). Characterisation of lead-induced stress molecular biomarkers in *Medicago sativa* plants. *Environ. Exp. Bot.* 123: 1–12.

He, L.L., Shen, H., Wang, Y. et al. (2014). Analysis of genomic DNA methylation level in radish under lead stress. *J. Nucl. Agr. Sci.* 29: 1278–1284.

He, Q., Zhu, S., and Zhang, B. (2014). MicroRNA–target gene responses to lead-induced stress in cotton (*Gossypium hirsutum* L.). *Funct. Integr. Genomics.* 14: 507–515.

Henderson, I.R. and Jacobsen, S.E. (2007). Epigenetic inheritance in plants. *Nature* 447: 418–424.

Herbette, S., Taconnat, L., Hugouvieux, V. et al. (2006). Genome-wide transcriptome profiling of the early cadmium response of *Arabidopsis* roots and shoots. *Biochimie* 88: 1751–1765.

Hernández, I., Alegre, L., Van Breusegem, F., and Munné-Bosch, S. (2009). How relevant are flavonoids as antioxidants in plants? *Trends Plant Sci.* 14: 125–132.

Hirata, K., Tsuji, N., and Miyamoto, K. (2005). Biosynthetic regulation of phytochelatins, heavy metal-binding peptides. *J. Biosci. Bioeng.* 100: 593–599.

Hogge, L.R., Reed, D.W., Underhill, E.W., and Haughn, G.W. (1988). HPLC separation of glucosinolates from leaves and seeds of *Arabidopsis thaliana* and their identification using thermospray liquid chramatography/mass spectrometry. *J. Chromatogr. Sci.* 26: 551–556.

Hou, J., Liu, X., Wang, J. et al. (2015). Microarray-based analysis of gene expression in *Lycopersicon esculentum* seedling roots in response to cadmium, chromium, mercury, and lead. *Environ. Sci. Technol.* 49: 1834–1841.

Hu, C.A., Delauney, A.J., and Verma, D.P. (1992). A bifunctional enzyme (delta 1-pyrroline-5-carboxylate synthetase) catalyzes the first two steps in proline biosynthesis in plants. *Proc. Natl. Acad. Sci. U.S.A.* 89: 9354–9358.

Hu, X., Gao, Y., and Fang, Z. (2016). Integrating metabolic analysis with biological endpoints provides insight into nanotoxicological mechanisms of graphene oxide: from effect onset to cessation. *Carbon* 109: 65–73.

Huang, T.L. and Huang, H.J. (2008). ROS and CDPK-like kinase-mediated activation of MAP kinase in rice roots exposed to lead. *Chemosphere* 71: 1377–1385.

Huang, H., Li, T., Tian, S. et al. (2008). Role of EDTA in alleviating lead toxicity in accumulator species of *Sedum alfredii* H. *Bioresour. Technol.* 99: 6088–6096.

Izbiańska, K., Arasimowicz-Jelonek, M., and Deckert, J. (2014). Phenylpropanoid pathway metabolites promote tolerance response of lupine roots to lead stress. *Ecotoxicol. Environ. Saf.* 110: 61–67.

Jalmi, S.K. and Sinha, A.K. (2015). ROS mediated MAPK signaling in abiotic and biotic stress-striking similarities and differences. *Front. Plant Sci.* 6: 769.

Jalmi, S.K., Bhagat, P.K., Verma, D. et al. (2018). Traversing the links between heavy metal stress and plant signaling. *Front. Plant Sci.* 9: 12.

Janicka-Russak, M. (2011). Plant plasma membrane H+-ATPase in adaptation of plants to abiotic stresses. In: *Abiotic Stress Response in Plants-Physiological, Biochemical and Genetic Perspectives*, vol. 1 (ed. A. Shanker and B. Venkateswarlu), 197–218. IntechOpen.

Jiang, L., Wang, W., Chen, Z. et al. (2017). A role for APX1 gene in lead tolerance in *Arabidopsis thaliana*. *Plant Sci.* 256: 94–102.

Jonak, C., Ökrész, L., Bögre, L., and Hirt, H. (2002). Complexity, cross talk and integration of plant MAP kinase signalling. *Curr. Opin. Plant Biol.* 5: 415–424.

Jones, P.G., Mitta, M., Kim, Y. et al. (1996). Cold shock induces a major ribosomal-associated protein that unwinds double-stranded RNA in *Escherichia coli*. *Proc. Natl. Acad. Sci. U.S.A.* 93: 76–80.

Karanja, B.K., Fan, L., Xu, L. et al. (2017). Genome-wide characterization of the WRKY gene family in radish (*Raphanus sativus* L.) reveals its critical functions under different abiotic stresses. *Plant Cell Rep.* 36: 1757–1773.

Keunen, E.L.S., Peshev, D., Vangronsveld, J. et al. (2013). Plant sugars are crucial players in the oxidative challenge during abiotic stress: extending the traditional concept. *Plant Cell Environ.* 36: 1242–1255.

Khalid, N., Aqeel, M., and Noman, A. (2019). System biology of metal tolerance in plants: an integrated view of genomics, transcriptomics, metabolomics, and phenomics. In: *Plant Metallomics and Functional Omics* (ed. G. Sablok), 107–144. Springer.

Kieffer, P., Dommes, J., Hoffmann, L. et al. (2008). Quantitative changes in protein expression of cadmium-exposed poplar plants. *Proteomics* 8: 2514–2530.

Kim, D.Y., Bovet, L., Kushnir, S. et al. (2006). AtATM3 is involved in heavy metal resistance in *Arabidopsis*. *Plant Physiol.* 140: 922–932.

Kim, D.Y., Bovet, L., Maeshima, M. et al. (2007). The ABC transporter AtPDR8 is a cadmium extrusion pump conferring heavy metal resistance. *Plant J.* 50: 207–218.

Köhler, C., Merkle, T., and Neuhaus, G. (1999). Characterisation of a novel gene family of putative cyclic nucleotide-and calmodulin-regulated ion channels in *Arabidopsis thaliana*. *Plant J.* 18: 97–104.

Krämer, U., Talke, I.N., and andHanikenne, M. (2007). Transition metal transport. *FEBS Lett.* 581: 2263–2272.

Kumar, A. and Majeti, N.V.P. (2014). Proteomic responses to lead-induced oxidative stress in *Talinum triangulare* Jacq. (Willd.) roots: identification of key biomarkers related to glutathione metabolisms. *Environ. Sci. Pollut. Res.* 21: 8750–8764.

Kumar, A. and Prasad, M.N.V. (2018). Plant-lead interactions: transport, toxicity, tolerance, and detoxification mechanisms. *Ecotoxicol. Environ. Saf.* 166: 401–418.

Kumar, P.N., Dushenkov, V., Motto, H., and Raskin, I. (1995). Phytoextraction: the use of plants to remove heavy metals from soils. *Environ. Sci. Technol.* 29: 1232–1238.

Kushwaha, A., Hans, N., Kumar, S., and Rani, R. (2018). A critical review on speciation, mobilization and toxicity of lead in soil-microbe-plant system and bioremediation strategies. *Ecotoxicol. Environ. Saf.* 147: 1035–1045.

Laidlaw, M.A. and Filippelli, G.M. (2008). Resuspension of urban soils as a persistent source of lead poisoning in children: a review and new directions. *Appl. Geochem.* 23: 2021–2039.

Lebrun, M., Macri, C., Miard, F. et al. (2017). Effect of biochar amendments on As and Pb mobility and phytoavailability in contaminated mine technosolsphytoremediated by Salix. *J. Geochem. Explor.* 182: 149–156.

Lee, M., Lee, K., Lee, J. et al. (2005). AtPDR12 contributes to lead resistance in Arabidopsis. *Plant Physiol.* 138: 827–836.

Leopold, I., Günther, D., Schmidt, J., and Neumann, D. (1999). Phytochelatins and heavy metal tolerance. *Phytochemistry* 50: 1323–1328.

Li, Z.S., Lu, Y.P., Zhen, R.G. et al. (1997). A new pathway for vacuolar cadmium sequestration in *Saccharomyces cerevisiae*: YCF1-catalyzed transport of bis (glutathionato) cadmium. *Proc. Natl. Acad. Sci. U.S.A.* 94: 42–47.

Li, H.W., Zang, B.S., Deng, X.W., and Wang, X.P. (2011). Overexpression of the trehalose-6-phosphate synthase gene OsTPS1 enhances abiotic stress tolerance in rice. *Planta* 234: 1007–1018.

Li, K., Wang, X., Pidatala, V.R. et al. (2014). Novel quantitative metabolomic approach for the study of stress responses of plant root metabolism. *J. Proteome Res.* 13: 5879–5887.

Li, G.K., Gao, J., Peng, H. et al. (2016). Proteomic changes in maize as a response to heavy metal (lead) stress revealed by iTRAQ quantitative proteomics. *Genet. Mol. Res.* 15: 1–14.

Li, H., Hu, T., Amombo, E., and Fu, J. (2017). Transcriptome profilings of two tall fescue (*Festuca arundinacea*) cultivars in response to lead (Pb) stress. *BMC Genom.* 18: 1–16.

Li, S., Han, X., Lu, Z. et al. (2022). MAPK cascades and transcriptional factors: regulation of heavy metal tolerance in plants. *Int. J. Mol. Sci.* 23: 4463.

Lin, Y.F. and Aarts, M.G. (2012). The molecular mechanism of zinc and cadmium stress response in plants. *Cell. Mol. Life Sci.* 69: 3187–3206.

Liu, T., Liu, S., Guan, H. et al. (2009). Transcriptional profiling of *Arabidopsis* seedlings in response to heavy metal lead (Pb). *Environ. Exp. Bot.* 67: 377–386.

Liu, X., Wu, H., Ji, C. et al. (2013). An integrated proteomic and metabolomic study on the chronic effects of mercury in *Suaeda salsa* under an environmentally relevant salinity. *PLoS One* 8: e64041.

Liu, X., Shen, X., Lai, Y. et al. (2016). Toxicological proteomic responses of halophyte *Suaeda salsa* to lead and zinc. *Ecotoxicol. Environ. Saf.* 134: 163–171.

Lukens, L.N. and Zhan, S. (2007). The plant genome's methylation status and response to stress: implications for plant improvement. *Curr. Opin. Plant Biol.* 10: 317–322.

Lyu, G., Li, D., Li, S. et al. (2020). Genotoxic effects and proteomic analysis on *Allium cepa* var. agrogarum L. root cells under Pb stress. *Ecotoxicology* 29: 959–972.

Ma, Y., Egodawatta, P., McGree, J. et al. (2016). Human health risk assessment of heavy metals in urban stormwater. *Sci. Total Environ.* 557: 764–772.

Macovei, A., Vaid, N., Tula, S., and Tuteja, N. (2012). A new DEAD-box helicase ATP-binding protein (OsABP) from rice is responsive to abiotic stress. *Plant Signal. Behav.* 7: 1138–1143.

Maestri, E., Marmiroli, M., Visioli, G., and Marmiroli, N. (2010). Metal tolerance and hyperaccumulation: costs and trade-offs between traits and environment. *Environ. Exp. Bot.* 68: 1–13.

Mahdavian, K., Ghaderian, S.M., and Schat, H. (2016). Pb accumulation, Pb tolerance, antioxidants, thiols, and organic acids in metallicolous and non-metallicolous *Peganum harmala* L. under Pb exposure. *Environ. Exp. Bot.* 126: 21–31.

Malik, B. and Pirzadah, T.B. (2020). Lead induced changes in biomarkers and proteome map of Chicory (*Cichorium intybus* L.). *Curr. Bot.* 11: 111–116.

Małkowski, E., Sitko, K., Zieleźnik-Rusinowska, P. et al. (2019). Heavy metal toxicity: physiological implications of metal toxicity in plants. In: *Plant Metallomics and Functional Omics* (ed. G. Sablok), 253–301. Cham: Springer.

Maruyama-Nakashita, A., Inoue, E., Watanabe-Takahashi, A. et al. (2003). Transcriptome profiling of sulfur-responsive genes in *Arabidopsis* reveals global effects of sulfur nutrition on multiple metabolic pathways. *Plant Physiol.* 132: 597–605.

McBride, M.B. (1994). *Environmental Chemistry of Soils*. New York: Oxford Press.

Meng, L., Yang, Y., Ma, Z. et al. (2022). Integrated physiological, transcriptomic and metabolomic analysis of the response of *Trifolium pratense* L. to Pb toxicity. *J. Hazard. Mater.* 436: 129128.

Mishra, S., Srivastava, S., Tripathi, R.D. et al. (2006). Lead detoxification by coontail (*Ceratophyllum demersum* L.) involves induction of phytochelatins and antioxidant system in response to its accumulation. *Chemosphere* 65: 1027–1039.

Mishra, B., Sangwan, R.S., Mishra, S. et al. (2014). Effect of cadmium stress on inductive enzymatic and nonenzymatic responses of ROS and sugar metabolism in multiple shoot cultures of Ashwagandha (*Withania somnifera* Dunal). *Protoplasma* 251: 1031–1045.

Mohr, S., Stryker, J.M., and Lambowitz, A.M. (2002). A DEAD-box protein functions as an ATP-dependent RNA chaperone in group I intron splicing. *Cell* 109: 769–779.

Morel, M., Crouzet, J., Gravot, A. et al. (2009). AtHMA3, a P1B-ATPase allowing Cd/Zn/co/Pb vacuolar storage in Arabidopsis. *Plant Physiol.* 149: 894–904.

Muller, B., Pantin, F., Génard, M. et al. (2011). Water deficits uncouple growth from photosynthesis, increase C content, and modify the relationships between C and growth in sink organs. *J. Exp. Bot.* 62: 1715–1729.

Navari-Izzo, F., Cestone, B., Cavallini, A. et al. (2006). Copper excess triggers phospholipase D activity in wheat roots. *Phytochemistry* 67: 1232–1242.

Nield, J., Orlova, E.V., Morris, E.P. et al. (2000). 3D map of the plant photosystem II supercomplex obtained by cryoelectron microscopy and single particle analysis. *Nat. Struct. Biol.* 7: 44–47.

Nishikawa, K., Onodera, A., and Tominaga, N. (2006). Phytochelatins do not correlate with the level of Cd accumulation in *Chlamydomonas* spp. *Chemosphere* 63: 1553–1559.

Norton, G.J., Deacon, C.M., Xiong, L. et al. (2010). Genetic mapping of the rice ionome in leaves and grain: identification of QTLs for 17 elements including arsenic, cadmium, iron and selenium. *Plant Soil* 329: 139–153.

Obiora, S.C., Chukwu, A., Toteu, S.F., and Davies, T.C. (2016). Assessment of heavy metal contamination in soils around lead (Pb)-zinc (Zn) mining areas in *Enyigba, southeastern* Nigeria. *J. Geol. Soc. India* 87 (4): 453–462.

Opdenakker, K., Remans, T., Vangronsveld, J., and Cuypers, A. (2012). Mitogen-activated protein (MAP) kinases in plant metal stress: regulation and responses in comparison to other biotic and abiotic stresses. *Int. J. Mol. Sci.* 13: 7828–7853.

Ortega, A., Garrido, I., Casimiro, I., and Espinosa, F. (2017). Effects of antimony on redox activities and antioxidant defence systems in sunflower (*Helianthus annuus* L.) plants. *PLoS One* 12: e0183991.

Pawlak-Sprada, S., Arasimowicz-Jelonek, M., Podgórska, M., and Deckert, J. (2011a). Activation of phenylpropanoid pathway in legume plants exposed to heavy metals. Part I. Effects of cadmium and lead on phenylalanine ammonia-lyase gene expression, enzyme activity and lignin content. *Acta Biochim. Pol.* 58: 211–216.

Pawlak-Sprada, S., Stobiecki, M., and Deckert, J. (2011b). Activation of phenylpropanoid pathway in legume plants exposed to heavy metals. Part II. Profiling of isoflavonoids and their glycoconjugates induced in roots of lupine (*Lupinus luteus*) seedlings treated with cadmium and lead. *Acta Biochim. Pol.* 58 (2): 217–223.

Pelletier, M.K. and Shirley, B.W. (1996). Analysis of flavanone 3-hydroxylase in *Arabidopsis* seedlings (coordinate regulation with chalcone synthase and chalcone isomerase). *Plant Physiol.* 111: 339–345.

Perrone, A. and Martinelli, F. (2020). Plant stress biology in epigenomic era. *Plant Sci.* 294: 110376.

Peters, K., Worrich, A., Weinhold, A. et al. (2018). Current challenges in plant eco-metabolomics. *Int. J. Mol. Sci.* 19: 1385.

Pidatala, V.R., Li, K., Sarkar, D. et al. (2016). Identification of biochemical pathways associated with lead tolerance and detoxification in *Chrysopogon zizanioides* L. Nash (Vetiver) by metabolic profiling. *Environ. Sci. Technol.* 50: 2530–2537.

Pidatala, V.R., Li, K., Sarkar, D. et al. (2018). Comparative metabolic profiling of vetiver (*Chrysopogon zizanioides*) and maize (*Zea mays*) under lead stress. *Chemosphere* 193: 903–911.

Piechalak, A., Tomaszewska, B., Baralkiewicz, D., and Malecka, A. (2002). Accumulation and detoxification of lead ions in legumes. *Phytochemistry* 60: 153–162.

Pilon-Smits, E. (2005). Phytoremediation. *Annu. Rev. Plant Biol.* 56: 15–39.

Podazza, G., Arias, M., and Prado, F.E. (2012). Cadmium accumulation and strategies to avoid its toxicity in roots of the citrus rootstock Citrumelo. *J. Hazard. Mater.* 215: 83–89.

Pourrut, B., Perchet, G., Silvestre, J. et al. (2008). Potential role of NADPH-oxidase in early steps of lead-induced oxidative burst in *Vicia faba* roots. *J. Plant Physiol.* 165: 571–579.

Pourrut, B., Shahid, M., Dumat, C. et al. (2011). Lead uptake, toxicity, and detoxification in plants. *Rev. Environ. Contam. Toxicol.* 213: 113–136.

Putri, S.P., Yamamoto, S., Tsugawa, H., and Fukusaki, E. (2013). Current metabolomics: technological advances. *J. Biosci. Bioeng.* 116: 9–16.

Quintana, G.C., Rocha, G.J., Gonçalves, A.R., and Velásquez, J.A. (2008). Evaluation of heavy metal removal by oxidized lignins in acid media from various sources. *BioResources* 3: 1092–1102.

Rai, V.K. (2002). Role of amino acids in plant responses to stresses. *Biol. Plant.* 45: 481–487.

Reddy, A.M., Kumar, S.G., Jyothsnakumari, G. et al. (2005). Lead induced changes in antioxidant metabolism of horsegram (*Macrotyloma uniflorum* (Lam.) Verdc.) and bengalgram (*Cicer arietinum* L.). *Chemosphere* 60: 97–104.

Reeves, R.D., Baker, A.J., Jaffré, T. et al. (2018). A global database for plants that hyperaccumulate metal and metalloid trace elements. *New Phytol.* 218: 407–411.

Reis, G.S.M., de Almeida, A.A.F., de Almeida, N.M. et al. (2015). Molecular, biochemical and ultrastructural changes induced by Pb toxicity in seedlings of *Theobroma cacao* L. *PLoS One* 10: e0129696.

Rodriguez, E., da Conceição Santos, M., Azevedo, R. et al. (2015). Photosynthesis light-independent reactions are sensitive biomarkers to monitor lead phytotoxicity in a Pb-tolerant *Pisum sativum* cultivar. *Environ. Sci. Pollut. Res.* 22: 574–585.

Romeo, S., Trupiano, D., Ariani, A. et al. (2014). Proteomic analysis of *Populus*× *euramericana* (clone I-214) roots to identify key factors involved in zinc stress response. *J. Plant Physiol.* 171: 1054–1063.

Rosa, M., Prado, C., Podazza, G. et al. (2009). Soluble sugars: metabolism, sensing and abiotic stress: a complex network in the life of plants. *Plant Signal. Behav.* 4: 388–393.

Sahoo, J.P., Behera, L., Sharma, S.S. et al. (2020). Omics studies and systems biology perspective towards abiotic stress response in plants. *Am. J. Plant Sci.* 11: 2172.

Saito, K., Yokoyama, H., Noji, M., and Murakoshi, I. (1995). Molecular cloning and characterization of a plant serine acetyltransferase playing a regulatory role in cysteine biosynthesis from watermelon. *J. Biol. Chem.* 270: 16321–16326.

Sazuka, T., Keta, S., Shiratake, K. et al. (2004). A proteomic approach to identification of transmembrane proteins and membrane-anchored proteins of *Arabidopsis thaliana* by peptide sequencing. *DNA Res.* 11: 101–113.

Sengar, R.S., Gautam, M., Sengar, R.S. et al. (2008). Lead stress effects on physiobiochemical activities of higher plants. *Rev. Environ. Contam. Toxicol.* 196: 73–93.

Shafiq, S., Zeb, Q., Ali, A. et al. (2019). Lead, cadmium and zinc phytotoxicity alter DNA methylation levels to confer heavy metal tolerance in wheat. *Int. J. Mol. Sci.* 20: 4676.

Shafiq, S., Ali, A., Sajjad, Y. et al. (2020). The interplay between toxic and essential metals for their uptake and translocation is likely governed by DNA methylation and histone deacetylation in maize. *Int. J. Mol. Sci.* 21: 6959.

Sharma, S.S. and Dietz, K.J. (2006). The significance of amino acids and amino acid-derived molecules in plant responses and adaptation to heavy metal stress. *J. Exp. Bot.* 57: 711–726.

Sharma, P. and Dubey, R.S. (2005). Lead toxicity in plants. *Braz. J. Plant Physiol.* 17: 35–52.

Sheen, J. (1996). Ca^{2+}-dependent protein kinases and stress signal transduction in plants. *Science* 274: 1900–1902.

Shen, Y., Zhang, Y., Chen, J. et al. (2013). Genome expression profile analysis reveals important transcripts in maize roots responding to the stress of heavy metal Pb. *Physiol. Plant.* 147: 270–282.

Shen, C.C., Chen, M.X., Xiao, T. et al. (2021). Global proteome response to Pb(II) toxicity in poplar using SWATH-MS-based quantitative proteomics investigation. *Ecotoxicol. Environ. Saf.* 220: 112410.

Shi, S., Li, S., Asim, M. et al. (2018). The *Arabidopsis* calcium-dependent protein kinases (CDPKs) and their roles in plant growth regulation and abiotic stress responses. *Int. J. Mol. Sci.* 19: 1900.

Simiele, M., Lebrun, M., Del Cioppo, G. et al. (2020). Evaluation of different amendment combinations associated with *Trifolium repens* to stabilize Pb and As in a mine-contaminated soil. *Water Air Soil Pollut.* 231: 1–15.

Simiele, M., Sferra, G., Lebrun, M. et al. (2021). In-depth study to decipher mechanisms underlying *Arabidopsis thaliana* tolerance to metal (loid) soil contamination in association with biochar and/or bacteria. *Environ. Exp. Bot.* 182: 104335.

Singh, R., Tripathi, R.D., Dwivedi, S. et al. (2010). Lead bioaccumulation potential of an aquatic macrophyte *Najas indica* are related to antioxidant system. *Bioresour. Technol.* 101: 3025–3032.

Smeets, K., Opdenakker, K., Remans, T. et al. (2013). The role of the kinase OXI1 in cadmium- and copper-induced molecular responses in *Arabidopsis thaliana*. *Plant Cell Environ.* 36: 1228–1238.

Song, W.Y., Ju Sohn, E., Martinoia, E. et al. (2003). Engineering tolerance and accumulation of lead and cadmium in transgenic plants. *Nat. Biotech.* 21: 914–919.

Sun, Q., Ye, Z.H., Wang, X.R., and Wong, M.H. (2005). Increase of glutathione in mine population of *Sedum alfredii*: a Zn hyperaccumulator and Pb accumulator. *Phytochem* 66:2549–2556.

Sunkar, R., Kapoor, A., and Zhu, J.K. (2006). Posttranscriptional induction of two Cu/Zn superoxide dismutase genes in *Arabidopsis* is mediated by downregulation of miR398 and important for oxidative stress tolerance. *Plant Cell* 18: 2051–2065.

Szuba, A., Marczak, Ł., and andKozłowski, R. (2020). Role of the proteome in providing phenotypic stability in control and ectomycorrhizal poplar plants exposed to chronic mild Pb stress. *Environ. Pollut.* 264: 114585.

Takahashi, F., Mizoguchi, T., Yoshida, R. et al. (2011). Calmodulin-dependent activation of MAP kinase for ROS homeostasis in *Arabidopsis*. *Mol. Cell* 41: 649–660.

Tang, M., Xu, L., Wang, Y. et al. (2019). Genome-wide characterization and evolutionary analysis of heat shock transcription factors (HSFs) to reveal their potential role under abiotic stresses in radish (*Raphanus sativus* L.). *BMC Genom.* 20: 1–13.

Tong, Y.P., Kneer, R., and Zhu, Y.G. (2004). Vacuolar compartmentalization: a second-generation approach to engineering plants for phytoremediation. *Trends Plant Sci.* 9: 7–9.

Turchi, A., Tamantini, I., Camussi, A.M., and Racchi, M.L. (2012). Expression of a metallothionein A1 gene of *Pisum sativum* in white poplar enhances tolerance and accumulation of zinc and copper. *Plant Sci.* 183: 50–56.

Urano, K., Kurihara, Y., Seki, M., and Shinozaki, K. (2010). 'Omics' analyses of regulatory networks in plant abiotic stress responses. *Curr. Opin. Plant Biol.* 13: 132–138.

Uzu, G., Sobanska, S., Aliouane, Y. et al. (2009). Study of lead phytoavailability for atmospheric industrial micronic and sub-micronic particles in relation with lead speciation. *Environ. Pollut.* 157: 1178–1185.

Van den Ende, W. and Valluru, R. (2009). Sucrose, sucrosyl oligosaccharides, and oxidative stress: scavenging and salvaging? *J. Exp. Bot.* 60: 9–18.

Verbruggen, N., Hermans, C., and Schat, H. (2009). Molecular mechanisms of metal hyperaccumulation in plants. *New Phytol.* 181: 759–776.

Verret, F., Gravot, A., Auroy, P. et al. (2004). Overexpression of AtHMA4 enhances root-to-shoot translocation of zinc and cadmium and plant metal tolerance. *FEBS Lett.* 576: 306–312.

Visvalingam, S., Hønsi, T.G., and Bones, A.M. (1998). Sulphate and micronutrients can modulate the expression levels of myrosinases in *Sinapis alba* plants. *Physiol. Plant.* 104: 30–37.

Walliwalagedara, C., Atkinson, I., van Keulen, H. et al. (2010). Differential expression of proteins induced by lead in the Dwarf Sunflower *Helianthus annuus*. *Phytochemistry* 71: 1460–1465.

Wang, H.H., Shan, X.Q., Wen, B. et al. (2007). Effect of indole-3-acetic acid on lead accumulation in maize (*Zea mays* L.) seedlings and the relevant antioxidant response. *Environ. Exp. Bot.* 61: 246–253.

Wang, P., Zhang, S., Wang, C., and Lu, J. (2012). Effects of Pb on the oxidative stress and antioxidant response in a Pb bioaccumulator plant *Vallisneria natans*. *Ecotoxicol. Environ. Saf.* 78: 28–34.

Wang, Y., Xu, L., Chen, Y. et al. (2013). Transcriptome profiling of radish (*Raphanus sativus* L.) root and identification of genes involved in response to lead (Pb) stress with next generation sequencing. *PLoS One* 8: e66539.

Wang, L., Yang, H., Liu, R., and Fan, G. (2015a). Detoxification strategies and regulation of oxygen production and flowering of *Platanus acerifolia* under lead (Pb) stress by transcriptome analysis. *Environ. Sci. Pollut. Res.* 22: 12747–12758.

Wang, Y., Zhao, Z., Deng, M. et al. (2015b). Identification and functional analysis of microRNAs and their targets in *Platanus acerifolia* under lead (Pb) stress. *Int. J. Mol. Sci.* 16: 7098–7111.

Wang, Y., Xu, L., Shen, H. et al. (2015c). Metabolomic analysis with GC-MS to reveal potential metabolites and biological pathways involved in Pb and Cd stress response of radish roots. *Sci. Rep.* 5: 1–13.

Wang, Y., Xu, L., Tang, M. et al. (2016). Functional and integrative analysis of the proteomic profile of radish root under Pb exposure. *Front. Plant Sci.* 7: 1871.

Wojas, S., Ruszczyńska, A., Bulska, E. et al. (2007). Ca^{2+}-dependent plant response to Pb^{2+} is regulated by LCT1. *Environ. Pollut.* 147: 584–592.

Wu, H., Liu, X., Zhao, J. et al. (2012). Toxicological effects of environmentally relevant lead and zinc in halophyte *Suaeda salsa* by NMR-based metabolomics. *Ecotoxicology* 21: 2363–2371.

Xia, X.J., Zhou, Y.H., Shi, K. et al. (2015). Interplay between reactive oxygen species and hormones in the control of plant development and stress tolerance. *J. Exp. Bot.* 66: 2839–2856.

Xia, C., Hong, L., Yang, Y. et al. (2019). Protein changes in response to lead stress of lead-tolerant and lead-sensitive industrial hemp using SWATH technology. *Genes* 10: 396.

Xiao, S., Gao, W., Chen, Q.F. et al. (2008). Overexpression of membrane-associated acyl-CoA-binding protein ACBP1 enhances lead tolerance in *Arabidopsis*. *Plant J.* 54: 141–151.

Xu, Y., Zhou, G., Zhou, L. et al. (2007). Expression patterns of the rice class I metallothionein gene family in response to lead stress in rice seedlings and functional complementation of its members in lead-sensitive yeast cells. *Chin. Sci. Bull.* 52: 2203–2209.

Xu, B., Wang, Y., Zhang, S. et al. (2017). Transcriptomic and physiological analyses of *Medicago sativa* L. roots in response to lead stress. *PLoS One* 12: e0175307.

Yadav, S.K. (2010). Heavy metals toxicity in plants: an overview on the role of glutathione and phytochelatins in heavy metal stress tolerance of plants. *S. Afr. J. Bot.* 76: 167–179.

Yakimova, E.T., Kapchina-Toteva, V.M., and Woltering, E.J. (2007). Signal transduction events in aluminum-induced cell death in tomato suspension cells. *J. Plant Physiol.* 164: 702–708.

Yan, X., Yu, D., Wang, H., and Wang, J. (2006). Response of submerged plant (*Vallisneria spinulosa*) clones to lead stress in the heterogenous soil. *Chemosphere* 63: 1459–1465.

Yan, H., Zou, D., Zhou, G. et al. (2021). Metabolomics of ginger based on ultra-performance liquid chromatography coupled with quadrupole time-of-flight mass spectrometry technology. *Food Qual. Saf.* 5: fyaa036.

Yang, Y.Y., Jung, J.Y., Song, W.Y. et al. (2000). Identification of rice varieties with high tolerance or sensitivity to lead and characterization of the mechanism of tolerance. *Plant Physiol.* 124: 1019–1026.

Yang, J.L., Liu, L.W., Gong, Y.Q. et al. (2007). Analysis of genomic DNA methylation level in radish under cadmium stress by methylation-sensitive amplified polymorphism technique. *J. Plant Physiol. Mol. Biol.* 33: 219–226.

Yusuf, D., Butland, S.L., Swanson, M.I. et al. (2012). The transcription factor encyclopedia. *Genome Biol.* 13: 1–25.

Zafari, S., Sharifi, M., Chashmi, N.A., and Mur, L.A. (2016). Modulation of Pb-induced stress in Prosopis shoots through an interconnected network of signaling molecules, phenolic compounds and amino acids. *Plant Physiol. Biochem.* 99: 11–20.

Zeng, H., Xu, L., Singh, A. et al. (2015). Involvement of calmodulin and calmodulin-like proteins in plant responses to abiotic stresses. *Front. Plant Sci.* 6: 600.

Zhang, Z., Gao, X., and andQiu, B. (2008). Detection of phytochelatins in the hyperaccumulator *Sedum alfredii* exposed to cadmium and lead. *Phytochemistry* 69: 911–918.

Zhang, Y., Ge, F., Hou, F. et al. (2017). Transcription factors responding to Pb stress in maize. *Genes* 8: 231.

Zhang, X., Yang, H., Schaufelberger, M. et al. (2020). Role of flavonol synthesized by nucleus FLS1 in *Arabidopsis* resistance to Pb stress. *J. Agric. Food Chem.* 68: 9646–9653.

Zhang, L., Xiang, Y., Chen, S. et al. (2022). Mechanisms of MicroRNA biogenesis and stability control in plants. *Front. Plant Sci.* 13: 844149.

Zhao, X., Lin, H., Zhang, Z. et al. (2014). Mapping of QTLs controlling Pb($^{2+}$) content in maize kernels under Pb($^{2+}$) stress. *Hereditas* 36: 821–826.

Zhao, X., Liu, Y., Wu, W. et al. (2018a). Genome-wide association analysis of lead accumulation in maize. *Mol. Genet. Genom.* 293: 615–622.

Zhao, X., Cao, Y., Li, Y. et al. (2018b). Identification of QTL and candidate genes for Pb accumulation in maize at maturity stage. *Maydica* 61: 9.

Zhu, F.Y., Li, L., Lam, P.Y. et al. (2013). Sorghum extracellular leucine-rich repeat protein SbLRR2 mediates lead tolerance in transgenic *Arabidopsis*. *Plant Cell Physiol.* 54: 1549–1559.

Zhu, F.Y., Chan, W.L., Chen, M.X. et al. (2016). SWATH-MS quantitative proteomic investigation reveals a role of jasmonic acid during lead response in *Arabidopsis*. *J. Proteome Res.* 15: 3528–3539.

Zimeri, A.M., Dhankher, O.P., McCaig, B., and Meagher, R.B. (2005). The plant MT1 metallothioneins are stabilized by binding cadmiums and are required for cadmium tolerance and accumulation. *Plant Mol.Biol.* 58: 839–855.

Zulfiqar, U., Farooq, M., Hussain, S. et al. (2019). Lead toxicity in plants: impacts and remediation. *J. Environ. Manage.* 250: 109557.

19

Interaction of Heavy Metal with Drought/Salinity Stress in Plants

Ziqian Li[1], Wentao Chen[1], Qianlong Tan[1], Yuanyuan Hou[1], Taimoor Hassan Farooq[2], Baber Iqbal[3], and Yong Li[1]

[1] *Life and Science Department, National Engineering Laboratory for Applied Technology of Forestry & Ecology in South China, Laboratory of Urban Forest Ecology of Hunan Province, College of Life Science and Technology, Central South University of Forestry and Technology, Changsha, Hunan, China*
[2] *Bangor College China, A Joint Unit of Bangor University and Central South University of Forestry and Technology, Changsha, Hunan, China*
[3] *School of Environment and Safety Engineering, Jiangsu University, Zhenjiang, China*

19.1 Introduction

With the continuous development of industry, heavy metal pollution has become the focus of environmental biology (Duruibe et al. 2007; Yang et al. 2018). When the heavy metal ions in the soil exceed their self-cleaning effect on heavy metal ions, they can affect plant growth and metabolism (Bradl 2004). Human activities such as vehicle emissions, sewage irrigation, pesticides, herbicides, and fertilizer use are sources of heavy metals (Shukla and Jain 2022). Heavy metals enter the environment to participate in the water-soil-biological system cycle. They can accumulate in the roots, stems, leaves, and seeds of plants, and pose a great threat to the health of organisms and humans throughout the food chain. Certain metals such as zinc (Zn), copper (Cu), and nickel (Ni) are essential micronutrients for plants as cofactors for various enzymes (Ghori et al. 2019; Healoui et al. 2020; Nasiri et al. 2010). In addition, other metals such as cadmium (Cd) and lead (Pb) in pesticides do not have any beneficial effects (Rusin et al. 2021). When the concentration exceeds a certain level, the plants become toxic, resulting in slow growth, short stature plants, reduced yield, inhibited root elongation, and even death in severe cases (Haider et al. 2021; Zahra et al. 2021; Zulfiqar et al. 2019).

The toxic effects of heavy metals on plants are different due to different kinds of heavy metals. The plant cell membrane system (including tonoplast, plasma membrane, and organelle membrane) is the interface and barrier for material exchange and information transmission between plant cells and the external environment (Schindler 1989). Its stability is the basis for the normal physiological functions of the cells. Heavy metal stress can

Heavy Metal Toxicity and Tolerance in Plants: A Biological, Omics, and Genetic Engineering Approach, First Edition. Edited by Mohammad Anwar Hossain, AKM Zakir Hossain, Sylvain Bourgerie, Masayuki Fujita, Om Parkash Dhankher, and Parvez Haris.

damage plant cell membrane permeability (Feng et al. 2021; O'Lexy et al. 2018). In addition, heavy metals inhibited plant photosynthesis by affecting electron transport and damaging chloroplast integrity and showed significant concentration and time effects (Clijsters and Assche 1985; Emamverdian and Ding 2018; Küpper et al. 2003; Wang et al. 2013). Plant respiration is disrupted by heavy metal stress, and the energy available for normal life activities is reduced. Some energy is transferred to the adaptation process of heavy metal stress, such as the repair of damage and the formation of heavy metal complexes, which leads to the inhibition of plant growth and development (Hussain et al. 2017). Heavy metal stress can lead to inactivation, denaturation, destruction of enzyme activity, imbalance of carbohydrate synthesis metabolism, and nitrogen metabolism (Sharma et al. 2020). Plants, like other organisms, are frequently sensitive to the availability and scarcity of certain heavy metal ions as essential micronutrients. Studies are being conducted worldwide to determine the effects of heavy metal toxicities on plants. Thus, heavy metals have become an important ecological and environmental problem due to their potential adverse ecological effects leading to a decline in soil quality.

With global warming, drought and salt stress are widespread and intensifying (Trenberth et al. 2013). Drought and salt are important environmental factors affecting plant survival and growth (Liu et al. 2016). The effects of drought stress on plants are reflected in physiological structures such as organs, tissues, and cell functions (Liu et al. 2016). According to drought causes, drought stress is divided into atmospheric drought, soil drought, and physiological drought (Salehi-Lisar and Bakhshayeshan-Agdam 2016). Drought in the atmosphere and soil is caused by a lack of atmospheric and soil moisture, which is directly caused by plant root water consumption being greater than water absorption, resulting in a water deficit in plants, and inhibiting normal growth and metabolism. Physiological drought is due to external environmental factors, such as soil salinization, low temperature, and excessive fertilizer. Drought causes root water loss to exceed water uptake, decreases cell water potential and turgor pressure, and leads to leaf wilting flower and fruit drop and death (Farooq et al. 2009; Steudle 2000). In addition, protoplasts and cell walls are mechanically damaged, biofilm systems are damaged and, membrane permeability changes, stomatal opening decreases or closes, resulting in a decrease in photosynthetic rate (Küpper et al. 2003; Wang et al. 2013).

Soil salt stress will increase the osmotic pressure inside the soil, and the soil permeability and water permeability will decrease to varying degrees (Nedjimi 2013). Plants cannot grow normally, causing death and community degradation under salt stress (Singh et al. 2015). Research on the harms of salt stress to plants and vegetation's physiological and biochemical changes in response to stress has become the focus of current research. Damages caused by salt stress to plants are divided into primary salt damage caused by osmotic stress and secondary salt damage caused by nutrient deficiency caused by ion stress (Yildiz et al. 2020). Primary salt damage is divided into direct salt damage and indirect salt damage. Direct salt damage is the change of plasma membrane permeability caused by ion toxicity caused by an unbalanced soil solution. Indirect salt damage is the disorder of plant physiological metabolism caused by membrane lipid peroxidation induced by oxidative stress *in vivo*. Secondary salt damage, on the one hand, due to osmotic stress, leads to plant physiological drought; on the other hand, due to lack of nutrients, it affects plant growth and development. Under salt stress, the root system got first injured by osmotic stress, and then a series of physiological and biochemical reactions occurred, which affected the growth and development of aboveground parts (Yang et al. 2010). Relative growth and photosynthetic parameters were the comprehensive performance of plants to salt stress (Redondo-Gómez et al. 2007). There are many reports on the effects of drought/salinity and heavy metals on plants. This chapter discusses the effects of drought/salinity and heavy metal stress on plants to provide more favorable environmental conditions for plant growth in practical work.

19.2 Plant Physiology and Biochemistry

Plants face the stress that is nonunique in the natural environment. When plants are subjected to multiple stresses, the growth and development of plants will be more limited (Iqbal et al. 2022). Studies have shown that when plants are under multiple stresses, they show a common response to individual and interactive stresses (Hamed et al. 2013). This section summarizes the effects of the interaction between three typical heavy metals and drought stress on plant physiology and biochemistry.

19.2.1 Zinc (Zn)

Zn is an essential micronutrient for plants and is involved in a variety of physiological functions, including protein and carbohydrate synthesis (Mousavi et al. 2013). Seed germination is the first stage and is considered the basis for plant growth and productivity mainly affected by drought stress. In crop management, changes in emergence rates caused by environmental factors are crucial. Therefore, seed germination and seedling emergence highly affect plant final productivity. Moreover, the interaction between environmental factors and the internal mechanisms of seeds determines the level of seed germination and emergence under certain conditions. Studies have found that zinc can increase the germination rate and yield of maize, wheat, and chickpeas (Potarzycki and Grzebisz 2009; Reis et al. 2018; Seyedi et al. 2012). In addition, drought can inhibit crop growth by fluctuating the physiological mechanism inside the crop body. Studies have found that zinc application can improve the adaptability of crops to drought. Cakmak (2008) applied Zn under drought conditions increasing the synthesis of IAA and GA3, and increasing the length and weight of the embryo under drought stress. Paul et al. (2016) spraying Zn on wheat leaves under drought conditions increased wheat yield. In addition, drought can lead to a decrease in the ability of plants to absorb zinc, which limits root growth (Gupta et al. 2020). Hajiboland and Amirazad (2010) studied the effects of zinc deficiency on red cabbage plants under normal water supply or drought conditions. Compared with the normal water supply conditions, zinc deficiency under drought conditions caused greater damage to plant growth and led to a sharp decline in Zn content (Hajiboland and Amirazad 2010). Adequate zinc nutrition improves alfalfa growth and drought tolerance by promoting root growth and root water content (Grewal and Williams 2000). Maintaining good leaf water status is the result of protecting cells from being properly controlled under different conditions, thereby promoting growth and improving the viability of errors under stress conditions (Steudle 2000). From this point of view, the application of zinc can improve the adaptability of crops to drought.

Zinc deficiency is now considered to be one of the most serious micronutrient deficiencies for plants growing in high-pH calcareous, saline-alkaline, and acidic soils. Alpaslan et al. (1999) and Shahriaripour et al. (2010) found that zinc application could alleviate the salt toxicity of pistachio and tomato seedlings under NaCl stress by inhibiting the absorption or transport of Na and Cl in salt-affected areas. It was reported that a similar phenomenon was found in rice and barley crops. The application of Zn inhibited Na transport in plants and promoted plant growth (Iqbal et al. 2000; Muhammad et al. 2020).

19.2.2 Cadmium (Cd)

Cadmium is an extremely harmful pollutant that enters the environment through natural and human activities. Domestic sewage, atmospheric deposition, and industrial discharge are important sources of cadmium (Benavides et al. 2005). Of all the nonessential heavy metals, Cd is of greatest concern because it is a highly toxic element to humans and can be transferred to humans through plant and animal consumption. Cadmium also has a high solubility in water and high mobility in soil–plant systems (Pinto et al. 2004). Due to its high solubility, it is easily absorbed through the roots of the plant and then transferred to the twigs (Das et al. 1997). Moreover, cadmium also interacts with the microplastics present in the soil by diminishing the growth and yield of the crop (Iqbal et al. 2022). Cadmium is generally 2–20 times more toxic than similar concentrations of other heavy metals (Maria et al. 2013; Shah and Dubey 1995). Cadmium can affect many morphological, physiological, and biochemical changes in plants (Azevedo et al. 2005; Hammami et al. 2004; Turgut et al. 2005). In addition, when plants were stressed by the interaction of Cd and drought, their responses were different from those under single stress and closely related to plant varieties and developmental stages. El-Mahdy et al. (2021) found that the combined stress of Cd and drought inhibited the growth of bananas (*Musa acuminata* L.) and detected yellowing and visible toxic symptoms in plant leaves (El-Mahdy et al. 2021). Bashir et al. (2019) exposed four soybean varieties (*Glycine max* L.) to Cd and drought stress. The results showed that the reduction of root length, root area, and root diameter of drought-tolerant varieties (Shenlong 20 and Liaodou 32) was not as significant as that of drought-resistant varieties (Liaodou 3 and Liaodou 10). Cadmium stress and PEG-induced drought stress significantly inhibited the germination and early seedling growth of black mung bean (*Vigna mungo* L.) under single and combined stresses (Swapna and Gopal 2015). In combined treatment, the early seedling growth stage was more sensitive than the germination stage (Swapna and Gopal 2015). Drought and Cd stress reduced the growth of mung bean (*Vigna radiata* L.) seedlings because seed germination rate and shoot and root length decreased significantly with the increase of Cd concentration and drought intensity. The interaction between Cd and drought stress on dry weight and root length was also significant, but the interaction between the two stresses was less than the additive effect.

In addition, drought played different roles in Cd translocation and redistribution at different developmental stages of plants. Xia et al. (2015) found that drought reduced the root cadmium concentration in seedlings of two high cadmium accumulation varieties (Haihua 1 and Zhenghong 3). However, for mature peanut plants, drought stress increased cadmium accumulation in roots, pod walls, and seeds. Liu et al. (2016) also found the same rule.

At the seedling stage of peanuts, drought reduced the cadmium concentration in the aboveground parts of two varieties (Fenghua 1 and Silihong), while at the mature stage, drought increased the cadmium content in the aboveground parts and seeds (Liu et al. 2016). The Cd concentration in seeds was negatively correlated with the biomass of buds and pods but positively correlated with the Cd concentration in buds and pods (Liu et al. 2016). It is speculated that the increase of Cd concentration in seeds under drought stress may be due to the concentration effect caused by drought-induced plant growth decline (Liu et al. 2016). Liu et al. (2016) reported that drought stress mainly concentrated Cd in the fine roots of *Spartina alterniflora* Loisel, which may be because root endodermis absorbed and hindered the transport of Cd from roots to photosynthetic sites (Hasan et al. 2009). Under PEG-induced drought stress, the concentration and uptake of Cd in castor decreased in the middle, while increased in other plants (Liu et al. 2016; Rizwan et al. 2018). In a comparative study of two genotypes of Peruvian tomato (*Lycopersicon peruvianum* L.), Ünyayar et al. (2005) found that the accumulation of Cd in drought-tolerant lines was lower than that in drought-sensitive tomatoes. This result indicates that drought tolerance can regulate cadmium uptake due to water-limited conditions in the soil and emphasizes that PEG can reduce cadmium uptake due to water-limited conditions in the medium.

19.2.3 Aluminium (Al)

Aluminum toxicity and drought stress are two major constraints in the crop production worldwide, especially in tropical regions. With global climate change, changes in rainfall distribution and long drought periods during crop growth in most tropical areas are becoming increasingly important yield-limiting factors. Aluminum toxicity is the most important factor restricting plant growth in acid-mineral soils (Carver and Ownby 1995; Keyser and Munns 1979). A typical and most sensitive symptom of Al toxicity is rapid inhibition of root elongation, leading to root reduction and damage, limiting the absorption of nutrients and water (Delhaize and Ryan 1995; Horst et al. 1992; Kochian et al. 2004). In addition, Al toxicity in the soil exacerbates drought stress, leading to reduced root elongation, limiting plant roots to explore and absorb water into acid substratum reducing plant resistance to drought stress (Yang et al. 2013). Yang et al. (2012) studied the interaction between aluminum toxicity and drought stress on root tip growth in acidic and aluminum-toxic soils. The results showed that both aluminum and drought inhibited root growth, and the combined stress inhibited root elongation more seriously. They believe that the decrease in soil moisture during drought may increase the concentration of aluminum in soil solutions, increasing aluminum toxicity. In Scots and Norway spruce, drought stress significantly reduced root elongation rate and the number of growing roots in high aluminum soils compared to low aluminum soils (Bartsch 1987). Krizek and Foy (1988) and Krizek et al. (1988) observed that drought exacerbated aluminum toxicity in plants while increasing soil moisture reduced aluminum toxicity in barley (*Hordeum vulgare* L.) and sunflower (*Helianthus annuus* L.). Drought-induced low Al accumulation in plants was reported in Norway spruce seedlings, evergreen Mediterranean oak leaves, and common beans (Sardans and Peñuelas 2007; Slugeňová et al. 2011; Yang et al. 2010). Yang et al. (2010) used polyethylene glycol (PEG 6000) to simulate drought stress in hydroponics and found that PEG 6000 strongly prevented the accumulation of aluminum in root tips, thereby reducing the toxic inhibition of aluminum on root elongation of common bean. It is speculated that the removal of aluminum is mainly caused by the change of CW porosity caused by PEG 6000-induced dehydration of root apoplast (Yang et al. 2010).

19.2.4 Other Metals

Cr is an essential trace element for plants, but excessive Cr can also damage plants and inhibit their growth and development (Ali et al. 2015; Bukhari et al. 2016a, b). Biomass loss in plant growth is the most prominent effect of drought and heavy metals on plants (Osakabe et al. 2014; Silva et al. 2012; Tauqeer et al. 2016). Naz et al. (2021) found that the combined effects of drought and Cr stress resulted in dwarfism and reduced biomass in maize, and the effects of the two stresses were additive. Cr toxicity leads to decreases in maize growth and biomass, chlorophyll, and carotenoids (Habiba et al. 2019). It also causes a decrease in nutrient uptake, which ultimately leads to a decrease in plant biomass (Tauqeer et al. 2016). The harmful effects of drought and chromium toxicity may be related to increased chromium uptake or plant inability to obtain essential nutrients (Singh et al. 2015). Mercury (Hg) is ubiquitous and occurs in trace amounts throughout the lithosphere, hydrosphere, atmosphere and biosphere, and igneous rocks (Azevedo and Rodriguez 2012; Mercury 1970). Hg is a dangerous pollutant that can be rapidly absorbed and accumulated in plants and animals and spread rapidly through its consumption in ecosystems, with toxic effects on humans. Hg has higher phytotoxicity than other heavy metals (Gautam et al. 2010; Munzuroglu and Geckil 2002). Different physiological processes, such as plant water uptake, transpiration, chlorophyll synthesis, and photosynthesis, were observed to decrease significantly after mercury exposure. Both forms of mercury (organic/inorganic composite) were found to cause loss of other elements such as Fe, Mn, Mg, and K (Bonnet et al. 2000). Lead (Pb) is the main environmental heavy metal in the soil. The agency for toxic substances and disease registry lists it as the second of all harmful substances. Lead is naturally found on the surface of the earth, from where it is released into the environment. Some human activities also lead to the distribution of Pb in the environment. Lead may exist in the form of free metal ions and may also form complexes with inorganic components (such as Cl_2, SO_4^{2-}, CO_3^{2-}, and HCO_3^-). It can also be found in soils as organic ligands such as amino acids, fulvic acids, and humic acids (Sammut et al. 2010; Tabelin and Igarashi 2009; Uzu et al. 2009; Vega et al. 2010). Since the twentieth century, the increase in the world's lead ore processing has significantly increased air-lead (Air-Pb) emissions and soil-lead (Soil-Pb) pollution (Bi et al. 2006a, b, 2007; Feng et al. 2004, 2006). In addition, air-Pb and soil-Pb can enter the food chain, posing a health risk to humans and animals (Canales-Pastrana and Paredes 2013; Fahr et al. 2013). Severe changes in nutrient cycling or energy flow due to air-lead and soil-lead due to increased human activity (Grantz et al. 2003). These effects may make plants more vulnerable to additional stresses such as drought, which is caused by the reduced ability of smaller roots to absorb water, possible blockage of water channels, and reduced water use efficiency (WUE) (Arduini et al. 2010; Ryser and Emerson 2007).

19.3 Photosynthesis

Photosynthesis is the most basic function of plants by triggering the conversion of light energy into chemical energy, which provides fuel for a variety of metabolic processes. Due to their sessile nature, plants are often exposed to a variety of abiotic stresses, which in turn affect the efficiency of photosynthesis (Wuana and Okieimen 2011). Heavy metal and drought/salt stress reduce plant photosynthetic efficiency mainly by affecting pigment concentration, electron transport system, and gas exchange mechanism (Xu et al. 2020).

Chlorophyll, the main pigment in plant photosynthesis, is located in the thylakoid membrane and plays a central role in light absorption. Results showed that the combined effects of heavy metals and drought stress would lead to the decrease of chlorophyll content in red maple (Silva et al. 2012). Chromium (Cr) can reduce plant uptake of magnesium (Mg) and nitrogen (N) and affect the synthesis of chlorophyll content. Ma et al. (2017) observed that under the combined stress conditions (poly metals such as Cr and drought), the pigment content of plants decreased the most, which greatly limited plant photosynthesis. The heavy metal Cd is also one of the factors leading to chlorophyll degradation in plants (Kapoor et al. 2019). It is well known that Zinc (Zn) is an important trace element that plays a crucial role in photosynthesis (Rehman et al. 2018). However, excess Zn can lead to reduced photosynthetic pigment synthesis and impair photosynthetic machinery (Paunov et al. 2018). However, small amounts of Zn could increase wheat yield and mitigate the negative effects of drought (Farooq et al. 2015). In the combination of Cu and drought stress, thrips could tolerate drought stress well, indicating the positive effect of Cu on morphological traits and chlorophyll content under drought stress; therefore, trace Cu nutrients could improve drought stress in plants (None 1948). At higher Cu concentrations, the total chlorophyll content decreased sharply, showing a combined effect of Cu and drought stress.

The electron transport mechanism occurs in the light reaction of photosynthesis and is responsible for converting light energy into chemical energy, which is further used in the dark reaction (Hasni et al. 2015). This mechanism occurs in the thylakoid membrane of chloroplasts, where photosystem II is one of the major components. During photosynthesis, an extremely complex biochemical reaction occurs on the thylakoid membrane, among which an intermediate called oxygen evolution complex (OEC) is formed, which is highly sensitive to heavy metals and drought/salt stress (Debus 1992; Win et al. 2018). Cd has been reported to alter the donor site of photosystem II by displacing Ca in the OEC (Cristina et al. 2006). Similarly, Pb dissociates extrinsic peptides at the donor site of photosystem II and can displace chloride ions (Cl^-) and manganese (Mn), as well as Ca from the OEC, thereby hindering the oxygen release process (Rashid et al. 1991). Overall, several studies of Copper(Cu), Mercury(Hg), Ni, Zn, and aluminum (Al) have shown that these heavy metals lead to the release of three extrinsic peptides of OEC, which are indispensable peptides for photosynthesis (Sharma et al. 2020). The abnormal formation of oxygen evolution complex caused by heavy metals and water loss caused by drought/salt stress affect the damage of PS II reaction center, which affects the photosynthetic electron transport activity and reduces the photochemical efficiency, leading to the dissipation of excess light energy in the form of heat through non-photochemical pathways (Vishnupradeep et al. 2022). For example, when exposed to the combined stresses of Cr and drought, photosynthetic Fv/Fm values ranged from 0.45 to 0.71 (Gururani et al. 2013), indicating severe changes in the integrity and functionality of PSII reaction centers, resulting in decreased ΦPSII (Xiao et al. 2019). Maize is highly susceptible to Al and drought treatment stress, greatly reducing the photosynthetic rate of maize seedlings and hindering the accumulation of organic compounds, gradually leading to the cell death of maize seedlings (Muqadas et al. 2020).

The gas exchange mechanism, which generally reduces CO_2 assimilation by altering RubisCO activity and reducing mesophyll conductance, as well as other non-stomatal factors, thus increasing the limitation of stomatal regulation (stomatal closure and conduction) caused by Ci reduces CO_2 utilization by chloroplasts. Finally, the photosynthetic rate was inhibited and the vulnerability of photosystems (PS I and PS II) was increased (Federico et al. 2019). Studies have shown that transpiration rate and stomatal conductance increase after Cd exposure, and drought stress aggravates stress; therefore, the first response of all plants under heavy metal and drought/salt stress is to reduce transpiration by closing stomata and reducing the frequency of photosynthesis (Pankovic 2000). Under the interaction, not all plants will be affected. Some scholars have found that under the stress of lead and drought, male pine shows greater plasticity in photosynthetic capacity than female (Han et al. 2013). Tea plants exposed to zinc sulfate ($ZnSO_4$) and drought stress decreased the dry weight of tea leaves and increased Zn uptake in leaves, indicating that Zn could regulate drought-mediated photosynthesis in tea plants (Upadhyaya et al. 2013). In general, photosynthesis in plants can be hindered under conditions such as heavy metals and drought/salt stress, affecting the normal physiological functions of plants.

19.4 Antioxidant System

Plants that suffer from heavy metal or drought/salt stress will lead to excessive production of reactive oxygen species (ROS), such as superoxide (O^{2-}), hydroxyl (–OH), hydrogen peroxide (H_2O_2), and other very active small molecules. These small molecules can react with a large number of cellular components, oxidation of various macromolecules, such as nucleic acids, proteins, sugars, and fats, causing oxidative stress in plant cells, leading to cell membrane damage, and malondialdehyde (MDA) accumulation and cell death (Aganchich et al. 2007; Guo et al. 2007; Manchanda and Garg 2011). The activities of catalase (CAT), peroxidase (POD), and superoxide dismutase (SOD) have long been considered important physiological indicators for determining the resistance or tolerance of plants to biotic and abiotic stresses (Ippolito et al. 2011; Minutolo et al. 2012; Song et al. 2012). The accumulation of ROS is mainly cleared by the activity of antioxidant enzymes such as catalase (CAT), superoxide dismutase (SOD), and peroxidase (POD). Peroxidase plays an important role in scavenging ROS by catalyzing H_2O_2 reduction to protect plant cells from oxidative stress damage (Choudhury et al. 2013; Reddy et al. 2004). The response of plant antioxidant enzymes to heavy metal and drought combined stress is more complex than that under single stress. But all showed that appropriate interactive stress was beneficial to improve the activity of antioxidant enzymes in plants. Song et al. (2014) treated maize seedlings with Cd^{2+}, Hg^{2+}, $Pb^{2+,}$ and PEG-induced drought stress. The results showed that heavy metals and drought stress could increase the activities of CAT, POD, and SOD, and the content of MDA had a certain concentration effect on heavy metals and drought stress. Bashir et al. (2019) found that the expression of the soybean peroxidase gene (U51191.1) was upregulated under Cd and PEG-mediated drought, stress and the expression level in tolerant varieties (Liaodou 32 and Shennong 20) was higher than that in sensitive varieties (Liaodou 3 and Liaodou 10). The expression levels of SOD, POD, and CAT genes in drought-tolerant varieties were higher than those in drought-sensitive varieties (Bhardwaj and Yadav 2012; Chutipaijit 2016; Yang et al. 2017). The ability of plants to improve oxidative stress depends on the induction of these genes. Bashir et al. (2019) found that the activity of antioxidant enzymes under the combined stress of drought and Cd was higher than that of Cd or drought alone. In contrast, Battana and Ghanta (2014) showed that drought stress induced more antioxidants than heavy metals, and the combined effects of drought and heavy metal stress were less than additive effects. In addition, drought stress decreased the activities of antioxidant enzymes in plants, resulting in increased malondialdehyde (MDA) content and ion leakage (Smirnoff 2010). Zinc application increased the activities of antioxidant enzymes such as SOD and POD and reduced the effects of oxidative stress, MDA accumulation, and electrolyte leakage (Shemi et al. 2021). Therefore, Zn protects cell membranes from drought stress damage (Ibrahim et al. 2017; Sakya et al. 2018; Ullah et al. 2019; Vazin 2012).

The research on combined stress of heavy metals and drought is rare, but the heavy metal pollution caused by industrial wastes is becoming more and more serious. The lack of water resources has led to increasing drought, and the combined stress of drought and heavy metals is facing increasingly serious problems.

19.5 Conclusions and Prospects

Heavy metal is an important environmental pollutant. It has long been accumulated in soils through human activities, such as vehicle exhaust emissions, sewage irrigation, and the use of pesticides, herbicides, and fertilizers. These activities lead to the entry of heavy metals into the environment and participate in the circulation of water–soil–biological systems. Heavy metals accumulate in the roots, stems, leaves, and seeds of plants through plant absorption and can pose a great threat to organisms and human health through the accumulation of food chains. In addition, drought/salt stress is also an abiotic stress that severely limits global crop yield. It has various adverse effects on plants, including reduced seed germination, seedling growth, biomass and grain yield, and nutrient and water uptake. In addition to the effects of heavy metals and drought on plants alone, the effects of combined stress on plant growth and yield parameters are very different from those of single stress. Therefore, it is necessary to observe the interaction between these two abiotic stresses and plants to better understand the effects of their interaction on plants. Researchers in the laboratory are extensively studying the effects of different abiotic stress factors on plant growth in the field in an individual manner. However, studies on the combined effects of metals and drought stress are limited to very few reports.

By studying the changes of SOD, CAT, POD, and other antioxidant enzyme activities and MDA content in plants, it is of great significance to explore the mechanism of plant resistance to heavy metal stress in arid areas, to accelerate the selection of heavy metal stress and hyperaccumulators, to make full use of land resources, to alleviate the ecological crisis of heavy metal pollution, and to ensure ecological security. At the same time, it can provide a theoretical basis for the toxicological evaluation of heavy metal pollution in arid areas and ecological restoration of abandoned land in mining areas. Furthermore, our understanding of the biochemical and molecular mechanisms that govern plant responses to these two stress combinations is insufficient in the existing literature. Therefore, it is necessary to conduct extensive research on these mechanisms. The assessment of multiple stresses is also important for obtaining a realistic view of the effects of dynamic environmental conditions.

Acknowledgments

This study was supported by the Science and Technology Innovation Program of Hunan Province (2022RC1118), the Scientific Research Fund of Hunan Provincial Education Department (18B166, 18B171), the Key Research and Development Project of Hunan Province (2020NK2022), the Training Program for Excellent Young Innovators of Changsha (kq1905065), and the Hunan Graduate Research Innovation Project (CX20220714).

References

Aganchich, B., Tahi, H., Wahbi, S. et al. (2007). Growth, water relations and antioxidant defence mechanisms of olive (*Olea europaea* L.) subjected to partial root drying (PRD) and regulated deficit irrigation (RDI). *Plant Biosyst.* 141: 252–264.

Ali, B., Gill, R.A., Yang, S. et al. (2015). Regulation of cadmium-induced proteomic and metabolic changes by 5-aminolevulinic acid in leaves of *Brassica napus* L. *PloS One* 10: e0123328.

Alpaslan, M., Inal, A., Gunes, A. et al. (1999). Effect of zinc treatment on the alleviation of sodium and chloride injury in tomato (*Lycopersicon esculentum* L.) grown under salinity. *Turk. J. Bot.* 23: 1–6.

Arduini, I., Godbold, D.L., and Onnis, A. (2010). Cadmium and copper change root growth and morphology of *Pinus pinea* and *Pinus pinaster* seedlings. *Physiol. Plant.* 92: 675–680.

Azevedo, R. and Rodriguez, E. (2012). Phytotoxicity of mercury in plants: a review. *J. Bot.* 2012: 1–6.

Azevedo, H., Pinto, C.G.G., Fernandes, J. et al. (2005). Cadmium effects on sunflower growth and photosynthesis. *J. Plant Nutr.* 28: 2211–2220.

Bartsch, N. (1987). Responses of root systems of young *Pinus sylvestris* and *Picea abies* plants to water deficits and soil acidity. *Can. J. For. Res.* 17: 805–812.

Bashir, W., Anwar, S., Zhao, Q. et al. (2019). Interactive effect of drought and cadmium stress on soybean root morphology and gene expression. *Ecotoxicol. Environ. Saf.* 175: 90–101.

Battana, S. and Ghanta, R.G. (2014). Antioxidative enzyme responses under single and combined effect of water and heavy metal stress in two *Pigeon pea* cultivars. *Int. J. Eng. Sci.* 3: 72–80.

Benavides, M.P., Susana, M., and Tomaro, M. (2005). Cadmium toxicity in plants. *Braz. J. Plant Physiol.* 17: 2134.

Bhardwaj, J. and Yadav, S.K. (2012). Comparative study on biochemical parameters and antioxidant enzymes in drought tolerant and sensitive variety of horsegram (*Macrotyloma uniflorum*) under drought stress. *Am. J. Plant Physiol.* 7: 17–29.

Bi, X.Y., Feng, X.B., Yang, Y.G. et al. (2006a). Quantitative assessment of cadmium emission from zinc smelting and its influences on the surfacesoils and mosses in Hezhang County, Southwestern China. *Atmos. Environ.* 40: 4228–4233.

Bi, X.Y., Feng, X.B., Yang, Y.G. et al. (2006b). Environmental contamination of heavy metals from zinc smelting areas in Hezhang County, western Guizhou, China. *Environ. Int.* 32: 883–890.

Bi, X.Y., Feng, X.B., Yang, Y.G. et al. (2007). Heavy metals in an impacted wetland systems: a typical case from southwestern China. *Sci. Total Environ.* 387: 257–268.

Bonnet, M., Camares, O., and Veisseire, P. (2000). Effects of zinc and influence of *Acremonium lolii* on growth parameters, chlorophyll a fluorescence and antioxidant enzyme activities of ryegrass (*Lolium perenne* L. cv Apollo). *J. Exp. Bot.* 51: 945–953.

Bradl, H.B. (2004). Adsorption of heavy metal ions on soils and soils constituents. *J. Colloid Interface Sci.* 277: 1–18.

Bukhari, S.A.H., Zheng, W., Xie, L. et al. (2016a). Cr-induced changes in leaf protein profile, ultrastructure and photosynthetic traits in the two contrasting tobacco genotypes. *Plant Growth Regul.* 79: 147–156.

Bukhari, S.A.H., Wang, R., Wang, W. et al. (2016b). Genotype-dependent effect of exogenous 24-epibrassinolide on chromium-induced changes in ultrastructure and physicochemical traits in tobacco seedlings. *Environ. Sci. Pollut.* 23: 18229–18238.

Cakmak, I. (2008). Enrichment of cereal grains with zinc: agronomic or genetic biofortification. *Plant Soil* 302: 1–17.

Canales-Pastrana, R.R. and Paredes, M. (2013). Phytoremediation dynamic model as an assessment tool in the environmental management. *Open J. Appl. Sci.* 3: 208.

Carver, B.F. and Ownby, J.D. (1995). Acid soil tolerance in wheat. *Adv. Agron.* 54: 117–173.

Choudhury, S., Panda, P., Sahoo, L. et al. (2013). Reactive oxygen species signaling in plants under abiotic stress. *Plant Signal. Behav.* 8: e23681.

Chutipaijit, S. (2016). Changes in physiological and antioxidant activity of *indica* rice seedlings in response to mannitol-induced osmotic stress. *Chil. J. Agric. Res.* 76: 455–462.

Clijsters, H. and Assche, F. (1985). Inhibition of photosynthesis by heavy metals. *Photosyn. Res.* 7: 31–40.

Cristina, P., Marco, R., Francesca, D.V. et al. (2006). Evidence for PSII donor-side damage and photoinhibition induced by cadmium treatment on rice (*Oryza sativa* L.). *J. Photochem. Photobiol. B Biol.* 8: 70–78.

Das, P., Samantaray, S., and Rout, G.R. (1997). Studies on cadmium toxicity in plants: a review. *Environ. Pollut.* 97: 110–113.

Debus, R.J. (1992). The manganese and calcium ions of photosynthetic oxygen evolution. *Biochim. Biophys. Acta* 1102: 269–352.

Delhaize, E. and Ryan, P.R. (1995). Aluminum toxicity and tolerance in plants. *Plant Physiol.* 107: 315–321.

Duruibe, J.O., Ogwuegbu, M.O.C., and Egwurugwu, J.N. (2007). Heavy metal pollution and human biotoxic effects. *Int. J. Phys. Sci.* 2: 112–118.

El-Mahdy, M.T., Abdel-Wahab, D.A., and Youssef, M. (2021). *In vitro* morpho-physiological performance and DNA stability of banana under cadmium and drought stresses. *In Vitro Cell. Dev. Biol. Plant* 57: 460–469.

Emamverdian, A. and Ding, Y. (2018). Effects of inhibition threshold in heavy metals on the growth and development of *Arundinaria fortunei*. *Pak. J. Bot.* 50: 887–892.

Fahr, M., Laplaze, L., Bendao, N. et al. (2013). Effect of lead on root growth. *Front. Plant Sci.* 4: 175.

Farooq, M., Wahid, A., Kobayashi, N. et al. (2009). Plant drought stress: effects, mechanisms and management. *Agron. Sustain. Dev.* 29: 185–212.

Farooq, M., Ahmad, R., Rehman, A. et al. (2015). Seed priming with zinc improves the germination and early seedling growth of wheat. *Seed Sci. Technol.* 43: 262–268.

Federico, B., Susanna, P., Aida, R. et al. (2019). Root colonization by *Pseudomonas chlororaphis* primes tomato (*Lycopersicum esculentum*) plants for enhanced tolerance to water stress. *J. Plant Physiol.* 232: 82–93.

Feng, X., Li, G., and Qiu, G. (2004). A preliminary study on mercury contamination to the environment from artisanal zinc smelting using indigenous method in Hezhang County, Guizhou, China: part 1. Mercury emission from zinc smelting and its influences on the surface waters. *Atmos. Environ.* 38: 6223–6230.

Feng, X., Li, G., and Qiu, G. (2006). A preliminary study on mercury contamination to the environment from artisanal zinc smelting using indigenous method in Hezhang County, Guizhou, China: part 2. Mercury contaminations to soil and crop. *Sci. Total Environ.* 36: 47–55.

Feng, Z., Ji, S.Y., Ping, J.F., and Cui, D. (2021). Recent advances in metabolomics for studying heavy metal stress in plants. *Trends Analyt. Chem.* 143: 116402.

Gautam, M., Sengar, R.S., Chaudhary, R. et al. (2010). Possible cause of inhibition of seed germination in two rice cultivars by heavy metals Pb^{2+} and Hg^{2+}. *Toxicol. Environ. Chem. Rev.* 92: 1111–1119.

Ghori, N.H., Ghori, T., Hayat, M.Q. et al. (2019). Heavy metal stress and responses in plants. *Int. J. Environ. Sci. Technol.* 16: 1807–1828.

Grantz, D.A., Garner, J.H.B., and Johnson, D.W. (2003). Ecological effects of particulate matter. *Environ. Int.* 29: 213–219.

Grewal, H.S. and Williams, R. (2000). Zinc nutrition affects alfalfa responses to water stress and excessive moisture. *J. Plant Nutr.* 23: 949–962.

Guo, B., Liang, Y.C., and Zhu, Y.G. (2007). Role of salicylic acid in alleviating oxidative damage in rice roots (*Oryza sativa*) subjected to cadmium stress. *Environ. Pollut.* 147: 734–739.

Gupta, A., Rico-Medina, A., and Cao-Delgado, A.I. (2020). The physiology of plant responses to drought. *Science* 368: 266–269.

Gururani, M.A., Upadhyaya, C.P., Venkidasamy, B. et al. (2013). Plant growth-promoting rhizobacteria enhance abiotic stress tolerance in *Solanum tuberosum* through inducing changes in the expression of ROS-scavenging enzymes and improved photosynthetic performance. *J. Plant Growth Regul.* 32: 245–258.

Habiba, U., Ali, S., Rizwan, M. et al. (2019). Alleviative role of exogenously applied mannitol in maize cultivars differing in chromium stress tolerance. *Environ. Sci. Pollut. Res.* 26: 5111–5121.

Haider, F.U., Cai, L., Coulter, J.A. et al. (2021). Cadmium toxicity in plants: impacts and remediation strategies. *Ecotoxicol. Environ. Saf.* 211: 111887.

Hajiboland, R. and Amirazad, H. (2010). Drought tolerance in Zn-deficient red cabbage (*Brassica oleracea* L. var. *capitata* f. *rubra*) plants. *Hort. Sci.* 37: 88–98.

Hamed, K.B., Ellouzi, H., Talbi, O.Z. et al. (2013). Physiological response of halophytes to multiple stresses. *Funct. Plant Biol.* 40: 883–896.

Hammami, S.S., Chaffai, R., and Ferjiani, E.E. (2004). Effect of cadmium on sunflower growth, leaf pigment and photosynthetic enzymes. *Pak. J. Biol. Sci.* 7: 1419–1426.

Han, Y., Wang, L., Zhang, X. et al. (2013). Sexual differences in photosynthetic activity, ultrastructure and phytoremediation potential of *Populus cathayana* exposed to lead and drought. *Tree Physiol.* 33: 1043–1060.

Hasan, S.A., Fariduddin, Q., Ali, B. et al. (2009). Cadmium: toxicity and tolerance in plants. *J. Environ. Biol.* 30: 165–174.

Hasni, I., Yaakoubi, H., Hamdani, S. et al. (2015). Mechanism of interaction of Al^{3+} with the proteins composition of photosystem II. *PLoS One* 10: e0120876.

Healoui, S., Boughattas, I., Hattab, S. et al. (2020). Effects of nickel on growth and the reproductive organs of *Vicia faba* plants. *Braz. J. Biol. Sci.* 7: 305–317.

Horst, W.J., Asher, C.J., Cakmak, J. et al. (1992). Short-term responses of soybean roots to aluminium. *J. Plant Physiol.* 140: 174–178.

Hussain, M.M., Hina, A., Saeed, A. et al. (2017). Impact of heavy metals on plants and animals in relation to sewage water-a review. *Sci. Technol. Soc.* 36: 215–226.

Ibrahim, S.A., Desoky, E., and Elrys, A.S. (2017). In fluencing of water stress and micronutrients on, physio-chemical attributes, yield and anatomical features of common bean plants (*Phaseolus vulgaris* L.). *Egypt. J. Agron.* 39: 251–265.

Ippolito, M.P., Fasciano, C., Aquino, L.D., and Tommasi, F. (2011). Responses of antioxidant systems to lanthanum nitrate treatments in tomato plants during drought stress. *Plant Biosyst.* 145: 248–252.

Iqbal, M., Aslam, M., Ranjha, A.M., and Akhtar, J. (2000). Salinity tolerance of rice affected by zinc application. *Pak. J. Biol. Sci.* 3: 2055–2057.

Iqbal, B., Zhao, T.T., Yin, W. et al. (2022). Impacts of soil microplastics on the crops: a review. *Agric. Ecosyst. Environ. Appl. Soil Ecol.* 181: 104680.

Kapoor, D., Singh, M.P., Kaur, S. et al. (2019). Modulation of the functional components of growth, photosynthesis, and anti-oxidant stress markers in cadmium exposed *Brassica juncea* L. *Plants* 8: 260.

Keyser, H.H. and Munns, D.N. (1979). Effects of calcium, manganese, and aluminum on growth of rhizobia in acid media. *Soil Sci. Soc. Am. J.* 43: 500.

Kochian, L.V., Hoekenga, O.A., and Piñeros, M.A. (2004). How do crop plants tolerate acid soils? Mechanisms of aluminum tolerance and phosphorous efficiency. *Annu. Rev. Plant Biol.* 55: 459–493.

Krizek, D.T. and Foy, C.D. (1988). Role of water stress in differential aluminum tolerance of two barley cultivars grown in an acid soil. *J. Plant Nutr.* 11: 351–367.

Krizek, D.T., Foy, C.D., and Wergin, W.P. (1988). Role of water stress in differential aluminum tolerance of six sunflower cultivars grown in an acid soil. *J. Plant Nutr.* 11: 387–408.

Küpper, H., Setlik, I., Setlikova, E. et al. (2003). Copper-induced inhibition of photosynthesis: limiting steps of *in vivo* copper chlorophyll formation in *Scenedesmus quadricauda*. *Funct. Plant Biol.* 30: 1187–1196.

Liu, C.F., Yu, R.G., and Shi, G. (2016). Effects of drought on the accumulation and redistribution of cadmium in peanuts at different developmental stages. *Arch. Agron. Soil Sci.* 63: 1–9.

Ma, Y., Rajkumar, M., Moreno, A. et al. (2017). Serpentine endophytic bacterium *Pseudomonas azotoformans* ASS1 accelerates phytoremediation of soil metals under drought stress. *Chemosphere* 185: 75–85.

Manchanda, G. and Garg, N. (2011). Alleviation of salt-induced ionic, osmotic and oxidative stresses in *Cajanus cajan* nodules by AM inoculation. *G. Bot. Ital.* 145: 88–97.

Maria, S.D., Puschenreiter, M., and Rivelli, A.R. (2013). Cadmium accumulation and physiological response of sunflower plants to Cd during the vegetative growing cycle. *Plant Soil Environ.* 59: 254–261.

Mercury, S. (1970). Mercury in the environment. *BMJ-Brit. Med. J.* 4: 606.

Minutolo, M., Caruso, I., Caruso, G. et al. (2012). Establishment of *Aster sedifolius* and *Aster caucasicus* callus as apotential source of antioxidants. *G. Bot. Ital.* 146: 41–46.

Mousavi, S.R., Galavi, M., and Rezaei, M. (2013). Zinc (Zn) importance for crop production – a review. *Int. J. Agron. Plant Prod.* 4: 64–68.

Muhammad, U.H., Muhammad, A., Muhammad, U.C. et al. (2020). The critical role of zinc in plants facing the drought stress. *Agriculture* 10: 396.

Munzuroglu, O. and Geckil, H. (2002). Effects of metals on seed germination, root elongation, and coleoptile and hypocotyls growth in *Triticum aestivum* and *Cucumis sativus*. *Arch. Environ. Contam. Toxicol.* 43: 203–213.

Muqadas, S., Ali, Q., and Malik, A. (2020). Genetic association among seedling traits of *Zea mays* under multiple stresses of salts, heavy metals and drought. *Biol. Clin. Sci. Res. J.* 2020: 26.

Nasiri, Y., Zehtab-Salmasi, S., Nasrullahzadeh, S. et al. (2010). Effects of foliar application of micronutrients (Fe and Zn) on flower yield and essential oil of chamomile (*Matricaria chamomilla* L.). *J. Med. Plant Res.* 4: 1733–1737.

Naz, R., Sarfraz, A., Anwar, Z. et al. (2021). Combined ability of salicylic acid and spermidine to mitigate the individual and interactive effects of drought and chromium stress in maize (*Zea mays* L.). *Plant Physiol. Biochem.* 159: 285–300.

Nedjimi, B. (2013). Effects of salinity on growth, membrane permeability and root hydraulic conductivity in three saltbush species. *Biochem. Syst. Ecol.* 52: 4–13.

None (1948). Trace elements in plant physiology. *Nature* 161: 364–365.

O'Lexy, R., Kasai, K., Clark, N. et al. (2018). Exposure to heavy metal stress triggers changes in plasmodesmatal permeability via deposition and breakdown of callose. *J. Exp. Bot.* 69: 3715–3728.

Osakabe, Y., Osakabe, K., Shinozaki, K. et al. (2014). Response of plants to water stress. *Front. Plant Sci.* 5: 86.

Pankovic, D. (2000). Effects of nitrogen nutrition on photosynthesis in Cd-treated sunflower plants. *Ann. Bot.* 86: 841–847.

Paul, A.K., Bala, T.K., Shahriar, S. et al. (2016). Effect of foliar application of zinc on yield of wheat grown under water stress condition. *Int. J. Bio-Resour. Stress Manag.* 7: 1025–1031.

Paunov, M., Koleva, L., Vassilev, A. et al. (2018). Effects of different metals on photosynthesis: cadmium and zinc affect chlorophyll fluorescence in durum wheat. *Int. J. Mol. Sci.* 19: 787.

Pinto, A.P., Mota, A.M., and Pinto, F.C. (2004). Influence of organic matter on the uptake of cadmium, zinc, copper and iron by Sorghum plants. *J. Environ. Sci. (China)* 326: 239–247.

Potarzycki, J. and Grzebisz, W. (2009). Effect of zinc foliar application on grain yield of maize and its yielding components. *Plant Soil Environ.* 55: 519–527.

Rashid, A., Bernier, M., Pazdernick, L. et al. (1991). Interaction of Zn^{2+} with the donor side of photosystem II. *Photosyn. Res.* 30: 123–130.

Reddy, A.R., Chaitanya, K.V., and Vivekanandan, M. (2004). Drought-induced responses of photosynthesis and antioxidant metabolism in higher plants. *J. Plant Physiol.* 161: 1189–1202.

Redondo-Gómez, S., Redondo-Gómez, E., Davy, A.J. et al. (2007). Growth and photosynthetic responses to salinity of the salt-marsh shrub *Atriplex portulacoides*. *Ann. Bot.* 100: 555–563.

Rehman, A., Farooq, M., Naveed, M. et al. (2018). *Pseudomonas*-aided zinc application improves the productivity and biofortification of bread wheat. *Crop Pasture Sci.* 69: 659–672.

Reis, S., Pavia, I., Carvalho, A. et al. (2018). Seed priming with iron and zinc in bread wheat: effects in germination, mitosis and grain yield. *Protoplasma* 255: 1179–1194.

Rizwan, M., Ali, S., Abbas, T. et al. (2018). Residual effects of biochar on growth, photosynthesis and cadmium uptake in rice (*Oryza sativa* L.) under Cd stress with different water conditions. *J. Environ. Manage.* 206: 676–683.

Rusin, M., Domagalska, J., Rogala, D. et al. (2021). Concentration of cadmium and lead in vegetables and fruits. *Sci. Rep.* 11: 11913.

Ryser, P. and Emerson, P. (2007). Growth, root and leaf structure, and biomass allocation in *Leucanthemum vulgare* Lam. (Asteraceae) as influenced by heavy-metal containing slag. *Plant Soil* 59: 2461–2467.

Sakya, A.T., Sulistyaningsih, E., Indradewa, D. et al. (2018). Stomata character and chlorophyll content of tomato in response to Zn application under drought condition. *IOP Conf. Series Earth Environ. Sci.* 142: 12–33.

Salehi-Lisar, S.Y. and Bakhshayeshan-Agdam, H. (2016). Drought stress in plants: causes, consequences, and tolerance. *Environ. Sci.* 1: 1–16.

Sammut, M., Noack, Y., Rose, J. et al. (2010). Speciation of Cd and Pb in dust emitted from sinter plant. *Chemosphere* 78: 445–450.

Sardans, J. and Peñuelas, J. (2007). Drought changes the dynamics of trace element accumulation in a Mediterranean *Quercus ilex* forest. *Environ. Pollut.* 147: 567–583.

Schindler, M. (1989). RGD-dependent linkage between plant cell wall and plasma membrane: consequences for growth. *J. Cell Biol.* 108: 1955–1965.

Seyedi, M., Hamzei, J., Fathi, H. et al. (2012). Effect of seed priming with zinc sulfate on germination characteristics and seedling growth of chickpea (*Cicer arietinum* L.) under salinity stress. *Int. J. Agric. Innov. Res.* 2: 108–114.

Shah, K. and Dubey, S. (1995). Effect of cadmium on RNA level as well as activity and forms of ribonuclease in growing rice seedlings. *Plant Physiol. Biochem.* 33: 577–584.

Shahriaripour, R., Pour, A.T., Mozaffari, V. et al. (2010). Effect of salinity and soil zinc application on growth and chemical composition of pistachio seedlings. *J. Plant Nutr.* 33: 1166–1179.

Sharma, A., Kumar, V., Shahzad, B. et al. (2020). Photosynthetic response of plants under different abiotic stresses: a review. *J. Plant Growth Regul.* 39: 509–531.

Shemi, R., Wang, R., Gheith, E. et al. (2021). Role of exogenous-applied salicylic acid, zinc and glycine betaine to improve drought-tolerance in wheat during reproductive growth stages. *BMC Plant Biol.* 21: 1–15.

Shukla, L. and Jain, N. (2022). A review on soil heavy metals contamination: effects, sources and remedies. *Appl. Ecol. Environ. Res.* 10: 15–18.

Silva, D., Cholewa, E., and Ryser, P. (2012). Effects of combined drought and heavy metal stresses on xylem structure and hydraulic conductivity in red maple (*Acer rubrum* L.). *J. Exp. Bot.* 63: 5957–5966.

Singh, S., Srivastava, P.K., Kumar, D. et al. (2015). Morpho-anatomical and biochemical adapting strategies of maize (*Zea mays* L.) seedlings against lead and chromium stresses. *Biocatal. Agric. Biotechnol.* 4: 286–295.

Slugeňová, K., Ditmarová, Ľ., Kurjak, D. et al. (2011). Drought and aluminium as stress factors in Norway spruce (*Picea abies* L. Karst) seedlings. *J. For. Sci.* 57: 547–554.

Smirnoff, N. (2010). The role of active oxygen in the response of plants to water deficit and desiccation. *New phytol.* 125: 27–58.

Song, W.Y., Zheng, A.Z., Shao, H.B. et al. (2012). The alleviative effect of salicylic acid on the physiological indices of the seedling leaves in six different wheat genotypes under lead stress. *Plant Omics.* 5: 486–493.

Song, W.Y., Peng, S.P., Shao, C.Y. et al. (2014). Ethylene glycol tetra-acetic acid and salicylic acid improve anti-oxidative ability of maize seedling leaves under heavy-metal and polyethylene glycol 6000-simulated drought stress. *Plant Biosyst.* 148: 96–108.

Steudle, E. (2000). Water uptake by roots: effects of water deficit. *J. Exp. Bot.* 51: 1531–1542.

Swapna, B. and Gopal, R. (2015). Germination and early growth responses of *Vigna mungo* L. Hepper cultivars to water stress and heavy metals. *Am. J. Med. Sci.* 1: 169–174.

Tabelin, C. and Igarashi, T. (2009). Mechanisms of arsenic and lead release from hydrothermally altered rock. *J. Hazard. Mater.* 169: 980–990.

Tauqeer, H.M., Ali, S., Rizwan, M. et al. (2016). Phytoremediation of heavy metals by *Alternanthera bettzickiana*: growth and physiological response. *Ecotoxicol. Environ. Saf.* 126: 138–146.

Trenberth, K.E., Dai, A., Gerard, V. et al. (2013). Global warming and changes in drought. *Nat. Clim. Change* 4: 17–22.

Turgut, C., Pepe, M.K., and Cutright, T.J. (2005). The effect of EDTA on *Helianthus annuus* uptake, selectivity and translocation of heavy metals when grown in Ohio, New Mexico and Colombia soils. *Chemosphere* 58: 1087–1095.

Ullah, A., Romdhane, L., Rehman, A. et al. (2019). Adequate zinc nutrition improves the tolerance against drought and heat stresses in chickpea. *Plant Physiol. Biochem.* 143: 11–18.

Ünyayar, S., Keleş, Y., and Çekiç, F.Ö. (2005). The antioxidative response of two tomato species with different drought tolerances as a result of drought and cadmium stress combinations. *Plant Soil Environ.* 51: 57–64.

Upadhyaya, H., Dutta, B.K., and Panda, S.K. (2013). Zinc modulates drought-induced biochemical damages in tea [*Camellia sinensis* (L) O Kuntze]. *J. Agric. Food Chem.* 61: 6660–6670.

Uzu, G., Sobanska, S., Aliouane, Y. et al. (2009). Study of lead phytoavailability for atmospheric industrial micronic and sub-micronic particles in relation with lead speciation. *Environ. Pollut.* 157: 1178–1185.

Vazin, F. (2012). Effect of zinc sulfate on quantitative and qualitative characteristics of corn (*Zea mays*) in drought stress. *Cercet. Agron. Mold.* 45: 15–24.

Vega, F., Andrade, M., and Covelo, E. (2010). Influence of soil properties on the sorption and retention of cadmium, copper and lead, separately and together, by 20 soil horizons: comparison of linear regression and tree regression analyses. *J. Hazard. Mater.* 174: 522–533.

Vishnupradeep, R., Benedict, L.B., Zarin, T. et al. (2022). Plant growth promoting bacteria improve growth and phytostabilization potential of *Zea mays* under chromium and drought stress by altering photosynthetic and antioxidant responses. *Environ. Technol. Innov.* 25: 102154.

Wang, X.H., Song, Y.P., Ren, J. et al. (2013). Response of biomass and photosynthesis in non-heading Chinese cabbage to excess copper. *J. Anim. Plant Sci.* 23: 1659–1665.

Win, K.T., Tanaka, F., Okazaki, K. et al. (2018). The ACC deaminase expressing endophyte *Pseudomonas* spp. enhances NaCl stress tolerance by reducing stress-related ethylene production, resulting in improved growth, photosynthetic performance, and ionic balance in tomato plants. *Plant Physiol. Biochem.* 127: 599–607.

Wuana, R. and Okieimen, F. (2011). Heavy metals in contaminated soils: a review of sources, chemistry, risks and best available strategies for remediation. *Int. Schol. Res. Not.* 2011: 1–20.

Xia, S., Wang, X., Su, G. et al. (2015). Effects of drought on cadmium accumulation in peanuts grown in a contaminated calcareous soil. *Environ. Sci. Pollut. Res.* 22: 18707–18717.

Xiao, M.H., Li, Y.Y., Wang, J.W. et al. (2019). Study on the law of nitrogen transfer and conversion and use of fertilizer nitrogen in paddy fields under water-saving irrigation mode. *Water* 11: 218.

Xu, Q., Ma, X., Lv, T. et al. (2020). Effects of water stress on fluorescence parameters and photosynthetic characteristics of drip irrigation in rice. *Water* 12: 289.

Yang, Z.B., Eticha, D., Rao, I.M. et al. (2010). Alteration of cellwall porosity is involved in osmotic stress-induced enhancement of aluminium resistance in common bean (*Phaseolus vulgaris* L.). *J. Exp. Bot.* 61: 3245–3258.

Yang, Z.B., Eticha, D., Albacete, A. et al. (2012). Physiological and molecular analysis of the interaction between aluminium toxicity and drought stress in common bean (*Phaseolus vulgaris*). *J. Exp. Bot.* 63: 3109–3125.

Yang, Z.B., Rao, I.M., and Horst, W.J. (2013). Interaction of aluminium and drought stress on root growth and crop yield on acid soils. *Plant Soil* 372: 3–25.

Yang, H., Zhao, L., Zhao, S. et al. (2017). Biochemical and transcriptomicanalyses of drought stress responses of LY1306 tobacco strain. *Sci. Rep.* 7: 17442.

Yang, Q., Li, Z., Lu, X. et al. (2018). A review of soil heavy metal pollution from industrial and agricultural regions in china: pollution and risk assessment. *Sci. Total Environ.* 642: 690–700.

Yildiz, M., Poyraz, R., Avdar, A. et al. (2020). Plant responses to salt stress. In: *Plant Breeding - Current and Future Views*. London, UK: IntechOpen.

Zahra, N., Hafeez, M.B., Shaukat, K. et al. (2021). Iron toxicity in plants: impacts and remediation. *Physiol. Plant.* 173: 201–222.

Zulfiqar, U., Farooq, M., Hussain, S. et al. (2019). Lead toxicity in plants: impacts and remediation. *J. Environ. Manage.* 250: 109557.

20

Hormonal Regulation of Heavy Metal Toxicity and Tolerance in Crop Plants

Éderson Akio Kido, Gizele de Andrade Luz, Valquíria da Silva, Maria Fernanda da Costa Gomes, and José Ribamar Costa Ferreira Neto

Department of Genetics, Biosciences Center, Federal University of Pernambuco (UFPE), Recife, Pernambuco, Brazil

20.1 Introduction

In nature, heavy metals (HMs) (Figure 20.1), such as arsenic (As), cadmium (Cd), chromium (Cr), copper (Cu), iron (Fe), lead (Pb), mercury (Hg), molybdenum (Mo), nickel (Ni), and zinc (Zn), accumulated on air, soil, or water, threaten crop yield, food safety, and human health. Usually, HMs derive from natural and anthropogenic sources, such as industrial and agricultural activities, and these critical contributors increase their availability in the ecosystems (Edelstein and Ben-Hur 2018; Ali et al. 2019). Since they are not biodegradable, HMs remain in the environment, being transferred to living organisms and accumulating at different trophic levels (Ali et al. 2019).

The HM absorption by plants is mainly by roots via apoplastic (passive diffusion) and symplastic (active transport across the plasma membrane) pathways, with the participation of metals channels/transporters (Jamla et al. 2021; Thakur et al. 2022). Concerning plants, some of these HM are essential for plant metabolism, as the micronutrients performing relevant functions or constituting enzymes/cofactors, such as B, Cu, Fe, Mn, Mo, Ni, and Zn, or those nonessential, including As, Cr, Pb, and Hg. However, even for those essential for plants, the requirement is trace amounts (ppm level), while reaching toxic levels disrupt the plant integrity, causing biochemical, physiological, and morphological abnormalities (Andresen et al. 2018; Jamla et al. 2021) (Figure 20.1).

When plants sense the HM stress, they activate stress signal transduction pathways involving signaling molecules, such as MAPK (mitogen-activated protein kinases), plant hormones, and calcium ions (Ca^{2+}), to induce HM-responsive genes, triggering plant strategies to perform an adequate plant stress-response (Jalmi et al. 2018). Among the signaling molecules, phytohormones (PHYs) promote regulatory networks, playing crucial roles in cellular processes during plant growth and development (PG&D) (Wani et al. 2016). Also, in plants under HM stress, the PHYs stimulating the respective signaling pathway and

Heavy Metal Toxicity and Tolerance in Plants: A Biological, Omics, and Genetic Engineering Approach, First Edition. Edited by Mohammad Anwar Hossain, AKM Zakir Hossain, Sylvain Bourgerie, Masayuki Fujita, Om Parkash Dhankher, and Parvez Haris.

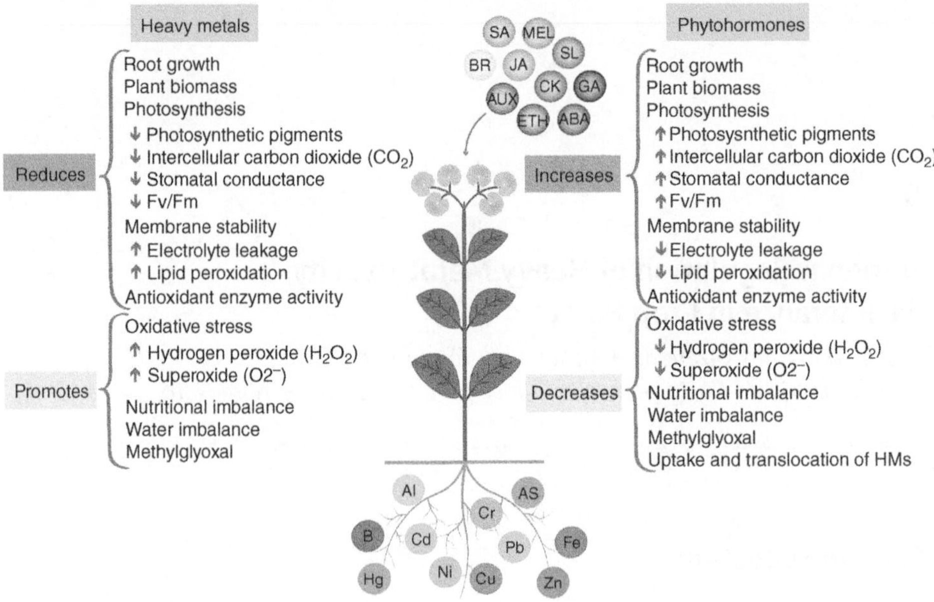

Figure 20.1 Overview of phytohormones alleviating physiological and biochemical imbalances caused by heavy metals in plants. Phytohormones: SA, salicylic acid; MEL, melatonin; SL, strigolactone; GA, gibberellin; JA, jasmonate; CK, cytokinin; ABA, abscisic acid; BR, brassinosteroid; AUX, auxin; ETH, ethylene; heavy metals: Al, aluminum; B, boron; Hg, mercury; Cd, cadmium; Ni, nickel; Cu, copper; Cr, chromium; Pb, lead; As, arsenic; Fe, iron; Zn, zinc; Fv/Fm, maximum quantum yield of photosystem II.

inducing the expected responsive genes will relieve the HM toxicity, alleviating the HM-induced adverse effects, including PG&D inhibition.

In this way, understanding plant hormonal responses to HM stress could provide alternatives for plant breeders to develop tolerant plants for growing in HM-contaminated soils as an "eco-friendly" alternative to improve bioremediation processes. Thus, this chapter will cover relevant aspects of the hormonal signaling pathways of plants under HM stress, highlighting their molecular strategies for stress alleviation.

20.2 General Aspects of Plants Under HM Stress

Plants exposed to HM stress limit the potential PG&D as a result of biochemical and physiological dysfunctions, generally involving (Figure 20.1):

a) Reactive oxygen species (ROS): in cells, at a basal level, ROS are signaling molecules controlling processes during the cell cycle and the PG&D, also coordinating plant responses to environmental signals, while excess ROS are harmful to cell membranes and macromolecules, such as lipids, proteins, and nucleic acids. This is the case of plants exposed to HMs with excessive ROS production, such as hydrogen peroxide (H_2O_2) and superoxide (O_2^-), as a result of direct biological redox reactions or indirectly (Shahid et al. 2014);

b) Oxidative stress: increased ROS deteriorates the antioxidant defense systems and disrupts the cellular redox homeostasis, enhancing the oxidative stress (Shahid et al. 2014);
c) Lipid peroxidation (LP) and electrolyte leakage (EL): increased ROS induces the LP, which represents oxidative degradation of membranes, affecting fluidity and stabilities, and also damaging membrane proteins, and inactivating receptors and ion channels; the malondialdehyde (MDA) is one of the end products from the LP product degradation; some LP by-products interfere with the photosynthesis performance. In turn, the EL is a landmark of cells death, triggering the programmed cell death (PCD); these processes affect the structures of stomata and chloroplasts (Hashem 2014; Guedes et al. 2021);
d) Methylglyoxal (MG) generation: the MG is a by-product of several metabolic pathways, including the glycolytic metabolism, acting at a basal level as a signaling metabolite in several reactions, but the MG accumulated at a toxic level is harmful to membranes, lipids, and proteins (Hoque et al. 2016; Kamran et al. 2021; Kaushik et al. 2022);
e) Photosynthetic capacity: plants under HM stress inhibit their photosynthetic efficiency as a consequence of damages in the photosystem II (PSII), impairing the electron transport; furthermore, the increased activities of chlorophyllase (Chlase) and Mg-dechelatase (MDCase) degrade photosynthetic pigments (chlorophyll and carotenoids) (Salavati et al. 2021).
f) Plant water relations: in plants under HM stress, the HMs promote osmotic imbalances and disturb the plant water relations as they reduce the root absorption area and water uptake. Also, HMs decrease root hydraulic conductivity and vessel size, reducing water transfer (apoplastic and symplastic routes) and water delivered to shoots, which is also due to transpiration inhibition since HMs reduce leaf size, affecting stomata density and stomatal closure (Rucińska-Sobkowiak 2016; Jan et al. 2020).
g) Nutritional balance: plants under HM stress can present a mineral deficiency, affecting the PG&D; the HMs inhibiting the root system also alter the uptake of mineral nutrients from the soil since some HMs presenting similarities with nutrient cations compete with them for absorption by roots, like arsenate (AsO_4^{3-}) and phosphate (PO_4^{3-}) due to similar acid dissociation constants or Cd with Zn, as a consequence of nonspecific ionic transporters, affecting also their translocation and the PG&D (DalCorso et al. 2013; Maia et al. 2022); in addition, antagonistic and synergistic interactions between HMs and micronutrients were reported, concerning their uptake by plants (Singh et al. 2016).

To minimize those mentioned HM-induced adverse effects, plants evolved defense mechanisms, some of them triggered by plant hormones.

20.3 Phytohormone-Mediating Plant Response to HM Stress

Plants under distinct environmental growing conditions face (a)biotic stresses during their life cycles. Throughout evolution, plants developed mechanisms sensing stress signals (Verma et al. 2016). Also, aiming to achieve effective cellular communication while performing an adequate stress response, plants use phytohormones (PHYs) as signaling molecules to promote complex regulatory networks during PG&D (Wani et al. 2016). At low concentrations in plant cells, PHYs regulate cellular processes acting locally, at the site of synthesis, or along the vegetal (Santner et al. 2009; Umezawa et al. 2010; Zwanenburg et al. 2016). Several PHY families are known (Figure 20.1), including abscisic acid (ABA), auxin (AUX), brassinosteroid (BR), cytokinin (CK), ethylene (ETH), gibberellin (GA), jasmonate (JA), melatonin (MEL), salicylic acid (SA), and strigolactone (SL) (Wang et al. 2015; Arnao and Hernández-Ruiz 2019).

Plants under HM stress explore signaling pathways, including the hormonal, activating PHY-responsive genes and triggering regulatory networks to perform the appropriated plant stress response (Figure 20.1). An HM-induced plant response, alleviating the HM stress and attenuating its toxicity, results of combined plant strategies described below (Figure 20.2):

a) Antioxidant defense systems: the HM-induced oxidative stress reduction involves induction of antioxidant enzyme activities of peroxidases (PODs), catalases (CATs), glutathione peroxidases (GPxs), superoxide dismutases (SODs), among others (Zhou et al. 2018); furthermore, the ROS scavenging process is enhanced by stimulation of the

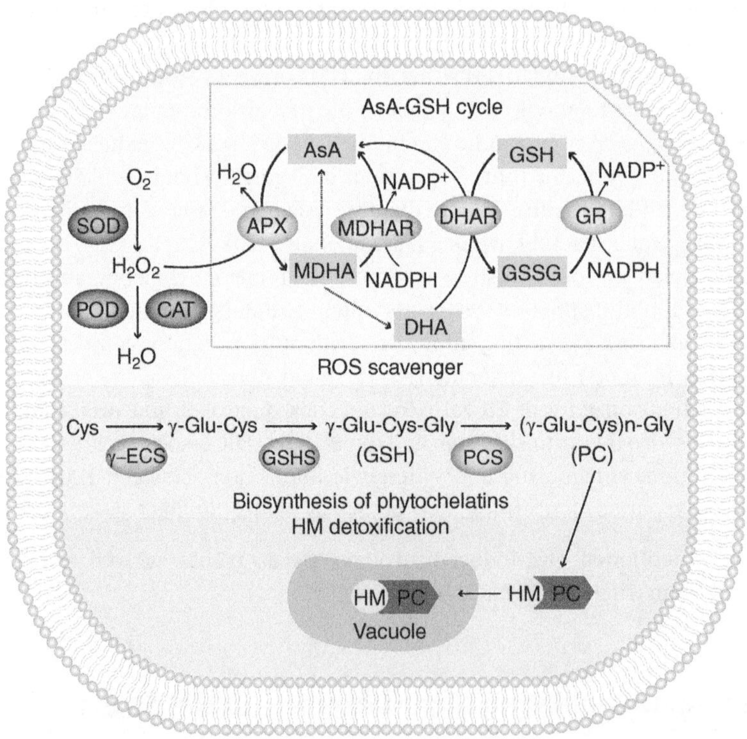

Figure 20.2 Antioxidant defense and heavy metal detoxification in plants. HM, heavy metal; SOD, superoxide dismutase; POD, peroxidase; CAT, catalase; AsA, ascorbate; GSH, glutathione; APX, ascorbate peroxidase; MDHAR, monodehydroascorbate reductase; DHAR, dehydroascorbate reductase; GR, glutathione reductase; MDHA, monodehydroascorbate; GSSG, oxidized glutathione; γ-ECS, γ-glutamylcysteine synthetase; GSHS, glutathione synthase; PCS, phytochelatin synthase; PC, phytochelatin.

ascorbate-glutathione (AsA-GSH) cycle, with involvement of nonenzymatic antioxidant compounds, such as AsA and GSH, besides the enzymes ascorbate peroxidase (APX), monodehydroascorbate reductase (MDHAR), dehydroascorbate reductase (DHAR), and glutathione reductase (GR) (Kaya et al. 2020; Mostofa et al. 2020). Several PHY-mediated responses enhance antioxidant defense systems, including ABA (Leng et al. 2021), AUX (Gong et al. 2019), BR (Soares et al. 2020), CK (Sameena and Puthur 2021), ETH (Chahel et al. 2020), GA (Alatawi et al. 2023), JA (Kamran et al. 2021), MEL (Kaya et al. 2019), SA (Naeem et al. 2020), and SL (Tai et al. 2017).

b) Methylglyoxal (MG) detoxification: MG is a by-product of many metabolic reactions, and, at basal levels, MG signals aspects of plant metabolism, including cellular homeostasis and cell proliferation, but at toxic levels, MG generates ROS and inhibits cell growth (Hoque et al. 2016; Mostofa et al. 2018a); some PHY-mediated response (JA, MEL) stimulates the glyoxalase system, and the increased activity and expression of glyoxalases reduce the MG level (Kaya et al. 2020; Kamran et al. 2021; Kaushik et al. 2022).

c) Photosynthetic performance: increased ROS promotes photosynthetic damage inhibiting the photosystem II (PSII) damage repair cycle, and CKs act in this repairing process (Cortleven et al. 2014); furthermore, besides protecting chloroplasts and photosynthetic pigments, PHYs (AUX, ABA, BR, and SA) also induce changes to various photosynthetic components and regulate plastidial and nuclear genes involved in photosynthesis, improving the photosynthetic capacity (Gururani et al. 2015; Cunha et al. 2021).

d) Plant water relations: since HMs reduce the root absorption area, affecting the water uptake and translocation, also due to transpiration inhibition affected by the stomatal closure, PHYs actions, including ABA-mediated signals, can induce aquaporin (*AQP*) genes, and some AQPs can mediate the stomatal closure and improve plant hydraulic properties, increasing the plant water status (Rucińska-Sobkowiak 2016). Also, PHY (SA) interferes with the plant root architecture, enhancing the root-absorbing area and the water uptake, contributing to the water status in the plant (Mostofa et al. 2020).

e) Mineral and metal homeostasis: since HMs interfere with the root absorption and root-to-shoot translocation of nutrients and micronutrients, the PHY-mediated response of plants under HM stress contributes to the mineral/metal homeostasis combining pleiotropic actions, covering changes in the root architecture, limiting the HM absorption or its translocation, applying HM-chelation or vacuolar HM-sequestration/detoxification, or performing extracellular exclusion by transporter proteins presented in tonoplast or plasma membranes, such as P(1B)-type ATPases transporting HMs (Cu^{+}, Cu^{2+}, Zn^{2+}, Co^{2+}, Cd^{2+}, Pb^{2+}) across membranes (Singh et al. 2016; Shi et al. 2019; Salavati et al. 2021).

f) Quelation, sequestration, and detoxification: since chelating molecules binding HM ions reduce their toxicities, some PHY-mediated response of plants under HM stress improves the synthesis of secondary metabolites, relevant for plant-environment interactions; thus, some PHYs (SA, JA, MEL) improve the biosynthesis of chelators [GSH, phytochelatin (PC), and metallothionein (MT)], induce roots to release chelating exudates, induce HM immobilization in pectins (polysaccharide), besides HM accumulation in vacuoles and deposition in root cell walls (Huda et al. 2016; Cao et al. 2019).

g) Gasotransmitter: like ROS, the gasotransmitter nitric oxide (NO) is a reactive nitrogen species (RNS) and a signaling molecule; in plants under HM stress, some PHY (SA) induced the synthesis of NO and hydrogen sulfide (H_2S), another gasotransmitter, which interactions with SA increased the activity of glyoxalases and enzymes from the AsA-GSH cycle, minimizing the oxidative stress and enhancing the plant HM-tolerance (Kaya et al. 2020).

Thus, exploring the PHYs influences on plants under HM stress is relevant since potential PHY treatment (spraying or supplementation) applied to plants growing in HM-contaminated soils is an alternative to improve the phytoremediation process or cultivation of HM-tolerant crops without risks to the environment or human health. However, the benefits of such PHYs effectively restoring plant physiological processes depend on several factors, including the PHY (type, presentation, formula, concentration, dosage, application method), the HM stress (chemical element, bioavailability, toxicity, area), and the plant (species, life cycle, developmental stage). Also, although PHYs control several developmental stages of plants facing (a)biotic stress, including HMs, each PHY presents some particularities described below.

20.3.1 Abscisic Acid

The abscisic acid (ABA) is crucial to PG&D, embracing bud dormancy, seed germination, leaf senescence, passing through the stomatal opening, and growth inhibition (Nambara et al. 2010; Kuromori et al. 2011). The ABA synthesized in roots reaches by the xylem, the aerial parts of the plant, despite ABA being also synthesized in leaves. In turn, the transport system covering the influx and efflux of ABA involves proteins, some of them tissue-specific, for instance, the proteins ABCG25 (ATP-binding cassette G25) (Kuromori et al. 2011) and AIT (abscisic acid-importing transporter) (Kanno et al. 2012) are mainly expressed in vascular tissues, and the PDR (Pleiotropic drug resistance) (Kang et al. 2010) proteins expressed in guard cells.

Concerning plant (a)biotic stress responses, the ABA is a central regulator, increasing its biosynthesis with the accumulated ABA triggering a cascade of signals, inducing ABA-responsive genes and promoting the stress response. According to the ABA signal transduction pathway (Figure 20.3a), during the absence of ABA, the PP2C (Protein phosphatase 2C) phosphatases inactivated SnRK2 (Serine/threonine-protein kinase srk2) kinases, inhibiting the ABA-mediated responses. When ABA is present, the phytohormone binds to its receptors (PYR/PYL, pyrabactin resistance/pyrabactin resistance-like) and disrupts the interaction of PP2C and SnRK2. After SnRK2 autophosphorylation, the kinases activate ABI5 or ABF (abscisic acid-responsive element-binding factors) TFs, which will regulate the ABA-responsive genes, performing the ABA-mediated response (Umezawa et al. 2010; Ng et al. 2014).

Applications of ABA treatment to plants under HM stress induce antioxidant defenses, alleviating the HM-induced oxidative stress, and reducing ROS and MDA levels, which minimize LP and EL processes and consequently damage to membranes and macromolecules (Leng et al. 2021; Deng et al. 2022). An ABA supplementation to *Artemisia annua* L. plants under Cu stress prevented chloroplast damage, protected photosynthetic pigments, and increased activities of photosynthesis-related enzymes, including RuBisCo (Ribulose-1,5-bisphosphate carboxylase/oxygenase) (Zehra et al. 2020). Also, spraying

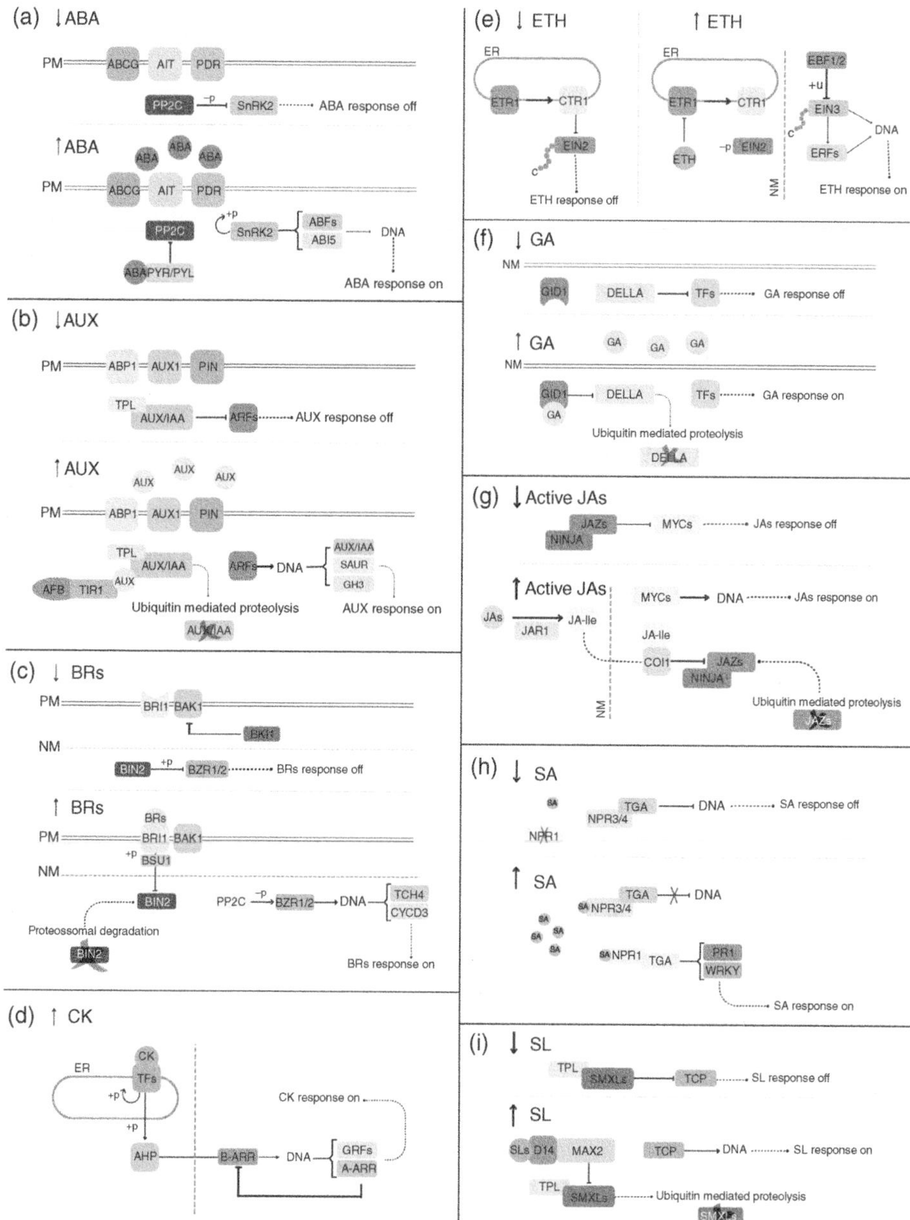

Figure 20.3 Schematic diagrams of nine major phytohormones (PHYs) signaling pathways: ABA signaling (a), AUX signaling (b), BR signaling (c), CK signaling (d), ETH signaling (e), GA signaling (f), JA signaling (g), SA signaling (h), and SL signaling (i). Arrows represent PHY levels during signaling pathways. The components highlighted in the PHY signaling pathways are described in the text. PM, plasma membrane; ER, endoplasmic reticulum; MN, nuclear membrane.

ABA (1–20 μmol l^{-1}) on lettuce (*Lactuca sativa*) plants under Cd stress improved gas exchange and photosynthesis capacity (Tang et al. 2020).

Furthermore, spraying ABA (5 μM) on *Malus hupehensis* Rehd leaves restricted root Cd^{2+} influx and root-to-shoot translocation, and since the transpiration rate drives the absorption and transport of inorganic salts in plants, with ABA decreasing the transpiration rate, this reduces the potential of (Cd^{2+})-transport to stems and leaves (Deng et al. 2022). Also, ABA supplementation induced the AIT1 transporter (ABA-importing transporter), favoring ABA uptake by roots while inhibiting the expression of the IRT1 transporter, a broad metal spectrum transporter (including Zn, Mn, Cd, and Co), minimizing the HM (Cd) accumulated in cells (Pan et al. 2020).

In addition, ABA supplementation also influences genes involved in plant HM-tolerance. Supplementation of ABA (10 μM) treatment in grapevine (*Vitis vinifera* L.) plants reduced the Zn stress (10 mM Zn^{2+}), inducing Zn uptake-related genes (from the ZIP family of metal transporters) and HM detoxification-related genes (Song et al. 2019). In the same way, ABA (10 μM in Hoagland nutrient solution) supplemented to poplar (*Populus × canescens*) plants exposed to Pb^{2+} (3 mM $PbCl_2$) induced genes related to Pb uptake, translocation, and detoxification (Shi et al. 2019). This enhanced Pb movement induced by ABA in a fast-growing plant with expressive biomass highlights the *Populus* species as relevant candidates for phytoremediation of Pb-contaminated soils (Shi et al. 2019). In general, ABA applications (0.5–10 μM) contributes to reversing the HM-induced adverse effects, including the PG&D inhibition (Pan et al. 2020; Leng et al. 2021).

20.3.2 Auxin

The auxin (AUX) is a growth-regulating plant hormone influencing several physiological processes, including embryonic development, tropisms, apical dominance, lateral root formation, cell differentiation, leaf abscission, and parthenocarpy, among others (Pandey et al. 2019). Its biosynthesis is tryptophan (Trp)-independent or -dependent and based on the well-defined TAA/YUC pathway, the tryptophan aminotransferase of *Arabidopsis* (TAA) and then flavin monooxygenase (YUCCA) converts Trp to AUX (Singh et al. 2021).

The AUX signaling pathway (Figure 20.3b) is a regulatory complex of plant–environment interactions whose actions culminate in the repression or activation of AUX-responsive genes through the degradation of transcriptional repressors. When cells present low AUX levels, AUX/IAA (Auxin-responsive protein IAA) transcriptional repressors inhibit the activity of ARFs (Auxin response factors) TFs, inactivating AUX-mediated responses. When increasing AUX levels in cells, interactions between TIR1 (transport inhibitor response 1) proteins and AUX/IAA transcriptional repressors drive AUX/IAA to degradation via the proteasome, releasing ARF TFs to regulate gene expression of AUX-responsive genes (Lavy and Estelle 2016).

In addition to controlling usual physiological functions, the AUX influences plants' responses to (a)biotic stresses, including HMs stresses. Thus, plants under HM stress increase AUX levels by inducing genes related to the AUX-biosynthetic pathway, including the YUCCA family, and those encoding AUX signaling and transport, such as ABCB (ATP-binding transporter), and PIN (PIN-FORMED) to ensure the AUX accumulation (Kaur et al. 2021). A higher level of AUXs activates the respective signaling pathway, inducing

AUX-mediated responses. Those responses will reduce the HM-induced oxidative stress, limit the HM uptake and root-to-shoot translocation, reducing the HM toxicity and the adverse effects while restoring the water content, mineral homeostasis, photosynthesis performance, and the PG&D (Emamverdian et al. 2021).

The activation of antioxidant defense systems (enzymatic and nonenzymatic) reduces excessive ROS production and oxidative stress, minimizing LP and EL events and consecutively preserving membranes and macromolecules (Hashem 2014). Exogenous IAA applications highlight these positive effects. Application of IAA ($10\,\mathrm{mg\,kg^{-1}}$) in *Cinnamomum camphora* under Cd stress reduced MDA levels, alleviating the HM-induced oxidative stress by increasing antioxidant enzymes, soluble sugars, and proline, a stress-related signal (Zhou et al. 2020). An IAA ($60\,\mathrm{mg\,l^{-1}}$) treatment in spinach plants (*Spinacia oleracea* L.) exposed to Cu stress-induced antioxidant enzymes (SOD, APX and GR), reducing the overproduced ROS, also reducing the HM toxicity and the Cu level in roots and leaves (Gong et al. 2019). Additionally, the application of IBA (indole-3-butyric acid, AUX precursor) in barley (*Hordeum vulgare* L.) under Cd stress minimized the oxidative stress by inducing GPx enzymes, which contributed to reducing organic peroxides levels (Demecsová et al. 2020).

In turn, spraying IAA ($20\,\mathrm{mg\,l^{-1}}$) on rice (*Oryza sativa* L.) plants under As stress stimulated the hemicellulose synthesis, reducing the As translocation to rice grains (He et al. 2022). Also, an exogenously applied IAA ($10\,\mu\mathrm{M}$) treatment in tea [*Camellia sinensis* (L.) Kuntze] plants reduced the Cd uptake and the root-to-shoot translocation (Zhang et al. 2020). In addition, plants under HM stress also reduce HM toxicity by performing a vacuolar HM compartmentalization (Singh et al. 2021), as demonstrated in eggplant (*Solanum melongena* L.) plants exposed to As stress supplemented with IAA ($5\,\mu\mathrm{M}$).

Furthermore, plants under HM stress increase the gasotransmitter (such as NO and H_2S) production, and these stress-signaling molecules contribute to the plant's HM tolerance (Kaya et al. 2020). The IAA ($5\,\mu\mathrm{M}$) applied to tomato (*Lycopersicum esculentum* Mill., synonym *Solanum lycopersicum* L.) under Cd stress restricted the Cd accumulation and reduced HM toxicity while increasing the NO accumulated in roots (Khan et al. 2019). Similarly, the AUX-precursor IBA applied on barley roots under Cd stress increased NO levels (Demecsová et al. 2020).

Finally, considering that plants under HM stress presented inhibited PG&D, and exogenously applied AUX treatment improves PG&D. The application of artificial IAA (10^{-6} M) and IAA from two rhizospheric bacteria (*Lysinibacillus varians* and *Pseudomonas putida*) in mustard (*Brassica nigra* L.) plants under Cd stress increased the chlorophyll content and improved the PG&D (Pal et al. 2019). The IBA treatment (10^{-9} M) in corn (*Zea mays* L.) roots exposed to Cd stress increased the nutrient uptake, enhancing the PG&D (Siposová et al. 2021). Sometimes, these positive effects are IAA concentration-dependent; in tamarillo [*Cyphomandra betacea* (Cav.) Sendtn.] plants, the treatments with lower IAA concentrations ($30–60\,\mathrm{mg\,l^{-1}}$) were more efficient, concerning increasing root biomass, photosynthetic pigments, and soluble sugar content (Li et al. 2020). Thus, these AUX-mediated responses uncovered by plants under HM stress are strategies to be explored by breeders developing HM-tolerant plants or considered in bioremediation processes.

20.3.3 Brassinosteroid

The brassinosteroid (BR) is a class of steroid hormones coordinating broad biological processes in plants, including cell division and elongation, vascular differentiation, seed germination, flowers/ fruits development, plant immunity, and PG&D (Cukor et al. 2018; Hong et al. 2018; Lee et al. 2019; Furio et al. 2019). Also, BRs induce the biosynthesis of other PHYs to favor the PG&D (Gruszka et al. 2016). Several studies associated BRs with multiple effects (direct and indirect) on plants under (a)biotic stress, including HM stress (Gruszka et al. 2016; Sharma et al. 2018; Maia et al. 2022).

During the BR signal transduction pathway (Figure 20.3c), at low BR levels, negative BRI1 (brassinosteroid insensitive1) regulator proteins inactivate positive regulator proteins BKI1 (brassinosteroid intensive1 kinase inhibitor1), BSK (brassinosteroid signaling kinase), and CDG1 (Constitutive differential growth1). Also, the negative regulators BIN2 (brassinosteroid insensitive 2) phosphorylate and inactivate BZR1/2 (brassinazole resistant1/2) TF proteins, which are retained by GRF (general regulatory factor) proteins, inhibiting the BR-mediated response. During higher BRs levels, BRI1 receptors disassociated from BKI1 (BRI1 kinase inhibitors) proteins are activated. Then, BRI1 proteins phosphorylate and activate BSK, CDG1, and BSU1 proteins, together inactivating BIN2 proteins. In turn, PP2A (phosphatase protein2A) proteins dephosphorylate BZR1/2 TF proteins, triggering the expression of BR-responsive genes and performing the BR-mediated response (She et al. 2011; Ye et al. 2011).

Plants under HM stress trigger the BR-signaling pathways and induce the responsive genes to present a BR-mediated response involving alleviation of the HM toxicity, reduction of the HM-induced oxidative stress, stimulation of the redox mechanism and ionic homeostasis, protection of photosynthetic pigments and apparatus, trying to circumvent the HM-adverse effects, including the PG&D inhibition (Li et al. 2015a; Pociecha et al. 2016; Shu et al. 2016; Sharma et al. 2018).

In this way, an exogenous application of EBL, also named EBR (24-epibrassinolide), the most bioactive plant steroid form, has been explored. Thus, supplementation of EBL (10^{-8} M) on Indian mustard seeds [*Brassica juncea* (L.) Czern.] under Pb stress reduced the oxidative stress by increasing antioxidant enzymes (CAT and POX) and reducing the ROS levels, enhancing seed germination and the PG&D (Soares et al. 2020). Also, pretreatment with EBR ($0.10\,mg\,l^{-1}$) in grapevine under Cu stress alleviated oxidative damages by improving antioxidant defense systems [enzymes (CAT, POD, SOD, APX, MDHAR, GR, and DHAR) and antioxidant compounds (AsA and DHA)], reducing the levels of MDA, H_2O_2, and O_2^-, and inducing proline-biosynthesis-related genes (Zhou et al. 2018). Additionally, EBR (100 nM) supplemented on tomato leaves attenuated Ni stress by inducing antioxidant enzymes, reducing levels of ROS [O_2^-, H_2O_2] and MDA, also minimizing EL events (Maia et al. 2022). Similarly, rice plants exposed to Pb stress after EBR (100 nM) treatment induced antioxidant enzymes, reducing MDA levels and EL events (Guedes et al. 2021).

Furthermore, the BRs influence the plant's photosynthetic capacity. A supplemented EBR (1 μM) treatment in pepper (*Capsicum annuum* L.) seedlings under Cr stress increased chlorophyll and carotenoid contents and induced chlorophyll-biosynthesis-related genes (Mumtaz et al. 2022). Spraying EBR (100 nM) treatment on rice plants exposed to Pb stress increased chlorophyll levels (a, b, and total) and the Mg content, as necessary for the chlorophyll synthesis. Also, an EBR treatment restored the efficiency of the photosystem II (PSII), improving related parameters [ΦPSII (effective quantum yield of PSII photochemistry), qP (photochemical quenching), and ETR (electron transport rate)] (Guedes et al. 2021). Another EBR (100 nM) treatment applied to *Eucalyptus urophylla* S.T. Blake plants under Cd stress improved gas exchange and photosynthetic capacity, as reinforced by values of Fm (maximal fluorescence yield of the dark-adapted state), Fv/Fm (maximal quantum yield of PSII photochemistry), ΦPSII, qP, and ETR (Cunha et al. 2021).

Concerning the plant water status and the BR influence, an exogenous application of EBR (10^{-7} M) in tomato plants exposed to Cr stress restored the relative water content (RWC), also increasing the osmolytes proline and betaine glycine (Jan et al. 2020). Also, an EBR treatment (100 nM) on soybean plants [*Glycine max* (L.) Merr] exposed to excess Ni^{2+} mitigated anatomical disturbances in leaf and root structures. In roots, the applied EBR improved the epidermis (physical barrier), as reflected by the parameters RET (root epidermis thickness), RDT (root endodermis thickness), RCT (root cortex thickness), VCD (vascular cylinder diameter) and RMD (root metaxylem diameter). In leaves, improvements in stomatal density and functionality facilitated the carbon fixation in mesophyll cells (Saraiva et al. 2021). Still, as EBR plays a relevant role in stomatal development, the pretreatment with EBR (100 nM) on tomato leaves under Ni stress stimulated stomatal performance, as reinforced by the parameters g_s (stomatal conductance), SD (stomatal density), and EDS (equatorial diameter of the stomata), and SI (stomatal index) (Maia et al. 2022).

Furthermore, regarding the plant mineral nutrition, an exogenously applied ERB treatment (100 nM) in rice plants (exposed to Pb stress) induced ionic homeostasis by re-establishing the ratios of essential nutrients (Ca^{+2}/Pb^{+2} and Mg^{+2}/Pb^{+2}) in roots and shoots (Guedes et al. 2021). Also, an EBR treatment (100 nM) on tomato leaves under Ni stress improved the nutrient absorption in roots (P, S, Ca, Mn, Cu, and Zn) and leaves (Ca, Mn, Cu, and Zn) by enhancing the permeability of membranes (Maia et al. 2022). Also, an EBR (100 nM) treatment in *E. urophylla* S.T. Blake plants under Cd stress positively regulated the uptake of nutrients (Mg, K, Ca, Cu, Zn, and Fe), reducing the Cd^{2+} uptake and accumulation (Cunha et al. 2021). All these BR-mediated responses showed by plants under HM stress are strategies to be explored in bioremediation processes or developing HM-tolerant plants.

20.3.4 Cytokinin

The natural cytokinin (CK) is a plant hormone derived from adenine, carrying an isoprenoid or aromatic side chain at the adenine ring's N6 position (Hwang and Sakakibara 2006). Isoprenoid-type CKs are the most abundant in the plant kingdom. The CK controls several physiological processes involved in PG&D, including root architecture, root nodule formation, meristem maintenance, and vascular development (Reid et al. 2016; Wybouw and De Rybel 2019).

During the CK signal transduction pathway (Figure 20.3d), the CK receptor AHK (*Arabidopsis* Histidine Kinase) autophosphorylates when bound to CK. Subsequently, the phosphate group is transferred to protein AHP (*Arabidopsis* Histidine phosphotransfer protein), which carries the phosphate group to the nucleus. Then, AHP phosphorylates type-B response regulator (RRB) transcription factor (TF), and this TF activates CK-responsive gene, triggering the CK-mediated response (Cortleven and Schmülling 2015).

The CK also mediates plants responding to (a)biotic stresses (Cortleven et al. 2019), and its effects regarding the endogenous CK levels under HM stress appear controversial. The green alga (*Acutodesmus obliquus*) (Piotrowska-niczyporuk et al. 2020) and soybean (Hashem 2014), when exposed to HM (Pb or Cd), decreased the pool of active CKs concerning the contents of zeatin and zeatin riboside, respectively. In the green alga, the Pb stress favors the glycosylation of CKs; however, exogenous CK (and AUX) applications induce changes in the CK endogenous levels, adjusting the CK homeostasis (Piotrowska-niczyporuk et al. 2020). In turn, pepper plants exposed to Ag nanoparticles increased total CK concentrations in leaves (Vinković et al. 2017), while *Arabidopsis thaliana* plants exposed to Cd increased in tips of roots the CK-mediated signaling (Bruno et al. 2017).

Considering the CK-mediated responses, some transgenic tobacco (*Nicotiana tabacum* L.) plants expressing an agrobacterial *IPT* (isopentyl transferase) gene highlighted the relevance of CK-biosynthesis-related genes in plants under HM stress. Such transgenic plants under Cu stress (100–200 μM) exhibited lower LP of membranes and higher Cu tolerance than the wild-type plants (Ermoshin et al. 2019). Also, transgenic tobacco plants under Zn stress (250–750 mg kg^{-1} soil) overexpressing an *IPT* gene increased the CK content and the HM tolerance. Those lines also improved photosynthesis while reducing the transpiration rate (Pavlíková et al. 2014). Additionally, a rice *OsHIPP17* gene overexpressed in *A. thaliana* plants under Cu stress (up to 50 μM $CuSO_4$) negatively regulated the *AtCKX1* gene, which is responsible for CK degradation, helping to regulate the CK homeostasis in plants under stress (Huang et al. 2022a).

Also improving photosynthesis and PG&D, an exogenous application of CK *trans*-zeatin (tZR) in wheat (*Triticum aestivum* L.) plants alleviated the As stress (50–100 μM; Zaheer et al. 2022). Additionally, spraying CK treatment (*trans*-zeatin riboside, 10 μM) in plants [*Kosteletzkya pentacarps* (L.) Ledeb.] under Cd (10 μM) or Zn (100 μM) stress mitigated HM-adverse effects by reducing the oxidative stress and also by increasing g_s, net photosynthesis (P_N), and PG&D. In the same way, a supplemented CK kinetin (KN, 10–50 μM) improved the PG&D of fenugreek (*Trigonella foenum-graecum* L.) under Cd stress (0–9 mg Cd kg^{-1} of soil) inducing the AsA-GSH cycle, reducing H_2O_2, and improving the photosynthetic capacity (Bashri et al. 2021). Another exogenous application of kinetin (KN) and 6-benzylaminopurine (BAP) in castor bean (*Ricinus communis* L.) plants reduced the Cu-uptake and the Cu stress after improved antioxidant enzymes and secondary metabolites with antioxidant properties, including phenolic compounds and flavonoids (Sameena and Puthur 2021). Similarly, a CK treatment (KN 50 μM) applied in mung bean [*Vigna angularis* (Willd.) Ohwi & H. Ohashi] plants increased phenolic and flavonoid contents, enhancing the ROS scavenging and alleviating the Cd stress (100 μM of $CdCl_2$) (Ahanger et al. 2020).

The CK also influences the root architecture by inhibiting root elongation and branching. Therefore, reducing CK levels could improve the root system (Rivas et al. 2022). Thus, CK-deficient plants could HM-accumulate higher amounts (such as Cd and Zn) since more developed roots exploit the HM-contaminated area more efficiently (Nehnevajova et al. 2019). In turn, a CK treatment applied to *Cannabis sativa* L. plants increased the phytoremediation of As-contaminated soil. However, not by regulating the root system but by increasing the HM translocated to the aerial parts of the plant (Grifoni et al. 2021). Thus, the potential for HM mitigation depends on the CK strategy to be selected and enhanced in the phytoremediation process.

20.3.5 Ethylene

The ethylene (ETH), a gaseous plant hormone, induces and coordinates a range of physiological processes, including the formation of root hair and nodules, seed germination, and fruit development, including maturation, ripening, and senescence (Iqbal et al. 2017).

During the ETH signal transduction pathway (Figure 20.3e), in the ETH absence, ETR1 (ethylene response 1) receptor proteins induce the kinase activity of CTR1 (constitutive triple 1) proteins, and these act in the inactivation and degradation of the positive regulator EIN2 (ethylene intensive 2) proteins. In the presence of ETH, the EIN2 proteins accumulate due to inactivated CTR1 proteins. The C-terminal domains of EIN2 proteins are then transported to the nucleus, stabilizing EIN3 (ethylene intensive 3) TF proteins, which, in turn, induce genes encoding ERF (ethylene-responsive transcription factor) FTs, thus activating ETH-responsive genes involved in the ETH-mediated response (Ju et al. 2012; Ji and Guo 2013).

After the initial stress perception, the ETH signaling pathway stimulates ROS production, and the H_2O_2 signaling the stress modulates the defense system. During the ETH signaling pathway, the EIN2 family members play relevant roles, helping to control the ROS levels. Plants (*A. thaliana*) with a mutation in the *EIN2-5* gene exposed to Hg stress presented lower H_2O_2 accumulation than wild-type plants; this gene was relevant to HM tolerance (Montero-Palmero et al. 2014). In *A. thaliana*, the *EIN2* gene also modulated the antioxidant defense cascade since the mutant line (*EIN2-1* gene), which is sensible to Pb stress, presented a negative regulation of the *GSH1* (*gamma-glutamylcysteine synthetase 1*) gene, which encodes the first enzyme of the GSH biosynthetic pathway (Cao et al. 2009).

The main active components of the ETH biosynthesis pathway in plants under non-stressful conditions are the enzymes ACC (1-amino cyclopropane-1-carboxylic acid) synthase (ACS) and ACC oxidase (ACO). The ACS converts SAM (*S*-adenosyl-methionine) into ACC, whose oxidation by ACO produces ETH, CO_2, and cyanide (Xu and Zhang 2015). The genes *ACS* and *ACO* are also responsive to HM stress, and plants under HM stress increase ETH biosynthesis after the ROS signaling. In this case, the MAPK3/6 phosphorylation cascade phosphorylates ACS enzymes, making them more stable and active than native ACSs (Xu and Zhang 2014).

In *A. thaliana*, the gene expression of *ACS* and *ACO* (*AtACS2, AtACS6, AtACO2,* and *AtACO4*) increased after the Cd stress (Schellingen et al. 2014). Also, several *ACS* genes overexpressed in rice roots exposed to As stress (Huang et al. 2012), with similar results in rice roots after Cr exposure (Trinh et al. 2014). Analysis in different tissues (*A. thaliana*) pointed to the highest ETH level induced by HMs (Cu/Cd) in root tips (Arteca and Arteca 2007). Also, Ni and Zn stresses tend to increase ETH levels in plants (*B. juncea* L.) after *ACS* gene overexpression (Khan and Khan 2014).

In plants under abiotic stress, the ERF TFs mediate the integration of hormonal and redox signaling pathways (Dietz et al. 2010; Abiri et al. 2017). The overexpression of *ERF* genes in plants under HM stress is widely reported in crops and model plants, including *A. thaliana* (Schellingen et al. 2014), alfalfa (*Medicago sativa* L.) (Montero-Palmero et al. 2014), rice (Trinh et al. 2014), and barley (Lopes et al. 2013).

Also, *ERF* genes have been considered good transgenes. Transgenic seedlings (*Glycyrrhiza uralensis* Fisch.) under Cd stress overexpressing the *ERF061* gene increased antioxidant enzymes (SOD, CAT, and POD) and the Cd tolerance (Chahel et al. 2020). Also, transgenic plants (tobacco) overexpressing the transgene *LchERF* presented Cd tolerance after inducing GSH-biosynthesis-related genes and the antioxidant defense systems (Guan et al. 2015). Similarly, transgenic tobacco seedlings overexpressing the *TdSHN1* gene (encoding ERF from *Triticum turgidum* L.) showed higher HM tolerance (Cd, Cu, and Zn) than the wild-type plants under HM stress. These transgenic lines enhanced antioxidant defense systems and ROS scavenging, besides improved roots, biomass accumulation, and chlorophyll content (Djemal and Khoudi 2022). All these ETH-mediated responses presented by plants under HM stress are strategies to be explored in bioremediation processes or developing HM-tolerant plants.

20.3.6 Gibberellin

The gibberellin (GA) comprises a class with more than 136 GAs, but few are biologically active, these standing out GA1, GA3, and GA4 in higher plants (Giacomelli et al. 2013; Gao and Chu 2020). The GAs participates in several processes in PG&D, such as seed germination, stem elongation, leaf expansion, flowering, fruit ripening, and responses to environmental stresses (Xu et al. 2016; Penfield 2017; Gao and Chu 2020). Also, GAs interacts positively and negatively with other PHYs (Weiss and Ori 2007; Huang et al. 2021).

In the GA signal transduction pathway (Figure 20.3f), during the absence of GA, DELLA proteins (DELLA-contained protein) inhibit the activity of PIF transcription factors (PTIs; phytochrome-interacting factor), suppressing GA-mediated responses. When GA levels increase in cells, the bioactive forms change the conformation of the GID1 receptors; then, the resulting GA-GID1 complexes bind to DELLA proteins and drive them to degradation via the proteasome. Thus, the released PIF TFs activate GA-responsive genes, performing the GA-mediated response (Li et al. 2016a).

Considering adverse effects induced by the HM stress, the GA-mediated responses relieve some of these effects, reducing the HM toxicity and improving the PG&D (Emamverdian et al. 2020). A GA3 treatment ($100\,mg\,l^{-1}$) applied to pea plants under Cu stress enhanced antioxidant enzyme activities (SOD, POD, and CAT), reduced the levels of H_2O_2, MDA, EL, and increased the photosynthetic capacity [chlorophyll content, P_N, intercellular CO_2 concentration (C_i)] (Javed et al. 2021). Similarly, a GA treatment ($100\,mg\,l^{-1}$) applied to *Corchorus capsularis* L. under Cu stress enhanced antioxidant defense systems by increasing enzymes and non-enzymatic compounds (phenolics, flavonoids, anthocyanins, ascorbic acid, and proline). Also, the treatment improved photosynthesis performance and mineral nutrition (Mg and Ca) (Alatawi et al. 2023). Still in *C. capsularis* under Cu stress, a GA3 ($10–100\,mg\,l^{-1}$) treatment improved the photosynthetic capacity (P_N, g_s and C_i), increased antioxidant enzymes, and reduced the levels of MDA, H_2O_2, and EL (Saleem et al. 2020).

A GA treatment also affects the HM uptake and translocation in plants under HM stress. The GA3 treatment ($10\,\mu mol\,l^{-1}$) in lettuce (*L. sativa* L.) plants conferred Cd tolerance and reduced the Cd-uptake and root-to-shoot translocation by repressing genes related to its transport and translocation [*IRT1* (*iron-regulated transporter 1*), *NRAMP1* (*natural resistance-associated macrophage protein 1*), *HMA2* (*heavy metal ATPase 2*), and *HMA4* (*heavy metal ATPase 4*)] (Chen et al. 2021). A GA treatment (5 μM) applied to *Arabidopsis* under Cd stress repressed a gene encoding IRT1 transporter, attenuating Cd uptake and accumulation in roots (Zhu et al. 2012).

In turn, a GA3 treatment (0.5 μM) applied on barley seeds under Cd (also Mo) stress increased (during the seed germination) amino acids, soluble sugars, and hydrolytic enzymes (α-amylase, and β-amylase; Amri et al. 2016). Also, a GA treatment (10 μM) in peas (*Pisum sativum* L.) plants under Cr stress promoted antioxidant defenses and increased the N assimilation and the protein content (Gangwar et al. 2011). Furthermore, a GA treatment ($300\,mg\,l^{-1}$) in tamarillo plants under Cd stress increased antioxidant enzymes, soluble proteins and sugars, photosynthesis performance, and biomass, also reducing the Cd accumulated in different tissues (Yang et al. 2021). Similarly, a GA treatment (10^{-4} M) applied to *Vigna radiata* L. plants under Ni stress reduced the Ni accumulation in roots and shoots, improving the PG&D (Ali et al. 2015). However, GA treatments also increased the HM accumulation in different parts of the plant (Saleem et al. 2020; Javed et al. 2021). Thus, plants (candidates) for phytoremediation of HM-contaminated soils could explore the mentioned plant strategies.

20.3.7 Jasmonate

The jasmonic acid or jasmonate (JA) coordinates several physiological processes on PG&D, including seed germination, leaf growth, and senescence (Linkies and Leubner-Metzger 2012; Jiang et al. 2014; Pluskota et al. 2019).

During the JA signaling pathway (Figure 20.3g), in the absence of JA bioactive forms, the proteins JAZs (jasmonate zim-domain proteins), NINJAs (novel interactor of jasmonate zim-domain proteins), and TPLs (topless proteins) work together to inactivate MYC (MYC-related transcriptional activator) TFs. With JA bioactive forms, JAs bind to COI1 (coronatine intensive 1) receptors. In turn, COI1 proteins lead JAZ proteins to degradation via the proteasome, releasing MYC TFs to activate JA-responsive genes, performing the JA-mediated response (Pauwels et al. 2010; Sheard et al. 2010).

Plants responding to (a)biotic stresses also perform JA-mediated responses (Falk et al. 2014; Kazan 2015; Balfagón et al. 2019). Thus, JA induces antioxidant defense systems (enzymatic and non-enzymatic compounds) and the glyoxalase system. The induced glyoxalases (Gly I and Gly II) improve the GSH regeneration, helping to reduce ROS and MG levels and alleviating the HM-induced oxidative stress (Kamran et al. 2021; Kaushik et al. 2022).

In this way, in plants under HM stress, JA-mediated responses stimulate protective effects on cells, reducing the overproduced ROS, minimizing LP and EL events, and, consequently, membrane damages, favoring cellular ionic homeostasis (Verma et al. 2020). The protection of cell membranes inhibits the entry of HMs into the cytoplasm, decreasing the HM uptake.

In addition, JA induces metal-transport proteins, such as P1B-type heavy-metal ATPases. These ATPases, besides their contribution to transporting HMs (Cu^{+}, Cu^{2+}, Zn^{2+}, Co^{2+},

Cd^{2+}, Pb^{2+}) across membranes, also play a key role in metal homeostasis (Salavati et al. 2021). The methyl jasmonate (MeJA), a methyl ester of JA, besides inhibiting the synthesis of endogenous JA, also improved foliar concentrations of mineral nutrients (P, Mg, K, Ca, and Mn), enhancing the PG&D (Qiu et al. 2021). In turn, JA stimulates the production of pectin, hemicelluloses, and lignin, which favor the binding of HMs to the root cell walls, and these natural barriers restrict the transport of HMs through the plasma membrane from roots to shoots, reducing potential damages to the photosynthetic apparatus (Kamran et al. 2021; Salavati et al. 2021).

All these JA-mediated responses shown by plants under HM stress are also strategies to be explored in bioremediation processes or developing HM-tolerant plants.

20.3.8 Melatonin (MT)

The melatonin (MEL, *N*-acetyl-5-methoxy-tryptamine) is a molecule naturally occurring in several kingdoms, including animal and vegetal. The MEL influences several biological processes, including seed germination, vegetative growth, photosynthesis, flowering, and senescence (Arnao and Hernández-Ruiz 2019). As an antioxidant compound, MEL also emerges as a relevant molecule for plants responding to (a)biotic stresses (Moustafa-Farag et al. 2020; Nawaz et al. 2021).

In plants, the MEL is a multi-regulatory molecule derived from the amino acid tryptophan (Back et al. 2016). The discovery of a MEL-receptor (CAND2/PMTR1) in *A. thaliana* (Wei et al. 2018) favored MEL as a PHY or at least a master regulator of PG&D based on several multi-functionalities, which actions are receptor-dependent or -independent (Arnao and Hernández-Ruiz 2019).

In *Arabidopsis* plants, the MEL signaling mediated by the receptor CAND2/PMTR1 regulates stomatal closure through H_2O_2 and Ca^{2+} signals (Wei et al. 2018). Plants exogenously supplemented with MEL also modulate MEL-biosynthesis-related genes and their signaling (Jahan et al. 2020; Xu et al. 2020). The MEL treatment (50 μM) in radish (*Raphanus sativus* L.) plants under Pb stress increased the expression of genes related to signal transduction, including *CBL* (*calcineurin B-like protein*), *CaM* (*calmodulin*), *CML* (*CaM-like protein*), *CDPK* (*Calcium-dependent protein kinase*), *MAPK* (*mitogen-activated protein kinase*), and *MAPKKK* (*MAP kinase kinase kinase*) (Tang et al. 2021). Similarly, in radish plants exposed to Cd stress, the MEL treatment induced genes related to calcium signaling [CaM-like protein (CML)] and MAPK signaling (mitogen-activated protein kinase) pathways (Xu et al. 2020). In tomato plants exposed to Cd stress, the overexpression of the gene *HsfA1a* (*heat-shock factor A1a*), encoding a TF that induces the gene *COMT1* (*caffeic acid O-methyltransferase 1*) from the MEL-biosynthesis pathway leads to the MEL accumulation, which partially regulates the expression of HSPs and the plant Cd tolerance (Cai et al. 2017).

Arabidopsis transgenic plants overexpressing *MsSNAT* (*serotonin N-acetyltransferase* from alfalfa) gene, an MEL-synthesis-related gene, increased endogenous MEL levels, showing the roots a higher Cd tolerance and a lower Cd content, reflecting the modulation of Cd transporters. The authors highlighted the heavy metal transporter *HMA4* gene (Gu et al. 2017) and the induced ABC transporter *PDR8* gene, already reported to reduce Cd content in plants (Kim et al. 2007).

Concerning MEL stimulating antioxidant defense systems, spraying MEL (50–100 μM) on wheat leaves promoted Cd tolerance by inducing antioxidant enzyme activities (CAT, SOD, and POD), reducing H_2O_2, MDA, and EL levels and stimulating the NO production (Kaya et al. 2019). An MEL (50–100 μM) treatment in *Brassica napus* L. under Al or Cd stress increased endogenous MEL levels and attenuated the oxidative stress by increasing antioxidant enzymes; the treatment also increased H^+-ATPase activity and improved the plasma membrane stability (Sami et al. 2020). An MEL treatment (0.05–0.10 mM) applied in corn under Pb stress alleviated the oxidative stress by increasing antioxidant enzymes, with reduction of H_2O_2, MDA, and EL levels, besides enhancing the NO accumulation (Okant and Kaya 2019). Also, wheat seeds soaked with MEL (100 μM) induced genes encoding antioxidant enzymes and repressed genes (*TaRbohD* and *TaRbohF*) encoding NADPH oxidases (ROS-generating enzymes); the applied treatment improved the seed germination under Cr stress (Lei et al. 2021). Similarly, the supplementation of MEL (50 μM) and Ca^{2+} (30 mM) to fava beans (*Vicia faba* L.) under As stress synergistically reduced ROS, increasing antioxidant enzyme activities and reducing activities of ROS-producing enzyme (Siddiqui et al. 2020). In turn, an MEL treatment (50 μM) applied to wheat under Cd stress, besides increased antioxidant compounds levels (AsA and GSH), also highlighted in roots the induction of *GR*, *APX*, and *GST* genes, with repression of *GSHS* (glutathione synthase), *DHAR*, and *MDHAR* genes. About shoots, the induced genes were *GR*, *APX*, *DHAR*, *MDHAR*, and *GST* (Li et al. 2022).

Regarding MEL stimulating the photosynthesis performance, a MEL treatment (100 μM) supplemented on leaves (wheat) exposed to excess B increased chlorophyll content by increasing the activity of δ-aminolevulinic acid dehydratase, a chlorophyll-biosynthesis-related enzyme, and by reducing the chlorophyllase activity. The treatment also increased the enzyme activities of RuBisCO and carbonic anhydrase and improved photosynthesis performance (P_N and g_s; Al-Huqail et al. 2020). Also, an MEL (100 μM) treatment applied to cucumber under excess Fe improved photosynthetic capacity (P_N, g_s, and transpiration rate) under low ($3\,mg\,l^{-1}$) and high ($90\,mg\,l^{-1}$) Fe concentration (Ahammed et al. 2020). In turn, a combined treatment [MEL (50 μM) and Ca^{2+} (30 mM)] applied to fava beans under As stress synergistically improved the chlorophyll content, decreasing chlorophyllase activity and increasing the activity of δ-aminolevulinic acid dehydratase. The treatment also increased RuBisCO and carbonic anhydrase enzymatic activities and some photosynthetic parameters (P_N, g_s, and C_i; Siddiqui et al. 2020).

MEL treatments also modulated plant metabolism of sulfur (S), carbon (C), nitrogen (N), and secondary metabolism. An exogenous MEL (100 μM) treatment on tomato plants under Cd stress improved sulfur (S) acquisition and enhanced the Cd tolerance by increasing PCs and promoting Cd-chelation (Hasan et al. 2019). The treatment also increased the activities of S-uptake-related enzymes, including ATPS (ATP sulfurylase), APSR (adenosine 5′-phosphosulfate reductase), SiR (sulfite reductase), and OASTL (*O*-acetylserine (thiol)lyase), as well as the expression of respective genes and those encoding sulfate transporters (*SUT1*, *SUT2*; Hasan et al. 2019). Another MEL treatment (100 μM) covering *B. napus* under Se stress (50 μM) increased levels of nonprotein thiols (NPTs) and PCs, also increasing the activities of two related enzymes [γ-ECS (γ-glutamylcysteine synthetase) and PCS (phytochelatins synthase)] (Ulhassan et al. 2019). Also, an MEL ($10\,nmol\,l^{-1}$) treatment applied to cucumber plants under Cu stress activated the C metabolism, presenting modulation of genes from glycolysis and pentose phosphate pathways (Cao et al. 2019).

Concerning secondary metabolism, a MEL (100 μM) treatment in cucumber plants under excess Fe increased PAL and PPO (polyphenol oxidase) enzymes, culminating with enhancement (in leaves) of phenols and flavonoids levels under lower and higher Fe availability (Ahammed et al. 2020). Also, an MEL treatment (100 μM) in tomato plants under Ni stress increased phenols, flavonoids, and anthocyanin contents (Jahan et al. 2020). Additionally, a foliar application of MEL (100 μM) in tea under As stress increased the anthocyanin content, with modulation of anthocyanin-biosynthesis-related genes [*CsCHS* (*chalcone synthase*) and *CsANS* (*anthocyanidin synthase*)] (Li et al. 2021).

All these MEL-mediated responses showed by plants under HM stress are strategies to be explored in bioremediation processes or developing HM-tolerant plants.

20.3.9 Salicylic Acid (SA)

Salicylic acid is a phenolic compound relevant in PG&D controlling several aspects of plant life cycles, such as seed germination, root nodulation/nitrogen metabolism, photosynthesis, and flower and fruit development (Lebeis et al. 2015; Arif et al. 2020). In plants, two distinct metabolic routes synthesize the SA: (i) from the chorismate (shikimic acid pathway), which conversion to isochorismate in plastids, and after conjugation with GSH in the cytosol generates isochorismate-9-glutamate, and that converted (spontaneously or enzymatically) to SA; (ii) from phenylalanine, being converted to *trans*-cinnamic acid, and subsequently to benzoic acid, and then to SA (Peng et al. 2021).

During the SA signal transduction pathway (Figure 20.3h), at lower SA levels, NPR1 (Nonexpressor of pathogenesis-related gene1) proteins (transcriptional activators) are inactivated, and NPR3 and NPR4 proteins (transcriptional repressors) inhibit the expression of SA-responsive genes. However, at higher SA levels, some SA binding to NPR1 proteins promotes their activation as transcriptional activators while binding to NPR3/NPR4 proteins release target genes from their repression. Then, NPR1 proteins activate TF members of the families WRKY and TGA (TGACG sequence-specific binding proteins), inducing responsive genes to perform the SA-mediated responses (Zhang and Li 2019; Peng et al. 2021). In this way, SA plays a relevant role in plants responding to (a)biotic stresses (Khan et al. 2015).

Regarding biotic stresses, SA is a central regulator of the systemic acquired resistance of plants against pathogens (Peng et al. 2021). About abiotic stresses, particularly HM stress, SA-mediated responses involve physiological and biochemical changes. Primarily, the SA attenuates HM toxicity in plants by alleviating the HM-induced oxidative stress. The SA increasing the H_2O_2 level triggers the SA-signaling pathway, activating related genes. Then, the induced plant antioxidant defense systems protect membranes and macromolecules from oxidative damage, reducing ROS, LP, and EL (Emamverdian et al. 2020). Also, by improving mineral nutrition, plant water status, and photosynthetic performance, plants alleviate the HM stress and the PG&D inhibition.

In this way, an SA treatment (10^{-6} M) applied to *A. annua* plants under As stress increased antioxidant enzyme activities, reducing H_2O_2 levels (Naeem et al. 2020). An SA treatment (0.1 mM) applied in rice plants (under Se stress) induced genes encoding antioxidant enzymes also increased the activity of enzymes of the AsA-GSH cycle and decreased the ROS levels (O_2^- and H_2O_2; Mostofa et al. 2020). Additionally, an SA (0.5 mM) treatment applied to pepper plants under Pb stress induced the H_2S gasotransmitter; its interaction increased the activity of enzymes from the AsA-GSH cycle and improved the Pb tolerance (Kaya 2021). Likewise, treatment with SA in corn plants under As stress increased the NO gasotransmitter, and its interaction improved glyoxalases and AsA-GSH cycle enzymes, conferring tolerance to As (Kaya et al. 2020).

In plants under HM stress, the SA also restores the plant water status and the photosynthesis performance. An SA treatment (0.1 mM) applied to rice plants under Se stress increased the photosynthetic pigments, relative leaf water content, and osmoprotectant compounds (soluble sugars and proteins) (Mostofa et al. 2020). Additionally, an SA treatment (10^{-5} M) in Indian mustard plants under Ni stress increased the relative water content, water use efficiency, chlorophyll content, g_s, C_i, and P_N (Zaid et al. 2019). Another SA treatment (0.5 mM) applied to pepper plants under Pb stress, in addition to increasing the gasotransmitter H_2S, also improved the chlorophyll content and photosynthetic capacity (Kaya 2021).

Furthermore, concerning plants under HM stress, the SA application restricts the HM uptake and root-to-shoot translocation. An SA treatment applied to rice plants under Cd stress increased the expression of genes encoding metal transporter proteins, such as *OsNRAMP5* (*Mn/Cd transporter*) and *OsHMA3* (*vacuolar transporter*). Those metal transporters increasing Fe and Mn levels will inhibit Cd uptake and transport (by competition), reducing the Cd content and its toxicity (Huang et al. 2022b). The treatment also inducted some genes: *OsICS1* (*isochorismate synthase 1; an SA-biosynthesis-related gene*) and *OsWRKY6*, which encodes a TF regulating the *OsICS1* gene (Huang et al. 2022b). Another SA treatment (100 μmol l^{-1}) on rice plants under Cd stress reduced Cd accumulation in roots and shoots, promoting its deposition in cell walls. To prevent the Cd-uptake by root cells, the authors stood out the strengthening of cell walls by increasing lignin, as reinforced by induced lignin-biosynthesis-related genes [*OsPAL* (*phenylalanine ammonia-lyase*) and *Os4CL3* (*4-coumarate-CoA ligase*)]. Also, the induced *OsPME29* (*pectin methylesterase*) gene and the increase in pectin content pointed to the relevance of pectin esterification in cell walls leading to HM tolerance (Pan et al. 2021). In addition, an SA treatment in rice under Cd stress reduced the Cd accumulated in grains and increased the Cd tolerance. The results highlighted SA-signaling-related genes: as those induced [regulator protein *OsNPR1* (*O. sativa non-expressor of PR1*) and the *OsWRKY45 TF*] and those repressed (*OsWRKY7* and *OsWRKY70*), both genes encoding TFs negatively correlated with the SA content. Also, SA-treated plants showed a reduction of Cd translocated to stem, leaves, and panicles, reducing the HM in grains (Wang et al. 2021).

Furthermore, an SA treatment (100 μM) in rice plants under Cd stress increased the expression of genes encoding chelating agents (PC and MT), such as *OsPCS1* (*phytochelatin synthase*) and *OsMT1* (*metallothionein*), in addition to a gene encoding a vacuolar transporter (*OsHMA3*), suggesting the Cd sequestration (Huda et al. 2016).

All these SA-mediated responses showed by plants under HM stress are strategies to be explored by plant breeders developing HM-tolerant plants or considered in bioremediation processes.

20.3.10 Strigolactone (SL)

Strigolactones (SLs) are multifunctional metabolites derived from carotenoids that occur naturally in plants as allelochemicals in root exudates. Discovered in the mid-1960s (Cook et al. 1966), SL was first associated with inducing seed germination of root-parasitic weeds and attraction of symbiotic arbuscular mycorrhizal fungi (Torres-Vera et al. 2016; Kobae et al. 2018). The SLs influence several plants' biological processes, including root development, stem/shoot growth, and leaf senescence (Rameau et al. 2019).

During the SL signal transduction pathway (Figure 20.3i), at lower SL levels, D53 (SMXLs) proteins interact with TPL (TOPLESS) TF proteins, inhibiting activities of TCP (teosinte branched 1/cycloidea/proliferating cell factor) TF proteins, and inhibiting the SL-mediated response. When at higher SL levels, proteins D14 (DWARF14) together with D3 (MAX2) drive the D53/SMXLs to degradation, releasing TCP TFs to regulate the expression of responsive genes, performing the SL-mediated response (Zhou et al. 2013).

Regarding HM stress, some physiological and biochemical changes highlight SL-induced positive effects on PG&D. Thus, exogenous application of GR24 (1 μM), a synthetic SL analog of strigol, to barley plants under Cd stress (10 μM) mitigated the HM toxicity by improving antioxidant defense system, reducing MDA levels, also increasing mineral nutritional, chlorophyll content, and photosynthetic capacity (Qiu et al. 2021). Also, a GR24 treatment (1 μM) applied to rice seedlings under Cd ($2\,mg\,l^{-1}$) stress reduced the Cd accumulated in the aerial parts of the rice plant (Xu et al. 2019). In addition, a GR24 treatment ($1\,\mu mol\,l^{-1}$) supplemented on grass (*Panicum virgatum* L.) seedlings under Cd stress ($10\,\mu mol\,l^{-1}$) increased antioxidant enzymes, mitigating the oxidative stress, also decreased the Cd uptake (increasing the absorption of Zn and Fe) and increased the chlorophyll content, photosynthetic capacity, and PG&D (Tai et al. 2017).

In plants under stress, the SLs also regulate the architecture and development of roots. The SLs in plants under higher Pi (inorganic P form) levels inhibit lateral roots and root hair density, reducing root absorption (Ruyter-Spira et al. 2011). In addition, some HMs, including As, are absorbed via the root system by phosphate transporter (PT) proteins (Li et al. 2016b). This strategy (phytoextraction) could be explored considering the remediation of HM-contaminated soils.

Abiotic stresses also influence the biosynthesis of SLs. In rice seedlings under As stress, some SL-defective mutant lines, covering two SL-biosynthesis-related genes (*D10* and *D17*), accumulated higher As levels in roots and increased the As-sensitivity concerning the wild-type plants. These mutants increased oxidative stress, repressed genes encoding antioxidant enzymes (SOD, APX, and CAT), and decreased *phosphate transporter genes* (*OsPT1*, *OsPT2*, *OsPT4*, and *OsPT8*) in roots (Mostofa et al. 2021).

Although the mentioned studies highlighted SL's beneficial actions in plants facing HM stress, the development of bioactive optimized synthetic SLs could accelerate the knowledge acquisition since the SL-bioavailability and high diversification still limit the studies (Screpanti et al. 2016; Mostofa et al. 2018b). However, some mentioned SL-mediated responses showed by plants under HM stress are strategies to be explored in bioremediation processes or developing HM-tolerant plants.

20.4 Crosstalk of Phytohormones in Plants Responding to Heavy Metals

In HM-contaminated soils, the HM uptake into root cells induces complex changes at genetic, biochemical, and physiological levels. Considering the cellular condition, the PHYs interact with each other in a synergistic (additive interactions) or antagonistic (negative interactions) way to regulate PG&D in a mechanism known as crosstalk (Khan et al. 2020). The hormonal crosstalk in plants under HM stress coordinates actions to circumvent HM-induced adverse effects, including the PG&D inhibition, as described below.

A hormonal crosstalk (AUX-CK) influenced the root growth and pattern in *Arabidopsis* plants under Cd stress (Bruno et al. 2017). The two hormones are synergistic in controlling the shoot stem cell niche, but they act antagonistically to control the size of the root meristem. In root morphogenesis, AUX promotes a pivotal role; however, the Cd stress (25–50 μM) inhibits primary root growth (despite a root branching increase) due to an imbalance of accumulated AUX at the root tips reducing the root apical meristem (RAM) (Bruno et al. 2017). The Cd stress triggering hormonal signaling pathways prevent the AUX distribution at the root apex by decreasing the expression of *PIN* (*PIN-FORMED*) genes and promoting a mislocalization of PIN proteins, affecting the meristem pattern. At a short distance, the AUX distribution is cell to cell, involving membrane transporters, such as the PIN proteins (AUX-transport facilitator, as efflux carriers), which polar distribution on plasma membranes drives the AUX flow direction. The excess Cd provokes PIN mislocalization. Also, the Cd stress deregulates the expression of *SCR* (*SCARECROW*) genes. These genes encode TF members of the GRAS family with involvement in the AUX-CK crosstalk by controlling the RAM activity and maintenance (Bruno et al. 2017).

Another crosstalk (ETH-AUX) involved *Arabidopsis* plants exposed to Cr(VI) (100 μM) with the ETH mediating the Cr(VI)-induced inhibition of primary root elongation by increasing AUX accumulation and polar transport after inducing the *AUX1* gene (Wakeel et al. 2018). Plants under Cr stress reduce seed germination, primary root elongation, leaf number, and biomass. Previous results associated the reduced root length with repressed meristematic cell division. Also predicted an association of *repressed cell cycle-related genes* [*CYCA2-1*, *CYCB1-1*, *CYCD1-1*, *CYCD2-1*] with induced *E2Fb* gene, which encoded TF negatively regulates cell division. Thus, the Cr stress triggers the ETH signaling and biosynthesis, while the ETH mediates *AUX1* gene expression, increasing AUX accumulation, leading the primary root elongation inhibition (Wakeel et al. 2018). The synergistic effect involves the AUX enhancing the ETH biosynthesis and inducing the ETH signaling pathway; in turn, the ETH induces AUX-biosynthesis-related genes, thereby inducing accumulation, distribution, and polar transport of AUXs, which mediate the root growth inhibition.

Furthermore, another ETH-AUX crosstalk involved plants (under excess Fe) inhibiting the lateral root (LR) formation. During the LR development, the AUX acts as a positive regulator. In plants under excess Fe, the ETH measures the tolerance of LR formation through AUX transporter proteins (Li et al. 2015b). The excess Fe inhibits the LR formation downregulating *PIN2* genes (encoding *AUX efflux carrier proteins*) and upregulating *ETH-biosynthesis-related genes*. With higher ETH levels, the ETH attenuates the inhibiting effects of excess Fe partially through AUX1, the AUX influx carrier proteins (Li et al. 2015b). Thus, the ETH-mediated LR formation is dependent on the AUX signaling pathway (Ivanchenko et al. 2008; Lewis et al. 2011). This fine-tuning adjustment of the root system architecture restricts excessive Fe uptake (avoiding its toxicity), with the ETH counterbalancing the AUX1 and preventing massive LR reduction to maintain the nutrient uptake (Heil and Baldwin 2002).

Exogenous PHY applications on plants under HM stress also modulate hormonal crosstalks. BR-AUX crosstalk provided by 24-EBL (active BR form) supplementation improved PG&D, reduced the Cd toxicity, and induced differentially expressed *AUX-signaling-related genes*, including those encoding ARF TFs and Aux/IAA proteins, reinforcing the crosstalk (Mumtaz et al. 2022). According to Sun et al. (2020), the BRs control the nuclear AUX input and the AUX output signaling since the BR-signaling modulates AUX transport facilitator PILS (PIN-LIKES) proteins, accumulation of which in the endoplasmic reticulum provides the AUXs the nuclear abundance and the signaling.

Another crosstalk (MEL-JA) involved a MEL pretreatment (300 μmol l^{-1}) on melon (*Cucumis melo* L.) seeds with the promotion of the root development under Cu stress (300 μmol l^{-1} $CuSO_4$) by stimulation of the lipoxygenase pathway and suppression of the JA biosynthesis. The pretreatment improved antioxidant defense systems and the root vigor. The analysis highlighted four *LOX* (*lipoxygenase*) genes (all repressed) and two metabolites (linoleic acid and lecithin) as differentially expressed candidates. The LOX enzyme is central in lipid degradation and JA biosynthesis. In turn, the lecithinase enzyme converts the lecithin to linoleic acid, while the LOX enzyme converts linoleic acid to JA (Feussner and Wasternack 2002). The MEL treatment results showed increased lecithin and decreased linoleic acid contents, and these results, together with the repression of LOX-related genes, reinforced the JA biosynthesis inhibition (Hu et al. 2020).

In turn, another crosstalk (AUX-SA) involved the root morphology, cellular homeostasis, and molecular signaling in rice plants responding to multiple (a)biotic stresses, including excess Cr(VI) (Tiwari et al. 2020). The analysis highlighted MYB TFs (R1-type) controlling the crosstalk. Plants under Cr(VI) exposure [150–300 μM Cr(VI) ($K_2Cr_2O_7$)] upregulating the *OsMyb-R1* gene showed Cr tolerance and increased the LR formation. The Cr(VI)-tolerant transgenic lines overexpressing *OsMYB-R1* accumulated AUX and triggered the SA-signaling pathway inducing responsive genes [*calcium-dependent protein kinase* (*CDPK*), *POD*, and *GST*]. The treatment increased antioxidant enzymes, reduced oxidative stress, and improved vacuolar Cr-sequestration, which reduced free Cr(VI) translocated to shoots. The enhanced HM-compartmentalization into root cells prevented its translocation and accumulation in shoots (usually used for fodder) and grains (for human nutrition) (Tiwari et al. 2020).

An exogenous SA treatment (0.5 mmol l^{-1}) applied on sunflower (*Helianthus annuus* L.) seeds under zinc stress (0.3 mmol l^{-1} Zn^{2+}) highlighted crosstalk (SA-GA-ABA) attenuating Zn-associated adverse effects on seed germination by modulation of GA and ABA levels (Huang et al. 2021). Both phytohormones regulate the seed germination process, with GA inducing the seed germination (activating hydrolases), while the ABA accumulation enforces the seed dormancy (Rajjou et al. 2012). The SA treatment reduced the Zn^{2+} level by decreasing ABA and increasing levels of GA and endogenous SA contents, improving seed germination. The treatment also induced Zn-transporter-related genes from the families MTP (metal tolerance protein), known as CDF (cation diffusion facilitator) and HMA (heavy metal ATPase) (Huang et al. 2021). These families comprise Zn-eliminating proteins, mainly HMA members involved in transmembrane transport of divalent cations, and MTP members involved in HM-detoxification processes, including discharge of excess Zn^{2+} from the cytoplasm or vacuolar HM-compartmentation (Fasani et al. 2017). The SA treatment also downregulated the ABA content improving the seed germination. Usually, a seed germination process reduces the ABA level, but during the HM stress, this ABA level increases. Also, during seed germination, the endogenous GA content increases; however, the HM stress reduces the GA content. Thus, the SA treatment increased the GA content, improving the seed germination by a synergistic effect. Furthermore, regular seed germination gradually increases the endogenous SA content, but the HM stress reduces this SA accumulation. In turn, the applied SA treatment supplemented the endogenous SA deficiency improving the seed germination (Huang et al. 2021).

Indian mustard plants under As ($24\,mg\,kg^{-1}$) supplemented with Se ($2\,mg\,kg^{-1}$ soil) + Ethephon® (2-chloroethylphosphonic acid, $200\,\mu l\,l^{-1}$), an ETH source, revealed crosstalk (ETH-ABA) improving photosynthetic capacity and reducing oxidative stress, ABA content, and excessive ETH production (Sehar et al. 2022). The ETH and ABA act antagonistically in the stomatal regulation since the ETH induces stomatal opening (by controlling redox homeostasis) while inhibiting ABA-induced stomatal closure. The treatment (ETH + Se) modulated in the plants the ABA levels, alleviating the As toxicity since applications of NBD, an ETH action inhibitor, on plants under As stress increased oxidative stress and ABA levels, reducing the photosynthetic capacity. These results highlighted the relevance of ETH in reducing the ABA levels to promote plant protection (Sehar et al. 2022).

In some cases, rather than a single PHY application, the co-application of PHYs may be more beneficial concerning alleviating HM-associated adverse effects. The co-application (pre-sowing of seeds) of 24-EBL ($10^{-7}\,M$) and SA (1 mM) treatment in Indian mustard plants exposed to Pb (0.25–0.75 mM) alleviated the HM-associated adverse effects by enhancing antioxidant defenses and osmolyte contents (proline, trehalose, and glycine betaine), while reduced the HM uptake and accumulation (Kohli et al. 2018). The treated plants under HM stress enhanced adaptive mechanisms compared to individual treatments, such as improving plant acclimatization probably due to membranes with stabilized electrical properties, blocking the HM uptake, minimizing the root-to-shoot transportation, and its accumulation in shoots (Kohli et al. 2018).

Another example of combination included JA (1 nM) and GA3 (10–6 M), as seed priming followed by foliar spray, respectively, in chickpea (*Cicer arietinum* L.) plants exposed to Cd stress (150 µM $CdSO_4{\cdot}8H_2O$). This treatment also enhanced osmolyte accumulation and improved ROS scavenging (Ahmad et al. 2021). Thus, combining PHYs and exploring crosstalk responses is a valuable alternative to suppress some HM-associated PG&D inhibition.

20.5 Final Considerations

The future scenario of the human species points to a growing population living worldwide in places with limited natural resources. However, besides this scenario, the next could be worst since the rising population will increase waste, and waste not adequately managed or recycled, will contaminate soil, water, and air. This growing population increasingly demand food, ordering larger quantity and better quality. However, crop productivity does not always follow the demands. Traditional plant breeding programs require time, money, and expertise. The omics technology is shortening some requirements. In turn, non-observation of good agricultural practices deteriorates soil and water quality, restricting lands. Soil salinization is an example, making areas increasingly unproductive. In addition, world climate change is accelerating, becoming more frequent and intense. These events will condemn cultivars provided by crop breeding programs since they will not respond to the new environmental conditions. At the same time, the rising urban population increasing their activities will increase waste, requiring adequate disposal and treatment to avoid contamination of natural reservoirs and even farmlands. Concerning the residues, since HMs are not recycled and presented in several products, such as the technological ones, HM is a current and future problem, contaminating ecosystems. Plants constantly face environmental stresses during their life cycles, and during their evolution, performing stress responses, new genes playing adaptive and physiological responses allow them to finish their life cycles. Thus, the genetic variability is the substrate to the evolutionary processes and also for selections applied by breeders. Phytohormones of different classes trigger many of these plant responses. Understanding how phytohormones signal HM stresses and regulate genes and products mediating the HM-tolerant response is relevant to plant breeders. The knowledge of intrinsic hormonal plant response to HM stress could provide strategies to be optimized by biotechnology, aiming to develop HM-tolerant materials. Furthermore, plants efficiently performing those strategies are promising candidates for HM-phytoextraction in the phytoremediation of HM-contaminated soils, representing an economically sustainable alternative. We hope this chapter helps students and professionals deal with this challenge surrounding us.

References

Abiri, R., Shaharuddin, N.A., Maziah, M. et al. (2017). Role of ethylene and the APETALA 2/ethylene response factor superfamily in rice under various abiotic and biotic stress conditions. *Environmental and Experimental Botany* 134: 33–44.

Ahammed, G.J., Wu, M., Wang, Y. et al. (2020). Melatonin alleviates iron stress by improving iron homeostasis, antioxidant defense and secondary metabolism in cucumber. *Scientia Horticulturae* 265: 109205.

Ahanger, M.A., Aziz, U., Alsahli, A.A. et al. (2020). Combined kinetin and spermidine treatments ameliorate growth and photosynthetic inhibition in *Vigna angularis* by up-regulating antioxidant and nitrogen metabolism under cadmium stress. *Biomolecules* 10: 147.

Ahmad, P., Raja, V., Ashraf, M. et al. (2021). Jasmonic acid (JA) and gibberellic acid (GA3) mitigated Cd-toxicity in chickpea plants through restricted cd uptake and oxidative stress management. *Scientific Reports* 11: 1–17.

Alatawi, A., Wang, X., Saleem, M.H. et al. (2023). Individual and synergic effects of phosphorus and gibberellic acid on organic acids exudation pattern, ultra-structure of chloroplast and stress response gene expression in Cu-stressed jute (*Corchorus Capsularis* L.). *Journal of Plant Growth Regulation* 42: 1186–1211.

Al-Huqail, A.A., Khan, M.N., Ali, H.M. et al. (2020). Exogenous melatonin mitigates boron toxicity in wheat. *Ecotoxicology and Environmental Safety* 201: 110822.

Ali, M.A., Asghar, H.N., Khan, M.Y. et al. (2015). Alleviation of nickel-induced stress in mungbean through application of gibberellic acid. *International Journal of Agriculture and Biology* 17: 990–994.

Ali, H., Khan, E., and Ilahi, I. (2019). Environmental chemistry and ecotoxicology of hazardous heavy metals: environmental persistence, toxicity, and bioaccumulation. *Journal of Chemistry* 2019: 1–14.

Amri, B., Khamassi, K., Ali, M.B. et al. (2016). Effects of gibberellic acid on the process of organic reserve mobilization in barley grains germinated in the presence of cadmium and molybdenum. *South African Journal of Botany* 106: 35–40.

Andresen, E., Peiter, E., and Küpper, H. (2018). Trace metal metabolism in plants. *Journal of Experimental Botany* 69: 909–954.

Arif, Y., Sami, F., Siddiqui, H. et al. (2020). Salicylic acid in relation to other phytohormones in plant: a study towards physiology and signal transduction under challenging environment. *Environmental and Experimental Botany* 175: 104040.

Arnao, M.B. and Hernández-Ruiz, J. (2019). Melatonin: a new plant hormone and/or a plant master regulator? *Trends in Plant Science* 24: 38–48.

Arteca, R.N. and Arteca, J.M. (2007). Heavy-metal-induced ethylene production in *Arabidopsis thaliana*. *Journal of Plant Physiology* 164: 1480–1488.

Back, K., Tan, D.X., and Reiter, R.J. (2016). Melatonin biosynthesis in plants: multiple pathways catalyze tryptophan to melatonin in the cytoplasm or chloroplasts. *Journal of Pineal Research* 61: 426–437.

Balfagón, D., Sengupta, S., Gómez-Cadenas, A. et al. (2019). Jasmonic acid is required for plant acclimation to a combination of high light and heat stress. *Plant Physiology* 181: 1668–1682.

Bashri, G., Singh, S., Prasad, S.M. et al. (2021). Kinetin mitigates Cd-induced damages to growth, photosynthesis and PS II photochemistry of Trigonella seedlings by up-regulating ascorbate-glutathione cycle. *PLoS One* 16: e0249230.

Bruno, L., Pacenza, M., Forgione, I. et al. (2017). In *Arabidopsis thaliana* cadmium impact on the growth of primary root by altering SCR expression and auxin-cytokinin cross-talk. *Frontiers in Plant Science* 8: 1323.

Cai, S.Y., Zhang, Y., Xu, Y.P. et al. (2017). HsfA1a upregulates melatonin biosynthesis to confer cadmium tolerance in tomato plants. *Journal of Pineal Research* 62: e12387.

Cao, S., Chen, Z., Liu, G. et al. (2009). The Arabidopsis ethylene-insensitive 2 gene is required for lead resistance. *Plant Physiology and Biochemistry* 47: 308–312.

Cao, Y.Y., Qi, C.D., Li, S. et al. (2019). Melatonin alleviates copper toxicity via improving copper sequestration and ROS scavenging in cucumber. *Plant and Cell Physiology* 60: 562–574.

Chahel, A.A., Yousaf, Z., Zeng, S. et al. (2020). Growth and physiological alterations related to root-specific gene function of LrERF061-OE in *Glycyrrhiza uralensis* Fisch. hairy root clones under cadmium stress. *Plant Cell, Tissue and Organ Culture* 140: 115–127.

Chen, H., Yang, R., Zhang, X. et al. (2021). Foliar application of gibberellin inhibits the cadmium uptake and xylem transport in lettuce (*Lactuca sativa* L.). *Scientia Horticulturae* 288: 110410.

Cook, C.E., Whichard, L.P., Turner, B. et al. (1966). Germination of witchweed (*Striga lutea* Lour.): isolation and properties of a potent stimulant. *Science* 154: 1189–1190.

Cortleven, A. and Schmülling, T. (2015). Regulation of chloroplast development and function by cytokinin. *Journal of Experimental Botany* 66: 4999–5013.

Cortleven, A., Nitschke, S., Klaumünzer, M. et al. (2014). A novel protective function for cytokinin in the light stress response is mediated by the ARABIDOPSIS HISTIDINE KINASE2 and ARABIDOPSIS HISTIDINE KINASE3 receptors. *Plant Physiology* 164: 1470–1483.

Cortleven, A., Leuendorf, J.E., Frank, M. et al. (2019). Cytokinin action in response to abiotic and biotic stresses in plants. *Plant, Cell and Environment* 42: 998–1018.

Cukor, J., Rasakova, N.A.M., Linda, R. et al. (2018). Effects of brassinosteroid application on seed germination of scots pine under standard and heat stress conditions. *Baltic Forestry* 24: 60–67.

Cunha, L.F.S., Oliveira, V.P., Nascimento, A.W.S. et al. (2021). Leaf application of 24-epibrassinolide mitigates cadmium toxicity in young *Eucalyptus urophylla* plants by modulating leaf anatomy and gas exchange. *Physiologia Plantarum* 173: 67–87.

DalCorso, G., Manara, A., and Furini, A. (2013). An overview of heavy metal challenge in plants: from roots to shoots. *Metallomics* 5: 1117–1132.

Demecsová, L., Zelinová, V., Liptáková, Ľ. et al. (2020). Indole-3-butyric acid priming reduced cadmium toxicity in barley root tip via NO generation and enhanced glutathione peroxidase activity. *Planta* 252: 1–16.

Deng, B., Zhang, W., and Yang, H. (2022). Abscisic acid decreases cell death in *Malus hupehensis* Rehd. under cd stress by reducing root Cd^{2+} influx and leaf transpiration. *Journal of Plant Growth Regulation* 41: 639–646.

Dietz, K.J., Vogel, M.O., and Viehhauser, A. (2010). AP2/EREBP transcription factors are part of gene regulatory networks and integrate metabolic, hormonal and environmental signals in stress acclimation and retrograde signalling. *Protoplasma* 245: 3–14.

Djemal, R. and Khoudi, H. (2022). The ethylene-responsive transcription factor of durum wheat, TdSHN1, confers cadmium, copper, and zinc tolerance to yeast and transgenic tobacco plants. *Protoplasma* 259: 19–31.

Edelstein, M. and Ben-Hur, M. (2018). Heavy metals and metalloids: sources, risks and strategies to reduce their accumulation in horticultural crops. *Scientia Horticulturae* 234: 431–444.

Emamverdian, A., Ding, Y., and Mokhberdoran, F. (2020). The role of salicylic acid and gibberellin signaling in plant responses to abiotic stress with an emphasis on heavy metals. *Plant Signaling and Behavior* 15: 1777372.

Emamverdian, A., Ding, Y., Mokhberdoran, F., and Ahmad, Z. (2021). Mechanisms of selected plant hormones under heavy metal stress. *Polish Journal of Environmental Studies* 30: 497–508.

Ermoshin, A.A., Kiseleva, I.S., and Shatunova, S.A. (2019). Agrobacterial cytokinin synthesis gene *ipt* increases plant resistance to high doses of copper ions. *AIP Conference Proceedings* 2063: 030006.

Falk, K.L., Kästner, J., Bodenhausen, N. et al. (2014). The role of glucosinolates and the jasmonic acid pathway in resistance of *Arabidopsis thaliana* against molluscan herbivores. *Molecular Ecology* 23: 1188–1203.

Fasani, E., DalCorso, G., Varotto, C. et al. (2017). The *MTP1* promoters from *Arabidopsis halleri* reveal cis-regulating elements for the evolution of metal tolerance. *New Phytologist* 214: 1614–1630.

Feussner, I. and Wasternack, C. (2002). The lipoxygenase pathway. *Annual Review of Plant Biology* 53: 275–297.

Furio, R.N., Albornoz, P.L., Coll, Y. et al. (2019). Effect of natural and synthetic brassinosteroids on strawberry immune response against *Colletotrichum acutatum*. *European Journal of Plant Pathology* 153: 167–181.

Gangwar, S., Singh, V.P., Srivastava, P.K., and Maurya, J.N. (2011). Modification of chromium (VI) phytotoxicity by exogenous gibberellic acid application in *Pisum sativum* (L.) seedlings. *Acta Physiologiae Plantarum* 33: 1385–1397.

Gao, S. and Chu, C. (2020). Gibberellin metabolism and signaling: targets for improving agronomic performance of crops. *Plant and Cell Physiology* 61: 1902–1911.

Giacomelli, L., Rota-Stabelli, O., Masuero, D. et al. (2013). Gibberellin metabolism in *Vitis vinifera* L. during bloom and fruit-set: functional characterization and evolution of grapevine gibberellin oxidases. *Journal of Experimental Botany* 64: 4403–4419.

Gong, Q., Li, Z., Wang, L. et al. (2019). Exogenous of indole-3-acetic acid application alleviates copper toxicity in spinach seedlings by enhancing antioxidant systems and nitrogen metabolism. *Toxics* 8: 1–12.

Grifoni, M., Rosellini, I., Petruzzelli, G. et al. (2021). Application of sulphate and cytokinin in assisted arsenic phytoextraction by industrial *Cannabis sativa* L. *Environmental Science and Pollution Research* 28: 47294–47305.

Gruszka, D., Janeczko, A., Dziurka, M. et al. (2016). Barley brassinosteroid mutants provide an insight into phytohormonal homeostasis in plant reaction to drought stress. *Frontiers in Plant Science* 7: 1–14.

Gu, Q., Chen, Z., Yu, X. et al. (2017). Melatonin confers plant tolerance against cadmium stress via the decrease of cadmium accumulation and reestablishment of microRNA-mediated redox homeostasis. *Plant Science* 261: 28–37.

Guan, C., Ji, J., Wu, D. et al. (2015). The glutathione synthesis may be regulated by cadmium-induced endogenous ethylene in *Lyciumchinense*, and overexpression of an ethylene responsive transcription factor gene enhances tolerance to cadmium stress in tobacco. *Molecular Breeding* 35: 1–13.

Guedes, F.R.C.M., Maia, C.F., Silva, B.R.S. et al. (2021). Exogenous 24-Epibrassinolide stimulates root protection, and leaf antioxidant enzymes in lead stressed rice plants: central roles to minimize Pb content and oxidative stress. *Environmental Pollution* 280: 116992.

Gururani, M.A., Mohanta, T.K., and Bae, H. (2015). Current understanding of the interplay between phytohormones and photosynthesis under environmental stress. *International Journal of Molecular Sciences* 16: 19055–19085.

Hasan, M.K., Ahammed, G.J., Sun, S. et al. (2019). Melatonin inhibits cadmium translocation and enhances plant tolerance by regulating sulfur uptake and assimilation in *Solanum lycopersicum* L. *Journal of Agricultural and Food Chemistry* 67: 10563–10576.

Hashem, H.A. (2014). Cadmium toxicity induces lipid peroxidation and alters cytokinin content and antioxidant enzyme activities in soybean. *Botany* 92: 1–7.

He, Y., Zhang, T., Sun, Y. et al. (2022). Exogenous IAA alleviates arsenic toxicity to rice and reduces arsenic accumulation in rice grains. *Journal of Plant Growth Regulation* 41: 734–741.

Heil, M. and Baldwin, I.T. (2002). Fitness costs of induced resistance: emerging experimental support for a slippery concept. *Trends in Plant Science* 7: 61–67.

Hong, J., Sung, J., and Ryu, H. (2018). Brassinosteroids-mediated regulation of ABI3 is involved in high-temperature induced early flowering in plants. *Journal of Plant Biotechnology* 45: 83–89.

Hoque, T.S., Hossain, M.A., Mostofa, M.G. et al. (2016). Methylglyoxal: an emerging signaling molecule in plant abiotic stress responses and tolerance. *Frontiers in Plant Science* 7: 327–345.

Hu, Z., Fu, Q., Zheng, J. et al. (2020). Transcriptomic and metabolomic analyses reveal that melatonin promotes melon root development under copper stress by inhibiting jasmonic acid biosynthesis. *Horticulture Research* 7: 79.

Huang, T.L., Nguyen, Q.T.T., Fu, S.F. et al. (2012). Transcriptomic changes and signalling pathways induced by arsenic stress in rice roots. *Plant Molecular Biology* 80: 587–608.

Huang, Y.T., Cai, S.Y., Ruan, X.L. et al. (2021). Salicylic acid enhances sunflower seed germination under Zn^{2+} stress via involvement in Zn^{2+} metabolic balance and phytohormone interactions. *Scientia Horticulturae* 275: 109702.

Huang, J., Shi, Y., Wang, M. et al. (2022a). Oshipp17 is Involved in Plant Heavy Metal Toxic Response Via Affecting Cytokinin Homeostasis. SSRN 4045413.

Huang, Q., Xu, R., Zhang, Y. et al. (2022b). Salicylic acid ameliorates cadmium toxicity by increasing nutrients uptake and upregulating antioxidant enzyme activity and uptake/transport-related genes in *Oryza sativa* L. *indica*. *Journal of Plant Growth Regulation* 42: 1–13.

Huda, A.K.M., Swaraz, A.M., Reza, M.A. et al. (2016). Remediation of chromium toxicity through exogenous salicylic acid in rice (*Oryza sativa* L.). *Water, Air, and Soil Pollution* 227: 1–11.

Hwang, I. and Sakakibara, H. (2006). Cytokinin biosynthesis and perception. *Physiologia Plantarum* 126: 528–538.

Iqbal, N., Khan, N.A., Ferrante, A. et al. (2017). Ethylene role in plant growth, development and senescence: interaction with other phytohormones. *Frontiers in Plant Science* 8: 475.

Ivanchenko, M.G., Muday, G.K., and Dubrovsky, J.G. (2008). Ethylene-auxin interactions regulate lateral root initiation and emergence in *Arabidopsis thaliana*. *The Plant Journal* 55: 335–347.

Jahan, M.S., Guo, S., Baloch, A.R. et al. (2020). Melatonin alleviates nickel phytotoxicity by improving photosynthesis, secondary metabolism and oxidative stress tolerance in tomato seedlings. *Ecotoxicology and Environmental Safety* 197: 110593.

Jalmi, S.K., Bhagat, P.K., Verma, D. et al. (2018). Traversing the links between heavy metal stress and plant signaling. *Frontiers in Plant Science* 9: 12.

Jamla, M., Khare, T., Joshi, S. et al. (2021). Omics approaches for understanding heavy metal responses and tolerance in plants. *Current Plant Biology* 27: 100213.

Jan, S., Noman, A., Kaya, C. et al. (2020). 24-Epibrassinolide alleviates the injurious effects of Cr(VI) toxicity in tomato plants: insights into growth, physio-biochemical attributes, antioxidant activity and regulation of Ascorbate–glutathione and Glyoxalase cycles. *Journal of Plant Growth Regulation* 39: 1587–1604.

Javed, T., Ali, M.M., Shabbir, R. et al. (2021). Alleviation of copper-induced stress in pea (*Pisum sativum* L.) through foliar application of gibberellic acid. *Biology* 10: 120.

Ji, Y. and Guo, H. (2013). From endoplasmic reticulum (ER) to nucleus: EIN2 bridges the gap in ethylene signaling. *Molecular Plant* 6: 11–14.

Jiang, Y., Liang, G., Yang, S., and Yu, D. (2014). Arabidopsis WRKY57 functions as a node of convergence for jasmonic acid–and auxin-mediated signaling in jasmonic acid–induced leaf senescence. *The Plant Cell* 26: 230–245.

Ju, C., Yoon, G.M., Shemansky, J.M. et al. (2012). CTR1 phosphorylates the central regulator EIN2 to control ethylene hormone signaling from the ER membrane to the nucleus in *Arabidopsis*. *Proceedings of the National Academy of Sciences* 109: 19486–19491.

Kamran, M., Wang, D., Alhaithloul, H.A.S. et al. (2021). Jasmonic acid-mediated enhanced regulation of oxidative, glyoxalase defense system and reduced chromium uptake contributes to alleviation of chromium (VI) toxicity in choysum (*Brassica parachinensis* L.). *Ecotoxicology and Environmental Safety* 208: 111758.

Kang, J., Hwang, J.U., Lee, M. et al. (2010). PDR-type ABC transporter mediates cellular uptake of the phytohormone abscisic acid. *Proceedings of the National Academy of Sciences* 107: 2355–2360.

Kanno, Y., Hanada, A., Chiba, Y. et al. (2012). Identification of an abscisic acid transporter by functional screening using the receptor complex as a sensor. *Proceedings of the National Academy of Sciences* 109: 9653–9658.

Kaur, R., Das, S., Bansal, S. et al. (2021). Heavy metal stress in rice: uptake, transport, signaling, and tolerance mechanisms. *Physiologia Plantarum* 173: 430–448.

Kaushik, S., Sharma, P., Kaur, G. et al. (2022). Seed priming with methyl jasmonate mitigates copper and cadmium toxicity by modifying biochemical attributes and antioxidants in *Cajanus cajan*. *Saudi Journal of Biological Sciences* 29: 721–729.

Kaya, C. (2021). Salicylic acid-induced hydrogen sulphide improves lead stress tolerance in pepper plants by upraising the ascorbate-glutathione cycle. *Physiologia Plantarum* 173: 8–19.

Kaya, C., Okant, M., Ugurlar, F. et al. (2019). Melatonin-mediated nitric oxide improves tolerance to cadmium toxicity by reducing oxidative stress in wheat plants. *Chemosphere* 225: 627–638.

Kaya, C., Ashraf, M., Alyemeni, M.N. et al. (2020). Salicylic acid-induced nitric oxide enhances arsenic toxicity tolerance in maize plants by upregulating the ascorbate-glutathione cycle and glyoxalase system. *Journal of Hazardous Materials* 399: 123020.

Kazan, K. (2015). Diverse roles of jasmonates and ethylene in abiotic stress tolerance. *Trends in Plant Science* 20: 219–229.

Khan, M.I.R. and Khan, N.A. (2014). Ethylene reverses photosynthetic inhibition by nickel and zinc in mustard through changes in PS II activity, photosynthetic nitrogen use efficiency, and antioxidant metabolism. *Protoplasma* 251: 1007–1019.

Khan, M.I.R., Fatma, M., Per, T.S. et al. (2015). Salicylic acid-induced abiotic stress tolerance and underlying mechanisms in plants. *Frontiers in Plant Science* 6: 462.

Khan, M.Y., Prakash, V., Yadav, V. et al. (2019). Regulation of cadmium toxicity in roots of tomato by indole acetic acid with special emphasis on reactive oxygen species production and their scavenging. *Plant Physiology and Biochemistry* 142: 193–201.

Khan, N., Bano, A., Ali, S., and Babar, M. (2020). Crosstalk amongst phytohormones from planta and PGPR under biotic and abiotic stresses. *Plant Growth Regulation* 90: 189–203.

Kim, D.Y., Bovet, L., Maeshima, M. et al. (2007). The ABC transporter AtPDR8 is a cadmium extrusion pump conferring heavy metal resistance. *The Plant Journal* 50: 207–218.

Kobae, Y., Kameoka, H., Sugimura, Y. et al. (2018). Strigolactone biosynthesis genes of rice are required for the punctual entry of arbuscular mycorrhizal fungi into the roots. *Plant and Cell Physiology* 59: 544–553.

Kohli, S.K., Handa, N., Bali, S. et al. (2018). Modulation of antioxidative defense expression and osmolyte content by co-application of 24-epibrassinolide and salicylic acid in Pb exposed Indian mustard plants. *Ecotoxicology and Environmental Safety* 147: 382–393.

Kuromori, T., Sugimoto, E., and Shinozaki, K. (2011). *Arabidopsis* mutants of AtABCG22, an ABC transporter gene, increase water transpiration and drought susceptibility. *The Plant Journal* 67: 885–894.

Lavy, M. and Estelle, M. (2016). Mechanisms of auxin signaling. *Development* 143: 3226–3229.

Lebeis, S.L., Paredes, S.H., Lundberg, D.S. et al. (2015). Salicylic acid modulates colonization of the root microbiome by specific bacterial taxa. *Science* 349: 860–864.

Lee, J., Han, S., Lee, H.Y. et al. (2019). Brassinosteroids facilitate xylem differentiation and wood formation in tomato. *Planta* 249: 1391–1403.

Lei, K., Sun, S., Zhong, K. et al. (2021). Seed soaking with melatonin promotes seed germination under chromium stress via enhancing reserve mobilization and antioxidant metabolism in wheat. *Ecotoxicology and Environmental Safety* 220: 112241.

Leng, Y., Li, Y., Ma, Y.H. et al. (2021). Abscisic acid modulates differential physiological and biochemical responses of roots, stems, and leaves in mung bean seedlings to cadmium stress. *Environmental Science and Pollution Research* 28: 6030–6043.

Lewis, D.R., Negi, S., Sukumar, P., and Muday, G.K. (2011). Ethylene inhibits lateral root development, increases IAA transport and expression of PIN3 and PIN7 auxin efflux carriers. *Development* 138: 3485–3495.

Li, J., Yang, P., Gan, Y. et al. (2015a). Brassinosteroid alleviates chilling-induced oxidative stress in pepper by enhancing antioxidation systems and maintenance of photosystem II. *Acta Physiologiae Plantarum* 37: 1–11.

Li, G., Song, H., Li, B. et al. (2015b). Auxin resistant1 and PIN-FORMED2 protect lateral root formation in *Arabidopsis* under iron stress. *Plant Physiology* 169: 2608–2623.

Li, K., Yu, R., Fan, L.M. et al. (2016a). DELLA-mediated PIF degradation contributes to coordination of light and gibberellin signalling in *Arabidopsis*. *Nature Communications* 7: 1–11.

Li, N., Wang, J., and Song, W.Y. (2016b). Arsenic uptake and translocation in plants. *Plant and Cell Physiology* 57: 4–13.

Li, Z., Zhu, J., Wang, Y. et al. (2020). Effects of exogenous indole acetic acid on growth and cadmium accumulation of *Cyphomandra betacea* seedlings. *International Journal of Environmental Analytical Chemistry* 102: 771–779.

Li, X., Ahammed, G.J., Zhang, X.N. et al. (2021). Melatonin-mediated regulation of anthocyanin biosynthesis and antioxidant defense confer tolerance to arsenic stress in *Camellia sinensis* L. *Journal of Hazardous Materials* 403: 123922.

Li, G.Z., Wang, Y.Y., Liu, J. et al. (2022). Exogenous melatonin mitigates cadmium toxicity through ascorbic acid and glutathione pathway in wheat. *Ecotoxicology and Environmental Safety* 237: 113533.

Linkies, A. and Leubner-Metzger, G. (2012). Beyond gibberellins and abscisic acid: how ethylene and jasmonates control seed germination. *Plant Cell Reports* 31: 253–270.

Lopes, M.S., Iglesia-Turiño, S., Cabrera-Bosquet, L. et al. (2013). Molecular and physiological mechanisms associated with root exposure to mercury in barley. *Metallomics* 5: 1305–1315.

Maia, C.F., Pereira, Y.C., Silva, B.R.S. et al. (2022). Exogenously applied 24-epibrassinolide favours stomatal performance, ROS detoxification and nutritional balance, alleviating oxidative damage against the photosynthetic apparatus in tomato leaves under nickel stress. *Journal of Plant Growth Regulation* 1–16. https://doi.org/10.1007/s00344-022-10693-3.

Montero-Palmero, M.B., Martín-Barranco, A., Escobar, C., and Hernández, L.E. (2014). Early transcriptional responses to mercury: a role for ethylene in mercury-induced stress. *New Phytologist* 201: 116–130.

Mostofa, M.G., Ghosh, A., Li, Z.G. et al. (2018a). Methylglyoxal–a signaling molecule in plant abiotic stress responses. *Free Radical Biology and Medicine* 122: 96–109.

Mostofa, M.G., Li, W., Nguyen, K.H. et al. (2018b). Strigolactones in plant adaptation to abiotic stresses: an emerging avenue of plant research. *Plant, Cell and Environment* 41: 2227–2243.

Mostofa, M.G., Rahman, M.M., Siddiqui, M.N. et al. (2020). Salicylic acid antagonizes selenium phytotoxicity in rice: selenium homeostasis, oxidative stress metabolism and methylglyoxal detoxification. *Journal of Hazardous Materials* 394: 122572.

Mostofa, M.G., Rahman, M.M., Nguyen, K.H. et al. (2021). Strigolactones regulate arsenate uptake, vacuolar-sequestration and antioxidant defense responses to resist arsenic toxicity in rice roots. *Journal of Hazardous Materials* 415: 125589.

Moustafa-Farag, M., Elkelish, A., Dafea, M. et al. (2020). Role of melatonin in plant tolerance to soil stressors: salinity, pH and heavy metals. *Molecules* 25: 5359.

Mumtaz, M.A., Hao, Y., Mehmood, S. et al. (2022). Physiological and transcriptomic analysis provide molecular insight into 24-epibrassinolide mediated Cr(VI)-toxicity tolerance in pepper plants. *Environmental Pollution* 306: 119375.

Naeem, M., Sadiq, Y., Jahan, A. et al. (2020). Salicylic acid restrains arsenic induced oxidative burst in two varieties of *Artemisia annua* L. by modulating antioxidant defence system and artemisinin production. *Ecotoxicology and Environmental Safety* 202: 110851.

Nambara, E., Okamoto, M., Tatematsu, K. et al. (2010). Abscisic acid and the control of seed dormancy and germination. *Seed Science Research* 20: 55–67.

Nawaz, K., Chaudhary, R., Sarwar, A. et al. (2021). Melatonin as master regulator in plant growth, development and stress alleviator for sustainable agricultural production: current status and future perspectives. *Sustainability* 13: 294.

Nehnevajova, E., Ramireddy, E., Stolz, A. et al. (2019). Root enhancement in cytokinin-deficient oilseed rape causes leaf mineral enrichment, increases the chlorophyll concentration under nutrient limitation and enhances the phytoremediation capacity. *BMC Plant Biology* 19: 1–15.

Ng, L.M., Melcher, K., Teh, B.T., and Xu, H.E. (2014). Abscisic acid perception and signaling: structural mechanisms and applications. *Acta Pharmacologica Sinica* 35: 567–584.

Okant, M. and Kaya, C. (2019). The role of endogenous nitric oxide in melatonin-improved tolerance to lead toxicity in maize plants. *Environmental Science and Pollution Research* 26: 11864–11874.

Pal, A.K., Mandal, S., and Sengupta, C. (2019). Exploitation of IAA producing PGPR on mustard (*Brassica nigra* L.) seedling growth under cadmium stress condition in comparison with exogenous IAA application. *Plant Science Today* 6: 22–30.

Pan, W., You, Y., Shentu, J.L. et al. (2020). Abscisic acid (ABA)-importing transporter 1 (AIT1) contributes to the inhibition of Cd accumulation via exogenous ABA application in *Arabidopsis*. *Journal of Hazardous Materials* 391: 122189.

Pan, J., Guan, M., Xu, P. et al. (2021). Salicylic acid reduces cadmium (Cd) accumulation in rice (*Oryza sativa* L.) by regulating root cell wall composition via nitric oxide signaling. *Science of the Total Environment* 797: 149202.

Pandey, V., Bhatt, I.D., and Nandi, S.K. (2019). Role and regulation of auxin signaling in abiotic stress tolerance. In: *Plant Signaling Molecules* (ed. M.I.R. Khan, P.S. Reddy, A. Ferrante, and N.A. Khan), 319–331. Woodhead Publishing.

Pauwels, L., Barbero, G.F., Geerinck, J. et al. (2010). NINJA connects the co-repressor TOPLESS to jasmonatesignalling. *Nature* 464: 788–791.

Pavlíková, D., Pavlík, M., Procházková, D. et al. (2014). Nitrogen metabolism and gas exchange parameters associated with zinc stress in tobacco expressing an ipt gene for cytokinin synthesis. *Journal of Plant Physiology* 171: 559–564.

Penfield, S. (2017). Seed dormancy and germination. *Current Biology* 27: R874–R878.

Peng, Y., Yang, J., Li, X., and Zhang, Y. (2021). Salicylic acid: biosynthesis and signaling. *Annual Review of Plant Biology* 72: 761–791.

Piotrowska-Niczyporuk, A., Bajguz, A., Kotowska, U. et al. (2020). Auxins and cytokinins regulate phytohormone homeostasis and thiol-mediated detoxification in the green alga *Acutodesmus obliquus* exposed to lead stress. *Scientific Reports* 10: 1–14.

Pluskota, W.E., Pupel, P., Głowacka, K. et al. (2019). Jasmonic acid and ethylene are involved in the accumulation of osmotin in germinating tomato seeds. *Journal of Plant Physiology* 232: 74–81.

Pociecha, E., Dziurka, M., Oklestkova, J., and Janeczko, A. (2016). Brassinosteroids increase winter survival of winter rye (*Secale cereale* L.) by affecting photosynthetic capacity and carbohydrate metabolism during the cold acclimation process. *Plant Growth Regulation* 80: 127–135.

Qiu, C.W., Zhang, C., Wang, N.H. et al. (2021). Strigolactone GR24 improves cadmium tolerance by regulating cadmium uptake, nitric oxide signaling and antioxidant metabolism in barley (*Hordeum vulgare* L.). *Environmental Pollution* 273: 116486.

Rajjou, L., Duval, M., Gallardo, K. et al. (2012). Seed germination and vigor. *Annual Review of Plant Biology* 63: 507–533.

Rameau, C., Goormachtig, S., Cardinale, F. et al. (2019). Strigolactones as plant hormones. *Strigolactones-Biology and Applications* 5: 47–84.

Reid, D.E., Heckmann, A.B., Novák, O. et al. (2016). CYTOKININ OXIDASE/DEHYDROGENASE3 maintains cytokinin homeostasis during root and nodule development in *Lotus japonicus*. *Plant Physiology* 170: 1060–1074.

Rivas, M.Á., Friero, I., Alarcón, M.V., and Salguero, J. (2022). Auxin-cytokinin balance shapes maize root architecture by controlling primary root elongation and lateral root development. *Frontiers in Plant Science* 13: 836592.

Rucińska-Sobkowiak, R. (2016). Water relations in plants subjected to heavy metal stresses. *Acta Physiologiae Plantarum* 38: 257.

Ruyter-Spira, C., Kohlen, W., Charnikhova, T. et al. (2011). Physiological effects of the synthetic strigolactone analog GR24 on root system architecture in *Arabidopsis*: another belowground role for strigolactones? *Plant Physiology* 155: 721–734.

Salavati, J., Fallah, H., Niknejad, Y., and Tari, D.B. (2021). Methyl jasmonate ameliorates lead toxicity in *Oryza sativa* by modulating chlorophyll metabolism, antioxidative capacity and metal translocation. *Physiology and Molecular Biology of Plants* 27: 1089–1104.

Saleem, M.H., Fahad, S., Adnan, M. et al. (2020). Foliar application of gibberellic acid endorsed phytoextraction of copper and alleviates oxidative stress in jute (*Corchorus capsularis* L.) plant grown in highly copper-contaminated soil of China. *Environmental Science and Pollution Research* 27: 37121–37133.

Sameena, P.P. and Puthur, J.T. (2021). Exogenous application of cytokinins confers copper stress tolerance in *Ricinus communis* L. seedlings. *Journal of Plant Growth Regulation* 41: 1–15.

Sami, A., Shah, F.A., Abdullah, M. et al. (2020). Melatonin mitigates cadmium and aluminium toxicity through modulation of antioxidant potential in *Brassica napus* L. *Plant Biology* 22: 679–690.

Santner, A., Calderon-Villalobos, L.I.A., and Estelle, M. (2009). Plant hormones are versatile chemical regulators of plant growth. *Nature Chemical Biology* 5: 301–307.

Saraiva, M.P., Maia, C.F., Silva, B.R.S. et al. (2021). 24-Epibrassinolide induces protection against nickel excess in soybean plants: anatomical evidences. *Brazilian Journal of Botany* 44: 197–205.

Schellingen, K., Straeten, D.V.D., Vandenbussche, F. et al. (2014). Cadmium-induced ethylene production and responses in *Arabidopsis thaliana* rely on ACS2 and ACS6 gene expression. *BMC Plant Biology* 14: 1–14.

Screpanti, C., Fonné-Pfister, R., Lumbroso, A. et al. (2016). Strigolactone derivatives for potential crop enhancement applications. *Bioorganic and Medicinal Chemistry Letters* 26: 2392–2400.

Sehar, Z., Iqbal, N., Fatma, M. et al. (2022). Ethylene suppresses abscisic acid, modulates antioxidant system to counteract arsenic-inhibited photosynthetic performance in the presence of selenium in mustard. *Frontiers in Plant Science* 13: 852704.

Shahid, M., Pourrut, B., Dumat, C. et al. (2014). Heavy-metal-induced reactive oxygen species: phytotoxicity and physicochemical changes in plants. *Reviews of Environmental Contamination and Toxicology* 232: 1–44.

Sharma, I., Sharma, A., Pati, P., and Bhardwaj, R. (2018). Brassinosteroids reciprocates heavy metals induced oxidative stress in radish by regulating the expression of key antioxidant enzyme genes. *Brazilian Archives of Biology and Technology* 61: 1–13.

She, J., Han, Z., Kim, T.W. et al. (2011). Structural insight into brassinosteroid perception by BRI1. *Nature* 474: 472–476.

Sheard, L.B., Tan, X., Mao, H. et al. (2010). Jasmonate perception by inositol-phosphate-potentiated COI1–JAZ co-receptor. *Nature* 468: 400–405.

Shi, W.G., Liu, W., Yu, W. et al. (2019). Abscisic acid enhances lead translocation from the roots to the leaves and alleviates its toxicity in *Populus* × *canescens*. *Journal of Hazardous Materials* 362: 275–285.

Shu, K., Liu, X.D., Xie, Q., and He, Z.H. (2016). Two faces of one seed: hormonal regulation of dormancy and germination. *Molecular Plant* 9: 34–45.

Siddiqui, M.H., Alamri, S., Khan, M.N. et al. (2020). Melatonin and calcium function synergistically to promote the resilience through ROS metabolism under arsenic-induced stress. *Journal of Hazardous Materials* 398: 122882.

Singh, S., Parihar, P., Singh, R. et al. (2016). Heavy metal tolerance in plants: role of transcriptomics, proteomics, metabolomics, and ionomics. *Frontiers in Plant Science* 6: 1143.

Singh, H., Bhat, J.A., Singh, V.P. et al. (2021). Auxin metabolic network regulates the plant response to metalloids stress. *Journal of Hazardous Materials* 405: 124250.

Siposová, K., Labancová, E., Kučerová, D. et al. (2021). Effects of exogenous application of indole-3-butyric acid on maize plants cultivated in the presence or absence of cadmium. *Plants* 10: 2503.

Soares, T.F.S.N., Dias, D.C.F.S., Oliveira, A.M.S. et al. (2020). Exogenous brassinosteroids increase lead stress tolerance in seed germination and seedling growth of *Brassica juncea* L. *Ecotoxicology and Environmental Safety* 193: 110296.

Song, C., Yan, Y., Rosado, A. et al. (2019). ABA alleviates uptake and accumulation of zinc in grapevine (*Vitis vinifera* L.) by inducing expression of ZIP and detoxification-related genes. *Frontiers in Plant Science* 10: 872.

Sun, L., Feraru, E., Feraru, M.I. et al. (2020). PIN-LIKES coordinate brassinosteroid signaling with nuclear auxin input in *Arabidopsis thaliana*. *Current Biology* 30: 1579–1588.

Tai, Z., Yin, X., Fang, Z. et al. (2017). Exogenous GR24 alleviates cadmium toxicity by reducing cadmium uptake in switchgrass (*Panicum virgatum*) seedlings. *International Journal of Environmental Research and Public Health* 14: 852.

Tang, Y., Wang, L., Xie, Y. et al. (2020). Effects of exogenous abscisic acid on the growth and cadmium accumulation of lettuce under cadmium-stress conditions. *International Journal of Environmental Analytical Chemistry* 100: 720–731.

Tang, M., Xu, L., Wang, Y. et al. (2021). Melatonin-induced DNA demethylation of metal transporters and antioxidant genes alleviates lead stress in radish plants. *Horticulture Research* 8: 1–14.

Thakur, M., Praveen, S., Divte, P.R. et al. (2022). Metal tolerance in plants: molecular and physicochemical interface determines the "not so heavy effect" of heavy metals. *Chemosphere* 287: 131957.

Tiwari, P., Indoliya, Y., Chauhan, A.S. et al. (2020). Auxin-salicylic acid cross-talk ameliorates *OsMYB–R1* mediated defense towards heavy metal, drought and fungal stress. *Journal of Hazardous Materials* 399: 122811.

Torres-Vera, R., García, J.M., Pozo, M.J., and López-Ráez, J.A. (2016). Expression of molecular markers associated to defense signaling pathways and strigolactone biosynthesis during the early interaction tomato-*Phelipanche ramosa*. *Physiological and Molecular Plant Pathology* 94: 100–107.

Trinh, N.N., Huang, T.L., Chi, W.C. et al. (2014). Chromium stress response effect on signal transduction and expression of signaling genes in rice. *Physiologia Plantarum* 150: 205–224.

Ulhassan, Z., Huang, Q., Gill, R.A. et al. (2019). Protective mechanisms of melatonin against selenium toxicity in *Brassica napus*: insights into physiological traits, thiol biosynthesis and antioxidant machinery. *BMC Plant Biology* 19: 1–16.

Umezawa, T., Nakashima, K., Miyakawa, T. et al. (2010). Molecular basis of the core regulatory network in ABA responses: sensing, signaling and transport. *Plant and Cell Physiology* 51: 1821–1839.

Verma, V., Ravindran, P., and Kumar, P.P. (2016). Plant hormone-mediated regulation of stress responses. *BMC Plant Biology* 16: 1–10.

Verma, G., Srivastava, D., Narayan, S. et al. (2020). Exogenous application of methyl jasmonate alleviates arsenic toxicity by modulating its uptake and translocation in rice (*Oryza sativa* L.). *Ecotoxicology and Environmental Safety* 201: 110735.

Vinković, T., Novák, O., Strnad, M. et al. (2017). Cytokinin response in pepper plants (*Capsicum annuum* L.) exposed to silver nanoparticles. *Environmental Research* 156: 10–18.

Wakeel, A., Ali, I., Upreti, S. et al. (2018). Ethylene mediates dichromate-induced inhibition of primary root growth by altering *AUX1* expression and auxin accumulation in *Arabidopsis thaliana*. *Plant, Cell and Environment* 41: 1453–1467.

Wang, C., Liu, Y., Li, S.S., and Han, G.Z. (2015). Insights into the origin and evolution of the plant hormone signaling machinery. *Plant Physiology* 167: 872–886.

Wang, F., Tan, H., Zhang, Y. et al. (2021). Salicylic acid application alleviates cadmium accumulation in brown rice by modulating its shoot to grain translocation in rice. *Chemosphere* 263: 128034.

Wani, S.H., Kumar, V., Shriram, V., and Sah, S.K. (2016). Phytohormones and their metabolic engineering for abiotic stress tolerance in crop plants. *The Crop Journal* 4: 162–176.

Wei, J., Li, D.X., Zhang, J.R. et al. (2018). Phytomelatonin receptor PMTR1-mediated signaling regulates stomatal closure in *Arabidopsis thaliana*. *Journal of Pineal Research* 65: e12500.

Weiss, D. and Ori, N. (2007). Mechanisms of cross talk between gibberellin and other hormones. *Plant Physiology* 144: 1240–1246.

Wybouw, B. and De Rybel, B. (2019). Cytokinin–a developing story. *Trends in Plant Science* 24: 177–185.

Xu, J. and Zhang, S. (2014). Regulation of ethylene biosynthesis and signaling by protein kinases and phosphatases. *Molecular Plant* 7: 939–942.

Xu, J. and Zhang, S. (2015). Ethylene biosynthesis and regulation in plants. *Ethylene in Plants* 5: 1–25.

Xu, Q., Krishnan, S., Merewitz, E. et al. (2016). Gibberellin-regulation and genetic variations in leaf elongation for tall fescue in association with differential gene expression controlling cell expansion. *Scientific Reports* 6: 1–12.

Xu, B., Yu, J., Zhong, Y. et al. (2019). Influence of Br24 and Gr24 on the accumulation and uptake of Cd and As by rice seedlings grown in nutrient solution. *Regulation of Plant Growth and Development* 19: 20.

Xu, L., Zhang, F., Tang, M. et al. (2020). Melatonin confers cadmium tolerance by modulating critical heavy metal chelators and transporters in radish plants. *Journal of Pineal Research* 69: e12659.

Yang, L., Huan, Y., Sun, J. et al. (2021). Effects of exogenous gibberellic acid on growth and cadmium accumulation in *Cyphomandra betacea* seedlings. *Environmental Progress and Sustainable Energy* 40: e13655.

Ye, H., Li, L., and Yin, Y. (2011). Recent advances in the regulation of brassinosteroid signaling and biosynthesis pathways F. *Journal of Integrative Plant Biology* 53: 455–468.

Zaheer, M.S., Ali, H.H., Erinle, K.O. et al. (2022). Inoculation of *Azospirillum brasilense* and exogenous application of trans-zeatin riboside alleviates arsenic induced physiological damages in wheat (*Triticum aestivum*). *Environmental Science and Pollution Research* 29: 33909–33919.

Zaid, A., Mohammad, F., Wani, S.H., and Siddique, K.M. (2019). Salicylic acid enhances nickel stress tolerance by up-regulating antioxidant defense and glyoxalase systems in mustard plants. *Ecotoxicology and Environmental Safety* 180: 575–587.

Zehra, A., Choudhary, S., Wani, K.I. et al. (2020). Exogenous abscisic acid mediates ROS homeostasis and maintains glandular trichome to enhance artemisinin biosynthesis in *Artemisia annua* under copper toxicity. *Plant Physiology and Biochemistry* 156: 125–134.

Zhang, Y. and Li, X. (2019). Salicylic acid: biosynthesis, perception, and contributions to plant immunity. *Current Opinion in Plant Biology* 50: 29–36.

Zhang, C., He, Q., Wang, M. et al. (2020). Exogenous indole acetic acid alleviates Cd toxicity in tea (*Camellia sinensis*). *Ecotoxicology and Environmental Safety* 190: 110090.

Zhou, F., Lin, Q., Zhu, L. et al. (2013). D14–SCFD3-dependent degradation of D53 regulates strigolactone signalling. *Nature* 504: 406–410.

Zhou, Y.L., Huo, S.F., Wang, L.T. et al. (2018). Exogenous 24-epibrassinolide alleviates oxidative damage from copper stress in grape (*Vitis vinifera* L.) cuttings. *Plant Physiology and Biochemistry* 130: 555–565.

Zhou, J., Cheng, K., Huang, G. et al. (2020). Effects of exogenous 3-indoleacetic acid and cadmium stress on the physiological and biochemical characteristics of *Cinnamomum camphora*. *Ecotoxicology and Environmental Safety* 191: 109998.

Zhu, X.F., Jiang, T., Wang, Z.W. et al. (2012). Gibberellic acid alleviates cadmium toxicity by reducing nitric oxide accumulation and expression of *IRT1* in *Arabidopsis thaliana*. *Journal of Hazardous Materials* 239: 302–307.

Zwanenburg, B., Pospíšil, T., and ĆavarZeljković, S. (2016). Strigolactones: new plant hormones in action. *Planta* 243: 1311–1326.

21

Heavy-Metal-Induced Reactive Oxygen Species and Methylglyoxal Formation and Detoxification in Crop Plants

Modulation of Tolerance by Exogenous Chemical Compounds

Beatrycze Nowicka[1], Tahsina Sharmin Hoque[2], Sheikh Mahfuja Khatun[3], Jannatul Naim[3], Ahmed Khairul Hasan[4], and Mohammad Anwar Hossain[3]

[1] *Department of Plant Physiology and Biochemistry, Faculty of Biochemistry, Biophysics and Biotechnology, Jagiellonian University, Kraków, Poland*
[2] *Department of Soil Science, Bangladesh Agricultural University, Mymensingh, Bangladesh*
[3] *Department of Genetics and Plant Breeding, Bangladesh Agricultural University, Mymensingh, Bangladesh*
[4] *Department of Agronomy, Bangladesh Agricultural University, Mymensingh, Bangladesh*

21.1 Introduction

The term "heavy metals" (HMs) most often refers to metals and metalloids with the density higher than $5\,g\,cm^{-3}$. The threshold $4\,g\,cm^{-3}$ is also used (Nagajyoti et al. 2010). In some publications concerning HM toxicity and tolerance, lighter metal elements, in particular Al, are also taken into consideration (Hossain et al. 2012). According to the density criterion, 53 of 90 naturally occurring elements belong to HMs. However, only 17 [silver (Ag), arsenic (As), cadmium (Cd), cobalt (Co), chromium (Cr), copper (Cu), iron (Fe), mercury (Hg), manganese (Mn), molybdenum (Mo), nickel (Ni), lead (Pb), antimony (Sb), uranium (U), vanadium (V), tungsten (W), zinc (Zn)] are bioavailable in physiological conditions, while the rest occur only in very low amounts or their compounds are insoluble in water. Among above-mentioned elements, there are essential micronutrients displaying toxicity only when their concentration in the environment exceeds the optimal range for given species. The others (Ag, As, Cd, Hg, Pb, Sb, and U) do not play physiological roles in living organisms (Nowicka 2022). Considering higher plants, the essential HMs are as follows: Co, Cu, Fe, Mn, Mo, Ni, and Zn (Nagajyoti et al. 2010). Among them, Fe, Mn, and Mo display rather low toxicity, whereas the others are potentially more harmful. Higher plants are known to be especially vulnerable to Cu, which causes metabolic disturbances at intracellular concentration only slightly higher than the optimal level. Very toxic HMs nonessential for plants are Ag, Cd, Cr, and Hg (Nowicka 2022).

HMs occur in the Earth crust and natural processes, especially rock weathering and volcanic activity, result in the enrichment of soils and waters in these elements. However, in

Heavy Metal Toxicity and Tolerance in Plants: A Biological, Omics, and Genetic Engineering Approach, First Edition. Edited by Mohammad Anwar Hossain, AKM Zakir Hossain, Sylvain Bourgerie, Masayuki Fujita, Om Parkash Dhankher, and Parvez Haris.

the recent centuries, anthropogenic sources have been playing a significant role in HMs release (Nowicka 2022). Rapid industrialization, urbanization, and usage of agricultural amendments have resulted in the contamination of the environment with HM ions (Hasan et al. 2017). Industrial sources of HMs include mining and smelting of metal ores, fossil fuel combustion, production of electronics, textiles, paper, plastic, wood preservatives, and paints. In agriculture, the production and application of fertilizers and plant protection products lead to the increase in HMs concentration in soil and water. Other important sources of HMs are transport and corrosion of waste products (Pinto et al. 2003; Nagajyoti et al. 2010). Contamination with HMs leads to the decrease in crop productivity and poses a threat to human health as a result of biomagnification of HMs in the food chains (Nagajyoti et al. 2010).

The toxicity of HMs is a complex phenomenon due to their pleiotropic effects. On the biochemical level, we may distinguish four main modes of action. First, HM cations may compete with and displace the essential metal cations. This effect is observed both for highly toxic HMs and HMs with lower toxicity, the latter display it when present in high concentrations. For example, Cd is able to displace Zn from active sites of enzymes, it may also interfere with Ca^{2+}-based signaling. Secondly, some HMs have high affinity to functional groups of proteins and low-molecular compounds. The most important of such groups are thiol groups of proteins and low-molecular-weight compounds including glutathione (GSH). High affinity to thiols is important mode of action of HMs displaying high toxicity, such as Ag, Hg, Cd, and Pb. The other targets are carboxyl groups of proteins and histidyl groups, the latter are known to bind Ni (Nowicka 2022). The oxyanions of As and V are similar to phosphate group and therefore can inhibit or disturb phosphate-metabolizing systems (DalCorso 2012; Larsson et al. 2013). Finally, some HMs are redox active, which means that they react with reactive oxygen species (ROS), which leads to the conversion of less harmful ROS into more active ones. This mode of action will be described in more details in the next section.

All the above-mentioned processes lead to the inhibition of enzymes, pumps, and ion channels. They also result in the malfunctioning of cellular signaling cascades. This, in turn, causes disturbance of metabolism and transport (Küpper and Andresen 2016). In such a case, various toxic metabolites, such as methylglyoxal (MG), are formed and provoke secondary damage. HMs also cause structural damage, both at the level of macromolecules and organelles. This results from the improper protein folding caused by ion substitution, thiol binding, or oxidation. Another reason is lipid peroxidation in membranes. HMs exert genotoxic action. Known effects of the exposure to these pollutants are mutations, DNA damage such as strand breaks, and chromosome aberrations (Hossain et al. 2012; Nowicka 2022). On the whole organism level, the observed results of HM toxicity to plants are growth inhibition, chloroses and necroses, nutrient deficiency, disturbed water balance, decreased yield, and death (Nagajyoti et al. 2010).

The toxicity of HMs depends on many factors including the type of HM, its oxidative state, the certain form (cation, oxyanion, organic compound as it is possible in the case of As and Hg) and the concentration of given HM. The soil properties, such as pH, presence and concentration of other ions, and the amount of mineral and organic compounds able to bind metal ions, are also crucial. Important feature affecting the toxicity of HMs is their mobility, which in turn depends on the solubility of certain HM salts in water and the affinity of metal cations to chelating compounds (Nowicka 2022). Cd is very dangerous HM due to its high mobility, as its salts are relatively well soluble and its cations are bound less strongly to soil organic matter than Pb, Hg, or Cu ions (Kalaivanan and Ganeshamurthy 2016). Of course, the tolerance to certain HMs differs between plant species and cultivars (Hossain et al. 2012). The other organisms present in soil, such as mycorrhizal fungi, are also important for the overall effect of HM exposure (Schützendübel and Polle 2002; Singh et al. 2020).

Being sessile organisms, plants are routinely exposed to various abiotic and biotic stresses. To protect themselves, they evolved both constitutive and adaptive mechanisms. The general stress-response strategies can be divided into "avoidance" and "tolerance." The first is a strategy aimed at preventing the stress factor from affecting the plant cells. In the case of HMs, "avoidance" means that plant limits uptake and transport of HM ions from the soil to the shoots. The second strategy is aimed at withstanding the stress, in this case – surviving high internal concentration of HM ions (Hossain et al. 2012). Plants have evolved many mechanisms to protect them from HM-induced stress. The exudation of chelating compounds, such as organic acids (mainly malate, citrate, oxalate) that are able to bind HMs leads to the decrease in their mobility in soil (Anjum et al. 2015). HM ions can be also bound to the cell walls or to excreted mucilaginous substances. Another way to limit mobility is bioconversion of metal ions leading to the formation of insoluble salts and oxides (Hossain et al. 2012; Pasricha et al. 2021). The restricted uptake of HMs results from the suppression or enhanced degradation of certain transporters (Manara 2012; Hasan et al. 2017). The active efflux of HMs enables to decrease their concentration inside the cells (Ismael et al. 2019). Plants can regulate transport and distribution of HM ions. Reduced translocation of HM ions to the shoots is a widespread tolerance mechanism. On the other hand, there are known species of so-called hyperaccumulator plants, which efficiently take HMs from the soil and maintain high concentrations of HM ions in their green tissues (Hossain et al. 2012). HM ions present inside the cells are often transported and sequestered in the vacuoles. Very important protective mechanism is chelation of HM ions by high-affinity ligands. Plants synthesize two types of peptides that able to bind HMs: phytochelatins and metallothioneins. Phytochelatins are small, enzymatically produced, thiol-containing peptides. They complexate HM ions in the cytosol, then such complexes may be exported into the vacuole. Metallothioneins are metal-binding polypeptides synthesized on ribosomes and occur in cytosol (Hossain et al. 2012; Hasan et al. 2017). Higher plants are also known to chelate HMs inside the cell (mainly in vacuole) by organic acids and phytate. In some cases, chemical modification (usually via redox reactions) enables transformation of more toxic HM ions into less toxic ones (Hossain et al. 2012; Sharma et al. 2016). To counteract the effects of HM toxicity, HM-exposed plants enhance their antioxidant defense system and glyoxalase system and mechanisms of repair, resynthesis, and degradation of damaged biomolecules (Hossain et al. 2012). The exposure of plants to toxic concentrations of HM ions triggers signaling cascades, which leads to the induction of protective mechanisms (Sytar et al. 2013). The current chapter focuses on the ROS and MG metabolism in plant cells in response to HM stress. The role of exogenously applied chemical compounds (signaling molecules, phytohormones, antioxidants, osmoprotectants, chelators, and nutrients) in providing tolerance to HM stress in crop plants will also be discussed.

21.2 Heavy-Metal-Induced ROS and Methylglyoxal Production in Plant Cells

Atomic oxygen in its ground state (triplet oxygen, $^3O_2$) is diradical having two unpaired electrons with parallel spins in two antibonding orbitals $\pi * 2p$. This makes $^3O_2$ less reactive because the majority of chemical compounds have paired antiparallel electrons in their molecular orbitals. The excitation or partial reduction of $^3O_2$ leads to the formation of ROS, which may then undergo further interconversions and reactions with organic and inorganic compounds (Nowicka 2022). Major ROS formed in plant cells are shown in Table 21.1. Overproduction of ROS is dangerous situation leading to the oxidation of proteins, nucleic acids, and lipids. Protein damage results in the loss of activity, degradation, and aggregation. Oxidation of lipids, mainly due to radical-mediated or non-radical lipid peroxidation, causes decrease in membrane fluidity and increase in its permeability, which results in the disturbance of membrane structure and functions (Nowicka 2022). On the other hand, in plants ROS play also beneficial roles in signaling, cell wall synthesis and modification, and during pathogen response (Gechev et al. 2006; Janku et al. 2019).

ROS are inevitable by-products of aerobic metabolism. In photosynthetic organisms they are generated in side reactions occurring in electron transfer chains, as a result of activity of various enzymes, and due to photooxidizing action of pigments, mainly chlorophylls (Chls) (Table 21.2). In plant green tissues, the highest amounts of ROS are formed in chloroplasts due to the functioning of photosynthetic electron transfer chain and increased O_2 concentration in these organelles. ROS formation in chloroplasts is increased in the case of over-reduction of photosynthetic electron transfer chain. Such situation often takes place under exposure to the stress factors that slow down later reactions of the light phase and dark phase reactions (Edreva 2005). The other two most important ROS producers are peroxisomes and mitochondria. The share of mitochondria in overall ROS production is higher in non-photosynthetic tissues and in the dark, whereas ROS production in peroxisomes is increased in conditions that enhance photorespiration (Foyer and Noctor 2003; Nowicka 2022).

A situation when there is overproduction of ROS in cells is called oxidative stress. This type of stress is the result of the exposure of plants to various stress factors including HMs (Sytar et al. 2013). HMs may induce oxidative stress both directly and indirectly. Redox-active HMs, such as Cu, Cr, Fe, and V, may occur in cells in multiple oxidation states. If the redox potential of certain heavy metal ion fits the range −420 to +800 mV, this ion may undergo redox reactions in cells. Such metal ions may be reduced in cells by $O_2^{\bullet-}$ or reducing compounds (such as reduced nucleotides or ascorbate) and then react with H_2O_2, which results in the formation of extremely dangerous $OH^{\bullet}$:

i) $Me^{n+1} + O_2^{\bullet-} \rightarrow Me^n + O_2$
ii) $Me^n + H_2O_2 \rightarrow Me^{n+1} + OH^- + OH^{\bullet}$

Table 21.1 Major reactive oxygen species formed in plant cells.

Reactive oxygen species	Formation	Activity and selected properties
Radical forms		
Superoxide anion $O_2^{\bullet-}$	One-electron reduction of O_2	• damages Fe–S centres • oxidizes thiol groups of Cys • oxidizes His, Met, Trp • reacts with compounds containing unsaturated bonds, which leads to hydroperoxide formation • reduces redox-active metal ions • reacts with NO, which leads to $ONOO^-$ formation
Hydroperoxyl radical $HO_2^{\bullet}$	Protonated form of superoxide, more abundant in low pH	• similar to that of $O_2^{\bullet-}$ • being uncharged, it may penetrate lipid membranes and initiate lipid peroxidation
Hydroxyl radical $OH^{\bullet}$	Three-electron reduction of O_2	• the most reactive ROS, reacts at a high rate with most inorganic and organic molecules
Lipid alkoxyl $LO^{\bullet}$ and peroxyl $LOO^{\bullet}$ radicals	Formed during radical-initiated lipid peroxidation	• propagate radical peroxidation reaction
Non-radical forms		
Singlet oxygen $^1O_2$	Excitation of $^3O_2$	• oxidizes compounds containing unsaturated bonds including chlorophyll and lipids • oxidizes proteins, especially Cys, Trp, Tyr, His, Met residues • oxidizes DNA, mainly guanine
Hydrogen peroxide H_2O_2	Two-electron reduction of O_2	• less reactive, further range of diffusion • uncharged, may diffuse through membranes • oxidizes thiol groups • oxidizes Trp, His, Met • damages Mn cluster in PS II • damages heme groups
Lipid hydroperoxide LOOH	Product of lipid peroxidation	• relatively easily undergoes reactions leading to $LO^{\bullet}$ and $LOO^{\bullet}$ formation

Source: From Edreva (2005), Triantaphylidès and Havaux (2009), Bhattacharjee (2019), and Nowicka (2022)
PS II, photosystem II.

In the case of Fe ions, the second reaction is called Fenton reaction and whole cycle of $OH^{\bullet}$ generation in the presence of Fe ions and H_2O_2 is known as the Haber–Weiss reaction. The term "Haber–Weiss cycling" is often used in general to describe such type of reactions catalyzed also by other redox-active HMs (Nowicka 2022). There is experimental evidence

Table 21.2 ROS formation in plant cells.

Site of ROS generation	Process or enzyme responsible	ROS formed
Chloroplasts	Energy transfer between triplet excited state of chlorophyll (^{3}P680*) and $^3O_2$ in PS II	$^1O_2$
	Electron leakage at PS II acceptor site	$O_2^{\bullet-}$
	Energy transfer between triplet excited state of chlorophyll and $^3O_2$ in antennae	$^1O_2$
	Cytochrome b_6f	$^1O_2$
	Electron leakage at Fe–S clusters in PS I	$O_2^{\bullet-}$
	Reduction of O_2 by reduced ferredoxin (Mehler reaction)	$O_2^{\bullet-}$
Peroxisomes and glyoxysomes	Glycolate oxidation in photorespiratory pathway	H_2O_2
	β-oxidation of fatty acids	H_2O_2
	Urate oxidase	H_2O_2
	Flavin-containing oxidases, i.e. sarcosine and polyamine oxidase	H_2O_2
	Sulfite detoxification by sulfite oxidase	H_2O_2
	Peroxisomal membrane polypeptides	$O_2^{\bullet-}$
	Xanthine oxidase	$O_2^{\bullet-}$
Mitochondria	Electron leakage at complex I	$O_2^{\bullet-}$
	Nonenzymatic UQH_2 oxidation	$O_2^{\bullet-}$
	Electron leakage at complex III	$O_2^{\bullet-}$
Endoplasmic reticulum and cytosol	Detoxification reactions carried out by cytochromes	$O_2^{\bullet-}$
	Aldehyde oxidases	$O_2^{\bullet-}$ and H_2O_2
	Xanthine oxidase	$O_2^{\bullet-}$
Cell membrane	Plasma membrane NADPH-dependent oxidase	$O_2^{\bullet-}$
Apoplast	Cell wall peroxidases	H_2O_2
	Amine oxidases	H_2O_2
	Oxalate oxidases	H_2O_2

Source: From Bhattacharjee (2005), Edreva (2005), Krieger-Liszkay (2005), Del Río et al. (2006), Hossain et al. (2011), Tripathy and Oelmüller (2012), Corpas et al. (2015), and Janku et al. (2019).
PS I, photosystem I, PS II, photosystem II.

supporting redox cycling of Cr ions in plant cells (Balasaraswathi et al. 2017). The induction of oxidative stress is thought to be the main mode of the toxicity of redox-active HMs. Elements with filled orbital *d*, such as Cd, Hg, Pb, and Zn, belong to the redox-inactive ones (Nowicka 2022).

HMs, both redox-active and redox-inactive, are also able to cause oxidative stress indirectly. They disturb ROS-detoxifying mechanisms via binding to thiol groups of GSH and other low-molecular thiols. They can also inhibit antioxidant enzymes, for example Cd is known to displace Zn in Cu/Zn superoxide dismutase (Cu/ZnSOD) (Stoiber et al. 2013;

Küpper and Andresen 2016). The inhibition of photosynthetic and respiratory electron transfer chains and the inhibition of Calvin cycle enzymes by HMs also enhance ROS production due to over-reduction of the chains, which leads to electron leakage. In the case of photosynthesis inhibition, another adverse effect is an increase in the duration of ^{1}P680 excited state that facilitates interconversion to ^{3}P680* and $^1O_2$ formation (Krieger-Liszkay 2005; Nowicka 2022). Many HMs display such toxicity mechanisms. Ag^+ is known to disturb light phase of photosynthesis via substitution of Cu in plastocyanin and to inhibit mitochondrial electron transfer due to reaction with thiol groups of proteins (Marambio-Jones and Hoek 2010; Yan and Chen 2019). Cd disturbs the action of both photosystems and inhibits some dark phase enzymes including Rubisco, where it replaces Mg^{2+} ions crucial for the reaction. What is more, Cd causes stomata closing, which limits CO_2 availability leading to the enhancement of photorespiration (DalCorso 2012; Küpper and Andresen 2016). Cu acts at many sites in PS II disturbing its function, it also inhibits Rubisco and phosphoenolpyruvate carboxylase (PEPC) crucial for CO_2 assimilation in C_4 and CAM plants (Burda et al. 2003; Yruela 2005; DalCorso 2012). When applied in high concentration, Co inhibits PS II and PEPC (Palit et al. 1994). Cr (+6) inhibits mitochondrial complexes I, III, and IV, and both photosystems (Singh et al. 2013). Hg inhibits mitochondrial electron transfer, as well as both light and dark phases of photosynthesis (Nagajyoti et al. 2010; Kalaivanan and Ganeshamurthy 2016). Ni is known to inhibit the activity of both photosystems and some enzymes of Calvin cycle including Rubisco. Considering the last enzyme mentioned, the mechanism is similar to that of Cd and is a result of displacement of Mg^{2+} ions by Ni^{2+} (DalCorso 2012; Shahzad et al. 2018). It is thought that Pb ions may replace Mn in Mn cluster in PS II. Pb also inhibits enzymes of dark phase of photosynthesis, e.g. Rubisco, PEPC, ribulose-5-phosphate kinase (DalCorso 2012). The replacement of crucial divalent cations is thought to be the reason of inhibitory action of Zn to PS II and Rubisco (Küpper and Andresen 2016). Sometimes, exposure to HMs results in the activation of enzymatic ROS production, i.e. Cd, Cu, Ni, and Pb were shown to induce $O_2^{\bullet-}$ formation by plasma membrane-bound NADPH oxidases (Hossain et al. 2014). The inhibitory actions of HMs on various other enzymes, transporters, as well as genotoxic action of HMs and HM-induced structural damage lead to the overall disturbance of cell metabolism and signaling, which further enhances oxidative stress.

Malonyldialdehyde (MDA) and other thiobarbituric acid reactive substances formed during lipid peroxidation process are oxidative stress markers often measured in organisms exposed to various stress factors. The occurrence of lipid peroxidation in HM-treated plants and green algae is widely documented (DalCorso 2012; Sytar et al. 2013; Tiwari and Sarangi 2015; Nowicka 2022). The ROS production in HM-exposed plants and algae was also confirmed using fluorescent probes, NBT assay for $O_2^{\bullet-}$ staining, and assays for H_2O_2 determination (Mourato et al. 2012; Stoiber et al. 2013; Tiwari and Sarangi 2015; Nowicka et al. 2020). Interestingly, the hydroxy derivatives of plastoquinone (PQ) and plastochromanol (PC-OH), which are considered to be the characteristic products of oxidation of above-mentioned compounds by $^1O_2$, were not observed in Cd and Cr-exposed green microalga *Chlamydomonas reinhardtii* (Szymańska et al. 2014; Nowicka et al. 2020).

Apart from ROS, the toxic action is displayed by various low-molecular-weight compounds formed during stress. Secondary products of lipid peroxidation, such as aldehydes, hydroxydialdehydes, ketones, cyclic peroxides, epoxides, ethers, and isoprostanes, may cause damage to proteins and DNA. A well-known reaction is cross-linking of proteins by MDA (Nowicka 2022). Another toxic aldehyde is MG. This highly reactive αβ-dicarbonyl ketoaldehyde is formed in plants spontaneously under physiological conditions, by nonenzymatic or enzymatic β-elimination of phosphate from 1,2-enediol form of glyceraldehyde-3-phosphate and dihydroxyacetone phosphate. These phosphotrioses are intermediates of glycolysis and dark phase of photosynthesis; therefore, the disturbance of these processes during exposure to stress factors leads to the enhanced formation of MG (Hossain et al. 2012). MG has mutagenic properties and induces DNA strand breaks. It also causes irreversible modifications and cross-links in protein. At high concentrations, MG inhibits cell proliferation (Hossain et al. 2012). The increased generation of MG was observed in Zn-exposed tobacco (Singla-Pareek et al. 2006), Cd-exposed pumpkin, mung bean, and *Lepidium sativum* (Hossain et al. 2009; Gómez Ojeda et al. 2013; Nahar et al. 2016), Ni-exposed and Cu-exposed rice (Mostofa et al. 2015a; Hasanuzzaman et al. 2019), and Cr-exposed maize and tomato (Kharbech et al. 2020a, b; Alam et al. 2021).

21.3 Detoxification of ROS and Methylglyoxal in Plant Cells

To protect themselves from oxidative stress, living organisms have evolved protective mechanisms based on low-molecular-weight antioxidants and ROS-detoxifying enzymes (Sytar et al. 2013). In higher plants, these mechanisms form complex and multilevel network. The concentration of important antioxidants, such as ascorbate (AsA) and tocopherol (Toc) is usually higher in green tissues of land plants than in algae. What is more, genomes of higher plants contain more genes encoding antioxidant enzymes than algal genomes (Nowicka 2022). Each type of ROS can be detoxified in many ways, and every cell compartment contains its antioxidant system (Gechev et al. 2006). There is redundancy between cellular antioxidant mechanisms, for example shortage of α-Toc can be compensated by the increased content of other low-molecular-weight compounds and increased activity of ROS-detoxifying enzymes (Nowicka et al. 2021). Elaborate signaling network enables plants to monitor redox status of their cells and regulate antioxidant defense accordingly (Ahmad et al. 2010; Sytar et al. 2013; Dumanović et al. 2021).

The low-molecular-weight compounds can be subdivided into hydrophilic and lipophilic ones (Table 21.3). The major hydrophilic antioxidants of higher plants are AsA and GSH. The former is more abundant, in leaves its concentration may exceed that of GSH by 5–10 times (Smirnoff 2005). Particularly high AsA levels occur in the leaves of alpine plants, where AsA may constitute up to 19% of the leaf carbon pool (Gest et al. 2013). Chloroplasts contain 30–40% of cellular AsA pool and the concentration of this compound may reach 300 mM (Triantaphylidès and Havaux 2009; Ahmad et al. 2010). AsA is able to directly scavenge ROS, it is also a cofactor of H_2O_2-detoxifying enzyme ascorbate peroxidase (APX). What is more, this compound participates in nonenzymatic regeneration of other antioxidants and plays a role in redox signaling (Nowicka 2022). AsA oxidation results in monodehydroascorbate (MDHA) formation. MDHA can disproportionate into AsA and dehydroascorbate (DHA). Both MDHA and DHA can be re-reduced enzymatically (Dumanović et al. 2021). The AsA content was shown to increase in plants in response

Table 21.3 Major low-molecular-weight antioxidants of higher plants.

Compound	Antioxidant activity	Localization
Hydrophilic compounds		
Ascorbate, AsA	• scavenges $O_2^{\bullet-}$, H_2O_2, $^1O_2$, organic radicals • plays a role in non-enzymatic regeneration of carotenoids, tocopherols and phenolics • cofactor of H_2O_2-detoxifying enzyme ascorbate peroxidase	Plastids, mitochondria, peroxisomes, cytosol, nucleus, vacuoles, apoplast
Glutathione, GSH	• scavenges $O_2^{\bullet-}$, H_2O_2, $^1O_2$, organic radicals • cofactor of enzymes responsible for H_2O_2 and ROOH detoxification: glutathione peroxidases and glutaredoxins • cofactor of enzymes re-reducing oxidized thiol groups • cofactor of dehydroascorbate reductase participating in regeneration of Asc	Plastids, mitochondria, peroxisomes, cytosol, endoplasmic reticulum, nucleus, vacuoles, apoplast
Phenolic compounds	• scavenge $O_2^{\bullet-}$ and organic radicals • quench and scavenge $^1O_2$ • play a role in nonenzymatic tocopherol regeneration • cofactors of peroxidases participating in H_2O_2 and ROOH detoxification	Vacuoles, cytosol, endoplasmic reticulum, plastids, nucleus, apoplast
Proline, Pro	• scavenges $^1O_2$ and organic radicals	Plastids, mitochondria, cytosol
Lipophilic compounds		
Carotenoids, Car	• scavenge $O_2^{\bullet-}$ and lipid radicals • quench and scavenge $^1O_2$ • quench chlorophyll excited states preventing $^1O_2$ formation	Plastids
Chromanols, mainly α-tocopherol, α-Toc	• scavenge $O_2^{\bullet-}$ and lipid radicals • quench and scavenge $^1O_2$	Plastids, oleosomes (in seeds)
Isoprenoid quinols, mainly plastoquinol, PQH_2, and ubiquinol, UQH_2	• scavenge $O_2^{\bullet-}$, lipid radicals, and perferryl radical • quench and scavenge $^1O_2$ • play a role in nonenzymatic tocopherol regeneration	Plastids (PQH_2), mitochondria (UQH_2)

Source: From Dumanović et al. (2021) and Nowicka (2022).

to HMs (Zengin and Munzuroglu 2005; Collin et al. 2008; Štolfa et al. 2015). A tripeptide GSH (γ-Glu-Cys-Gly) is the major low-molecular-weight thiol in both eukaryotes and prokaryotes. This compound is crucial for plant stress responses due to its multiple functions. It is a cofactor of ROS- and MG-detoxifying enzymes, as well as glutathione-*S*-transferases (GSTs), which are crucial for detoxification of xenobiotics. Being a cofactor of dehydroascorbate reductase (DHAR), GSH is necessary for AsA recycling in plant cells. GSH may also scavenge ROS directly. It is a substrate for phytochelatin synthesis and plays a role in redox signaling (Hossain et al. 2012; Nowicka 2022). Under physiological conditions, GSH predominantly exists in a reduced form. Under oxidizing conditions it forms glutathione disulfide (GSSG), which can be re-reduced by glutathione reductase (GR). Similarly to AsA, GSH is the most abundant in chloroplasts, where its concentration is about 1–4 mM (Nowicka 2022). The role of GSH in response to HMs was reviewed in Hossain et al. (2012).

Antioxidant function is also displayed by other compounds, including phenolics, amino acids, and sugars (Dumanović et al. 2021). Phenolic compounds are large and diverse group of secondary metabolites playing various roles in plant cells. The antioxidant capacity of certain compound depends on the number and the arrangement of hydroxyl groups in its molecule. Some phenolics were shown to be more effective in radical scavenging than AsA and Toc. These compounds can also chelate HM ions (Nowicka 2022). Considering amino acids, proline is known to be an osmoprotectant playing a role in the response to drought and salt stress, but it displays antioxidant properties as well (Siripornadulsil et al. 2002). The content of this compound often increases in HM-exposed plants (Štolfa et al. 2015).

Lipophilic antioxidants are crucial for protection of the membranes and stored lipids from lipid peroxidation. In plants, the major antioxidants belonging to this group are isoprenoid chromanols, isoprenoid quinols, and carotenoids. The latter are accessory photosynthetic pigments, but they are also crucial for efficient photoprotection and ROS scavenging in thylakoid membranes (Latowski et al. 2014). The molecules of isoprenoid chromanols and quinols comprise of prenyl side-chain and hydrophilic head group. The most important isoprenoid chromanols are Tocs and tocotrienols, the latter containing unsaturated side chain. The α-, β-, γ-, and δ-forms are distinguished on the basis of the substituent configuration in the chromanol ring. In plant green tissues, α-Toc is the most abundant isoprenoid chromanol, whereas other members of this group often occur in oil seeds. The increase in α-Toc was observed in HM-exposed plants when compared to control (Zengin and Munzuroglu 2005; Gajewska and Skłodowska 2007; Collin et al. 2008; Kumar et al. 2012). Isoprenoid quinones play a role of electron carriers in photosynthetic and respiratory electron transfer chains, they also participate in signaling and are enzyme cofactors (Nowicka and Kruk 2010). Their reduced quinol forms are potent antioxidants. Quinone forms show some antioxidant properties (detoxification of $O_2^{\bullet-}$ and $^1O_2$), but they are less pronounced than those of quinols (Nowicka et al. 2021). One has to remember that under certain conditions, low-molecular-weight antioxidants may act as pro-oxidants (Nowicka 2022).

The activity and cellular localization of the major enzymes playing a role in antioxidant defense in plants are summarized in Table 21.4. They include enzymes participating in ROS detoxification, regeneration of AsA and GSH, and re-reduction of oxidized thiol groups of proteins. Superoxide dismutases (SODs) are the only plant enzymes able to

Table 21.4 The most important enzymes playing a role in antioxidant defense in plants.

Enzyme	Antioxidant activity	Localization
Detoxification of ROS and reduction of oxidized proteins		
Superoxide dismutases, SODs	• dismutation of $O_2^{\bullet-}$	Cu/ZnSOD: cytosol, chloroplasts, peroxisomes MnSOD: mitochondria, peroxisomes FeSOD: chloroplasts
Catalase, CAT	• dismutation of H_2O_2	Peroxisomes, glyoxysomes, mitochondria, apoplast
Ascorbate peroxidase, APX	• reduction of H_2O_2 to H_2O using Asc as a cofactor	Chloroplasts, mitochondria, peroxisomes, glyoxysomes, cytosol, apoplast
Glutathione peroxidases, GPXs	• reduction of H_2O_2 to H_2O and/or ROOH to ROH using either GSH or thioredoxins as reductants	Chloroplasts, mitochondria, peroxisomes, cytosol endoplasmic reticulum, apoplast
Glutathione-*S*-transferases, GSTs	• glutathionylation/deglutathionylation playing a role in the detoxification of xenobiotics, transport, and regulation • some may detoxify ROOH • some may reduce DHA	Chloroplasts, mitochondria, cytosol, nucleus, peroxisomes, apoplast
Thioredoxins, TRXs	• reduction of disulfide bridges in proteins • electron donors for peroxiredoxins	Chloroplasts, mitochondria, cytosol, nucleus, plasma membrane
Glutaredoxins, GRXs	• reduction of disulfide bridges in proteins • reduction ROOH to ROH using GSH or thioredoxin reductases as electron donors • reduction of dehydroascorbate to ascorbate • electron donors for peroxiredoxins	Cytosol, chloroplasts, mitochondria, endoplasmic reticulum, nucleus, plasma membrane
Peroxiredoxins, PRXs	• reduction of H_2O_2 to H_2O and ROOH to ROH • reduction of $ONOO^-$	Chloroplasts, mitochondria, cytosol, nucleus
Guaiacol peroxidases, POXs	• reduction of H_2O_2 to H_2O using various substrates as a reductants • can also produce ROS	Apoplast, cytosol, vacuoles, mitochondria

(*Continued*)

Table 21.4 (Continued)

Enzyme	Antioxidant activity	Localization
Ascorbate and glutathione recycling		
Monodehydroascorbate reductase, MDHAR	• reduction of monodehydroascorbate using NAD(P)H as electron donor	Chloroplasts, mitochondria, peroxisomes, cytosol, plasma membrane
Ferredoxin	• reduction of monodehydroascorbate	Chloroplasts
Dehydroascorbate reductase, DHAR	• reduction of dehydroascorbate using GSH as electron donor	Chloroplasts, mitochondria, cytosol
Glutathione reductase, GR	• reduction of oxidized glutathione (GSSG) using NADPH as electron donor	Chloroplasts, mitochondria, peroxisomes, cytosol, nucleus

Source: From Gechev et al. (2006), Noctor et al. (2018), Estévez and Hernández (2020), and Nowicka (2022).
ROOH, organic peroxides.

detoxify $O_2^{\bullet-}$. They are a group of metalloisoenzymes classified depending on the metal ion(s) in their active center into MnSOD, FeSOD, Cu/ZnSOD, CuSOD, and NiSOD. Higher plants contain Cu/ZnSOD, MnSOD, and FeSOD, the last is not present in all species. FeSOD and MnSOD are structurally related; they are thought to be more ancient than Cu/ZnSOD. Different SODs are targeted to different cell compartments (Nowicka 2022). Catalases (CATs) are the only enzymes able to detoxify H_2O_2 without a need for any additional reductant. They have high turnover numbers and V_{max}, but their affinity to the substrate is low when compared to APX or peroxiredoxins (PRXs). Therefore, CAT activity plays a significant role in the case of high H_2O_2 generation (Nowicka 2022). APX is considered to be the main H_2O_2-detoxifying enzyme in chloroplasts. APX family consists of at least five different isoforms including both soluble and membrane-bound forms targeted to various cell compartments. Due to its high affinity to H_2O_2, APX is thought to be involved in the modulation of ROS signaling (Ahmad et al. 2010). Glutathione peroxidases (GPXs) are a family of multiple isozymes catalyzing reduction of peroxides using GSH or thioredoxins (TRXs) as electron donors. They are classified into three types: selenium-dependent GPX, nonselenium-dependent phospholipid hydroperoxide GPX, and GSTs showing glutathione peroxidase activity. In plants, these enzymes were postulated to play role also in redox signaling (Nowicka 2022). TRXs, glutaredoxins (GRXs), and PRXs are phylogenetically related family of proteins that cooperate in re-reduction of oxidized thiol groups of proteins and in ROS detoxification. They also play important regulatory roles. Genomes of higher plants contain several genes encoding these proteins. The reductases of MDHA, DHA, and GSH are crucial for the efficient recycling of AsA and GSH. In plant green tissues, more than 80% of GR activity was reported to be of chloroplastic isoform, which is another example of the importance of efficient ROS detoxification in these organelles (Nowicka 2022).

The exposure to HMs causes changes in the activity of antioxidant enzymes. The exact pattern of these changes depends on many factors including HM type and concentration applied, time of exposure, and the degree of tolerance of certain species or cultivar to given HM. The increase in the activity of ROS-detoxifying enzymes is often observed, whereas more pronounced stress may cause the decrease in their activity. Sometimes, in HM-treated plants, some antioxidant enzymes are induced, whereas the activity of the others is decreased (Schützendübel and Polle 2002; Mourato et al. 2012; Sytar et al. 2013; Štolfa et al. 2015).

The pathway of MG detoxification operates in cytosol, mitochondria, and chloroplasts. It consists of two enzymes: glyoxalase I (lactoylglutathione lyase, Gly I) and glyoxalase II (hydroxyacylglutahione hydrolase, Gly II) (Hossain et al. 2011; Hoque et al. 2016; Hasanuzzaman et al. 2017a). Hemithioacetal product of spontaneous reaction of MG and GSH is converted by Gly I to S-D-lactoylglutathione. The latter is then hydrolyzed to D-lactate and GSH by Gly II. Apart from MG, Gly I and II may convert other 2-oxoaldehydes to 2-hydroxyacids (Hossain et al. 2012). Recently, the glyoxalase III able to convert MG to D-lactate directly has been discovered (Hasanuzzaman et al. 2017a). HM-induced stress had an impact on Gly I and II activity. The observed effect depended on dose, duration of the stress, plant species and HM applied (Hasanuzzaman et al. 2017a). Both increase (Singla-Pareek et al. 2006; Hasanuzzaman et al. 2019) and decrease in Gly I and II activity was observed in plants after HM treatment (Hossain and Fujita 2009). Sometimes the activity of one enzyme was increased, while the other was decreased (Mostofa et al. 2015a; Nahar et al. 2016; Kharbech et al. 2020b; Alam et al. 2021). Transgenic tobacco over-expressing *glyoxalase I* and *glyoxalase II* displayed enhanced tolerance to Zn (Singla-Pareek et al. 2006).

21.4 Exogenous Chemical-Compounds-Mediated Heavy Metal/Metalloid Tolerance in Crop Plants

Plenty of recent studies (Table 21.5) have shown that the application of exogenous protectants such as singaling compounds (nitric oxide [NO], hydrogen sulfide [H_2S], H_2O_2, etc.), plant hormones (abscisic acid [ABA], salicylic acid [SA], brassinosteroids etc.), osmoprotectants (proline [Pro], glycinebetaine [GB], trehalose [Tre], polyamines), antioxidants (GSH and AsA), nutrient elements (selenium [Se], calcium [Ca], etc.) and chelating compounds like ethylenediaminetetraacetic acid (EDTA) and hemin enhanced HM or metalloid stress tolerance through the modulation of nonenzymatic antioxidants contents (Hossain et al. 2010; Mostofa et al. 2014, 2015a, b; Rady and Hemida 2015; Rahman et al. 2015, 2016; Hasanuzzaman et al. 2017b, c; Ahmad et al. 2021; Ben Massoud et al. 2022; Vega et al. 2022), the activities of antioxidant enzymes (Hossain et al. 2010; Mostofa et al. 2014, 2015a, b; Rady and Hemida 2015; Rahman et al. 2015, 2016; Hasanuzzaman et al. 2017b, c; Ahmad et al. 2020, 2021; Fang et al. 2020; Kharbech et al. 2020a, 2020b; Altaf et al. 2022), and glyoxalase pathway enzymes (Hossain et al. 2010; Hasanuzzaman et al. 2012, 2017b, c, 2018; Islam et al. 2015; Mostofa et al. 2015a; Rahman et al. 2015, 2016; Mahmud et al. 2019; Ahmad et al. 2021; Altaf et al. 2022), Pro synthesis (Mostofa et al. 2015a; Noreen et al. 2018), and by maintaining the lower level of ROS

Table 21.5 Protective effects of exogenous compounds to the ROS and MG detoxification system in several plant species under metal/metalloid toxicity.

Crop species	Type of protectants	Doses of protectants	Metal or metalloid doses	Major effects of ROS and MG detoxification	References
Rice (*Oryza sativa* L.)	Signaling compounds such as hydrogen sulfide (H_2S), nitric oxide (NO), and hydrogen peroxide	H_2S; 100 μM sodium hydrosulfide (NaHS)	Cd; 250 and 500 μM $CdCl_2$	Cd stress resulted in a significant reduction in leaf Chl, water content, essential minerals, and enzymatic and nonenzymatic antioxidants leading to severe oxidative stress as evidenced by increased levels of $O_2^{\bullet-}$, H_2O_2, MG, and MDA. H_2S-pretreamtent of rice seedlings modulated the activities of ROS and MG detoxification system and reduced Cd-induced oxidative damage as indicated by lower level of ROS, MG and MDA. H_2S-treatment also reduced Cd accumulation and content in root and leaf tissues	Mostofa et al. (2015b)
Maize (*Zea mays*)		H_2S; 500 μM NaHS	Cr; 200 μM $K_2Cr_2O_7$	NaHS-derived H_2S restricted nitro-oxidative stress induced by Cr partly through suppression of SNO (*S*-nitrosothiols) accumulation. Mitigation of Cr effects appears to involve GSH-mediated reduction of MG and maintaining AsA-GSH homeostasis in epicotyls	Kharbech et al. (2020a)
Alfalfa (*Medicago sativa* L.)		H_2S; 500 μM NaHS	Pb; 217 $mg\,kg^{-1}$ Cd; 4.95 $mg\,kg^{-1}$	Exogenous H_2S increased antioxidant activities of POD, CAT, SOD, and APX, reduced oxidative damage caused by ROS and MDA, and decreased the metal ion absorption	Fang et al. (2020)
Cauliflower (*Brassica oleracea* L.)		H_2S; 200 μM NaHS	Cr; 10, 100, 200 μM $K_2Cr_2O_7$	Exogenous H_2S restricted oxidative stress damage caused by electrolyte leakage, H_2O_2 and MDA and enhanced antioxidant activities, viz. SOD, CAT, POD, and APX in leaves and root tissues	Ahmad et al. (2020)

Rice (*Oryza sativa* L.)	H_2S; 200 μM NaHS	Hg; 100 μM $HgCl_2$	H_2S acted as an antioxidant to inhibit or scavenge ROS production for maintaining the lower MDA and H_2O_2 levels, and thus prevented oxidative damage	Chen et al. (2016)
Rice (*Oryza sativa* L.)	NO, GSH; 200 μM NO as SNP or 200 μM GSH or 200 μM SNP + 200 μM GSH	Cu; 100 mM $CuSO_4$	Exogenous SNP and GSH decreased Cu-induced ROS and MDA levels, showing substantial protection against oxidative stress. Reduced Cu uptake, elevated levels of AsA, AsA, and GSH redox ratios and higher activities of CAT, GPX, GST, and DHAR play an important role in SNP- and GSH-mediated alleviation of Cu-induced oxidative stress	Mostofa et al. (2014)
Tomato (*Solanum lycopersicum* L.)	NO; 100 μM NO as SNP	Cd; 150 μM $CdCl_2$	NO prevented oxidative damage of Cu-stressed seedlings by upregulating antioxidant enzymes, viz. SOD, CAT, APX, GR, MDHAR, DHAR, and other enzymatic activities of AsA-GSH cycle and promoting a significant decline in ROS-induced lipid peroxidation and electrolyte leakage	Ahmad et al. (2018)
Rapeseed (*Brassica napus* L.).	H_2O_2; 50 μM H_2O_2	Cd; 0.5 and 1.0 mM $CdCl_2$	Cd stress impacted ROS and MG detoxification systems leading to abrupt increase in ROS and ROS-induced damage (lipid peroxidation). H_2O_2-pretreatment counteracted the negative impact of Cd stress on plants by maintaining or increasing the activities of APX, MDHAR, DHAR, GR, GST, GPX, CAT, Gly I, and Gly II of Cd-affected seedlings that resulted in decreased levels of H_2O_2 and MDA	Hasanuzzaman et al. (2017b)
Barley (*Hordeum vulgare* L.)	NO; 500 μM SNP $Na_2(Fe(CN)_5NO){\cdot}2H_2O$	Cu; 200 μM $CuCl_2$	NO alleviated Cu toxicity by affecting the antioxidant defense system and maintaining the GSH-AsA cycle status, suggesting a protection of proteins against oxidation by regulating the cellular redox homeostasis	Ben Massoud et al. (2022)

(*Continued*)

Table 21.5 (Continued)

Crop species	Type of protectants	Doses of protectants	Metal or metalloid doses	Major effects of ROS and MG detoxification	References
Rice (*Oryza sativa* L.)	Plant hormones	SA; 200 μM SA	V; 35 mg L^{-1}	SA pretreatment increased SOD, CAT, GPX, GST, and AsA-GSH pathway enzymes and enhanced AsA and GSH levels, and minimized ROS in rice seedlings by regulating the gene expression of antioxidant enzymes (*OsCuZnSOD1, OsCaTB, OsAPX1, OsGR2, OsGSTU37*). SA reduced MG toxicity and enhanced glyoxalase enzyme activity by upregulating the gene expression of glyoxalase genes (*OsGlyI-1* and *OsGlyII-2*) under V stress	Altaf et al. (2022)
Mustard (*Brassica juncea* L.)		SA; 100 μM H_2S + 0.5 mM SA	Cd; 50 μM $CdCl_2$	Under Cd stress, dual treatment with H_2S and SA increased APX activity by 105%, MDHAR 83%, GR 196%, DHAR 85%, SOD 98%, and GST 91% compared to Cd-stressed plants. Both H_2S and SA function synergistically mitigate the detrimental effects of Cd on nutrient homeostasis and boost antioxidant pathway	Kaur et al. (2021)
Milk thistle (*Silybum marianum* L. Gaertn)		SA; 0.25 μM SA + 10 μM H_2O_2	Cd; 500 μM	Priming and foliar spray of SA and H_2O_2 suppressed the negative impact of Cd^{2+} and decreased the oxidative damage parameters, i.e. MDA and H_2O_2, thereby protecting membrane integrity under Cd^{2+} stress	Nizar et al. (2022)
Pepper (*Capsicum annuum* L.)		SA; 0.5 mM SA + 0.2 mM NaHS	Pb; 0.1 mM $PdCl_2$	Both SA and NaHS upregulated AsA-GSH cycle enzyme activities and alleviated Pb-induced phytotoxicity	Kaya (2020)
Castor (*Ricinus communis* L.)		Cytokinins; 15 μM cytokinins, viz. kinetin (KIN) and 6-benzylaminopurine (BAP)	Cu; 80 and 160 μM $CuSO_4$	Exogenous application of cytokinins efficiently alleviated Cu stress effects via effective modulation of antioxidant mechanisms and accumulation of bioactive compounds (phenolic and flavonoid)	Sameena and Putur (2021)

Chickpea (*Cicer arietinum*)		JA, GA; 1 nM JA or 10^{-6} M GA_3	Cd; 150 μM $CdSO_4 \cdot 8H_2O$	Combined application of JA and GA_3 modulated the AsA-GSH enzyme cycle differently, through altering the activities of APX, MDHAR, and DHAR to a varying level and causing an enhancement in GR activity. Besides, JA and GA_3 treatment played a pivotal role in maintaining the AsA/DHA and GSH/GSSG ratios that also helped to reduce the oxidative damage through regulation of the GSH-based Gly systems to detoxify MG	Ahmad et al. (2021)
Tomato (*Solanum lycopersicum* L.)		BRs; 10^{-8} M 24-EBL (24-epibrassinolide) and 0.1 mM H_2O_2	Ni; 100 mg kg^{-1} $NiCl_2 \cdot 7H_2O$	Exogenous BR and H_2O_2 restricted the Ni uptake and minimized the Ni-induced oxidative damage as indicated by lower ROS levels through the modulation of CAT and SOD activities. The treated plants also sustained higher Chl content, net photosynthetic rate, protein content, proline, shoot length, fresh and dry weight, nitrate reductase, and beta-carotenoids	Nazir et al. (2019)
Mustard (*Brassica juncea* L.)		BRs; 10^{-6}, 10^{-8} and 10^{-10} M 24-EBL (24-epibrassinolide)	Pb; 3 mM $(CH_3CO_2)_2Pb$	Mustard seeds pre-soaking with 24-EBL activated the antioxidant defense system as indicated by increased activities of SOD, CAT, APX, and POX to overcome the toxic effects of Pb stress	Soares et al. (2020)
Mung bean (*Vigna radiata*)	Osmoprotectants	GB, Pro; 5 mM GB or 5 mM Pro	Cd; 1 mM $CdCl_2$	Exogenous application of Pro or GB caused an increase in GSH and AsA content, maintained a high GSH/GSSG ratio and increased the activities of APX, DHAR, MDHAR, GR, GST, GPX, CAT, Gly I, and Gly II involved in ROS and MG detoxification system as compared to the control and Cd-stressed plants, with a concomitant decrease in GSSG content, H_2O_2 and lipid peroxidation level	Hossain et al. (2010)

(*Continued*)

Table 21.5 (Continued)

Crop species	Type of protectants	Doses of protectants	Metal or metalloid doses	Major effects of ROS and MG detoxification	References
Wheat (*Triticum aestivum* L.)		Pro, 80 mM Pro	Cu; 400 µM $CuSO_4 \cdot 5H_2O$	The exogenous application of Pro resulted in the reduced generation of ROS by increasing SOD, POD and CAT activities, and enhanced accumulation of Pro and protein contents in plants under Cu stress	Noreen et al. (2018)
Tobacco (*Nicotiana tabacum* L.)		Pro, GB; 1 and 10 mM Pro; 1 and 10 mM GB	Cd; 100 µM $CdCl_2 \cdot 2.5H_2O$	Exogenous application of Pro or GB to the Cd-stressed cells resulted in a decrease in ROS, MG, and carbonyl contents. In Cd-stressed cells, exogenously applied Pro or GB led to an increase in GPX and Gly I activities that are involved in the detoxification of ROS and MG, respectively	Islam et al. (2015)
Wheat (*Triticum aestivum* L.)		GB; 100 mM GB	Cr: 0.25 and 0.5 mM $K_2Cr_2O_7$	Foliar GB application alleviated Cr toxicity in wheat plants by preventing Cr accumulation and enhancing antioxidant capacity of CAT and APX in shoots and roots	Ali et al. (2015)
Cotton (*Gossypium hirsutum* L.)		GB; 1 mM GB	Cd; 1 and 5 µM $CdCl_2$	Addition of GB markedly reduced Cd-induced MDA, H_2O_2 accumulation and electrolyte leakage. GB also increased the activities of key antioxidant enzymes such as APX, SOD, POD, and CAT in roots and shoots as compared to respective Cd-only treatments	Farooq et al. (2016)

Rice (*Oryza sativa* L.)		Tre; 10 mM Tre	Cu; 100 μM $CuSO_4$	Tre pretreatment improved Cu tolerance in rice plants by inhibiting Cu uptake, regulating the antioxidant and glyoxalase systems and decreasing Cu-induced oxidative damage by lowering the accumulation of ROS and MDA. Tre counteracted the Cu-induced increase in Pro and GSH content, but significantly improved AsA content and redox status. Increased activities of glyoxalases I and II correlated with reduced levels of MG in Tre-pretreated Cu-stressed rice plants	Mostofa et al. (2015a)
Wheat (*Triticum aestivum* L.)		Polyamines; 2 mM spermine (Spm) or 2 mM spermidine (Spd)	Cd; 1 mM $CdCl_2$	Exogenous Spd or Spm increased the activities of SOD and CAT and the concentrations of nonenzymatic antioxidants, viz. AsA, total GSH, Spd and Spm in wheat seedlings grown under Cd stress	Rady and Hemida (2015)
Wheat (*Triticum aestivum* L.)	Antioxidants	AsA; 0.2 and 0.6 mM AsA	Pb; 2 mM Pb	Exogenous AsA application mitigated the Pb toxicity-induced oxidative damage by enhancing the activities of antioxidant enzymes, such as SOD, CAT, and GR. Improved activity of antioxidant enzymes suppressed the formation of MDA and H_2O_2 and improved growth characteristics	Alamri et al. (2018)
Wheat (*Triticum aestivum* L.)		GSH; 1 mM GSH	Pb; 0.5 and 1 mM $Pb(NO_3)_2$	GSH supplementation under Pb stress increased antioxidant redox pool, i.e. AsA and GSH pool and augmented the activities of antioxidant enzymes such as MDHAR, DHAR, GR, CAT, and GPX and decreased ROS production. Increased GSH level and the higher activities of Gly I and Gly II helped in MG detoxification	Hasanuzzaman et al. (2018)

(*Continued*)

Table 21.5 (Continued)

Crop species	Type of protectants	Doses of protectants	Metal or metalloid doses	Major effects of ROS and MG detoxification	References
Barley (*Hordeum vulgare* L.)		GSH; 20 mg L^{-1} GSH	Cd; 5 μM $CdCl_2$	Exogenous GSH decreased Cd-induced ROS and MDA accumulation and facilitated substantial protection against oxidative injury in responses to Cd stress. Glutathione stimulated APX (especially cAPX, sAPX, and tAPX isoenzymes) and CAT (CAT1 and CAT2) activities playing an important role in alleviation of oxidative stress	Chen et al. (2010)
Rice (*Oryza sativa* L.)		GSH; 50 and 100 mg kg^{-1} GSH	As; 15 μM $NaAsO_2$	Exogenous GSH application in As-treated seedlings reduced As-induced oxidative stress, improved antioxidant defense systems by maintaining antioxidant and/or redox enzyme homeostasis, and increased the AsA and GSH contents. The increase in GSH redox state was linked to an increase in the AsA redox ratio via the induction of the AsA–GSH cycle	Jung et al. (2019)
Mustard (*Brassica juncea* L.)	Chelating compounds	EDTA; 0.5 mM EDTA	Cd; 0.5 and 1 mM $CdCl_2$	EDTA supplemented to Cd-stressed seedlings improved the constituents of the AsA-GSH cycle and upregulated the activities of APX, MDHAR, DHAR, GR, GPX, SOD, CAT, and notably enhanced Gly I activity. Addition of EDTA to the Cd-stressed seedlings notably decreased H_2O_2 and $O_2{\bullet}^-$ generation, lipid peroxidation, and MG content in the leaves	Mahmud et al. (2019)
Canola (*Brassica napus* L.)		EDTA: 2.5 mM EDTA	Cd; 10 and 50 μM $CdCl_2$	Application of EDTA alone and with combination of Cd increased the antioxidant enzymes activities of APX, SOD, POD, and CAT and reduced the electrolyte leakage and production of H_2O_2 and MDA	Farid et al. (2015)

Canola (*Brassica napus* L.)		EDTA; 2.5 mM EDTA	Pb; 50 and 100 µM lead nitrate ($Pb(NO_3)_2$)	Exogenous application of EDTA significantly increased the activities of SOD, CAT, POD, and APX in roots and leaves of the seedlings and caused a significant reduction in electrolyte leakage, H_2O_2 and MDA content at both levels of Pb stress	Kanwal et al. (2014)
Rice (*Oryza sativa* L.)	Nutrients: Ca	2.5 mM $CaCl_2$	Cd; 0.25 and 1 mM $CdCl_2$	Exogenous Ca regulated the antioxidant defense and glyoxalase systems, which decreased overproduction of ROS and MG and reduced Cd toxicity. Ca application significantly increased AsA content and the activities of SOD, CAT, GST, MDHAR, and DHAR in the antioxidant system and upregulated Gly I and Gly II activities in the glyoxalase system in rice seedlings exposed to both levels of Cd stress	Rahman et al. (2016)
Rice (*Oryza sativa* L.)		10 mM $CaCl_2$	As; 0.5 and 1 mM Na_2HAsO_4	In As-stressed rice seedlings, Ca supplementation reduced ROS production, increased AsA content, and the activities of MDHAR, DHAR, CAT, GPX, SOD, Gly I, and Gly II suggesting that Ca supplementation improves rice seedlings tolerance to As-induced oxidative stress by reducing As uptake, enhancing their antioxidant defense and glyoxalase systems	Rahman et al. (2015)
Lentil (*Lens culinaris*)		5 mM $Ca(NO_3)_2$	Cd; 20 and 40 µm $CdCl_2$	Ca priming modulated the antioxidant defense response in favor of plant growth under elevated Cd concentrations, and H_2O_2 level, playing a pivotal role in the crosstalk between Ca and ROS signaling	Talukdar (2012)

(*Continued*)

Table 21.5 (Continued)

Crop species	Type of protectants	Doses of protectants	Metal or metalloid doses	Major effects of ROS and MG detoxification	References
Rapeseed (*Brassica napus*)	Nutrients: Se	50 and 100 μM Se as Na_2SeO_4	Cd; 0.5 and 1.0 mM $CdCl_2$	Cd stress significantly impacted the ROS and MG detoxification system leading to oxidative damage as indicated by stringent increase in H_2O_2 and MDA levels. Exogenous Se modulated the APX, MDHAR, DHAR, GR, GPX, CAT, Gly I and Gly II activities and conferred low oxidative damage by maintaining low level of ROS	Hasanuzzaman et al. (2012)
Sunflower (*Helianthus annuus*)		5, 10 and 20 μM Se as $Na_2SeO_4{\cdot}10H_2O$	Cd; 20 μM $CdCl_2$	Pre-soaking of seeds with Se alleviated the negative effect of Cd stress on growth and led to a decrease in oxidative injuries caused by Cd. Furthermore, Se enhanced the activities of CAT, APX, and GR, but lowered that of SOD and POD	Saidi et al. (2014)
Oilseed rape (*Brassica napus* L.)		Se (0, 5 and 10 μM as sodium selenite, Na_2SeO_3)	Cr; 0, 100 and 200 μM potassium dichromate ($K_2Cr_2O_7$)	Exogenous application of Se under Cr stress modulated the photosynthesis-related parameters, enhanced the nutrient uptake and improved the essential amino acid content. Se also activated the antioxidative and glyoxalase pathway enzymes and redistricted the excessive accumulation of ROS and MG	Ulhassan et al. (2019)
Wheat (*Triticum aestivum* L.)	Nutrients: Si	1.5 mM Na_2SiO_3	Cd, 0, 25, 50 and 75 μM $CdCl_2$	Addition of Si to the growth medium caused an increase in leaf Pro concentration, photosynthetic pigments, and the activity of SOD, and altered the uptake, translocation, and accumulation of Cd in both shoots and grains in wheat cultivars. Exogenous application of Si at the boot stage decreased accumulation of Cd in wheat grains	Hussain et al. (2015)

Wheat (*Triticum aestivum* L.)		1.5 and 3 mM Na_2SiO_3	Cd; 10 and 25 mg kg^{-1} $CdCl_2$	Si application increased the growth-related attributes, photosynthetic pigments, essential nutrient uptake, and also enhanced the activities of various antioxidant enzymes, viz. SOD, POD, APX, and CAT leading to a decrease in the reduction of oxidative stress indicators, like MDA, H_2O_2, and electrolyte leakage and Cd uptake in root and shoot of wheat cultivars	Thind et al. (2021)
Rapeseed (*Brassica napus* L.)		1 mM SiO_2	Cd; 0.5 and 1 mM $CdCl_2$	Exogenous Si attenuated oxidative damage in plants to make them more tolerant under Cd stress through augmentation of different antioxidant components and MG detoxification system. Si reduced H_2O_2 and MDA contents and improved antioxidant defense mechanism through increasing the AsA and GSH pools and activities of AsA-GSH cycle (APX, MDHAR, DHAR, and GR) and glyoxalase system (Gly I and Gly II) enzymes and CAT	Hasanuzzaman et al. (2017c)
Tartary buckwheat (*Fagopyrum tararicum*)	Nutrients: S	400 mg l^{-1} $(NH_4)_2SO_4$	Cd; 100 mg L^{-1} $CdCl_2 \cdot 2.5H_2O$	Exogenous S significantly mitigated Cd-induced oxidative stress with the aids of antioxidant enzymes and agents both in leaves and roots, including POD, APX, GPX, GST, AsA, and GSH	Lu et al. (2019)
Mung bean (*Vigna radiata*)	Nutrients: Fe	0.5 mM Hemin (ferroprotoporphyrin IX)	Cd; 50 μM $CdCl_2$	Hemin supplementation under Cd stress improved plant growth by enhancing the harvestable biomass and photosynthetic pigments, increasing antioxidant activities (SOD, APX, POD, Heme oxygenase-1 (HO-1) and Pro), lowering oxidative damage and increasing Cd tolerance in plants	Mahawar et al. (2021)
Rice (*Oryza sativa* L.)		1 and 5 μM Hemin	Zn; 100 μM $ZnSO_4$ Pb; 100 μM $Pb(NO_3)_2$ Cr; 80 μM $K_2Cr_2O_7$	Hemin reduced heavy-metal-induced toxicity in rice seedlings by upregulating activities of SOD, APX, and GR; promoting accumulation of AsA and GSH and downregulating production of ROS and inhibiting metal accumulation in roots	Chen et al. (2017)

(Hossain et al. 2010; Hasanuzzaman et al. 2012, 2017b, c, 2018; Farid et al. 2015; Islam et al. 2015; Rahman et al. 2015, 2016; Chen et al. 2016; Alamri et al. 2018; Ahmad et al. 2020; Kharbech et al. 2020a, 2020b; Kaur et al. 2021; Thind et al. 2021; Nizar et al. 2022) and MG (Islam et al. 2015; Mostofa et al. 2015a; Rahman et al. 2015; Mahmud et al. 2019; Ulhassan et al. 2019), lipid peroxidation levels (Hossain et al. 2010; Hasanuzzaman et al. 2012, 2017b, c, 2018; Mostofa et al. 2014; Farid et al. 2015; Islam et al. 2015; Rahman et al. 2015, 2016; Chen et al. 2016; Ahmad et al. 2018, 2020; Alamri et al. 2018; Mahmud et al. 2019; Fang et al. 2020; Thind et al. 2021; Nizar et al. 2022) (Table 21.5). Additionally, application of exogenous chemical compounds was also found to reduce the uptake, accumulation, and translocation of HMs from root to shoot (Hussain et al. 2015; Mostofa et al. 2015a; Rahman et al. 2015; Fang et al. 2020; Thind et al. 2021). Few studies also showed that exogenous application of these compounds enhanced the expression of antioxidative and glyoxalase pathway genes (Altaf et al. 2022) (Table 21.5).

21.5 Conclusions and Future Perspectives

Extensive research carried out during last decades lead us to understand the phenomenon of HM toxicity in plants. Protective mechanisms important for providing tolerance to this type of stress have been also examined, including the antioxidant defense system and glyoxalase system. Plenty of studies have shown that co-ordinated induction of ROS detoxifying antioxidant defense system and MG detoxifying glyoxalase system conferred better tolerance to abiotic stresses including HM/metalloid stresses through the regulation of multiple stress responsive pathways (summarized in Figure 21.1). Knowing the role of oxidative and MG stress in HM/metalloid toxicity, as well as the action of ROS- and MG-detoxifying enzymes, low-molecular compounds and signaling cascades aimed at induction or enhancement of antioxidant defense, we may try to modulate these processes. One of the ways of such modulation is the application of exogenous chemical compounds. The results presented above showed that such an approach has a potential (Table 21.5). Hence, the application of exogenous chemical compounds is a promising strategy to enhance plant tolerance to HMs. Such technique does not require genetic modification of the plants, therefore it would be faster to introduce into agricultural practices. However, further studies are required to dissect the roles of ROS and MG metabolism and signaling in orchestrating HM/metalloid stress tolerance through the regulation of multiple stress-responsive pathways and genes.

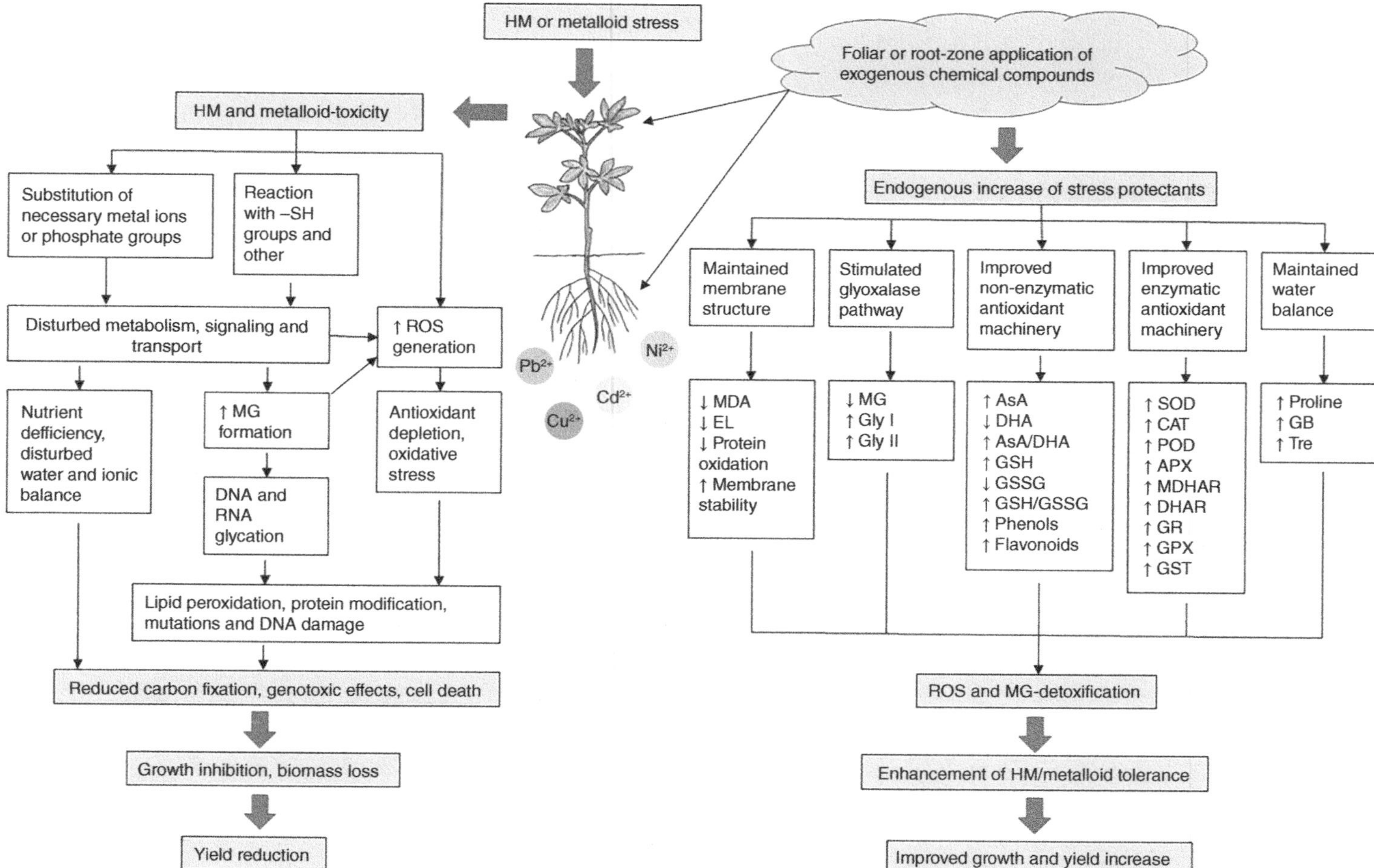

Figure 21.1 A schematic diagram representing HM/metalloid-induced toxicity and protective mechanisms of exogenous chemical compound-mediated tolerance. APX, ascorbate peroxidase; AsA, ascorbate; CAT, catalase; DHA, dehydroascorbate; DHAR, dehydroascorbate reductase; EL, electrolyte leakage; GB, glycine betaine; Gly I, glyoxalase I; Gly II, glyoxalase II; GPX, glutathione peroxidase; GR, glutathione reductase; GSH, glutathione; GSSG, oxidized glutathione; GST, glutathione-*S*-transferase; HM, heavy metal; MDA, malonyldialdehyde; MDHAR, monodehydroascorbate reductase; MG, methylglyoxal; ROS, reactive oxygen species; POD, peroxidase; SOD, superoxide oxidase; Tre, trehalose.

References

Ahmad, P., Jaleel, C.A., Salem, M.A. et al. (2010). Roles of enzymatic and nonenzymatic antioxidants in plants during abiotic stress. *Crit. Rev. Biotechnol.* 30: 161–175.

Ahmad, P., Ahanger, M.A., Alyemeni, M.N. et al. (2018). Exogenous application of nitric oxide modulates osmolyte metabolism, antioxidants, enzymes of ascorbate-glutathione cycle and promotes growth under cadmium stress in tomato. *Protoplasma* 255: 79–93.

Ahmad, R., Ali, S., Rizwan, M. et al. (2020). Hydrogen sulfide alleviates chromium stress on cauliflower by restricting its uptake and enhancing antioxidative system. *Physiol. Plant.* 168: 289–300.

Ahmad, P., Raja, V., Ashraf, M. et al. (2021). Jasmonic acid (JA) and gibberellic acid (GA_3) mitigated Cd-toxicity in chickpea plants through restricted Cd uptake and oxidative stress management. *Sci. Rep.* 11: 19768.

Alam, P., Balawi, T.H., Altalayan, F.H. et al. (2021). Silicon attenuates the negative effects of chromium stress in tomato plants by modifying antioxidant enzyme activities, ascorbate–glutathione cycle and glyoxalase system. *Acta Physiol. Plant.* 43: 110.

Alamri, S.A., Siddiqui, M.H., Al-Khaishany, M.Y.Y. et al. (2018). Ascorbic acid improves the tolerance of wheat plants to lead toxicity. *J. Plant Interact.* 13 (1): 409–419.

Ali, S., Chaudhary, A., Rizwan, M. et al. (2015). Alleviation of chromium toxicity by glycinebetaine is related to elevated antioxidant enzymes and suppressed chromium uptake and oxidative stress in wheat (*Triticum aestivum* L.). *Environ. Sci. Pollut. Res.* 22: 10669–10678.

Altaf, M.M., Diao, X.P., Wang, H. et al. (2022). Salicylic acid induces vanadium stress tolerance in rice by regulating the AsA-GSH cycle and glyoxalase system. *J. Soil Sci. Plant Nutr.* https://doi.org/10.1007/s42729-022-00788-x.

Anjum, N.A., Hasanuzzaman, M., Hossain, M.A. et al. (2015). Jacks of metal/metalloid chelation trade in plants—an overview. *Front. Plant Sci.* **6**: 192. https://doi.org/10.3389/fpls.2015.00192.

Balasaraswathi, K., Jayaveni, S., Sridevi, J. et al. (2017). Cr–induced cellular injury and necrosis in *Glycine max* L.: biochemical mechanism of oxidative damage in chloroplast. *Plant Physiol. Biochem.* 118: 653–666.

Ben Massoud, M., Kharbech, O., Mahjoubi, Y. et al. (2022). Effect of exogenous treatment with nitric oxide (NO) on redox homeostasis in barley seedlings (*Hordeum vulgare* L.) under copper stress. *J. Soil Sci. Plant Nutr.* https://doi.org/10.1007/s42729-021-00757-w.

Bhattacharjee, S. (2005). Reactive oxygen species and oxidative burst: roles in stress, senescence and signal transduction in plants. *Curr. Sci.* 89: 1113–1121.

Bhattacharjee, S. (2019). ROS and oxidative stress: origin and implication. In: *Reactive Oxygen Species in Plant Biology*, 1–31. New Delhi: Springer.

Burda, K., Kruk, J., Schmid, G.H., and Strzalka, K. (2003). Inhibition of oxygen evolution in Photosystem II by Cu(II) ions is associated with oxidation of cytochrome b_{559}. *Biochem. J.* 371: 597–601.

Chen, F., Wang, F., Wu, F. et al. (2010). Modulation of exogenous glutathione in antioxidant defense system against Cd stress in the two barley genotypes differing in Cd tolerance. *Plant Physiol. Biol. Chem.* 48 (8): 663–672.

Chen, Z., Chen, M., and Jiang, M. (2016). Hydrogen sulfide alleviates mercury toxicity by sequestering it in roots or regulating reactive oxygen species productions in rice seedlings. *Plant Physiol. Biochem.* 111: 179–192.

Chen, O., Zhang, X., Liu, Y. et al. (2017). Hemin-mediated alleviation of zinc, lead and chromium toxicity is associated with elevated photosynthesis, antioxidative capacity, suppressed metal uptake and oxidative stress in rice seedlings. *Plant Growth Regul.* 81: 253–264.

Collin, V.C., Eymery, F., Genty, B. et al. (2008). Vitamin E is essential for the tolerance of *Arabidopsis thaliana* to metal-induced oxidative stress. *Plant Cell Environ.* 31: 244–257.

Corpas, F.J., Gupta, D.K., and Palma, J.M. (2015). Production sites of reactive oxygen species (ROS) in organelles from plant cells. In: *Reactive Oxygen Species and Oxidative Damage in Plants Under Stress* (ed. F.J. Corpas, D.K. Gupta, and J.M. Palma), 1–22. Springer International.

DalCorso, G. (2012). Heavy metal toxicity in plants. In: *Plants and Heavy Metals* (ed. A. Furini), 1–25. Springer.

Del Río, L.A., Sandalio, L.M., Corpas, F.J. et al. (2006). Reactive oxygen species and reactive nitrogen species in peroxisomes. Production, scavenging, and role in cell signaling. *Plant Physiol.* 141: 330–335.

Dumanović, J., Nepovimova, E., Natić, M. et al. (2021). The significance of reactive oxygen species and antioxidant defense system in plants: a concise overview. *Front. Plant Sci.* 11: 552969.

Edreva, A. (2005). Generation and scavenging of reactive oxygen species in chloroplasts: a submolecular approach. *Agric. Ecosyst. Environ.* 106: 119–133.

Estévez, I.H. and Hernández, M.R. (2020). Plant glutathione *S*-transferases: an overview. *Plant Gene* 23: 100233.

Fang, L., Ju, W., Yang, C. et al. (2020). Exogenous application of signaling molecules to enhance the resistance of legume-rhizobium symbiosis in Pb/Cd-contaminated soils. *Environ. Pollut.* 265: 114744.

Farid, M., Ali, S., Ishaque, W. et al. (2015). Exogenous application of ethylenediamminetetraacetic acid enhanced phytoremediation of cadmium by *Brassica napus* L. *Int. J. Environ. Sci. Technol.* 12: 3981–3992.

Farooq, M.A., Ali, S., Hameed, A. et al. (2016). Cadmium stress in cotton seedlings: physiological, photosynthesis and oxidative damages alleviated by glycine betaine. *South Afr. J Bot.* 104: 61–68.

Foyer, C.H. and Noctor, G. (2003). Redox sensing and signalling associated with reactive oxygen in chloroplasts, peroxisomes and mitochondria. *Physiol. Plant.* 119: 355–364.

Gajewska, E. and Skłodowska, M. (2007). Relations between tocopherol, chlorophyll and lipid peroxides contents in shoots of Ni-treated wheat. *J. Plant Physiol.* 164: 364–366.

Gechev, T.S., Van Breusegem, F., Stone, J.M. et al. (2006). Reactive oxygen species as signals that modulate plant stress responses and programmed cell death. *BioEssays* 28: 1091–1101.

Gest, N., Gautier, H., and Stevens, R. (2013). Ascorbate as seen through plant evolution: the rise of a successful molecule? *J. Exp. Bot.* 64: 33–53.

Gómez Ojeda, A., Corrales Escobosa, A.R., Wrobel, K. et al. (2013). Effect of Cd(II) and Se(IV) exposure on cellular distribution of both elements and concentration levels of glyoxal and methylglyoxal in *Lepidium sativum*. *Metallomics* 5: 1254–1261.

Hasan, M.K., Cheng, Y., Kanwar, M.K. et al. (2017). Responses of plant proteins to heavy metal stress – a review. *Front. Plant Sci.* 8: 1492.

Hasanuzzaman, M., Hossain, M.A., and Fujita, M. (2012). Exogenous selenium pretreatment protects rapeseed seedlings from cadmium-induced oxidative stress by upregulating antioxidant defense and methylglyoxal detoxification systems. *Biol. Trace Elem. Res.* 149: 248–261.

Hasanuzzaman, M., Nahar, K., Hossain, M.S. et al. (2017a). Coordinated actions of glyoxalase and antioxidant defense systems in conferring abiotic stress tolerance in plants. *Int. J. Mol. Sci.* 18: 200.

Hasanuzzaman, M., Nahar, K., Gill, S.S. et al. (2017b). Hydrogen peroxide pretreatment mitigates cadmium-induced oxidative stress in *Brassica napus* L.: an intrinsic study on antioxidant defense and glyoxalase systems. *Front. Plant Sci.* 8: 115.

Hasanuzzaman, M., Nahar, K., Anee, T.I., and Fujita, M. (2017c). Exogenous silicon attenuates cadmium-induced oxidative stress in *Brassica napus* L. by modulating AsA-GSH pathway and glyoxalase system. *Front. Plant Sci.* 8: 1061.

Hasanuzzaman, M., Nahar, K., Rahman, A. et al. (2018). Exogenous glutathione attenuates lead-induced oxidative stress in wheat by improving antioxidant defense and physiological mechanisms. *J. Plant Interact.* 13 (1): 203–212.

Hasanuzzaman, M., Alam, M.M., Nahar, K. et al. (2019). Silicon-induced antioxidant defense and methylglyoxal detoxification works coordinately in alleviating nickel toxicity in *Oryza sativa* L. *Ecotoxicology* 28: 261–276.

Hoque, T.S., Hossain, M.A., Mostofa, M.G. et al. (2016). Methylglyoxal: an emerging signaling molecule in plant abiotic stress responses and tolerance. *Front. Plant Sci.* 7: 1341.

Hossain, M.A. and Fujita, M. (2009). Purification of glyoxalase I from onion bulbs and molecular cloning of its cDNA. *Biosci. Biotechnol. Biochem.* 73: 2007–2013.

Hossain, M.A., Hossain, M.Z., and Fujita, M. (2009). Stress-induced changes of methylglyoxal level and glyoxalase I activity in pumpkin seedlings and cDNA cloning of glyoxalase I gene. *Aust. J. Crop Sci.* 3: 53–64.

Hossain, M.A., Hasanuzzaman, M., and Fujita, M. (2010). Up-regulation of antioxidant and glyoxalase systems by exogenous glycinebetaine and proline in mung bean confer tolerance to cadmium stress. *Physiol. Mol. Biol. Plants* 16: 259–272.

Hossain, M.A., Teixeira da Silva, J.A., and Fujita, M. (2011). Glyoxalase system and reactive oxygen species detoxification system in plant abiotic stress response and tolerance: an intimate relationship. In: *Abiotic Stress in Plants—Mechanisms and Adaptations* (ed. A. Shanker and B. Venkateswarlu), 235–266. Rijeka, Croatia: Intech Open.

Hossain, M.A., Piyatida, P., da Silva, J.A.T., and Fujita, M. (2012). Molecular mechanism of heavy metal toxicity and tolerance in plants: central role of glutathione in detoxification of reactive oxygen species and methylglyoxal and in heavy metal chelation. *J. Bot.* 2012: 1–37.

Hossain, M.A., Hoque, M.A., Burritt, D.J., and Fujita, M. (2014). Proline protects plants against abiotic oxidative stress: biochemical and molecular mechanisms. In: *Oxidative Damage to Plants: Antioxidant Networks and Signaling* (ed. P. Ahmad), 477–522. Academic Press.

Hussain, I., Ashraf, M.A., Rasheed, R. et al. (2015). Exogenous application of silicon at the boot stage decreases accumulation of cadmium in wheat (*Triticum aestivum* L.) grains. *Braz. J. Bot.* 38: 223–234.

Islam, M.M., Okuma, E., Jannat, R. et al. (2015). Exogenous proline and glycinebetaine enhance glutathione-dependent defense mechanisms and mitigate cadmium toxicity in cultured tobacco cells. *Ann. Bangladesh Agric.* 19: 67–80.

Ismael, M.A., Elyamine, A.M., Moussa, M.G. et al. (2019). Cadmim in plants: uptake, toxicity, and its interactions with selenium fertilizers. *Metallomics* 11: 255–277.

Janku, M., Luhová, L., and Petrivalský, M. (2019). On the origin and fate of reactive oxygen species in plant cell compartments. *Antioxidants* 8: 105.

Jung, H., Kong, M.-S., Lee, B.-R. et al. (2019). Exogenous glutathione increases arsenic translocation into shoots and alleviates arsenic-induced oxidative stress by sustaining ascorbate–glutathione homeostasis in rice seedlings. *Front. Plant Sci.* 10: 1089.

Kalaivanan, D. and Ganeshamurthy, A.N. (2016). Mechanisms of heavy metal toxicity in plants. In: *Abiotic Stress Physiology of Horticultural Crops* (ed. N.K. Srinivasa Rao, K.S. Shivashankara, and R.H. Laxman), 85–102. Springer, India.

Kanwal, U., Ali, S., Shakoor, M.B. et al. (2014). EDTA ameliorates phytoextraction of lead and plant growth by reducing morphological and biochemical injuries in *Brassica napus* L. under lead stress. *Environ Sci. Pollut. Res.* 21: 9899–9910.

Kaur, H., Hussain, S.J., Al-Huqail, A.A. et al. (2021). Hydrogen sulphide and salicylic acid regulate antioxidant pathway and nutrient balance in mustard plants under cadmium stress. *Plant Biol.* https://doi.org/10.1111/plb.13322.

Kaya, C. (2020). Salicylic acid-induced hydrogen sulphide improves lead stress tolerance in pepper plants by upraising the ascorbate-glutathione cycle. *Physiol. Plant* 173: 8–19.

Kharbech, O., Ben Massoud, M., Sakouhi, L. et al. (2020a). Exogenous application of hydrogen sulfide reduces chromium toxicity in maize seedlings by suppressing NADPH oxidase activities and methylglyoxal accumulation. *Plant Physiol. Biochem.* 154: 646–656.

Kharbech, O., Sakouhi, L., Ben Massoud, M. et al. (2020b). Nitric oxide and hydrogen sulfide protect plasma membrane integrity and mitigate chromium-induced methylglyoxal toxicity in maize seedlings. *Plant Physiol. Biochem.* 157: 244–255.

Krieger-Liszkay, A. (2005). Singlet oxygen production in photosynthesis. *J. Exp. Bot.* 56: 337–346.

Kumar, A., Prasad, M.N.V., and Sytar, O. (2012). Lead toxicity, defense strategies and associated indicative biomarkers in *Talinum triangulare* grown hydroponically. *Chemosphere* 89: 1056–1065.

Küpper, H. and Andresen, E. (2016). Mechanisms of metal toxicity in plants. *Metallomics* 8: 269–285.

Larsson, M.A., Baken, S., Gustafsson, J.P. et al. (2013). Vanadium bioavailability and toxicity to soil microorganisms and plants. *Environ. Toxicol. Chem.* 32: 2266–2273.

Latowski, D., Szymańska, R., and Strzałka, K. (2014). Carotenoids involved in antioxidant system of chloroplasts. In: *Oxidative Damage to Plants: Antioxidant Networks and Signaling* (ed. P. Ahmad), 289–319. Academic Press.

Lu, Y., Wang, Q., Li, J. et al. (2019). Effects of exogenous sulfur on alleviating cadmium stress in tartary buckwheat. *Sci. Rep.* 9: 7397.

Mahawar, L., Popek, R., Shekhawat, G.S. et al. (2021). Exogenous hemin improves Cd^{2+} tolerance and remediation potential in *Vigna radiata* by intensifying the HO-1 mediated antioxidant defence system. *Sci. Rep.* 11: 2811.

Mahmud, J.A., Hasanuzzaman, M., Nahar, K. et al. (2019). EDTA reduces cadmium toxicity in mustard (*Brassica juncea* L.) by enhancing metal chelation, antioxidant defense and glyoxalase systems. *Acta Agrobot.* 72: 1722.

Manara, A. (2012). Plant responses to heavy metal toxicity. In: *Plants and Heavy Metals* (ed. A. Furini), 27–53. Dordrecht: Springer.

Marambio-Jones, C. and Hoek, E.M.V. (2010). A review of the antibacterial effects of silver nanomaterials and potential implications for human health and the environment. *J. Nanopart. Res.* 12: 1531–1551.

Mostofa, M.G., Seraj, Z.I., and Fujita, M. (2014). Exogenous sodium nitroprusside and glutathione alleviate copper toxicity by reducing copper uptake and oxidative damage in rice (*Oryza sativa* L.) seedlings. *Protoplasma* 251: 1373–1386.

Mostofa, M.G., Hossain, M.A., Fujita, M., and Tran, L.S.P. (2015a). Physiological and biochemical mechanisms associated with trehalose-induced copper-stress tolerance in rice. *Sci. Rep.* 5: 11433.

Mostofa, M., Rahman, A., Ansary, M.M.U. et al. (2015b). Hydrogen sulfide modulates cadmium-induced physiological and biochemical responses to alleviate cadmium toxicity in rice. *Sci. Rep.* 5: 14078.

Mourato, M., Reis, R., and Martins, L.L. (2012). Characterization of plant antioxidative system in response to abiotic stresses: a focus on heavy metal toxicity. In: *Advances in Selected Plant Physiology Aspects* (ed. G. Montanaro and B. Dichio), 23–44. Intech Open.

Nagajyoti, P.C., Lee, K.D., and Sreekanth, T.V.M. (2010). Heavy metals, occurrence and toxicity for plants: a review. *Environ. Chem. Lett.* 8: 199–216.

Nahar, K., Hasanuzzaman, M., Alam, M.M. et al. (2016). Polyamine and nitric oxide crosstalk: antagonistic effects on cadmium toxicity in mung bean plants through upregulating the metal detoxification, antioxidant defense and methylglyoxal detoxification systems. *Ecotoxicol. Environ. Saf.* 126: 245–255.

Nazir, F., Hussain, A., and Fariduddin, Q. (2019). Interactive role of epibrassinolide and hydrogen peroxide in regulating stomatal physiology, root morphology, photosynthetic and growth traits in *Solanum lycopersicum* L. under nickel stress. *Environ. Exp. Bot.* 162: 479–495.

Nizar, M., Shaukat, K., Zahra, N. et al. (2022). Exogenous application of salicylic acid and hydrogen peroxide ameliorate cadmium stress in milk thistle by enhancing morpho-physiological attributes grown at two different altitudes. *Front. Plant Sci.* 12: 809183.

Noctor, G., Reichheld, J.P., and Foyer, C.H. (2018). ROS-related redox regulation and signaling in plants. *Semin. Cell Dev. Biol.* 80: 3–12.

Noreen, S., Akhter, M.S., Yaamin, T., and Arfan, M. (2018). The ameliorative effects of exogenously applied proline on physiological and biochemical parameters of wheat (*Triticum aestivum* L.) crop under copper stress condition. *J. Plant Interact.* 13: 221–230.

Nowicka, B. (2022). Heavy metal–induced stress in eukaryotic algae—mechanisms of heavy metal toxicity and tolerance with particular emphasis on oxidative stress in exposed cells and the role of antioxidant response. *Environ. Sci. Pollut. Res.* 29: 16860–16911.

Nowicka, B. and Kruk, J. (2010). Occurrence, biosynthesis and function of isoprenoid quinones. *Biochim. Biophys. Acta Bioenerg.* 1797: 1587–1605.

Nowicka, B., Fesenko, T., Walczak, J., and Kruk, J. (2020). The inhibitor-evoked shortage of tocopherol and plastoquinol is compensated by other antioxidant mechanisms in

Chlamydomonas reinhardtii exposed to toxic concentrations of cadmium and chromium ions. *Ecotoxicol. Environ. Saf.* 191: 110241.

Nowicka, B., Trela-Makowej, A., Latowski, D. et al. (2021). Antioxidant and signaling role of plastid-derived isoprenoid quinones and chromanols. *Int. J. Mol. Sci.* 22: 2950.

Palit, S., Sharma, A., and Talukder, G. (1994). Effects of cobalt on plants. *Bot. Rev.* 60: 149–181.

Pasricha, S., Mathur, V., Garg, A. et al. (2021). Molecular mechanisms underlying heavy metal uptake, translocation and tolerance in hyperaccumulators-an analysis: heavy metal tolerance in hyperaccumulators. *Environ. Chall.* 4: 100197.

Pinto, E., Sigaud-Kutner, T.C.S., Leitão, M.A.S. et al. (2003). Heavy metal-induced oxidative stress in algae. *J. Phycol.* 39: 1008–1018.

Rady, M.M. and Hemida, K.A. (2015). Modulation of cadmium toxicity and enhancing cadmium-tolerance in wheat seedlings by exogenous application of polyamines. *Ecotoxicol. Environ. Saf.* 119: 178–185.

Rahman, A., Mostofa, M.G., Alam, M.M. et al. (2015). Calcium mitigates arsenic toxicity in rice seedlings by reducing arsenic uptake and modulating the antioxidant defense and glyoxalase systems and stress markers. *Biomed. Res. Int.* 2015: 340812.

Rahman, A., Mostofa, M.G., Nahar, K. et al. (2016). Exogenous calcium alleviates cadmium-induced oxidative stress in rice (*Oryza sativa* L.) seedlings by regulating the antioxidant defense and glyoxalase systems. *Braz. J. Bot.* 39: 393–407.

Saidi, I., Chtourou, Y., and Djebali, W. (2014). Selenium alleviates cadmium toxicity by preventing oxidative stress in sunflower (*Helianthus annuus*) seedlings. *J. Plant Physiol.* 17: 85–91.

Sameena, P.P. and Putur, J.T. (2021). Exogenous application of cytokinins confers copper stress tolerance in *Ricinus communis* L. seedlings. *J. Plant Growth Regul.* https://doi.org/10.1007/s00344-021-10522-z.

Schützendübel, A. and Polle, A. (2002). Plant responses to abiotic stresses: heavy metal-induced oxidative stress and protection by mycorrhization. *J. Exp. Bot.* 53: 1351–1365.

Shahzad, B., Tanveer, M., Rehman, A. et al. (2018). Nickel; whether toxic or essential for plants and environment – a review. *Plant Physiol. Biochem.* 132: 641–651.

Sharma, S.S., Dietz, K.J., and Mimura, T. (2016). Vacuolar compartmentalization as indispensable component of heavy metal detoxification in plants. *Plant Cell Environ.* 39: 1112–1126.

Singh, H.P., Mahajan, P., Kaur, S. et al. (2013). Chromium toxicity and tolerance in plants. *Environ. Chem. Lett.* 11: 229–254.

Singh, S., Yadav, V., Arif, N. et al. (2020). Heavy metal stress and plant life: uptake mechanisms, toxicity, and alleviation. *Plant Life Under Changing Environ.* 271–287.

Singla-Pareek, S.L., Yadav, S.K., Pareek, A. et al. (2006). Transgenic tobacco overexpressing glyoxalase pathway enzymes grow and set viable seeds in zinc-spiked soils. *Plant Physiol.* 140: 613–623.

Siripornadulsil, S., Traina, S., Verma, D.P.S., and Sayre, R.T. (2002). Molecular mechanisms of proline-mediated tolerance to toxic heavy metals in transgenic microalgae. *Plant Cell* 14: 2837–2847.

Smirnoff, N. (2005). Ascorbate, tocopherol and carotenoids: metabolism, pathway engineering and functions. In: *Antioxidants and Reactive Oxygen Species in Plants* (ed. N. Smirnoff), 53–86. Blackwell Publishing Ltd.

Soares, T.F.S.N., Dias, D.C.F.d.S., Oliveira, A.M.S. et al. (2020). Exogenous brassinosteroids increase lead stress tolerance in seed germination and seedling growth of *Brassica juncea* L. *Ecotoxicol. Environ. Saf.* 193: 110296.

Stoiber, T.L., Shafer, M.M., and Armstrong, D.E. (2013). Induction of reactive oxygen species in *Chlamydomonas reinhardtii* in response to contrasting trace metal exposures. *Environ. Toxicol.* 28: 516–523.

Štolfa, I., Pfeiffer, T.Ž., Špoljarić, D. et al. (2015). Heavy metal-induced oxidative stress in plants: response of the antioxidative system. In: *Reactive Oxygen Species and Oxidative Damage in Plants Under Stress* (ed. F.J. Corpas, D.K. Gupta, and J.M. Palma), 127–163. Springer International Publishing.

Sytar, O., Kumar, A., Latowski, D. et al. (2013). Heavy metal-induced oxidative damage, defense reactions, and detoxification mechanisms in plants. *Acta Physiol. Plant.* 35: 985–999.

Szymańska, R., Nowicka, B., and Kruk, J. (2014). Hydroxy-plastochromanol and plastoquinone-C as singlet oxygen products during photo-oxidative stress in Arabidopsis. *Plant Cell Environ.* 37: 1464–1473.

Talukdar, D. (2012). Exogenous calcium alleviates the impact of cadmium-induced oxidative stress in *Lens culinaris* Medic. seedlings through modulation of antioxidant enzyme activities. *J. Crop Sci. Biotechnol.* 15: 325–334.

Thind, S., Hussain, I., Ali, S. et al. (2021). Silicon application modulates growth, physio-chemicals, and antioxidants in wheat (*Triticum aestivum* L.) exposed to different cadmium regimes. *Dose-Response* 19: 1–15. https://doi.org/10.1177/15593258211014646.

Tiwari, S. and Sarangi, B.K. (2015). Arsenic and chromium-induced oxidative stress in metal accumulator and nonaccumulator plants and detoxification mechanisms. In: *Reactive Oxygen Species and Oxidative Damage in Plants Under Stress* (ed. F.J. Corpas, D.K. Gupta, and J.M. Palma), 165–189. Springer International Publishing.

Triantaphylidès, C. and Havaux, M. (2009). Singlet oxygen in plants: production, detoxification and signaling. *Trends Plant Sci.* 14: 219–228.

Tripathy, B.C. and Oelmüller, R. (2012). Reactive oxygen species generation and signaling in plants. *Plant Signal. Behav.* 7: 1621–1633.

Ulhassan, Z., Gill, R.A., Huang, H. et al. (2019). Selenium mitigates the chromium toxicity in *Brassicca napus* L. by ameliorating nutrients uptake, amino acids metabolism and antioxidant defense system. *Plant Physiol. Biochem.* 145: 142–152.

Vega, A., Delgado, N., and Handford, M. (2022). Increasing heavy metal tolerance by the exogenous application of organic acids. *Int. J. Mol. Sci.* 23: 5438.

Yan, A. and Chen, Z. (2019). Impacts of silver nanoparticles on plants: a focus on the phytotoxicity and underlying mechanism. *Int. J. Mol. Sci.* 20: 1003.

Yruela, I. (2005). Coper in plants. *Braz. J. Plant Physiol.* 17: 145–156.

Zengin, F.K. and Munzuroglu, O. (2005). Effects of some heavy metals on content of chlorophyll, proline and some antioxidant chemicals in bean (*Phaseolus vulgaris* L.) seedlings. *Acta Biol. Cracov. Ser. Bot.* 47: 157–164.

22

Biochar Amendments in Soils and Heavy Metal Tolerance in Crop Plants

Agnieszka Medyńska-Juraszek and Bhakti Jadhav

Institute of Soil Science, Plant Nutrition and Environmental Protection, Wroclaw University of Environmental and Life Sciences, Wroclaw, Poland

22.1 Introduction

Increase of human activity related with agriculture has improved the living standard of human beings, but also created serious environmental problems related to soil degradation and accumulation of potentially toxic elements (PTEs) in plants. Rapid increase of organic (sewage sludge, manure and compost) and mineral fertilization, but also chemical plant protection in food production becomes an emerging trend consequently contributing to the reduction of soil productivity. Soil contamination with heavy metals is one of the most serious anthropogenic environmental stressors determining inhibition of plant growth. When plants receive heavy metals in small amounts, mechanisms of avoidance, e.g. translocation, complexation, and sequestration, are sufficient to overcome the negative impact (Juknys et al. 2012). Excessive accumulation of PTEs in agricultural soils results in soil contamination and has consequences for food quality and safety. Food is the major intake source of toxic metals by human beings. Among food system, vegetables and cereals are the most exposed food to environmental pollution (Jolly et al. 2013). Some of the potential toxic elements, e.g. zinc (Zn), cobalt (Co), copper (Cu), and manganese (Mn), are biologically essential micronutrients required in trace amounts for plants and animal growth. Cadmium (Cd), chromium (Cr), arsenic (As), and lead (Pb) can be toxic to living organism even at low concentrations (Kabata-Pendias and Pendias 2001). Unlike organic contaminants, potentially toxic elements (PTEs) do not degrade easily through either microbial or chemical degradation, resulting in contaminant accumulation in soil environment (Mohamed et al. 2017). In plants, excess of PTE(s) causes physiological, biochemical, and morphological alterations leading to decreased growth and biomass and even death of plants (Adrees et al. 2015a). General metal toxicity mechanisms are interactions with essential functional groups of biomolecules, disruption of the integrity of biomembranes, and production of

Heavy Metal Toxicity and Tolerance in Plants: A Biological, Omics, and Genetic Engineering Approach, First Edition. Edited by Mohammad Anwar Hossain, AKM Zakir Hossain, Sylvain Bourgerie, Masayuki Fujita, Om Parkash Dhankher, and Parvez Haris.

reactive oxygen species (ROS) (Choudhury et al. 2017). Metals bind to many functional groups of cellular macromolecules that form essential cell structures, such as lipids, polysaccharides, and proteins. HMs can damage DNA, denaturate proteins, and inhibit enzyme activity (Nguyen et al. 2021). Presence of PTEs in elevated concentrations not only affects plants but also symbiotic organisms like plant-growth-promoting rhizobacteria (PGPRs) (Giller et al. 2009). PGPRs are a heterogeneous group of bacteria that can be found in the rhizosphere and are capable of improving the extent or quality of plant growth directly and/or indirectly (Glick et al. 1998).

In recent years, much attention has been given to developing strategies to alleviate HM toxicity in crops and agricultural systems to fulfill the global food demand. Physicochemical techniques employed for the clean-up of contaminated sites include excavation, stabilization, or *in situ* fixation (i.e. by adding chemicals to change the metal structure so that it does not get absorbed by plants) (Syta et al. 2019). The principle of soil remediation strategies is to reduce the toxicity by reducing the bioavailability of trace elements for plants and soil biota. As plants acquire PTEs from soil solution through their root system, controlling of metal bioavailability in soil results in decreased uptake of toxic elements by plants. The sorption characteristics of heavy metals in soil systems are important with respect to plant nutrition and environmental pollution, and this sorption process is influenced by (Nguyen et al. 2021) the presence of organic matter. Among potential amendments reducing the risk of potentially toxic elements transfer to plants biochar attracts a lot of attention of researchers and has been considered as a universal amendment improving soil physical, chemical, and biological properties (Park et al. 2011; Kim et al. 2013; Cao et al. 2018; Medyńska-Juraszek and Ćwieląg-Piasecka 2020). Biochar is a fine-grained, carbon-rich, extremely porous by-product of the thermochemical conversion of plant biomass (pyrolysis) at wide range of temperatures (300–800 °C) in an oxygen-poor atmosphere (Amonette and Joseph 2009). Its properties can be designed for better adsorption of trace elements and ameliorating effects in soils. Biochar application to soil brings benefits, improving the physical, chemical, biological, and microbiological properties and consequently, improving soil fertility and productivity over time. Biochar has distinct properties such as high surface area and sorption capacity, presence of oxygen functional groups on its surface and alkaline nature that can contribute to the direct and indirect changes of trace element mobility and bioavailability in soil. Biochar addition dose not result in the decrease of the total heavy metal content of the soil; however reduces the bioavailability of heavy metals to plants and soil biota. Combining biochar application as part of phytoremediation strategy has a good potential. Biochar can improve soil properties (CEC, pH, increase bioavailability of nutrients) and improve soil physical properties crucial for biological reclamation and promotion of mycorrhizal fungi and plant root development. As a consequence of heavy metal immobilization, biochar can reduce the phytotoxicity of polluted soils (Paz-Ferreiro et al. 2014) playing important role in soil bioremediation and protecting crops from PTE accumulation. This chapter presents current understanding regarding the effects of biochar application to soil, enhanced plant growth and heavy metal tolerance in crop plants including mechanisms related to changes of heavy metal uptake conditions and direct changes in plant physiology, reducing the metal stress and negative plant respond to elevated concentrations of toxic elements in soil.

22.2 Heavy Metal Immobilization Mechanisms on Biochar

The physicochemical properties such as surface area, surface charge, pore spaces, and BC production parameters, such as pyrolysis retention time, pyrolysis temperature, and feedstock type, are major variables that determine the adsorption capacity of PTEs of biochar (Uchimiya et al. 2010; Park et al. 2011; Haider et al. 2017). Some materials are better suited than others to remove various heavy metals due to the large range of features that biochar may possess. Therefore, scientists should be conscious of both the features of the biochar and the soil type when choosing a biochar for remediation purposes. Additionally, it should be taken into considaration that pretreatments can modify crucial biochar features including surface area, pH, ash, and carbonates contents, which improves biochar capacity to immobilize heavy metals (Lima et al. 2014). It is important to understand how biochar affects the bioavailable portion of soil heavy metals and can also lessen their leachability before discussing the mechanisms underlying the association between biochar and heavy metals. The high specific surface area of biochar, which has a large capacity for toxic substance adsorption on its surface may cause trace elements interaction with negatively charged carbon surfaces or by ion exchange between trace elements cations and acidic carbon surfaces (Sohi et al. 2010; Uchimiya et al. 2011a, 2012).

Numerous studies have employed scanning electron microscopy to show how heavy metals bind to the surface of biochar (Beesley and Marmiroli 2011; Lu et al. 2012). The creation of complexes between the heavy metals and the biochar functional groups, the exchange of the heavy metal with the cations K^+, Na^+, Mg^{2+}, and Ca^{2+} (Uchimiya et al. 2011b), and physical adsorption are possible causes of this sorption (Lu et al. 2012). Trakal et al. (2014) found that physical sorption on the surface of the biochar was the predominant cause of Cu^{2+} sorption. Other substances found in the ash, such as carbonates, phosphates, or sulfates (Cao et al. 2009; Karami et al. 2011; Park et al. 2013) can also aid in stabilizing heavy metals through precipitation of these substances with the pollutants (Figure 22.1). Xu et al. (2013) reported that biochar made from dairy manure (350 °C) increased Zn, Cd, and Cu, sorption primarily through precipitation, with PO_4^{3-} (less than 10% of total precipitated) or

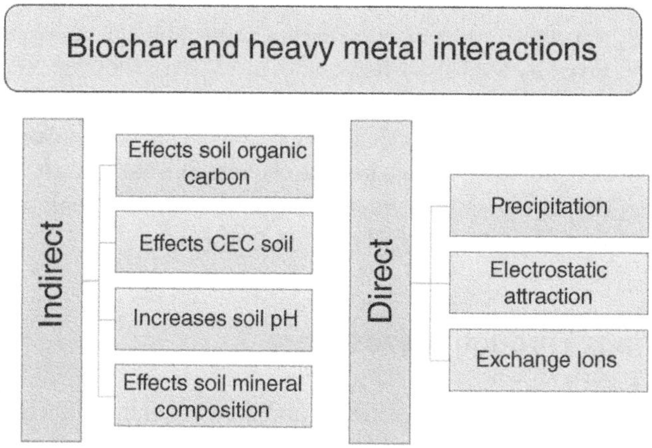

Figure 22.1 Biochar and heavy metal interactions in soil.

CO_3^{2-} (more than 90%) coming from biochar, and less than 25% through surface adsorption through chelation of trace elements by phenolic –OH site.

22.2.1 Heavy Metal Immobilization Through Soil pH Modification

The bioavailability of PTE(s) like Pb, Zn, Cr, Ni, and Cd in the soil is often reduced when soil pH increases (Beesley et al. 2014; Hu et al. 2014; Rees et al. 2014; Houben and Sonnet 2015; Lim et al. 2015; Rees et al. 2015; Venegas et al. 2015; Rinklebe et al. 2016). Depending on the feedstock, processing temperature, and ash content in biochar, pH can range from mildly acidic to alkaline (Cantrell et al. 2012; Wu et al. 2012). In general, as pyrolysis temperatures rise, biochar pH values rise as well (Ehsan et al. 2014; Jiang et al. 2014). Generally, by adding biochar and raising the soil pH, the competition for sorption sites among H^+ and metal cations (Mn^+) is reduced, which decreases the soil mobility and availability of HMs (Gomez-Eyles et al. 2013). Precipitation of HMs as insoluble hydroxides, phosphates, carbonates, and other mineral species result from a raised biochar soil pH and is a common explanation for reductions in HM mobilization rates (Figure 22.1). The relationship between soil pH and HM mobility is dependent on soil properties, quantity of applied biochar, and the duration of its application (Fellet et al. 2011; Kim et al. 2015; Lu et al. 2015a). According to Lim et al. (2015), after 21 days of incubation, soil treated with 10% (*w/v*) biochar made from poultry dung improved its pH from 6.94 (control) to 7.24 and 7.88 at 300 °C and 700 °C, respectively. Zheng et al. (2015) described that application of bean and rice straw biochar increased pH of paddy soil from 6.2 to 6.7 and 7.4, respectively.

Better effects of biochar application in reducing metal uptake by plants are usually observed on sandy, acidic soils (Rees et al. 2015; Medyńska-Juraszek et al. 2020a). Many experiments recorded reduced mobility of HMs in the leachate due to induced decline in the acid-soluble HM fractions (Taraqqi-A-Kamal et al. 2021). In general, higher pH biochars resulted in greater reductions in plant Cd concentrations. Plant tissue Cd reductions were 18%, 35%, and 49%, respectively, for biochars grouped as having pH of <8.5, 8.6–10, and >10 (Chen et al. 2018). The effects of biochar on plant heavy metal uptake is generally dependent on soil pH but were inconsistent and varied for different metals (Chen et al. 2018) (Figure 22.1). Houben et al. (2013) reported that biochar applied at 5% and 10% reduced the Cd and Pb content of shoots by 33% and 75%, respectively, as well as decreasing extractable Cd and Pb in the soil. Similar findings were observed by Medyńska-Juraszek et al. (2020a), showing 20% reduced content of bioavailable Cd, as well as decrease of extractable Cu and Pb in contaminated soil from copper mining site. Increased surface area for adsorption of cations, increased charge density per unit surface, or a combination of both leads to elevated biochar CECs (Cao et al. 2018). This biochar properties has a great impact on plant growth improvement, but mainly accumulation of trace elements in plant tissues (Shen et al. 2016).

22.3 Biochar Interactions Through Rhizosphere

Roots are the first point of contact between biochar particles and growing plants. Biochar may affect root growth, and therefore plant performance, through two mechanisms: (i) as a direct nutrient source and (ii) through impacts on nutrient availability

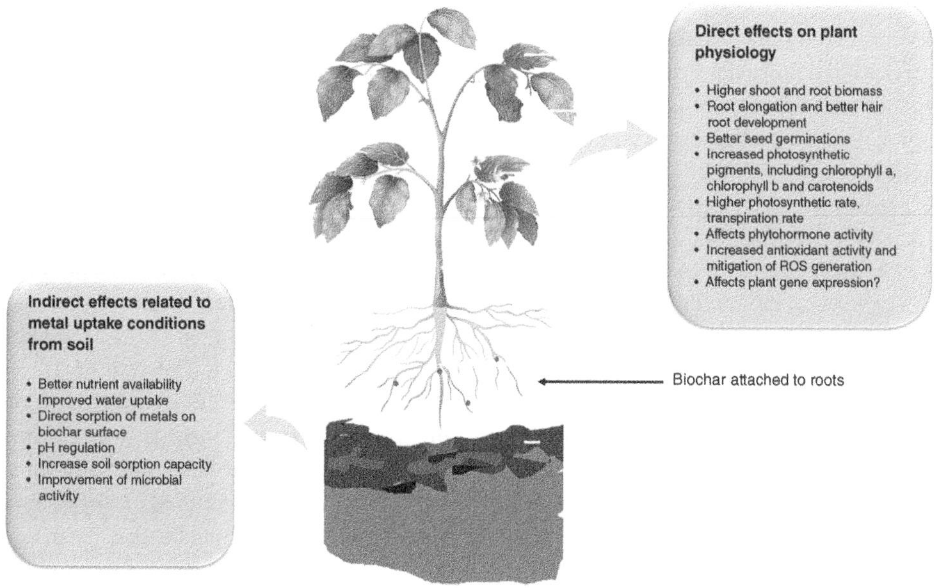

Figure 22.2 Influence of biochar on physiological factors related to heavy metal uptake and stress tolerance.

(Prendergast-Miller et al. 2014). Root activity creates distinct root–soil zones. Thus, biochar addition can affect root–soil interactions, for example nitrate retention in the wheat (*Triticum aestivum* L.) rhizosphere in biochar-amended soils (Prendergast-Miller et al. 2011). The mechanisms of root–biochar interactions coincide with biochar impact on soil: pH, bulk density, aeration and water holding capacity, nutrient content, and availability (Figure 22.2). Biochar has also an influence on microbial communities in soil and on particles signaling molecules in the soil, by absorbing or releasing them (Głuszek et al. 2017). Effects of biochar application on properties of rhizosphere soil might be distinct from the effects in bulk soil. Houben and Sonnet (2015) described that in the bulk soil, the application of 5% biochar significantly reduced the exchangeable pool of metals primarily due to a liming effect, which subsequently promoted the metal shift into the carbonate-bound pool. However, metals were re-mobilized in the rhizosphere of both *Agrostis capillaris* and *Lupinus albus* due to root-induced acidification, which counteracted the liming effect of biochar. Similar findings were observed by Medyńska-Juraszek et al. (2020c), describing that root-induced acidification of the rhizosphere also occurs in biochar-amended soil; however, Cu, Zn, Pb, and Cd extractability and plant uptake of metals were significantly reduced even due to higher accumulation of elements in rhizosphere soil.

22.3.1 Effect on Plant Root Development

Biochar may affect the development of the root system and heavy metal uptake by modification of soil physical properties. High porosity and light density of biochar impacts soil bulk density and porosity, causing direct changes of soil water holding capacity, increasing the content of water and amounts of trace elements in forms easily available for plants

(Głuszek et al. 2017). According to Andrenelli et al. (2016), soil treated with pelletized biochar had a bulk density that was almost 10% lower than control soil. Additionally, it increases the soil's capacity to retain water. This improvement may be attributed to the biochar natural retention ability as well as a more functional arrangement of the soil aggregates and particles when biochar pellets are present. Due to lower bulk density and higher porosity in biochar amended soils, plant root formation can be affected (Medyńska-Juraszek et al. 2021). Biochar application may alter root growth and traits and thereby affect plant performance (Figure 22.2). The biochar-induced increases in root length (+52%) and number of root tips (+17%) were much larger than the increase in root diameter (+9.9%). This result suggests that biochar application benefits root morphological development to alleviate plant nutrient and water deficiency (Xiang et al. 2017). Kamran et al. (2020) described that application of wood-chip-derived biochar at rate of 2% to heavy metal polluted soil enhanced shoot fresh biomass (41% and 72%), root fresh biomass (34% and 62%), respectively. Study of Prendergast-Miller et al. (2014) showed that spring barley (*Hordeum vulgare* L.) developed larger rhizosphere zones when grown on biochar amended soil than the control treatment. The rhizosphere contained more biochar particles than the bulk soil, indicating that roots preferred soil-containing biochar particles. Biochar particles retained more soil nitrogen (N) in the form of nitrates but also supplied plants with phosphorus (P).

22.3.2 Changes in Elements Uptake from Rhizosphere

One of the most important strategy of heavy metal immobilization in soil and reduction of toxicity is to transfer the PTEs into residual (not bioavailable for plants) forms, easily by cation binding in solid phase, e.g. on organic matter or clay minerals (Van Poucke et al. 2018). Depending on particle size distribution, soil organic matter content, and cation exchange capacity (CEC), soils will have different ability to bind metals in sorption complex; however, this properties can be modified by application of organic amendments (Głąb et al. 2016). By improving soil properties, biochar improves the plant adsorption of macro- and micronutrients, e.g. Cu, Zn, Mg, Fe, K, P, Ca, and N (Figure 22.2). When biochar is applied to soil, soluble and mineral compounds can be leached from biochar by soil water, which affects soil chemical properties increasing the content of dissolved organic carbon (DOC), content of cations and anions present in soil solution (Joseph et al. 2021). The direct effects of biochar includes fertlization effect and addition of nitrogen (N), potassium (K), and phosphorus (P) in to the soil, which in turn increases the growth of plants (Karami et al. 2011). Due to high sorption capacity and presence of exchangeable ions on its surface, biochar can release P, K, Ca, Mg, and trace elements into soil solution increasing directly the availability of elements for plants (Różyło et al. 2017). Biochar is usually poor in available nitrogen; however, its application to soil and increase of water retention decreases the nitrate leaching and increases indirectly N availability for crops (Bagheri et al. 2020). Biochar as a carrier of Mg^{2+}, K^{+}, and Ca^{2+} may act as amorphous buffers embedded in the cell walls of plant tissues, reducing the adsorption of PTEs and uptake from rhizosphere to plants (Haider et al. 2022). As plants can competitively absorb cations, divalent metal cations can compete for exchange sites in soil with biochar-derived ions decreasing the adsorption of heavy metals. Biochar adsorption of HM ions induces the liberation of H ions and of Na, K, Ca, Mg cations into the soil solution (Uchimiya et al. 2010) (Figure 22.2). Essential nutrients can affect heavy metal uptake and thereby reduce the concentration of toxic elements in plant tissues. In general, biochar reduces the average concentrations of heavy metals in plant tissue as compared to plants grown in soils without biochar (Chen et al. 2018). Across studies, biochar has been shown to immobilize and reduce the phytoavailability (Kim et al. 2013) and plant uptake of PTEs (Xu et al. 2013), including Pb, Al, Cr, Mn, Fe, Co, Ni, Cu, Zn, Cd, and Tl (Beesley et al. 2010; Fellet et al. 2011; Saletnik et al. 2017; Bogusz and Oleszczuk 2018; Lee and Park 2020; Medyńska-Juraszek et al. 2020a). Production of biochar in high temperature develops more aromatic structure and process of biochar aging causes oxidation upon exposure to air, creating oxygen-containing negatively charged functional groups (phenol, hydroxyl, carboxyl, and carbonyl groups) over the biochar surface. Aging of biochar also may enhance its ability to stabilize heavy metals in soil. Biochar undergoes oxidation over time, which promotes increase in carboxyl groups and the net negative charge that generates its cation exchange capacity (Bian et al. 2014).

22.4 Biochar-Induced Plant Respond to Metal Stress

22.4.1 Biochar Induces Changes in Photosynthetic Activity

Certain biochemical changes, such as a decrease in chlorophyll content and photosynthetic activity, occur when plants are cultivated under PTE(s) stress (Arshad et al. 2015; Rizwan et al. 2015). Reduced chlorophyll synthesis, which results from the direct inhibition of enzymes and/or a lack of some crucial nutrients under PTE(s) stress is blamed for the fall in photosynthetic activity (Adrees et al. 2015a, b; Ali et al. 2015). Spinach and fenugreek were subjected to Cd and Ni stress and application of biochar-increased photosynthetic pigments, including chlorophyll a, chlorophyll b, and carotenoids, as well as gas exchange attributes, including photosynthetic rate, transpiration rate, and sub stomatal CO_2 concentration (Younis et al. 2015a, b, c). Similar findings were observed by Fiaz et al. (2014) studying the effect of biochar derived from poultry manure on Cd and Pb uptake by mustard greens and showed that application of biochar increased concentrations of photosynthetic compounds like chlorophyll a and b, carotenoids, as well as accessory pigments like anthocyanin and lycopene in plants under PTE(s) stress (Figure 22.2). Waqas et al. (2014) described that during Zn stress, soybean showed an increase in photosynthesis that was mediated by biochar, while Buss et al. (2012) reported increased net photosynthetic activity in quinoa cultivated under Cu stress and biochar amendment. Mechanisms of biochar-induced changes of photosynthetic activity are unknown and this topic needs further research.

22.4.2 Biochar Induces Changes in Antioxidant and Phytohormone Activity

To combat PTE(s) stress, plants have, however, adapted a variety of strategies, including increased antioxidant activity (lipid peroxidation, cysteine content, ascorbic acid, and non-protein thiol content) (Singh et al. 2004; Paz-Ferreiro et al. 2014; Mohamed et al. 2017). The response of plants to the adverse effects of PTEs is regulated by different external and internal factors. Among the internal plant factors, phytohormones have important role in stress tolerance and adaption. Plant development is regulated and coordinated by activity of several phytohormones like abscisic acid (ABA), gibberellins (GAs), ethylene, auxin (indole-3-acetic acid [IAA]), cytokinins (CKs), salicylic acid (SA), jasmonic acid (JA), and brassinosteroids (BRs), which controls many physiological and biochemical processes in the plant (Zhang et al. 2014; Fahad et al. 2015; Hmid et al. 2015). There is evidence that biochar application can stimulate gibberellin, auxin, and brassinosteroid regulation promoting plant growth (Racioppi et al. 2019) (Figure 22.2). Increased content of these substances in plant tissues can be a good indicator of abiotic stress and recently the abscisic acid (ABA) has been proposed as a mediator and the major internal signal in the plant response to abiotic stress (Farhangi-Abriz and Torabian 2018). ABA controls plant development and growth, including embryo development, seed dormancy, inhibition of lateral root formation, transpiration, synthesis of proteins, and photosynthesis. It has been documented that ABA can induce the expression of antioxidant genes and enhance the capacity of antioxidant defense systems, including enzymatic and nonenzymatic constituents (Wang et al. 2013). Exposure to HMs such as copper, mercury, cadmium, or lead induces expression of ABA biosynthetic genes and in turn increases the endogenous levels of ABA, which regulates up to 10% of protein-encoding genes (Syta et al. 2019). Plants respond to biotic and abiotic stress factors by increasing the amount of active (free) form of the hormone, which takes place primarily by hydrolysis of bound ABA (Nguyen et al. 2021). Pretreatment with abscisic acid has been reported as method of increasing rice tolerance to cadmium (Hsu and Kao 2003). Wang et al. (2013) reported that low concentration of exogenous ABA-mediated reduction of Pb-induced toxicity to the Chinese herb *Atractylodes macrocephala* is also associated with the increase of antioxidative defense systems and increase in accumulation of soluble sugars, soluble protein, and chlorophyll. On the contrary, a higher concentration of ABA increased oxidative stress, causing tissue damage, thereby decreasing the growth of Pb-stressed plants. Biochar may reduce oxidative stress by mitigating Reactive Oxygen Species (ROS) generation and improving antioxidant activity (Fahad et al. 2016). When antioxidant activity rises in response to BC, it aids in the prevention of the start or expansion of oxidizing chain reactions, resulting in inhabitation or oxidation delays in plant cells, lipids, proteins, and other essential components (Abbas et al. 2018). Haider et al. (2022) showed that application of biochar and thiourea mitigated the harmful effects of Cd impact on lipid peroxidation and cell membrane permeability. Application of BC decreased electrolyte leakage (EL) and lipid peroxidation (MDA) contents by 69% and 9% respectively, compared with Cd stress alone. Similar effects on MDA and activities of antioxidants, i.e. catalase, superoxide dismutase, and ascorbate peroxides, content in quinoa plants cultivated on Ni-contaminated soil were described by Ramzani et al. (2017). In the same publication, authors described that protein, fat, fiber, starch, and ascorbic acid can be significantly reduced, when crops are cultivated in stressed soils. Biochar can also boost superoxide dismutase (SOD), peroxidase (POD), and catalase (CAT) activities in rice flag leaves while decreasing MDA concentrations (Zhang et al. 2014). In bean plants, biochar mitigated Zn stress by inducing malic enzymes, glutamate dehydrogenase, isocitrate dehydrogenase, and POD activity (Hmid et al. 2015). According to recent reports, the application of biochar improved tomato resistance to *Botrytis cinerea* via influencing signaling pathways like hydrogen oxide and jasmonic acid buildup (Mehari et al. 2015).

French and Iyer-Pascuzzi (2018) observed that biochar promotes growth partially through stimulation of the gibberellins (GA) pathway. Viger et al. (2015) found evidence for upregulation of auxin and brassinosteroid pathways in *Arabidopsis* plants grown in biochar, though this may be due to differences in species response or the type of biochar used. In the same study significant differences in 1076 genes expression were revealed in plants grown in biochar soil comparing control plants without BC. Copley et al. (2017) presented that soybean plants grown in 5% biochar exhibited upregulated transcription of β-1,4-glucanase involved in cell wall expansion, which is a hallmark of GA-mediated growth. Racioppi et al. (2019) observed similar effect of biochar application on GA pathway in wheat. In wheat grown on biochar amended soil, ABA and 3-IAA concentrations were significantly lower than those of GA. These results demonstrate that, at a molecular level, the biochar soil amendment was able to markedly improve GA content of wheat seeds, thereby increasing their germination ability. One of the possible mechanisms of altered cell signaling can be related to biochar actions in soil, mainly increase of pH in rhizosphere and increased water holding capacity leading to improved transport of water and nutrients to plant tissues. Some results also confirm that biochar upregulates genes related to water and sugar transport such as TIP and PIP (Szenthe et al. 2007). However knowledge about biochar-induced gene expression and modifications are very limited.

22.4.3 Biochar as a Source of Specific Chemical Compounds Affecting Heavy Metal Uptake By Plants

Biochar contains also bioactive compounds, mainly karrikins and humic substance products, which have been shown to have hormone-like effects on plant growth, including GA-like responses (Nelson et al. 2009; French and Iyer-Pascuzzi 2018). Karrikins are small organic compounds generated during burning of plant materials, typically for charcoal or biochar. Karrikins are very stable in soil however decay rapidly in UV light or can be easily leached from soil to groundwaters (Flematti et al. 2015). Karrikins affect seedling growth and development. They are reported to cause more rapid or vigorous growth of some seedlings, including maize and tomato, and in *Arabidopsis thaliana* karrikins influence seedling photomorphogenesis, causing smaller stature seedlings with larger seed leaves (Waters et al. 2014). Flematti et al. (2015) showed potential pathway of plant respond to karrikins directly reacting with *KAI2* proteins, targeting the genes responsible for plant growth.

Dissolved organic carbon plays an important role in both carbon cycling and chemistry of soils, through its influence on acidity, nutrient availability, metal toxicity, and transport. Dissolved organic carbon (DOC), the most mobile and active component of biochar, that can be rapidly released into soil water via soil pore water (Sherene 2009). It can immediately modify the physicochemical characteristics of soil, increase the organic matter content of soil, impact the microbial community structure, and change the soil organic pollutants and heavy metal migration (Novak et al. 2009; Liu et al. 2022). The effect of DOC on heavy metal mobility and bioavailability is related to formation of DOC–metal complexes (Mancinelli et al. 2018). The organic acids present in the DOC can act as a chelating agent and thereby enhance the mobilization of toxic elements. Content of DOC in biochar depends on the feedstock type, woody biomass generally releases less DOC compared to non-woody biomass (Zimmerman et al. 2011) and decreases with the increase in pyrolysis temperature (Gaskin et al. 2008). Part of biochar-derived DOC consists of low-molecular-weight organic compounds (LMOC), mainly organic acids (oxalic, malic, acetic, and citric acid) typically derived from decomposition of soil organic matter or plant root exudates. The main role of LMOC is to increase the bioavailability of trace elements in soil. Part of biochar-DOC are humic substances and fulvic acids. Content of components taking active part in trace element transformations on biochar surface depends mainly on the temperature of pyrolysis. In low temperature, biochar fulvic acids are dominant and play very important role in metal binding (Ali et al. 2015), while higher than 550 °C, they generate more humic substances (Zimmerman et al. 2011). The effect of humic substances in the mitigation of abiotic stress effects in plants is well known and generally described as result of increase of enzymatic and nonenzymatic antioxidant defense, increase in compatible solutes production, and changes in ion balance (Canellas et al. 2020). Humic acids are able to interact with metal cations in soil solution, and this interaction increases with pH (Spark et al. 1997), which can explain enhanced sorption of PTEs in soil amended with alkaline biochar. Moreover, biochar can become a source of humic substances in soil. So-called biochar-derived humic acids (BDHA) can be released from biochar during biotic and abiotic process leading to biochar oxidation. Jin et al. (2018) demonstrated that BDHA have higher sorption capacity and distinct sorption mechanism of pollutants compared to soil-derived humic acids. This finding is important to understand the fate, transport, and bioavailability of PTEs in biochar-amended soils.

DOC derived from biochar plays also another important role stimulating microbial activity and increasing microbial abundance (Lin et al. 2012). Biochar can impact soil organisms in different ways and many different interactions from inhibiting and toxic effect (mainly due to volatile organic compounds) of biochar on soil microbes to stimulating effects and increased diversity of microbial functional groups in the presence of biochar (Głąb et al. 2016; Van Poucke et al. 2018). Plant growth-promoting rhizobacteria (PGPR) can fix atmospheric nitrogen and supply it to plants. They can synthesize low-molecular-weight siderophores, which can solubilize and sequester iron from the soil, and then provide it to plant cells; they can synthesize several different phytohormones, notably indole acetic acid (IAA), which can act to enhance various stages of plant growth; finally they may facilitate the solubilization of minerals such as phosphorus from the soil, which then become more readily available for plant growth; and they may synthesize some enzymes that modulate plant growth and development by influencing the levels of plant hormones (Glick et al. 1998).

Recently also attention has been paid to ethylene, which regulates several growth and cellular defense mechanisms in response to environmental stress (Van de Poel et al. 2015). PGPR can effectively lower plant ethylene levels by sequestering the ethylene precursor 1-aminocyclopropane-1-carboxy late (ACC) from the plant and then hydrolyzing it through the action of the enzyme ACC deaminase (Glick et al. 1998). Abiotic stress induces ethylene production and this is directly related to the concentration of ACC in plant tissues, ensuring that the level of ethylene does not turn out to be high from the point where root growth is impaired (Fahad et al. 2015). ACC is the immediate precursor of ethylene that may be exuded from plant roots. For many plants, a burst of ethylene is required to break seed dormancy, but after seed germination a sustained level of ethylene inhibits root elongation. It has been reported that bacteria contain an enzyme ACC-deaminase that hydrolyses ACC into ammonia and α-ketobutyrate on the surface of roots and eliminates the potential inhibitory effects of higher ethylene concentrations and facilitates formation of longer roots (Noreen et al. 2012). Biochar can become a source of ethylene for plants. Rates of ethylene production vary across biochars depending mainly on feedstock type and temperature of biochar production. High-temperature biochars with higher surface area, mainly used in soil remediation, contain less ethylene compared to materials obtained during torrefaction. Spokas et al. (2010) suggested that release of ethylene from biochar can affect plants, but also soil microbes.

22.5 Effect of Biochar on Heavy Metal Concentrations in Different Crops

The reductions in plant heavy metal concentrations achieved with biochar application varied significantly for different types of plants. On average, biochar reduced Cd concentrations in rice (40%), wheat (42%), maize (36%), vegetables (41%), grass (40%), and hyperaccumulating plants (42%) (Cui et al. 2012; Younis et al. 2016; Chen et al. 2018; Medyńska-Juraszek et al. 2020b). Pb concentrations were reduced in crops in each of the eight categories, with the effect size ranging between 14% and 75%. The greatest decrease in Pb concentration occurred in maize, followed by vegetables, while the smallest decrease was in wheat (Houben and Sonnet 2015; Shen et al. 2016). Cu concentrations were reduced by biochar in rice (32%), wheat (46%), maize (24%), vegetables (26%), and grass (15%). As for changes in plant Zn concentrations, biochar uses led to declines only for vegetable crops, legumes, and hyperaccumulating plants, while it had no effects in reducing Zn concentrations in rice, wheat, maize, and grasses (Chen et al. 2018; Medyńska-Juraszek et al. 2020a).

22.6 Effect of Biochar Type on Heavy Metal Immobilization

The amount of trace elements a plant can absorb depends on the type of biochar (Fellet et al. 2014; Luo et al. 2014; Kloss et al. 2015; Tan et al. 2015). According to Khan et al. (2015), peanut shell biochar modification resulted in a decreased uptake of specific TE(s) (Cd, Cu, Pb, and Zn) by turnip (*Brassica rapa*) compared to other biochar (sewage sludge, rice, and soybean straw). The uptake of Pb by rice (*Oryza sativa*) was reported to be reduced by bamboo biochar (2% *w/v*), compared to coconut shell biochar, pine wood shavings biochar, and sugarcane bagasse biochar under similar conditions (Tan et al. 2015). The fact that bamboo biochar has a lower acid-soluble Pb content in the soil than other biochars, may be it is the cause of reduced Pb uptake in rice. In a contaminated shooting range soil, Ahmad et al. (2016) found that soybean stover-derived biochar was a more effective Pb and Cu immobilizer than pine needle biochar. Brennan et al. (2014) research revealed that the amount of As that maize absorbed varied depending on the biochar applied, declining with pine wood chip biochar while dramatically increasing with olive tree pruning biochar. On the other hand, the application of both biochars considerably reduced the Cu concentration in maize. Red amaranth (*Amaranthus tricolor*) had a lower cadmium concentration in its root, stem, and leaves after soil application of poultry litter biochar as opposed to eucalyptus biochar (Lu et al. 2014a). It has been noted that *Sedum plumbizincicola* cadmium and zinc hyperaccumulator plants treated with rice straw biochar as opposed to bamboo biochar showed lower amounts of Cu and Pb and higher concentrations of Cd (Lu et al. 2014b).

According to Zheng et al. (2013), as compared to rice husk and bran biochar both made at 350 °C, rice straw BC lowered Cd, Zn, and As concentrations while increasing Pb concentrations in wheat (*T. aestivum*) shoots. Similar to this, Zheng et al. (2012) found that rice straw biochar was superior to rice bran and husk in lowering PTE(s), Cd, Zn, and Pb in rice seedlings. According to Fellet et al. (2014), Cd and Pb concentrations in the shoots of *Noccearo tundifolium* (L.) and *Anthyllis vulneraria* increased when biochar was obtained from wood, whereas biochar derived from manure pellets and pruning residues decreased Cd and Pb concentrations. Opposite effects were observed for arsenic and according to Gregory et al. (2014), applying willow biochar As-contaminated soil boosted As uptake by rye grass, while study of Beesley et al. (2013) showed that As concentrations were reduced by BC shoots and roots of tomato plant. We may infer from these researches that a careful selection of material is a crucial stage in the preparation of biochar, particularly for application in the field for agronomic crop growth.

All described in the chapter functionalities of biochar makes this material a promising tool in mitigating the problem of soil chemical degradation related to intensive use of arable soil for food production. Improvements in plant yield after biochar addition are often attributed to changes of soil physicochemical properties, mainly water and nutrient retention, which also benefits in microbiological life improvement and potential toxic elements immobilization due to increase of soil pH. All these functionalities are crucial for crop protection against metal stress and fundamental in soil remediation and renewal of conditions for plant development. An increase in the demand for food causes a decrease in its quality due to the use of degraded soils and low-quality fertilizers for cultivation. The negative footprint of food production and agricultural activity is increasing rapidly. Unexpectedly, all human efforts related to maintaining soil productivity have opposite effects, causing a decrease in plant yield, mainly due to soil microbiological life degradation and high toxicity of substances introduced to soil. The risk of toxic compounds transfer from polluted soil is high, bringing higher risk of adverse effects on human health and animal welfare. Against humans is also the changing climate. More severe droughts and water shortage meets with high consumption of water by plants and only sustainable land management and putting back to soil carbon and organic matter can suppress this negative changes. Even if the cost of biochar application is high, the benefits of its use can be paid back in a short time. Thus, actions related to a wide use of biochar in agriculture should be promoted by policy-makers.

References

Abbas, T., Rizwan, M., Ali, S. et al. (2018). Effect of biochar on alleviation of cadmium toxicity in wheat (*Triticum aestivum* L.) grown on Cd-contaminated saline soil. *Environ Sci Pollut Res* 25 (26): 25668–25680.

Adrees, M., Ali, S., Rizwan, M. et al. (2015a). Mechanisms of silicon-mediated alleviation of heavy metal toxicity in plants: a review. *Ecotoxicol Environ Saf* 119: 186–197.

Adrees, M., Ali, S., Rizwan, M. et al. (2015b). The effect of excess copper on growth and physiology of important food crops: a review. *Environ Sci Pollut Res* 22 (11): 8148–8162.

Ahmad, M., Ok, Y., Rajapaksha, A. et al. (2016). Lead and copper immobilization in a shooting range soil using soybean stover- and pine needle-derived biochars: chemical, microbial and spectroscopic assessments. *J Hazard Mater* 301: 179–186.

Ali, S., Bharwana, S., Rizwan, M. et al. (2015). Fulvic acid mediates chromium (Cr) tolerance in wheat (*Triticum aestivum* L.) through lowering of Cr uptake and improved antioxidant defense system. *Environ Sci Pollut Res* 22 (14): 10601–10609.

Amonette, J. and Joseph, S. (2009). Characteristics of biochar: microchemical properties. In: *Biochar for Environmental Management* (ed. J. Lehmann and S. Joseph). Routledge. P-20 9781849770552.

Andrenelli, M., Maienza, A., Genesio, L. et al. (2016). Field application of pelletized biochar: short term effect on the hydrological properties of a silty clay loam soil. *Agric Water Manag* 163: 190–196.

Arshad, M., Ali, S., Noman, A. et al. (2015). Phosphorus amendment decreased cadmium (Cd) uptake and ameliorates chlorophyll contents, gas exchange attributes, antioxidants, and mineral nutrients in wheat (*Triticum aestivum* L.) under Cd stress. *Arc Agron Soil Sci* 62 (4): 533–546.

Bagheri, S., Hassandokht, M., Mirsoleimani, A., and Mousavi, A. (2020). Effects of palm leaf biochar on the availability of soil nutrients, leaf nutrient concentration, and physiological characteristics of melon plants (*Cucumis melo* L.) under drought stress. *Acta Agrobot* 73 (1): 13.

Beesley, L. and Marmiroli, M. (2011). The immobilization and retention of soluble arsenic, cadmium and zinc by biochar. *Environ Pollut* 159 (2): 474–480.

Beesley, L., Moreno-Jiménez, E., and Gomez-Eyles, J. (2010). Effects of biochar and greenwaste compost amendments on mobility, bioavailability and toxicity of inorganic and organic contaminants in a multi-element polluted soil. *Environ Pollut* 158 (6): 2282–2287.

Beesley, L., Marmiroli, M., Pagano, L. et al. (2013). Biochar addition to an arsenic contaminated soil increases arsenic concentrations in the pore water but reduces uptake to tomato plants (*Solanum lycopersicum* L.). *Sci Total Environ* 454-455: 598–603.

Beesley, L., Inneh, O., Norton, G. et al. (2014). Assessing the influence of compost and biochar amendments on the mobility and toxicity of metals and arsenic in a naturally contaminated mine soil. *Environ Pollut* 186: 195–202.

Bian, R., Joseph, S., Cui, L. et al. (2014). A three-year experiment confirms continuous immobilization of cadmium and lead in contaminated paddy field with biochar amendment. *J Hazard Mater* 272: 121–128.

Bogusz, A. and Oleszczuk, P. (2018). Sequential extraction of nickel and zinc in sewage sludge- or biochar/sewage sludge-amended soil. *Sci Total Environ* 636: 927–935.

Brennan, A., Jiménez, E., Puschenreiter, M. et al. (2014). Effects of biochar amendment on root traits and contaminant availability of maize plants in a copper and arsenic impacted soil. *Plant Soil* 379 (1–2): 351–360.

Buss, W., Kammann, C., and Koyro, H. (2012). Biochar reduces copper toxicity in *Chenopodium* quinoa wild in a Sandy Soil. *J Environ Qual* 41 (4): 1157–1165.

Canellas, L., Canellas, N., da Sirineu, L. et al. (2020). Plant chemical priming by humic acids. *Chem Biol Technol Agric* 7 (1): 12.

Cantrell, K., Hunt, P., Uchimiya, M. et al. (2012). Impact of pyrolysis temperature and manure source on physicochemical characteristics of biochar. *Bioresour Technol* 107: 419–428.

Cao, X., Ma, L., Gao, B., and Harris, W. (2009). Dairy-manure derived biochar effectively sorbs lead and atrazine. *Environ Sci Technol* 43 (9): 3285–3291.

Cao, Y., Gao, Y., Qi, Y., and Li, J. (2018). Biochar-enhanced composts reduce the potential leaching of nutrients and heavy metals and suppress plant-parasitic nematodes in excessively fertilized cucumber soils. *Environ Sci Pollut Res* 25 (8): 7589–7599.

Chen, D., Liu, X., Bian, R. et al. (2018). Effects of biochar on availability and plant uptake of heavy metals A meta-analysis. *J Environ Manag* 222: 76–85.

Choudhury, F., Rivero, R., Blumwald, E., and Mittler, R. (2017). Reactive oxygen species, abiotic stress and stress combination. *Plant J* 90 (5): 856–867.

Copley, T., Bayen, S., and Jabaji, S. (2017). Biochar amendment modifies expression of soybean and rhizoctoniasolani genes leading to increased severity of *Rhizoctonia* foliar blight. *Front. Plant Sci.* 8: 221.

Cui, L., Pan, G., Li, L. et al. (2012). The reduction of wheat Cd uptake in contaminated soil via biochar amendment: a two-year field experiment. *BioResour* 7 (4): 5666–5676.

Ehsan, M., Barakat, M., Husein, D., and Ismail, S. (2014). Immobilization of Ni and Cd in soil by biochar derived from unfertilized dates. *Water Air Soil Pollut* 225 (11): 2123.

Fahad, S., Hussain, S., Bano, A. et al. (2015). Potential role of phytohormones and plant growth-promoting rhizobacteria in abiotic stresses: consequences for changing environment. *Environ Sci Pollut Res* 22 (7): 4907–4921.

Fahad, S., Hussain, S., Saud, S. et al. (2016). A combined application of biochar and phosphorus alleviates heat-induced adversities on physiological, agronomical and quality attributes of rice. *Plant Physiol Biochem* 103: 191–198.

Farhangi-Abriz, S. and Torabian, S. (2018). Effect of biochar on growth and ion contents of bean plant under saline condition. *Environ Sci Pollut Res* 25 (12): 11556–11564.

Fellet, G., Marchiol, L., DelleVedove, G., and Peressotti, A. (2011). Application of biochar on mine tailings: effects and perspectives for land reclamation. *Chemosphere* 83 (9): 1262–1267.

Fellet, G., Marmiroli, M., and Marchiol, L. (2014). Elements uptake by metal accumulator species grown on mine tailings amended with three types of biochar. *Sci Total Environ* 468–469: 598–608.

Fiaz, K., Danish, S., Younis, U. et al. (2014). Drought impact on Pb/Cd toxicity remediated by biochar in *Brassica campestris*. *J Plant Nutr Soil Sci* 14 (4): 845–854.

Flematti, G., Dixon, K., and Smith, S. (2015). What are karrikins and how were they 'discovered' by plants? *BMC Biol* 13 (1): 108.

French, E. and Iyer-Pascuzzi, A. (2018). A role for the gibberellin pathway in biochar-mediated growth promotion. *Sci Rep* 8 (1): 538.

Gaskin, J.W., Steiner, C., Harris, K. et al. (2008). Effect of low-temperature pyrolysis conditions on biochar for agricultural use. *Trans ASABE* 51 (6): 2061–2069.

Giller, K., Witter, E., and McGrath, S. (2009). Heavy metals and soil microbes. *Soil Biol Biochem* 41 (10): 2031–2037.

Głąb, T., Palmowska, J., Zaleski, T., and Gondek, K. (2016). Effect of biochar application on soil hydrological properties and physical quality of sandy soil. *Geoderma* 281: 11–20.

Glick, B., Penrose, D., and Li, J. (1998). A model for the lowering of plant ethylene concentrations by plant growth-promoting bacteria. *J Theor Biol* 190 (1): 63–68.

Głuszek, S., Sas-Paszt, L., Sumorok, B., and Kozera, R. (2017). Biochar-rhizosphere interactions – a review. *Pol J Microbiol* 66 (2): 151–161.

Gomez-Eyles, J., Beesley, L., Moreno-Jimenez, E. et al. (2013). The potential of biochar amendments to remediate contaminated soils. In: *Biochar and Soil Biota* (ed. N. Ladygina and F. Rineau), 100–133. CRC Press.

Gregory, S.J., Anderson, C.W.N., Camps Arbestain, M., and McManus, M. (2014). Response of plant and soil microbes to biochar amendment of an arsenic-contaminated soil. *Agric Ecosyst Environ* 191: 133–141.

Haider, G., Steffens, D., Moser, G. et al. (2017). Biochar reduced nitrate leaching and improved soil moisture content without yield improvements in a four-year field study. *Agric Ecosyst Environ* 237: 80–94.

Haider, F., Virk, A., Rehmani, M. et al. (2022). Integrated application of thiourea and biochar improves maize growth, antioxidant activity and reduces cadmium bioavailability in cadmium-contaminated soil. *Front Plant Sci* 12: 809322.

Hmid, A., Al Chami, Z., Sillen, W. et al. (2015). Olive mill waste biochar: a promising soil amendment for metal immobilization in contaminated soils. *Environ Sci Pollut* 22 (2): 1444–1456.

Houben, D. and Sonnet, P. (2015). Impact of biochar and root-induced changes on metal dynamics in the rhizosphere of *Agrostis capillaris* and *Lupinus albus*. *Chemosphere* 139: 644–651.

Houben, D., Evrard, L., and Sonnet, P. (2013). Beneficial effects of biochar application to contaminated soils on the bioavailability of Cd, Pb and Zn and the biomass production of rapeseed (*Brassica napus* L.). *Biomass Bioenergy* 57: 196–204.

Hsu, Y. and Kao, C. (2003). Role of abscisic acid in cadmium tolerance of rice (*Oryza sativa* L.) seedlings: ABA and Cd tolerance of rice seedlings. *Plant Cell Environ* 26 (6): 867–874.

Hu, J., Wu, F., Wu, S. et al. (2014). Biochar and *Glomus caledonium* influence Cd accumulation of upland kangkong (*Ipomoea aquatica* Forsk.) intercropped with alfredstonecrop (*Sedum alfredii Hance*). *Sci Rep* 4 (1): 4671.

Jiang, T., Xu, R., Gu, T., and Jiang, J. (2014). Effect of crop-straw derived biochars on Pb(II) adsorption in two variable charge soils. *J Integr Agric* 13 (3): 507–516.

Jin, J., Sun, K., Yang, Y. et al. (2018). Comparison between soil- and biochar-derived humic acids: composition, conformation, and phenanthrene sorption. *Environ Sci Technol* 52 (4): 1880–1888.

Jolly, Y.N., Islam, A., and Akbar, S. (2013). Transfer of metals from soil to vegetables and possible health risk assessment. *Springerplus* 2 (1): 385.

Joseph, S., Cowie, A., Van Zwieten, L. et al. (2021). How biochar works, and when it doesn't: a review of mechanisms controlling soil and plant responses to biochar. *Glob Change Biol Bioenergy* 13 (11): 1731–1764.

Juknys, R., Vitkauskaitė, G., Račaitė, M., and Venclovienė, J. (2012). The impacts of heavy metals on oxidative stress and growth of spring barley. *Open Life Sci* 7 (2): https://doi.org/10.2478/s11535-012-0012-9.

Kabata-Pendias, A. and Pendias, H. (2001). *Trace Elements in Soils and Plants*, 548. Boca Raton: CRC Press.

Kamran, M., Malik, Z., Parveen, A. et al. (2020). Ameliorative effects of biochar on rapeseed (*Brassica napus* L.) growth and heavy metal immobilization in soil irrigated with untreated wastewater. *J Plant Growth Regul* 39 (1): 266–281.

Karami, N., Clemente, R., Moreno-Jiménez, E. et al. (2011). Efficiency of green waste compost and biochar soil amendments for reducing lead and copper mobility and uptake to ryegrass. *J Hazard Mater* 191 (1–3): 41–48.

Khan, S., Waqas, M., Ding, F. et al. (2015). The influence of various biochars on the bioaccessibility and bioaccumulation of PAHs and potentially toxic elements to turnips (*Brassica rapa* L.). *J Hazard Mater* 300: 243–253.

Kim, W.K., Shim, T., Kim, Y.S. et al. (2013). Characterization of cadmium removal from aqueous solution by biochar produced from a giant *Miscanthus* at different pyrolytic temperatures. *Bioresour Technol* 138: 266–270.

Kim, B.S., Lee, H.W., Park, S.H. et al. (2015). Removal of Cu^{2+} by biochars derived from green macroalgae. *Environ Sci Pollut Res* 23 (2): 985–994.

Kloss, S., Zehetner, F., Buecker, J. et al. (2015). Trace element biogeochemistry in the soil-water-plant system of a temperate agricultural soil amended with different biochars. *Environ Sci Pollut Res* 22 (6): 4513–4526.

Lee, J. and Park, Y. (2020). Applications of modified biochar-based materials for the removal of environment pollutants: a mini review. *Sustainability* 12 (15): 6112.

Lim, J., Lee, S., and Ok, Y. (2015). Efficiency of poultry manure biochar for stabilization of metals in contaminated soil. *J Appl Biol Chem* 58 (1): 39–50.

Lima, I.M., Boykin, D.L., Thomas Klasson, K., and Uchimiya, M. (2014). Influence of post-treatment strategies on the properties of activated chars from broiler manure. *Chemosphere* 95: 96–104.

Lin, Y., Munroe, P., Joseph, S. et al. (2012). Water extractable organic carbon in untreated and chemical treated biochars. *Chemosphere* 87 (2): 151–157.

Liu, H., Zhao, B., Zhang, X. et al. (2022). Investigating biochar-derived dissolved organic carbon (DOC) components extracted using a sequential extraction protocol. *Materials* 15 (11): 3865.

Lu, H., Zhang, W., Yang, Y. et al. (2012). Relative distribution of Pb^{2+} sorption mechanisms by sludge-derived biochar. *Water Res* 46 (3): 854–862.

Lu, H., Li, Z., Fu, S. et al. (2014a). Can biochar and phytoextractors be jointly used for cadmium remediation? *PLoS One* 9 (4): e95218.

Lu, K., Yang, X., Shen, J. et al. (2014b). Effect of bamboo and rice straw biochars on the bioavailability of Cd, Cu, Pb and Zn to *Sedum plumbizincicola*. *Agric Ecosyst Environ* 191: 124–132.

Lu, H., Li, Z., Fu, S. et al. (2015). Combining phytoextraction and biochar addition improves soil biochemical properties in a soil contaminated with Cd. *Chemosphere* 119: 209–216.

Luo, F., Song, J., Xia, W. et al. (2014). Characterization of contaminants and evaluation of the suitability for land application of maize and sludge biochars. *Environ Sci Pollut Res* 21 (14): 8707–8717.

Mancinelli, E., Baltrėnaitė, E., Baltrėnas, P. et al. (2018). Correction to: influence of biochar from slow pyrolysis on dissolved organic carbon and total dissolved nitrogen levels of urban storm-water runoff. *Water Air Soil Pollut* 229 (8): 265.

Medyńska-Juraszek, A. and Ćwieląg-Piasecka, I. (2020). Effect of biochar application on heavy metal mobility in soils impacted by copper smelting processes. *Pol. J. Environ. Stud.* 29 (2): 1749–1757.

Medyńska-Juraszek, A., Bednik, M., and Chohura, P. (2020a). Assessing the influence of compost and biochar amendments on the mobility and uptake of heavy metals by green leafy vegetables. *IJERPH* 17 (21): 7861.

Medyńska-Juraszek, A., Ćwieląg-Piasecka, I., Jerzykiewicz, M., and Trynda, J. (2020b). Wheat straw biochar as a specific sorbent of cobalt in soil. *Materials* 13 (11): 2462.

Medyńska-Juraszek, A., Rivier, P., Rasse, D., and Joner, E. (2020c). Biochar affects heavy metal uptake in plants through interactions in the rhizosphere. *Appl Sci* 10 (15): 5105.

Medyńska-Juraszek, A., Latawiec, A., Królczyk, J. et al. (2021). Biochar improves maize growth but has a limited effect on soil properties: evidence from a three-year field experiment. *Sustainability* 13 (7): 3617.

Mehari, Z., Elad, Y., Rav-David, D. et al. (2015). Induced systemic resistance in tomato (*Solanum lycopersicum*) against *Botrytis cinerea* by biochar amendment involves jasmonic acid signaling. *Plant Soil* 395 (1–2): 31–44.

Mohamed, B., Ellis, N., Kim, C., and Bi, X. (2017). The role of tailored biochar in increasing plant growth, and reducing bioavailability, phytotoxicity, and uptake of heavy metals in contaminated soil. *Environ Pollut* 230: 329–338.

Nelson, D., Riseborough, J., Flematti, G. et al. (2009). Karrikins discovered in smoke trigger Arabidopsis seed germination by a mechanism requiring gibberellic acid synthesis and light. *Plant Physiol* 149 (2): 863–873.

Nguyen, T., Sesin, V., Kisiala, A., and Emery, R. (2021). Phytohormonal roles in plant responses to heavy metal stress: implications for using macrophytes in phytoremediation of aquatic ecosystems. *Environ Toxicol Chem* 40 (1): 7–22.

Noreen, S., Ali, B., and Hasnain, S. (2012). Growth promotion of *Vigna mungo* (L.) by pseudomonas spp. exhibiting auxin production and ACC-deaminase activity. *Ann Microbiol* 62 (1): 411–417.

Novak, J.M., Lima, I., Xing, B. et al. (2009). Characterization of designer biochar produced at different temperatures and their effects on a loamy sand. *Ann Environ Sci* 3: 195–206.

Park, J., Choppala, G., Bolan, N. et al. (2011). Biochar reduces the bioavailability and phytotoxicity of heavy metals. *Plant Soil* 348 (1–2): 439–451.

Park, J.H., Choppala, G., Lee, S. et al. (2013). Comparative sorption of Pb and Cd by biochars and its implication for metal immobilization in soils. *Water Air Soil Pollut* 224 (12): 1711.

Paz-Ferreiro, J., Lu, H., Fu, S. et al. (2014). Use of phytoremediation and biochar to remediate heavy metal polluted soils: a review. *Solid Earth* 5 (1): 65–75.

Prendergast-Miller, M., Duvall, M., and Sohi, S. (2011). Localisation of nitrate in the rhizosphere of biochar-amended soils. *Soil Biol Biochem* 43 (11): 2243–2246.

Prendergast-Miller, M., Duval, M., and Sohi, S. (2014). Biochar-root interactions are mediated by biochar nutrient content and impacts on soil nutrient availability: biochar-root interactions. *Eur J Soil Sci* 65 (1): 173–185.

Racioppi, M., Tartaglia, M., De la Rosa, J. et al. (2019). Response of ancient and modern wheat varieties to biochar application: effect on hormone and gene expression involved in germination and growth. *Agronomy* 10 (1): 5.

Ramzani, P., Shan, L., Anjum, S. et al. (2017). Improved quinoa growth, physiological response, and seed nutritional quality in three soils having different stresses by the application of acidified biochar and compost. *Plant Physiol Biochem* 116: 127–138.

Rees, F., Simonnot, M., and Morel, J. (2014). Short-term effects of biochar on soil heavy metal mobility are controlled by intra-particle diffusion and soil pH increase. *Eur J Soil Sci* 65 (1): 149–161.

Rees, F., Germain, C., Sterckeman, T., and Morel, J. (2015). Plant growth and metal uptake by a non-hyperaccumulating species (*Lolium perenne*) and a Cd-Zn hyperaccumulator (*Noccaea caerulescens*) in contaminated soils amended with biochar. *Plant Soil* 395 (1–2): 57–73.

Rinklebe, J., Shaheen, S., and Frohne, T. (2016). Amendment of biochar reduces the release of toxic elements under dynamic redox conditions in a contaminated floodplain soil. *Chemosphere* 142: 41–47.

Rizwan, M., Meunier, J., Davidian, J. et al. (2015). Silicon alleviates Cd stress of wheat seedlings (*Triticum turgidum* L. cv. Claudio) grown in hydroponics. *Environl Sci Pollut Res* 23 (2): 1414–1427.

Różyło, K., Świeca, M., Gawlik-Dziki, U. et al. (2017). The potential of biochar for reducing the negative effects of soil contamination on the phytochemical properties and heavy metal accumulation in wheat grain. *A F Sci* 26 (1): 34.

Saletnik, B., Zaguła, G., Grabek-Lejko, D. et al. (2017). Biosorption of cadmium(II), lead(II) and cobalt(II) from aqueous solution by biochar from cones of larch (*Larix decidua* Mill. subsp. decidua) and spruce (*Picea abies* L. H. Karst). *Environ Earth Sci* 76 (16): 574.

Shen, X., Huang, D., Ren, X. et al. (2016). Phytoavailability of Cd and Pb in crop straw biochar-amended soil is related to the heavy metal content of both biochar and soil. *J Environ Manag* 168: 245–251.

Sherene, T. (2009). Effect of dissolved organic carbon (DOC) on heavy metal mobility in soils. *Nat Environ Pollut Technol* 8 (4): 5.

Singh, S., Sinha, S., Saxena, R. et al. (2004). Translocation of metals and its effects in the tomato plants grown on various amendments of tannery waste: evidence for involvement of antioxidants. *Chemosphere* 57 (2): 91–99.

Sohi, S., Krull, E., Lopez-Capel, E., and Bol, R. (2010). A review of biochar and its use and function in soil. *Adv Agron* 105: 47–82.

Spark, K., Wells, J., and Johnson, B. (1997). The interaction of a humic acid with heavy metals. *Aust J Soil Res* 35 (1): 89.

Spokas, K., Baker, J., and Reicosky, D. (2010). Ethylene: potential key for biochar amendment impacts. *Plant Soil* 333 (1–2): 443–452.

Syta, O., Kumari, P., Yadav, S. et al. (2019). Phytohormone priming: regulator for heavy metal stress in plants. *J Plant Growth Regul* 38 (2): 739–752.

Szenthe, A., Schäfer, H., Hauf, J. et al. (2007). Characterisation and expression of monosaccharide transporters in lupins, *Lupinus polyphyllus* and *L. albus*. *J Plant Res* 120 (6): 697–705.

Tan, X., Liu, Y., Gu, Y. et al. (2015). Immobilization of Cd(II) in acid soil amended with different biochars with a long term of incubation. *Environ Sci Pollut Res* 22 (16): 12597–12604.

Taraqqi-A-Kamal, A., Atkinson, C.J., Khan, A. et al. (2021). Biochar remediation of soil: linking biochar production with function in heavy metal contaminated soils. *Plant Soil Environ* 67 (4): 183–201.

Trakal, L., Šigut, R., Šillerová, H. et al. (2014). Copper removal from aqueous solution using biochar: effect of chemical activation. *Arab J Chem* 7 (1): 43–52.

Uchimiya, M., Lima, I., Klasson, K., and Wartelle, L. (2010). Contaminant immobilization and nutrient release by biochar soil amendment: roles of natural organic matter. *Chemosphere* 80 (8): 935–940.

Uchimiya, M., Chang, S., and Klasson, K. (2011a). Screening biochars for heavy metal retention in soil: role of oxygen functional groups. *J Hazard Mater* 190 (1–3): 432–441.

Uchimiya, M., Klasson, K., Wartelle, L., and Lima, I. (2011b). Influence of soil properties on heavy metal sequestration by biochar amendment: 1. Copper sorption isotherms and the release of cations. *Chemosphere* 82 (10): 1431–1437.

Uchimiya, M., Bannon, D., and Wartelle, L. (2012). Retention of heavy metals by carboxyl functional groups of biochars in small arms range soil. *J Agric Food Chem* 60 (7): 1798–1809.

Van de Poel, B., Smet, D., and Van Der Straeten, D. (2015). Ethylene and hormonal cross talk in vegetative growth and development. *Plant Physiol* 169 (1): 61–72.

Van Poucke, R., Ainsworth, J., Maeseele, M. et al. (2018). Chemical stabilization of Cd-contaminated soil using biochar. *Appl Geochem* 88: 122–130.

Venegas, A., Rigol, A., and Vidal, M. (2015). Viability of organic wastes and biochars as amendments for the remediation of heavy metal-contaminated soils. *Chemosphere* 119: 190–198.

Viger, M., Hancock, R., Miglietta, F., and Taylor, G. (2015). More plant growth but less plant defence? First global gene expression data for plants grown in soil amended with biochar. *Glob Change Biol Bioenergy* 7 (4): 658–672.

Wang, J., Chen, J., and Pan, K. (2013). Effect of exogenous abscisic acid on the level of antioxidants in *Atractylodes macrocephala* Koidz under lead stress. *Environ Sci Pollut Res* 20 (3): 1441–1449.

Waqas, M., Khan, A.L., Kang, S. et al. (2014). Phytohormone-producing fungal endophytes and hardwood-derived biochar interact to ameliorate heavy metalstress in soybeans. *Biol Fertil Soils* 50 (7): 1155–1167.

Waters, M., Scaffidi, A., Sun, Y. et al. (2014). The karrikin response system of Arabidopsis. *Plant J* 79 (4): 623–631.

Wu, W., Yang, M., Feng, Q. et al. (2012). Chemical characterization of rice straw-derived biochar for soil amendment. *Biomass Bioenergy* 47: 268–276.

Xiang, Y., Deng, Q., Duan, H., and Guo, Y. (2017). Effects of biochar application on root traits: a meta-analysis. *Glob Change Biol Bioenergy* 9 (10): 1563–1572.

Xu, X., Cao, X., Zhao, L. et al. (2013). Removal of Cu, Zn, and Cd from aqueous solutions by the dairy manure-derived biochar. *Environ Sci Pollut Res* 20 (1): 358–368.

Younis, U., Athar, M., Malik, S.A. et al. (2015a). Biochar impact on physiological and biochemical attributes of Spinach (*Spinacia oleracea* L.) in nickel contaminated soil. *Global J Environ Sci Manag* 1: 245–254.

Younis, U., Malik, S., Qayyum, M. et al. (2015b). Biochar affects growth and biochemical activities of fenu-greek (*Trigonella corniculata*) in cadmium polluted soil. *J App Bot Food Qual* 88: 29–33.

Younis, U., Qayyum, M., Shah, M. et al. (2015c). Growth, survival, and heavy metal (Cd and Ni) uptake of spinach (*Spinacia oleracea*) and fenugreek (*Trigonella corniculata*) in a biochar-amended sewage-irrigated contaminated soil. *J Plant Nutr Soil Sci* 178 (2): 209–217.

Younis, U., Malik, S.A., Rizwan, M. et al. (2016). Biochar enhances the cadmium tolerance in spinach (*Spinacia oleracea*) through modification of Cd uptake and physiological and biochemical attributes. *Environ Sci Pollut Res* 23 (21): 21385–21394.

Zhang, Z., Meng, J., Dang, S., and Chen, W. (2014). Effect of biochar on relieving cadmium stress and reducing accumulation in super japonica rice. *J Integr Agri1* 3 (3): 547–553.

Zheng, R., Cai, C., Liang, J. et al. (2012). The effects of biochars from rice residue on the formation of iron plaque and the accumulation of Cd, Zn, Pb, As in rice (*Oryza sativa* L.) seedlings. *Chemosphere* 89 (7): 856–862.

Zheng, R., Chen, Z., Cai, C. et al. (2013). Effect of biochars from rice husk, bran, and straw on heavy metal uptake by pot-grown wheat seedling in a historically contaminated soil. *Bio Resour* 8 (4): 5965–5982.

Zheng, R., Chen, Z., Cai, C. et al. (2015). Mitigating heavy metal accumulation into rice (*Oryza sativa* L.) using biochar amendment -a field experiment in Hunan, China. *Environ Sci Pollut Res* 22 (14): 11097–11108.

Zimmerman, A., Gao, B., and Ahn, M. (2011). Positive and negative carbon mineralization priming effects among a variety of biochar-amended soils. *Soil Biol Biochem* 43 (6): 1169–1179.

23

Plant-Growth-Promoting Rhizobacteria and Their Metabolites

Clean and Green Approaches to Deal with Heavy Metal Toxicity

Imtinen Sghaier[1,2], Ameur Cherif[1], and Mohamed Neifar[1,3]

[1] BVBGR-LR11ES31, Department of Biotechnology, Higher Institute of Biotechnology of Sidi Thabet (ISBST), University of Manouba, Ariana, Tunisia
[2] Department of Biology, Faculty of Sciences of Tunis, University of Tunis El Manar, Tunis, Tunisia
[3] APVA-LR16ES20, Department of Biology, National School of Engineers of Sfax (ENIS), University of Sfax, Sfax, Tunisia

23.1 Introduction

With increasing global food demand, industrialization, and long-term application of synthetic fertilizers, the world has facing to considerable concerns. These threats include pollution, environmental deterioration, climate change, public health damage, soil contamination, etc. Thus, excessive use of chemical fertilizers has led to heavy metal pollution (such as cadmium [Cd], lead [Pb], zinc [Zn], and arsenic [As]), accumulation of various toxic xenobiotics on soil, higher pH, less availability of essential nutrients, lower organic matter content, and inadequate soil structure. Heavy metals accumulated in soils could be absorbed by plants and transmitted to animals and human by soil–food chain. Also, application of high concentrations of synthetic pesticides and fungicides has extremely damaged soil fertility and ecosystem balance stimulating pathogen resistance and appearance of new diseases. As consequence, negative impacts on soil qualities have been emerged threatening agriculture sustainability and food security (Song et al. 2017; Chandini et al. 2019; Carne et al. 2021). To deal with these concerns, scientists have given importance to biological processes. Application of microorganisms such as plant growth promoting rhizobacteria (PGPR) has emerged as an eco-friendly and cost-effective strategy to improve sustainability. PGPR correspond to bacteria that colonize around or in/on the root system (rhizosphere) of plants. They coexist or compete with other microorganisms and efficiently interact with plant. Thus, they possess various benefits improving plant growth and development (Vejan et al. 2016; Khatoon et al. 2020). Plant–microbe interactions determine the rhizosphere metabolome diversity and types of metabolites secreted by either plant or microorganism. PGPR enhance plant growth via multiple mechanisms directly and indirectly. Direct traits include essentially (i) nutrient uptake enhancement such as solubilization of phosphorus (P) and potassium (K), nitrogen (N) fixation and improved mobilization

Heavy Metal Toxicity and Tolerance in Plants: A Biological, Omics, and Genetic Engineering Approach, First Edition. Edited by Mohammad Anwar Hossain, AKM Zakir Hossain, Sylvain Bourgerie, Masayuki Fujita, Om Parkash Dhankher, and Parvez Haris.

of micronutrients (iron, Fe; zinc, Zn; manganese, Mn, etc.), (ii) secretion of growth regulators or phytohormones, (iii) production of siderophores, (iv) production of amino 1-aminocyclopropane-1-carboxylic acid (ACC) deaminase. For indirect mechanisms, PGPR produce different metabolites such as antibiotics, exoplysaccharides, volatile organic compounds (VOCs), lytic enzymes, lipopeptides (LPs), quorum sensing molecules such as *N*-acyl homoserine lactones (AHLs), and other secondary metabolites etc.; and stimulate induced systemic resistance (ISR) in plants (Aloo et al. 2020; Ríos-Ruiz et al. 2020). The most bacterial genera showing PGP traits are *Pseudomonas*, *Bacillus*, *Xanthomonas*, *Rhizobium*, *Arthrobacter*, *Azospirillum*, *Azobacter*, and *Agrobacterium* (Rani et al. 2022). *Pseudomonas* and *Bacillus* genera have been largely investigated for plant growth promotion (Israr et al. 2016; Setiawati et al. 2022). Briefly, PGPR are able with versatile traits to control plant resistance to phytopathogens and damaged diseases. Added to that, PGPR are able to improve plant resistance against abiotic stress conditions (drought, salinity, contaminants, etc.) and boost crop productivity (Hassen et al. 2018; Sunita et al. 2020; Sati et al. 2021). Among current severe concerns, soil contamination by heavy metals has been largely reported by researchers using PGPR as metal bioremediators. Thus, PGPR exert different effective strategies to deal with hazardous metals and clean up metal-contaminated soils. PGPR could improve plant growth and the phytoremediation of elements from damaged soils. The versatile plant-growth-promoting (PGP) activities mentioned above give to PGPR the potentialities of biofertilization, biostimultion, biocontrol, and bioremediation. Figure 23.1 summarizes the different mechanisms that are adopted by PGPR. This chapter provides an overview on PGPR and their different potentialities focusing on heavy metal bioremediation capacity.

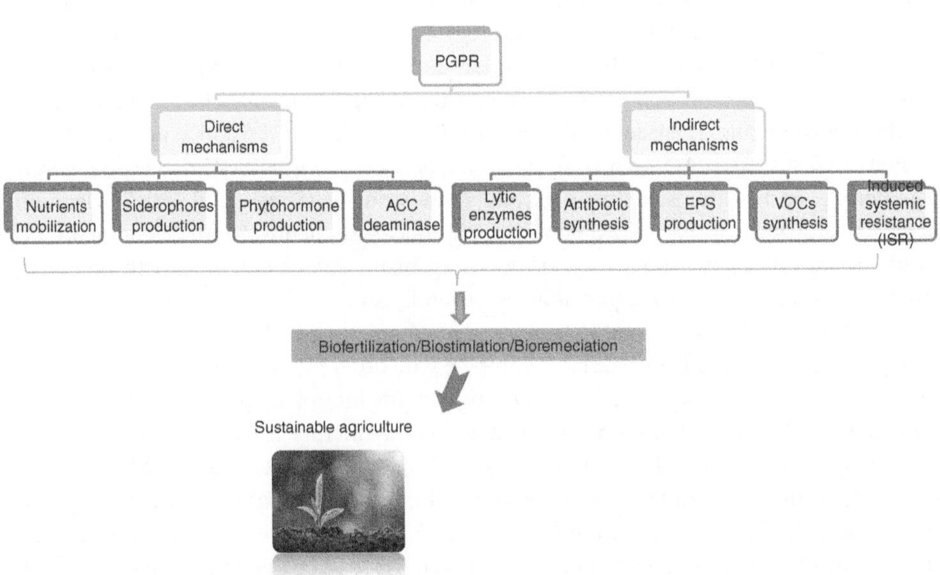

Figure 23.1 Overview of the different mechanisms adopted by PGPR. *Source:* Lovelyday12/ Adobe stock.

23.2 Chemical Fertilizers and Their Impacts

The rapid demographic increase in world has led to famine, poverty, and food insecurity. To meet strong food demand, farmers have been obliged to intensify crop productivity mainly by the supply of chemical fertilizers. Synthetic phosphate fertilizers such as triple super phosphate or phosphoric acid and nitrogen fertilizers provide essential nutrients indispensable in plant growth. They are highly soluble, and rapidly and easily absorbed by plant. As a result, they increase harvest yield and agricultural productivity. Meanwhile, excessive use of chemical fertilizers has caused different concerns in the ecosystem. Heavy metal contamination is one of the main of these threats (Carne et al. 2021).

23.2.1 Impacts of Chemical Fertilizers on Atmospheric Ecosystem

The intensification of chemical fertilizer production and usage has lead to air pollution. Agro-chemical industries are sources of toxic emissions such as nitrogen oxides (NO, N_2O, NO_2), chloro-fluoro hydrocarbons, carbon dioxide fixation, methane, and hydrogen sulfide (H_2S). These emissions have caused greenhouse effect, global warming, and climate change. The emission of ammonia destruct life cycle and vegetation. Ammonia, nitric, and sulfuric acid lead to acid rain, which could damage plant animals and human life. Added to that, they favorite the appearance of allergies and public health risks (Tosi et al. 2020; Gao et al. 2022a).

23.2.2 Impacts of Chemical Fertilizers on Aquatic Ecosystem

Chemical fertilizer industries release huge amounts of wastewater effluents, without or with inadequate treatment into natural ecosystem. Effluents highly charged by different pollutants could lead to eutrophication and groundwater contamination. In addition to that, application of enormous quantities of synthetic phosphate fertilizers is useless, especially that plant use only about 20–30% of P from applied fertilizer. The rest is rapidly converted into insoluble complex compounds due to its interaction with calcium and magnesium in calcareous soils, or aluminum and iron in acid soils. As a consequence, serious risks have been reported affecting aquatic life and human health by the development of dental and bone fluorosis (Attallah et al. 2019; Ikhajiagbe and Ogwu 2020; Srivastav 2020).

23.2.3 Impacts of Chemical Fertilizers on Soil

Phosphate industries, their wastewaters effluents, and the over use of synthetic fertilizers are the main activities responsible for soil contamination. Synthetic phosphate fertilizers derived mainly from rock phosphate include significant amounts of toxic impurities such as trace metals and radionucleides (Chaabouni et al. 2014; Mussarat et al. 2021). At high concentrations, heavy metals accumulate in various soil strata and disturb its normal properties. Soil contamination depends on different factors such as soil physical and chemical properties, fertilization competition among plant species, plant type, accessibility of soil components or leaf deposits, etc. (Guo et al. 2020). Altansuvd et al. (2014) demonstrated that long-term chemical phosphatic fertilizers have led to highly selenium (Se) accumulation in fields. Similarly, Carne et al. (2021) reported that phosphatic fertilizers are the main source of Cd accumulation in croplands, knowing toxic health risk of Cd even at low amounts. Besides that Wang et al. (2020) reported that chemical fertilizers strongly affected soil microbial communities, restricting microbial diversity. The strong increase in food demand leading to usage intensification of agrochemicals has posed serious repercussions. Soil remediation has become an urgent to meet current food requirements without damaging the soil, environment, and public health.

23.2.4 Impacts of Chemical Fertilizers on Plants

Trace elements are involved in plant metabolism and act as cofactors of various enzymes. Elements such as Fe, cobalt (Co), copper (Cu), molybdenum (Mo), Zn, and nickel (Ni) are essential for plant and may inhibit its growth if they are insufficient, but if in high amounts, they will have toxic impacts on crop growth. Primarily, trace metals are taken up by plant roots, mobilized into the aerial parts, and then into vascular tissues. Finally, they accumulate into intercellular spaces and vacuole. Chemical fertilizers are considered as the main source of metal toxicity in plants. Heavy metals directly or indirectly affect plant metabolism, physiological, biochemical and molecular processes, reducing the crop development and agricultural yields (Berni et al. 2019). Also, metals reduce nutritional value and quality of crops (Manzoor et al. 2018). High concentrations of lead (Pb), cadmium (Cd), arsenic (As), Ni, and Zn caused leaf chlorosis, root development, and pigment production inhibition resulting in plant senescence (Vinogradov and Zubkova 2022). A long-term and an over use of phosphate fertilizers led to the accumulation of many heavy metals in potato field. As a consequence, high concentrations of As, Cd, Cr, Cu, Pb, and Zn were found in potato tubers (Cheraghi et al. 2013). Similarly, Ugulu et al. (2021) demonstrated that the health risk index of metals is higher in wheat supplemented with chemical fertilizers. Taki et al. (2018) reported that chemical fertilizer increased the ozone layer depletion, photochemical oxidation, abiotic depletion, and acidification rates. Also, chemical fertilizer treatments had the most energy consumption and the main hotspots in wheat production.

23.3 PGPR and Biofertilization Traits

23.3.1 Acquisition of Nutrients

PGPR promote plant growth by increasing mobilization and uptake of nutrients by plants. PGPR enhance availability of the three major essential macromolecules for plant growth: N, P, and K. Thus, PGPR are able to enhance N plant uptake via plant-associated symbiotic bacteria, which are able to stimulate roots to produce nodule. It is a result of mutualistic relationship between bacteria and leguminous/nonleguminous plants. *Rhizobium* genera are the most studied symbiotic PGPR strains (Tewari et al. 2020). Other bacteria and cyanobacteria adopt the biological nitrogen fixation (BNF) process to convert atmospheric N_2 into forms available by plants using nitrogenase enzyme. Both symbiotic and nonsymbiotic PGPR have been reported to fix N_2. *Rhizobium, Bradyrhizobium, Pantoea, Klebsiella, Mesorhizobium, Sinorhizobium, Beijerinckia,* and *Azoarcus* are the most widely reported N_2

fixers in the soil (Khatoon et al. 2020). The use of nitrogen fixing bacteria can minimize the applications of N chemical fertilizers and reduce production cost. PGPR with nitrogen fixing capacity can provide more soil organic nitrogen and other essential nutrients in agricultural soil and increasing the availability of nutrients (Ríos-Ruiz et al. 2020). The second major limiting nutrient for plants is P. Despite its abundance, it is usually in complex forms that are not available for plant uptake. PGPR, especially phosphorus solubilizing bacteria (PSB), are able to solubilize P either organic or inorganic form. They transform phosphate complex compounds into soluble forms. To do it, PSB adopt different mechanisms such as production of enzymes (like phosphatase and glucose dehydrogenase) and organic acid production (Rasul et al. 2019). Other micronutrients, such as Zn and Fe, are also abundant in soils but unavailable to plants and can limit crop yields. Several strains of Zn-mobilizing bacteria have been shown to increase Zn uptake, and thus increase crops yield. Zn solubilization mechanism is not largely studied, but it is likely similar to P solubilization. Zn mobilization is namely due to the production of chelating agents and organic acids. Dinesh et al. (2018) characterized a potent Zn-solubilizing bacteria *Bacillus megaterium*. Zn solubilization was accompanied by remarkable decrease in pH owing to enhanced gluconic acid production. Additionally, many PGPR strains increase the availability of Fe through the production of organic acids and siderophores (Patel et al. 2020).

23.3.2 Production of Siderophores

Iron (Fe) is essential, in defined concentration, for plant growth. But it is unavailable for plants. It is present in nature in the form of ferric iron (Fe^{3+}), which is highly insoluble. Fe deficiency could inhibit chlorophyll synthesis leading to chlorosis, decreases photosynthetic level of young leaves, and could even inhibit plant development and reduce crop yield. Iron deficiency has become a major limiting factor affecting crop yields worldwide. PGPR could solve this problem. Thus, PGPR produce low-molecular-weight iron binding protein compounds called siderophores. Siderophores are capable to chelate Fe^{3+} (Tariq et al. 2017). Under iron deficiency, PGPR produce different types of siderophores like pyoverdines, bacillibactins, cephalosporins, etc. Under Fe deficiency, PGPR-producing siderophores chelate Fe^{3+} and facilitate its uptake by plants (Ferreira et al. 2019). Gao et al. (2022b) reported that under Fe limiting conditions, *Pseudomonas* sp. strain SP3 can effectively improve apple rootstock growth and enhance plant iron availability by secreting siderophore type pyoverdine. Under stress and/or calcareous conditions, Fe is almost deficient in plants. For example, Fe bioavailability is very low in saline soils. In this context, Sultana et al. (2021) demonstrated that siderophore-producing PGPR *Bacillus aryabhattai* MS3 could enhance effectively plant growth under high salinity and Fe deficiency. The strain promoted the expression of *entD* gene involved in siderophore biosynthesis.

23.3.3 Production of Exopolysaccharides

Exopolysaccharides (EPS) are carbohydrate polymers that are synthesized by PGPR. They are constituted by a variable number of monomers (typically >10 monomers; oligosaccharides, ≤ 10 monomers) linked together by glycoside bonds, and therefore, they are characterized by a high molecular weight. EPS structure can be simple with no branches or complicated with different branches. They were classified as homopolysaccharides, if composed of one type of monomers, or heteropolysaccharides, when formed by two types of monomers at least. They have significant diversity on their structure. They differ by monosaccharide conformation, sugar constituents, alpha and beta binding, noncarbohydrate substituent, charge, MW, and physicochemical properties (Sati et al. 2021). The structural diversity provides a variety of functions to EPS, which could have a large scale of industrial applications. Thus, EPS have vital roles in different physicochemical and biological processes such as formation of biofilm, gelling ability, plasma substituting capacity, protection of bacterial cell from desiccation, harsh conditions, antibacterial activity against pathogenetic agents, pollutant degradation kinetics, maintaining cell metabolic functions and bioremediation potential (Chouchane et al. 2018, 2021; Guesmi et al. 2019). The literature review concluded that EPS are mainly produced under harsh conditions. PGPR could produce EPS with homo or heteropolysaccharides (count protein, mucopolysaccharides, and humic substances), which could be linked to their cell wall as a bound capsule layer or released in to cell surrounding and extracellular parts as exudates. Therefore, formation of biofilm is due to these bacterial exudates that play a significant role in toxic heavy metal complexion and in reducing their mobility and availability in soils and plant parts (Ansari and Ahmad 2018). Dar et al. 2021 reported that EPS-producing PGPR *Rhizobium phaseoli* and *Pseudomonas bathysetes* have the capacity to promote physiology, yield, and growth of maize culture. They increased plant height, root length, root fresh and dry weight, root length density, SPAD value, leaf areas index, photosynthesis rate, transpiration, and stomata conductance. Thus, EPS produced by PGPR strain increased organic matter content in plant rhizosphere soil resulting in the enhancement of plant-available soil water contents in soil. Also, a higher biomass production was observed. Added to that, they enhanced physical soil properties such as porosity, bulk density, and soil aggregate stability. Thus, EPS, which is a gummy material, could join soil particles forming soil aggregates.

23.4 Resistance to Abiotic Stress

A wide range of secondary metabolites and volatile organic compounds (VOCs) could be produced by PGPR to resist toward abiotic stress conditions. Different PGPR genera have shown significance capacity to resist under extreme conditions such as drought, salinity, and alkalinity (Cherif et al. 2015; Soussi et al. 2016). Zahedi and Samira Abbas (2015) proved that under drought stress, inoculation of three PGPR *Rhizobium japonicum*, *Azotobacter chroococcum*, and *Azospirillum brasilense* promotes soybean growth. To do that, PGPR strains stimulate production of phytohormones (cytokinin, gibberellin, and auxin) and polyamines (putrescine and spermidine). Asghari et al. (2019) proved that application of PGPR strains *A. chroococcum* and *A. brasilense* enhances pennyroyal *Mentha pulegium* growth under drought stress. Thus, the two strains increased antioxidant osmoprotectant accumulation, membrane lipid peroxidation, enzymes activity, DPPH radical scavenging activity, and the production of abscisic acid (ABA), phenolic, flavonoid, oxygenated monoterpenes and essential oil contents. Similarly, Nadeem et al. (2021) confirmed that the combination of the two PGP traits, ACC deaminase and EPS production give plant significant resistance toward drought stress. As a result, maize inoculation with ACC deaminase and EPS-producing PGPR bacteria enhanced photosynthesis rate, stomatal conductance, vapor pressure, water-use efficiency, and transpiration rate in maize culture. To resist toward salinity, multiple secondary metabolites such as osmoprotectants/compatible solutes, exopolysaccharides (EPS), and volatile organic compounds (VOCs) produced by halotolerant PGPR strains have been reported (Sunita et al. 2020). Naseem et al. (2018) concluded that the application of EPS-producing PGPR is considered as an efficient eco-friendly approach to face drought stress conditions and improve crop productivity. In fact, the survival of PGPR under low moisture is due to the formation of nodules and especially the production of EPS. Thus, EPS surrounded plant roots forming a biofilm and protecting the plant from desiccation, microbial aggregation, surface attachment, bioremediation, etc. Plants inoculated with EPS-producing PGPR showed higher accumulation of sugars, proline, and free amino acids under drought stress. As well, Banerjee et al. (2019) discussed the potential of PGPR in the resistance against extreme and harsh environmental conditions. It was reported that PGPR are distinguished by high genetic diversity and physiological adaptations. The main mechanisms that are significantly related to stress tolerance are the growth in biofilm consortia, the synthesis of compatible solutes, production of EPS, and extremozymes. This report talked also about salt-tolerant PGPR, which effectively alleviate soil salinity stress during plant growth. Thus, under salinity stress, salt-tolerant PGPR induce biofilm formation and EPS production. Thereby, EPS improve bacterial colonization of plant roots and enhance soil structure and particles. Also, EPS has a significant capacity to mitigate salinity stress by reducing the content of Na^+ available for plant uptake.

23.5 Biostimulation Potential and PGPR

PGPR can produce phytostimulators and phytohormones, which are key players in plant growth regulation. These phytostimulators are able to enhance or inhibit plant development. PGPR can produce auxins, cytokinins, gibberellins, and inhibitors of ethylene production (Maheshwari et al. 2015; Khan et al. 2020). Indole-3-acetic acid (IAA) is the most studied auxin produced by PGPR. It enhances root proliferation and architecture, growth rate, cell division, geotropisms, phototropism, etc. Tryptophan is the main precursor molecule for biosynthesis of IAA in bacteria and its conversion is an important step. It is an amino acid found in root exudates. Meanwhile, Dashti et al. (2021) characterized PGPR isolates that were capable to produce IAA both in the presence and absence of tryptophan substrate. But, there was a strong increase in IAA synthesis when tryptophan was supplemented in bacterial culture media. Also, IAA enhances induced systemic tolerance (IST) against various abiotic stresses and induced systemic resistance (ISR) against different phytopathogenic diseases and provides resistance to plant under abiotic stress conditions (Yousef 2018; Meena et al. 2020). Cytokinins promote cell division, vascular cambium sensitivity, vascular differentiation and control root proliferation by inhibiting primary root elongation and lateral root initiation and stimulating root hair development (Saqlain Zaheer 2018). Meanwhile, gibberellins stimulate the development of stem tissue, root elongation, and lateral root extension. Also, it is involved in seed germination and emergence, floral induction, fruit development, and steam and leaf growth. However, the most dominant physiological effect of gibberellins is shoot elongation (Sharma et al. 2018).

1-Aminocyclopropane-1-carboxylic acid (ACC) deaminase is an enzyme that can inhibit ethylene production, which increases under biotic and abiotic stress. At low concentration, ethylene is a key phytohormone, which can promote plant growth. It plays important role in the stimulation of root initiation, seed germination, fruit ripening, and leaf abscission and in the synthesis of other plant hormones. At high concentration, it is considered as plant growth inhibitor. It induces defoliation and other cellular processes leading to senescence (Patel et al. 2020). Renoud et al. (2022) reported that *Azospirillum* inoculant enhanced maize growth by producing ACC deaminase, which modulate ethylene metabolism and stimulate root growth, and 2,4-diacetylphloroglucinol, which is an auxinic signal enhancing also root development.

23.6 Biocontrol Potential and PGPR

PGPR could be an alternative approach in plant control disease, functioning as efficient biological control agents (Vaish et al. 2021). PGPR promote plant growth through the control of biotic stress conditions caused by phytopathogenic agents, which could be bacteria, fungi, virus pathogens, and pests (David et al. 2018; Mhatre et al. 2019; Wang et al. 2021). PGPR adopt different mechanisms to fight against phytopathogenic diseases. Different metabolites produced are used as defense systems, mainly antifungal compounds and antibiotics (Mishra and Arora 2018; Kenawy et al. 2019). PGPR produce lytic enzymes, such as chitinase and glucanase, to inhibit fungal growth by destruction of fungal cell wall composed by chitin and beta-glucan. Added to lytic enzymes, PGPR produce hydrogen cyanide (HCN), which inhibits the growth of fungal pathogens (Kesaulya et al. 2015). Metabolites such as siderophores, VOCs, jasmonic acid (JA), QS molecules, lipopeptides (LPs), and MAMPs have also been involved (Wu et al. 2017. VOCs and JA have been shown to have an important role in the elicitation of induce systemic resistance (ISR) and are further responsible for regulating the cascade of defense gene expression (Mashabela et al. 2022). Fatima et al. (2022) reported that inoculation of four different PGPR strains (*Serratia* sp. IN-1, *Serratia* sp. IS-1, *Enterobacter* sp. IN-2, *Enterobacter* sp. IN-6) showed a suppression of chickpea wilt disease caused by *Fusarium oxysporum* f. sp. *ciceris* (Foc). PGPR strains produced amylase, protease, cellulase, chitinase, indole-3-acetic acid, and HCN. Also, they induced the expression of other various key defense enzymes, like β-1,3-glucanase, peroxidase, phenylalanine ammonia lyase, and polyphenol peroxidase, which might have helped plant to fight against FOC attack. Using confocal laser scanning microscopy, strains showed root colonization capacity and elimination of broken Foc mycelia and hyphae from roots. Also, they promote root development and proliferation resulting to crop productivity enhancement. Many different PGPR genera produce multiple secondary metabolites to protect against pathogenic diseases. *Azotobacter salinestris* showed an antifungal capacity against *Fusarium* species proving significant promotion cereals growth by the production of vitamins, amino acids, HCN, lytic enzymes, siderophotes, etc. (Chennappa et al. 2018). Some of the largest studied antibiotics produced by PGPR, particularly *Pseudomonas* genus, are phenazines. It is a large family, which inhibits several bacterial, fungal, and oomycete plant pathogens (Zboralski et al. 2020). Xia et al. 2020 characterized a multiresistant and potent biocontrol PGPR strain *Bacillus xiamenensis* PM14. The strain showed *in vitro* tolerance toward salinity, drought, high temperature, and heavy metal toxicity. Also, it was tested positive for all *in vitro* PGP activities except HCN production. It showed also remarkable inhibition of six phytopathogenic fungi, which are *Colletotrichum falcatum*, *F. oxysporum*, *Fusarium moniliforme*, *Rhizoctonia solani*, *Macrophomina phaseolina*, and *Phythium splendens*. *Bacillus subtilis* and *Pseudomonas koreensis* are two PGPR, which could be efficient biological control candidates of late wilt disease caused by *Cephalosporium maydis* in maize plant, reported Ghazy and El-Nahrawy (2021). The strain induced the production of antioxidant enzymes such as catalase (CAT), peroxidase (POX), and polyphenol oxidase (PPO), increased total chlorophyll and carotenoids content, and the thickness of the sclerenchymatous sheath layer surrounding the vascular bundles in maize stem, resulting to maize development, growth promotion, and higher crop yield. One of the first mechanisms also described to inhibit phytopathogens is the production of siderophores. *Pseudomonas aeruginosa* FP6, a siderophore-production PGPR reported by Sasirekha and Srividya (2016), showed biocontrol potential of chilli fungal diseases caused by *R. solani* and *C. gloesporioides*.

23.7 PGPR and Heavy Metal Bioremediation

Heavy metals (such as Cd, Pb, As, Zn, Cu, and Ni) could be classified as essential, toxic, and very toxic according to their toxicity, solubility, and chemical properties. Metals persist for long time in any ecosystem (water, atmosphere, soil) and they are difficult to degradable. The native or natural metals usually exist in soils in low amounts and in highly immobile forms. Meanwhile, heavy metals obtained from anthropogenic activities often appear in active forms, mobile, and plant-available. After contact with soil, they are progressively converted to insoluble forms and incorporate into soil through different chemical reactions and mechanisms. Heavy metal bioavailability in soils is a complex process, which is affected by different factors like soil pH, ionic composition, oxides, texture, organic matter content, calcium carbonate equivalent, etc. The chemical forms of heavy metals in soils are defined by the sequential extraction procedure. Then, the toxicity and the availability of metal are determined after its incorporation into soil and its reaction with soil species. About their affinity with soil, metals could be exchangeable, water soluble, bound to carbonate, bound to iron-manganese, bound to oxides, bound to organic matter, and residual (Lock and Janssen 2003). High concentrations of metals could affect essential nutrient uptake like P and K leading to plant growth inhibition. They could also cause an oxidative stress in plants inducing disequilibrium of cellular redox balance and formation of reactive oxygen species (ROS). Plants could tolerate harsh environmental conditions using multiple defense strategies including enzymes machinery (peroxidase, superoxide dismutase, and catalase) and metabolite production like amino acids, organic acids, osmolytes, bio-regulators, antioxidants. The inoculation of PGPR could strongly consolidate plant defense strategy and its phytoremediation potential (Abbaszadeh-Dahaji et al. 2016). Removal of heavy metals using chemical conventional methods costs high. Also, it requires labor intensity and generates enormous quantities of wastes highly charged of secondary pollutants and sludge. However, the removal of heavy metals using bioremediation has given hope to soil decontamination and sustainability. Bioremediation is a biological process using microorganisms, in their survival adequate conditions, to remediate contaminated sites and treat target pollutants. PGPR inoculants have been considered as the most suitable candidates for reducing metal toxicity. They reduce the bioavailability of toxic metals by the detoxification of contaminants and nutrient cycling maintaining soil structure and fertility (Kumar et al. 2018; Mandal and Basu 2021). Table 23.1 summarizes some of PGPR strains that have previously showed heavy metal bioremediation potential. Thus, PGPR strains that have bioremediation potency are commonly highly resistant to metal toxicity via the development of different strategies to tolerate this metallic stress conditions. They adopt different mechanisms, mainly two strategies in heavy metal bioremediation, which are bioleaching and bioaccumulation. Bioleaching corresponds to the extraction of metals from metal-contaminated sites and ecosystems. It is based on solubilization and transformation of metallic complex compounds to soluble forms using potent PGPR strains. In this first strategy, we could also find the terms bio-oxidation and bio-mining, which enhances metal recovery. The second one is metal immobilization corresponding to bioaccumulation and biosorption of metals. Bioaccumulation is an active process, which is mediated by microbial metabolism, irreversible, and complex. Meanwhile, biosorption is a passive process, rapid, and reversible. It could be occurred via intracellular accumulation, extracellular accumulation, or cell surface adsorption. The two strategies include various ways such as precipitation, chelation, acidification, complexation, and transport across cell membrane, ion exchange, methylation, volatilization, EPS sequestration, enzymatic detoxification, and production of siderophores, organic acids, and other different metabolites (Manoj et al. 2020; Upadhyay et al. 2021). For example, phosphorus-solubilizing bacteria (PSB)

Table 23.1 Some examples of PGPR strains possessing heavy metal bioremediation potential.

PGPR strain(s)	Heavy metal	Plant	Mechanism(s)/ Metabolite(s)	References
Pseudomonas gessardii and *Pseudomonas fluorescens*	Pb	Sunflower	Immobilization, antioxidant enzymes	Saleem et al. (2018)
Pseudomonas aeruginosa and *Burkholderia gladioli*	Cd	*Lycopersicon esculentum*	Organic acids, antioxidant enzymes, osmolytes	Khanna et al. (2019)
Pseudomonas fluorescens and *Bacillus subtilis*	Cd	*Vigna radiata* L.	Bioaccumulation siderophores, antioxidant enzymes	Rajendran and Sundaram (2020)
Stenotrophomonas maltophilia sp. and *Agrobacterium* sp.	As	*Arundo donax* L.	Volatilization, bioaccumulation	Guarino et al. (2020)
Klebsiella huaxiensis, *Pantoea cypripedii*	Zn, Cd	Napier grass	Chelation, auxins	Sumranwanich et al. (2022)
Serratia liquefaciens and *Bacillus thuringiensis*	Pb, Cd	Potato	Immobilization, antioxidant enzymes	Cheng et al. (2020)
Bacillus sp.	Cr	Wheat	Antioxidant enzymes	Mazhar et al. (2020)
Bacillus halotolerans and *Enterobacter hormaeche*	Zn	Jujube plant	Solubilization, IAA	Fahsi et al. (2021)

produce organic acids that reduce soil pH and stimulate metal ion sequestration. PSB are able to solubilize phosphorus and increase its availability mainly through this soil acidification, ion exchange reactions, and mineralization of organic phosphate compounds. Teng et al. (2019) characterized two PSB strains, *Leclercia adecarboxylata* and *Pseudomonas putida*, which are potential in Pb immobilization. Thus, these strains were able to grow at Pb concentration greater than 4 mM. Through phosphate solubilization and biomineralization, the two strains transformed lead ions into lead hydroxyapatite and pyromorphite. Additionally, PGPR produce inorganic acids, which could reduce metal stress by precipitation forming metallic insoluble precipitates. Like P, iron is one of the most abundant elements in nature but it is not available for plants as it is mostly found in an insoluble form. Contaminated, salt, calcareous, and stressed soils are usually iron deficient. In such condition, siderophore-producing PGPR can play a significant role to mitigate iron deficiency in plants. Many research reports confirmed that plants treated with siderophore-producing PGPR improve chlorophyll concentration as well as other growth parameters in plants under metal-stressed condition. It has been observed that siderophore forms complexes with heavy metals and reduces the inhibitory effects of heavy metals in IAA biosynthesis and regulates IAA level in plants under heavy metal-stressed conditions. Siderophores also enhance peroxidase activity and mitigate oxidative damage under metal-stressed condition. About metal biosorption strategy, Kour et al. (2019) reported zinc biosorption by PGPR zinc-tolerant bacteria. These strains showed an arsenal of PGP activities. They were tested positive for IAA, GA3 and ammonia production, ACC deaminase and phytase activities, siderophore and HCN production, and P solubilization. Zinc is essential for plant nutrition at low levels. However, it will highly phytotoxic at high concentrations. PGPR zinc-tolerant bacteria showed zinc solubilization of 2% insoluble ZnO in tested medium. Liu et al. (2022) reported that rice (*Oryza sativa*) treatment with the PGPR strain *Bacillus* sp. showed more efficiency in phytoremediation. The strain improved Zn immobilization, Cd extraction; maintained ion homeostasis in rice; and enhanced soil microbial diversity and soil fertility. Ke et al. (2021) showed that PGPR could be an eco-friendly promising tool to enhance the phytostabilization efficiency of Cu and Cd in HM-contaminated soils. The strains adopting metal adsorption revealed that PGPR inoculation increased shoot and root biomasses of ryegrass in co-contaminated soil with Cu and Cd. Also, a decrease in metal availability and translocation were observed in inoculated ryegrass due to the decrease of diethylenetriamine pentaacetic acid-extractable metal content and the increase of residual metal-fraction percentage. Additionally, lower malondialdehyde content and superoxide dismutase activity were noticed in inoculated ryegrass than those of un-inoculated. EPS play a role in trace metal adsorption EPS-producing activity for improving maize growth under drought stress these strains were also significantly better in improving the plant physiological parameters including photosynthesis rate, stomata conductance, vapor pressure, water-use efficiency and transpiration rate. In this context, Zainab et al. (2020) reported that PGPR strains *Bacillus gibsonii* PM11 and *B. xiamenensis* PM14 adopted to produce mainly EPS, IAA, and ACC-deaminase to increase metal phytoremediation by flax plant (*Linum usitatissimum* L.), PGPR have the potential to enhance plant heavy metal tolerance and get more efficient phyoremediation. Phytohormones are very important for plants and their health, which regulate different morphological and physiological functions. Thus, IAA has a significant role in the development and proliferation of surface area, length, and architecture root, giving to plant a greater access to extract metals and other nutrients from the soil. Ahmad et al. (2016) reported that PGPR strains *Bacillus* sp., *Klebsiella* sp., *Leifsonia* sp., *Enterobacter* sp. improve phytoaccumulation of Cd by *Zea mays* L. via mainly IAA production and also by EPS and antioxidant enzyme production and phosphate solubilization. Ethylene is another crucial plant hormone that regulates cellular elongation and metabolism. Meanwhile, ethylene overproduction could lead to plant growth inhibition. It disturbs some morphological and physiological processes in plants inducing premature senescence. Under harsh conditions (climate change, pathogen infections, pollutants, etc.), plants produced ACC, which is the primary precursor for ethylene synthesis. Here, PGPR enhance the expression of ACC deaminase enzyme and converts the ACC precursor to α-ketobutyrate and ammonia, reducing ethylene level and increasing plant growth (Murali et al. 2021; Chandwani and Amaresan 2022). In this context, an ACC deaminase-producing bacterial strain *Achromobacter* sp. A1 was evaluated by Sun et al. (2022) for Cd immobilization and maize growth promotion. The strain significantly increased the maize dry weights, chlorophyll content, invertase, urease, and catalase activity with an important decrease of ethylene concentration. Also, it reduced tissue Cd uptake. Thus, the expression levels of Cd transporter genes *HMA3* and *Nramp5* were significantly decreased in root and shoot. Similarly, Zafar-ul-hye et al. (2018) concluded that *Agrobacterium fabrum* and *Stenotrophomonas maltophilia* are two PGPR strains that are able to promote wheat growth under Cd toxicity and reduce Cd plant uptake through ACC deaminase production. Kumar et al. (2022) discussed the role and constraints of arsenic-resistant PGPR in reducing the arsenic toxicity in food crops. As-resistant PGPR could reduce arsenic contamination increasing phytoremediation efficiency as hyperaccumulator plants. Abdollahi et al. (2022) reported that application of PGPR strain *B. megaterium* in heavy-metal-contaminated soil enhanced phytoremediation process. Thus, it increased the metal phytoavailability, mobility, translocation factor and biomass production, and the Pb and Cd concentrations in the roots and shoots. Ju et al. (2022) applied PGPR to reduce Cu and nitrate leaching risk. This study showed that PGPR combined with the biodegradable chelator (*S*,*S*)-)-*N*,*N*′-ethylenediaminedisuccinic acid (EDDS) are effective in enhancing metal mobility and ammonium release. Meanwhile, the risk of metal and/or nitrate residues within EDDS biodegradation has some limitations such as its short-term application. Thus, EDDS supplemented with PGPR showed significant increase in Cu phytoextraction and soil nitrate content, important plant biomass, and decrease in soil Cu and nitrate levels.

23.8 Conclusion and Future Prospects

It is concluded that industrialization and chemical fertilizers are one of the main sources of heavy metal contamination, soil damage, and public health concerns. Soil metal contamination is a severe threat, which strongly affects soil fertility, crop productivity, and agriculture sustainability. Keeping in view the above-given beneficial effects of PGPR, it is amply clear that the PGPR application could give hope to sustainable agriculture. It is really an eco-friendly attractive and economic approach. Using their arsenal of traits, PGPR develop multiple potentialities in biofertilization, biostimulation, biocontrol, and bioremediation. They are capable to reduce metals from contaminated sites adopting various mechanisms.

To get a potent metal bioremediator PGPR strain, effective studies should be carried out under field conditions with different climate and stressful conditions. To finish, we should insist that bioremediation of contaminated soils has become more important. Thus, more work will be required to deal with environmental and health concerns. It is time to explore eco-friendly approaches giving hope to a sustainable agriculture and safe environment.

Acknowledgments

This project is carried out under the MOBIDOC scheme, funded by The Ministry of Higher Education and Scientific Research through the PromEssE project and managed by the ANPR.

References

Abbaszadeh-Dahaji, P., Omidvari, M., and Ghorbanpour, M. (2016). Increasing phytoremediation efficiency of heavy metal-contaminated soil using PGPR for sustainable agriculture. In: *Plant-Microbe Interaction: An Approach to Sustainable Agriculture* (ed. D.K. Choudhary, A. Varma, and N. Tuteja), 187–204. Singapore: Springer.

Abdollahi, S., Golchin, A., Shahryari, F., and Alamdari, P. (2022). PGPR inoculation of a contaminated soil affects plant growth and phytoavailability of Cd and Pb. *Archives of Agronomy and Soil Science* 68 (5): 579–596.

Ahmad, I., Akhtar, M.J., Asghar, H.N. et al. (2016). Differential effects of plant growth-promoting rhizobacteria on maize growth and cadmium uptake. *Journal of Plant Growth Regulation 35* (2): 303–315.

Aloo, B.N., Makumba, B.A., and Mbega, E.R. (2020). Plant growth promoting rhizobacterial biofertilizers for sustainable crop production: the past, present, and future. *Preprints* **2020**: 2020090650. https://doi.org/10.20944/preprints202009.0650.v1.

Altansuvd, J., Nakamaru, Y.M., Kasajima, S. et al. (2014). Effect of long-term phosphorus fertilization on soil Se and transfer of soil Se to crops in northern Japan. *Chemosphere* 107: 7–12.

Ansari, F.A. and Ahmad, I. (2018). Biofilm development, plant growth promoting traits and rhizosphere colonization by *Pseudomonas entomophila* FAP1: a promising PGPR. *Advances in Microbiology* 8 (03): 235.

Asghari, B., Khademian, R., & Sedaghati, B. (2020). Plant growth promoting rhizobacteria (PGPR) confer drought resistance and stimulate biosynthesis of secondary metabolites in pennyroyal (Mentha pulegium L.) under water shortage condition. *Scientia Horticulturae*, 263, 109132.

Attallah, M.F., Metwally, S.S., Moussa, S.I., and Soliman, M.A. (2019). Environmental impact assessment of phosphate fertilizers and phosphogypsum waste: elemental and radiological effects. *Microchemical Journal* 146: 789–797.

Banerjee, A., Sarkar, S., Cuadros-Orellana, S., and Bandopadhyay, R. (2019). Exopolysaccharides and biofilms in mitigating salinity stress: the biotechnological potential of halophilic and soil-inhabiting PGPR microorganisms. In: *Microorganisms in Saline Environments: Strategies and Functions*, 133–153. Cham: Springer.

Berni, R., Luyckx, M., Xu, X. et al. (2019). Reactive oxygen species and heavy metal stress in plants: impact on the cell wall and secondary metabolism. *Environmental and Experimental Botany* 161: 98–106.

Carne, G., Leconte, S., Sirot, V. et al. (2021). Mass balance approach to assess the impact of cadmium decrease in mineral phosphate fertilizers on health risk: the case-study of French agricultural soils. *Science of the Total Environment* 760: 143374.

Chaabouni, A., Chtara, C., Nzihou, A., and El Feki, H. (2014). Study the nature and the effects of the impurities of phosphate rock in the plants of production of phosphoric acid. *Journal of Advances in Chemistry* 7 (2): 1296–1299.

Chandwani, S. and Amaresan, N. (2022). Role of ACC deaminase producing bacteria for abiotic stress management and sustainable agriculture production. *Environmental Science and Pollution Research* 29: 22843–22859.

Cheng, D., Jiang, C., Xu, J. et al. (2020). Characteristics and applications of alginate lyases: a review. *International Journal of Biological Macromolecules* 164: 1304–1320.

Chennappa, G., Sreenivasa, M.Y., and Nagaraja, H. (2018). *Azotobacter salinestris*: a novel pesticide-degrading and prominent biocontrol PGPR bacteria. In: *Microorganisms for Green Revolution* (ed. D. Panpatte, Y. Jhala, H. Shelat, and R. Vyas), 23–43. Singapore: Springer.

Cheraghi, M., Lorestani, B., Merrikhpour, H., and Rouniasi, N. (2013). Heavy metal risk assessment for potatoes grown in overused phosphate-fertilized soils. *Environmental Monitoring and Assessment* 185 (2): 1825–1831.

Cherif, H., Marasco, R., Rolli, E. et al. (2015). Oasis desert farming selects environment-specific date palm root endophytic communities and cultivable bacteria that promote resistance to drought. *Environmental Microbiology Reports* 7 (4): 668–678.

Chouchane, H., Mahjoubi, M., Ettoumi, B. et al. (2018). A novel thermally stable heteropolysaccharide-based bioflocculant from hydrocarbonoclastic strain *Kocuria rosea* BU22S and its application in dye removal. *Environmental Technology* 39 (7): 859–872.

Chouchane, H., Boutiti, S., Ouertani, A. et al. (2021). Effect of gamma irradiation on enhanced biological activities of exopolysaccharide from *Halomonas desertis* G11: biochemical and genomic insights. *Polymers* 13 (21): 3798.

Dar, A., Zahir, Z.A., Iqbal, M. et al. (2021). Efficacy of rhizobacterial exopolysaccharides in improving plant growth, physiology, and soil properties. *Environmental Monitoring and Assessment* 193 (8): 1–15.

Dashti, N.H., Al-Sarraf, N.Y.A., Cherian, V.M., and Montasser, M.S. (2021). Isolation and characterization of novel plant growth-promoting rhizobacteria (PGPR) isolates from tomato (*Solanum lycopersicum* L.) rhizospherical soil: a novel IAA producing bacteria. *Kuwait Journal of Science* 48 (2): 1–17.

David, B.V., Chandrasehar, G., and Selvam, P.N. (2018). *Pseudomonas fluorescens*: a plant-growth-promoting rhizobacterium (PGPR) with potential role in biocontrol of pests of crops. In: *Crop Improvement Through Microbial Biotechnology* (ed. K. Goyal, T. Kumar, P. Sharma, et al.), 221–243. Elsevier.

Dinesh, R., Srinivasan, V., Hamza, S. et al. (2018). Isolation and characterization of potential Zn solubilizing bacteria from soil and its effects on soil Zn release rates, soil available Zn and plant Zn content. *Geoderma* 321: 173–186.

Fahsi, N., Mahdi, I., Mesfioui, A. et al. (2021). Plant growth-promoting rhizobacteria isolated from the Jujube (*Ziziphus lotus*) plant enhance wheat growth, Zn uptake, and heavy metal tolerance. *Agriculture* 11 (4): 316.

Fatima, I., Hakim, S., Imran, A. et al. (2022). Exploring biocontrol and growth-promoting potential of multifaceted PGPR isolated from natural suppressive soil against the causal agent of chickpea wilt. *Microbiological Research* 260: 127015.

Ferreira, M.J., Silva, H., and Cunha, A. (2019). Siderophore-producing rhizobacteria as a promising tool for empowering plants to cope with iron limitation in saline soils: a review. *Pedosphere* 29 (4): 409–420.

Gao, Y., Fang, Z., Van Zwieten, L. et al. (2022a). A critical review of biochar-based nitrogen fertilizers and their effects on crop production and the environment. *Biochar* 4 (1): 1–19.

Gao, B., Chai, X., Huang, Y. et al. (2022b). Siderophore production in *Pseudomonas* sp. strain SP3 enhances iron acquisition in apple rootstock. *Journal of Applied Microbiology* 133 (2): 720–732.

Ghazy, N. and El-Nahrawy, S. (2021). Siderophore production by *Bacillus subtilis* MF497446 and *Pseudomonas koreensis* MG209738 and their efficacy in controlling *Cephalosporium maydis* in maize plant. *Archives of Microbiology* 203 (3): 1195–1209.

Guarino, F., Miranda, A., Castiglione, S., and Cicatelli, A. (2020). Arsenic phytovolatilization and epigenetic modifications in *Arundo donax* L. assisted by a PGPR consortium. *Chemosphere* 251: 126310.

Guesmi, S., Chouchane, H., Neifar, M. et al. (2019). Radiation-inducible radioprotective exopolysaccharides of *Bacillus siamensis* CV5 from irradiated roots of *Cistanche violacea* to decrease free radical damage produced by ionizing radiation. *International Journal of Radiation Biology* 95 (11): 1552–1563.

Guo, J., Muhammad, H., Lv, X. et al. (2020). Prospects and applications of plant growth promoting rhizobacteria to mitigate soil metal contamination: a review. *Chemosphere* 246: 125823.

Hassen, W., Neifar, M., Cherif, H. et al. (2018). *Pseudomonas rhizophila* S211, a new plant growth-promoting rhizobacterium with potential in pesticide-bioremediation. *Frontiers in Microbiology* 9: 34.

Ikhajiagbe, B. and Ogwu, M.C. (2020). Hazard quotient, microbial diversity, and plant composition of spent crude oil-polluted soil. *Beni-Suef University Journal of Basic and Applied Sciences* 9 (1): 1–9.

Israr, D., Mustafa, G., Khan, K.S. et al. (2016). Interactive effects of phosphorus and *Pseudomonas putida* on chickpea (*Cicer arietinum* L.) growth, nutrient uptake, antioxidant enzymes and organic acids exudation. *Plant Physiology and Biochemistry* 108: 304–312.

Ju, W., Duan, C., Liu, L. et al. (2022). Reduction of Cu and nitrate leaching risk associated with EDDS-enhanced phytoextraction process by exogenous inoculation of plant growth promoting rhizobacteria. *Chemosphere* 287: 132288.

Ke, T., Guo, G., Liu, J. et al. (2021). Improvement of the Cu and Cd phytostabilization efficiency of perennial ryegrass through the inoculation of three metal-resistant PGPR strains. *Environmental Pollution 271*: 116314.

Kenawy, A., Dailin, D.J., Abo-Zaid, G.A. et al. (2019). Biosynthesis of antibiotics by PGPR and their roles in biocontrol of plant diseases. In: *Plant Growth Promoting Rhizobacteria for Sustainable Stress Management* (ed. R.Z. Sayyed, N.K. Arora, and M.S. Reddy), 1–35. Singapore: Springer.

Kesaulya, H., Zakaria, B., and Syaiful, S.A. (2015). Isolation and physiological characterization of PGPR from potato plant rhizosphere in medium land of Buru Island. *Procedia Food Science* 3: 190–199.

Khan, N., Bano, A., Ali, S., and Babar, M. (2020). Crosstalk amongst phytohormones from planta and PGPR under biotic and abiotic stresses. *Plant Growth Regulation* 90 (2): 189–203.

Khanna, K., Jamwal, V.L., Gandhi, S.G. et al. (2019). Metal resistant PGPR lowered Cd uptake and expression of metal transporter genes with improved growth and photosynthetic pigments in *Lycopersicon esculentum* under metal toxicity. *Scientific Reports* 9 (1): 1–14.

Khatoon, Z., Huang, S., Rafique, M. et al. (2020). Unlocking the potential of plant growth-promoting rhizobacteria on soil health and the sustainability of agricultural systems. *Journal of Environmental Management* 273: 111118.

Kour, R., Jain, D., Bhojiya, A.A. et al. (2019). Zinc biosorption, biochemical and molecular characterization of plant growth-promoting zinc-tolerant bacteria. *3 Biotech* 9 (11): 1–17.

Kumar, S., Kumar, M., Singh, R. et al. (2018). Plant–microbe symbiosis A synergistic approach for heavy-metal bioremediation. In: *Recent Advances in Environmental Management* (ed. R.N. Bharagava), 299–315. CRC Press.

Chandini, R.K., Kumar, R., and Om, P. (2019). The impact of chemical fertilizers on our environment and ecosystem. In: *Research Trends in Environmental Sciences*, 2e, 71–86. https://www.researchgate.net/publication/331132826.

Kumar, S., Choudhary, A.K., Suyal, D.C. et al. (2022). Leveraging arsenic resistant plant growth-promoting rhizobacteria for arsenic abatement in crops. *Journal of Hazardous Materials* 425: 127965.

Liu, A., Wang, W., Zheng, X. et al. (2022). Improvement of the Cd and Zn phytoremediation efficiency of rice (*Oryza sativa*) through the inoculation of a metal-resistant PGPR strain. *Chemosphere* 302: 134900.

Lock, K. and Janssen, C.R. (2003). Influence of aging on metal availability in soils. *Reviews of Environmental Contamination and Toxicology* 178: 1–21.

Maheshwari, D.K., Dheeman, S., and Agarwal, M. (2015). Phytohormone-producing PGPR for sustainable agriculture. In: *Bacterial Metabolites in Sustainable Agroecosystem* (ed. D.K. Maheshwari), 159–182. Cham: Springer.

Mandal, D. and Basu, A. (2021). Role of heavy-metal resistant bacteria isolated from rhizosphere in bioremediation and plant development. Eds., Soumya Mukherjee, František Baluška In: *Rhizobiology: Molecular Physiology of Plant Roots*, 411–435. Cham: Springer.

Manoj, S.R., Karthik, C., Kadirvelu, K. et al. (2020). Understanding the molecular mechanisms for the enhanced phytoremediation of heavy metals through plant growth promoting rhizobacteria: a review. *Journal of Environmental Management 254*: 109779.

Manzoor, J., Sharma, M., and Wani, K.A. (2018). Heavy metals in vegetables and their impact on the nutrient quality of vegetables: a review. *Journal of Plant Nutrition* 41 (13): 1744–1763.

Mashabela, M. D., Piater, L. A., Dubery, I. A., Tugizimana, F., & Mhlongo, M. I. (2022). Rhizosphere tripartite interactions and PGPR-mediated metabolic reprogramming towards ISR and plant priming: A metabolomics review. *Biology*, 11(3), 346.

Mazhar, R., Ilyas, N., Arshad, M. et al. (2020). Isolation of heavy metal-tolerant PGPR strains and amelioration of chromium effect in wheat in combination with biochar. *Iranian Journal of Science and Technology, Transactions A: Science* 44 (1): 1–12.

Meena, M., Swapnil, P., Divyanshu, K. et al. (2020). PGPR-mediated induction of systemic resistance and physiochemical alterations in plants against the pathogens: current perspectives. *Journal of Basic Microbiology* 60 (10): 828–861.

Mhatre, P.H., Karthik, C., Kadirvelu, K. et al. (2019). Plant growth promoting rhizobacteria (PGPR): a potential alternative tool for nematodes bio-control. *Biocatalysis and Agricultural Biotechnology* 17: 119–128.

Mishra, J. and Arora, N.K. (2018). Secondary metabolites of fluorescent *pseudomonads* in biocontrol of phytopathogens for sustainable agriculture. *Applied Soil Ecology 125*: 35–45.

Murali, M., Gowtham, H.G., Singh, S.B. et al. (2021). Bio-prospecting of ACC deaminase producing rhizobacteria towards sustainable agriculture: a special emphasis on abiotic stress in plants. *Applied Soil Ecology* 168: 104142.

Mussarat, M., Ali, H., Muhammad, D. et al. (2021). Comparing the phosphorus use efficiency of pre-treated (organically) rock phosphate with soluble P fertilizers in maize under calcareous soils. *PeerJ* 9: e11452.

Nadeem, S.M., Ahmad, M., Tufail, M.A. et al. (2021). Appraising the potential of EPS-producing rhizobacteria with ACC-deaminase activity to improve growth and physiology of maize under drought stress. *Physiologia Plantarum* 172 (2): 463–476.

Naseem, H., Ahsan, M., Shahid, M.A., and Khan, N. (2018). Exopolysaccharides producing rhizobacteria and their role in plant growth and drought tolerance. *Journal of Basic Microbiology* 58 (12): 1009–1022.

Patel, J.S., Yadav, S.K., Bajpai, R. et al. (2020). PGPR secondary metabolites: an active syrup for improvement of plant health. In: *Molecular Aspects of Plant Beneficial Microbes in Agriculture* (ed. V. Sharma, R. Salwan, and L.K.T. Al-Ani), 195–208. Academic Press.

Rajendran, S.K. and Sundaram, L. (2020). Degradation of heavy metal contaminated soil using plant growth promoting rhizobacteria (PGPR): assess their remediation potential and growth influence of *Vigna radiata* L. *International Journal of Agricultural Technology* 16 (2): 365–376.

Rani, K., Tigga, P., Roy, A. et al. (2022). Nutrient availability and plant productivity through PGPR: mechanisms, potential and constraints. In: *Advances in Agricultural Biotechnology*, vol. 5 (ed. K.C. Mamata and A. Lamichhane), 138. AkiNik Publications.

Rasul, M., Yasmin, S., Suleman, M. et al. (2019). Glucose dehydrogenase gene containing phosphobacteria for biofortification of phosphorus with growth promotion of rice. *Microbiological Research* 223: 1–12.

Renoud, S., Abrouk, D., Prigent-Combaret, C. et al. (2022). Effect of inoculation level on the impact of the PGPR *Azospirillum lipoferum* CRT1 on selected microbial functional groups in the rhizosphere of field maize. *Microorganisms* 10 (2): 325.

Ríos-Ruiz, W.F., Torres-Chávez, E.E., Torres-Delgado, J. et al. (2020). Inoculation of bacterial consortium increases rice yield (*Oryza sativa* L.) reducing applications of nitrogen fertilizer in San Martin region, Peru. *Rhizosphere* 14: 100200.

Saleem, M., Asghar, H.N., Zahir, Z.A., and Shahid, M. (2018). Impact of lead tolerant plant growth promoting rhizobacteria on growth, physiology, antioxidant activities, yield and lead content in sunflower in lead contaminated soil. *Chemosphere* 195: 606–614.

Saqlain Zaheer, M. (2018). Effect of plant growth promoting rhizobacteria (PGPR) and cytokinins on the growth and yield of wheat (*Triticum Aestivum* L.) under drought. Doctoral dissertation. http://prr.hec.gov.pk/jspui/handle/123456789/10592.

Sasirekha, B. and Srividya, S. (2016). Siderophore production by *Pseudomonas aeruginosa* FP6, a biocontrol strain for *Rhizoctonia solani* and *Colletotrichum gloeosporioides* causing diseases in chilli. *Agriculture and Natural Resources* 50 (4): 250–256.

Sati, D.; Pande, V.; Pandey, S.C.; Samant, M. Recent advances in pgpr and molecular mechanisms involved in drought stress tolerance. Preprints.org 2021, 2021050331. https://doi.org/10.20944/preprints202105.0331.v1.

Setiawati, T.C., Erwin, D., Mandala, M., and Hidayatulah, A. (2022). Use of *Bacillus* as a plant growth-promoting rhizobacteria to improve phosphate and potassium availability in acidic and saline soils. *KnE Life Sciences* 541–558. https://doi.org/10.18502/kls.v7i3.11160.

Sharma, S., Sharma, A., and Kaur, M. (2018). Extraction and evaluation of gibberellic acid from *Pseudomonas* sp.: plant growth promoting rhizobacteria. *Journal of Pharmacognosy and Phytochemistry* 7 (1): 2790–2795.

Song, K., Xue, Y., Zheng, X. et al. (2017). Effects of the continuous use of organic manure and chemical fertilizer on soil inorganic phosphorus fractions in calcareous soil. *Scientific Reports* 7 (1): 1–9.

Soussi, A., Ferjani, R., Marasco, R. et al. (2016). Plant-associated microbiomes in arid lands: diversity, ecology and biotechnological potential. *Plant and Soil* 405 (1): 357–370.

Srivastav, A.L. (2020). Chemical fertilizers and pesticides: role in groundwater contamination. In: *Agrochemicals Detection, Treatment and Remediation* (ed. M.N.V. Prasad), 143–159. Butterworth-Heinemann.

Sultana, S., Alam, S., and Karim, M.M. (2021). Screening of siderophore-producing salt-tolerant rhizobacteria suitable for supporting plant growth in saline soils with iron limitation. *Journal of Agriculture and Food Research* 4: 100150.

Sumranwanich, T., Leartsiwawinyu, W., Meeinkuirt, W., and Chayapan, P. (2022). Application of plant growth-promoting rhizobacteria (PGPR) associated with energy plant, Pennisetum purpurenum, in cadmium and zinc contaminated soil. *Research Square* https://doi.org/10.21203/rs.3.rs-1401400/v1.

Sun, L., Zhang, X., Ouyang, W. et al. (2022). Lowered Cd toxicity, uptake and expression of metal transporter genes in maize plant by ACC deaminase-producing bacteria *Achromobacter* sp. *Journal of Hazardous Materials* 423: 127036.

Sunita, K., Mishra, I., Mishra, J. et al. (2020). Secondary metabolites from halotolerant plant growth promoting rhizobacteria for ameliorating salinity stress in plants. *Frontiers in Microbiology* 11: 567768.

Taki, M., Soheili-Fard, F., Rohani, A. et al. (2018). Life cycle assessment to compare the environmental impacts of different wheat production systems. *Journal of Cleaner Production* 197: 195–207.

Tariq, M., Noman, M., Ahmed, T. et al. (2017). Antagonistic features displayed by plant growth promoting rhizobacteria (PGPR): a review. *Journal of Plant Science and Phytopathology* 1 (1): 038–043.

Teng, Z., Shao, W., Zhang, K. et al. (2019). Characterization of phosphate solubilizing bacteria isolated from heavy metal contaminated soils and their potential for lead immobilization. *Journal of Environmental Management* 231: 189–197.

Tewari, S., Pooniya, V., and Sharma, S. (2020). Next generation bioformulation prepared by amalgamating *Bradyrhizobium*, cell free culture supernatant, and exopolysaccharides enhances the indigenous rhizospheric rhizobial population, nodulation, and productivity of pigeon pea. *Applied Soil Ecology* 147: 103363.

Tosi, M., Brown, S., Machado, P.V.F. et al. (2020). Short-term response of soil N-cycling genes and transcripts to fertilization with nitrification and urease inhibitors, and relationship with field-scale N_2O emissions. *Soil Biology and Biochemistry* 142: 107703.

Ugulu, I., Ahmad, K., Khan, Z.I. et al. (2021). Effects of organic and chemical fertilizers on the growth, heavy metal/metalloid accumulation, and human health risk of wheat (*Triticum aestivum* L.). *Environmental Science and Pollution Research* 28 (10): 12533–12545.

Upadhyay, K.H., Vaishnav, A.M., Tipre, D.R., and Dave, S.R. (2021). Metal bioremediation, mechanisms, kinetics and role of marine bacteria in the bioremediation technology. In: *Biotechnology for Sustainable Environment* (ed. S.J. Joshi, A. Deshmukh, and H. Sarma), 173–199. Singapore: Springer.

Vaish, S., Garg, N., and Ahmad, I.Z. (2021). Bioprospecting of microbial isolates from biodynamic preparations for PGPR and biocontrol properties. *Journal of Environmental Biology* 42 (3): 644–651.

Vejan, P., Abdullah, R., Khadiran, T. et al. (2016). Role of plant growth promoting rhizobacteria in agricultural sustainability—a review. *Molecules* 21 (5): 573.

Vinogradov, D.V. and Zubkova, T.V. (2022). Accumulation of heavy metals by soil and agricultural plants in the zone of technogenic impact. *Indian Journal of Agricultural Research* 56 (2): 201–207.

Wang, F., Han, W., Chen, S. et al. (2020). Fifteen-year application of manure and chemical fertilizers differently impacts soil ARGs and microbial community structure. *Frontiers in Microbiology* 11: 62.

Wang, H., Liu, R., You, M.P. et al. (2021). Pathogen biocontrol using plant growth-promoting bacteria (PGPR): role of bacterial diversity. *Microorganisms* 9 (9): 1988.

Wu, Y., Lin, H., Lin, Y. et al. (2017). Effects of biocontrol bacteria *Bacillus amyloliquefaciens* LY-1 culture broth on quality attributes and storability of harvested litchi fruit. *Postharvest Biology and Technology* 132: 81–87.

Xia, Y., Farooq, M.A., Javed, M.T. et al. (2020). Multi-stress tolerant PGPR *Bacillus xiamenensis* PM14 activating sugarcane (*Saccharum officinarum* L.) red rot disease resistance. *Plant Physiology and Biochemistry 151*: 640–649.

Yousef, N.M. (2018). Capability of plant growth-promoting rhizobacteria (PGPR) for producing indole acetic acid (IAA) under extreme conditions. *European Journal of Biological Research* 8 (4): 174–182.

Zafar-Ul-Hye, M., Shahjahan, A., Danish, S. et al. (2018). Mitigation of cadmium toxicity induced stress in wheat by ACC-deaminase containing PGPR isolated from cadmium polluted wheat rhizosphere. *Pakistan Journal of Botany* 50 (5): 1727–1734.

Zainab, N., Din, B.U., Javed, M.T. et al. (2020). Deciphering metal toxicity responses of flax (*Linum usitatissimum* L.) with exopolysaccharide and ACC-deaminase producing bacteria in industrially contaminated soils. *Plant Physiology and Biochemistry 152*: 90–99.

Zahedi, H., & Abbasi, S. (2015). Effect of plant growth promoting rhizobacteria (PGPR) and water stress on phytohormones and polyamines of soybean. *Indian Journal of Agricultural Research*, 49 (5), 427–431.

Zboralski, A., Biessy, A., Savoie, M.C. et al. (2020). Metabolic and genomic traits of phytobeneficial phenazine-producing *Pseudomonas* spp. are linked to rhizosphere colonization in *Arabidopsis thaliana* and *Solanum tuberosum*. *Applied and Environmental Microbiology* 86 (4): e02443–e02419.

24

Applications of Nanotechnology for Improving Heavy Metal Stress Tolerance in Crop Plants

Meng Jiang[1,2,3], *Yue Song*[1,2,3], *Mukesh Kumar Kanwar*[1,4], *and Jie Zhou*[1,4,5]

[1] *College of Agriculture and Biotechnology, Zhejiang University, Hangzhou, PR China*
[2] *Hainan Institute, Zhejiang University, Sanya, PR China*
[3] *Institute of Crop Sciences, National Key Laboratory of Rice Biology, Zhejiang University, Hangzhou, PR China*
[4] *Department of Horticulture, Zhejiang Provincial Key Laboratory of Horticultural Plant Integrative Biology, Zhejiang University, Hangzhou, PR China*
[5] *Shandong (Linyi) Institute of Modern Agriculture, Zhejiang University, Linyi, PR China*

24.1 Introduction

Heavy metals (HMs) are naturally occurring elements with a high density and atomic weight, and are lethal to plant cells once the concentrations reach a suitable verge (Tchounwou et al. 2012). Toxic heavy metals like lead (Pb), mercury (Hg), chromium (Cr), cadmium (Cd), arsenic (As), and thallium (Tl) are acknowledged to be nonessential for living animals or plants (Rahman and Singh 2019). In addition, there are some essential HMs, e.g. selenium (Se), molybdenum (Mo), manganese (Mn), boron (B), iron (Fe), copper (Cu), and zinc (Zn) (Yılmaz et al. 2017). Because they exhibit characteristics of both metals and nonmetals, several of these elements are referred to as metalloids. Plant growth and development are also subjected to their hazardous and positive qualities (Uchimiya et al. 2020). For instance, metalloids, e.g. Si and B, help in cell conformation and cellular functions (Bienert and Chaumont 2011); whereas other metalloids like Cd and As are poisonous for metabolism along with growth of plants (Chandrakar et al. 2016; Jiang et al. 2020; Mawia et al. 2021). HMs bring the reduction of functional moieties like sulfhydryl groups of glutathione and free ions leading to the production of reactive oxygen species (ROS), which causes cells oxidative toxicity in plants (Stohs and Bagchi 1995). Beyond a certain verge, all metals, including the essential ones, reduce crop yield and inhibit plant growth. Plants uptake and transport both non-essential and necessary HMs in a variety of ways, which includes both inactive and active passage (Shahid et al. 2017). The essential metals after being absorbed translocated to sub-cellular location or tissue in plants and perform their vital role, e.g. being cofactors (Kaur et al. 2021). The excess number of

Heavy Metal Toxicity and Tolerance in Plants: A Biological, Omics, and Genetic Engineering Approach, First Edition. Edited by Mohammad Anwar Hossain, AKM Zakir Hossain, Sylvain Bourgerie, Masayuki Fujita, Om Parkash Dhankher, and Parvez Haris.

essential HMs or the occurrence of nonessential HMs in the plant cells can trigger signaling pathways *via* several intermediate trails, e.g. MAPK (mitogen-activated protein kinase) family signal, Ca^{2+} ion signal, hormone signal, which activates defense mechanisms against HM damage and ultimately get resistant toward toxicity of HM owing to plant tolerance (Mani and Sankaranarayanan 2018). The pollution of HM is one of the main factors threatening food security and limiting crop production (Irshad et al. 2020; Javaid 2020). More significantly, the world population is predicted to rise to 9.73 billion people by the mid-twenty-first century, implying that the present production of food would have to double to fulfill the global population's requirements (Mueller et al. 2012). Consequently, it is essential to popularize and improve HM lessen/removal expertise in agricultural systems.

In the last decade, fields of cosmetics, agriculture, industry, and medicine were improved intensely by the applications of nanoparticles (NPs) (Adeel et al. 2019; Jiang et al. 2021). Materials that have one dimension as a minimum than 100 nm are referred to as NPs. Along with the criteria, NPs with diverse functions, geometries, and particle sizes can be created (Adeel et al. 2020). In comparison with the normal materials, NPs have numerous superior aspects, e.g. special magnetic characters, additional surface reaction sites, good surface activity, and high catalytic efficiency (Yang et al. 2017, 2018; Wang et al. 2019a). Previous research has looked at the influences of NPs on the environment. For example, Song et al. reported that the lower doses of ZnO-NPs (50 and 100 $mg\,l^{-1}$) alleviated chilling stress in rice (*Oryza sativa* L.) by regulating the transcription factors of chilling-response and inducing the antioxidative system (Song et al. 2021a); whereas the plant growth, tetrapyrrole biosynthesis, and amino acid metabolism of rice were severely inhibited at higher doses of ZnO-NPs (500 and 1000 $mg\,kg^{-1}$) (Song et al. 2021b). Exposure to 50 $mg\,kg^{-1}$ Ag-NPs harmed the quality and biomass of peanuts, according to Rui et al. (2017).

NPs have been now effectively used in environmental processes and agricultural systems. Numerous researches have exposed that NPs enhanced crop yield, photosynthesis, resistance to oxidative stresses, and seed germination (Li et al. 2016; Palchoudhury et al. 2018; Kah et al. 2019). Moreover, NPs are operated as nano-pesticides (Hao et al. 2019; Zhao et al. 2018; Adeel et al. 2021) or nano-fertilizers (Rui et al. 2016, 2018; Li et al. 2020), which are gradually released compared to regular ones and have the benefits of being effortlessly assimilated by crops (Lowry et al. 2019). However, the possible environmental and ecological hazards related to NPs still require detailed studies (Lombi et al. 2019). NPs (e.g. Mn_3O_4, CeO_2, and TiO_2 NPs) can improve the antioxidant enzyme activity that can decrease the content of ROS in crops, mitigating toxicity and enhancing the crop yield and quality (Usman et al. 2020; Wang et al. 2020a). Due to their special characteristics, NPs can perform new roles and are more repairable than regular materials (Moharem et al. 2019). Foliar application, in addition to soil application, is a suitable strategy to enhance the resilience of HMs in plants (Toumey 2020; Usman et al. 2020). For instance, when it comes to reducing Cd stress in maize, soil application of TiO_2-NPs is less active than foliar treatment (Lian et al. 2020). Si and Se-NPs mitigated Pb and Cd toxicity in rice through foliar spraying (Hussain et al. 2020).

In view of increasing HM pollution at the global level, this chapter will discuss the impacts of HM toxicity, the probable mechanisms of interaction between HMs and NPs, and the applications of diverse NPs to crops under HM toxicity, with a prime focus on the recent research carried on HM in the past decade.

24.2 Impacts of NPs on the HM Stress in Plants

24.2.1 Silicon

Silicon (Si) is the second most plentiful element in the earth's crust. It is not a crucial component for plant growth, but it is inseparable related to higher plants, particularly for *Gramineae* species (e.g. rice) (Gaur et al. 2020). Numerous researches have reported that Si may help crops to withstand abiotic stress (Mateos-Naranjo et al. 2015; Bhat et al. 2019). Until now, the broadly known mitigation of HM toxicity by Si has concluded the following underlying mechanisms: (i) regulation of HM transport genes expression (Cao et al. 2020), (ii) effecting as a nutrient source (Su et al. 2019), (iii) complexation and immobilization of HM that can decrease their biological effects (Liang et al. 2007), (iv) stimulation of the biochemical and physiological defense systems to enhance the ability of ROS scavenging (Emamverdian et al. 2018). For instance, foliar application of Si-NPs (2.5 mM) to rice seedlings grown in hydroponic solution supplemented with $CdCl_2$ (20 M) revealed to have smaller amounts of MDA and Cd content in rice shoots, signifying that Si-NPs could decrease Cd toxicity in rice seedlings (Wang et al. 2015). In a rice cell culture experiment, the amounts of living cells under Cd (40 M) treatment were induced significantly after the application of Si-NPs (1.0 mM) (Cui et al. 2017). This increase might be correlated with the reduction in the expression of Cd transport and uptake genes (*OsNramp5* and *OsLCT1*) followed by the induction of the Cd-vacuole transport gene (*OsHMA3*). The stress of Al was mitigated by the application of 4 $mg\,kg^{-1}$ Si-NPs in maize via triggering the antioxidant system; while the Al content did not change (de Sousa et al. 2019). Foliar spraying of 10 and 20 $mg\,l^{-1}$ Si-NPs efficiently improved the crop yield and reduced the accumulation of Cd in grains of rice (Hussain et al. 2020). Similarly, the addition of Si-NPs mitigated Cd stress (1.21 $mg\,kg^{-1}$) in wheat and enhanced growth (Khan et al. 2020).

Additionally, SiO_2-NPs also perform vital functions in mitigating HM stress in plants. For example, foliar spraying of 5 mM SiO_2-NPs reduced As content in rice grains by 40% and 28% in the open field and laboratory experiments, respectively (Liu et al. 2014). The application of SiO_2-NPs (0.1–10 mM) decreased the stress of As and the expression levels of gene *Lsi2* and *Lsi1* in suspension cells of rice (Cui et al. 2020).

24.2.2 Selenium

Selenium (Se) is a vital element for animals' growth and development, which has numerous significant biological functions, e.g. anticancer, antioxidant, antiviral, and immune regulation (Fairweather-Tait et al. 2011; Rayman 2012; Lv et al. 2014). Se is chiefly assimilated through the consumption of cereal crops; however, the amount of Se in most rice-producing nations is low (Dinh et al. 2018). Hence, the application of Se fertilizer to induce the fortification of Se in crops is important for humans.

Se, as a vital active component of glutathione peroxidase, may effectively eliminate the ROS accumulation, activating the activity of antioxidant enzymes, and thus efficiently improve the stress resistance of plants. For instance, it has been reported that Se-NPs can decrease Cd toxicity and absorption in rice (Lin et al. 2012; Chang et al. 2013; Hu et al. 2014). The influence of Se on rice Cd absorption may be linked to the production of iron filming on rice root surface (Huang et al. 2020). Se greatly decreased the content of Cd and facilitated the production of the iron film on rice root surface with the application of Fe^{2+}. The phytic acid content of brown rice decreased, and biomass, yield, and Se accumulation were induced in Pb and Cd polluted soil after the application of Se-NPs (Hussain et al. 2018). Both, foliar spraying with 0.5 and 1.0 $mmol\,l^{-1}$ concentrations of Se-NPs and soil addition with 0.5 and 2.5 $mg\,kg^{-1}$ concentrations of Se-NPs enhanced the biochemical attributes of rice in Cd-polluted soil, with soil application having worse influences on decreasing the Cd accumulation and improving the yield and chlorophyll in rice than foliar spraying (Li 2018). Under Cd toxicity (10 and 20 $mg\,kg^{-1}$), the MDA and Cd contents of leaves were decreased, and the height, biomass, chlorophyll, GSH-Px, and SOD contents were enhanced in Chinese cabbage after foliar application of 1.25 and 12.5 $mg\,l^{-1}$ Se-NPs (Zhang 2019).

24.2.3 Iron

Iron (Fe) is an important component of the development of plants and animals (Li et al. 2020). It is involved in a variety of biochemical and physiological processes, e.g. photosynthesis, respiration, and cell metabolism (Lin et al. 2019). For instance, Fe is an enzyme cofactor that is strictly linked to enzyme function (Briat et al. 2007). The mitigation mechanisms of Fe-NPs chiefly scavenge the ROS, stimulation of oxidative defense system, and adsorption of HM (Konat et al. 2017). Numerous researches have shown that Fe-NPs have a great possible application as fertilizers. For instance, the growth of legume root was improved by ~3.7 times by using low-dose pre-soaking Fe_2O_3-NPs (Palchoudhury et al. 2018). The starvation of Fe leads to reduced photosynthetic capacity and yellowing leaves, and the particular protein-chlorophyll multiplexes synthesis needs Fe in chloroplasts (Konate et al. 2018).

In sunflowers, nanoscale zero-valent iron (nZVI) has been found to enhance plant growth, induce POD and SOD activity in leaves, and decrease the HM content (Michálková et al. 2017). After the application of 100 $mg\,l^{-1}$ nZVI, gene expression of Fe transporters involved for both Cd and Fe uptake (*YSL2*, *YSL15*, *IRT1*, and *IRT2*) were meaningfully downregulated in rice plants. While the gene expression of *OsCAX4* and *OsVIT1* was upregulated in vacuoles and causes inhibition of Cd transport (Guha et al. 2020).

Due to the special electronic and structure characteristics, Fe-NPs are mostly powerful adsorbents. A magnetic field may easily separate magnetic NPs (Fe_3O_4 and Fe_2O_3 NPs) from the adsorbing medium (Lin et al. 2019). Fe_2O_3-NPs have been planned as a Fe-comprising fertilizer for peanuts while there is still a lot more work to be done to fully understand the possible threats. (Rui et al. 2016). Foliar and soil applications of 5, 10, 15, and 20 ppm Fe_2O_3-NPs under Cd stress (0.93 $mg\,kg^{-1}$) improved the dry weight and antioxidant system, and reduced the Cd content and leaf electrolyte leakage rate in wheat (Hussain et al. 2019). In addition, they reported that soil application of Fe-NPs is worse than foliar application because the absorption of Fe can be influenced by several features in soil, e.g. the mixture with other minerals and pH-guided absorption. When biochar was added to the soil, the effects of foliar spraying Fe-NPs on Cd toxicity were improved in rice (Rizwan et al. 2019a). Fe_3O_4 and nZVI (0–4 $mg\,l^{-1}$) mixed with hydroponic growth media decreased As translocation from roots to the shoots (Huang et al. 2018).

24.2.4 Zinc Oxide

Zinc (Zn) is an indispensable nutrient for both humans and plants. Zinc oxide (ZnO) NPs are among the most formed and engineered NPs, broadly utilized in paints, coatings, and personal care products. The mitigation mechanisms of ZnO-NPs contain improvement of biochemical and physiological roles, offering nutritional value and initiation of the oxidative stress system in plants. For instance, ZnO-NPs relieved the drought stress by inducing the potassium and nitrogen absorption in sorghum (Dimkpa et al. 2019). ZnO-NPs application enhanced the antioxidant system and decreased the leaves' electrolyte leakage rate in wheat leaves (Ali et al. 2019). ZnO-NPs ($50\,mg\,l^{-1}$) seed priming enhanced germination, root growth, cell wall biosynthesis, and embryo glycolytic metabolism (Sun et al. 2020).

For alleviation of Cd stress, only the ZnO-NPs application reduced the Cd content, while the accumulation of Cd was induced by Zn^{2+} in rice shoot. The content of As in rice roots and shoots was greatly decreased by ZnO-NPs and Zn^{2+} (Ma et al. 2020). The applications of 25, 50, and $100\,mg\,l^{-1}$ ZnO-NPs by soil addition and foliar spraying enhanced wheat photosynthesis, growth, and grain yield under Cd toxicity (Hussain et al. 2018). But the soil addition was less effective than foliar spraying in decreasing Cd accumulation in wheat grains and inducing grain yield. A suitable content of ZnO-NPs to develop the maize seeds improved maize resistance to the stress of Cd (Rizwan et al. 2019b). The application of ZnO-NPs alleviated the Cd toxicity with increasing yield, biomass, and photosynthesis in wheat (Khan et al. 2019). ZnO-NPs ($100\,mg\,l^{-1}$) mixed with hydroponic growth media reduced the As accumulation by 72% and 68% in rice roots, when treated with As(V) and As(III) in the laboratory, respectively (Wang et al. 2018); $100\,mg\,l^{-1}$ ZnO-NPs decreased the concentration of As in shoots (−60.2% and −80.0%) and roots (−39.5% and −83.3%) of rice in greenhouse (Ma et al. 2020). The application of $10–200\,mg\,l^{-1}$ ZnO-NPs decreased the As content by 10.2–56.6% and 8.4–72.3% in rice shoots and roots, respectively (Wu et al. 2020).

The combined application of biochar (soil addition) and ZnO-NPs (foliar application) efficiently mitigated the toxicity of Cd, which was more effective than the addition of biochar or ZnO-NPs only (Ali et al. 2019; Rizwan et al. 2019b; Bashir et al. 2020). These results propose that the joint addition of NPs with other supplements can serve as an effective method in HM decrease of plants.

24.2.5 Titanium Dioxide

Titanium dioxide (TiO_2) NPs, common metal oxide NPs, are extensively utilized in coatings, cosmetics, and pigments due to their favorable optical properties, high stability, low toxicity, and strong adhesive properties (Mueller and Nowack 2008). The chief mechanisms of TiO_2-NPs in mitigating HM toxicity rely on the stimulation of the oxidative defense system and HM adsorption. Numerous researches have reported that TiO_2-NPs addition decreases the plants' oxidative toxicity. For instance, TiO_2-NPs induced the activity of antioxidant enzymes in corn (Wang et al. 2010). The photosynthetic rate and physiological parameters of soybean were induced after adding TiO_2-NPs into the soil, and efficiently restrain the Cd stress on soybean (Singh and Lee 2016). Consequently, TiO_2-NPs have boundless applications probable in mitigating oxidative toxicity triggered by HM in plants.

For mitigation of Cd toxicity, TiO_2-NPs had little influence on the biomass of rice plants in hydroponic solutions (Ji et al. 2017). But the antioxidant enzyme activity, hormone level, plant height, root length, and other physiological indices were enhanced after TiO_2-NPs addition, demonstrating that TiO_2-NPs application mitigated the Cd toxicity to rice plants. The applications of 100 and 250 mg l^{-1} TiO_2-NPs were utilized by foliar spraying and soil addition to maize planted in Cd-polluted soil (Lian et al. 2020). Soil application of TiO_2-NPs improved the uptake of Cd and significantly reduced maize biomass; however, foliar spraying blocked Cd transport in maize and induced its biomass. It was also revealed that foliar addition of TiO_2-NPs enhanced alanine, aspartame acid, galactose, and other metabolic pathways and prompted the activities of GST and SOD to mitigate the Cd toxicity to maize. These findings showed that the mode of application had an important effect on mitigation practices. In addition, for the alleviation of As-induced damage, 100 mg l^{-1} TiO_2-NPs treatment reduced the As bioaccumulation by 40–90% and subsequently decreased its toxicity in rice plants (Wu et al. 2020).

The effect of sodium dodecylbenzene sulfonate (SDBS)-coated TiO_2-NPs and bare TiO_2-NPs were investigated in Cd-exposed wheat by Dai et al. (2019). The hydrodynamic diameter of TiO_2-NPs was decreased, and the dispersion stability of TiO_2-NPs was enhanced by electrostatic and spatial repulsion after SDBS coating, thus inducing the adsorption ability of Cd^{2+} by exposing more accessible adsorption sites. They also reported that the influence of TiO_2-NPs with SDBS on alleviating the stress damage of Cd was greatly better than those of uncoated TiO_2-NPs. These results showed that surfactants or other materials-coated NPs can influence HM toxicity and bioavailability in a better way.

TiO_2-NPs addition efficiently induced the 133Cs content in soybean in 133Cs-polluted soil (Singh and Lee 2018). It proposes that NPs can be used for phytoremediation while lasting open field-depended research is required.

24.2.6 Cerium Dioxide

CeO_2-NPs have valuable properties and general utilization in agriculture systems and are light rare earth elements (Zhang et al. 2017). CeO_2-NPs can mitigate the toxicity of HM because of the dangling Ce^{4+} and Ce^{3+} links on the surface, which diminish the oxidative stress and scavenge ROS produced by HM (Dutta et al. 2006; Wu et al. 2017). For instance, when compared with the positively charged CeO_2-NPs, negatively charged CeO_2-NPs efficiently eliminated ROS accumulated in the chloroplasts of *Arabidopsis*, but the removal efficiency was greatly influenced by the Ce^{4+}/Ce^{3+} ratio (Wu et al. 2017). According to the experts, the modest accumulation of CeO_2-NPs is useful to plant development. For instance, 100 mg kg^{-1} CeO_2-NPs with polyvinylpyrrolidone (PVP) (coating or uncoating) enhanced plant growth and induced the photosynthesis rate, while the higher concentration of CeO_2-NPs (500 mg kg^{-1}) decreased the net photosynthesis rate in soybeans (Cao et al. 2017). The application of 10 and 100 mg l^{-1} bulk CeO_2-NPs enhanced *Brassica rapa* biomass, but CeO_2-NPs showed no significant influence at the same concentration (Ma et al. 2014). These effects might be explained in terms of appropriate doses of CeO_2-NPs and their formulation of aggregates. These results signposted that there were still numerous unknowns concerning the consequence of NP surface alterations that powerfully influenced plant growth via the mobility and stability of NPs.

For the alleviation of Cd damage, foliar spraying of 500 mg kg^{-1} CeO_2-NPs did not influence on the transport of Cd, but greatly enhanced the Ce content in soybean tissues grown in Cd-polluted soil (0.25 and 1.0 mg kg^{-1}) (Rossi et al. 2017). This was partially produced by the co-culture of CeO_2-NPs and Cd that affected the apoplastic barriers of the root. The addition of 200 mg l^{-1} CeO_2-NPs effectively blocked the Cd uptake and triggered the activity of antioxidant enzymes in rice seedlings planted in 50 M $CdCl_2$ hydroponic solutions (Wang et al. 2019b). For mitigation of As stress, 100 mg l^{-1} CeO_2-NPs had no significant change in total As(III) and As(V) contents in hydroponically grown rice (Wang et al. 2018). In summary, we advocate that more in-depth research is required to comprehend the mitigation mechanisms of NPs in plants, as some contradictory results were also obtained by the research raising concern about their potential use in sustainable agriculture.

24.3 Mechanisms of NPs to Mitigate the Toxicity of HM

The biochemistry, physiology, and morphology of plants have been revealed to be affected by the toxicity of HM. There are numerous approaches utilized to mitigate the toxicity of HM in plants (Figure 24.1). For instance, decreasing the accumulation of bioavailable HM in the hydroponics or soils (Moharem et al. 2019), improving the antioxidant systems and other physiological attributes in plants (Wang et al. 2020a), enhancing the creation of protection agents (e.g. phytochelatins, root exudates, and organic acids) (Cao et al. 2020; Lian et al. 2020; Sun et al. 2020), and controlling the translocation of HM

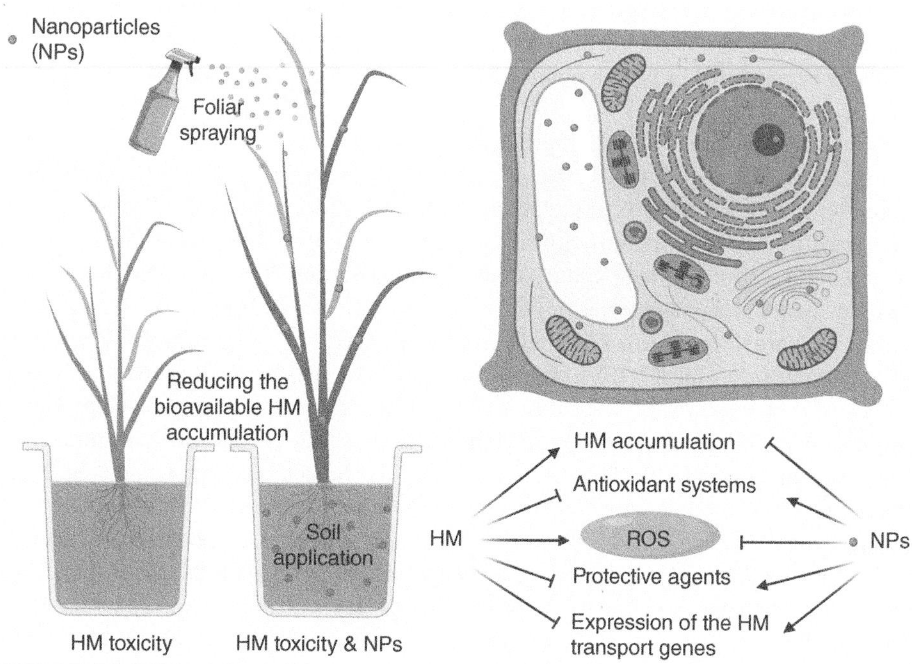

Figure 24.1 Numerous approaches are utilized to mitigate the toxicity of HM in plants.

uptake and identification of associated genes in crops (Sun et al. 2020). For a more intuitive understanding of the effects of NPs on HM (As and Cd) toxicity and accumulation in plants, we choose rice (*O. sativa* L.) as an example to summarize HM effects and presented them in Table 24.1.

The HM can be transformed and absorbed by the NPs in the soil by decreasing their bioavailability and mobility. For instance, the fluidity of Cd and other HMs was reduced by the application of Fe_3O_4-NPs in the soil (Sebastian et al. 2019). The formation of Cd was transformed by mercapto Si-NPs into a steadier constituent after 36 months in the open-field experiment (Wang et al. 2020b). In addition, diverse NPs enhance the properties of soil, e.g. hydroxyapatite NPs can induce soil pH and release phosphate, which decreases the unfavorable influences of HMs in soil (Cui et al. 2018).

A huge number of NPs adsorb to the cell walls and bind with HM, rendering them unusable by forming complexes. These complexes turned out to be accumulated on the cell surfaces (Wang et al. 2015; Cui et al. 2017), thus the biological activity and movement of HM are decreased in plants. Furthermore, the organic acids accumulating in the plant cell walls may be chelated with HM to minimize their toxicity to plants. NPs have been demonstrated to boost the creation of structural protective agents, e.g. exogenous Si-NPs encouraged the synthesis of organic acids, thus decreasing Cd stress in plants.

The biological role of apoplastic barriers on plant roots is to monitor the passage of ions, oxygen, and water while also protecting the plants' development (Yang et al. 2013). NPs affected the development of apoplastic barriers that reduces the accumulation of HM in the roots (Rossi et al. 2017). But the modification of apoplastic barriers cannot be a positive approach to decrease the toxicity of HMs, as plant root includes numerous transporter proteins and ions in the plasma membrane of the root cell, which can significantly translocate HM at the same time. Furthermore, the expression level of several metal transport genes can be regulated by NPs in plants, thus improving the extracellular barriers of plants to interrupt the transport of HM.

Stimulating the plant's oxidation protective mechanism is another way to decrease HM toxicity. Plants create ROS under various metabolic processes (Wu et al. 2017). For instance, plants continually create ROS in the chloroplasts and other portions of cells during metabolic activities including photosynthesis and respiration. ROS function as chemical messengers in protection, growth, and development at low levels. Increased ROS buildup, on the other hand, is damaging the proteins, cell membranes, and other biological components, which are further provoked under HM stress (Wu et al. 2017). Antioxidant enzymes, e.g. glutathione peroxidase (GPX), glutathione reductase (GR), ascorbate peroxidase (APX), catalase (CAT), superoxide peroxidase (POD), and dismutase (SOD), scavenge ROS in plants. Nonenzymatic low-molecular-weight metabolites, e.g. polyphenols, vitamin E, and vitamin C, can also scavenge ROS (Zhang et al. 2017; Zhang et al. 2018; Ullah et al. 2020). Under stress, metabolic pathways associated with ROS elimination will be activated. For example, the oxidative stress of plant can be mitigated by the metabolism of ascorbate, aspartic acid, alanine, and galactose and the biosynthesis of shikimate-phenylpropanoid (Zhang et al. 2018; Lian et al. 2020). In general, using NPs with antioxidant enzyme activity (e.g. C_{60}, Mn_3O_4 NPs, Fe_3O_4 NPs, and CeO_2 NPs) enhance their capacity to decrease ROS and reduce yield and growth losses in crops (Konat et al. 2017; Wu et al. 2017; Yao et al. 2018; Wang et al. 2019a).

Table 24.1 Effect of nanoparticles on Cd and As toxicity and accumulation in rice plant.

HM stress	Nanoparticle	Concentration	Experimental procedure	Growth medium	Applicatio mode	Consequences	References
Cd	Si	2.5 mM	Laboratory	Hydroponic	Foliar spraying	Increased mineral elements, GSH content, and POD and SOD activities, decreased Cd contents, and CAT activity in shoots	Wang et al. (2015)
Cd	Se	0.5–2.5 $mg\,kg^{-1}$	Greenhouse	Soil	Soil addition	Induced the biomass, stomatal conductance, and chlorophyll, reduced the accumulation of Cd in grains and leaves and the exchangeable Cd in soil	Li (2018)
Cd	ZnO	0–100 $mg\,l^{-1}$	Laboratory	Hydroponic	Foliar spraying	Induced root fresh weight, root-shoot length, and the modulations in physiological and biochemical attributes	Li et al. (2021)
Cd	TiO_2	10–1000 $mg\,l^{-1}$	Laboratory	Hydroponic	Mixed with growth media	Increased net photosynthetic rate, chlorophyll, and the activities of SOD and POD, reduced MDA and Cd accumulation in roots and leaves	Ji et al. (2017)
Cd	CeO_2	200 $mg\,l^{-1}$	Laboratory	Hydroponic	Mixed with growth media	Improved the rice growth indicators and photosynthesis, reduced the activities of SOD and POD	Wang et al. (2019b)
As	SiO_2	5 mM	Greenhouse and open field	Hydroponic and soil	Foliar spraying	Decreased total As in the grains by 28% and 40% for pot and field experiments, respectively	Liu et al. (2014)
As	SiO_2	0.1–10 mM	Laboratory	Soil	Cell suspension	Decreased As toxicity in rice suspension cells, decreased expression levels of Lsi1 and Lsi2 genes	Cui et al. (2020)

(*Continued*)

Table 24.1 (Continued)

HM stress	Nanoparticle	Concentration	Experimental procedure	Growth medium	Applicatio mode	Consequences	References
As	Fe_3O_4 and zero-valent iron	0–4 mg l^{-1}	Laboratory	Hydroponic	Mixed with growth media	Decreased As transport from the roots to the shoots	Huang et al. (2018)
As	ZnO	100 mg l^{-1}	Laboratory	Hydroponic	Mixed with growth media	Reduced total As content in rice roots by 72% and 68% after exposure to As(III) and As(V), respectively	Wang et al. (2018)
As	ZnO	10–200 mg l^{-1}	Laboratory	Hydroponic	Mixed with growth media	Decreased total As concentrations reduced by 8.4–72.3% and 10.2–56.6%, respectively, in rice roots and shoots	Wu et al. (2020)
As	ZnO	100 mg l^{-1}	Greenhouse	Hydroponic	Mixed with growth media	Decreased total As accumulation in rice roots (−39.5% and −83.3%) and shoots (−60.2% and −80.0%)	Ma et al. (2020)
As	TiO_2	0–100 mg l^{-1}	Laboratory	Hydroponic	Mixed with growth media	Reduced As toxicity and decreased arsenic bioaccumulation in rice by 40-90%	Wu et al. (2020)
As	CeO_2	100 mg l^{-1}	Laboratory	Hydroponic	Mixed with growth media	No significant changed in total As content in rice	Wang et al. (2018)
As	CuO	0.1–100 mg l^{-1}	Greenhouse	Soil	Soil addition	Decreased total As accumulation by 35% in dehusked grains	Liu et al. (2018)

24.4 Summary and Prospect

The first research on the application of NPs to remediate HM-polluted soil appeared in 2010. The early study mostly focused on the utilization of NPs as a supplement for HM recovery in the soil, as well as their adsorption ability. Cd-related studies were initially published in 2016. HM contamination in soil occurs in the complex formulation of numerous metals, not just a single element, in real-world situations. As a result, in past few years, experts have begun to analyze the compound contamination of HM to acquire a better understanding of the true environmental problem. Thus, more comprehensive approaches are required to handle HM contamination in a holistic way to pinpoint the real-world problem. Later, research in biochemistry and physiology of plants grew and discovered that the application of NPs might boost antioxidant enzyme activity of plants, lowering HM levels and oxidative toxicity, and improving plant growth. The efficacy of biochar, phytoremediation, and or other combination treatments, can be improved with the application of NPs, as they considerably boost the adsorption of HMs and improve resistance toward HMs in plants. Furthermore, research into the use of NPs in combination with microbes, and other novel materials has exploded in past decades, demonstrating a brilliance of NP in real-world problems.

This chapter evaluated the studies advocating the utilization of several NPs to increase the resistance to HM damage by foliar application, seed priming, solution application, and soil addition in plants. Foliar spraying and soil additions allow NPs to get into plant leaves and roots via the pathways of apoplast and symplast, respectively (Figure 24.2).

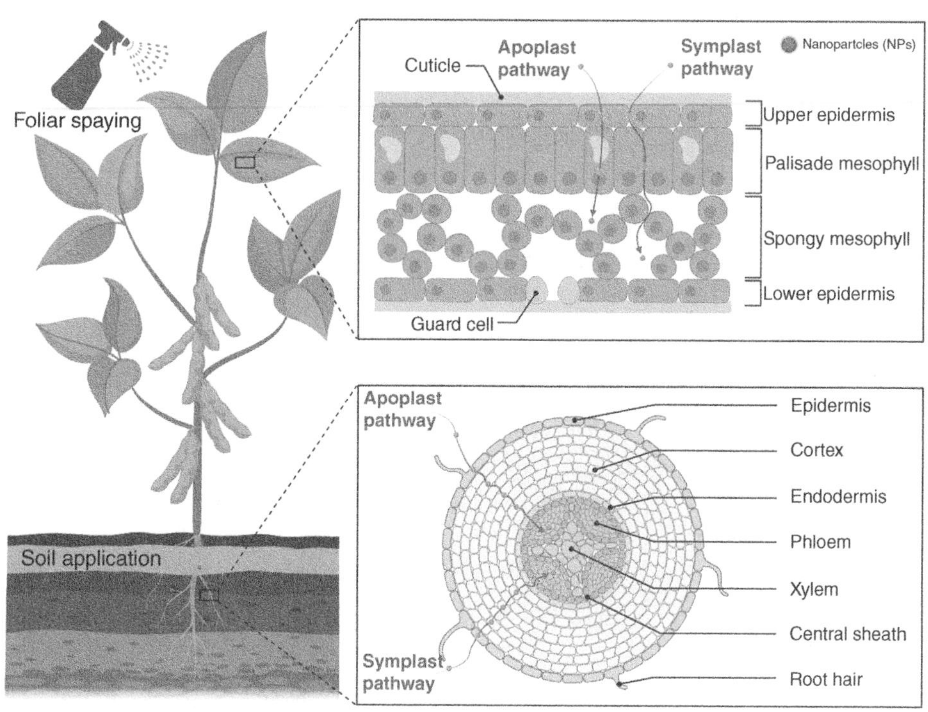

Figure 24.2 Nanoparticles enter plants in different ways.

HM stress has a major impact on plant biochemistry, morphology, and physiology. Inducing the capability of apoplastic barriers to block HMs, decreasing the amount of bio-available HM in soil, supplying more nutrition, improving the biological function of organs, enhancing the capability of antioxidant enzymes, increasing the production of structural protective agents, and regulating the expression of HM transport genes in plants are all important approaches for enhancing HM resistance (e.g. organic acids, phytochelatins, and root exudates). We also created a schematic model of putative NP pathways to reduce HM stress in plant cells (Figure 24.3).

Diverse levels of NPs exposure and application techniques have varying influences on plant development in studies on alleviating HM toxicity. The low dose has no meaningful impact, and a high dosage would have detrimental effects and rise superfluous expenditures on plants, which is a serious concern for modern agriculture. Furthermore, there are still a lot of questions about how to apply NPs in the soil environment that need to be answered. As a result, more research into the causes and therapeutic effects of various modalities is required. Additionally, combining various types of NPs with other materials is unquestionably a novel study avenue. Application of NPs with other alleviators such as auxin, biochar, nitric oxide, hydrogen peroxide, and HM-resistant microorganisms, for example, should be studied.

Numerous forms of NPs, containing surface-coatings or mixed with other materials, are becoming accessible in the market as technology and science improve. These NPs can enrich and increase their present functionalities. Furthermore, the physicochemical feature of NPs (charge or size) that govern how they interact with HM in the soil with crop plants need to be investigated further. Since plant cell walls have a porosity of 15 nm, some

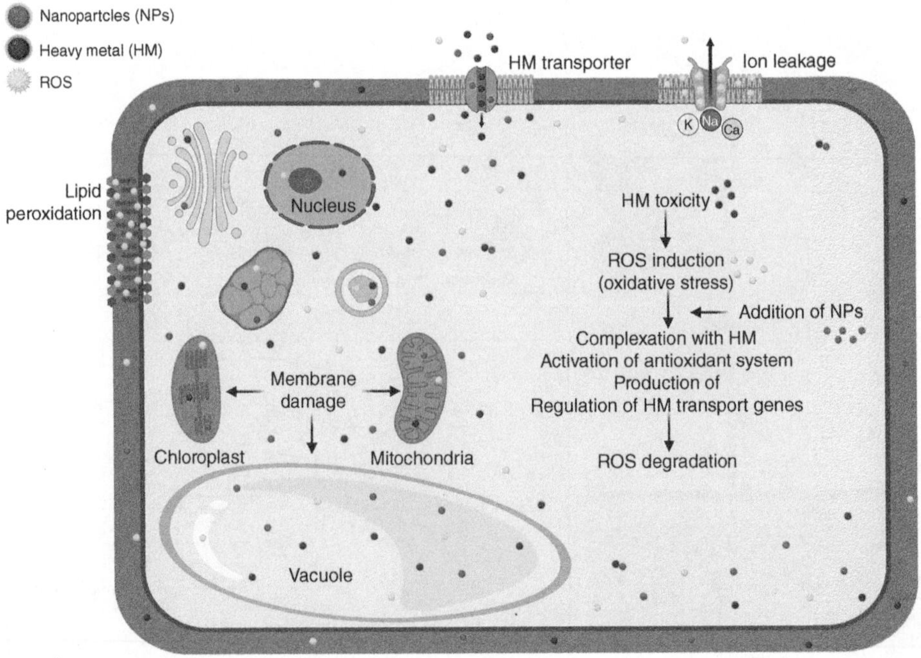

Figure 24.3 A schematic model of putative NP pathways to reduce HM stress in plant cells.

NPs may pass through (Wu et al. 2017). On the other hand, neutral or positive NPs can permeate the membrane less efficiently than negatively charged NPs (Wong et al. 2016). Various studies on the negative effects of NPs in plants and animals have been published, but a large amount of them have been conducted with single NPs, and further deeper research on complex formulation of NPs is required.

Environmental science and other disciplines are already incorporating artificial intelligence and big data technologies. For instance, artificial intelligence was employed to forecast the level of Cd and CeO_2-NPs in *Brassica napus* L. based on physiological factors of the plant (Rossi et al. 2019). With the advancement of technology and the collection of experimental data, it is expected that the content of HM in plants, as well as many physiological characteristics of plants, can be predicted depending on NPs properties, soil properties, and plant types. It is feasible to choose the right levels and kinds of NPs for maximizing the positive influence and decreasing the needless expenditures.

At last, one of the most important constraints limiting NPs' broader implementation in the agricultural system is their cost. As a result, effectively manufacturing NPs using green technologies at low cost is the essential aspect of nanotechnology's future growth. Investigators must undertake further studies to appreciate how NPs might increase plants' resistance, along with the consideration of NPs effects on the terrestrial environment. Furthermore, combining NPs with computer oriented programs/algorithms will allow us to more precisely assess the environmental influence of NPs, which will further be allowing us to use NPs as a tool for monitoring HM toxicity.

References

Adeel, M., Lee, J.Y., Zain, M. et al. (2019). Cryptic footprints of rare earth elements on natural resources and living organisms. *Environ Int* 127: 785–800.

Adeel, M., Jin, T., Hussain, T. et al. (2020). Bioaccumulation of ytterbium oxide nanoparticles insinuate oxidative stress, inflammatory, and pathological lesions in ICR mice. *Environ Sci Pollut Res* 27: 32944–32953.

Adeel, M., Farooq, T., White, J. et al. (2021). Carbon-based nanomaterials suppress Tobacco Mosaic Virus (TMV) infection and induce resistance in *Nicotiana benthamian*. *J Hazard Mater* 404: 124167.

Ali, S., Rizwan, M., Noureen, S. et al. (2019). Combined use of biochar and zinc oxide nanoparticle foliar spray improved the plant growth and decreased the cadmium accumulation in rice (*Oryza sativa* L.) plant. *Environ Sci Pollut Res* 26: 11288–11299.

Bashir, A., Rizwan, M., Ali, S. et al. (2020). Effect of composted organic amendments and zinc oxide nanoparticles on growth and cadmium accumulation by wheat; a life cycle study. *Environ Sci Pollut Res* 27: 23926–23936.

Bhat JA, Shivaraj SM, Singh P, Navadagi DB, Tripathi DK, Dash PK, Solank A.U, Sonah H, Deshmukh R. (2019) Role of silicon in mitigation of heavy metal stresses in crop plants. *Plants* 8: 71

Bienert GP, Chaumont F. (2011) Plant aquaporins: roles in water homeostasis, nutrition, and signaling processes. In: Markus Geisler, Kees Venema Transporters and Pumps in Plant Signaling. 7 Berlin, Heidelberg: Springer 3–36. doi: https://doi.org/10.1007/978-3-642-14369-4_1

Briat, J., Curie, C., and de Ric, G.F. (2007). Iron utilization and metabolism in plants. *Curr Opin Plant Biol* 10: 276–282.

Cao, Z., Stowers, C., Rossi, L. et al. (2017). Physiological effects of cerium oxide nanoparticles on the photosynthesis and water use efficiency of soybean (*Glycine max* (L.) Merr.). *Environ Sci-Nano* 4: 1086–1094.

Cao, F., Dai, H., Hao, P.F., and Wu, F. (2020). Silicon regulates the expression of vacuolar H(+)-pyrophosphatase 1 and decreases cadmium accumulation in rice (*Oryza sativa* L.). *Chemosphere* 240: 124907.

Chandrakar, V., Naithani, S.C., and Keshavkant, S. (2016). Arsenic-induced metabolic disturbances and their mitigation mechanisms in crop plants: a review. *Biologia* 71 (4): 367–377.

Chang, H., Zhou, X., Wang, W. et al. (2013). Effects of selenium application in soil on formation of iron plaque outside roots and cadmium uptake by rice plants. In: *Advanced Materials Research*, vol. 750 (ed. J. Bu and Y.H. Kim), 1573–1576. Zurich, Switzerland: Trans Tech Publications Ltd.

Cui, J., Liu, T., Li, F. et al. (2017). Silica nanoparticles alleviate cadmium toxicity in rice cells: mechanisms and size effects. *Environ Pollut* 228: 363–369.

Cui, H., Shi, Y., Zhou, J. et al. (2018). Effect of different grain sizes of hydroxyapatite on soil heavy metal bioavailability and microbial community composition. *Agric Ecosyst Environ* 267: 165–173.

Cui, J., Li, Y., Jin, Q., and Li, F. (2020). Silica nanoparticles inhibit arsenic uptake into rice suspension cells via improving pectin synthesis and the mechanical force of the cell wall. *Environ Sci Nano* 7 (1): 162–171.

Dai, C., Shen, H., Duan, Y. et al. (2019). TiO_2 and SiO_2 nanoparticles combined with surfactants mitigate the toxicity of Cd^{2+} to wheat seedlings. *Water Air Soil Pollut* 230: 232.

Dimkpa, C.O., Singh, U., Bindraban, P.S. et al. (2019). Zinc oxide nanoparticles alleviate drought-induced alterations in sorghum performance, nutrient acquisition, and grain fortification. *Sci Total Environ* 688: 926–934.

Dinh, Q.T., Cui, Z., Huang, J. et al. (2018). Selenium distribution in the Chinese environment and its relationship with human health: a review. *Environ Int* 112: 294–309.

Dutta, P., Pal, S., Seehra, M.S. et al. (2006). Concentration of Ce^{3+} and oxygen vacancies in cerium oxide nanoparticles. *Chem Mater* 18: 5144–5146.

Emamverdian, A., Ding, Y., Xie, Y., and Sangari, S. (2018). Silicon mechanisms to ameliorate heavy metal stress in plants. *BioMed Res Int* 10: 8492898.

Fairweather-Tait, S.J., Bao, Y., Broadley, M.R. et al. (2011). Selenium in human health and disease. *Antioxid Redox Signal* 14: 1337–1383.

Gaur, S., Kumar, J., Kumar, D. et al. (2020). Fascinating impact of silicon and silicon transporters in plants: a review. *Ecotoxicol Environ Saf* 202: 110885.

Guha, T., Barman, S., Mukherjee, A., and Kundu, R. (2020). Nano-scale zero valent iron modulates Fe/Cd transporters and immobilizes soil Cd for production of Cd free rice. *Chemosphere* 260: 127533.

Hao, Y., Fang, P., Ma, C. et al. (2019). Engineered nanomaterials inhibit *Podosphaera pannosa* infection on rose leaves by regulating phytohormones. *Environ Res* 170: 1–6.

Hu, Y., Norton, G.J., Duan, G. et al. (2014). Effect of selenium fertilization on the accumulation of cadmium and lead in rice plants. *Plant Soil* 384: 131–140.

Huang, Q., Liu, Q., Lin, L. et al. (2018). Reduction of arsenic toxicity in two rice cultivar seedlings by different nanoparticles. *Ecotoxicol Environ Saf* 159: 261–271.

Huang, G., Ding, C., Li, Y. et al. (2020). Selenium enhances iron plaque formation by elevating the radial oxygen loss of roots to reduce cadmium accumulation in rice (*Oryza sativa* L.). *J Hazard Mater* 398: 122860.

Hussain, A., Ali, S., Rizwan, M. et al. (2018). Zinc oxide nanoparticles alter the wheat physiological response and reduce the cadmium uptake by plants. *Environ Pollut* 242: 1518–1526.

Hussain, A., Ali, S., Rizwan, M. et al. (2019). Responses of wheat (*Triticum aestivum*) plants grown in a Cd contaminated soil to the application of iron oxide nanoparticles. *Ecotoxicol Environ Saf* 173: 156–164.

Hussain, B., Lin, Q., Hamid, Y. et al. (2020). Foliage application of selenium and silicon nanoparticles alleviates Cd and Pb toxicity in rice (*Oryza sativa* L.). *Sci Total Environ* 712: 136497.

Irshad, H.M.K., Noman, A., Alhaithloul, H. et al. (2020). Goethite-modified biochar ameliorates the growth of rice (*Oryza sativa* L.) plants by suppressing Cd and As-induced oxidative stress in Cd and As co-contaminated paddy soil. *Sci Total Environ* 717: 137086.

Javaid, S. (2020). Heavy metals stress, mechanism and remediation techniques in rice (*Oryza sativa* L.): a review. *Pure Appl Biol* 9 (1): 403–426.

Ji, Y., Zhou, Y., Ma, C. et al. (2017). Jointed toxicity of TiO_2 NPs and Cd to rice seedlings: NPs alleviated Cd toxicity and Cd promoted NPs uptake. *Plant Physiol Biochem* 110: 82.

Jiang, M., Jiang, J., Li, S. et al. (2020). Glutamate alleviates cadmium toxicity in rice via suppressing cadmium uptake and translocation. *J Hazard Mater* 384: 121319.

Jiang, M., Song, Y., Kanwar, M.K. et al. (2021). Phytonanotechnology applications in modern agriculture. *J Nanobiotech* 19 (1): 430.

Kah, M., Tufenkj, N., and White, J.C. (2019). Nano-enabled strategies to enhance crop nutrition and protection. *Nat Nanotechnol* 14: 532–540.

Kaur, R., Das, S., Bansal, S. et al. (2021). Heavy metal stress in rice: uptake, transport, signaling, and tolerance mechanisms. *Physiol Plant* 173 (1): 430–448.

Khan, Z.S., Rizwan, M., Hafeez, M. et al. (2019). The accumulation of cadmium in wheat (*Triticum aestivum*) as influenced by zinc oxide nanoparticles and soil moisture conditions. *Environ Sci Pollut Res* 26: 19859–19870.

Khan, Z.S., Rizwan, M., Hafeez, M. et al. (2020). Effects of silicon nanoparticles on growth and physiology of wheat in cadmium contaminated soil under different soil moisture levels. *Environ Sci Pollut Res* 27: 4958–4968.

Konat, A., He, X., Zhang, Z. et al. (2017). Magnetic (Fe_3O_4) Nanoparticles reduce heavy metals uptake and mitigate their toxicity in wheat seedling. *Sustainability* 9: 790.

Konate, A., Wang, Y., He, X. et al. (2018). Comparative effects of nano and bulk-Fe_3O_4 on the growth of cucumber (*Cucumis sativus*). *Ecotoxicol Environ Saf* 165: 547–554.

Li, M. (2018). Effect of Selenium application on Cadmium accumulation and in Rice (*Oryza sativa*). Master's Thesis. Guangxi University, Nanning, China.

Li, J., Hu, J., Ma, C. et al. (2016). Uptake, translocation and physiological effects of magnetic iron oxide (–Fe_2O_3) nanoparticles in corn (*Zea mays* L.). *Chemosphere* 159: 326–334.

Li, M., Adeel, M., Peng, Z., and Yukui, R. (2020). Physiological impacts of zero valent iron, Fe_3O_4 and Fe_2O_3 nanoparticles in rice plants and their potential as Fe fertilizers. *Environ Pollut* 269: 116134.

Li, Y., Liang, L., Li, W. et al. (2021). ZnO nanoparticle-based seed priming modulates early growth and enhances physio-biochemical and metabolic profiles of fragrant rice against cadmium toxicity. *J Nanobiotech* 19 (1): 75.

Lian, J., Zhao, L., Wu, J. et al. (2020). Foliar spray of TiO_2 nanoparticles prevails over root application in reducing Cd accumulation and mitigating Cd-induced phytotoxicity in maize (*Zea mays* L.). *Chemosphere* 239: 124794.

Liang, Y., Sun, W., Zhu, Y., and Christie, P. (2007). Mechanisms of silicon-mediated alleviation of abiotic stresses in higher plants: a review. *Environ Pollut* 147: 422–428.

Lin, L., Zhou, W., Dai, H. et al. (2012). Selenium reduces cadmium uptake and mitigates cadmium toxicity in rice. *J Hazard Mater* 235: 343–351.

Lin, J., He, F., Su, B. et al. (2019). The stabilizing mechanism of cadmium in contaminated soil using green synthesized iron oxide nanoparticles under long-term incubation. *J Hazard Mater* 379: 120832.

Liu, C., Wei, L., Zhang, S. et al. (2014). Effects of nanoscale silica sol foliar application on arsenic uptake, distribution, and oxidative damage defense in rice (*Oryza sativa* L.) under arsenic stress. *RSC Adv* 4: 57227–57234.

Liu, J., Simms, M., Song, S. et al. (2018). Physiological effects of copper oxide nanoparticles and arsenic on the growth and life cycle of rice (*Oryza sativa* japonica 'Koshihikari'). *Environ Sci Technol* 52 (23): 13728–13737.

Lombi, E., Donner, E., Dusinska, M., and Wickson, F. (2019). A One Health approach to managing the applications and implications of nanotechnologies in agriculture. *Nat Nanotechnol* 14: 523–531.

Lowry, G.V., Avellan, A., and Gilbertson, L.M. (2019). Opportunities and challenges for nanotechnology in the agri-tech revolution. *Nat Nanotechnol* 14: 517–522.

Lv, Y., Yu, T., Yang, Z. et al. (2014). Constraint on selenium bioavailability caused by its geochemical behavior in typical Kaschin-Beck disease areas in Aba, Sichuan Province of China. *Sci Total Environ* 493: 737–749.

Ma, Y., Zhang, P., Zhang, Z. et al. (2014). Origin of the different phytotoxicity and biotransformation of cerium and lanthanum oxide nanoparticles in cucumber. *Nanotoxicology* 9: 262–270.

Ma, X., Sharifan, H., Dou, F., and Sun, W. (2020). Simultaneous reduction of arsenic (As) and cadmium (Cd) accumulation in rice by zinc oxide nanoparticles. *Chem Eng J* 384: 123802.

Mani A, Sankaranarayanan K. (2018) Heavy metal and mineral element induced abiotic stress in rice plant. In: Farooq Shah, Zafar Hayat Khan and Amjad Iqbal Rice Crop: Current Developments, p. London, UK: Intech Open 149. doi: https://doi.org/10.5772/intechopen.76080

Mateos-Naranjo, E., Gallé, A., Florez-Sarasa, I. et al. (2015). Assessment of the role of silicon in the Cu-tolerance of the C4 grass *Spartina densiflora*. *J Plant Physiol* 178: 74–83.

Mawia, A.M., Hui, S., Zhou, L. et al. (2021). Inorganic arsenic toxicity and alleviation strategies in rice. *J Hazard Mater* 408: 124751.

Michálková, Z., Fernández, D.M., and Komárek, M. (2017). Interactions of two novel stabilizing amendments with sunflower plants grown in a contaminated soil. *Chemosphere* 186: 374–380.

Moharem, M., Elkhatib, E., and Mesalem, M. (2019). Remediation of chromium and mercury polluted calcareous soils using nanoparticles: sorption-desorption kinetics, speciation and fractionation. *Environ Res* 170: 366–373.

Mueller, N.C. and Nowack, B. (2008). Exposure modeling of engineered nanoparticles in the environment. *Environ Sci Technol* 42: 4447–4453.

Mueller, N.D., Gerber, J.S., Johnston, M. et al. (2012). Closing yield gaps through nutrient and water management. *Nature* 490: 254–257.

Palchoudhury, S., Jungjohann, K.L., Weerasena, L. et al. (2018). Enhanced legume root growth with pre-soaking in Fe_2O_3 nanoparticle fertilizer. *RSC Adv* 8: 24075–24083.

Rahman, Z. and Singh, V.P. (2019). The relative impact of toxic heavy metals (THMs) (arsenic (As), cadmium (Cd), chromium (Cr)(VI), mercury (Hg), and lead (Pb)) on the total environment: an overview. *Environ Monit Assess* 191 (7): 419.

Rayman, M.P. (2012). Selenium and human health. *Lancet* 379: 1256–1268.

Rizwan, M., Noureen, S., Ali, S. et al. (2019a). Influence of biochar amendment and foliar application of iron oxide nanoparticles on growth, photosynthesis, and cadmium accumulation in rice biomass. *J Soils Sediments* 19: 3749–3759.

Rizwan, M., Ali, S., Rehman, M.Z.U. et al. (2019b). Alleviation of cadmium accumulation in maize (Zea mays L.) by foliar spray of zinc oxide nanoparticles and biochar to contaminated soil. *Environ Pollut* 248: 358–367.

Rossi, L., Zhang, W., Schwab, A.P., and Ma, X. (2017). Uptake, uptake, accumulation, and in planta distribution of coexisting cerium oxide nanoparticles and cadmium in *Glycine max* (L.) Merr. *Environ Sci Technol* 51: 12815–12824.

Rossi, L., Bagheri, M., Zhang, W. et al. (2019). Using artificial neural network to investigate physiological changes and cerium oxide nanoparticles and cadmium uptake by *Brassica napus* plants. *Environ Pollut* 246: 381–389.

Rui, M., Ma, C., Hao, Y. et al. (2016). Iron oxide nanoparticles as a potential iron fertilizer for peanut (*Arachis hypogaea*). *Front Plant Sci* 7: 815.

Rui, M., Ma, C., Tang, X. et al. (2017). Phytotoxicity of silver nanoparticles to peanut (*Arachis hypogaea* L.): physiological responses and food safety. *ACS Sustain Chem Eng* 5: 6557–6567.

Rui, M., Ma, C., White, J.C. et al. (2018). Metal oxide nanoparticles alter peanut (*Arachis hypogaea* L.) physiological response and reduce nutritional quality: a life cycle study. *Environ Sci Nano* 5: 2088–2102.

Sebastian, A., Nangia, A., and Prasad, M.N.V. (2019). Cadmium and sodium adsorption properties of magnetite nanoparticles synthesized from *Hevea brasiliensis* Muell. Arg. bark: relevance in amelioration of metal stress in rice. *J Hazard Mater* 371: 261–272.

Shahid, M., Dumat, C., Khalid, S. et al. (2017). Foliar heavy metal uptake, toxicity and detoxification in plants: a comparison of foliar and root metal uptake. *J Hazard Mater* 325: 36–58.

Singh, J. and Lee, B. (2016). Influence of nano-TiO_2 particles on the bioaccumulation of Cd in soybean plants (*Glycine max*): a possible mechanism for the removal of Cd from the contaminated soil. *J Environ Manag* 170: 88–96.

Singh, J. and Lee, B. (2018). Effects of Nano-TiO_2 particles on bioaccumulation of Cs-133 from the contaminated soil by soybean (*Glycine max*). *Process Saf Environ* 116: 301–311.

Song, Y., Jiang, M., Zhang, H., and Li, R. (2021a). Zinc oxide nanoparticles alleviate chilling stress in rice (*Oryza sativa* L.) by regulating antioxidative system and chilling response transcription factors. *Molecules* 26 (8): 2196.

Song, Y., Wang, B., Qiu, D. et al. (2021b). Melatonin enhances metallic oxide nanoparticle stress tolerance in rice via inducing tetrapyrrole biosynthesis and amino acid metabolism. *Environ Sci: Nano* 8: 2310–2323.

de Sousa, A., Saleh, A.M., Habeeb, T.H. et al. (2019). Silicon dioxide nanoparticles ameliorate the phytotoxic hazards of aluminum in maize grown on acidic soil. *Sci Total Environ* 693: 133636.

Stohs, S.J. and Bagchi, D. (1995). Oxidative mechanisms in the toxicity of metal ions. *Free Radical Bio Med* 18 (2): 321–336.

Su, M., Xu, J., Zhang, L. et al. (2019). Research progress on effects of silicon on plant growth under cadmium stress. *Guangzhou Chem Ind* 47: 41–43.

Sun, L., Song, F., Li, X. et al. (2020). Effects of ZnO nanoparticles on seed germination and root carbon metabolism in maize (*Zea mays* L.). *Soils Crop* 9: 40–49.

Tchounwou, P.B., Yedjou, C.G., Patlolla, A.K., and Sutton, D.J. (2012). Heavy metal toxicity and the environment. *Exp Suppl* 2101: 133–164.

Toumey, C. (2020). Notes on environmental nanoscience. *Nat Nanotechnol* 15: 250–251.

Uchimiya, M., Bannon, D., Nakanishi, H. et al. (2020). Chemical speciation, plant uptake, and toxicity of heavy metals in agricultural soils. *J Agric Food Chem* 68 (46): 12856–12869.

Ullah, S., Adeel, M., Zain, M. et al. (2020). Physiological and biochemical response of wheat (*Triticum aestivum*) to TiO_2 nanoparticles in phosphorous amended soil: a full life cycle study. *J Environ Manag* 263: 110365.

Usman, M., Farooq, M., Wakeel, A. et al. (2020). Nanotechnology in agriculture: current status, challenges and future opportunities. *Sci Total Environ* 721: 137778.

Wang, Z.Y., Yu, X.L., Gao, D.M. et al. (2010). Effect of nano-rutile TiO_2 and multiwalled carbon nanotubes on the growth of maize (*Zea mays* L.) seedlings and the relevant antioxidant response. *Environ Sci* 31: 480–487.

Wang, S., Wang, F., and Gao, S. (2015). Foliar application with nano-silicon alleviates Cd toxicity in rice seedlings. *Environ Sci Pollut Res* 22: 2837–2845.

Wang, X., Sun, W., Zhang, S. et al. (2018). Elucidating the effects of cerium oxide nanoparticles and zinc oxide nanoparticles on arsenic uptake and speciation in rice (*Oryza sativa*) in a hydroponic system. *Environ Sci Technol* 52 (17): 10040–10047.

Wang, Y., Jiang, F., Ma, C. et al. (2019a). Effect of metal oxide nanoparticles on amino acids in wheat grains (*Triticum aestivum*) in a life cycle study. *J Environ Manag* 241: 319–327.

Wang, Y., Wang, L., Ma, C. et al. (2019b). Effects of cerium oxide on rice seedlings as affected by co-exposure of cadmium and salt. *Environ Pollut* 252: 1087–1096.

Wang, Z., Yue, L., Dhankher, O.P., and Xing, B. (2020a). Nano-enabled improvements of growth and nutritional quality in food plants driven by rhizosphere processes. *Environ Int* 142: 105831.

Wang, Y., Liu, Y., Zhan, W. et al. (2020b). Long-term stabilization of Cd in agricultural soil using mercapto-functionalized nano-silica (MPTS/nano-silica): a three-year field study. *Ecotoxicol Environ Saf* 197: 110600.

Wong, M.H., Misra, R.P., Giraldo, J.P. et al. (2016). Lipid exchange envelope penetration (LEEP) of nanoparticles for plant engineering: a universal localization mechanism. *Nano Lett* 16: 1161–1172.

Wu, H., Tit, N., and Giraldo, J.P. (2017). Anionic cerium oxide nanoparticles protect plant photosynthesis from abiotic stress by scavenging reactive oxygen species. *ACS Nano* 11: 11283–11297.

Wu, X., Hu, J., Wu, F. et al. (2020). Application of TiO_2 nanoparticles to reduce bioaccumulation of arsenic in rice seedlings (*Oryza sativa* L.): a mechanistic study. *J Hazard Mater* 405: 124047.

Yang, C., Zhang, X., Liu, G. et al. (2013). Progress on the structure and physiological functions of apoplastic barriers in root. *Bull Bot Res* 33: 114–119.

Yang, J., Cao, W., and Rui, Y. (2017). Interactions between nanoparticles and plants: phytotoxicity and defense mechanisms. *J Plant Interact* 12: 158–169.

Yang, J., Jiang, F., Ma, C. et al. (2018). Alteration of crop yield and quality of wheat upon exposure to silver nanoparticles in a life cycle study. *J Agric Food Chem* 66: 2589–2597.

Yao, J., Cheng, Y., Zhou, M. et al. (2018). ROS scavenging Mn_3O_4 nanozymes for in vivo anti-inflammation. *Chem Sci* 9 (11): 2927–2933.

Yılmaz, A.B., Yanar, A., and Alkan, E.N. (2017). Review of heavy metal accumulation on aquatic environment in Northern East Mediterrenean Sea part I: some essential metals. *Rev Environ Health* 32: 119–163.

Zhang, S. (2019). Mechanism of migration and transformation of nano selenium and mitigates cadmium stress in plants. Master's Thesis. Shandong University, Jinan, China.

Zhang, P., Ma, Y., Liu, S. et al. (2017). Phytotoxicity, uptake and transformation of nano-CeO_2 in sand cultured romaine lettuce. *Environ Pollut* 220: 1400–1408.

Zhang, H., Du, W., Peralta-videa, J. et al. (2018). Metabolomics reveals how cucumber (*Cucumis sativus*) reprograms metabolites to cope with silver ions and silver nanoparticle-induced oxidative stress. *Environ Sci Technol* 52: 8016–8026.

Zhao, X., Jia, Y., Li, J. et al. (2018). Indole derivative-capped gold nanoparticles as an effective bactericide *in vivo*. *ACS Appl Mater Interfaces* 10: 29398–29406.

25

The Dynamics of Phytoremediation of Heavy Metals

Recent Progress and Future Perspective

Imran Haider Shamsi[1], *Xiaoli Jin*[1], *Xin Zhang*[2], *Qidong Feng*[1], *Zakir Ibrahim*[1,3], *Muhammad Faheem Adil*[1], *Muhammad Fazal Karim*[4], *and Najeeb Ullah*[5]

[1] *Department of Agronomy, College of Agriculture and Biotechnology, Key Laboratory of Crop Germplasm Resource, Zhejiang University, Hangzhou, China*
[2] *School of Marxism, Zhejiang University, Hangzhou, China*
[3] *Faculty of Agriculture, Lasbela University of Agriculture, Water and Marine Sciences, Uthal, Pakistan*
[4] *Department of Agronomy, PMAS Arid Agriculture University, Rawalpindi, Pakistan*
[5] *Agricultural Research Station, Office of VP for Research and Graduate Studies, Qatar University, Doha, Qatar*

25.1 Introduction

Global industrialization has led to widespread environmental pollution due to the constant addition of organic and inorganic wastes. Soils, surface- and groundwaters become contaminated with harmful compounds as a consequence of natural activities (such as geological erosion and salt infiltration), and human activities (such as industry, agriculture, wastewater treatment, construction and mining). Contaminants can be traced to specific sources, point sources, or originate from a broad array of non-point sources. Contaminants of concern are inorganic and organic compounds (heavy metals, radionuclides, nitrates, phosphates, inorganic acids, organic chemicals) (Sagonda et al. 2021). The biggest risk to the environment and to human health is posed by contaminated soil and water; however, new phytoremediation technologies can help to partially address this issue.

Phytoremediation is a biological process, which integrates various types of plants to eradicate, transport, stabilize and/or destroy soil and groundwater contaminants. The term "phytoremediation" was coined by Chaney (1983), the research involving the interactions among plants and environmental pollutants began quite early in the twentieth century. The exploitation of plants for the evacuation of xenobiotics and heavy metals from spillage sites, sewage waters, slops, and contaminated regions have turned into a significant exploratory and reasonable methodology throughout recent years. Information on the physiological and atomic systems of phytoremediation has started to arise along with natural and designed procedures intended to upgrade and further develop the phytoremediative interactions (Schwitzguebel 2000). A few plants, selected over the course of 20 years as

Heavy Metal Toxicity and Tolerance in Plants: A Biological, Omics, and Genetic Engineering Approach, First Edition. Edited by Mohammad Anwar Hossain, AKM Zakir Hossain, Sylvain Bourgerie, Masayuki Fujita, Om Parkash Dhankher, and Parvez Haris.

profoundly viable in retaining and gathering different poisonous minor components, are being assessed for their role in the phytoremediation of soils and waters polluted with minor components (Llao and Chang 2004). Phytoremediation is a creative approach/methodology that offers more biological advantages and a cost-effective alternative. Although it is a less expensive technique yet requires a specialized system, master project planners with field experience pick the legitimate species and cultivars for specific metals and districts (Mudgal et al. 2010). It presents many benefits when contrasted with other remediation procedures: it may be performed very well with insignificant natural unsettling influence; it encompasses a broad-spectrum of impurities, incorporating numerous metals with restricted or elective choices; perhaps less optional air or potentially water wastes are created than with customary strategies; natural poisons might be corrupted to CO_2 and H_2O, eliminating ecological harmfulness; it is practical for huge volumes of water having low convergences of pollutants; plant take-up of tainted groundwater can forestall off-site movement (Schwitzguebel 2000). Bioremediation can be classified into two kinds, *in situ* and *ex-situ*, and both (field treatments) have been used to remove contamination, most of which rely on physical and chemical processes. *In situ* methods include evaporation, surfactant washout, vitrification (thermal rehabilitation of contaminated soils at high temperatures via electricity), along with separation and containment employing barrier physics. Field techniques include extrication following heat treatment, chemical extraction, and/or solidification (packaging) prior to landfill disposal.

25.1.1 Types of Phytoremediation

Since utilizing plants to remove harmful metals and metalloids is a good alternative to physical and chemical remediation, the idea of using plants to sequester HMs is becoming more and more popular (Ali et al. 2013; Mapodzeke et al. 2021; Zeshan et al. 2021). It is important to talk about the following phytoremediative methods for removing heavy metals from media (environments) (Figure 25.1).

25.1.1.1 Phytostabilization

Phytostabilization or phyto-immobilization denotes the usage of plants that have the potential to either reduce the mobility and bioavailability of a metal or to avert its leaching into groundwater and entrance into the food chain. Several mechanisms include root adsorption, precipitation, and root complexation (Erakhrumen 2007).

25.1.1.2 Phytovolatalization

Another method is phytovolatalization, which involves converting metal into a volatile form and releasing it into the atmosphere via stomata (Ghosh and Singh 2005). This approach is most useful for mercury (Hg) because its ion is converted into a less-hazardous elemental form. Because precipitation can recycle the volatile form of Hg released into the atmosphere, this strategy provides a temporary solution to the problem.

25.1.1.3 Phytoextraction

Phytoextraction is the main phytoremediative approach for the deduction of metal(loid)s from polluted soils, water, bio-solids, sludge and silt (He et al. 2005; Seth 2012;

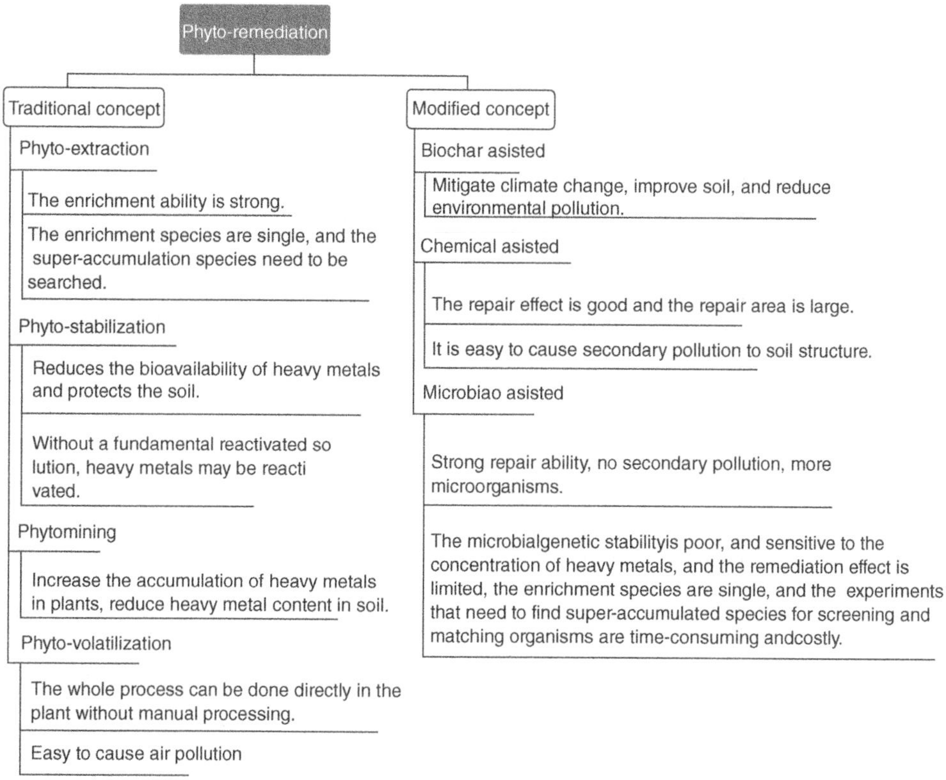

Figure 25.1 Various mechanisms involved in phytoremediation.

Ali et al. 2013), which is very much appropriate for business application when contrasted with other phytoremediation procedures (Sun et al. 2011). Various variables influence the proficiency of phytoextraction including soil properties, metal accessibility to plants, metal speciation, and concerned plant qualities. Plants chosen for phyto-extraction ought to have the qualities of; (i) quick development rate, (ii) more biomass creation, (iii) hyperaccumulate HMs, (iv) broadly conveyed, (v) move metal from root to shoot, (vi) endure the poisonous impacts of HM, (vii) impervious to microorganisms and nuisances, (viii) all around adjusted to fluctuating climatic circumstances, (ix) simple to propagate and harvest, and (x) have no appeal to herbivores in order to keep away from its entrance into the pecking order (Tong et al. 2004; Ali et al. 2013).

25.1.2 Modified Concept

Adulterated soil and water pose a major environmental and human health problem that can be partially solved by new phytoremediation technologies. Conventional phytoremediation approaches face certain limitations, such as (i) they require a longer time span to remediate the tainted soils, (ii) the phytoextractive capacity of hyperaccumulators is restricted owing to the low shoot biomass production, (iii) biology to adapt (the proportion of bioavailable metals is very low and their concentration varies with pH-values of soil, organic matter, competing cations, lime content, etc.), (iv) insufficient knowledge of agronomy, reproductive potential, pest and disease spectrum; and (v) improper management or neglect, which can lead to contamination of the food chain (Ali et al. 2013). These unavoidable limitations require researchers to modify traditional phytoremediation approaches to curtail these restraints and ensure the large-scale application of phytoremediation.

25.1.2.1 Chemical-Assisted Phytoremediation Employing Non-hyperaccumulator Plants

The decision of selecting the suitable species for heavy metal phytoremediation is most important. In general, species capable of accumulating high concentrations of HMs in their aerial parts are considered superior and are called hyperaccumulators. Baker et al. (1994) proposed the shoot-to-root metal concentration ratio as a criterion for determining whether a plant is a hyperaccumulator or not. A ratio above 1.0 in plant species, therefore, indicates that the metal is accumulating in the shoots rather than the roots, and such species can be utilized as hyperaccumulators. Krämer (2010) described that the above-ground HMs concentrations in hyperaccumulator plants vary between 1000 and 10,000 $mg\,kg^{-1}$ depending on the toxic content of the HMs, and even up to 100 $mg\,kg^{-1}$. To date, most plants classified as hyperaccumulators meet the above criteria. The main disadvantage of working with these plants is their very slow growth and little aboveground biomass. These limitations make such plants unsuitable for large-scale phytoremediation.

However, these plants can serve as a reference for producing genetically engineered plants with the potential to accrue high levels of metals and produce more biomass. Likewise, non-hyperaccumulator plants, which produce more biomass above-ground, should be evaluated for their phytoremediative prospective. This is because these plants sequester comparatively few heavy metals. To improve their extractability with synthetic or organic chelating agents, using a process known as chelation or induced plant extraction becomes more beneficial (Salt and Rauser 1995). The use of synthetic chelating agents, such as ethylenediaminetetraacetic acid (EDTA), diethylenetriaminepentaacetic acid (DTPA), and ethylene glycol tetraacetic acid (AGTA), increases metal bioavailability and improves uptake by plants (Pereira et al. 2010). However, this technique has certain limitations due to the risk of environmental pollution due to the high mobility of heavy metals. Additionally, up-taking large amounts of chelated HMs can turn out to be toxic for plants themselves.

25.1.2.2 Biochar-Assisted Phytoremediation

Biochar is a carbonaceous porous material synthesized by means of the pyrolysis of organic raw materials, such as plant material, organic fertilizers, and sludge. Charcoal, wood biochar, is the most widely used biochar and has been around for a long time. Biochar has some specific physicochemical properties, viz. high pH, large surface area for metal sorption, alkalinity, ash, carbon content, and ability to immobilize toxic heavy metals, which make it an excellent remediation agent for HMs. The large surface area available for sorption is attributable to the exchange of HMs cations with other metal cations (i.e., Na^+, K^+, Ca^{2+}, Mg^{2+}) at functional groups present in biochar, or physisorption. The high pH and alkalinity of biochar can reduce the bioavailability of metals and increase their precipitation in biochar-rich soils. Biochar pH has been reported to increase with pyrolysis temperature, possibly due to the increase in biochar ash content.

25.1.2.3 Microbial-Assisted Phytoremediation

Plant-associated soil microbes can influence the availability and uptake of HMs by plants in the rhizosphere. Some microorganisms live in association with the plant's roots, while others live freely. Mycorrhizal fungi are the key organisms in the root zone and live in communities with higher terrestrial plants in various forms, e.g. ectomycorrhizal, arbuscular mycorrhizal, orchid mycorrhizal, and mycorrhizal fungi of the Ericaceae family, most commonly arbuscular mycorrhizal fungi. These associations between fungi and plant roots are mediated in many ways, including other phytonutrients, such as N, P, K, Ca, S, Zn, Co, Ni, and Cu, available through extensive mycelial networks in plants. The associations of these fungi alter soil pH as well as the chemical composition of root exudates, which consequently alters the bioavailability of HMs in soil. Plant growth-promoting bacteria (PGPR) are another important factor in microbial community that helps plants to remediate soil spiked with HMs (Seth 2012). These microorganisms can be distinguished as commensal bacteria and free-living rhizobia. They can boost plant growth through a variety of mechanisms, including decreased ethylene production under stress, nitrogen fixation, and specific enzymatic activity.

25.2 Importance of Phytoremediation

Phytoremediation refers to a broad range of environmental remediation efforts and is becoming an important tool in the bioremediation of metal contaminants in abiotic sources. These plants are highly adapted to accumulate toxic trace metals while growing rapidly in contaminated soil (Ahmadpour et al. 2012). Their use in all forms of agriculture, from urban to rural areas, can prevent loss of important agricultural soil resources due to metal contamination, stop and avoid soil erosion/metal leaching by stabilizing HMs and reduce the risk of contamination. It also develops soil richness by releasing numerous organic matter that increases soil fertility. In recent years, tremendous amount of research has been carried out to identify the molecular mechanisms underlying heavy metal resistance and to advance tools and techniques to advance our proficiency in phytoremediation. This overview describes the mechanisms by which heavy metals are consumed and transported to the flora. Cleanup strategies (avoidance and tolerance) instilled in plants in response to heavy metals were also discussed. A major debate is overlooking current improvements in advanced phytoremediation technologies, including regimens to improve biological preparation, concentration, and tolerance of heavy metals (Wuana and Okieimen 2011). Economic development and an ever-growing world population are placing enormous strains on the food system. Feeding a population of nearly nine billion by 2050 will require a new vision to ensure food security, environmental sustainability, and economic opportunity through agriculture. Agricultural sustainability is essential to nourishing a growing population and preserving soil health. Sustainable agriculture requires returning more nutrients from food waste and sewage back into the soil. Field application of wastewater and compost derived from municipal solid waste can actually cause metal contamination. The availability of efficient phytoremediation techniques to keep soils fertile is becoming increasingly important (Belliturk 2011).

25.3 Role of Phytoremediation as a Sustainable Solution

Concentrations of harmful (heavy) metals in different ecological compartments remain a major risk for different biomes, surrounding environments, and biological concentrations of these harmful elements in successive layers of nutrients through the process of biological reactions increases the austerity of the problem (Ali et al. 2013). Compared to organic contaminants, primary metals/metals do not undergo any chemical or microbial decomposition or destruction, resulting in a constant increase in the amount of this deadly metal, leading to persistent soil toxicity (Adriano et al. 2004). The steady increase of these lethal elements in the ecosystems (aquatic and terrestrial) is a major problem globally (Sharma and Pandey 2014). As a result, the pursuit of innovative technological innovations has contributed to the enormity of this counterfeiting, which threatens the very existence of existing communities of organisms (Ebbs et al. 2016). The consequences are deterioration of water quality and human health risks, even at lower doses of heavy metals. Heavy metals such as lead, cadmium, chromium, mercury, aluminum, and copper are contaminants in the water created by human agents and natural events such as wildfires, volcanoes, and permeable rocks. Freshwater sources are largely concentrated by pathogens deposited by raw or untreated wastewater as well as by heavy metal by-products released from mining sites as well as industrial wastes (Caravanos et al. 2016).

The exclusion of HMs components from industrial wastewater is paramount due to the environmental problems associated with HM-contaminated soil and water. Tyler et al. (1989), reported that the release of biologically available heavy metals by human activity could damage or alter natural and man-made ecosystems. One of the effective and affordable technological solutions is the application of a plant treatment process, which is the application of plants to treat/remove contaminated soil, sediment, and water. Trees decompose, assimilate, metabolize or detoxify inorganic and organic contaminants from soil and water or make them harmless. It is an economical "green" technology contingent on the utilization of carefully chosen metal-concentrating plants to eliminate toxic metals from the environment (Garbisu and Alkorta 2001). Besides the advantage of low cost, extraction and concentration of a specific element from the environment, the harvested plant tissue, rich in accumulated pollutants, is easily and safely disposed of by drying, burning, or mixing. These metals can later be recovered from the ash, further reducing hazardous waste generation and generating recycling revenue. Conversely, regardless of these benefits, phytoremediation is not yet commonly practiced in developing countries. This is due to the incomplete awareness regarding this technology. Due to inadequate legal regulations, industrial waste mostly containing toxic HMs is being discharged into the environment untreated, contaminating soil, groundwater, and surface water.

The following are the after-effects:

i) Ecological destruction
ii) Destruction of marine biota
iii) Fishing and other aqua-cultural enterprises are hampered
iv) Seafood poisoning, pollution of surface and groundwater
v) Disturbances in soil function

However, plant productivity and survival chances are reduced. Some elements, such as Cu^{2+}, Zn^{2+} and Fe^{2+}, are crucial for plant's metabolism, but can be highly toxic when present in excess. Cadmium is one of the most toxic HMs, which till date is considered nonessential to life while zinc (Zn) is an essential and beneficial element for the human body and plants (Mapodzeke et al. 2021). Lead (Pb) alone has been reported as nontoxic to plants (Gupta 1980). Water tainted with HMs have historically been treated using ion exchange and activated carbon filtration processes. The drawbacks of these processes are that they are expensive and they are not selective in removing heavy metals. Thereby, phytoremediation, a cheap, effective, and efficient method, is now used. Soltan and Rasheed (2003) documented that *Eichhornia crassipes* (water hyacinth) is one of the highest yielding plants on earth and is considered the worst aquatic plant in the world as it forms dense mats that affect recreation, navigation, irrigation, and power generation. It also pollutes surface water, making it unsuitable for consumption. Plants accumulate toxic metals, such as Pb, Cd, and Zn, directly from the water they feed on. Phytoremediation has proven useful in many places around the world (Sumiahadi and Acar 2018). In Torch Lake, Michigan, USA, sunflower (*Helianthus annuus*) has been used to remove copper from mining-contaminated sediments (Li and Wusirika 2011).

25.4 Biophilic Design as Phytoremediation in Urban Sustainability

25.4.1 Eco-Design

The term eco-design also refers to environmental design, which is defined as "assimilating environmental considerations into product design and development with the purpose of plummeting the negative impact on the environment throughout the product life cycle". Environmentally friendly design (EFD) or eco-design has expanded its contribution to sustainable production in recent decades. Product design is recognized as an important life cycle phase that contributes to the environmental impact of the life cycle. An estimated 80% of environmental pollution is detected at this initial phase. Eco-design means a design that incorporates nature in consideration of the environment, modern applications of eco-design including the development of new techniques as a transition to environmental sustainability characterized as a shift from "high-tech" to "Eco-tech". Restoration, permaculture, green building, biotechnology, wind farms, and hybrid vehicles are examples of ecological design. Another area of ecological design has urban bio-network in mind; in addition the field of conservation biology is concerned with preserving the natural world. Designers take into account the natural world when designing landscapes and buildings or something that affects the interaction with wildlife.

25.4.2 Biophilic Design

A similar type of eco-design approach is biophilic design, where plants are incorporated into a built environment along with other natural elements (Figure 25.2). Biophilic design is a way of thinking about design that values systems and processes that are natural and this philosophy is often used in the design of buildings and other spaces (Kellert 2008).

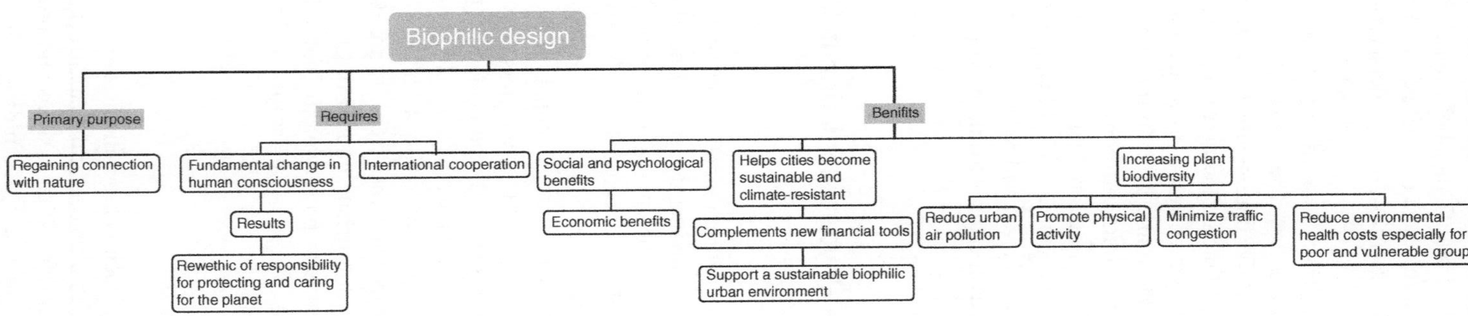

Figure 25.2 Illustration of biophilic design as phytoremediation in urban sustainability.

It is primarily founded on the hypothesis of biophilia, which asserts that humans are innately connected to nature (Wilson 1984), and therefore contact with mother nature is important to human welfare. However, modern societies often lack interaction between former and later (Kellert 2005), because of sociological tendencies in urbanization, architecture, and way of life. Subsequently, the goal of "biophilic design" is to integrate natural elements and systems into constructed spaces in order to give humans much-needed access to nature (Kellert 2008). The global building industry has recently shown a keen interest in biophilic design. Living Building Challenge (2015) clearly incorporates biophilic design into its two building rating systems, which were developed in the United States but are now promoted internationally, which combines them via biophilia imperative, and the new well building standard (WELL Building Standard Resources n.d.), which combines them via biophilia precondition, and biophilic optimization entails consultancy businesses, such as terrapin bright green (Terrapin 2015; Human Spaces 2015), which publishes several white papers on biophilic design, and interface parquet, which founded the human spaces website to stimulate conversation about biophilic design, have also incorporated the concept (Browning et al. 2014).

The use of natural occurrences and phenomena to inspire architectural design is encouraged by the biophilic design concept (Kellert 2008). This is based on the theory that exposure to natural settings and features positively impacts human health and well-being, which is supported by a wealth of research (Bowler et al. 2014). According to the biophilic hypothesis, these beneficial effects of nature exposure occur within the biological link concerning humans and nature (Wilson 1984). These ideas are supported by two of his ideas created from research in environmental psychology: attention recovery theory (Kaplan and Kaplan 1989) and stress recovery theory (Joye et al. 2012), which were taken further. According to both theories there are stressful and nonstressful environments, which help people to actively recover from stress and mental exhaustion. An environment that possesses the features of enticing people's attention without forcing them and helps them recover more rapidly and completely from mental exhaustion and stress is known as a recovery environment (Gifford and McGunn 2012). According to Kaplan (1995) and Ulrich et al. (1991), especially in the natural environment, there are elements that draw new attention by communicating absenteeism, charm, vastness, and compatibility, and includes elements that encourage survival and favorable evaluation (Hartig et al. 2003). On the contrary, urban areas frequently have a wide variety of features that are challenging, stressful, exciting, or boring. It suggests that the environment can be made more restful. Gifford and McGunn (2012) suggest that restorative design, which is a broader term, may include biophilic design. Studies on restorative environments that support the advantages of biophilic design for the well-being of humans are widely cited in the majority of peer-reviewed literature (Joye 2007; Gray and Birrell 2014; Ryan et al. 2014a). Despite the fact that it is a relatively new idea, significant investigations on nature and the restorative environment clearly support the potential benefits to one's health of incorporating its principles into the built environment. While there is much evidence supporting the positive impacts of exposure to nature, there is less support for other biophilic practices like using natural materials. The argument that biophilic design is advantageous can be supported by data from more than 30 years of research on how nature affects human health and well-being, especially in the scientific literature on biophilic design. They are still quite a few. One of the earliest studies on biophilic architecture is from Joye (2007), who has observed the use of empirical psychology research from diverse fields to biophilic design, and came to a conclusion that the realm of restoration environments supports these ideas.

25.4.2.1 Hypothesis of Biophilic

Erich Fromm, a psychologist, originally used the term biophilia in his 1973 book *Anatomy of Human Destruction*, in which he described "the ardent love of life and all that is living... whether in one person, a plant, an idea, or a social community" (Fromm 1973). The term has been extensively used and adapted to a number of different scientific fields. The hypothesis emphasizes that there is an inherent desire in humans to interact with other living organisms and nature since we have evolved to rely on them for our existence and fulfillment (Gullone 2000). People travel and spend money to explore nature reserves and rainforests, recline on beaches, climb for hours in the mountains, and visit national parks, all of which are relevant to the everyday lives of people. In addition, a lot of sports, like surfing, mountain biking, and skiing, are outside activities. Buyers are more than willing to spend 7% more on homes with fantastic views, 58% more on waterfront properties, and 127% more on riverfront properties. From a housing standpoint, customers are more likely to splurge more money on homes with views of nature. Animal companionship is also enjoyable for humans.

25.4.2.2 Dimensions of Biophilic Design

The pioneer of biophilic design, Stephen Kellert, developed a framework for integrating nature into built environments in a way that satisfies human needs (Kellert et al. 2011), and considers it an important eco-engineering design principle (Poff et al. 2016). These principles are adopted to honor, respect, and create a sustainable and inclusive urban environment. The parameters and characteristics that makeup Kellert's biophilic framework include (i) direct experience of nature, (ii) indirect experience of nature, and (iii) experience of space and place.

25.4.2.3 Direct Experience of Nature

It involves physical interaction with both living and non-living elements of the natural world, such as fire, light, water, plants, animals, weather, and the natural world. These characteristics have multiple effects on the environment. For example, light allows directionality about the time of day and season, orientation, and relaxation. However, light may produce shadows, movement, and many natural patterns and shapes. This ensures the well-being and interest of residents. In addition, the air is important for ventilation, temperature, and humidity. Such conditions can be applied using most importantly, altering these elements can increase occupant comfort and productivity, which includes windows and other passive approaches. Due to its multisensory effect, water is used for movement, sound, touch, and sight in buildings. Water features, wetlands, and aquariums can all have designs. Water has a special meaning for humans, and using it can improve performance, well-being, and health while lowering stress. Similarly, plant diversity has a variety of applications, including the addition of phytoremediation types to the outside and interior areas of buildings to enhance a building's connection to and benefit from nature. This should be plentiful (i.e. use a green wall or lots of potted plants) and have some vegetation in bloom.

Plants have been shown to enhance physical health, performance, and productivity and reduce stress levels. Interaction with animals promotes conservation concerns and enhances the functioning, mental stimulation, and enjoyment of ecosystems and services (Stephen and Heerwagen 2008). The natural landscape of the constructed environment brings immeasurable joy by simulating human evolutionary behavior and contributing to overall health and well-being.

25.4.2.4 Indirect Experience of Nature

They are described as contact with images and expressions of nature. The residents have found this to be fulfilling, both emotionally and intellectually. Paintings, photos, sculptures, murals, movies, and other forms of art can all incorporate images of nature (Kellert et al. 2011). People prefer natural materials because of their mental stimulation factor; therefore, plants can be integrated into the building as a natural component of the structure of the interior design. Plant-based leather made from mushrooms, pineapple skins, or cacti is now considered a viable alternative. It can also be seen that feeling and bringing nature and animals closer is counterproductive and contrary to the hypothesis of biophilia, simply destroying them in the process.

25.4.2.5 Experience of Place and Space

This experience is a spatial relationship designed to improve well-being and includes perspective and sanctuary (the capacity of a structure to produce a cozy and nourishing interior), perspective being the horizon, movement, and highlighting the source of the hazard. This principle also includes an organized complexity that reflects the necessity of regulated variability. The building architecture is repeated, modified, and detailed in design to achieve this (Kellert 2008). Plants are an integral part of most biocompatible design principles, providing well-being, comfort, ecosystem services, and a sustainable environment for their residents. These experiences also include connections with places, both cultural and ecological. This is accomplished by including local geography and history in the design. The development of ecosystems that enable the utilization of local flora and fauna results in the construction of an ecological identity (Stephen and Heerwagen 2008). When using biophilia in a project, each of these experiences needs to be taken into account separately. In terms of construction types, there is no one right response. The architects and project owners of each building must collaborate to implement the biophilic principles they feel the best match their objectives and best serve their tenants.

25.4.2.6 Sustainable Biophilic Cities

Humans spend about 80% of their time indoors, such as in homes, offices, and cars, and come into contact with potentially toxic pollutants (Parhizkar and Elzeyadi 2020). Rapid urbanization directly correlates with reduced exposure to nature (Jim 2004; Zhao et al. 2013). Increasing environmental stressors such as road traffic noise and pollution (Maas et al. 2008). Meanwhile, urban areas are facing a dominant environmental crisis due to increased outdoor pollution, aggressive ventilation strategies, and a lack of green space. Consider that solutions based on nature can combat stress and improve overall environmental well-being (Maas et al. 2008). Over the last few years, more and more joint efforts have been undertaken around the world to advance urban green spaces such as trees, parks, and public gardens. In 286 cities in China, an increase has been reported in average green area from 17% to 37% between 1989 and 2009 (Chan et al. 2021). Meanwhile, in 202 cities in Europe, green space enlarged by an average of 0.54% per year during 2000 and 2006 (Kabisch and Haase 2013). Numerous investigations have demonstrated that biophilic design, defined as a response to humans' natural needs to interact with nature in the workplace, improves productivity, reduces sick leave, and improves user well-being (Maas et al. 2008; Elzeyadi 2012; Ayuso et al. 2018). According to scientists, the major objective of a sustainable or bio-friendly city is to establish a setting where citizens desire to actively engage in, conserve, and interact with the local natural environment. There are established approaches to achieving this through frameworks of substructure, authority, awareness, and action. These dimensions can also be indicators of preexisting biophilic features already present in modern metropolises (Timothy 2011).

Biophilic Settings and Infrastructure The idea that there should always be a certain number of people near green spaces and parks. It creates an assimilated city's walking pathways and ecological network, designating vegetation and forests, green and biophilic building design capabilities, and specific land areas to take advantage of city-wide flora and fauna can be achieved by doing (Timothy 2011).

Biophilic Activities This includes staying outside for longer while visiting parks, while at school, taking better city walks, participating more actively in community gardens and greenhouse groups, and engaging in more volunteer work in the area (Timothy 2011).

Biophilic Attitudes and Knowledge Regions with elements of urban biophilic design will increase the number of residents who care about nature and can identify local native species. Residents' curiosity about local ecosystems is also growing (Timothy 2011).

Biophilic Establishments and Governance Activities related to nature and biocompatibility receive funding from local government entities. This indicator includes new laws demanding greener and more bio-friendly design principles, funding programs to inspire using nature and bio-friendliness, and engagement with natural history museums and academic applications, so that more and more conservation agencies and network agencies contribute (Timothy 2011).

Significance of Biophilic Designs There is no doubt that biophilic design will bring many benefits to the city surroundings and residents by using strengthening the connection with nature. For cities, many agree that the biggest supporters of this idea are their ability to be extra resilient to the environmental stresses they might also face (Timothy 2011).

25.4.3 Health Benefits

Studies show that factors such as natural melodies improved mental health 37% quicker than normal city noise after being exposed to stressors. According to the same study, when surgical patients were exposed to aromatherapy, they needed 45% less morphine overall and 56% fewer painkillers (Ryan et al. 2014b). Additional published literature shows that plant intake as indoor phytoremediation reduces stress and increases tolerance to pain. Using water elements and incorporating natural scenery also provides psychological relaxation to the resident (Gillis and Gatersleben 2015). According to Söderlund and Newman (2015), improving the visibility of the room reduces the patient's depression and pain and reduces the length of hospital stay from 3.67 to 2.6 days. Scholars also point out that biophilic cities have increased social connectivity and improved their ability to deal with life-threatening crises. This reduced the rate of violent and aggressive crime (Riggs 2012). According to the same study, the introduction of outdoor services such as the "Green Gym" allows people to eliminate overgrown vegetation, adapt to an active lifestyle and exercise more easily (walking, running, hiking, etc.). For that reason, it has proven to enhance physical activity, boost mental health, and improve quality of life. In addition, children raised in such areas have been found to have less asthma. In green areas, lower death rates and health inequalities among the rich and the poor have also been observed (Riggs 2012).

25.4.4 Biophilic as an Antidepressant in Urban Environment

A team of psychologists at NTU Singapore is using virtual reality to demonstrate that vertical green has a stress-buffering effect (harmful alleviation) and enhances the capacity to absorb the effects of stress in an urban setting (Chan et al. 2021). The study asked 111 participants to walk a virtual street for five minutes while using a virtual headset. Participants were arbitrarily assigned to streets lined with greenery (balconies, walls, building pillars, etc.) or streets with only buildings whose walls were painted green instead of green. Realistic experiences were created by emitting loud traffic noises as participants crossed a virtual street. Heart rate variability was continuously monitored with a portable electrocardiogram (ECG) device. The researchers found that those who saw only green buildings experienced a significant increase in stress, while those who saw buildings with green plants experienced changes in stress, as recorded by heart rate variability measurements. After the experiment, participants completed a questionnaire evaluating their positive emotions (e.g. curious, excited) and undesirable emotions (e.g. upset, hostile), and the anxiety they felt. Participants informed that strolling around the street with a building covered only with green walls made them feel worse, while a building covered with vegetation made them feel better.

The findings can impact the welfare of urban dwellers and guide urban greening efforts. The green walls help lower the ambient temperature and decrease the cooling system's energy usage, it can also help to minimize CO_2 emissions and the "urban heat island" impact (Wang and Akbari 2016). Due to the limited greenery and densely populated buildings, the temperature in the city center is much higher than in less populated areas (Wang and Akbari 2016). Vertical green spaces are often planted for these sustainability benefits and give city planners more impetus to adopt the "biophilic design" concept to contribute to the psychological well-being. This is a popular approach as a "Nature-based Solution" (NBS) in cities like Singapore, Wellington (NZ), and San Francisco (Wang and Akbari 2016). According to the study's investigators, with urbanization, more people are expected to reside in cities around the world in the future. Urban planning, therefore, has a direct impact on people's quality of life. Understanding the aspects that can facilitate a healthy life style is critical for urban planners and architects because of the possibilities.

By integrating and proving to be a feasible way to promote mental health, we can guide efforts toward green cities. Prior studies have found the effects of green vegetation, but the color green is merely a primordial factor leading to beneficial effects. Thanks to new technologies such as VR, we have been able to overcome this limitation and use control conditions that match vertical green and green colors in our studies (Chan et al. 2021).

25.4.5 Economic Benefits

Biophilia can have a slightly higher direct cost due to the addition of maintenance-requiring natural elements and expensive organic matter, but the perception of health and environmental benefits outweighs this. It is considered. Cities like New York can save almost $470 million from increased worker productivity and $1.7 billion from decreased crime expenses by implementing biophilic architecture and landscaping (Maas et al. 2008; Söderlund and Newman 2015). It was also discovered that businesses on streets with vegetation attracted customers who were more inclined to spend a 25% higher amount. According to the same survey, more sunlight from the store's skylights increased sales by 40–45%. Biophilic design properties also benefit from a 16% higher selling price than traditional buildings (Maas et al. 2008). According to Witters et al. (2012), phytoremediation is a competitive advantage over traditional restoration techniques for making sustainable, informed decisions about which crops to grow on contaminated land.

25.4.6 Sustainability and Resilience

At the urban level, the biophilic design enables cities to better adapt to climate change and, in turn, pressures created by the local environment (Timothy 2011). Researchers have devised a framework for bio-friendly cities, which permits us to discover the capability to amplify the resilience and sustainability of cities; it consists of three sections: bio affinity urbanism, physical bio affinity, and green measures that may be performed to promote urban resilience, community resilience as an outcome of these physical changes, and how adaptive and persistent behavior can occur when both of these steps are taken. The Biophilic Urbanism section points out that one-way cities can become more resilient is by following the biophysical route. Local effective vegetation increases the urban natural protection barrier. New Orleans, for example, is a city that overbuilt and flooded its natural wet plains. It is estimated that cities can save $23 billion annually in storm protection if they keep Bayeux intact (Beatley and Newman 2013). The adaptability section argues that the relationship with places and routes creates an exciting and interesting natural environment for residents. This strengthens ties and ties with neighbors and homes and increases the likelihood that citizens will change their place of residence (Timothy 2011; Beatley and Newman 2013). Moreover, these people are motivated to rebuild and assist their communities than to escape in times of shock and stress. It can also strengthen government actions to protect cities from future disasters.

The scientists believe that the most substantial resilient effect of this approach is greater population adaptability by establishing biophilic urbanism and adaptive capability. Citizens are healthier and more physically fit as steps leading to resilience encourage people to stay active, get out and participate in activities. People who walk in nature have been shown to be less depressed, less angry, and more empowered than those who walk indoors. Some argue that by way of including physical natural factors in the constructed surroundings alongside plants such as trees, rain gardens, and green roofs, constructions and cities can higher control storm-water runoff. To preserve these natural systems in a cost-efficient manner, extra waste water can be reused to irrigate plant life, and in green areas, polluted water is also reduced by green walls and rooftops because the vegetation serves as a biofilter. Reducing carbon emissions and boosting biodiversity are achieved by adding greenery. In order to minimize carbon, plants sequester carbon during photosynthesis. In contrast to asphalt and dark surfaces, high-albedo rooftop greening and vegetation that provides shade on roads and buildings can lessen heat absorption. By doing so, you will need to use 25% less heating and cooling energy and 50% less fluctuating temperature (Beatley and Newman 2013; Söderlund and Newman 2015). Moreover, the addition of a green facade can increase the biodiversity of the region when native species are planted. Singapore's Cootech Puat Hospital uses vegetation throughout the outer walls of the building, and as a result, 103 species of butterflies have been revived.

25.5 Conclusion

Unprecedented introduction of pollutants in every dimension of environment is a major concern to the overall well-being of human society and environmental health. These pollutants, ranging from HMs into the soil, water, air and food chain to aesthetics of modern world, need environment-friendly treatment approaches, such as phytoremediation. Phytoremediation technique is proved important tool to tackle the problem as other physical and chemical approaches to decontaminate the polluted soils seem economically unfeasible and time-consuming with less effective results. This ecofriendly remediation approach utilizes two components: one involves the root colonizing microbes and the other involves plant as a whole, rendering the toxic compounds to nontoxic metabolites. Various compounds, viz. synthetic compounds, xenobiotics, pesticides, hydrocarbon and heavy metals, are major contaminants that can be effectively remediated by plants (Figure 25.3). Phytoremediation encompasses several techniques, which utilize vegetation, related enzymes, and other intricate methods. Together, these methods are able to isolate, terminate, transport, or eradicate organic and inorganic pollutants from contaminated media.

Most widely used approach in modern urbanized world is biophilic design. Biophilic design effectuates regaining connection with nature and applying modern methodologies with natural elements to the design of the building environment. Therefore, its realization requires a fundamental change in human consciousness that results in a new ethic of responsibility for protecting and caring for the planet. The biophilic design approach offers significant benefits. This includes multiple social and psychological benefits for residents and the functional and economic benefits of the city as a whole. Cooperation at international level is required to facilitate knowledge sharing on research, policy improvement, biophilic planning, and smart innovation. This not only aids cities to become climate-resistant and sustainable but also complements new financial tools that support technology, smarter urban infrastructure, and a sustainable biophilic urban environment. Increasing investments in natural capital preservation and increasing plant biodiversity

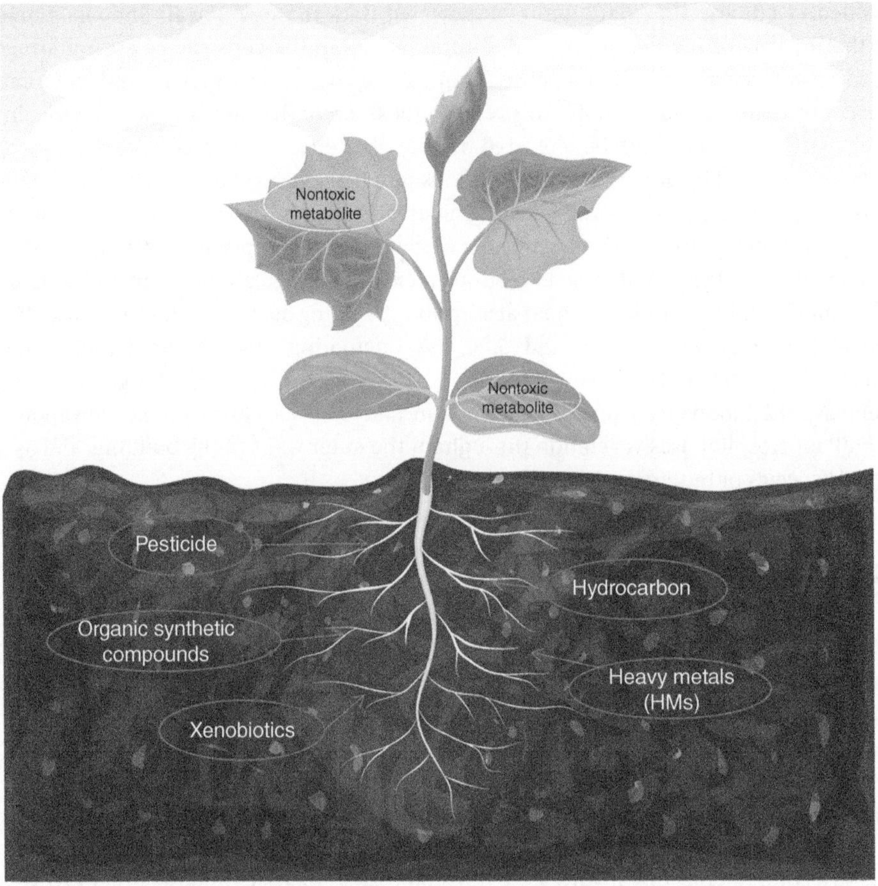

Figure 25.3 Schematic illustration of remediation by plants.

can reduce urban air pollution, promote physical activity, minimize traffic congestion, and reduce environmental health costs for particularly poor and vulnerable groups. Establishing a sustainable urban environment complements the design of biophilic cities.

25.6 Future Perspective

Traditional phytoremediation procedures are less cost-effective on a wide scale because naturally occurring hyperaccumulator plants are often slow growing and produce relatively less harvestable above ground biomass. Thus, using genetic engineering to create transgenic plants with high biomass output, increased metal accumulation, tolerance to metal toxicity, and adaptability to a broad range of environmental conditions may be more advantageous in this regard. Other contemporary phytoremediation techniques, such as chemical-assisted phyto-extraction and microbial assisted phytoremediation, could be utilized to detoxify polluted soils on a broader scale. Further research is necessary for the development of transgenic plants with improved phytoremediation abilities and to comprehend the underlying mechanisms and efficiency of phytoremediative methods in order to enhance their timesaving, and economically feasible features.

Acknowledgment

The authors gratefully acknowledge the Sino-Pakistan Project NSFC Grant No. 31961143008, National Natural Science Foundation of China, International (Regional) Cooperation and Exchange Program, Research fund for International young scientists Grant Nos. 31750110462, 32250410280, and Jiangsu Collaborative Innovation Center for Modern Crop Production (JCIC-MCP) China.

References

Adriano, D.C., Wenzel, W.W., Vangronsveld, J., and Bolan, N.S. (2004). Role of assisted natural remediation in environmental cleanup. *Geoderma* 122: 121–142.

Ahmadpour, P., Ahmadpour, F., Mahmud, T.M.M. et al. (2012). Phytoremediation of heavy metals: a green technology. *Afr. J. Biotechnol.* 11: 14036–14043.

Ali, H., Khan, E., and Sajad, M.A. (2013). Phytoremediation of heavy metals–concepts and applications. *Chemosphere* 91: 869–881.

Ayuso, S.J., Ikaga, T., and Vega-Sanchez, S. (2018). Quantitative improvement in workplace performance through biophilic design: a pilot experiment case study. *Energy Build* 177: 316–328.

Baker, A.J.M., Reeves, R.D., and Hajar, A.S.M. (1994). Heavy metal accumulation and tolerance in British populations of the metallophyte *Thlaspi caerulescens* J and C Presl (Brassicaceae). *New Phytol.* 127: 61–68.

Beatley, T. and Newman, P. (2013). Biophilic cities are sustainable, resilient cities. *Sustainability* 5 (8): 3328–3345.

Bellitürk, K. (2011). Determination of nutrient status of agricultural soils in Uzunkopru county of Edirne province. *J. Tekirdag Agric. Fac.* 8: 8–15.

Bowler, D., Buying-Ali, L., Knight, T., and Pullin, A. (2014). The importance of nature for health: is there a specific benefit of contact with green space? http://www.environmentalevidence.org/wp-content/uploads/2014/07/SR40.pdf (accessed 6 July 2015).

Browning, W.D., Ryan, C.O., and Clancy, J.O. (2014). *14 Patterns of Biophilic Design*. New York: Terrapin Bright Green, LLC.

Heerwagen, J. and Hase, B. (2001). Building biophilia: Connecting people to nature in building design. *Environ. Des. Constrn.* 3: 30–36.

Caravanos, J., Carrelli, J., Dowling, R. et al. (2016). Burden of disease resulting from lead exposure at toxic waste sites in Argentina, Mexico and Uruguay. *Environ. Health* 15: 1–9.

Chan, S.H.M., Qiu, L., Esposito, G., and Mai, K.P. (2021). Vertical greenery buffers against stress: evidence from psycho-physiological responses in virtual reality. *Landsc. Urban Plan.* 213: 104127.

Chaney, R.L. (1983). Food chain pathways for toxic metals and toxic organics in wastes. In: *Environment and Solid Wastes: Characterization, Treatment, and Disposal* (ed. C.W. Francis and S.I. Auerbach), 179–208. Boston, MA: Butterworths Publishers.

Ebbs, S.D., Bradfield, S.J., Kumar, P. et al. (2016). Accumulation of zinc, copper, or cerium in carrot (*Daucus carota*) exposed to metal oxide nanoparticles and metal ions. *Environ. Sci. Nano* 3: 114–126.

Elzeyadi, I.M.K. (2012). Quantifying the impacts of green schools on people and planet. *USGBC Green build Conf Expo.* 48–60.

Erakhrumen, A.A. (2007). Phytoremediation: an environmentally sound technology for pollution prevention, control and remediation in developing countries. *Educ. Res. Rev.* 2: 151–156.

Fromm, E. (1973). *The Anatomy of Human Destructiveness*. New York: Fawcett: Fawcett Crest.

Garbisu, C. and Alkorta, I. (2001). Phytoextraction: a cost effective plant-based technology for the removal of metals from the environment. *Bioresour. Technol.* 77 (3): 229–236.

Ghosh, M. and Singh, S.P. (2005). A review on phytoremediation of heavy metals and utilization of it's by products. *Asian. J. Energy Environ.* 6: 214–231.

Gifford, R. and McGunn, L.J. (2012). Appraisals of built environments and approaches to building design that pomote well-being and healthy behavior. In: *Environmental Psychology: An Introduction* (ed. L. Steg, A.E. van den Berg, and J.I.M. de Groot), 88–95. Hoboken, NJ: Wiley.

Gillis, K. and Gatersleben, B. (2015). A review of psychological literature on the health and wellbeing benefits of biophilic design. *Buildings* 5: 948–963.

Gray, T. and Birrell, C. (2014). Are biophilic-designed site office buildings linked to health benefits and high performing occupants? *Int. J. Environ. Res. Public Health* 11: 12204–12222.

Gullone, E. (2000). The biophilia hypothesis and life in the 21st century: increasing mental health or increasing pathology? *J. Happiness Stud.* 1: 293–321.

Gupta, G.C. (1980). Use of water hyacinth in wastewater treatment (a brief literature review). *J. Environ. Health* 43 (2): 80–82.

Hartig, T., Bringslimark, T., and Patil, G.G. (2008). Restorative environmental design: What, when, where, and for whom? In: *Biophilic Design: The Theory, Science, and Practice of Bringing Buildings to Life* (ed. S.R. Kellert, J. Heerwagen, and M. Mador), 385. Hoboken, NJ: Wiley.

He, Z.L., Yang, X.E., and StoVella, P.J. (2005). Trace elements in agroecosystems and impacts on the environment. *J. Trace Elem. Med. Biol.* 19: 125–140.

Human Spaces (2015). Biophilic Design in Hospitality. http://www.terrapinbrightgreen.com (accessed 23 August 2015).

Jim, C.Y. (2004). Green-space preservation and allocation for sustainable greening of compact cities. *Cities* 21 (4): 311–320.

Joye, Y. (2007). Architectural lessons from environmental psychology: the case of biophilic architecture. *Rev. Gen. Psychol.* 11: 305–328.

van den Berg, A.E. and Joye, Y. (2012). Restorative environments. In: *Environmental Psychology: An Introduction* (ed. A.E. van den Berg and L. Steg), 57–66. Wiley-Blackwell.

Kabisch, N. and Haase, D. (2013). Green spaces of European cities revisited for 1990–2006. *Landsc. Urban Plan.* 110 (1): 113–122.

Kaplan, S. (1995). The restorative benefits of nature: toward an integrative framework. *J. Environ. Psychol.* 15: 169–182.

Kaplan, R. and Kaplan, S. (1989). *The Experience of Nature: A Psychological Perspective*. Cambridge, UK: Cambridge University Press.

Kellert, S.R. (2005). *Building for Life: Designing and Understanding the Human-Nature Connection*. Washington, DC: Island Press.

Kellert, S.R. (2008). Dimensions, elements, and attributes of biophilic design. In: *Biophilic Design: The Theory, Science, and Practice of Bringing Buildings to Life* (ed. S.R. Kellert, J.H. Heerwagen, and M.L. Mador). Hoboken, NJ: Wiley.

Kellert, S.R., Heerwagen, J., and Mador, M. (2011). *Biophilic Design: The Theory, Science and Practice of Bringing Buildings to Life*. Wiley.

Krämer, U. (2010). Metal hyperaccumulation in plants. *Ann. Rev. Plant Biol.* 61: 517–534.

Li, K. and Wusirika, R. (2011). Effect of multiple metal resistant bacteria from contaminated lake sediments on metal accumulation and plant growth. *J. Hazard. Mater.* 189 (1–2): 531–539.

Living Building Challenge (2015). SM 3.0: A visionary path to a regenerative future. https://www.readkong.com/page/living-building-challengesm-3-0-living-future-institute-5328261 (accessed 5 August 2015).

Llao, S.W. and Chang., W.L. (2004). Heavy metal phytoremediation by water hyacinth at constructed wetlands in Taiwan. *J. Aquat. Plant Manage.* 42: 60–68.

Maas, J., Verheij, R.A., Spreeuwenberg, P., and Groenewegen, P.P. (2008). Physical activity as a possible mechanism behind the relationship between green space and health: a multilevel analysis. *BMC Public Health* 10 (8): 206.

Mapodzeke, J.M., Adil, M.F., Dongming, W. et al. (2021). Modulation of key physio-biochemical and ultrastructural attributes after synergistic application of zinc and silicon on rice under cadmium stress. *Plants* 10 (1): 87.

Mudgal, V., Madaan, N., and Mudgal, A. (2010). Heavy metals in plants: phytoremediation: plants used to remediate heavy metal pollution. *Agric. Biol. J. North Am.* 1 (1): 40–46.

Parhizkar, H. and Elzeyadi, I. (2020). Investigating the impact of plant phytoremediation on indoor air quality in work environments: a meta-analysis. *ASHRAE Trans.* 126: 109–119.

Pereira, B.F.F., De-Abreu, C.A., Herpin, U. et al. (2010). Phytoremediation of lead by jack beans on a rhodic hapludox amended with EDTA. *Sci. Agric.* 67: 308–318.

Poff, N.L., Brown, C.M., Grantham, T.E. et al. (2016). Sustainable water management under future uncertainty with eco-engineering decision scaling. *Nat. Clim. Chang.* 6: 25–34.

Riggs, W. (2012). Making healthy places: designing and building for health, well-being, and sustainability. *Berkeley Plan J.* 25 (1): https://doi.org/10.5070/BP325112307.

Ryan, C.O., Browning, W.D., Clancy, J.O. et al. (2014a). Biophilic design patterns emerging nature-based parameters for health and well-being in the built environment. *Int. J. Archit. Res.* 8 (2): 62–76.

Ryan, C.O., Browning, W.D., Clancy, J.O. et al. (2014b). Emerging nature-based parameters for health and well-being in the built environment. *Int. J. Archit. Res.* 8: 62–75.

Sagonda, T., Adil, M.F., Sehar, S. et al. (2021). Physio-ultrastructural footprints and iTRAQ-based proteomic approach unravel the role of *Piriformospora indica*-colonization in counteracting cadmium toxicity in rice. *Ecotoxicol. Environ. Saf.* 220 (3): 112390.

Salt, D.E. and Rauser, W.E. (1995). Mg–ATP dependent transport of phytochelatins across tonoplast of oat roots. *Plant Physiol.* 107: 1293–1301.

Schwitzguebel, J.P. (2000). Potential of pytoremediation, an emerging green technology. *Ecosystem Service and Sustainable Watershed Management in North China, International Conference*, Beijing, P.R., China (23–25 August 2000), 364–350.

Seth, C.S. (2012). A review on mechanisms of plant tolerance and role of transgenic plants in environmental clean-up. *Bot. Rev.* 78: 32–62.

Sharma, P. and Pandey, S. (2014). Status of phytoremediation in world scenario. *Int. J. Environ. Bioremediat. Biodegrad.* 2: 178–191.

Söderlund, J. and Newman, P. (2015). Biophilic architecture: a review of the rationale and outcomes. *Environ. Sci.* 950–969.

Soltan, M.E. and Rasheed, M.N. (2003). Laboratory study on the survival of water hyacinth under several conditions of heavy metal concentrations. *Adv. Environ. Res.* 7 (2): 321–334.

Stephen, R.K. and Heerwagen, J. (2008). *Biophilic Design: The Theory, Science and Practice of Bringing Buildings to Life*. Wiley.

Sumiahadi, A. and Acar, R. (2018). A review of phytoremediation technology: heavy metal update by plants. *IOP Conf. Ser. Earth Environ. Sci.* 142: 012023.

Sun, Y., Zhou, Q., Xu, Y. et al. (2011). The role of EDTA on cadmium phytoextraction in a cadmium–hyperaccumulator *Rorippa globosa*. *J. Environ. Chem. Ecotoxicol.* 3: 45–51.

Terrapin (2015). Terrapin collaborates with organizations to challenge assumptions and develop solutions that lead to improved environmental and financial performance through research, planning, guidelines, and product development. http://www.terrapinbrightgreen.com/ (accessed 20 June 2015).

Timothy, B. (2011). *Biophilic Cities: Integrating Nature into Urban Design and Planning*. Island Press.

Tong, Y.P., Kneer, R., and Zhu, Y.G. (2004). Vacuolar compartmentalization: a second generation approach to engineering plants for phytoremediation. *Trends. Plant Sci.* 9: 7–9.

Tyler, G., Pahlsson, A.M., Bengtsson, G. et al. (1989). Heavy metal ecology and terrestrial plants, microorganisms and invertebrates: a review. *Water Air Soil Pollut.* 47 (3–4): 189–215.

Ulrich, R.S., Simons, R.F., Losito, B.D. et al. (1991). Stress recovery during exposure to natural and urban environments. *J. Environ. Psychol.* 11: 201–230.

Wang, Y. and Akbari, H. (2016). Analysis of urban heat island phenomenon and mitigation solutions evaluation for Montreal. *Sustain. Cities Soc.* 26: 438–446.

WELL Building Standard Resources (2015). The WELL Building Standard, version 1.0. http://www.wellcertified.com/standard (accessed 5 August 2015).

Wilson, E.O. (1984). *Biophilia*. Cambridge, MS: Harvard University Press.

Witters, N., Mendelsohn, R., Van-Passel, S. et al. (2012). Phytoremediation, a sustainable remediation technology? II: Economic assessment of CO_2 abatement through the use of phytoremediation crops for renewable energy production. *Biomass Bioenergy* 39: 470–477.

Wuana, R.A. and Okieimen, F.E. (2011). Heavy metals in contaminated soils: a review of sources, chemistry, risks and best available strategies for remediation. *Int. Sch. Res. Notices* 402647. https://doi.org/10.5402/2011/402647.

Zeshan, A., Muhammad, A., Adil, M.F. et al. (2021). Improvement of morpho-physiological, ultrastructural and nutritional profiles in wheat seedlings through astaxanthin nanoparticles alleviating the cadmium toxicity. *J. Hazard. Mater.* 424 (Part C): 126511.

Zhao, J., Chen, S., Jiang, B. et al. (2013). Temporal trend of green space coverage in China and its relationship with urbanization over the last two decades. *Sci. Total Environ.* 442: 455–465.

26

Genetic Engineering for Heavy Metal/Metalloid Stress Tolerance in Plants

Mohammad Anwar Hossain[1], Md. Tahjib-Ul-Arif[2], Sopnil Ahmed Jahin[1], Abu Bakar Siddique[3], Mumtarin Haque Mim[1], Sharif-Ar-Raffi[1], Muhammad Javidul Haque Bhuiyan[2], and Beatrycze Nowicka[4]

[1] *Department of Genetics and Plant Breeding, Bangladesh Agricultural University, Mymensingh, Bangladesh*
[2] *Department of Biochemistry and Molecular Biology, Bangladesh Agricultural University, Mymensingh, Bangladesh*
[3] *Department of Plant Physiology, Umeå Plant Science Center (UPSC), Umeå University, Umeå, Sweden*
[4] *Department of Plant Physiology and Biochemistry, Faculty of Biochemistry, Biophysics and Biotechnology, Jagiellonian University, Kraków, Poland*

26.1 Introduction

Modern agricultural techniques, anthropogenic activities, fast industrialization and urbanization, and the use of fossil fuels all contribute to an increase in the agroecosystem's threshold level of various heavy metals (HMs) and metalloids (Oves et al. 2012; Stankovic and Stankovic 2013; Alengebawy et al. 2021; Gaur et al. 2021). Although many HMs occur naturally in the biosphere in specific places, the situation worsens due to their release in excess amounts into the soil environment due to the rising frequency of disasters and anthropogenic activities (Wuana and Okieimen 2011). Many HMs and metalloids are regarded as necessary in plants, including nickel (Ni), zinc (Zn), copper (Cu), iron (Fe), cobalt (Co), boron (B), and manganese (Mn) because they are required for fundamental and optimal plant metabolism and development (He et al. 2005; Vatansever et al. 2017). However, excessive concentrations of these critical HMs in the plant–soil continuum cause toxicity symptoms and the failure of several typical plant activities (He et al. 2005; Wani et al. 2018; Tahjib-Ul-Arif et al. 2021). Furthermore, metals such as cadmium (Cd), chromium (Cr), lead (Pb), aluminum (Al), and mercury (Hg), despite having no specific and recognized role in plant cell metabolism, are considered extremely dangerous even at low quantities (Balali-Mood et al. 2021; Jamla et al. 2021).

Heavy metals in the growth medium are hazardous to plants because they impair numerous critical physiological processes such as photosynthesis, respiration, reactive oxygen species (ROS) metabolism, and hormone control (Kalaivanan and Ganeshamurthy 2016; Hoque et al. 2021). Heavy metals have a deleterious impact on the plant's life cycle, from

Heavy Metal Toxicity and Tolerance in Plants: A Biological, Omics, and Genetic Engineering Approach, First Edition. Edited by Mohammad Anwar Hossain, AKM Zakir Hossain, Sylvain Bourgerie, Masayuki Fujita, Om Parkash Dhankher, and Parvez Haris.

germination to final production (Hoque et al. 2021). Overall, greater HM concentrations in plants impede essential nutrient absorption and transport pathways and disrupt other metabolic processes, affecting growth, development, and yield (Asati et al. 2016). Heavy metal pollution impacts agricultural output and quality and cause toxicity in the atmosphere and aquatic bodies. It can eventually build in the human body, creating significant health problems (Edelstein and Ben-Hur 2018; Hu et al. 2019; Mitra et al. 2022). Developing and cultivating HM-tolerant commercial or industrial crop might be a potential use of HM-affected areas. Moreover, developing HM-tolerant plants for effective phytoremediation might be a sustainable solution for the future use of HM-affected areas.

Recent advances imply that genetic engineering can be a viable strategy for enhancing HMs stress tolerance in plants. This chapter aims to offer a comprehensive review of current advances in genome transformation for improved plant tolerance to HMs/metalloids with special reference to the overexpression of genes encoding metal transporters, sulfur metabolites, metal chelators, and components of antioxidative system and glyoxalase pathways. In addition, we have discussed the potential of using CRISPR/Cas system in developing HM tolerance in plants.

26.2 Mechanisms of Heavy Metal/Metalloid Tolerance in Plants

During evolution, plants have acquired multiple mechanisms of protection from HM-induced stress (Figure 26.1). There are two strategies of stress response: avoidance and tolerance. The avoidance strategy is based on the restriction of both the uptake of HM ions by roots (especially for terrestrial plant species) and their transport to the shoots, whereas the tolerance strategy depends on intracellular detoxification of HMs (Singh et al. 2020). Avoidance mechanisms present in the roots are the first line of defense against HM toxicity. They include immobilization by mycorrhizal symbionts, complexation by root exudates, alterations of rhizosphere pH, binding in the cell wall, and redox reactions (Hossain et al. 2012; Singh et al. 2020). Plants exudate several chelating compounds, such as organic acids (mainly malate, citrate, oxalate) and other compounds (e.g. nicotianamine) (Chen et al. 2017). Change of pH in the rhizosphere results in the precipitation of HMs. Bioconversion of metal ions leads to the formation of insoluble oxides and salts. All these processes limit the mobility of HMs in the soil (Hossain et al. 2012; Ghori et al. 2019). Polysaccharides present in cell walls and in excreted mucilage efficiently bind metal cations (Ghori et al. 2019; Pasricha et al. 2021). Plants can also regulate HM transport via decreasing the number of the transporters involved in HM import and by active efflux of metal ions (Manara 2012; Hasan et al. 2017; Ismael et al. 2019). Regulation of transport inside the plant usually results in the limitation of HM translocation from roots to shoots. However, the species called hyperaccumulators maintain high concentrations of HMs in their green tissues (Hossain et al. 2012).

Tolerance mechanisms include chelation of HMs by high-affinity ligands, sequestration in the vacuoles, conversion of more toxic ions into less toxic ones, detoxification of ROS and other toxic compounds such as methylglyoxal (MG) that are formed during HM-induced stress, resynthesis, repair, and degradation of damaged biomolecules (Hossain et al. 2012; Sharma et al. 2016; Figure 26.1). Key chelating compounds are phytochelatins

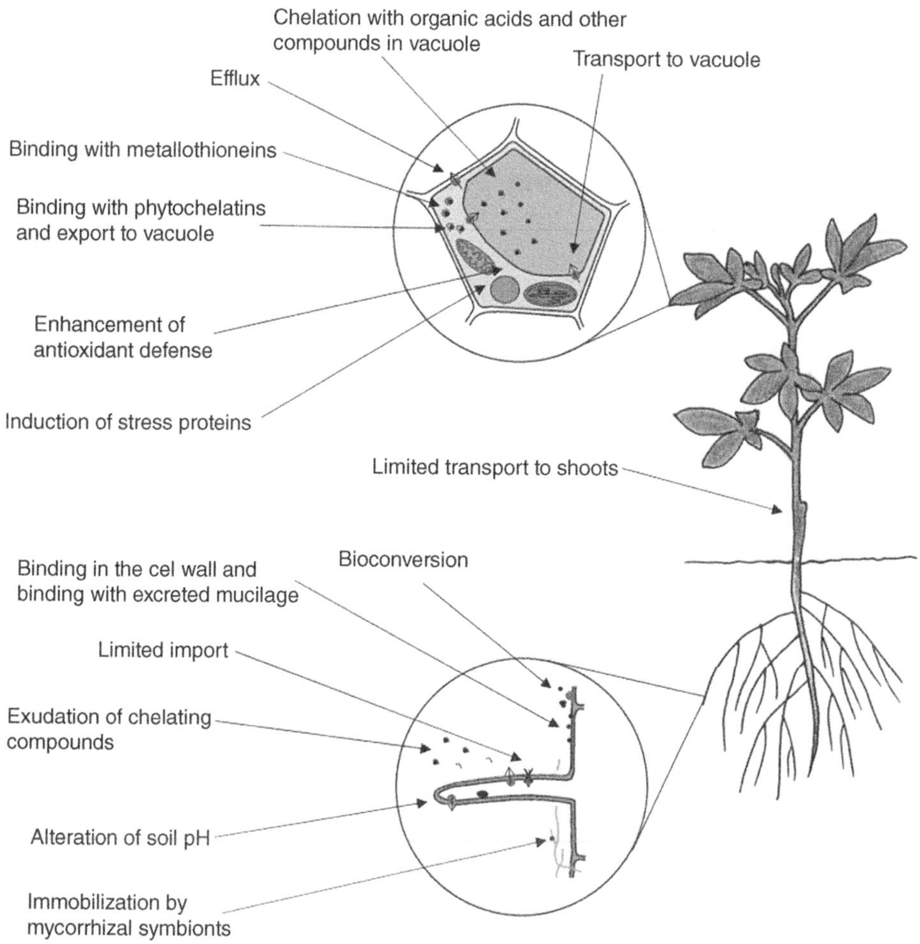

Figure 26.1 Mechanisms of heavy metal tolerance in plants.

(PCs) and metallothioneins (MTs). Phytochelatins are thiol-containing, short-chain peptides produced enzymatically. Metallothioneins are small HM-binding cytosolic proteins. Carboxylic and amino acids (citrate, malate, oxalate, histidine etc.) and phytate are also able to bind HM ions, mainly in the vacuole (Hossain et al. 2012; Hasan et al. 2017). Numerous transporter proteins are involved in the transport of HM ions into the vacuole (Ghori et al. 2019). Antioxidant response comprises both low-molecular-weight compounds able to quench and scavenge ROS, and ROS-detoxifying enzymes. The first group may be subdivided into hydrophilic (ascorbate, glutathione, phenolic compounds, proline) and lipophilic (carotenoids, isoprenoid chromanols and prenylquinones) compounds. Antioxidant enzymes include superoxide dismutases (SODs), catalase (CAT), ascorbate peroxidase (APX), glutathione peroxidases (GPXs), other peroxidases, thioredoxin–glutaredoxin–peroxiredoxin system and proteins involved in ascorbate and glutathione recycling (Nowicka 2022). Meanwhile, considering the repair of biomolecules, the protective role of heat shock proteins is important in the response to HMs (Hasan et al. 2017). Protective mechanisms include both constitutive and adaptive mechanisms. Exposure to HMs triggers signaling cascades leading to the induction and/or enhancement of protection (Hossain et al. 2012; Sytar et al. 2013).

26.3 Strategies for Improving Metal/Metalloid Stress Tolerance in Plants

The adjustment of ionic and osmotic equilibrium in plant cells is a fundamental physiological requirement for plant development. Plants that grow under HM stresses must adapt or acclimate to escape the significant consequences of HM-induced toxicity at the physiological, biochemical, and molecular levels (Rai et al. 2021). Researchers have proposed a number of strategies to improve the HM and metalloid stress tolerance in plants. In general, these strategies can be classified into three categories. The first approach is to screen HM-tolerant genotypes and develop HM-tolerant varieties using the classic breeding procedures. The screening and conventional breeding approach is lengthy process and limited to only specific plant species (Jackson et al. 1996).

The second approach is to create transgenic plants by manipulating several heavy metal transporter genes or transcription factors (TFs). Overexpression of *Arabidopsis thaliana* Heavy Metal ATPase 4 (*AtHMA4*) improved the growth and development of Arabidopsis plants under Zn, Cd, and Co stress conditions (Barabasz et al. 2016). Similarly, overexpression of Multidrug and Toxic Compound Extrusion (*MATE*) gene in *Medicago truncatula* and Aluminum-activated Malate Transporter 4 (*ALMT4*) gene in *Lotus japonicus* enhanced several HM stress tolerance (Wang et al. 2017). *Oryza sativa* Microsomal Triglyceride Transfer Protein 8.1 (*OsMTP8.1*) overexpression enhanced Mn accumulation and tolerance in yeast and rice (Chen et al. 2013). Additionally, the overexpression of the discovered genes that encode or regulate sulfur metabolites, metal chelators, and components of antioxidative system and glyoxalase pathway, and the encoded polypeptides responsible for HM absorption, translocation, and sequestration may enable the development of plants that may be successfully used for phytoremediation (Venegas-Rioseco et al. 2021). For example, transgenic plants overexpressing γ-glutamylcysteine synthetase (*GSH1* or *γ-ECS*), phytochelatin synthase (*PCS1*) (Bennett et al. 2003; Guo et al. 2008), glutathione synthetase (*GS*) (Bennett et al. 2003), and serine acetyl transferase (*SAT*) (Freeman et al. 2004) genes showed enhanced tolerance to several HMs. Moreover, overexpressing antioxidant-related genes such as those encoding CAT (Zhao et al. 2009), APX (Xu et al. 2008), SOD (Lee et al. 2007), and glutathione reductase (GR) (Li et al. 2022) improves tolerance to different HMs.

The third approach is to develop methods for reducing HM stress in plants by improving physiological adaptation, which includes avoiding oxidative damage, boosting photosynthesis, enhancing plant water status, and activating defensive genes. In this method, several chemical protectants or bio-stimulants, such as phytohormones, signaling molecules, minerals, osmolytes, nanoparticles, plant or algae extracts, were applied exogenously in seeds, growing media, or in leaves. This approach has been receiving a lot of attention in recent years because of its adaptability, quick results, and lower economic cost, which allows for the decrease of HM toxicity in a variety of plant species (Hua et al. 2021).

For example, exogenous nitric oxide (Terrón-Camero et al. 2019), ascorbic acid and citric acid (Tahjib-Ul-Arif et al. 2021), glycinebetaine (Bhatti et al. 2013), sulfur (Lu et al. 2019), cysteine (Cys) (Terzi and Yıldız 2021), 3-indoleacetic acid (Zhou et al. 2020), and salicylic acid (Alamri et al. 2018) were applied to improve plant tolerance to different HMs. In addition, plant growth promoting rhizobacteria (Singh et al. 2019) and arbuscular mycorrhizal fungi (Riaz et al. 2021) can effectively alleviate HM or metalloid stress in plants.

26.4 Transgenic Plants and Heavy Metal/Metalloid Stress Tolerance in Plants

26.4.1 Sulfur Metabolism Engineering and Heavy Metal Tolerance

Multi-functional role of sulfur-containing compounds in improving HM tolerance has been well documented in plants (Rajab et al. 2020). Sulfur-containing compounds, especially glutathione (GSH) and Cys, are vital in regulating HM-induced oxidative stress tolerance. Reactive oxygen species generated by HM or metalloid stress are directly detoxified by GSH (Schützendübel and Polle 2002), which enables maintaining redox homeostasis for metabolism, and plays a role in signal transduction and regulation of gene expression (Foyer and Noctor 2005). GSH is also a precursor of PC synthesis. Genes encoding enzymes involved in GSH biosynthesis, such as γ-ECS(*GSH1*) and GS(*GSH2*), and Cys biosynthesis (such as SAT and ATP sulfurylase, ATPS) were found to be upregulated in response to HM or metalloid stress (Rajab et al. 2020). A large number of studies (Table 26.1) have shown that transgenic plants overexpressing genes of GSH biosynthetic pathway or its analogues displayed enhanced HM or metalloid stress tolerance through the detoxification of ROS or PCs synthesis (Dhankher et al. 2002; Bennett et al. 2003; Freeman et al. 2004; Wangeline et al. 2004;Bittsanszky et al. 2005; Guo et al. 2008).

26.4.2 Glyoxalase Pathway Genes and Heavy Metal Stress Tolerance

Recent genetic engineering studies with glyoxalase pathway (composed of two enzymes, Gly I and Gly II able to detoxify MG to less toxic D-lactate) genes also conferred HM or metalloid stress tolerance through the regulation of multiple stress responsive pathways (Singla-Pareek et al. 2006; Lin et al. 2010; Zeng et al. 2016). Singla-Pareek et al. (2006) reported that the transgenic tobacco (*Nicotiana tabacum*) overexpressing both Gly I and Gly II showed enhanced HM stress tolerance as compared to single-gene transformants. Transgenic plants produced normal viable seeds even under high concentration of HM (5 mM $ZnCl_2$) and accumulated lesser amount of HM in the seeds. Biochemical studies showed that the transgenic plants (single or double gene transformants) accumulated lower amount of MG and displayed less pronounced lipid peroxidation as compared to WT plants. Their results also indicated that the increased level of PCs and maintenance of GSH homeostasis are the key players in formulating HM stress tolerance in transgenic plants. Lin et al. (2010) also showed that the transgenic plants overexpressing Gly I gene displayed improved tolerance to HM-induced stress ($ZnCl_2$). Similarly, Zeng et al. (2016) reported that transgenic rice plants overexpressing Gly I gene showed

Table 26.1 Sulfur metabolism engineering and heavy metal/metalloid stress tolerance.

Gene	Host plant	Findings/proposed mechanisms/reported phenotype	References
GSH1+AsPCS1	*Arabidopsis thaliana*	Transgenic plants overexpressing *GSH1* and/or *AsPCS1* showed improved root traits under Cd- or As-induced stress as compared to WT plants. The transgenic plants also accumulated higher As and Cd levels in above-ground tissues. Importantly, dual-gene transformants exhibited significantly higher tolerance and accumulated more Cd and As	Guo et al. (2008)
γ-ECS and *GS*	*Brassica juncea* (L.) Czern.	Transgenic plants overexpressing *γ-ECS* and *GS* accumulated 1.5-fold more Cd, 1.5 to 2-fold more Zn, and 2.4-fold to 3-fold more Cr, Cu, and Pb as compared to control. This study demonstrates the potential application of transgenic plants to remove HMs from the contaminated soil	Bennett et al. (2003)
γ-ECS	*Arabidopsis thaliana*	Transgenic plants overexpressing *γ-ECS* showed enhanced As tolerance and accumulation as compared to WT plants. However, the transgenic plants overexpressing both *γ-ECS* and arsenate reductase (*ArsC*) produced four to sevenfold more fresh weight and accumulated two to threefold more As per gram of tissue than WT plants or plants expressing *γ-ECS* or *ArsC* alone.	Dhankher et al. (2002)
APS	*Brassica juncea* (L.) Czern.	Transgenic plants overexpressing APS gene showed higher tolerance to As(III), As(V), Cd, Cu, Hg, and Zn, but they were less tolerant to Mo and V as compared to WT plants. Seedlings of transgenic line had up to 2.5-fold higher shoot concentrations of As(III), As(V), Hg, Mo, Pb, and V, and somewhat lower Cr levels	Wangeline et al. (2004)
SAT	*Arabidopsis thaliana*	Transgenic plants overexpressing SAT gene showed higher Ni stress tolerance as indicated by improved shoot (fivefold increase) and root (1.5-fold increase) growth. Importantly, differences in Ni resistance observed in the *SAT* overproducing lines also strongly correlated with total GSH concentrations in shoots of each line	Freeman et al. (2004)
GSH1	*Populus nigra*	Transgenic plants overexpressing *GSH1* gene either in cytosol or chloroplast showed improved tolerance to Zn	Bittsanszky et al. (2005)

APS, ATP sulfurylase; AsPCS1, phytochelatin synthase of *Allium cepa*; GS, glutathione synthetase; GSH1, bacterial γ-glutamylcysteine synthetase (γ-ECS); SAT, serine acetyl transferase.

enhanced tolerance to HM ($ZnCl_2$) as indicated by higher shoot height and shoot weight. The lower level of malonyldialdehyde (MDA) was also found in Gly I overexpressing plants. The transgenic plants showed higher Gly I activity and lower level of MG as compared to wild-type plants.

26.4.3 Enhanced Antioxidant Defense and Heavy Metal Tolerance

The increased formation of ROS and the occurrence of oxidative stress are intrinsic biomarkers of HM or metalloid stress. Although ROS at optimum level regulate a variety of stress responsive pathways, at higher concentration they are detrimental to cellular components leading to diminished growth or even death of plants (Hossain et al. 2012). Stringent regulation of ROS level is therefore vital for conferring tolerance to HM- or metalloid-induced oxidative stress. Large number of recent studies (Table 26.2) have shown that overexpression of genes encoding ROS-detoxifying enzymes in different plant species improved HM or metalloid stress tolerance through favorable modulation of the activities of multiple enzymes engaged with ROS detoxification or ROS-mediated signaling (Lee et al. 2007; Xu et al. 2008; Zhao et al. 2009; Dixit et al. 2011; Kumar et al. 2013; Liu et al. 2013; Srivastava et al. 2018; Li et al. 2022).

26.4.4 Phytochelatin and Metallothionein Genes and Heavy Metal Tolerance

Plants belonging to species of HM hyperaccumulators have distinct regulatory network to sequestrate and chelate HM ions (Talebi et al. 2019). There are specific metal-binding peptides such as PCs and MTs which are used by plants to immobilize, sequester and detoxify heavy metals (Shukla et al. 2013). Metallothioneins and PCs are well-defined heavy metal-binding ligands in plants (Joshi et al. 2016; Fan et al. 2018). The regulatory network maintains metal homeostasis through metal transport, chelation, trafficking, and sequestration activities. Both the PCs and MTs bind with metal ionsto produce stable metal complexes which are transferred to the vacuole by protein transporters (Cabral et al. 2018). These transporters maintain cellular homeostasis by efflux of the metal ions through the plasma membrane from the cytoplasm. Excessive generation of ROS is a common phenomenon in plants exposed to abiotic stresses, which damage the cells and cause ionic imbalance. Due to presence of thiols group in MTs, they bind and remove the excessive ROS from the cells. In the past few decades, a number of PCs and MTs genes have been identified and characterized along with their roles in improving the tolerance mechanisms against HM-induced stress (Fan et al. 2018; Su et al. 2020; Rono et al. 2021). Transgenic approaches to overexpress the PCs and MTs genes with their considerable potential in phytoremediation have been described in Tables 26.3 and 26.4, respectively.

26.4.5 Metal Ion Transporter Genes/Proteins and Heavy Metal Stress Tolerance

Maintaining metal homeostasis by metal uptake, trafficking, storage, chelation and efflux is considered very important mechanism enabling to avoid or counter effect to metal toxicity (Fasani et al. 2021). Transport of the metal ions alone or in a combined form with PCs and MTs to the vacuoles or other storage organs by a group of proteins belonging to various families of transporter proteins has been observed (Gao et al. 2020). As an example, the cation diffusion facilitator (CDF) family genes are well documented

Table 26.2 Transgenic plants overexpressing antioxidant genes and HM stress tolerance.

Gene	Host plant	Findings/proposed mechanisms/reported phenotype	References
GST	*Nicotiana tabacum*	GST overexpressing plants showed normal growth under drought, NaCl-, and Cd-induced stresses, and maintained higher GST activity and a significantly lower level of ROS	Liu et al. (2013)
GST + CAT	*Oryza sativa*	Transgenic plants under Cd stress and combined stress (Cd and heat) conditions showed a sharp increase in CAT, GST, APX, MDHAR, DHAR, GR activities, and increased oxidative stress tolerance	Zhao et al. (2009)
APX	*Arabidopsis thaliana*	Transgenic plants showed improved tolerance to Zn and Cd as indicated by lower ROS (H_2O_2) and MDA levels. The Zn and Cd content in the shoot tissues were higher in the transgenic plants as compared to WT plants	Xu et al. (2008)
CuZnSOD + APX	*Festuca arundinacea* Schreb.	Transgenic plants under Cd, Cu, and As stress showed lower H_2O_2 and MDA levels as compared to WT plants. Transgenic plants also displayed higher SOD and APX activities and maintained lower chlorophyll destruction and electrolyte leakage (EL) under HM stress	Lee et al. (2007)
GST	*Nicotiana tabacum*	Transgenic plants showed less pronounced lipid peroxidation under different levels of Cd stress. Under Cd stress transgenic plants also displayed enhanced GST, SOD, APX, GPX and CAT activities leading to HM tolerance. Cadmium accumulation in the plant biomass in transgenic plants was similar or lower than that of the WT plants	Dixit et al. (2011)
GR	*Arabidopsis thaliana*	Transgenic plants displayed increased root growth and biomass yield as compared to WT plants. Transgenic plants maintained higher GSH level and higher GSH/GSSG ratio. Under Cu stress, transgenic plants also maintained higher GR and POD activities and lower level of H_2O_2 conferring transgenic plants more tolerant to HM stress than WT plants	Li et al. (2022)
GST	*Arabidopsis thaliana*	Transgenic plants showed higher germination percentage and root length as compared to WT plants under Cd, Cr, and As stress. Interestingly, the transgenic plants also displayed enhanced tolerance to other abiotic stress factors such as cold, osmotic stress, and salt	Kumar et al. (2013)
GST	*Arabidopsis thaliana*	Transgenic plants displayed improved growth and yield attributing traits under Cr stress and drought stress conditions. They also showed higher proline content and lower MDA and EL under Cr stress and drought stress conditions	Srivastava et al. (2018)

APX, ascorbate peroxidase; CAT, catalase; DHAR, dehydroascorbate reductase; GR, glutathione reductase; GST, glutathione *S*-transferase; MDHAR, monodehydroascorbate reductase; POD, peroxidase; SOD, superoxide dismutase.

Table 26.3 Transgenic plants overexpressing PCS genes and HM stress tolerance.

Gene	Host plant	Findings/proposed mechanisms/reported phenotype	References
PCS1	*Arabidopsis thaliana*	Transgenic plants showed improved Cd tolerance with an increase in root length and better growth status during the seedling stage	Su et al. (2020)
PCS1	*Brassica napus*	Transgenic plants had significantly longer roots with higher root fresh weight compared to the WT plants. Transgenic plants also displayed enhanced activity of SOD and CAT, and increased PCs content under Cd stress. Leaf chlorophyll content was higher in transgenic plants when compared to WT plants	Bai et al. (2019)
PCS1	*Arabidopsis thaliana*	Transgenic plants showed improved growth with an increase in root length and PCs content under Cd stress, which was postulated to result from the enhanced detoxification of metal ions	Fan et al. (2018)
PCS1	*Oryza sativa*	Transgenic lines displayed enhanced tolerance to As and Cd. Disruption of *PCS1* in seedlings resulted in the sensitivity to Cd and As during the seedling stage. Mutant of *PCS1* showed weak phenotype with hypersensitive response to Cd and As stress	Uraguchi et al. (2017)
PCS1	*Oryza sativa*	Higher accumulation of PCs was observed in roots of transgenic lines with an unchanged PCs levels in shoots compared to WT plants. Developing seeds of transgenic lines showed higher amount of transcripts of *PCS1*. Mature seeds of these lines showed lower accumulation of Cd due to the high expression of *PCS1*	Das et al. (2017)
PCS1	*Brassica juncea*	Transgenic lines with moderate *PCS1* expression levels displayed significantly improved tolerance to Cd and Zn stress. However, transgenic lines with higher expression of *PCS1* did not exhibit enhanced Cd and Zn tolerance. Increased levels of PCs and thiol groups were observed in transgenic lines	Gasic and Korban (2007)
PCS1	*Nicotiana tabacum*	Transgenic lines showed increased accumulation of Cd in shoot and roots accompanied with higher content of PCs. Cd translocation towards the shoot was unaffected in transgenic plants at both seedling and adult stage. The degree of tolerance to Cd in transgenic plants with higher expression of *PCS1* showed dependence on the availability of GSH. Also, the overexpression of PCs did not enhance long distance root to shoot Cd transport	Pomponi et al. (2006)
PCS1	*Arabidopsis thaliana*	Transgenic plants showed normal growth at the shoot and root meristem level, whereas the WT plants showed roots with brown coloration when grown under the stress conditions. *PCS1* overexpression suppressed the HM sensitivity in mature plants. *PCS1* was postulated to be involved in the enhancement and acceleration of Cd transport in transgenic plants with lower accumulation of Cd in the roots by enhancing the long-distance root to shoot Cd transport	Gong et al. (2003)

(*Continued*)

Table 26.3 (Continued)

Gene	Host plant	Findings/proposed mechanisms/reported phenotype	References
PCS1	*Nicotiana tabacum*	Transgenic lines showed increased accumulation of PCs accompanied with increased accumulation of Cd in roots and shoots. Cd-exposed seedlings of transgenic plants showed increased root length when compared to WT. Stressed transgenic lines showed decreased MDA content and significantly increased activities of SOD, POD, and CAT	Liu et al. (2011)
PCS2	*Arabidopsis thaliana*	Overexpressing plants showed increased PCs accumulation with a distinct enzymatic difference between transgenic and WT plants. When exposed to Cd-induced stress, transgenic plants showed improved growth, whereas WT plants displayed leaf chlorosis along with a reduction in overall growth	Kuhnlenz et al. (2014)
PCS2	*Oryza sativa*	Transgenic plants showed unaltered accumulation of *PCS2* in shoots, whereas the up-regulation of *PCS2* under Cd stress was observed in these plants in developing seeds. Transgenic plants also displayed lower accumulation of Cd in mature seeds.	Das et al. (2017)

Table 26.4 Transgenic plants overexpressing MTs genes and HM stress tolerance.

Gene	Host plant	Findings/proposed mechanisms/reported phenotype	References
MT1	*Cicer arietinum*	Transgenic seedlings displayed improved growth under Pb stress compared to WT plants. Increased accumulation of proline and decreased accumulation of MDA was found in transgenic lines under Pb stress. Transgenic lines showed higher photosynthesis rate coupled with higher WUE under drought stress. The decreased level of H_2O_2 in transgenic lines under stress conditions showed that the improved stress tolerance resulted from the enhanced antioxidant activity. Also, the content of proline and L-cysteine in transgenic lines was increased compared to WT plants	Kumar et al. (2022)
MT1	*Oryza sativa*	Transgenic plants displayed enhanced tolerance to Cd stress due to improved detoxification of this metal. Transgenic lines showed increased shoot height and primary root length under Cd stress, as well as the increased root dry weight. Their leaves showed higher chlorophyll content when compared to stressed WT plants	Rono et al. (2021)
MT1	*Nicotiana tabacum*	Transgenic plants showed lack of chlorosis and improved root and shoot growth as compared to WT plants under higher level of Cu stress. Transgenic lines also displayed minimum reduction of root dry weight under stress conditions. Transgenic plants showed significantly increased Cu accumulation in roots and shoots accompanied with less-pronounced Cu-induced lipid peroxidation. POD activity was higher in transgenic plants; however, no significant differences in the activities of CAT, APX, SOD, and GR between the WT and transgenic plants were observed	Xia et al. (2012)

Table 26.4 (Continued)

Gene	Host plant	Findings/proposed mechanisms/reported phenotype	References
MT1	*Arabidopsis thaliana*	*MT1* knockout plants were hypersensitive to Cd and accumulated several fold lower levels of As, Cd and Zn compared to WT plants under Cd stress. Transgenic lines showed lower relative conductivity and lower MDA content indicating the less pronounced membrane damage	Zimeri et al. (2005)
MT2	*Nicotiana tabacum*	Transgenic lines showed lack of chlorosis and improved growth compared to WT plants under Zn stress. With the time, transgenic lines displayed improved floral organs with shorter life cycle and early maturity. Transgenic plants showed higher number of pods and total pods per plant under Zn stress. They also displayed improved photosynthetic parameters and higher photosynthetic efficiency compared to WT plants under stress conditions	Patel et al. (2021)
MT2	*Arabidopsis thaliana*	Transgenic plants showed significantly longer roots, larger sized leaves and greater biomass accumulation compared to the WT plants under salinity stress. Transgenic lines displayed the heavy branching of roots under H_2O_2 stress. Transgenic lines showed significantly lower Na^+ concentration under salinity stress accompanied with increased leaf chlorophyll content and higher photosynthetic activity. They also showed increased SOD activity and decreased MDA level under drought stress. Soil grown transgenic lines displayed improved tolerance to drought and salinity with lower accumulation of ROS	Patankar et al. (2019)
MT3	*Nicotiana tabacum*	Transgenic plants showed decreased accumulation of Cd in above-ground parts with similar content in lower parts when compared to WT plants. Transgenic plants performed better than WT plants in terms of leaf phenotypes. They also showed higher level of bound water, which increased the plant resistance	Zhi et al. (2020)
MT2 and *MT3*	*Vicia faba*	Transgenic plants were able to protect the chloroplasts of their guard cells from degradation upon exposure to Cd by maintaining lower level of ROS. Cd tolerance was suggested to be achieved not by the sequestration of this HM in the vacuole, but rather due to the reduction of ROS	Lee et al. (2004)

MTs, metallothioneins; WUE, water use efficiency.

with their efficient roles in efflux of the metal ions from the cytoplasm to either subcellular compartments or outside the cells to prevent their entry into cellular functions (Gustin et al. 2011; Gao et al. 2020). Transporters belonging to ABC protein family have been also described as detoxification proteins due to their ability to form a complex with heavy metals which are mostly functional in chloroplasts or cellular cytoplasm (Chen et al. 2021).Transgenic plants overexpressing the genes encoding transporter proteins have shown their roles in the enhancement of the plant tolerance to HM-induced stress. The results of the research on such transgenic lines were described in Table 26.5.

Table 26.5 Transgenic plants overexpressing metal ion transporter genes/proteins and HM stress tolerance.

Gene/ proteins	Host plant	Findings/proposed mechanisms/reported phenotype	References
ABCC3	*Arabidopsis thaliana*	Transgenic lines had longer roots than WT plants under Cd treatment, they also showed the increased vacuolar sequestration of Cd. However, ABCC3 was found to play a role in Cd sequestration under low concentration of this HM. ABCC3 is involved in the Cd detoxification by transporting the PC–Cd complexes into the vacuoles	Brunetti et al. (2015)
PDR8	*Arabidopsis thaliana*	Transgenic plants with *PDR8* knockout were more sensitive to Cd and Pb stress. Knocked out transgenic lines displayed the decreased shoot fresh weight, root length, and chlorophyll content when compared to WT plants under stress conditions. On the other hand, transgenic lines overexpressing *PDR8* displayed improved growth and resistance against Pb stress accompanied by a higher efflux of Cd^{2+}	Kim et al. (2007)
PDR12 (ABCG40)	*Arabidopsis thaliana*	Transgenic lines grew better with increased fresh weight and plant height than the WT plants under Pb stress. However, Pb resistance conferred by PDR12 was found to be independent of glutathione	Lee et al. (2005)
ZIP	*Oryza sativa*	Overexpression of *OsZIP1* in rice showed better growth under excess Zn, Cu, and Cd stress conditions as compared to wild-type plants. In contrast, both *oszip1* mutant and RNA interference (RNAi) lines showed a less accumulation of metal ions in roots rendering sensitivity to metal stress. These results suggest OsZIP1 is able to function as a metal exporter in rice under excess metal stress (Zn, Cu, and Cd) conditions	Liu et al. (2019)
HMA4	*Nicotiana tabacum var. Xanthi*	*AtHMA4* overexpressing tobacco plants showed a decrease in Cd uptake, root and shoot Cd concentration as well as root/shoot Cd concentration ratio as compared to wild type plants	Siemianowski et al. (2011)
MTP8	*Arabidopsis thaliana*	Transgenic lines were able to maintain high germination rate under Mn stress. They also showed the increased Mn accumulation in the seed coat. The expression of *MTP8* was correlated with the resistance of imbibing seeds to Mn stress	Höller et al. (2021)
MTP1 and *ZIP1*	*Manihot esculenta*	Transgenic plants overexpressing *ZIP1* and *MTP1* showed increased accumulation of Zn in their storage root parenchyma tissues compared to the WT plants. Enhanced accumulation of Zn in leaf tissues was found in transgenic plants overexpressing *ZIP1*	Gaitan-Solis et al. (2015)

ABCC3, ABC transporter C family member 3; HMA, *heavy metal* transporter *ATPase;* MTP, metal tolerance protein; PDR, pleiotropic drug resistance; ZIP, Zn-regulated transporter, iron-regulated transporter-like protein; ZIP, the Zrt/Irt-like protein.

26.5 CRISPR/Cas System and Heavy Metal Tolerance Development

Due to the immense improvement in plant molecular biology with genetic modifications, next plant breeding tools encompass a range of technologies to improve the stress tolerance in plants (Detain et al. 2022). Transgenic methods have been successfully used to improve the tolerance to HMs by enhancing the expression of genes related to metal ions transportation, sequestration, chelation and reducing the detrimental impact by activating the mechanisms needed to scavenge the HM-induced ROS (Hasanuzzaman et al. 2013). To find out and introduce the novel gain of functions of genes responsible for heavy metal tolerance, marker assisted selection is being used as a dominant method (Wang et al. 2016). CRISPR/Cas (Clustered Regulatory Interspaced Short Palindromic Repeat/CRISPR-associated protein) is very recent approach to improve the crop tolerance with or without GMOs. CRISPR/Cas systems offer the opportunity to overexpress, repress, knockout, and perform gene editing to help the plant to face the climate changes and the heavy metal stress (Kaur et al. 2019). Besides, the availability of sequence data of many important crops along with the higher support from the bioinformatics opened the new window to develop new crops having improved tolerance to climate change (Sarma et al. 2021).

CRISPR/Cas system has been exploited in genome editing and has been reported in different types of crops such as cereals, vegetables, fiber and fruit crops (Nakajima et al. 2017; Kim and Kang 2018; Miki et al. 2021). This method has the full potentiality in introgression of genes having a role in detoxification, bioaccumulation, chelation, and degradation mechanisms (Rai et al. 2021). CRISPR/Cas has been used to express the genes which products play a role in the formation of ligands (using PCs and MTs), in metal transport (HMA, CDF, MATE, and ZIP families), root exudation, as well as those responsible for an increase in the levels of phytohormones, which in turn shape the plant tolerance to heavy metals (Saxena et al. 2020; Rai et al. 2021). Phytoremediation based on CRISPR/Cas9 system can be used successfully to refine the soil profile, as well as to remove toxic metals, which in turn increases the crop productivity in HM-contaminated soils. Genome editing using the CRISPR/Cas system has been successfully applied in soybean to introduce several tolerance-related genes, which play role in developmental process and environmental cues (Cai et al. 2015). CRISPR/Cas9 has been used to develop the mutant lines for Cd tolerance in rice (Songmei et al. 2019).

However, CRISPR/Cas system has several pitfalls. Though marker free transgenic plants can be obtained using different techniques including several back crossing, it might cause off-target mutations, which can alter the gene expression and phenotypic traits. The elimination is difficult in the case of crops having higher heterozygosity or self-incompatibility. There are several factors needed to be considered in using the CRISPR/Cas9 techniques including the desired target sites, delivery vectors, and genotypes of plants (Agarwal and Rani 2022).

26.6 Conclusions and Future Prospects

Plant genetic alterations, including HM-responsive genes, TFs, and metal-binding proteins or ligands, showed significant improvements in plant tolerance to HM-induced stress and phytoremediation capabilities. By targeting specific genes via genome editing using site-specific nucleases researchers can improve abiotic stress tolerance in plants, including HM stress. In this regard, the CRISPR/Cas system has been recognized as an effective technique for plant genome engineering. Although the CRISPR/Cas9 technique has been extensively used for adapting plant species against abiotic or biotic stressors, its application in improving HM stress tolerance remains elusive. However, their full potential must be explored further by conducting future experiments utilizing a comprehensive approach to generate superior crop varieties with enhanced endurance to various abiotic stresses, including HM-induced stress. At the same time, researchers and policymakers should take initiatives to increase public awareness about GM technology to end the controversy surrounding transgenic crops.

Acknowledgment

Special thanks to Dr. Arifin Sandhi, Department of Biology and Environmental Science, Faculty of Health and Life Sciences, Linnaeus University, Kalmar 391 82, Sweden, for careful reading and critical comments on the chapter.

References

Agarwal, P. and Rani, R. (2022). Strategic management of contaminated water bodies: omics genome-editing and other recent advances phytoremediation. *Environ. Tech. Innov.* 27: 102463.

Alamri, S., Siddiqui, M.H., Al-Khaishany, M.Y. et al. (2018). Exogenous application of salicylic acid improves tolerance of wheat plants to lead stress. *Adv. Agric. Sci.* 6: 25–35.

Alengebawy, A., Abdelkhalek, S.T., Qureshi, S.R., and Wang, M.-Q. (2021). Heavy metals and pesticides toxicity in agricultural soil and plants: ecological risks and human health implications. *Toxics* 9: 42.

Asati, A., Pichhode, M., and Nikhil, K. (2016). Effect of heavy metals on plants: an overview. *Int. J. Appl. Innov. Eng. Manag.* 5: 56–66.

Bai, J., Wang, X., Wang, R. et al. (2019). Overexpression of three duplicated *BcPCS* genes enhanced Cd accumulation and translocation in *Arabidopsis thaliana* mutant *cad1-3*. *Bull. Environ. Contam. Toxicol.* 102: 146–152.

Balali-Mood, M., Naseri, K., Tahergorabi, Z. et al. (2021). Toxic mechanisms of five heavy metals: mercury, lead, chromium, cadmium, and arsenic. *Front. Pharmacol.* 12: 643972.

Barabasz, A., Klimecka, M., Kendziorek, M. et al. (2016). The ratio of Zn to Cd supply as a determinant of metal-homeostasis gene expression in tobacco and its modulation by overexpressing the metal exporter AtHMA4. *J. Exp. Bot.* 67: 6201–6214.

Bennett, L.E., Burkhead, J.L., Hale, K.L. et al. (2003). Analysis of transgenic Indian mustard plants for phytoremediation of metal-contaminated mine tailings. *J. Environ. Q.* 32: 432–440.

Bhatti, K.H., Anwar, S., Nawaz, K. et al. (2013). Effect of exogenous application of glycinebetaine on wheat (*Triticum aestivum* L.) under heavy metal stress. *Middle-East J. Sci. Res.* 14: 130–137.

Bittsanszky, A., Komives, T., Gullner, G. et al. (2005). Ability of transgenic poplars with elevated glutathione content to tolerate $Zn(^{2+})$ stress. *Environ. Int.* 31: 251–254.

Brunetti, P., Zanella, L., De Paolis, A. et al. (2015). Cadmium-inducible expression of the ABC-type transporter AtABCC3 increases phytochelatin-mediated cadmium tolerance in Arabidopsis. *J. Exp. Bot.* 66: 3815–3829.

Cabral, A., Jakovleska, J., Deb, A. et al. (2018). Further insights into the metal ion binding abilities and the metalation pathway of a plant metallothionein from *Musa acuminata*. *J. Biol. Inorg. Chem.* 23: 91–107.

Cai, Y., Chen, L., Liu, X. et al. (2015). CRISPR/Cas9 mediated genome editing in soybean hairy roots. *PLoS One* 10: e0136064.

Chen, Z., Fujii, Y., Yamaji, N. et al. (2013). Mn tolerance in rice is mediated by MTP8.1, a member of the cation diffusion facilitator family. *J. Exp. Bot.* 64: 4375–4387.

Chen, Y.T., Wang, Y., and Yeh, K.C. (2017). Role of root exudates in metal acquisition and tolerance. *Curr. Opin. Plant Biol.* 39: 66–72.

Chen, Y., Li, G., Yang, J. et al. (2021). Role of *Nramp* transporter genes of *Spirodela polyrhiza* in cadmium accumulation. *Ecotoxicol. Environ. Saf.* 227: 112907.

Das, N., Bhattacharya, S., Bhattacharyya, S., and Maiti, M.K. (2017). Identification of alternatively spliced transcripts of rice phytochelatin synthase 2 gene *OsPCS2* involved in mitigation of cadmium and arsenic stresses. *Plant Mol. Biol.* 94: 167–183.

Detain, A., Bhowmik, P., Leborgne-Castel, N., and Ochatt, S. (2022). Latest biotechnology tools and targets for improving abiotic stress tolerance in protein legumes. *Environ. Exp. Bot.* 197: 104824.

Dhankher, O.P., Li, Y., Rosen, B.P. et al. (2002). Engineering tolerance and hyperaccumulation of arsenic in plants by combining arsenate reductase and γ-glutamylcysteine synthetase expression. *Nat. Biotechnol.* 20: 1140–1145.

Dixit, P., Mukherjee, P.K., Ramachandran, V., and Eapen, S. (2011). Glutathione transferase from *Trichoderma virens* enhances cadmium tolerance without enhancing its accumulation in transgenic *Nicotiana tabacum*. *PLoS One* 6: e16360.

Edelstein, M. and Ben-Hur, M. (2018). Heavy metals and metalloids: sources, risks and strategies to reduce their accumulation in horticultural crops. *Sci. Hortic.* 234: 431–444.

Fan, W., Guo, Q., Liu, C. et al. (2018). Two mulberry phytochelatin synthase genes confer zinc/cadmium tolerance and accumulation in transgenic Arabidopsis and tobacco. *Gene* 645: 95–104.

Fasani, E., DalCorso, G., Zorzi, G. et al. (2021). Overexpression of ZNT1 and NRAMP4 from the Ni hyperaccumulator *Noccaea caerulescens* population Monte Prinzera in *Arabidopsis thaliana* perturbs Fe, Mn, and Ni accumulation. *Int. J. Mol. Sci.* 22: 11896.

Foyer, C.H. and Noctor, G. (2005). Redox homeostasis and antioxidant signaling: a metabolic interface between stress perception and physiological responses. *Plant Cell* 17: 1866–1875.

Freeman, J.L., Persans, M.W., Nieman, K. et al. (2004). Increased glutathione biosynthesis plays a role in nickel tolerance in Thlaspi nickel hyperaccumulators. *Plant Cell* 16: 2176–2191.

Gaitan-Solis, E., Taylor, N.J., Siritunga, D. et al. (2015). Overexpression of the transporters *AtZIP1* and *AtMTP1* in cassava changes zinc accumulation and partitioning. *Font. Plant. Sci.* 6: 492.

Gao, J., Chen, B., Lin, H. et al. (2020). Identification and characterization of the glutathione S-transferase (GST) family in radish reveals a likely role in anthocyanin biosynthesis and heavy metal stress tolerance. *Gene* 743: 144484.

Gasic, K. and Korban, S.S. (2007). Transgenic Indian mustard (*Brassica juncea*) plants expressing an Arabidopsis phytochelatin synthase (*AtPCS1*) exhibit enhanced As and Cd tolerance. *Plant Mol. Biol.* 64: 361–369.

Gaur, V.K., Sharma, P., Gaur, P. et al. (2021). Sustainable mitigation of heavy metals from effluents: toxicity and fate with recent technological advancements. *Bioengineered* 12: 7297–7313.

Ghori, N.H., Ghori, T., Hayat, M.Q. et al. (2019). Heavy metal stress and responses in plants. *Int. J. Environ. Sci. Technol.* 16: 1807–1828.

Gong, J., Lee, D., and Schroeder, J.I. (2003). Long distance root-to-shoot transport of phytochelatins and cadmium in Arabidopsis. *Proc. Natl. Acad. Sci. U.S.A.* 100: 10118–10123.

Guo, J., Dai, X., Xu, W., and Ma, M. (2008). Over-expressing *gsh1* and *AsPCS1* simultaneously increases the tolerance and accumulation of cadmium and arsenic in *Arabidopsis thaliana*. *Chemosphere* 72: 1020–1026.

Gustin, J.L., Zanis, M.J., and Salt, D.E. (2011). Structure and evolution of the plant cation diffusion facilitator family of ion transporters. *BMC Evol. Biol.* 11: 76.

Hasan, M.K., Cheng, Y., Kanwar, M.K. et al. (2017). Responses of plant proteins to heavy metal stress – a review. *Front. Plant Sci.* 8: 1492.

Hasanuzzaman, M., Nahar, K., Alam, M.M. et al. (2013). Physiological, biochemical, and molecular mechanisms of heat stress tolerance in plants. *Int. J. Mol. Sci.* 14: 9643–9684.

He, Z.L., Yang, X.E., and Stoffella, P.J. (2005). Trace elements in agroecosystems and impacts on the environment. *J. Trace Elem. Med. Biol.* 19: 125–140.

Höller, S., Kupper, H., Brückner, D. et al. (2021). Overexpression of *METAL TOLERANCE PROTEIN8*reveals new aspects of metal transport in *Arabidopsis thaliana* seeds. *Plant Biol.* 24: 23–29.

Hoque, M., Tahjib-Ul-Arif, M., Hannan, A. et al. (2021). Melatonin modulates plant tolerance to heavy metal stress: morphological responses to molecular mechanisms. *Int. J. Mol. Sci.* 22: 11445.

Hossain, M.A., Piyatida, P., da Silva, J.A.T. et al. (2012). Molecular mechanism of heavy metal toxicity and tolerance in plants: central role of glutathione in detoxification of reactive oxygen species and methylglyoxal and in heavy metal chelation. *J. Bot.* 2012: 872875.

Hu, Z., Li, J., Wang, H. et al. (2019). Soil contamination with heavy metals and its impact on food security in China. *J. Geosci. Environ. Prot.* 7: 168.

Hua, T., Zhang, R., Sun, H., and Liu, C. (2021). Alleviation of boron toxicity in plants: mechanisms and approaches. *Crit. Rev. Environ. Sci. Technol.* 51: 2975–3015.

Ismael, M.A., Elyamine, A.M., Moussa, M.G. et al. (2019). Cadmium in plants: uptake, toxicity, and its interactions with selenium fertilizers. *Metallomics* 11: 255–277.

Jackson, P., Robertson, M., Cooper, M., and Hammer, G. (1996). The role of physiological understanding in plant breeding; from a breeding perspective. *Field Crops Res.* 49: 11–37.

Jamla, M., Khare, T., Joshi, S. et al. (2021). Omics approaches for understanding heavy metal responses and tolerance in plants. *Curr. Plant Biol.* 27: 100213.

Joshi, R., Pareek, A., and Pareek, S.L.S. (2016). Plant metallothioneins: classification, distribution, function and regulation. In: *Plant Metal Interaction* (ed. P. Ahmad), 239–261. Elsevier.

Kalaivanan, D. and Ganeshamurthy, A.N. (2016). Mechanisms of heavy metal toxicity in plants. In: *Abiotic Stress Physiology of Horticultural Crops* (ed. N.K. Srinivasa Rao, K.S. Shivashankara, and R.H. Laxman), 85–102. Springer India.

Kaur, R., Yadav, P., Kohli, S.K. et al. (2019). Emerging trends and tools in transgenic plant technology for phytoremediation of toxic metals and metalloids. In: *Transgenic Technology for Remediation of Toxic Metals and Metalloids* (ed. M.N.V. Prasad), 63–88. Academic Press.

Kim, Y. and Kang, H. (2018). Comparative expression analysis of genes encoding metallothioneins in response to heavy metals and abiotic stresses in rice (*Oryza sativa*) and *Arabidopsis thaliana*. *Biosci. Biotechnol. Biochem.* 82: 1656–1665.

Kim, D.Y., Bovet, L., Maeshima, M. et al. (2007). The ABC transporter AtPDR8 is a cadmium extrusion pump conferring heavy metal resistance. *Plant J.* 50: 207–218.

Kuhnlenz, T., Schmid, H., Uraguchi, S., and Clemens, S. (2014). *Arabidopsis thaliana* phytochelatin synthase 2 is constitutively active *in vivo* and can rescue the growth defect of the *PCS1* deficient *cad1-3* mutant on Cd-contaminated soil. *J. Exp. Bot.* 65: 4241–4253.

Kumar, S., Asif, M.H., Chakrabarty, D. et al. (2013). Expression of a rice Lambda class of glutathione S-transferase, *OsGSTL2*, in Arabidopsis provides tolerance to heavy metal and other abiotic stresses. *J. Hazard. Mater.* 248: 228–237.

Kumar, S., Yadav, A., Verma, R. et al. (2022). Metallothionein (MT1): a molecular stress marker in chickpea enhances drought and heavy metal stress adaptive efficacy in transgenic plants. *Environ. Exp. Bot.* 199: 104871.

Lee, J., Shim, D., Song, W.Y. et al. (2004). Arabidopsis metallothionein 2a and 3 enhance resistance to cadmium when expressed in *Vicia faba* guard cells. *Plant Mol. Biol.* 54: 805–815.

Lee, M., Lee, K., Lee, J. et al. (2005). AtPDR12 contributes to lead resistance in Arabidopsis. *Plant Physiol.* 138: 827–836.

Lee, S.H., Ahsan, N., Lee, K.W. et al. (2007). Simultaneous overexpression of both CuZn superoxide dismutase and ascorbate peroxidase in transgenic tall fescue plants confers increased tolerance to a wide range of abiotic stresses. *J. Plant Physiol.* 164: 1626–1638.

Li, Y., Huang, F., Tao, Y. et al. (2022). BcGR1.1, a cytoplasmic localized glutathione reductase, enhanced tolerance to copper stress in *Arabidopsis thaliana*. *Antioxidants* 11: 389.

Lin, F., Xu, J., Shi, J. et al. (2010). Molecular cloning and characterization of a novel glyoxalase I gene TaGly I in wheat (*Triticum aestivum* L.). *Mol. Biol. Rep.* 37: 729.

Liu, G.Y., Zhang, Y.X., and Chai, T.Y. (2011). Phytochelatin synthase of *Thlaspi caerulescens* enhanced tolerance and accumulation of heavy metals when expressed in yeast and tobacco. *Plant Cell Rep.* 47: 515–523.

Liu, D., Liu, Y., Rao, J. et al. (2013). Overexpression of the glutathione S-transferase gene from *Pyrus pyrifolia* fruit improves tolerance to abiotic stress in transgenic tobacco plants. *Mol. Biol.* 47: 515–523.

Liu, X.S., Feng, S.J., Zhang, B.Q. et al. (2019). OsZIP1 functions as a metal efflux transporter limiting excess zinc, copper and cadmium accumulation in rice. *BMC Plant Biol.* 19: 283.

Lu, Y., Wang, Q.F., Li, J. et al. (2019). Effects of exogenous sulfur on alleviating cadmium stress in tartary buckwheat. *Sci. Rep.* 9: 7397.

Manara, A. (2012). Plant responses to heavy metal toxicity. In: *Plants and Heavy Metals* (ed. A. Furini), 27–53. Dordrecht: Springer.

Miki, D., Zinta, G., Zhang, W. et al. (2021). CRISPR/Cas9 based genome editing toolbox for *Arabidopsis thaliana*. In: *Arabidopsis Protocols* (ed. J.J. Sanchez-Serrano and J. Salinas), 121–146. New York: Humana.

Mitra, S., Chakraborty, A.J., Tareq, A.M. et al. (2022). Impact of heavy metals on the environment and human health: novel therapeutic insights to counter the toxicity. *J. King. Saud. Univ. Sci.* 34: 101865.

Nakajima, I., Ban, Y., Azuma, A. et al. (2017). CRISPR/Cas9 mediated targeted mutagenesis in grape. *PLoS One* 12: e0177966.

Nowicka, B. (2022). Heavy metal–induced stress in eukaryotic algae—mechanisms of heavy metal toxicity and tolerance with particular emphasis on oxidative stress in exposed cells and the role of antioxidant response. *Environ. Sci. Pollut. Res.* 29: 16860–16911.

Oves, M., Khan, M.S., Zaidi, A., and Ahmad, E. (2012). Soil contamination, nutritive value, and human health risk assessment of heavy metals: an overview. In: *Toxicity of Heavy Metals to Legumes and Bioremediation* (ed. A. Zaidi, P.A. Wani, and M.S. Khan), 1–27. Springer.

Pasricha, S., Mathur, V., Garg, A. et al. (2021). Molecular mechanisms underlying heavy metal uptake, translocation and tolerance in hyperaccumulators-an analysis: heavy metal tolerance in hyperaccumulators. *Environ. Challenges* 4: 100197.

Patankar, H.V., Al-Harrasi, I., Kharusi, L. et al. (2019). Overexpression of a *Metallothionein 2A* gene from date palm confers abiotic stress tolerance to yeast and *Arabidopsis thaliana*. *Int. J. Mol. Sci.* 20: 2871.

Patel, M.K., Pandey, S., Patel, J. et al. (2021). A type 2 metallothionein (*SbMT2*) gene cloned from *Salicornia brachiate* confers enhanced Zn stress tolerance in transgenic tobacco by transporting Zn^{2+} and maintaining photosynthesis efficacy. *Environ. Exp. Bot.* 191: 104626.

Pomponi, M., Censi, V., Girolamo, V.G. et al. (2006). Overexpression of Arabidopsis phytochelatin synthase in tobacco plants enhances Cd^{2+} tolerance and accumulation but not translocation to the shoot. *Planta* 223: 180–190.

Rai, K.K., Pandey, N., Meena, R.P., and Rai, S.P. (2021). Biotechnological strategies for enhancing heavy metal tolerance in neglected and underutilized legume crops: a comprehensive review. *Ecotoxicol. Environ. Saf.* 208: 111750.

Rajab, H., Khan, M.S., Wirtz, M. et al. (2020). Sulfur metabolic engineering enhances cadmium stress tolerance and root to shoot iron translocation in *Brassica napus* L. *Plant Physiol. Biochem.* 152: 32–43.

Riaz, M., Kamran, M., Fang, Y. et al. (2021). *Arbuscular mycorrhizal* fungi-induced mitigation of heavy metal phytotoxicity in metal contaminated soils: a critical review. *J. Hazard. Mater.* 402: 123919.

Rono, J.K., Wang, L.L., Wu, Z.C. et al. (2021). Identification of a new function of metallothionein-like gene *OsMT1e* for cadmium detoxification and potential phytoremediation. *Chemosphere* 265: 129136.

Sarma, H., Islam, N.F., Prasad, R. et al. (2021). Enhancing the phytoremediation of hazardous metal(loid)s using genome engineering CRISPR/Cas9 technology. *J. Hazard. Mater.* 414: 125493.

Saxena, P., Singh, N.K., Harish et al. (2020). Recent advances in phytoremediation using genome engineering CRISPR–Cas9 technology. In: *Bioremediation of Pollutants: From Genetic Engineering to Genome Engineering* (ed. V.C. Pandey and V. Singh), 125–141. Elsevier.

Schützendübel, A. and Polle, A. (2002). Plant responses to abiotic stresses: heavy metal-induced oxidative stress and protection by mycorrhization. *J. Exp. Bot.* 53: 1351–1365.

Sharma, S.S., Dietz, K.J., and Mimura, T. (2016). Vacuolar compartmentalization as indispensable component of heavy metal detoxification in plants. *Plant Cell Environ.* 39: 1112–1126.

Shukla, D., Kesari, R., Tiwari, M. et al. (2013). Expression of *Ceratophyllum demersum* phytochelatin synthase, *CdPCS1*, in *Escherichia coli* and Arabidopsis enhances heavy metal(loid)s accumulation. *Protoplasma* 250: 1263–1272.

Siemianowski, O., Mills, R.F., Williams, L.E., and Antosiewicz, D.M. (2011). Expression of the P_{1B}-type ATPase *AtHMA4* in tobacco modifies Zn and Cd root to shoot partitioning and metal tolerance. *Plant J.* 9: 64–74.

Singh, S., Kumar, V., Sidhu, G.K. et al. (2019). Plant growth promoting rhizobacteria from heavy metal contaminated soil promote growth attributes of *Pisum sativum* L. *Biocatal. Agric. Biotechnol.* 17: 665–671.

Singh, S., Yadav, V., Arif, N. et al. (2020). Heavy metal stress and plant life: uptake mechanisms, toxicity, and alleviation. In: *Plant Life Under Changing Environment* (ed. D.K. Tripathi, V.P. Singh, D.K. Chauhan, et al.), 271–287. Academic Press.

Singla-Pareek, S.L., Yadav, S.K., Pareek, A. et al. (2006). Transgenic tobacco overexpressing glyoxalase pathway enzymes grow and set viable seeds in zinc-spiked soils. *Plant Physiol.* 140: 613–623.

Songmei, L., Jie, J., Yang, L. et al. (2019). Characterization and evaluation of *OsLCT1* and *OsNramp5* mutants generated through CRISPR/Cas9 mediated mutagenesis for breeding low Cd rice. *Rice Sci.* 26: 88–97.

Srivastava, D., Verma, G., Chauhan, S. et al. (2018). Rice (*Oryza sativa* L.) tau class glutathione *S*-transferase (*OsGSTU30*) overexpression in *Arabidopsis thaliana* modulates a regulatory network leading to heavy metal and drought stress tolerance. *Metallomics* 11: 375–389.

Stankovic, S. and Stankovic, A.R. (2013). Bioindicators of toxic metals. In: *Green Materials for Energy, Products and Depollution* (ed. E. Lichtfouse, J. Schwarzbauer, and D. Robert), 151–228. Dordrecht: Springer.

Su, H., Zou, T., Lin, R. et al. (2020). Characterization of a phytochelatin synthase gene from *Ipomea pes-caprae* involved in cadmium tolerance and accumulation in yeast and plants. *Plant Physiol. Biochem.* 155: 743–755.

Sytar, O., Kumar, A., Latowski, D. et al. (2013). Heavy metal-induced oxidative damage, defense reactions, and detoxification mechanisms in plants. *Acta Physiol. Plant.* 35: 985–999.

Tahjib-Ul-Arif, M., Sohag, A.A.M., Mostofa, M.G. et al. (2021). Comparative effects of ascobin and glutathione on copper homeostasis and oxidative stress metabolism in mitigation of copper toxicity in rice. *Plant. Biol.* 23: 162–169.

Talebi, M., Ebrahim, B., Tabatabaei, S., and Akbarzadeh, H. (2019). Hyperaccumulation of Cu, Zn, Ni, and Cd in *Azolla* species inducing expression of methallothionein and phytochelatin synthase genes. *Chemosphere* 230: 488–497.

Terrón-Camero, L.C., Peláez-Vico, M.Á., Del-Val, C. et al. (2019). Role of nitric oxide in plant responses to heavy metal stress: exogenous application versus endogenous production. *J. Exp. Bot.* 70: 4477–4488.

Terzi, H. and Yıldız, M. (2021). Proteomic analysis reveals the role of exogenous cysteine in alleviating chromium stress in maize seedlings. *Ecotoxicol. Environ. Saf.* 209: 111784.

Uraguchi, S., Tanaka, N., Hofmann, C. et al. (2017). Phytochelatin synthase has contrasting effect on cadmium and arsenic accumulation in rice grains. *Plant Cell Physiol.* 58: 1730–1742.

Vatansever, R., Ozyigit, I.I., and Filiz, E. (2017). Essential and beneficial trace elements in plants, and their transport in roots: a review. *Appl. Biochem. Biotechnol.* 181: 464–482.

Venegas-Rioseco, J., Ginocchio, R., and Ortiz-Calderón, C. (2021). Increase in phytoextraction potential by genome editing and transformation: a review. *Plants* 11: 86.

Wang, H., Wang, H., Shao, H., and Tang, X. (2016). Recent advances in utilizing transcription factors to improve the plant abiotic stress tolerance by transgenic technology. *Front. Plant Sci.* 7: 67.

Wang, J., Hou, Q., Li, P. et al. (2017). Diverse functions of multidrug and toxin extrusion (MATE) transporters in citric acid efflux and metal homeostasis in *Medicago truncatula*. *Plant J.* 90: 79–95.

Wangeline, A.L., Burkhead, J.L., Hale, K.L. et al. (2004). Overexpression of ATP sulfurylase in Indian mustard: effects on tolerance and accumulation of twelve metals. *J. Environ. Q.* 33: 54–60.

Wani, W., Masoodi, K.Z., Zaid, A. et al. (2018). Engineering plants for heavy metal stress tolerance. *Rend. Lincei Sci. Fis. Nat.* 29: 709–723.

Wuana, R.A. and Okieimen, F.E. (2011). Heavy metals in contaminated soils: a review of sources, chemistry, risks and best available strategies for remediation. *Int. Sch. Res. Notices* 2011: 402647.

Xia, Y., Qi, Y., Yuan, Y. et al. (2012). Overexpression of *Elsholtzia haichowensis* metallothionein 1 (EhMT1) in tobacco plants enhances the copper tolerance and accumulation in root cytoplasm and decreases hydrogen peroxide production. *J. Hazard. Mater.* 233–234: 65–71.

Xu, W., Shi, W., Liu, F. et al. (2008). Enhanced zinc and cadmium tolerance and accumulation in transgenic Arabidopsis plants constitutively overexpressing a barley gene (*HvAPX1*) that encodes a peroxisomal ascorbate peroxidase. *Botany* 86: 567.

Zeng, Z., Xiong, F., Yu, X. et al. (2016). Overexpression of a glyoxalase gene, *OsGly I*, improves abiotic stress tolerance and grain yield in rice (*Oryza sativa* L.). *Plant Physiol. Biochem.* 109: 62–71.

Zhao, F.Y., Liu, W., and Zhang, S.Y. (2009). Different responses of plant growth and antioxidant system to the combination of cadmium and heat stress in transgenic and non-transgenic rice. *J. Integr. Plant Biol.* 51: 942–950.

Zhi, J., Liu, X., Yang, R., and Xu, J. (2020). Overexpression of the metallothionein gene *PaMT3* from *Phytolacca americana* enhances the plant tolerance to cadmium. *Plant Cell, Tissue Organ Cult.* 143: 211–218.

Zhou, J., Cheng, K., Huang, G. et al. (2020). Effects of exogenous 3-indoleacetic acid and cadmium stress on the physiological and biochemical characteristics of *Cinnamomum camphora*. *Ecotoxicol. Environ. Saf.* 191: 109998.

Zimeri, A.M., Dhankher, O.P., McCaig, B. et al. (2005). The plant MTI metallothioneins are stabilized by binding cadmium and are required for cadmium tolerance and accumulation. *Plant Mol. Biol.* 58: 839–855.

Index

Note: Page numbers followed by *f* or *t* indicate material in figures or tables respectively

a

aberrant gene expression 294
abscisic acid (ABA) signaling 383
Agrobacterium fabrum 524
Agrostis capillaris 497
Allium cepa 386
aluminum (Al)
  stress tolerance
    ALMT1 87, 89
    antioxidant defense system 93
    ART1 89
    CAMTA2 89
    comparative transcriptomics 93–95
    expression level polymorphism (ELP) 92–93
    future aspects 99
    genetic architecture 91–92
    genome-wide association studies (GWAS) 92, 94*f*
    iTRAQ approach 97
    plant mechanisms 88
    proteomics 95–98, 97*t*–98*t*
    QTL mapping 89, 90
    reactive oxygen species (ROS) signaling 94
    sensitivity to proton rhizotoxicity 1 (STOP1) mutant 87, 88
    soybean variety 93
    stress duration 95
    stress intensity 95
    vacuoles and detoxification 87
  toxicity tolerance
    breeding populations 114
    gene expression under stress condition 113*t*
    glutathione metabolism 117
    overview of 105–106
    plant signaling 107
    QTL identified 110*t*, 111*t*
    rice genetic mapping 107–109, 110*t*
    root-apex transition zone (TZ) 106
    root transcriptome 109–114
    wheat genetic mapping 114, 115*f*, 116
    wheat mate gene family 116
    wheat proteomics 117–118
aluminum-activated malate transporter (ALMT) 87, 89, 91–93, 95, 99, 109, 116, 117
aluminum resistance transcription factor 1 (ART1) 89, 91, 108, 109, 111
Anthyllis vulneraria 504
antioxidant defense system 39
Arabidopsis 49, 87, 92
  calcium-alleviated Al toxicity 96
  genome-wide association studies (GWAS) 94*f*

Heavy Metal Toxicity and Tolerance in Plants: A Biological, Omics, and Genetic Engineering Approach, First Edition. Edited by Mohammad Anwar Hossain, AKM Zakir Hossain, Sylvain Bourgerie, Masayuki Fujita, Om Parkash Dhankher, and Parvez Haris.

Arabidopsis thaliana 129, 141, 170, 266, 280, 328–329, 356, 374, 501
Arbuscular mycorrhiza 10, 11
arsenic (As)
  accumulation 296
  ascorbate–glutathione cycle 300
  carbohydrate metabolism 302
  detoxification 296
  environment and exposure to human 61*f*
  food chain contamination 62
  future aspects 311
  gene modulation in *Oryza sativa* 305*t*–306*t*
  genomic profiling 304
  inorganic uptake 296–297
  metabolomic profiling 308
  methylated arsenic uptake 297–298
  methylation and volatilization 298
  miRNA genes in *Oryza sativa* 306*t*–307*t*
  molecular biotechnology, As-tolerant plants 308–311
  nitrogen metabolism 302–303
  occurrence and distribution 295
  omics approaches 294
  omics modulation under challenged environment 303–304
  overview of 60, 293–294
  oxidative states 293
  pharmacokinetic process 62
  photosynthesis 300–301
  phytotoxicity
    germination and growth 298–299
    nutrient uptake 299
    oxidative stress and antioxidative defense 299–300
  primary metabolites alteration 309*t*
  proteomic profiling 307–308
  remedial options 63
  respiration 301–302
  secondary metabolites alteration 310*t*
  sources and forms 60, 62
  toxicological concerns 294
  toxicology process 62–63
  transcriptomic profiling 304, 306–307
  translocation 297*f*
  transporter assisted uptake 297*f*
  uptake 296
  vacuolar sequestration 297*f*
ascorbate peroxidase (APX) 356
ATP sulfurylase (APS) 327, 329
Avena sativa 354
avoidance, HMs
  mycorrhizal association 10–11, 11*f*
  root exudates 11*f*, 12
Azospirillum brasilense 518
Azotobacter chroococcum 518
Azotobacter salinestris 520

b

Bacillus gibsonii 523
Bacillus subtilis 521
Bacillus xiamenensis 520
back-cross double-haploid (BC_3DH) 199
Bacopa monniera 279
base-excision repair (BER) pathway 212
Berkheya coddii 208
bioaccumulation 293
biochar amendments
  dissolved organic carbon (DOC) 501
  heavy metal concentrations in crops 503
  heavy metal immobilization 495–496, 495*f*, 503–504
  phytoremediation strategy 494
  plant respond to metal stress
    antioxidant and phytohormone activity 499–501
    heavy metal uptake by plants 501–503
    photosynthetic activity 499
  potentially toxic elements (PTEs) 493
  rhizosphere
    changes in elements 498–499
    physiological factors 497, 497*f*
    plant root development 497, 498
  soil pH modification 496
  sorption characteristics 494
biochar-derived humic acids (BDHA) 502
bioconcentration factor (BCF) 350
biofortification 328–329

biophilic design
  as antidepressant 565–566
  dimensions of 562
  direct experience of nature 562–563
  experience of place and space 563
  hypothesis of 562
  indirect experience of nature 563
  natural occurrences and phenomena 561
  phytoremediation in urban sustainability 560*f*
  sustainable biophilic cities 563–564
Botrytis cinerea 500
branched-chain-amino-acid aminotransferases (BCATs) 381

c

cadmium (Cd)
  environment and exposure to human 65*f*
  food chain contamination 64, 66
  overview of 63–64
  pharmacokinetic process 66
  remedial options 67
  sources and forms 64
  toxicology process 66–67
Cajanus cajan 279
calcium-dependent protein kinases (CDPK) 382
calmodulin binding transcription activator 2 (CAMTA2) 89
Calvin–Benson cycle 151, 153
Cannabis sativa 379
carbohydrate metabolism 302
Cardamine hupingshanensis 327
Cardamine violifolia 327
catalase (CAT) enzymes 356
cation exchange capacity (CEC) 253
Celosia argentea 154
Ceratophyllum demersum 386
chalcone synthase (CHS) gene 218
Chlamydomonas reinhardtii 218, 467
chloroplast 22, 355
chop polymerase chain reaction (Chop-PCR) 375
chromatin 212
chromium (Cr)
  bioavailability 126–127
  decontamination 136
  detoxification mechanisms 129–130
  environment and exposure to human 73*f*
  food chain contamination 74
  industrial application 126
  metal-tolerant microbial consortium 135
  omics approach
    chromium-induced miRNAs 132
    metabolomics 133
    proteomics 133–134
    transcriptomics 130–132
  overview of 72, 125–126
  pharmacokinetic process 74
  phytoremediation approach 134–135
  remedial options 75–76
  soil contamination 125
  sources and forms 72
  sources of 126–127
  sub-cellular distribution 127–128
  toxicology process 74–75
  translocation 127–128
  uptake 127–128
Chromolaena odorata 354
chromosome segment substitution lines (CSSLs) 198
Cicer arietinum 278
Cichorium intybus 386
Citrus grandis 263
clustered regularly interspaced short palindromic repeats (CRISPR) 284
cobalt (Co)
  contamination 207–208
  environmental causes 207
  future aspects 220–221
  omics approach 216–220
    metabolomics 218–219
    proteomics 219–220
    transcriptomic sequencing 216–218, 217*f*
  overview of 207–208
  plant antioxidant defense system 213–216
    enzymatic antioxidants 213–215
    nonenzymatic antioxidants 215–216
  plant response to stress 208–211, 209*f*

cobalt (Co) (*Cont'd*)
molecular architecture 210*f*
plant ultrastructure 210*f*
translocation 209–211
uptake 209–211
ROS generation 211–213
damage to genetic material 212–213
lipid peroxidation 211–212
cobalt-induced plant antioxidant defense system 213–216
enzymatic antioxidants 213–215
catalases (CAT) 213–214
glutathione peroxidases (GPX) 214
glutathione reductase (GR) 214–215
superoxide dismutase (SOD) 213
nonenzymatic antioxidants 215–216
ascorbic acid 215
reduced glutathione (GSH) 216
tocopherols 215–216
Colletotrichum falcatum 521
copper (Cu)
in agriculture 257*t*
deficiency 252
deficient, sufficient, and toxic levels 255–256
dependent enzymes 254*f*
fertilizers and fungicides 256–258, 257*t*, 258*f*
functions of 253–255
genes encoding transport 260
maintenance of 252*f*
omics approach
genomics 259
metabolomics 263–264
miRNAomics 264–266, 265*f*
proteomics 261–263, 262*f*
transcriptomics 259–261, 261*f*
overview of 251–253
oxidative stress induction 261*f*
physiological effects 254
plant metabolic process 251
remobilization mechanisms 255
source of 253
uptake, transport, distribution 255
CRISPR/Cas9 System 284–285
crop defense mechanisms, Fe
retention in roots and shoot translocation 176*f*, 177–179
roots exclusion 175–177, 176*f*
ROS detoxification 176*f*, 179–180
shoots compartmentalization 176*f*, 177–179
Cucumis sativus 263
cystathionine-γ-synthase (CGS) 327

d

detoxification 296
Diectomis fastigiata 135
differentially expressed genes (DEGs) 283
dissolved organic carbon (DOC) 501
DNA methylation 381
DNA methylation/demethylation process 39
downregulation 303
drought/salinity stress, HMs
aluminium (Al) 411–412
antioxidant system 414–415
cadmium (Cd) 410–411
chromium (Cr) 412
drought stress 408
future aspects 415–416
lead (Pb) 412
mercury (Hg) 412
overview of 407–409
photosynthesis 413–414
salt stress 408
toxic effects 407
zinc (Zn) 409–410

e

electron transport rates (ETRs) 279
energy metabolic pathway 388–389
environmental contamination 35
ethylene-responsive factor (ERF) 383
Eucalyptus maculata 277
Eucalyptus urophylla 277

f

Fusarium moniliforme 521
Fusarium oxysporum 520

g

gas chromatography-mass spectrometry (GCMS) 218
genetic engineering, HMs
  CRISPR/ Cas system 585
  mechanisms of 574–576, 575*f*
  overview 573–574
  strategies for 576–577
  transgenic plants
    enhanced antioxidant defense 579, 580*t*
    glyoxalase pathway genes 577
    metal ion transporter genes/proteins 579, 583, 584*t*
    metallothionein genes 579, 582*t*–583*t*
    phytochelatin genes 579, 581*t*–582*t*
    sulfur metabolism engineering 577–578, 578*t*
genome-wide association studies (GWAS) 41–42, 286
  in Arabidopsis 94*f*
  expression level polymorphism (ELP) 92
  SNP markers 92
genomics 281, 283, 284*f*
  copper (Cu) 259
  heavy metals (HMs) 15, 39
  profiling 304
  zinc (Zn) 281, 283, 284*f*
genomics-assisted breeding
  genome-wide association studies 41–42
  quantitative trait loci (QTL) mapping 39–41
genomic template stability (GTS) 294
glutathione reductase (GR) 106, 356
Gossia bidwillii 156
Gossypium herbaceum 354
GSH *S*-transferase (GST) 387

h

Haber-Weiss/Fenton reaction 251
heat shock transcription factors (HSF) 383
heavy metal ATPase (HMA) 335
heavy-metal-induced ROS
  detoxification of 468–473, 469*t*, 471*t*–472*t*
  exogenous chemical-compounds 473–484, 474*t*–483*t*
  methylglyoxal in plant cells 464–468, 465*t*–466*t*
  plant cell 464–468
heavy metals (HMs)
  avoidance 463
    mycorrhizal association 10–11, 11*f*
    root exudates 11*f*, 12
  biotechnological techniques 36
  cellular redox homeostasis 1
  detrimental effects 2
  drought/salinity stress in plants
    aluminium (Al) 411–412
    antioxidant system 414–415
    cadmium (Cd) 410–411
    chromium (Cr) 412
    drought stress 408
    future aspects 415–416
    lead (Pb) 412
    mercury (Hg) 412
    overview of 407–409
    photosynthesis 413–414
    salt stress 408
    toxic effects 407
    zinc (Zn) 409–410
  essential heavy metals 59
  exogenous chemical-compounds 473–484, 474*t*–483*t*
  future aspects 22–23
  genetic engineering
    CRISPR technology 20–21
    gene silencing 22
    plastid transformation 21–22
  hormonal regulation
    overview of 425–426
    phytohormone-mediating plant response 427–445
    phytohormones crosstalk 445–447
    plants under HM stress 426–427
  metallothioneins (MTs) 12
  metals and metalloids 35
  metal sequestration 12, 13*t*–14*t*, 14
  nonessential heavy metals 59
  occurrence 461–462
  omics approach
    advanced technologies 16*f*

heavy metals (HMs) (*Cont'd*)
genomics 15
ionomics 18–19
metabolomics 17–18, 46–47
metallomics 19–20
miRNAomics 19, 47–49
phenomics 49–50
proteomics 17, 44–45
transcriptomics 15–17, 42–44
overview of 1
plant-metal interaction 2–3
on plants
biochemical process 4*t*–6*t*, 8–9
molecular responses 4*t*–6*t*, 9–10
morphoanatomical responses 3, 4*t*–6*t*, 7
physiological responses 4*t*–6*t*, 8
polluted environments 36
restricted uptake 463
schematic diagram 485
stable metal–metabolite complexes 12
toxicity 2, 35, 462–463
Helianthus annuus 389
Hibiscus esculentus 280
high-performance liquid chromatography (HPLC) 380
Hirschfeldia incana 384
HM-induced toxicity
effects on plants 36
epigenetic regulatory features 38
genomics-assisted breeding
genome-wide association studies 41–42
quantitative trait loci (QTL) mapping 39–41
histological response 37
metabolomics 46–47
methylation 38
microRNAs (miRNAs) 47–49
omics approaches 40*f*
phenomics 49–50
physicochemical properties 37
physiological mechanisms 37
plant's oxidation protective mechanism 540
proteomics 44–45
putative schematic diagram 38*f*
transcriptomics 42–44
Hordeum vulgare 217
hormonal regulation, HMs
overview of 425–426
phytohormone-mediating plant response 427–445
abscisic acid (ABA) 430–432, 431*f*
antioxidant defense systems 428–429, 428*f*
auxin (AUX) 432–433
brassinosteroid (BR) 434–435
cytokinin (CK) 435–437
detoxification 429
ethylene (ETH) 437–438
gasotransmitter 429–430
gibberellin (GA) 438–439
jasmonic acid/ jasmonate (JA) 439–440
melatonin (MT) 440–442
metal homeostasis 429
methylglyoxal (MG) detoxification 429
mineral homeostasis 429
photosynthetic performance 429
plant water relations 429
quelation 429
salicylic acid (SA) 442–443
sequestration 429
strigolactones (SLs) 444
phytohormones
biochemical imbalances 426*f*
crosstalk 445–447
physiological imbalances 426*f*
plant absorption 425
plants under HM stress 426–427
electrolyte leakage (EL) 427
lipid peroxidation (LP) 427
methylglyoxal (MG) generation 427
nutritional balance 427
oxidative stress 427
photosynthetic capacity 427
plant water relations 427
reactive oxygen species (ROS) 426
hydroponics-based screening 112
hyperMn-accumulating mutants 151

i

inductively coupled plasma mass spectrometry (ICP-MS) 218

International Agency for Research on Cancer (IARC) 293
iron (Fe)
acquisition systems in plants 170–171
cis-regulatory elements (CREs) 182
future aspects 183
homeostatic regulation 181
ions solubility 174*t*
in living organisms 169–170
retention in roots and shoots translocation 176*f*, 177–179
roots exclusion
genes involved in 176–177
hypothetical model 176*f*
plaque formation 175
silicon application 176
ROS detoxification
enzymatic detoxification 179
genes involved in 180
Glyoxalase I (GLYI) 180
hypothetical model 176*f*
shoots compartmentalization
genes involved in 178–179
hypothetical model 176*f*
YSL4 and YSL6 transporters 179
toxicity
global agriculture 171–172
pH effects 172
plant growth 174–175
in soil 172–174
soil improvement methods 173
soil state 172
water and drainage effects 173–174, 173*f*
toxicity tolerance in rice
candidate genes 196–197
cellular homeostasis 192
genetic variants in germplasm 197
mechanism of 193–197, 196*f*
molecular breeding 197–200, 201*t*
overview of 191–192
photosystem 193
retention in roots 194
role of 192
roots exclusion 193–194
ROS detoxification 195–196
shoot compartmentalization 194–195
shoots translocation 194
tissue damage 192
transcription factors 181–182
translocation systems in plants 171
isoflavone reductase (IFR) genes 218

j

Jatropha curcas 354

k

Kocuria rhizophila 217

l

Lathyrus sativus 386
lead (Pb)
binding proteins/transporters, 384–385
cell signaling 382–384
energy metabolic pathway 388–389
environment and exposure to human 69*f*
epigenetics regulations 380–382
food chain contamination 68
future aspects 392
genetics regulations 380–382
methionine-recycling pathway 391*f*
omics approach 375–380, 376*t*–379*t*
organic form 374
overview of 67–68, 373–374
oxidative stress and antioxidative mechanisms 385–388
pharmacokinetic process 68, 69
phenylpropanoid pathway 389–390
plant cell wall 382–384
remedial options 71–72
schematic representation 393*f*
sources and forms 68
sugar/carbohydrate metabolic pathway 388–389
sulfur-related pathway and phytohormones 390–392
toxicology process 70–71
transduction 382–384
leaf bronzing index (LBI) 199
Leclercia adecarboxylata 523
Lemna minor 219
Lepidium sativum 468
Leucaena glauca 353
leucine-rice repeat proteins (LRR) 374

liquid chromatography-mass spectrometry (LCMS) 218
low-cadmium, rice cultivars
  Cd uptake 336–338
  encoding genes 337*t*
  future aspects 343–344
  genome-editing technology 340–341
  molecular marker-assisted breeding 342–343
  mutation breeding 341–342
  overview of 335
  radial transport 338
  shoot distribution 338–339
  traditional transgenic technology 339–340
  xylem loading 338
low-molecular-weight (LMW) 129
Lupinus albus 497
Lupinus luteus 390

m

Macrophomina phaseolina 521
Malondialdehyde (MDA) 386
Malonyldialdehyde (MDA) 467
manganese (Mn)
  breeding strategies
    auxin homeostasis maintenance 157
    limiting absorption 155–156
    sequestration 156
    silicon uptake reinforcement and distribution 157–158
  concentration 141, 145
  fertilization 142
  future aspects 159
  overview of 141–142
  oxide reduction 144
  oxygenic photosynthesis 142
  soil availability 143–144
  toxicity
    agricultural land 144–146, 146*f*
    apoplastic 147–150
    brownish patchy spots in leaves 147–150
    callose synthesis 149
    Calvin–Benson cycle 151, 153
    in crop plants 143*t*
    foliar chlorosis under excess Mn 147–150
    history of 146–147
    occurrence of 144–146
    symplastic 150–154
    threat of 154–158
MAPK kinase kinase (MAPKKKs) 382
MAPK-promoted induction of E*IL* 94
marker-assisted selection (MAS) 198
massively parallel signature sequencing (MPSS) 283
matrix-assisted laser desorption ionization (MALDI) 218
matrix-assisted laser desorption ionization–time-of-flight mass spectrometry (MALDi-TOF-MS) 380
Medicago sativa 356, 386
mercury (Hg)
  anatomical response 354
  cellular responses 354–355
  content in drinking water 350
  environment and exposure to human 78*f*
  enzymatic responses 355–356
  food chain contamination 77, 78
  global mercury pollution 350–351
  hormonal responses 356
  metabolic responses 355–356
  morphological responses 353–354
  overview of 76, 349–350
  oxidative responses 356–357
  pharmacokinetic process 79
  photosynthetic responses 355
  phytoremediation prospects 360–361
  reactive oxygen species 356–357
  regional global emission 351*t*
  remedial options 80–81
  sources and forms 76–77
  tolerance mechanisms
    chelation 357–358
    enzymatic and antioxidative tolerance 358–359
    hormonal regulations 359–360
    miRNA-mediated tolerance 360
  toxicology process 79–80
  uptake and toxicity 352–353

metabolic pathways, Pb
energy metabolic pathway 388–389
phenylpropanoid pathway 389–390
sugar/carbohydrate metabolic pathway 388–389
sulfur-related pathway and phytohormones 390–392
metabolomics
chromium (Cr) 133
cobalt (Co) 218–219
copper (Cu) 263–264
heavy metals (HMs) 17–18, 46–47
HM-induced toxicity 46–47
nickel (Ni) 239*f*, 240
profiling 308
zinc (Zn) 281, 283, 284*f*
metal chelators 12, 14
metallothionein (MT) 387, 463
metal-tolerant protein 11 (MTP11) 148
methionine-*R*-sulfoxide reductase 1 (MsrB1) 325
methylated arsenic uptake 297–298
methylated DNA immunoprecipitation-sequencing (MeDIP-seq) 375
methylation sensitive amplified polymorphism (MSAP) 375
methylglyoxal in plant cells
detoxification of ROS 468–473, 469*t*, 471*t*–472*t*
future aspects 484
heavy-metal-induced ROS 464–468, 465*t*–466*t*
mutagenic properties 468
miRNAomics 284–285, 284*f*
copper (Cu) 264–266, 265*f*
heavy metals (HMs) 19, 47–49
mitogen-activated protein kinases (MAPKs) 359, 382
multidrug and toxin efflux (MATE) 116, 117
mycorrhizal association 10–11, 11*f*

n

nanotechnology applications
accumulation in rice plant 541*t*–542*t*
Cd and As toxicity 541*t*–542*t*
foliar spraying and soil additions 543, 543*f*
HM stress in plants
cerium dioxide 538–539
iron 536
schematic model 544*f*
selenium 535–536
silicon 535
titanium dioxide 537–538
zinc oxide 537
implementation in agricultural system 545
overview 533–534
toxicity of HM in plants 539–540, 539*f*
natural resistance-related macrophage proteins (NRAMP) 335
near-isogenic line (NIL) 343
next-generation sequencing (NGS) 283
nickel (Ni)
accumulation 233–234
anthropogenic sources 233
essential micronutrient 231
natural sources 232
omics approach
metabolomics 239*f*, 240
proteomics 239–240, 239*f*
transcriptomics 238–239, 239*f*
overview of 231–232
oxidative stress 232
pollution 232
role in plants 233
tolerance mechanisms 238–239
toxicity in plants
growth inhibition 234–235, 235*f*
oxidative stress induction 235*f*, 236–237
photosynthesis inhibition 235*f*, 236
uptake 233–234
Nicotiana tabacum 385
Noccearo tundifolium (L.) 504
NRAMP aluminum transporter 1 (NRAT1) transporter 91, 117
nuclear magnetic resonance spectroscopy (NMR) 218, 380
Nyssa sylvatica 208

o

omics approach
  applications of 294–295
  under challenged environment 303–304
  cobalt (Co) 216–220
    metabolomics 218–219
    proteomics 219–220
    transcriptomic sequencing 216–218, 217*f*
  copper (Cu)
    genomics 259
    metabolomics 263–264
    miRNAomics 264–266, 265*f*
    proteomics 261–263, 262*f*
    transcriptomics 259–261, 261*f*
  heavy metals (HMs)
    advanced technologies 16*f*
    genomics 15, 39
    ionomics 18–19
    metabolomics 17–18, 46–47
    metallomics 19–20
    miRNAomics 19, 47–49
    phenomics 49–50
    proteomics 17, 44–45
    stress-responsive mechanisms 40*f*
    transcriptomics 15–17, 42–44
  lead (Pb) 375–380, 376*t*–379*t*
  nickel (Ni)
    metabolomics 239*f*, 240
    proteomics 239–240, 239*f*
    transcriptomics 238–239, 239*f*
  zinc (Zn)
    CRISPR/Cas9 System 284–285
    genome-wide association study 286
    genomics 281, 283, 284*f*
    metabolomics 281, 283, 284*f*
    miRNA Omics 284–285, 284*f*
    proteomics 283–284
    quantitative trait locus (QTL) mapping 286
    transcriptomics 283–284, 284*f*, 285*t*
Origanum vulgare 135
Oryza rufipogon 108
Oryza sativa 141, 219
  gene modulation 305*t*–306*t*
  miRNA genes 306*t*–307*t*
OsNRAMP1 336, 342
OsNRAMP5 336, 340, 341, 343, 344
oxygen-evolving complex (OEC) 142

p

Persea americana 353
Phaseolus mungo 279
Phaseolus vulgaris 4, 279
phenylalanine ammonia-lyase (PAL) 389
phenylpropanoid pathway 389–390
phosphate transporter (PHT) proteins 294
photosynthesis 300–301
Phythium splendens 521
phytochelatins (PCs) 387
phytohormone-mediating plant response 427–445
  abscisic acid (ABA) 430–432, 431*f*
  antioxidant defense systems 428–429, 428*f*
  auxin (AUX) 432–433
  brassinosteroid (BR) 434–435
  cytokinin (CK) 435–437
  detoxification 429
  ethylene (ETH) 437–438
  gasotransmitter 429–430
  gibberellin (GA) 438–439
  jasmonic acid/ jasmonate (JA) 439–440
  melatonin (MT) 440–442
  metal homeostasis 429
  methylglyoxal (MG) detoxification 429
  mineral homeostasis 429
  photosynthetic performance 429
  plant water relations 429
  quelation 429
  salicylic acid (SA) 442–443
  sequestration 429
  strigolactones (SLs) 444
phytohormones, HMs
  biochemical imbalances 426*f*
  crosstalk 445–447
  physiological imbalances 426*f*
phytoremediation
  biochar-assisted phytoremediation 556
  chemical-assisted phytoremediation 556
  environmental remediation efforts 557
  future aspects 568–569

mechanisms of 555*f*
microbial-assisted phytoremediation 557
modified concept 555–557
non-hyperaccumulator plants 556
overview 553–554
schematic representation 568*f*
sustainable solution 558–559
types of
phytoextraction 554–555
phytostabilization 554
phytovolatalization 554
in urban sustainability
biophilic as antidepressant 565–566
biophilic design 559–564, 560*f*
eco-design, 559
economic benefits 566
health benefits 564–565
sustainability and resilience 566–567
phytotoxicity, Ar
germination and growth 298–299
nutrient uptake 299
oxidative stress and antioxidative defense 299–300
plant antioxidant defense system 2
plant growth promoting rhizobacteria (PGPRs)
biocontrol 520–521
biofertilization traits
acquisition of nutrients 516–517
exopolysaccharides (EPS) production 517–518
siderophores production 517
EDDS biodegradation 524
heavy metal bioremediation 521–524, 522*t*
hydrogen cyanide (HCN) 520
impacts of chemical fertilizers
aquatic ecosystem 515
atmospheric ecosystem 515
on plants 516
on soil 515–516
inoculation of 521
overview of 513–514, 514*f*
phytopathogenic diseases 520
phytostimulators and phytohormones 519–520
resistance to abiotic stress 518–519
plants
effect of HMs
biochemical process 4*t*–6*t*, 8–9
molecular responses 4*t*–6*t*, 9–10
morphoanatomical responses 3, 4*t*–6*t*, 7
physiological responses 4*t*–6*t*, 8
putative schematic diagram 38*f*
HM-induced toxicity 36–39, 38*f*
under HM stress
electrolyte leakage (EL) 427
lipid peroxidation (LP) 427
methylglyoxal (MG) generation 427
nutritional balance 427
oxidative stress 427
photosynthetic capacity 427
plant water relations 427
reactive oxygen species (ROS) 426
peptide metal-binding ligands 37
respond to metal stress
antioxidant and phytohormone activity 499–501
heavy metal uptake by plants 501–503
photosynthetic activity 499
pleiotropic drug resistance (PDR1) 374
potentially toxic elements (PTEs) 493
primary salt damage 408
proteomic profiling 307–308
proteomics
aluminum (Al)
stress tolerance 95–98, 97*t*–98*t*
toxicity tolerance 117–118
chromium (Cr) 133–134
cobalt (Co) 219–220
copper (Cu) 261–263, 262*f*
heavy metals (HMs) 17, 44–45
HM-induced toxicity 44–45
nickel (Ni) 239–240, 239*f*
wheat 117–118
zinc (Zn) 283–284
Pseudocallymma alliaceum 353
Pseudomonas aeruginosa 521
Pseudomonas koreensis 521
Pseudomonas putida 523

q

quadrupole time-of-flight mass spectrometry (QTOF-MS) 379
quantitative trait locus (QTL)
  copper (Cu) 259
  genomics-assisted breeding 39–41
  heavy metals (HMs) 39–41
  rice 198–200
  stress tolerance, Al 89, 90
  toxicity tolerance, Al 110*t*, 111*t*
  wheat 114
  zinc (Zn) 286

r

random amplified polymorphic DNAs (RAPDs) 294
Raphanus sativus 379
reactive nitrogen species (RNS) 328
reactive oxygen species (ROS) 142, 353, 374, 414
  aluminum (Al) 94
  cobalt (Co) 211–213
    damage to genetic material 212–213
    lipid peroxidation 211–212
  crop defense mechanisms, Fe 176*f*, 179–180
  detoxification of 468–473, 469*t*, 471*t*–472*t*
  exogenous chemical-compounds 473–484, 474*t*–483*t*
  future aspects 484
  heavy-metal-induced ROS 464–468, 465*t*–466*t*
  hormonal regulation, HMs 426
  iron (Fe)
    enzymatic detoxification 179
    genes involved in 180
    glyoxalase I (GLYI) 180
    hypothetical model 176*f*
  methylglyoxal in plant cells 468–473, 469*t*, 471*t*–472*t*
  overproduction 356
  oxidative stress 391
  plants under HM stress 426
  rice, iron (Fe) 195–196
real-time PCR (RT-qPCR) 327
recombinant inbred lines (RIL) 198
respiration 301–302
Rhizobium japonicum 518
Rhizoctonia solani 521
rice
  cultivars, low-Cadmium
    Cd uptake 336–338
    encoding genes 337*t*
    future aspects 343–344
    genome-editing technology 340–341
    molecular marker-assisted breeding 342–343
    mutation breeding 341–342
    overview of 335
    radial transport 338
    shoot distribution 338–339
    traditional transgenic technology 339–340
    xylem loading 338
  genetic mapping
    association mapping 108–109, 110*t*
    linkage mapping 107–108
  iron (Fe)
    candidate genes 196–197
    cellular homeostasis 192
    genetic variants in germplasm 197
    molecular breeding 197–200, 201*t*
    overview of 191–192
    photosystem 193
    retention in roots 194
    role of 192
    roots exclusion 193–194
    ROS detoxification 195–196
    shoot compartmentalization 194–195
    shoots translocation 194
    tissue damage 192
    tolerance mechanisms 193–197, 196*f*

s

Salsola vermiculata 36
Salvinia biloba 354
secondary salt damage 408
SeCys methyltransferase (SMT) 329
selenium (Se)
  biofortification 328–329
  edible plants 323
  necessity of 324

overview of 323–324
protein malformation 324–325
ROS-induced se phytotoxicity 325–326
tolerance 326–328
sensitivity to proton rhizotoxicity 1 (STOP1) mutant 87, 88
serial analysis of gene expression (SAGE) 17, 283
Sesuvium portulacastrum 388
simple sequence repeats (SSRs) 283
single nucleotide polymorphisms (SNPs) 92, 283
sodium dodecyl sulfate-polyacrylamide gel electrophoresis (SDS-PAGE) 239
Spirodela polyrrhiza 135
Stenotrophomonas maltophilia 524
Streptococcus pyogenes 20
Stylosanthes guianensis 155
Stylosanthes hamata 156
Suaeda salsa 379
sugar/carbohydrate metabolic pathway 388–389
sulfur-related pathway and phytohormones 390–392
superoxide dismutase (SOD) 356
Synallaxis albescens 328, 329

t

TaMATEs 116
Target of Avrb Operation1 (TAO1) 95
Thlaspi caerulescens 276
toxicity tolerance mechanisms
iron (Fe)
candidate genes 196–197
genetic variants in germplasm 197
retention in roots 194
roots exclusion 193–194
ROS detoxification 195–196
shoot compartmentalization 194–195
shoots translocation 194
manganese (Mn)
agricultural land 144–146, 146*f*
apoplastic 147–150
brownish patchy spots in leaves 147–150
callose synthesis 149
Calvin–Benson cycle 151, 153
in crop plants 143*t*
foliar chlorosis under excess Mn 147–150
history of 146–147
occurrence of 144–146
symplastic 150–154
threat of 154–158
mercury (Hg)
chelation 357–358
enzymatic and antioxidative tolerance 358–359
hormonal regulations 359–360
miRNA-mediated tolerance 360
nickel (Ni) 237–238
rice, iron (Fe) 193–197, 196*f*
selenium (Se)
ROS-induced se phytotoxicity 325–326
Se-induced protein malformation 324–325
zinc (Zn)
multi-omnics approach 281–286
oxidative metabolism 279–280
physiology and biochemistry 280–281
seed germination and growth 278–279
stress adaptation 281
transcriptomics 238–239, 239*f*, 283–284, 284*f*, 285*t*
aluminum (Al) 93–95
chromium (Cr) 130–132
copper (Cu) 259–261, 261*f*
heavy metals (HMs) 15–17, 42–44
HM-induced toxicity 42–44
nickel (Ni) 238–239, 239*f*
profiling 304, 306–307
sequencing 216–218, 217*f*
zinc (Zn) 283–284, 284*f*, 285*t*
transgenic plants and HMs tolerance
enhanced antioxidant defense 579, 580*t*
glyoxalase pathway genes 577
metal ion transporter genes/proteins 579, 583, 584*t*
metallothionein genes 579, 582*t*–583*t*
phytochelatin genes 579, 581*t*–582*t*
sulfur metabolism engineering 577–578, 578*t*
translocation 297*f*

transporter assisted uptake 297*f*
Trigonella foenum-graccum 354
two-dimensional difference gel
electrophoresis (2-D DIGE) 379

u

US Environmental Protection Agency
(EPA) 293

v

vacuolar sequestration 297*f*
Vallisneria natans 386
Vallisneria spinulosa 386
Vernonia cinerea 135
vesicle trafficking process 132
Vigna radiata 354
volatilization 298

w

wheat
genetic mapping 114, 115*f*, 116
mate gene family 116
proteomics 117–118

z

zinc (Zn)
accumulator plants 282*t*
future aspects 286
homeostasis 277, 278*t*
mobility 275
overview of 275
physio-genetics aspects 277–281
pollutant 277
toxicity in plants
multi-omnics approach 281–286
oxidative metabolism 279–280
physiology and biochemistry 280–281
seed germination and growth 278–279
stress adaptation 281
transporters and metal-binding
compounds 277
uptake and translocation
mechanisms 275–277, 276*f*